AN INTRODUCTION TO MARINE BIOGEOCHEMISTRY

AN INTRODUCTION TO MARINE BIOGEOCHEMISTRY

Susan M. Libes
Department of Marine Science
University of South Carolina—Coastal Carolina College
Conway, SC 29526

John Wiley & Sons, Inc.
New York • Chichester • Brisbane • Toronto • Singapore

Cover photo by James M. Cribb/Bruce Coleman Inc.

The following are the *part opener photo credits*:

Part 1 Opener	Courtesy Wood Hole Oceanographic Institution.
Part 2 Opener	Courtesy NOAA.
Part 3 Opener	Courtesy Rod Catanach, Woods Hole Oceanographic Institution.
Part 4 Opener	Courtesy Woods Hole Oceanographic Institution.
Part 5 Opener	Courtesy NOAA.
Part 6 Opener	Courtesy Greenpeace.

Acquisitions Editor	Nedah Rose
Production Manager	Joe Ford
Designer	Laura Nicholls
Production Supervisor	Lucille Buonocore
Manufacturing Manager	Lorraine Fumoso
Copy Editor	Elizabeth Swain
Illustration	Sigmund Malinowski

Recognizing the importance of preserving what has been written, it is a policy of John Wiley & Sons, Inc. to have books of enduring value published in the United States printed on acid-free paper, and we exert our best efforts to that end.

Library of Congress Cataloging in Publication Data:

Libes, Susan M.
An Introduction to marine biogeochemistry / Susan M. Libes.
p. cm.
Includes bibliographical references and index.
ISBN 0-471-50946-9
1. Chemical oceanography. 2. Biogeochemistry. I. Title.
GC116.L53 1992
551.46'01—dc20 91-28299
CIP

Printed in the United States of America

10 9 8 7 6 5 4 3 2 1

To my dear family
Sol, Lennie, and Don
for always getting me going and keeping me at it

PREFACE

We truly live on a salty blue planet since seawater covers nearly 71 percent of its surface. National interest in utilizing and conserving this vast resource has given rise to many college degree programs in oceanography. Most of their curricula are built on a core of courses that include marine geology, biology, physics, and chemistry. The latter can be viewed as a "capstone" course, as it involves a synthesis and extension of the information provided in the rest of the core. Nevertheless, there are few textbooks suitable for an undergraduate course in marine chemistry. This is largely a result of the rapid growth of this field, as well as its great diversity and complexity.

To overcome these difficulties, this book focuses on the ocean's role in the global biogeochemical cycling of selected elements. The impact of humans on the transport of elements within these cycles is given special emphasis. In Part 1, the concept of a global geochemical cycle is developed using the rock and water cycles as examples. Also included is a discussion of the chemical composition of seawater from the dual perspectives of elemental speciation and the impact of solutes on the physical behavior of water. In Part 2, the biogeochemical cycling of the nutrients and trace metals is considered from the perspective of redox reactions, most of which are driven by the biologically mediated oxidation of organic matter. Steady-state models are used to explain their global oceanic distributions.

Part 3 is a discussion of the biogeochemical phenomena that control the accumulation and preservation of marine sediments. The role of the sediments in determining the chemical composition of seawater is discussed from both equilibrium and kinetic perspectives. The biogeochemistry of selected organic compounds is the topic of Part 4 and includes a discussion of the ocean's role in producing petroleum and moderating the anthropogenic greenhouse effect. In Part 5, the marine chemistry of the radioactive and stable isotopes is discussed. This information is central to the rest of the text in that the use of isotopes as tracers of marine processes has provided much of the data supporting the conclusions presented throughout the book. Because of the relative chemical and mathematical complexity, the details of this approach have been relegated to Chapters 28 and 29. As the largest impact of pollution is felt in the estuaries, marine pollution is discussed from the perspective of estuarine chemistry in Part 6.

By focusing on the "hot" areas of study in marine chemistry, I have attempted to communicate the sense of excitement and discovery that is an essential characteristic of this relatively young and growing science. This book also includes a few aspects of marine analytical chemistry. However, because this subject is so applied, it is best covered in a separate course that also stresses field techniques and methods of data reduction and presentation. Finally, I also briefly discuss some atmospheric and aquatic processes.

This text is designed for a one-semester course in marine chemistry. Prerequisites for such a course include a two-semester survey of college chemistry and a one-semester introduction to marine science. Facility in algebra is required, and, though not essential, some knowledge of calculus is desirable. While this book was designed for an advanced undergraduate course, it is also suitable for students pursuing the master's degree in marine science.

ACKNOWLEDGMENTS

Many others, besides myself, spent long hours working on this textbook. My heartfelt thanks go to them and to my students, who provided me with the necessary teaching experience and motivation. Other important contributors include my editors, D. Sawicki and N. Rose, and my eight enthusiastic (and understanding) reviewers: T. Bidleman (University of South Carolina—Columbia), W.G. MacIntyre (Virginia Institute of Marine Science), F.J. Millero (University of Miami), P.S. Sansgiry (University of South Carolina—Coastal Carolina College), T. Tisue (Clemson University), J.G. Windsor (College of William and Mary), and J. Willey (University of North Carolina—Wilmington). I also wish to acknowledge USC–Coastal Carolina College for its generous financial and administrative support.

My particular interest in marine biogeochemistry stems from my experiences in the early 1980s as a graduate student in the Massachusetts Institute of Technology/Woods Hole Oceanographic Institution Joint Program in Oceanography and Ocean Engineering where I had first-hand contact with many of the most active researchers in this field. My thanks and admiration go to them all.

I would also like to thank all the authors and publishers for granting permission to use their copyrighted figures and tables. Finally, my thanks go to L. Libes and P.S. Sansgiry for proofreading this manuscript.

A FEW WORDS OF ADVICE TO STUDENTS ON HOW TO USE THIS BOOK

The subject of marine chemistry is challenging for undergraduates, but it is very satisfying because you get to use many of the skills and concepts presented in your basic science and math courses. Students are forewarned to be patient. You will not master the information in this book in a single semester. Rather, this book should be considered a permanent resource—a text that can be referred to and learned from long after you graduate.

Many technical terms are used in oceanography. For help, look in the glossary. Make sure you understand all the major concepts found in the chapter summaries and the key words printed in boldface. To test your understanding, try some of the homework problems. Consult the appendices for useful information such as conversion factors and the values of physical constants. To obtain more information on a particular subject, either consult other sections of this book by using the index or read additional material selected from the list of recommended references.

TABLE OF CONTENTS

PART ONE

The waters of the world are sovereign powers.

Jan Morris, *Travels*, 1976

THE PHYSICAL CHEMISTRY OF SEAWATER

CHAPTER 1

THE CRUSTAL-OCEAN FACTORY

INTRODUCTION

The study of marine chemistry encompasses all chemical changes that occur in seawater and the sediments. Since the ocean is a place where biological, physical, geological, and chemical processes interact, the study of marine chemistry is very interdisciplinary. As a result, this field is often referred to as marine **biogeochemistry**.

Not only are all the fields of marine science interconnected, but the ocean itself cannot be studied without considering interactions with the atmosphere and the crust of the earth. For these reasons, this textbook covers topics that range far beyond the margins of the seashore and seafloor, as well as the boundaries of a classical study of chemistry.

WHY THE STUDY OF MARINE CHEMISTRY IS IMPORTANT

Most of the water on Earth's surface is in the ocean; relatively little is present in the atmosphere or on land. Because of its chemical and physical properties, this water has had a great influence on the continuing biogeochemical evolution of this planet. Most notably, water is an excellent solvent. As such, at least a little bit of almost every substance on this planet is present in dissolved form in the ocean. This increases their chance of undergoing chemical changes as reactions are more likely to occur if reactants are in solution, as compared to the gaseous or solid phases.

Another important characteristic of water is its ability to absorb a great deal of heat without undergoing much of an increase in temperature. As a result, the ocean acts as a huge heat absorber, thereby influencing weather and climate. For all of these reasons, life, both marine and terrestrial, depends on water; this relationship is profound. Most scientists support the hypothesis that life itself evolved in the ocean. In turn, biological activity has had important effects on the chemical evolution of the planet. For example, the photosynthetic metabolism of plants is responsible for the relatively high concentration of oxygen in our present atmosphere.

In studying the ocean, marine biogeochemists focus on the exchanges of energy and material between the crust, atmosphere, and ocean. These exchanges have also exerted a central influence on the continuing biogeochemical evolution of the earth. For example, the uptake and release of gases by the ocean has a large impact in determining the chemical composition of the atmosphere. This in turn affects various features of global climate, such as temperature and patterns of rainfall, all of which are important in determining the habitability of our planet. Exchanges of still other substances between the land and sea control the distribution of marine life. For example, marine organisms are most abundant in coastal waters due to the transport of nutrients from the nearby continents. Thus the study of marine chemistry has great practical significance in helping us learn how both to use and conserve the ocean's vast mineral and biological resources.

THE CRUSTAL-OCEAN FACTORY AND GLOBAL BIOGEOCHEMICAL CYCLES

As illustrated in Figure 1.1, the planet can be viewed as a giant chemical factory in which elements are transported from one location to another. Along the way, some undergo chemical transformations. These changes are promoted by the ubiquitous presence of liquid water, which is also the most important

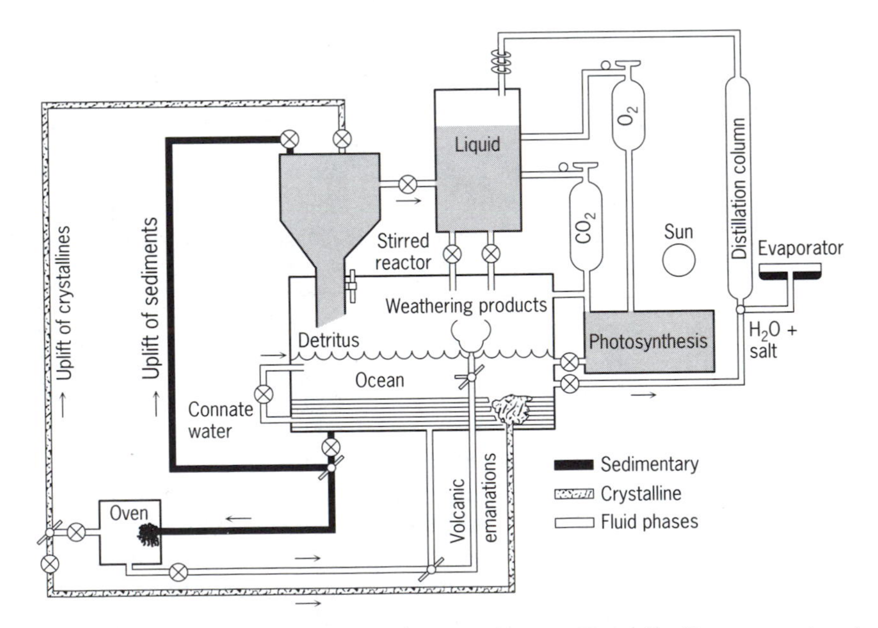

FIGURE 1.1. The crustal-ocean factory. *Source:* From R. Siever, reprinted with permission from *Sedimentology*, vol. 11, p. 21, copyright © 1968 by Elsevier Publishing Company, Amsterdam, The Netherlands.

transporting agent on Earth's surface. It carries dissolved and particulate chemicals from the land and the inner earth into the ocean via rivers and hydrothermal vents, respectively. Chemical changes that occur in the ocean cause most of these materials eventually to become buried as sediments. Geological processes uplift these sediments to locations where weathering occurs. The resulting products are then cycled back to the ocean. Thus the ocean can be thought of as a big stirred reactor, containing a great multitude of substances that have been brought into solution. Since these conditions promote chemical reactions, the ocean acts as a concentrator and settler of the products of crustal weathering.

Figure 1.1 is an example of a "model." Models provide a simplified representation of reality. Those that include mathematical representations of processes can be used to make quantitative predictions. Models are also used as a way of summarizing data and checking for internal consistency. Thus they tend to illuminate gaps in knowledge and hence suggest avenues for further study. In many cases, models constitute the final aim of a research project, since they can be used as a management tool.

The model illustrated in Figure 1.1 is a mechanistic one that emphasizes the flow of materials between various reservoirs. Because most material flows appear to follow closed circuits (if observed for long enough periods of time), the entire loop is referred to as a **biogeochemical cycle**. Such a cycle can be defined for any particular substance, whether it be an element, molecule, or solid. An example of the latter, the rock cycle, is given in Figure 1.2. This type of depiction is called a **box model** because each reservoir, or form that a substance occurs in, is symbolized by a box (e.g., sedimentary rock). The flow

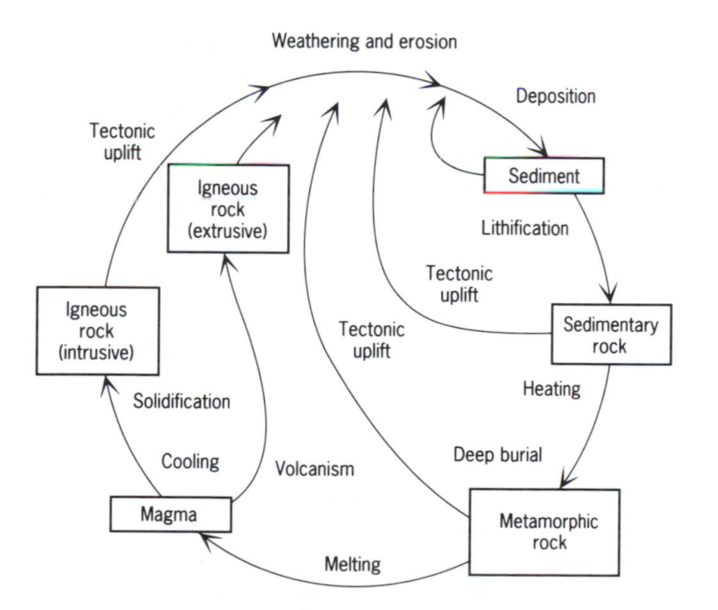

FIGURE 1.2. The global rock cycle.

of materials between reservoirs is indicated by arrows that point from the **source** of a substance to its **sink**. The magnitude of the exchange rates and sizes of the reservoirs are often included in these diagrams.

If the size of the reservoirs remains constant over time, the combined rates of input to each box must equal the combined output rates. This condition is referred to as a **steady state**. The average amount of time a substance spends in a particular reservoir before it is removed through some transport process(es) is called its **residence time**. This steady-state residence time is given by

$$\tau = \frac{\text{Total amount of a substance in a reservoir}}{\begin{array}{c}\text{Rate of supply or removal of the}\\ \text{substance to or from the reservoir}\end{array}} \qquad (1.1)$$

For example, since the total amount of sediment on the seafloor has a mass of 2.5×10^{24} g and the rate at which rivers bring particles into the ocean is 2.0×10^{16} g/y, the average particle entering the ocean spends 1.2×10^{8} y buried in the sediments before it is uplifted and weathered.

Despite differences in transport mechanisms, the mobility of all chemicals is strongly affected by partitioning at interfaces. These boundary zones include air–water and solid–water interfaces. In the ocean, these include the air–sea and sediment–water interfaces, as well as the contact zone between seawater and suspended or sinking particulate matter.

CONSIDERATION OF TIME AND SPACE SCALES

Conceptual models, such as those illustrated above, contain no information concerning variability in either the rates of material transfer or the geographic composition of the reservoirs. These types of variability are termed patchiness and heterogeneity, respectively. A classic example of the former is the **temporal variability** of plankton distributions. Blooms come and go on a seasonal, and even a daily, basis. An example of heterogeneity, or **spatial variability**, is the geographic difference in chemical composition among surface sediments.

As a result, large numbers of samples must be taken to ensure that a given data set is truly representative of a marine environment and not just an artifact of local variability. Thus sampling strategies are determined by the time and space scales over which the process of interest operates. As illustrated in Figure 1.3, marine phenomena occur over a wide range of time and space scales. Due to the complex nature of variability in the marine environment, statistical techniques are often used to help plan experimental design and aid in data analysis.

When considered on sufficiently long time scales, most elements in the ocean are present in steady state. External forces, such as changes in the rate of plate tectonics, can perturb these steady states. Despite this, the chemical composition of the **crustal-ocean factory** has remained relatively constant for the past few hundreds of millions of years. This long-term stability is the result

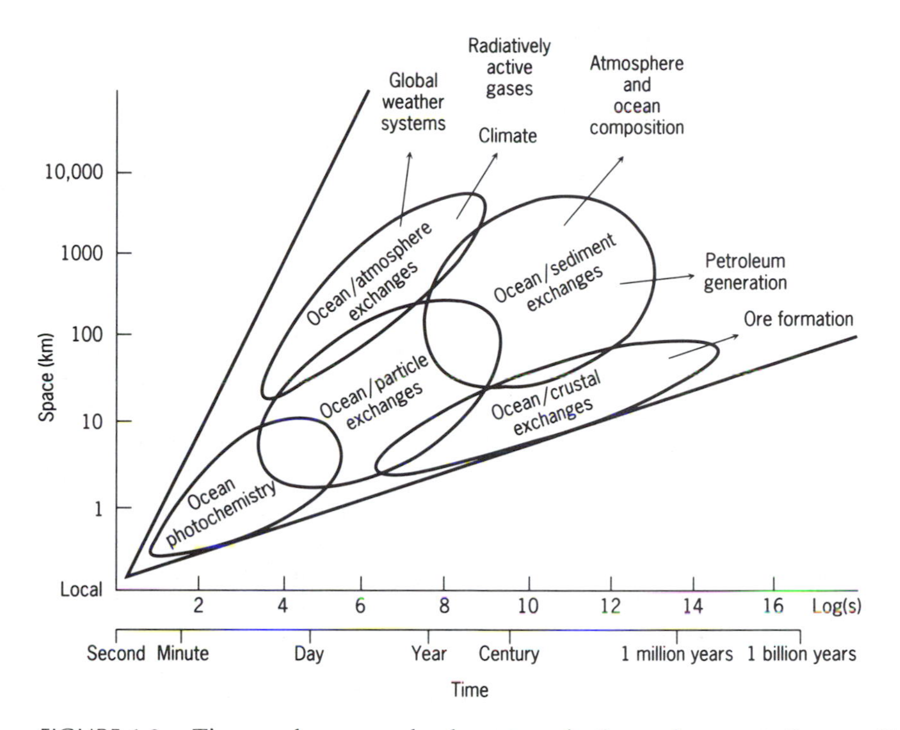

FIGURE 1.3. Time- and space-scale phenomena in the environment. *Source:* From A. Bard, E. D. Goldberg, and D. W. Spencer, reprinted with permission from *Applied Geochemistry,* vol. 3, p. 5, copyright © 1988 by Pergamon Press, Elmsford, NY.

of interactions among the transport processes that constitute the biogeochemical cycles. For example, a perturbation that causes an increase in the rate of supply of some element will be countered by an ensuing increase in the rate of its removal. In this way, the steady state is reestablished. This type of interconnected response is termed a negative **feedback loop** and is a feature common to many global biogeochemical cycles. In contrast, positive feedbacks amplify the effects of a perturbation.

Life appears to have evolved on this planet at a very early date. For life to have endured for billions of years requires Earth's environment to have been fairly stable for over both short and long time scales. As noted above, negative feedbacks among biogeochemical processes appear to stabilize some aspects of Earth's chemical composition and climate. This has led some scientists to suggest that Earth functions as a self-regulating organism in which the biota and their environment interact so as to maintain the atmosphere, land, and ocean in a steady state that favors the survival of life. This is termed the Gaia hypothesis. Though direct evidence for the existence of such a high level of organization has not yet been found, there is a significant amount of data supporting the existence of various negative feedbacks. Some are discussed in this text.

THE HISTORY OF THE STUDY OF MARINE CHEMISTRY

Marine chemistry became a formal subdiscipline of chemistry in the early 1900s, with the advent of scientists who focused all their research efforts in this field and with the development of doctoral degree programs. Prior to the 1900s, the study of marine chemistry was mostly restricted to investigations of the composition of the salts in seawater. The results of such work were first published in 1674 by the English chemist Robert Boyle, the discoverer of Boyle's Law, which describes the behavior of ideal gases.

In 1772, the French chemist Antoine Lavoisier published the first analysis of seawater based on a method of evaporation followed by solvent extraction. Twelve years later, the Swedish chemist Olaf Bergman also published results of the analysis of seawater. To make his measurements, Bergman developed the method of weighing precipitated salts. Through their efforts, the field of analytical chemistry was born.

Between 1824 and 1836, the technique of volumetric titrimetry was developed by Joseph Louis Gay-Lussac. Using this method of analysis, Gay-Lussac determined that the salt content of open-ocean seawater is nearly geographically constant. This conclusion was confirmed in 1818 by John Murray and in 1819–1822 by Alexander Marcet. These scientists discovered that highly precise and accurate measurements of the chemical composition of seawater could be accomplished by gravimetric analysis. This involved the selective precipitation of ions; that is, sulfate was precipitated as barium sulfate, calcium as calcium oxalate, chloride as silver muriate, and magnesium as magnesium pyrophosphate. From the mass of each precipitate, the amount of each ion was then inferred. In presenting his results, Marcet also proposed that seawater contained small quantities of all soluble substances and that the relative abundances of some were constant. The latter hypothesis is now known as Marcet's Principle.

By 1855, the elements that compose the bulk (99%) of sea salt had been discovered and subsequently detected in seawater. Two exceptions were fluorine, which was not isolated until 1886, and iodine. The latter was discovered in 1811 through isolation from the ashes of marine algae.

The concept of salinity was introduced by Georg Forchhammer in 1865. From extensive analyses of seawater samples, he was able to demonstrate the validity of Marcet's Principle for the most abundant of the salt ions: chlorine, sodium, calcium, potassium, magnesium, and sulfate. Thus, he recognized that the salinity of seawater could be inferred from the easily measurable chloride concentration or chlorinity. The details of this relationship were worked out by Martin Knudsen, Carl Forch, and S. P. L. Sorenson between 1899 and 1902. With the international acceptance of their equation relating salinity to chlorinity ($S ‰ = 1.805\ Cl ‰ + 0.030$), the standardization necessary for hydrographic research was provided. Further research into the chemical composition of seawater necessitated a slight revision in this equation ($S ‰ = 1.80655\ Cl ‰$), which was made in 1962 by international agreement.

The modern era of oceanography began in 1876 with the Challenger Expedition. This voyage of exploration was the first undertaken for purely scientific reasons. The results from the analysis of 77 seawater samples collected during this cruise were published by William Dittmar in 1884 and supported Marcet's Principle. During the remainder of the nineteenth century, progress was made in the development of analytical methods for the measurement of trace constituents, such as dissolved oxygen (O_2) and nutrients. With this information, attention shifted to the investigation of the chemical controls on marine life.

The study of oceanography grew increasingly more sophisticated during the period from 1925 to 1940, with the initiation of systematic and dynamic surveys. The most famous was done by the R/V *Meteor*, in which echo sounding was first used to map seafloor topography. Oceanography and the field of marine chemistry entered a new era in the 1940s, primarily as a result of submarine activity during World War II. This was a period of rapid development in technology and instrumentation. Analytical methods were developed for the measurement of isotopes and organic matter, with detection levels dropping to subnanomolar levels.

Modern oceanography is presently characterized by research projects that focus on global-scale phenomena. Due to the size of these projects and their high cost, most are supported by international efforts. This current era was initiated in 1958 with the International Geophysical Year, which was organized by the United Nations' UNESCO General Assembly. Unfortunately, the increasing cost of mounting large expeditions has continued to limit participation in international projects to the countries of Western Europe, North America, and the USSR. Much of their research focuses on predicting the impacts of human activities on the chemistry and biology of the ocean, particularly in coastal regions.

To understand the global role of the ocean in the crustal-ocean factory, marine chemists must obtain more comprehensive data sets. The most promising technique to satisfy this growing need is data collection by satellite. Marine navigation and global weather observations are already largely dependent on satellite information. Measurements of surface-water conditions, including temperature, chlorophyll content, and topography, have already been obtained by the *Seasat* and *Nimbus-7*, which were deployed in 1978. Oceanographers hope to improve their understanding of global-level phenomena, such as climate change, with the satellite work planned as part of NASA's "Mission to Planet Earth."

THE FUTURE OF MARINE CHEMISTRY

Though great progress has been made in the past two decades, many gaps remain in our understanding of the chemical processes that occur in the sea. There are several reasons for this. First, except for water and the six major ions, all the other substances in seawater are present at very low concentrations.

Combined with the interfering effects of the salts, this makes any type of analysis of trace constituents very difficult. To make matters worse, most elements are present in several different forms, or species, in seawater. The speciation of an element determines its reactivity. Thus the concentration of each species of an element must be known to understand the chemical behavior of that element. Fortunately, our ability to measure various substances in seawater is improving rapidly due to continuing technological advances in analytical chemistry.

Another difficulty arises from the great temporal and spatial variability over which most marine processes occur. Because of patchiness and heterogeneity, many samples must be taken to ensure that a representative data set has been collected. Large-scale, long-term sampling efforts are expensive and logistically difficult, but essential for the construction of global-level models of the biogeochemical cycles.

The last problem lies in our theoretical approach to the ocean. Traditionally, marine chemists have explained marine phenomena using equilibrium thermodynamics. Though simple to work with, this approach is limited as chemical reactions rarely attain equilibrium in the ocean. Attempts at kinetic descriptions of marine processes are mostly based on the assumption of first-order rate laws. For example, marine chemists commonly assume phytoplankton assimilate nutrients at rates that are directly proportional to the ambient nutrient concentrations. Such assumptions are made to simplify data interpretation. But it is far more likely that most marine processes involve higher-order kinetics. Indeed, nonlinear behavior is thought to be a source of stability in many biological systems. The application of recent advances in theoretical and analytical chemistry will enable a more sophisticated approach to the study of marine processes.

Marine chemistry has traditionally been divided into two fields. One seeks to understand the chemistry of organic substances in the ocean. The other investigates inorganic substances. Due to analytical difficulties, more is known about the latter than the former. Continuing methodological advances are causing this gap to close rapidly. Our growing recognition of the ubiquitous influence of marine organisms has also blurred the distinction between the two fields.

Another recent development has been the integration of marine chemistry with the other oceanographic subdisciplines in an effort to study the ocean's past. This new subdiscipline is called *paleoceanography*. The ocean covers most of Earth's surface, contains half the planet's biota, and controls our climate. Thus the story of the ocean's past is truly the story of Earth's past. Using information about the causes and behavior of such phenomena as ice ages and plate tectonics, paleoceanographers hope to predict the future of our ocean and planet. This goal is of more than academic interest. Humans have greatly accelerated the transport rates of some materials into the atmosphere and ocean. It is critical to our own continued existence on this planet that we predict the effects of our own actions.

SUMMARY

Marine chemists call themselves biogeochemists because in the ocean, chemical reactions interact with biological, geological, and physical processes. The study of marine chemistry is important because this science considers the mechanisms and rates by which materials are exchanged among the land, atmosphere, ocean, and sediments. Some of these exchanges influence the chemical composition of the atmosphere and climate. Thus they play an important role in making this planet habitable. In most cases, these material flows follow closed pathways, called **biogeochemical cycles**. The largest of these, the **crustal-ocean factory**, views the ocean as one of many material reservoirs on the planet.

A reservoir is said to be a **source** if it supplies materials to another reservoir, which is termed the **sink**. On certain time and space scales, the rates of material loss from a reservoir can equal the rates of supply. In this case, the size of the reservoir remains constant over time and is said to be in a **steady state**. The **residence time** of an average particle in that reservoir can be inferred from the size of the reservoir and either its removal or supply rate. **Box models** are commonly used to depict transport pathways and reservoirs in biogeochemical cycles.

On short time and space scales, the chemical composition of the ocean is relatively heterogeneous. But when considered on sufficiently long time scales, most of the elements in the ocean are present at steady state. Though external forces can perturb these steady states, they are reestablished through the stabilizing effect of **feedback loops**.

Marine chemistry is a relatively new science, having been recognized as a formal subdiscipline of chemistry only since the turn of the century. Since that time, significant progress has resulted from rapid improvements in analytical technology and the acquisition of large data sets. The latter are necessary due to the **temporal** and **spatial variability** of chemicals in the ocean.

CHAPTER 2

THE WATERS OF THE SEA

INTRODUCTION

What is the most abundant substance in the ocean? Water! Not only does water constitute approximately 97 percent of the mass of seawater, but it has some very unusual and important physical characteristics. Because water has a relatively high boiling point, it occurs mostly in the liquid phase. In fact, water is the most common liquid on our planet. Water is essential for life processes due largely to its unique ability to dissolve at least a little bit of virtually every substance. Water is also important because it plays a major role in controlling the distribution of heat on the planet. In this chapter, the physical and chemical features of water are discussed along with the processes by which this important substance is transported around our globe.

THE HYDROLOGICAL CYCLE

Among the planets of our solar system, Earth is unique in its great abundance of **free water**. Most of this water came from the degassing of the planet's interior during the early stages of Earth's formation. This free water is now transported between the land, atmosphere, and ocean through the **hydrological cycle**. As shown in Figure 2.1, the largest reservoir of free water is the ocean.

Water is supplied to the ocean through river and ground water runoff. Why then haven't the seas overflowed? The answer lies in Figure 2.1. Only 90 percent of the water that evaporates from the ocean is returned in the form of rainfall. The rest is transported over land, where it is rained out onto the continents. River runoff carries this missing 10 percent back to the sea. Since the rate at which water is removed from the sea is equal to the rate at which it is returned, the hydrological cycle is in a steady state. Thus the volume of water in the ocean does not change over time.

Water that precipitates on land participates in the rock cycle. The water erodes continental rock and then flows back into the ocean, carrying the solid and dissolved products of weathering. It takes 38,000 years for the average molecule of water to make the complete trip from the ocean to the continents and back again. In comparison, one mixing cycle of the ocean takes 1000 years.

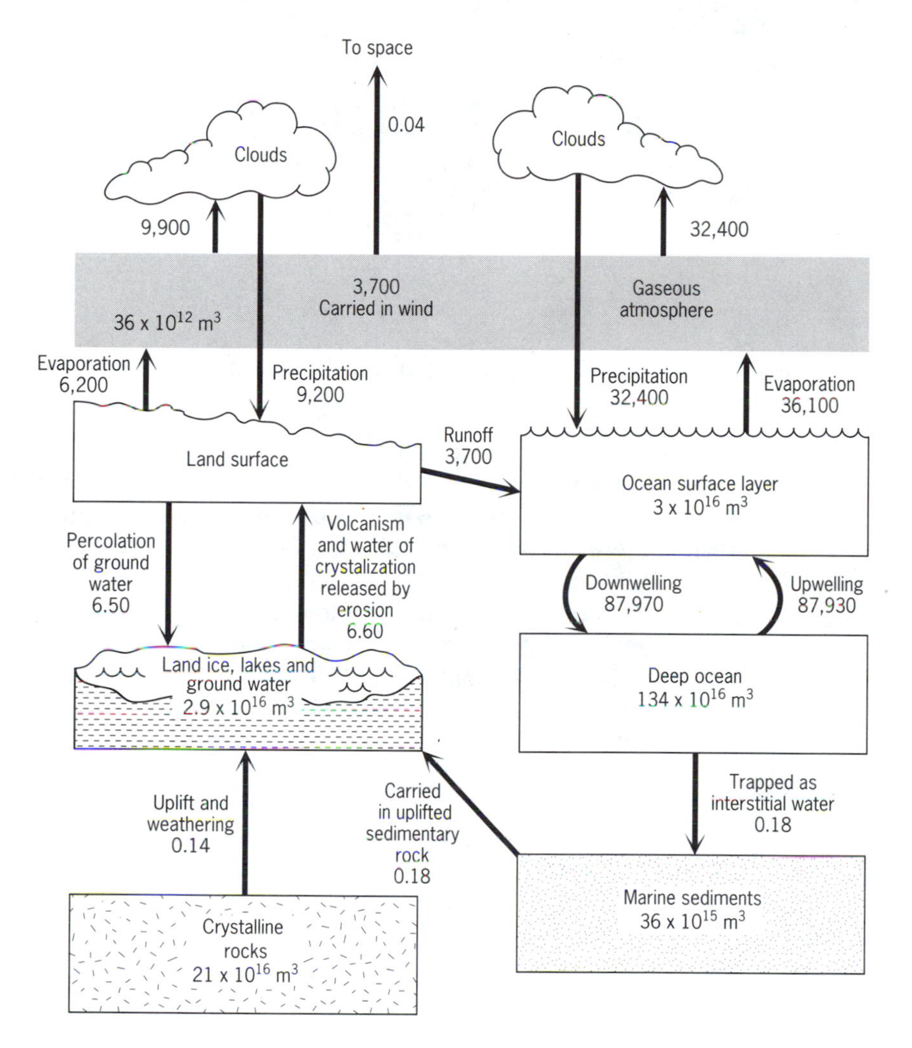

FIGURE 2.1. The hydrological cycle in 10^{10} m^3/y. *Source:* From *The World Ocean: An Introduction to Oceanography,* W. A. Anikouchine and R. W. Sternberg, copyright © 1981 by Prentice Hall, Inc., Englewood Cliffs, NJ, p. 126. Reprinted by permission.

WATER: A PHYSICALLY REMARKABLE LIQUID

The Molecular Structure of Water

The molecular structure of water is shown in Figure 2.2. Each atom of hydrogen is chemically bonded to the oxygen atom. The higher electronegativity of the oxygen atom and its two pairs of unshared electrons cause the oxygen end of

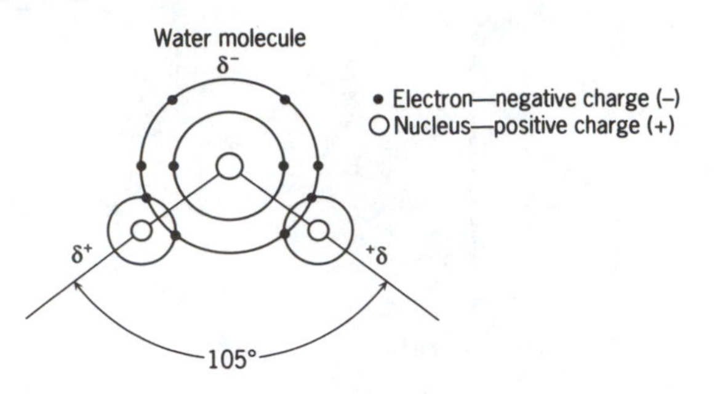

FIGURE 2.2. The Lewis structure and molecular geometry of the water molecule.

the molecule to have a small net negative charge. The hydrogen ends have small net positive charges. Because of their unequal charge distribution, these are termed **polar covalent bonds**. Since the net charges are significantly weaker than those associated with ions and ionic bonds, they are represented by the symbols δ^- and δ^+.

The oxygen atom possesses a total of four electron pairs, causing the water molecule to have approximately tetrahedral geometry. In ideal tetrahedral geometry, the bonding angles are 109.5°. The bond angles in water are slightly smaller (104.5°) because the lone electron pairs occupy a larger volume than the bonding pairs.

The Phases of Water

Water is one of the few substances on the planet that naturally occurs in three phases. Gaseous water is usually referred to as **steam** or water vapor. This phase is characterized by a relatively random arrangement of molecules. Like any gas, a quantity of steam has no definite shape or size. For example, one can put some gas in a balloon and then change the size and shape of the gas just by manipulating the size and shape of the balloon. Some gases, such as steam, are composed of molecules, while others, such as the noble gases, are composed of separate atoms. (The generic term *particle* is used to refer to either atoms or molecules.) In the gas phase, these particles of matter are less tightly packed together than in either the liquid or solid phases. The relative compactness of the phases of matter is shown in Figure 2.3.

The degree of compactness can be expressed as the **density** of a substance, which is defined as

$$\text{Density} = \frac{\text{Mass}}{\text{Volume}} \tag{2.1}$$

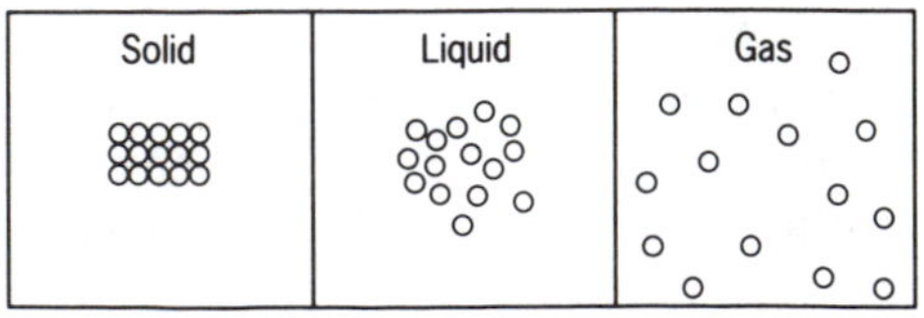

Solid | Liquid | Gas

FIGURE 2.3. Molecular distributions in the solid, liquid, and gaseous phases of matter. The volumes represented by each panel are equal.

The cgs units of density are g/cm^3. The density of pure **liquid water** at 4°C is exactly 1 g/cm^3. Thus a cube of liquid water, measuring exactly 1 cm on all sides, has a mass of exactly 1 gram. In fact, this is how the unit of a gram is defined. Density is an intrinsic property of matter. It remains constant regardless of the amount of substance being measured. For example, at 4°C both 1000 kg and 10 mg of pure water have a density of exactly 1 g/cm^3. The density of a substance gives important information on its behavior. For example, oil floats on liquid water because oil has a lower density than water. A rock will sink in liquid water because the rock has the higher density.

The liquid phase is denser than the gaseous phase and has a more orderly arrangement of particles. A liquid has a definite volume, but no definite shape. So a cup of liquid water can take on the shape of its container, whether it be a cylinder or a box.

Water in the solid phase is referred to as **ice**. Solids possess the most orderly arrangement of particles. As shown in Figure 2.4, crystalline solids possess such an orderly arrangement that the positions of the particles can be predicted over long distances. Due to this long-range order, solids are mechanically rigid. Thus solids have a definite size and shape, both of which are independent of their container.

The example given in Figure 2.4 represents part of a grain of table salt, or sodium chloride. The average grain of table salt (0.1 mm^3) contains about 10^{13} atoms of Na and Cl. But salt grains vary in size, depending on the environmental conditions under which they were formed. Thus it is not possible to write one molecular formula that describes crystalline sodium chloride. In such cases, chemists use an empirical formula to indicate the combining ratios of the atoms. For crystalline sodium chloride, this empirical formula is NaCl.

If the pressure on a substance is kept constant, its phase can be changed simply by adding or removing heat. For water, specific names are given to each phase change. The transition from solid to liquid state is termed **melting** and its reverse is **freezing**. If the water temperature is held at 0°C in a closed

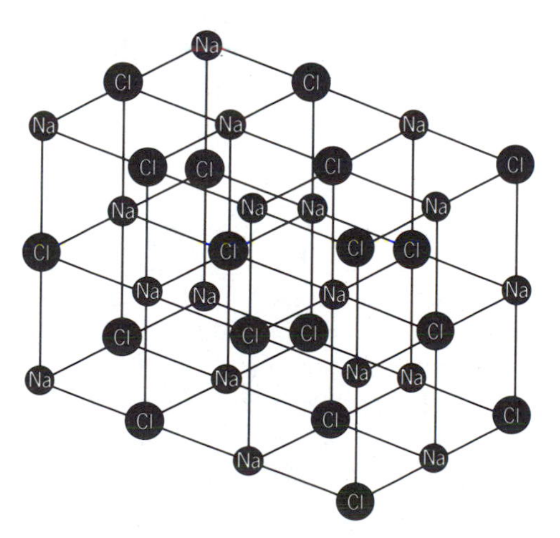

FIGURE 2.4. Crystal lattice of NaCl.

container containing 1 atm pressure, the two phases will coexist in the chemical equilibrium given in Eq. 2.2.

$$H_2O(s) \rightleftharpoons H_2O(l) \quad (2.2)$$

At equilibrium, the number of water molecules entering the solid state is exactly matched by the number entering the liquid state. The temperature at which this occurs is called the normal melting, or freezing, point of water. Note that no bonds are broken or formed, so no chemical reaction has occurred.

The transition from the liquid to the gaseous state is called **evaporation** or vaporization. The reverse is referred to as **condensation** or, in terms of rainfall, precipitation. If heated to 100°C in a closed container containing 1 atm pressure, the two phases will coexist in the equilibrium given in Eq. 2.3.

$$H_2O(l) \rightleftharpoons H_2O(g) \quad (2.3)$$

This temperature is called the normal boiling point of water. If the container were to be opened, some of the gas molecules would escape. To replace the missing water, the phase change represented by Eq. 2.3 would be driven to the right until all of the water evaporated. The direction transition from the solid to the gaseous phase is termed sublimation. Ice will sublime under arid conditions, especially in polar climates.

Heat transfer causes phase transitions by changing the average **kinetic energy** of the particles. When heated, particles move faster and, if unconfined, farther apart. In other words, the thermal energy is converted into kinetic energy. By driving the particles apart, the density of a substance is lowered. Thus, particle motion is fastest in the gaseous phase and slowest in the solid phase. When heat is removed from a substance, the particles slow down. In this lower energy state, the particles come closer together. Hence the density of the substance increases.

From this discussion, we would conclude that solids, which have less kinetic energy than liquids, should always be denser than their liquids. Why then does ice float in liquid water? Some force must be present in ice that keeps the water molecules apart, causing ice to have a lower density than liquid water. It is somewhat ironic that the most abundant and important of liquids on our planet is the only one to have this anomalous density behavior.

Hydrogen Bonding in Water

The force that influences the orientation of water molecules in ice is called **hydrogen bonding**. This is somewhat of a misnomer as hydrogen bonding is a relatively weak intermolecular force. It is caused by the electrostatic attraction of the negatively charged end of a water molecule for the positively charged end of a neighboring molecule. As shown in Figure 2.5, this attraction causes the unshared electron pairs on the oxygen end of each water molecule to orient themselves toward the hydrogen atoms of neighboring water molecules.

In ice, all of the water molecules have formed the maximum number of

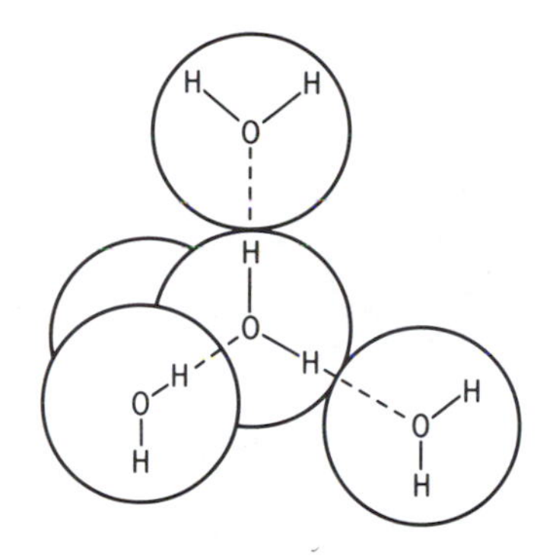

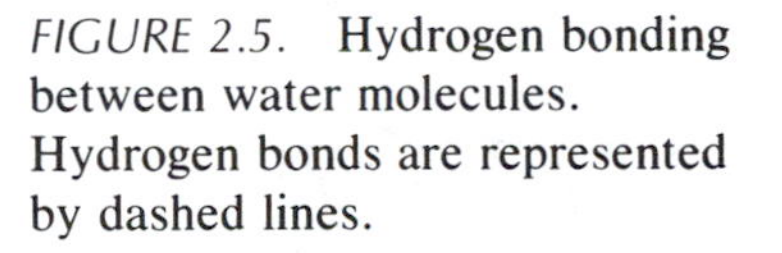

FIGURE 2.5. Hydrogen bonding between water molecules. Hydrogen bonds are represented by dashed lines.

hydrogen bonds, which is four per molecule. This creates the hexagonal pattern illustrated in Figure 2.6.

As shown in Figure 2.7, if hydrogen bonding is incomplete, order exists only over short distances. The less orderly the arrangement, the greater the liquid nature of the water. Though the details of the structure of liquid water are not well understood, this phase is thought to be composed of transitory clusters of molecules that result from the formation of multiple hydrogen bonds. Since the molecules have a high kinetic energy in the liquid state, these intermolecular bonds are rapidly broken and reformed.

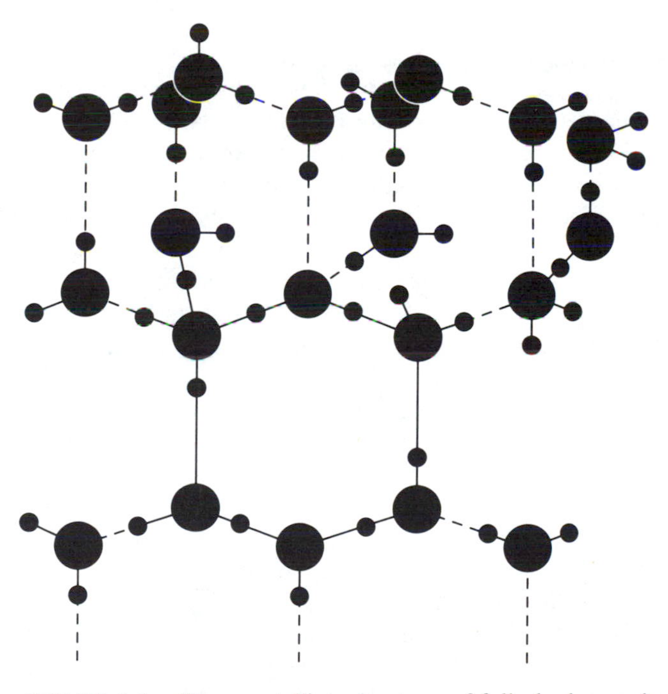

FIGURE 2.6. The crystalline structure of fully hydrogen-bonded water in ice. Hydrogen bonds are represented by dashed lines.

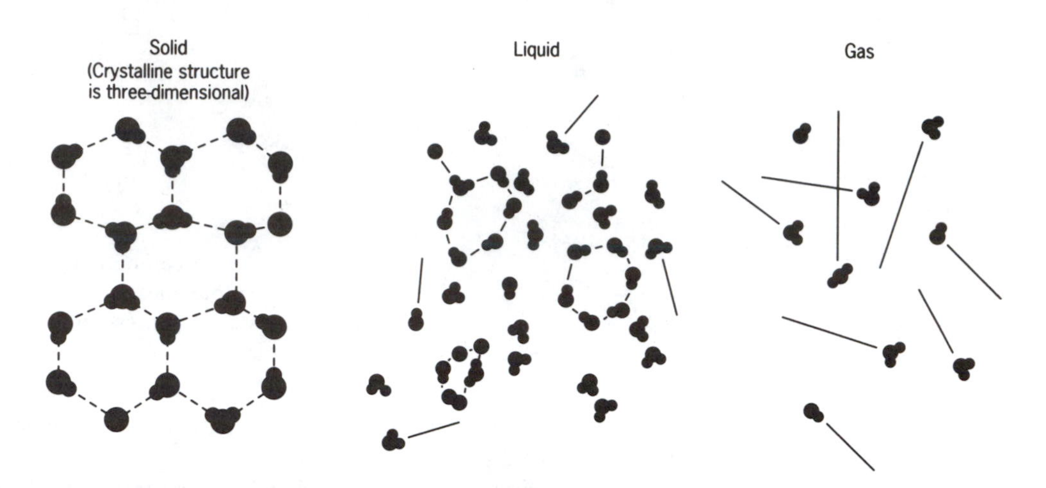

FIGURE 2.7. A comparison of hydrogen bonding in the solid, liquid, and gaseous phases of water. *Source:* From *Introductory Oceanography,* 5th ed., H. V. Thurman, copyright © 1988 by Merrill Publishing Company, Columbus, OH, p. 150. Reprinted by permission.

The Effect of Hydrogen Bonding on the Physical Behavior of Water

A few other hydrides, such as NH_3 and HF, have large enough dipole moments to induce hydrogen bonding. But these substances are gases at the temperatures and pressures usually encountered on this planet. Therefore hydrogen bonding is of little importance to their environmental chemistry, except when they are dissolved in water. Thus water is unique in the degree to which it engages in hydrogen bonding. This behavior is largely responsible for water's unusual physical characteristics, which are listed in Table 2.1 and discussed below.

As illustrated in Figure 2.8, extrapolation of the molecular weight trends established by the Group VIA hydrides suggests that the boiling and freezing points of H_2O should be −68°C and −90°C, respectively. The existence of hydrogen bonds between water molecules causes the liquid to have anomalously high boiling and freezing points.

Additional heat is required to reach the boiling point of water, because the hydrogen bonds must be disrupted before the liquid can be transformed into a gas. The anomalously high freezing point is caused by the formation of hydrogen bonds as water cools. This extra force helps organize the molecules into the long-range order necessary to produce a solid. Thus less heat removal is required to freeze water.

A closer look at the energetics associated with the phase transitions of water is provided in Figure 2.9. The temperature of 1 g of liquid water is increased by 1°C for every calorie of heat energy added. This is called the **heat capacity** of liquid water, which is exactly 1 cal $°C^{-1}$ g^{-1}. The heat capacities of ice and steam are 0.50 and 0.44 cal $°C^{-1}$ g^{-1}, respectively.

TABLE 2.1
Notable Physical Properties of Liquid Water

Property	*Comparison with Other Substances*	*Importance in Physical-Biological Environment*
Heat capacity	Highest of all solids and liquids except liquid NH_3	Prevents extreme ranges in temperature Heat transfer by water movements is very large Tends to maintain uniform body temperatures
Latent heat of fusion	Highest except NH_3	Thermostatic effect at freezing point owing to absorption or release of latent heat
Latent heat of evaporation	Highest of all substances	Large latent heat of evaporation extremely important for heat and water transfer in atmosphere
Thermal expansion	Temperature of maximum density decreases with increasing salinity; for pure water it is at 4°C	Fresh water and dilute seawater have their maximum density at temperatures above the freezing point; this property plays an important part in controlling temperature distribution and vertical circulation in lakes
Surface tension	Highest of all liquids	Important in physiology of the cell Controls certain surface phenomena and drop formation and behavior
Dissolving power	In general dissolves more substances and in greater quantities than any other liquid	Obvious implications in both physical and biological phenomena
Dielectric constant	Pure water has the highest of all liquids	Of utmost importance in behavior of inorganic dissolved substances because of resulting high dissociation
Electrolytic dissociation	Very small	A neutral substance, yet contains both H^+ and OH^- ions
Transparency	Relatively great	Absorption of radiant energy is large in infrared and ultraviolet; in visible portion of energy spectrum there is

(Table continues on following page)

TABLE 2.1 *continued*
Notable Physical Properties of Liquid Water

Property	*Comparison with Other Substances*	*Importance in Physical-Biological Environment*
		relatively little selective absorption, hence is "colorless"; characteristic absorption important in physical and biological phenomena
Conduction of heat	Highest of all liquids	Although important on small scale, as in living cells, the molecular processes are far outweighed by eddy diffusion

Source: From *The Oceans*, H. U. Sverdrup, M. W. Johnson, and R. H. Fleming, copyright © 1941 by Prentice Hall, Inc., Englewood Cliffs, NJ, p. 48. Reprinted by permission. Based on a table from *Recent Marine Sediments*, P. D. Trask (ed.), copyright © 1955 by the Society of Economic Petroleum Mineralogists, Tulsa, OK, p. 67. Reprinted by permission.

The cause of the relatively high heat capacity of liquid water is similar to that which produces the anomalously high boiling point. Due to the presence of hydrogen bonds, heat that would otherwise go to increasing the motion of the water molecules is instead absorbed by the hydrogen bonds. Once the hydrogen bonds have been disrupted, the added heat energy is expressed solely as an increase in molecular motion. It is this increased motion that is measured as a temperature rise by a thermometer.

You have probably experienced the high heat capacity of water for yourself during your last trip to the beach. Standing on the water's edge on a hot, sunny day, you can have one foot in the pleasantly cool water of the ocean and the other foot, just a few inches away, in the painfully hot sand. How can this be? Both the sand and the ocean have received the same amount of solar heat. The explanation, of course, is that the water has absorbed the heat without experiencing as large a rise in temperature as the sand.

The high heat capacity of water has important consequences for climate and life on this planet. During the summer, heat is stored by the ocean. This heat is radiated back to the atmosphere in the winter. Thus the ocean acts to moderate the climate, reducing the amplitude of seasonal temperature variations. This effect is most noticeable when comparing the moderate climate of the coastal zone with the hot summers and cold winters experienced at inland locations.

Water also has large **latent heats of fusion** and **evaporation**. The former is the amount of heat required to transform 1 g of ice into liquid water or the amount of heat that must be removed to transform 1 g of liquid water into ice. The heat of evaporation is analogous to the heat of fusion but refers to the

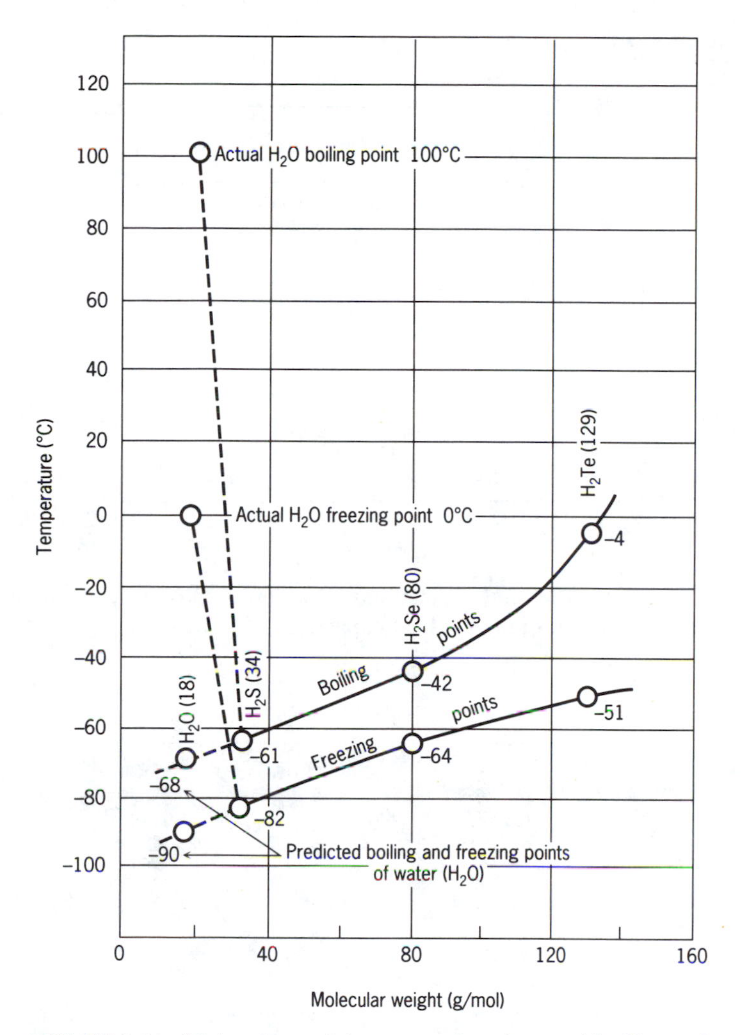

FIGURE 2.8. Molecular weight versus freezing- and boiling-point temperatures of the group VIA hydrides. *Source:* From *Introductory Oceanography,* 5th ed., H. V. Thurman, copyright © 1988 by Merrill Publishing Company, Columbus, OH, p. 148. Reprinted by permission.

liquid–gas phase transition. These relatively high latent heats are another consequence of hydrogen bonding. Before the water can undergo these phase transitions, heat is needed to disrupt the hydrogen bonds. More heat is required for the liquid-to-gas transition than for the solid-to-liquid transition because almost all the hydrogen bonds must be broken to reach the gaseous state.

The process by which water freezes is illustrated in Figure 2.10. Panels above the graph depict the arrangement of water molecules during various stages of cooling. Decreasing the temperature of water by removing heat slows the water molecules, bringing them closer together. In this way, the density

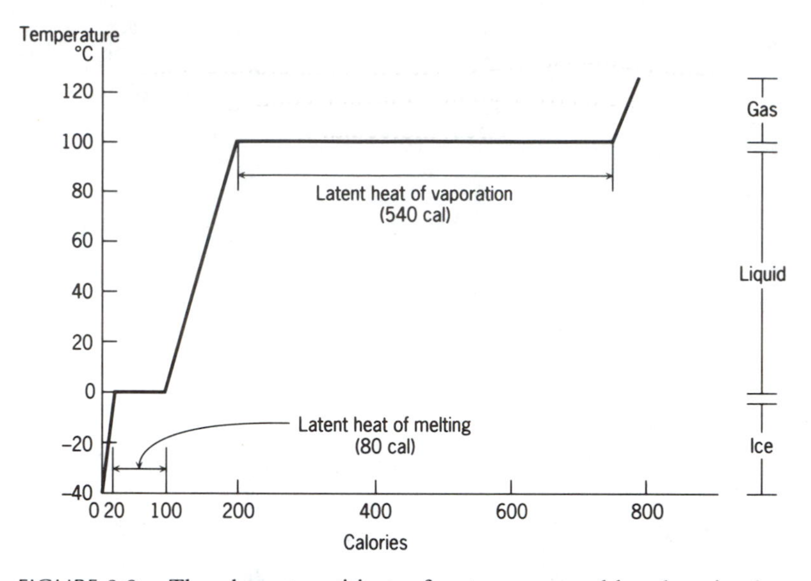

FIGURE 2.9. The phase transitions of water as caused by changing heat content. Slopes of the lines indicate heat capacity.

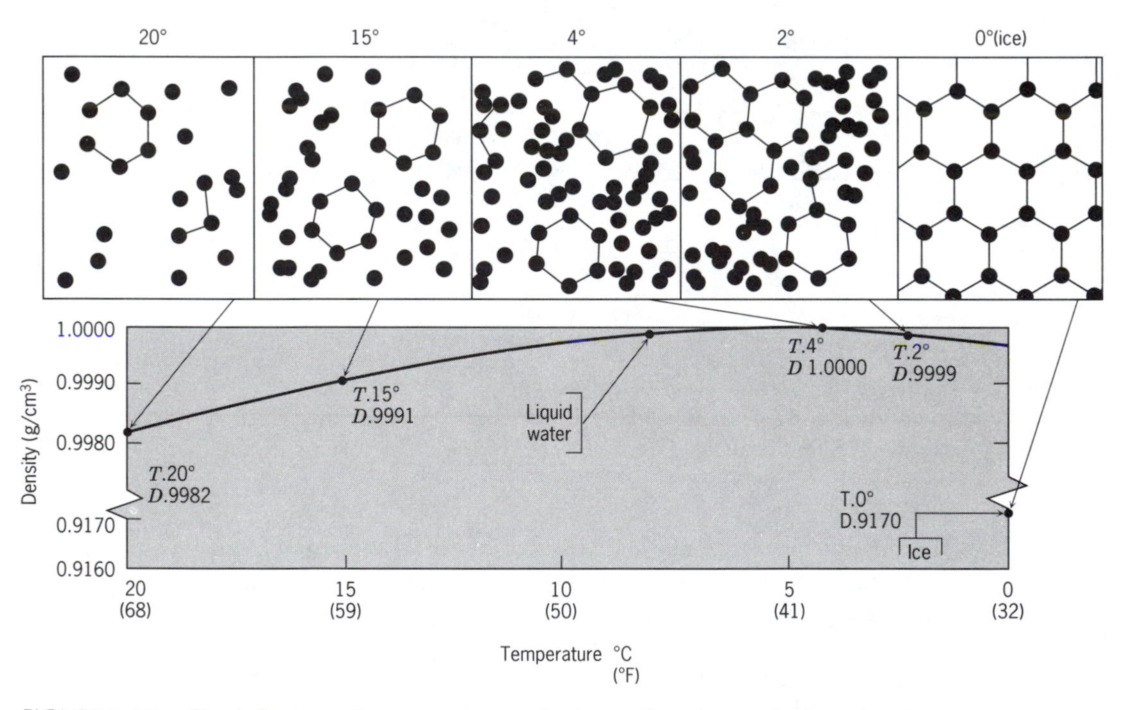

FIGURE 2.10. The influence of temperature on hydrogen bonding and water density. *Source:* From *Introductory Oceanography*, 5th ed., H. V. Thurman, copyright © 1988 by Merrill Publishing Company, Columbus, OH, p. 158. Reprinted by permission.

of water increases from 0.9982 g/cm^3 at 20°C to 0.9991 g/cm^3 at 15°C. This trend continues to 4°C. At this temperature, pure water reaches its maximum density, exactly 1 g/cm^3. Further cooling causes the density to decrease. At these low temperatures, molecular motion has been slowed such that hydrogen bonds form between enough molecules to create some hexagonal clusters. At 0°C, the water molecules are completely hydrogen bonded, forming the hexagonal **crystal lattice** that is ice.

As the diagram illustrates, ice has more open space than does liquid water. Thus water expands as it freezes, so ice floats in its liquid. You've probably experienced this expansion. Think what happens when you put a can of soda in the freezer. As the water freezes, its volume increases, making the can bulge.

The anomalous density behavior of water has important consequences for the survival of aquatic organisms at mid latitudes. As winter approaches, the surface waters of ponds and lakes cool. The ensuing increase in density causes this water to sink to the bottom of the water body. This process continues until water temperatures drop below 4°C. At lower temperatures, further cooling produces water of a lower density, so the sinking stops. If the atmospheric temperature reaches 0°C, ice will form at the surface and act as an insulating layer that isolates the underlying water from further atmospheric cooling. Thus ponds and lakes freeze from the top down. Since very long and cold winters are required to complete this process, frogs, fish, and other creatures that hibernate in the mud are protected from freezing.

Three other physical characteristics of water are affected by the presence of hydrogen bonds. These are the unusually high **surface tension** and **viscosity** of water and its relatively low **compressibility**. Hydrogen bonding creates such a highly interconnected web of water molecules that the surface of the liquid behaves as if it is covered by a thin skin. Because of this surface tension, a carefully filled cup of water forms a dome of the liquid above its rim. Viscosity is a measure of how much a fluid will resist changing its form when a force is exerted on it. Because hydrogen bonds hold water molecules together, they can not flow past each other as easily as molecules in fluids without hydrogen bonds.

Pressure increases by approximately 1 atm for every 10 m increase in water depth. Thus, at the average depth of the ocean, approximately 4000 m, the pressure is 400 times greater than at the sea surface. Due to the low compressibility of water, this great increase in pressure causes only a slight increase in density. This behavior is due to the presence of hydrogen bonds, as they restrict the degree to which the water molecules can be pushed together.

WATER AS THE UNIVERSAL SOLVENT

Water is called the **universal solvent** because of its ability to dissolve at least a little of virtually every substance. Water is a particularly good solvent for substances held together by polar or ionic bonds. Indeed, the most abundant substance dissolved in seawater is an ionic solid, sodium chloride. In comparison, only small amounts of nonpolar substances, such as hydrocarbon oils, will dissolve in water.

Ionic solids are also called **salts**. Salts are composed of atoms held together by ionic bonds. The bonds are the result of electrostatic attraction between positively charged ions (**cations**) and negatively charged ions (**anions**). The force of electrostatic attraction is inversely related to the square of the distance of separation of the ions.

When placed in water, salts, like sodium chloride, dissolve because the cations and anions are electrostatically attracted to the water molecules. The cations attract the oxygen ends of the water molecules, and the anions attract the hydrogen ends. When surrounded by water molecules, the ions are too far apart to exert any force of attraction on each other. Thus the ionic bond is broken and the ions are said to be dissolved or **hydrated**. When the surface ions become hydrated, the underlying salt ions are exposed to water and eventually become hydrated as well. This process is illustrated in Figure 2.11.

The chemical equation that describes the dissolution of NaCl is given in Eq. 2.4, where n and m equal the number of waters of hydration in direct contact with each ion.

$$NaCl(s) + (n + m)H_2O(l) \rightarrow Na(H_2O)_n^+ + Cl(H_2O)_m^- \qquad (2.4)$$

The number of waters of hydration is determined by an ion's charge and radius, as well as the presence of other solutes. In the case of Na^+ and Cl^-, this number ranges from four to six.

Since the only intramolecular bonds broken during the dissolution of salt are the ionic bonds, water is not truly a reactant in Eq. 2.4. Thus Eq. 2.4 is

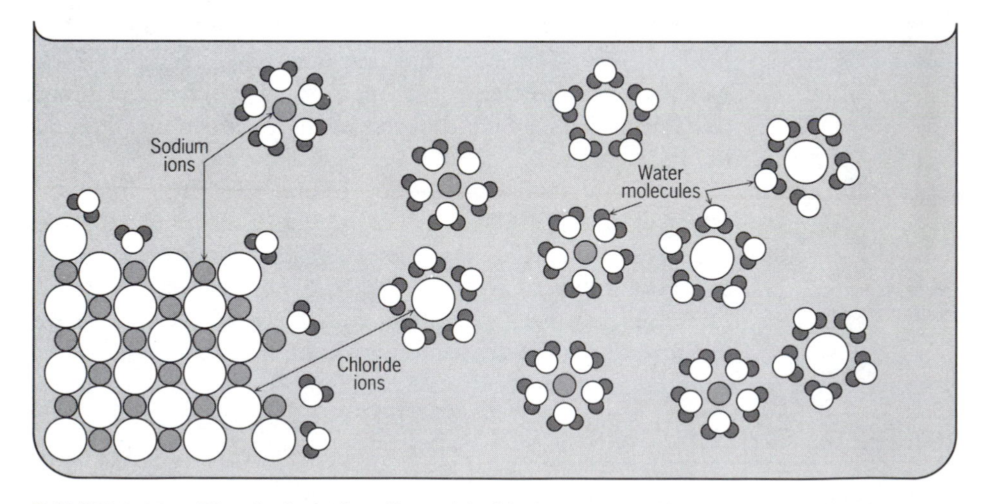

FIGURE 2.11. Dissolution of sodium chloride in water. *Source:* From *Chemistry and Chemical Reactivity*, J. C. Kotz and K. F. Purcell, copyright © 1987 by Saunders College Publishing Company, Philadelphia, p. 85. Reprinted by permission.

more commonly written as shown below,

$$NaCl(s) \rightarrow Na^+(aq) + Cl^-(aq) \tag{2.5}$$

where (aq), for aqueous, indicates that the ions are hydrated.

The ability of an ionic solid to dissolve depends on its lattice energy, as well as the degree to which its ions can become hydrated. The lattice energy of an ionic crystal is a measure of the strength of its three-dimensional network of bonds. If these interactions are weaker than the solute–solvent attractions, the ionic bonds will be easily disrupted by water molecules.

In the case of the dissolution of NaCl, the products are strongly favored at equilibrium. In other words, NaCl is very soluble in water, almost completely dissociating into $Na^+(aq)$ and $Cl^-(aq)$. Such salts are termed strong **electrolytes**. If an electromagnetic field is applied to a solution of strong electrolytes, the ions will migrate, producing an electric current. The ability of a solution to **conduct** an electrical current increases with increasing electrolyte concentration.

THE EFFECT OF SALT ON THE PHYSICAL PROPERTIES OF WATER

Adding salt to water increases the density of the solution. This effect is discussed in greater detail in the next chapter. As shown in Table 2.2, the presence of salt alters other physical properties.

TABLE 2.2
Comparison of Pure Water and Seawater Properties

Property	*Seawater, 35‰ S*	*Pure Water*
Density, g/cm^3, 25°C	1.02412	1.0029
Equivalent conductivity, 25°C, $cm^2\ ohm^{-1}\ equiv^{-1}$	—	—
Specific conductivity, 25°C, $ohm^{-1}\ cm^{-1}$	0.0532	—
Viscosity, 25°C, millipoise	9.02	8.90
Vapor pressure, mm Hg at 20°C	17.4	17.34
Isothermal compressibility, 0°C, unit vol/atm	46.4×10^{-6}	50.3×10^{-6}
Temperature of maximum density, °C	−3.52	+3.98
Freezing point, °C	−1.91	0.00
Surface tension, 25°C, dyne/cm	72.74	71.97
Velocity of sound, 0°C, m/s	1450	1407
Specific heat, 17.5°C, $J\ g^{-1}°C^{-1}$	3.898	4.182

Source: From *Marine Chemistry*, R. A. Horne, copyright © 1969 by John Wiley & Sons, Inc., New York, p. 57. Reprinted with permission.

For example, adding salt to water lowers the freezing point of the solution. Thus the freezing point of seawater is a function of its salt concentration, or salinity, as illustrated in Figure 2.12*c*. (A rigorous definition of salinity is given in Chapter 3.) At low enough temperatures, the water in the seawater will freeze. This increases the salt content of the remaining seawater, causing the freezing point of seawater to decline.

The temperature at which seawater reaches its maximum density also decreases with increasing salinity. Most seawater in the ocean has a salinity between 33‰ and 37‰. At salinities greater than 26‰, the freezing point of seawater is higher than the temperature at which it reaches its maximum density. Thus seawater never undergoes the anomalous density behavior of pure water. Instead, sea ice floats because it is mostly pure water.

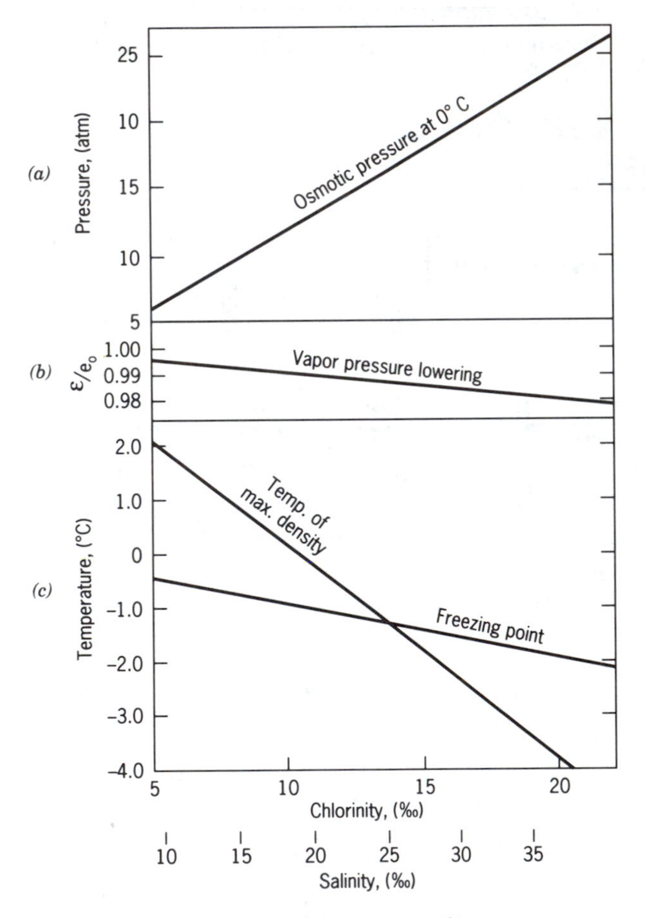

FIGURE 2.12. (*a*) Osmotic pressure, (*b*) vapor pressure relative to that of pure water, (*c*) freezing point and temperature of maximum density as a function of salinity. *Source:* From *The Oceans*, H. U. Sverdrup, M. W. Johnson, and R. H. Fleming, copyright © 1941 by Prentice Hall, Inc., Englewood Cliffs, NJ, p. 66. Reprinted by permission.

As shown in Figure 2.12*b*, the vapor pressure of seawater decreases with increasing salinity. Since more heat is required to raise the vapor pressure to atmospheric pressure, the boiling point of seawater increases with increasing salinity. There are few practical applications of this as seawater is rarely close to its boiling point. The hottest seawater is found in hydrothermal systems as a result of close contact with magma. This water reaches temperatures in excess of 400°C, but the high hydrostatic pressure lowers its vapor pressure and prevents boiling.

Because of the high concentrations of ions, seawater has a greater **osmotic pressure** than pure water. Differences in osmotic pressure cause water molecules to diffuse across semipermeable membranes from regions of low salt concentration to regions of higher concentration. Net diffusion ceases when the water transfer has equalized the salt concentrations across the membrane. The most common and important examples of natural semipermeable membranes are cell membranes. You have probably experienced osmosis if you've stayed in the ocean too long. Skin, especially around your hands, gets wrinkled. This is because water is diffusing out of your body against the salt concentration gradient. The salt content in the intracellular fluid of many marine organisms is close to that of seawater. This minimizes the amount of energy needed to maintain an internal salt content different from ambient seawater. The ion concentrations in the intracellular fluids of some marine organisms, humans, and seawater are given in Figure 2.13. The salt content of mammalian blood is quite close to that of seawater, approximately 3.5 percent wt/vol. This is not surprising, as life appears to have evolved in the ocean.

The addition of salt to water increases the viscosity of water. This is caused by the electrostatic attraction between the solutes and water. Due to slight spatial differences in salinity, the viscosity, and hence the speed of sound in seawater, is also geographically variable. This is of practical consequence as the operation of SONAR (sound navigation ranging) depends on a precise knowledge of the speed of sound in seawater. As described in Chapter 1, World War II made the need for accurate SONAR essential. This demand motivated the first detailed studies of the distribution of salinity and temperature in the ocean, which marked the beginning of the modern age of oceanography.

SUMMARY

Water is transported among the land, oceans, and atmosphere by the **hydrological cycle**. The largest reservoir of **free water** on the planet is the ocean. This reservoir is in a steady state because the rates of input via river runoff and rainfall are balanced by loss via evaporation. Earth is unique in having such a great abundance of free water on its surface. Water is unique on this planet for being the most abundant liquid and for naturally occurring in three phases, **steam, liquid water**, and **ice**. Transitions between these phases, such as **freezing, melting, evaporation**, and **condensation**, are commonly caused by the addition

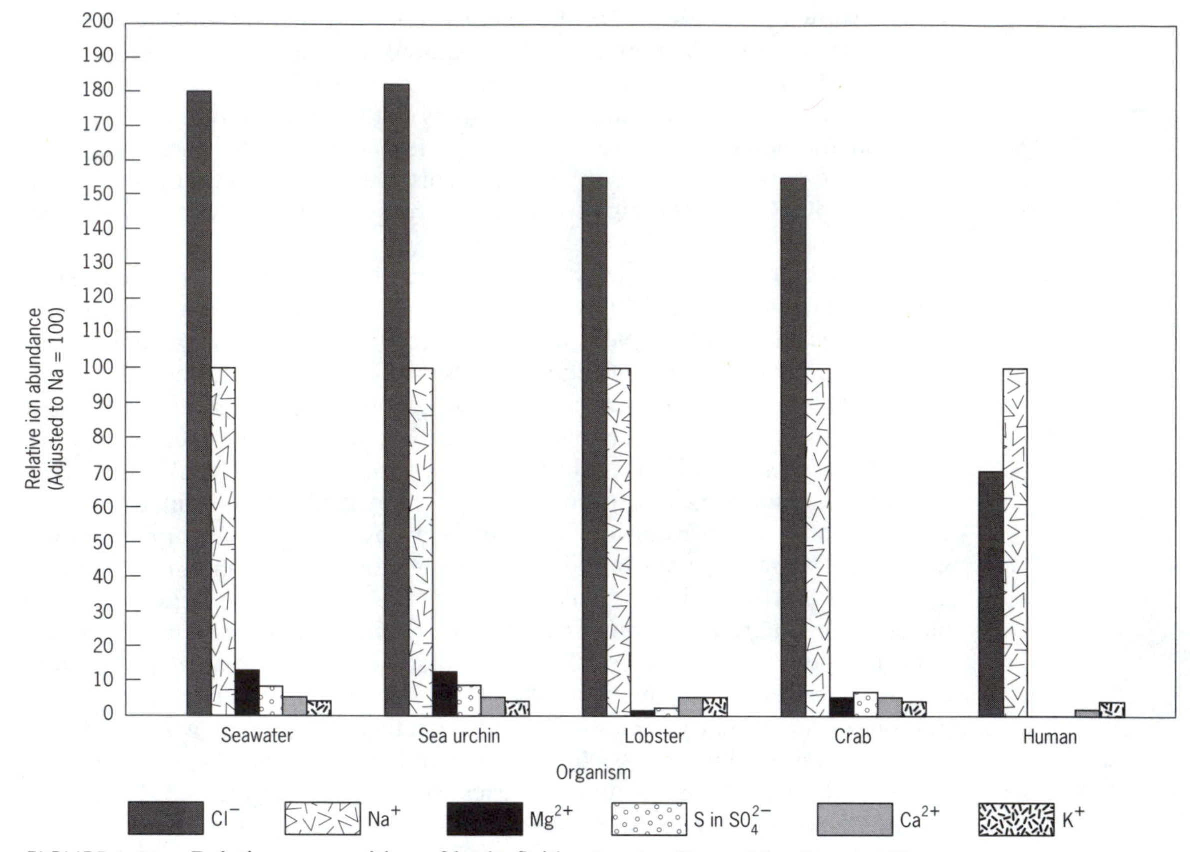

FIGURE 2.13. Relative composition of body fluids. *Source:* From *The Oceans,* H. U. Sverdrup, M. W. Johnson, and R. H. Fleming, copyright © 1941 by Prentice Hall, Inc., Englewood Cliffs, NJ, p. 48. Reprinted by permission. Based on data from J. D. Robertson, reprinted with permission from the *Journal of Experimental Biology,* vol. V.XVI, p. 394, copyright © 1939 by Company of Biologists, Ltd., Essex, England.

or removal of heat. These processes change the **kinetic energy** of the water molecules and hence their **density**.

Water has several unusual physical properties that are related to the nature of its chemical bonds. Because water has **polar covalent bonds**, slight electrical charge differences exist across the water molecule. This leads to **hydrogen bonding** between water molecules. The complete hydrogen bonding of water produces ice. As its **crystal lattice** has a very open structure, ice is not as dense as liquid water. This is why solid water floats in its liquid and ponds freeze from the top down. Hydrogen bonding also causes water to have relatively high **boiling** and **freezing points, heat capacity, latent heats of fusion** and **evaporation, surface tension, viscosity**, and to be nearly **incompressible**.

The polar covalent bonds also cause water to act as a **universal solvent**. **Ionic solids**, or **salts**, are very soluble in water because of the ability of water to **hydrate** the **cations** and **anions** that compose the salts. Adding salt to water increases its density, lowers its freezing point, elevates its boiling point, increases its electrical **conductivity**, increases its **osmotic pressure** and increases its viscosity. Thus the presence of salt in water has a very important impact on the physical behavior of seawater.

CHAPTER 3

SEASALT IS MORE THAN NaCl

INTRODUCTION

Transport through the hydrological cycle alters the chemical composition of natural waters as some substances dissolve and others precipitate or degas. This causes most natural waters to vary greatly in chemical composition. In contrast, the chemical composition of seawater is remarkably similar from sea surface to sea floor and from ocean to ocean. Seawater is also unusual for its relatively high concentrations of solutes.

The most abundant ions in seawater (chlorine, sodium, calcium, potassium, magnesium, and sulfate) are present in constant proportions because their concentrations are largely controlled by physical processes, such as the addition or removal of water. For this reason, these salts are termed **conservative ions**.

The rest of the substances in seawater are not present in constant proportions because their concentrations are altered by chemical reactions that occur in the ocean and sediment. These chemicals are termed **nonconservative**. Though most substances in seawater are nonconservative, they compose only a small fraction of the total mass of the ocean. The major classifications and average concentrations of these substances are given in Table 3.1.

The last two entries in Table 3.1 are not solutes. The theoretical size criteria used to differentiate the dissolved from the nondissolved substances are given in Table 3.2.

In practice, the concentration of solids is determined by filtering seawater through a membrane filter that has a nominal pore size of 0.4 μm. The particles retained by the filter are termed solids or **particulate matter**. The substances that pass are collectively referred to as dissolved matter and include the **colloids**. This is truly an operational definition as the separation is dependent on the conditions under which the filtration is done. In particular, clogging of the pores during filtration decreases the effective pore size of the filter, causing the retention of the smaller particles. As a result, the retention of the colloidal particles tends to be variable. This size separation of particles is also affected by the variable adsorption of solutes and colloids by the filter and the retained solids.

Due to their abundance, the major ions cause seawater to have a very high concentration of positive and negative charges. This causes chemical reactions in seawater to behave in a thermodynamically nonideal fashion. The major ions are also important as their conservative behavior enables them to be used

TABLE 3.1
Chemical Composition of Seawater in Order of Abundance

Category	*Examples*	*Concentration Range*
Major ions	Cl^-, Na^+, Mg^{2+}, SO_4^{2-}, Ca^{2+}, K^+	mM
Minor ions	HCO_3^-, Br^-, Sr^{2+}, F^-	μM
Gases	N_2, O_2, Ar, CO_2, N_2O, $(CH_3)_2S$, H_2S, H_2, CH_4	nM to mM
Nutrients	NO_3^-, NO_2^-, NH_4^+, PO_4^{3-}, H_4SiO_4	μM
Trace metals	Ni, Li, Fe, Mn, Zn, Pb, Cu, Co, U, Hg	<0.05 μM
Dissolved organic compounds	Amino acids, humic acids	ng/L to mg/L
Colloids	Sea foam, flocs	$\leq$mg/L
Particulate matter	Sand, clay, dead tissues, marine organisms, feces	μg/L to mg/L

as tracers of water mass motion. In this chapter, some of the physical and chemical aspects of the conservative ions are discussed.

MEASURING SALT CONTENT

Salinity

The most abundant solutes in seawater are Cl^-, Na^+, Mg^{2+}, So_4^{2-}, Ca^{2+}, and K^+. As mentioned above, they are referred to as the major or conservative ions. Taken together, they constitute over 99.8 percent of the mass of the solutes dissolved in seawater. The sodium and chlorine alone account for 86 percent.

Salinity is the most commonly used measure of the saltiness of seawater and is theoretically defined as the total number of grams of dissolved salt ions present in 1 kg of seawater. This can be represented mathematically as

$$S(‰) = \frac{\text{g of inorganic dissolved ions}}{\text{1 kg seawater}} \times 1000 \qquad (3.1)$$

TABLE 3.2
Classification by Particle Size

Class	*Particle Diameter (μm)*
Solids	≥ 0.1
Colloids	0.001 to 0.1
Solutes	≤ 0.001

Since the mass ratio is multiplied by 1000, the units of **salinity** are parts per thousand, symbolized by ‰. As shown in Figure 3.1, 99 percent of the seawater in the ocean has a salinity between 33‰ and 37‰. Note that the temperature of seawater is far more variable, ranging from −2° to 30°C. The average salinity of seawater is 35‰, which is equivalent to a 3.5 percent salt solution.

The most accurate and precise salinity measurements are made with an inductive **salinometer**, which measures the conductivity of a seawater sample. Due to their high concentrations, the major ions are responsible for most of the conductivity of seawater. Thus the higher their concentrations, the greater the conductivity of the sample. Other electrolytes, such as the minor ions, **trace metals, nutrients**, and **dissolved organic matter (DOM)** are present at such low concentrations that their contribution to the total conductivity of seawater is insignificant. The conductivity reading is converted to salinity using an empirical equation. The precision of a good inductive salinometer is ±0.001‰. Accuracy is insured by calibrating the salinometer on a routine basis against

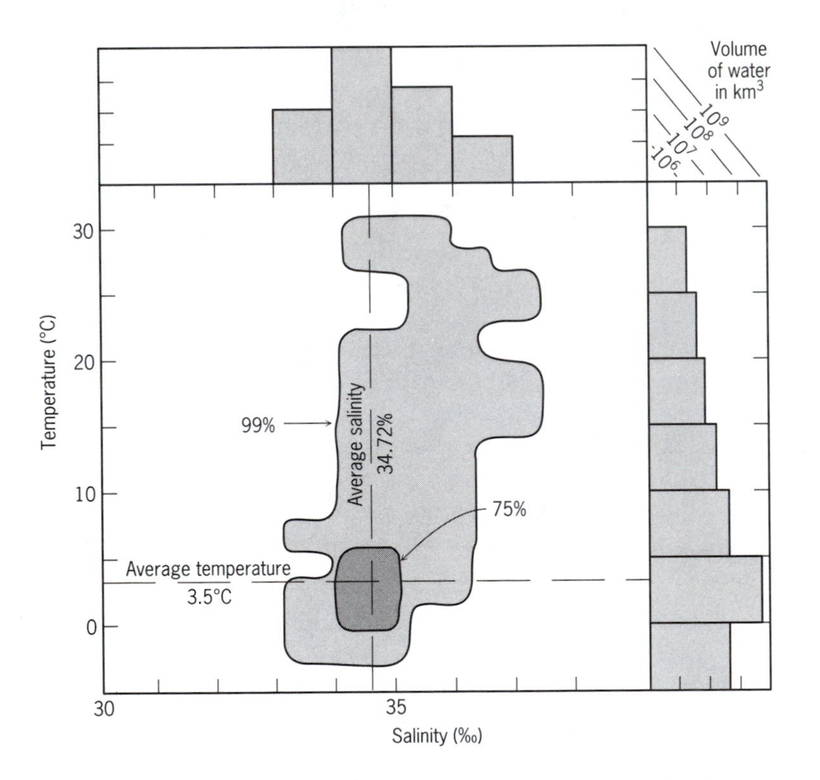

FIGURE 3.1. The temperature and salinity of 99% of ocean water are represented by points with the stippled area enclosed by the 99 percent contour. The shaded area represents the range of temperature and salinity of 75% of the water in the world ocean. *Source:* From *Oceanography: A View of the Earth,* 4th ed., M. G. Gross, copyright © 1987 by Prentice Hall, Inc., Englewood Cliffs, NJ, p. 164. Reprinted by permission.

IAPSO (International Association for Physical Sciences of the Ocean) Standard Sea Water, which is the internationally accepted salinity standard.

Chlorinity

Because the major ions are present in constant proportions, their concentrations can be used to infer the salinity of seawater. Before the invention of the salinometer, salinity was calculated from the chloride concentration of seawater because of the ease and accuracy with which this measurement can be made. The empirical equation that relates salinity to **chlorinity** is

$$\text{Salinity} = 1.80655 \times \text{Chlorinity} \tag{3.2}$$

where chlorinity is defined as the mass in grams of halides (expressed as chloride ions) that can be precipitated from 1000 g of seawater by Ag^+. Chlorinity is determined by titrating a sample of seawater with a standard solution of silver nitrate. The titration reaction is

$$3Ag^+(aq) + Cl^-(aq) + Br^-(aq) + I^-(aq) \rightarrow AgCl(s) + AgBr(s) + AgI(s) \tag{3.3}$$

Thus

$$\text{Chlorinity} = \left[\frac{\begin{matrix}\text{Atomic weight} \\ \text{of } Cl^-\end{matrix} \times \begin{matrix}\text{moles of } Ag^+ \text{ required to} \\ \text{precipitate } Cl^-, Br^-, \text{and } I^-\end{matrix}}{1000 \text{ g of seawater}}\right] \times 1000 \tag{3.4}$$

When done by a good analytical chemist, the titration produces results that are as precise and accurate as those from a salinometer. Unfortunately, minor deviations from the relationship given in Eq. 3.2 can occur. The causes for these anomalies are explained later in this chapter.

Density

The density of seawater is a measure of the total mass of seawater in a given unit volume. As with chlorinity, an empirical equation has been developed that relates density to salinity. This relationship is dependent on the in situ temperature and pressure and is called the **equation of state for seawater**. Due to the presence of nonconservative substances, this equation, like Eq. 3.2, is inexact.

Physical oceanographers calculate the rates of water circulation in the deep sea from density differences between adjacent water masses. Marine chemists use the concept of density to convert concentration units from mass per unit volume of seawater (e.g., mg/L) to mass per unit mass of seawater (e.g., mg/kg or ppm). Because density is not as easy to measure as salinity, oceanographers usually infer the density of a seawater sample from its salinity, using the empirical equation mentioned above. Since this information is frequently needed, tables have been constructed to provide the density of seawater as a function of in situ temperature and salinity. An example can be found in Appendix IX. In such tables, density is reported in a shorthand form, called

TABLE 3.3
Theoretical Calculation of the Density of 1000 g of Seawater with a Salinity of 35‰ and Temperature of 4°C

Substance	*Mass (g)*	*Density (g/cm³)*	*Computed Volume (mass/density, cm³)*
Water	965.00	1.0000	965.00
Salt	35.00	2.165[a]	16.17
Seawater	1000.00	1.0192[b]	981.17[c]

[a] For NaCl from 4 to 25°C.
[b] $\frac{\text{Mass}}{\text{Computed volume}} = (1000.00\ \text{g}/981.17\ \text{cm}^3)$.
[c] $\text{Volume}_{\text{water}} + \text{Volume}_{\text{salt}} = \text{Volume}_{\text{seawater}}$

sigma-t (σ_t), which is defined as

$$\sigma_t = (\rho - 1) \times 1000 \tag{3.5}$$

where ρ represents density in units of g/cm³.

If the salinity of a sample of seawater is 35‰ and its in situ temperature is 4°C, its density is

$$(27.81/1000) + 1 = 1.02781\ \text{g/cm}^3 \tag{3.6}$$

TABLE 3.4
Effect of Pressure on the Density of Seawater Having a Salinity of 35‰ and Temperature of 0°C

Pressure (db)	σ_t	*Percent Decrease in Volume*
0	28.13	0.000
100	28.60	0.046
200	29.08	0.093
500	30.50	0.231
1000	32.85	0.460
2000	37.47	0.909
3000	41.99	1.349
4000	46.40	1.778
5000	50.71	2.197
6000	54.94	2.609
7000	59.08	3.011
8000	63.14	3.406
9000	67.13	3.794
10,000	71.04	4.175

Source: From *Introduction to Marine Chemistry*, J. P. Riley and R. Chester, copyright © 1971 by Academic Press, Orlando, FL, p. 30. Reprinted by permission.

This value is about 1 percent higher than that predicted from summing the volumes of salt and water present in 1000 g of this seawater as shown in Table 3.3. The actual density is higher than the calculated value because water molecules cluster around salt ions forming pockets of denser solution. This process is called **electrostriction** and is illustrated in Figure 3.2.

Since seawater is nearly incompressible, pressure has a minor influence on the density of seawater. As shown in Table 3.4, the density of a water mass would be increased by 1.8 percent if it were lowered from the sea surface to 10,000 m. This effect is significant enough to impact the dynamic topography calculations of physical oceanographers. For this reason, empirical equations have also been developed relating density to pressure for the in situ temperatures and salinities commonly encountered in the deep sea.

THE CONSERVATIVE NATURE OF THE MAJOR IONS

The ions that compose the bulk of the dissolved salts are present in constant proportions to each other and to the total salt content of seawater. Their abundances relative to chloride are given in Table 3.5.

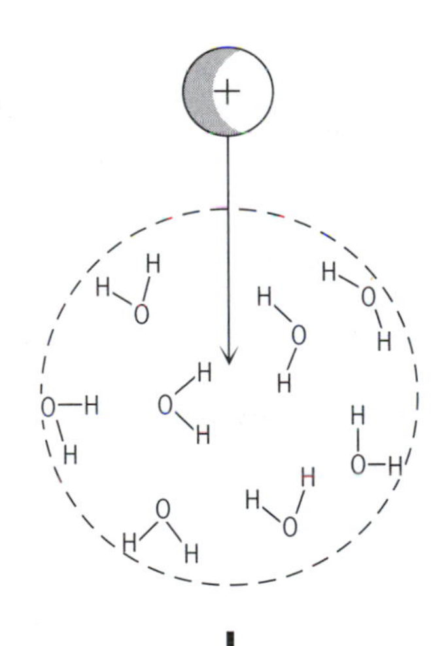

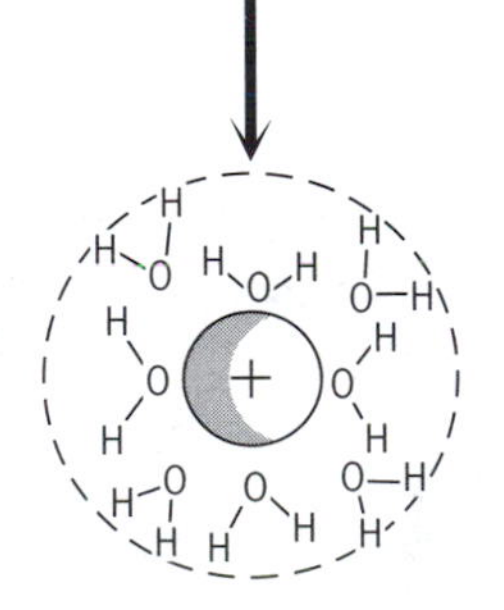

FIGURE 3.2. Electrostriction around a cation. *Source:* From *Survey of Progress in Chemistry*, vol. 4, R. A. Horne, copyright © 1968 by Academic Press, Orlando, FL, p. 15. Reprinted by permission.

TABLE 3.5
Composition of Sea Water

Ion	*g/kg of Water of Salinity 35‰*	*g/kg/Cl‰*
Chloride	19.344	—
Sodium	10.773	0.556
Sulfate	2.712	0.1400
Magnesium	1.294	0.0668
Calcium	0.412	0.02125
Potassium	0.399	0.02060
Bicarbonate	0.142	nonconservative
Bromide	0.0674	0.00348
Strontium	0.0079	0.00041
Boron	0.00445	0.00023
Fluoride	0.00128	6.67×10^{-5}

Source: Averaged results from F. Culkin and R. A. Cox, reprinted with permission from *Deep-Sea Research*, vol. 13, p. 801, copyright © 1966 by Pergamon Press, Elmsford, NY; and from J. P. Riley and M. Tongudai, reprinted with permission from *Chemical Geology*, vol. 2, p. 265, copyright © 1967 by Elsevier Science Publishers, Amsterdam, The Netherlands.

This constancy in relative ion concentration is known as **Marcet's Principle** or the **Rule of Constant Proportions**. Formally stated, it says that "regardless of how the salinity may vary from place to place, the ratios between the amounts of the major ions in the waters of the open ocean are nearly constant."

These proportions are constant because the rate at which water is moved through and within the ocean is much faster than any of the chemical processes that act to remove or supply the major ions. Since adding or removing water does not change the total amount of salt in the ocean, only the concentrations of the ions, and hence salinity, are altered. This is termed conservative behavior. The relatively slow chemical processes that determine the ion proportions and the total amount of salt in the ocean are described in Chapter 21.

The types of water transport that can alter the concentrations but not the relative abundances of the major ions are listed in Table 3.6. Note that these are all physical phenomena. In other words, the marine distributions of conservative substances are controlled by physical, rather than chemical, processes.

Pure liquid water is added to the ocean through rainfall (**precipitation**) and the melting of ice. The addition of pure water to seawater lowers its salinity. Since the concentrations of all the solutes are decreased by the same amount, the ion ratios remain constant. Likewise, the removal of water from the ocean, through the processes of evaporation and freezing, increases the salinity of

TABLE 3.6
Physical Factors That Can Alter the Salinity of Seawater

Evaporation
Precipitation
Freezing
Thawing
Molecular diffusion of ions between water masses of different salinity
Turbulent mixing between water masses of different salinity
Water-mass advection

seawater without altering the ion ratios. (This is not strictly true as small droplets of brine, enriched in some ions, can become occluded in the crystal lattice of sea ice.)

Table 3.6 contains one physical process that involves the movement of ions rather than water (i.e., **molecular diffusion**). As shown in Figure 3.3, solutes experience net transport when nonuniformly distributed in a solution. In the absence of external forces, solutes spontaneously undergo net diffusion from regions of higher concentration to lower concentration. This continues until a homogeneous distribution of the solute is achieved.

This net transfer of solutes is an example of a diffusive flux. As illustrated in Figure 3.4, a diffusive flux can be thought of as the amount of particles that move through a unit area (1 cm^2) in a unit time (1 s). If the concentration of the solute is expressed in the SI units of g/cm^3, then the units of diffusive flux of the solute are $g\ cm^{-2}\ s^{-1}$.

In the case where the solute is at steady state, its concentration gradient and flux are constant over time. Under these conditions, the diffusive flux of a solute C is given by Fick's First Law:

$$\text{Flux} = -D_c \frac{d[C]}{dz} = D_c \frac{([C]_{\text{high}} - [C]_{\text{low}})}{\Delta z} \tag{3.7}$$

which states that the flux is directly proportional to the concentration gradient, $d[C]/dz$, where z is the dimension along which the net flux takes place. This

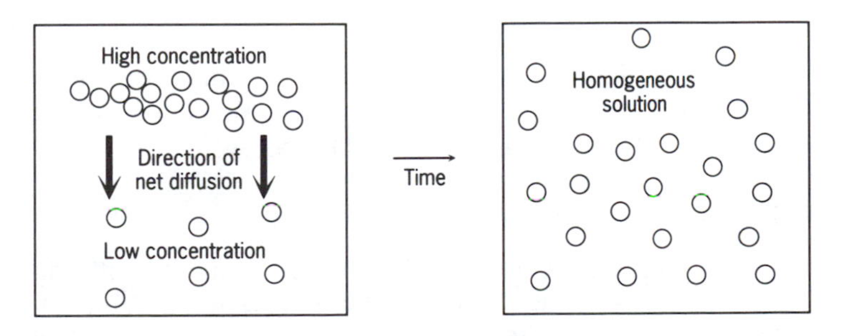

FIGURE 3.3. The process of molecular diffusion.

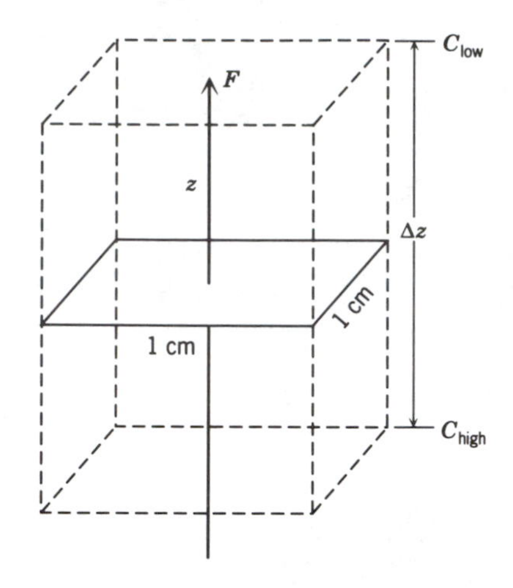

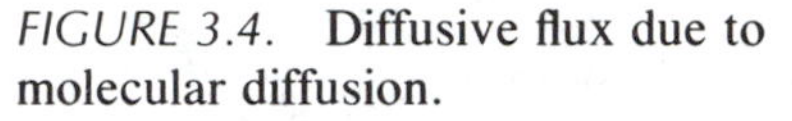

FIGURE 3.4. Diffusive flux due to molecular diffusion.

gradient can be approximated by the ratio of the concentration difference ($[C]_{high} - [C]_{low}$) to the distance, Δz, across which the concentration difference exists. Thus the greater the concentration difference or gradient, the larger the flux.

For solutes, the constant of proportionality, D_c, is called the molecular diffusivity coefficient. It is a function of solution temperature, as well as the radius and charge of the ionic solutes. In seawater, D_c ranges from 2×10^{-5} to 5×10^{-5} cm^2/s. As a result, the rate at which water molecules are transported in the ocean is usually much faster than the rate at which ions are moved by the process of molecular diffusion. Thus molecular diffusion is the dominant control on ionic distributions only in locations where water motion is relatively slow, such that the ions diffuse at slightly different rates.

The other physical processes that affect the salinity of seawater are the result of **water-mass advection** and **turbulent mixing**. These topics are the subject of Chapter 4.

EXCEPTIONS TO THE RULE OF CONSTANT PROPORTIONS

As noted earlier and shown in Table 3.7, small deviations in ion proportions have been observed in some locations. Some causes of exceptions to the Rule of Constant Proportions are described below.

Marginal Seas and Estuaries

The ion proportions in most river water is significantly different from that in seawater. As a result, **river runoff** can have a local impact on the ion ratios in coastal waters. This effect is most pronounced in marginal seas and estuaries where mixing with the open ocean is restricted and river input is relatively

TABLE 3.7
Major Constituent-to-Chlorinity Ratios for Various Oceans and Seas

Ocean or Sea	Na^+ / ‰ *Cl*	Mg^{2+} / ‰ *Cl*	K^+ / ‰ *Cl*	Ca^{2+} / ‰ *Cl*	Sr^{2+} / ‰ *Cl*	SO_4^{2-} / ‰ *Cl*	Br^- / ‰ *Cl*
N. Atlantic	—	—	0.02026	—	—	—	0.00337–0.00341
Atlantic	0.5544–0.5567	0.0667	0.01953–0.0263	0.02122–0.02126	0.000420	0.1393	0.00325–0.0038
N. Pacific	0.5553	0.06632–0.06695	0.02096	0.02154	—	0.1396–0.1397	0.00348
W. Pacific	0.5497–0.5561	0.06627–0.0676	0.02125	0.02058–0.02128	0.000413–0.000420	0.1399	0.0033
Indian	—	—	—	0.02099	0.000445	0.1399	0.0038
Mediterranean	0.5310–0.5528	0.06785	0.02008	—	—	0.1396	0.0034–0.0038
Baltic	0.5536	0.06693	—	0.02156	—	0.1414	0.00316–0.00344
Black	0.55184	—	0.0210	—	—	—	—
Irish	0.5573	—	—	—	—	0.1397	0.0033
Puget Sound	0.5495–0.5562	—	0.0191	—	—	—	—
Siberian	0.5484	—	0.0211	—	—	—	—
Antarctic	—	—	—	0.02120	0.000467	—	0.00347
Tokyo Bay	—	0.0676	—	0.02130	—	0.1394	—
Barents	—	0.06742	—	0.02085	—	—	—
Arctic	—	—	—	—	0.000424	—	—
Red	—	—	—	—	—	0.1395	0.0043
Japan	—	—	—	—	—	—	0.00327–0.00347
Bering	—	—	—	—	—	—	0.00341
Adriatic	—	—	—	—	—	—	0.00341

Source: From F. Culkin and R. A. Cox, reprinted with permission from *Deep-Sea Research*, vol. 13, p. 801, copyright © 1966 by Pergamon Press, Elmsford, NY.

large. The variable composition of river water and its impact on the chemical composition of seawater is discussed further in Chapter 21.

The impact of river input on the major ion ratios in seawater was not well understood at the time that the first analytical definitions of salinity were developed. As a matter of geographic convenience, water from the Baltic Sea was initially chosen as a salinity standard through which all laboratories could intercalibrate. A quick glance at Table 3.7 indicates that of all the areas listed, the Baltic Sea contains the major ions in ratios that deviate most widely from the average values given in Table 3.5. By the time this was recognized, a great deal of salinity data had been collected. Separate equations relating ocean salinity, chlorinity, and density to the peculiar composition of Baltic Sea water had to be developed to salvage that data.

Anoxic Basins

Deviations in the SO_4^{2-} ion ratios have also been observed in coastal areas, particularly in the sediments. This effect is due to bacterial reduction of sulfate to sulfide that occurs in waters devoid of dissolved oxygen. Environmental conditions that contribute to the depletion of dissolved oxygen include restricted water circulation and high rates of organic matter supply. This subject is discussed in Chapter 8.

Formation of Sea Ice

Small amounts of salt are commonly occluded in sea ice. Not all of the ions are incorporated to the same degree. This alters the ion ratios in the remaining brine, leading to deviations from the Rule of Constant Proportions.

Deposition of Calcareous Shells

Small deviations in the Ca^{2+} ion ratios are caused by the formation and dissolution of calcareous sea shells ($CaCO_3$). Despite the great abundance of shell-forming organisms, their activities produce a maximum variation of only 1 percent in Ca^{2+} ion ratios. The marine chemistry of carbonates is discussed in Chapter 15.

Hydrothermal Vents

Hydrothermal vents are another source of water entering the ocean. The vents are submarine hot-water geysers that are part of seafloor spreading centers. The hydrothermal fluids contain some major ions, such as magnesium and sulfate, in significantly different ratios than found in seawater. The importance of these areas in determining the chemical composition of the ocean is described in Chapter 19.

Evaporite Formation

Salt is removed from the sea when evaporation raises the major ion concentrations to levels that exceed the solubility of minerals such as NaCl (halite),

$CaCO_3$ (limestone), $CaMg(CO_3)_2$ (dolomite), and $CaSO_4 \cdot 2H_2O$ (anhydrite). These precipitates are deposited in stages, so the ions are removed at different times and rates. This causes the ion ratios in the remaining brine to deviate from those of average seawater. The formation of evaporites requires climatological conditions leading to high evaporation rates and restricted mixing with the open ocean. Because of the latter, most evaporites deposit in shallow-water environments. Evaporites represent a very large sedimentary reservoir of the major ions. Thus evaporite formation and dissolution play a major role in the marine geochemical cycle of the major ions. This subject is discussed in greater detail in Chapters 17 and 21.

Air–Sea Interface

The major ions are transported across the air–sea interface by the ejection of water droplets from the sea surface. Water turbulence at the sea surface causes microscopic bubbling. Some of these bubbles burst, ejecting seawater into the atmosphere. Since not all of the salt ions are ejected to the same degree, bursting bubbles can alter the ion ratios in the remaining water.

Once in the atmosphere, the water evaporates and some of the sea salt falls back to the sea surface. The rest is transported considerable distances by winds until it is washed out of the atmosphere by rainfall. The salts that are transported back to the continents by this process are termed cyclic salts. After having been rained out onto the continents, the salts are carried back into the ocean by river runoff. On short time scales, the marine geochemical cycles of chlorine and sodium are dominated by this process.

Interstitial Waters

A significant amount of seawater is trapped in the open spaces that exist between the particles in marine sediments. This fluid is termed pore, or interstitial, water. Marine sediments are the site of many chemical reactions, such as sulfate reduction, as well as mineral precipitation and dissolution, any of which can alter the major ion ratios. As a result, the chemical composition of pore water is usually quite different from that of seawater. The chemistry of marine sediments is the subject of Part 3.

SUMMARY

The chemical composition of seawater is unique as compared to other natural waters. First, it contains very high concentrations of solutes. Second, its chemical composition is not very geographically variable. Six inorganic salt ions compose more than 99% of the dissolved substances present in seawater. In order of abundance, the major ions are Cl^-, Na^+, Mg^{2+}, SO_4^{2-}, Ca^{2+}, and K^+. These ions are said to exhibit **conservative** behavior because their distributions in the sea are largely controlled by physical processes. This is due to their relatively slow rates of chemical reaction in seawater. The physical processes that affect the absolute concentrations of these ions include **precipitation**, evaporation, freezing, thawing, **molecular diffusion** of ions between water masses, **turbulent mixing** between water masses and **advection** of water masses.

These physical processes do not affect the relative proportions of the ions. Thus their absolute concentrations may change, but their relative ratios remain constant. This is a statement of **Marcet's Principle**, which is also called the **Rule of Constant Proportions**. Small deviations occur mostly in coastal areas due to such processes as **river runoff**, sulfate reduction, and evaporite formation.

Because the major ions are present in constant proportions, their concentrations are directly proportional to the total salt concentration, or **salinity**. Salinity is measured conductimetrically using a **salinometer**. The results are reported in parts per thousand (i.e., the total grams of dissolved major ions present in 1 kg of seawater). In the open ocean, the salinity of seawater ranges from 33‰ to 37‰. The salinity of seawater can also be inferred from its **chlorinity**. Chlorinity is measured titrimetrically by precipitating the halides as silver salts. The results are reported in parts per thousand (i.e., total grams of halides expressed as chloride present in 1 kg of seawater). The **density** of seawater can also be inferred from its salinity using an empirical **equation of state for seawater**. The density of seawater increases with decreasing temperature and increasing pressure and salinity. The latter effect is enhanced by the phenomenon of **electrostriction**. The units of density are g/cm^3, though a shorthand form, called **sigma-t**, is more commonly used by oceanographers.

Though most solutes in seawater are **nonconservative**, they constitute a relatively small fraction of the substances present in seawater. These substances are nonconservative because their distributions are controlled by both physical and chemical processes that occur in the ocean. The major classes of such compounds include the minor ions, gases, **nutrients, trace metals, particulate matter, colloids**, and organic compounds.

CHAPTER 4

SALINITY AS A CONSERVATIVE TRACER

INTRODUCTION

Local variations in the rates of water input and removal cause geographic and temporal variations in salinity. For example, during rainy seasons, increased rates of river runoff lower the salinity in estuaries. In the open ocean, latitudinal variations in the relative rates of precipitation and evaporation control the salinity of surface seawater.

The rates at which water is transported through the hydrological cycle are strongly influenced by the global heat cycle. In this chapter, the factors that control local variations in the global heat and water cycles are discussed. This information is used to explain the horizontal and vertical variations in salinity that occur in the open ocean.

Because of its conservative behavior, salinity can be used as a tracer of water motion. Salinity has also proven useful in studying the biogeochemical cycling of nonconservative substances. Both applications are discussed below.

GLOBAL HEAT AND WATER BALANCE

Solar radiation is a source of heat to the earth. Some of this incoming radiation is absorbed and some is reflected by the atmosphere. The remainder strikes the earth's surface and is referred to as **insolation**. Insolation that strikes the sea surface is either reflected or absorbed by water molecules. The latter supplies energy that can cause water to evaporate, thereby causing salinity to increase.

Insolation is not constant; it varies geographically and over time. For example, insolation decreases with increasing latitude as the angle of the sun's rays increases. This effect is shown in Figure 4.1. Since the sun is at it highest angle (90°) over the equator, this location receives the most insolation. The effect of the sun's angle is threefold. At low latitudes, a ray of solar radiation is spread over a smaller surface area. The ray also passes through a smaller thickness of atmosphere. Lastly, the higher the sun's angle, the less insolation is reflected from Earth's surface.

The geographic variation in insolation causes the temperature of the surface

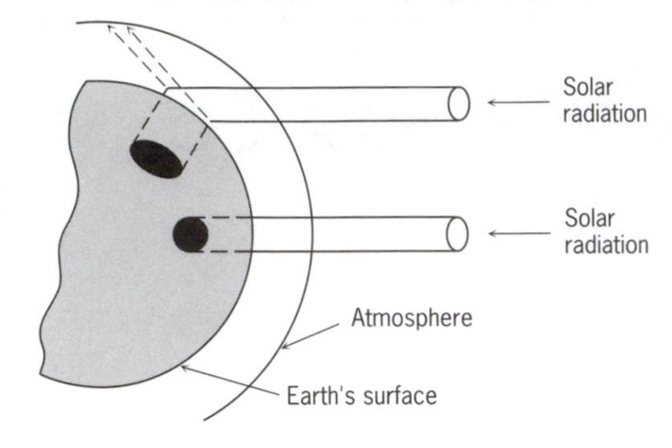

FIGURE 4.1. Latitudinal variations in insolation caused by differences in the angle at which the sun's radiation strikes Earth's surface. *Source:* From *Introductory Oceanography*, 5th ed., H. V. Thurman, copyright © 1988 by Merrill Publishing Company, Columbus, OH, p. 170. Reprinted by permission.

water to increase with decreasing latitude, as shown in Figure 4.2. This temperature distribution varies over time as a result of changes in insolation. Daily changes are caused by Earth's rotation. Seasonal fluctuations are the result of Earth's revolution around the sun and the tilt of its orbital axis, as illustrated in Figure 4.3. For a given hemisphere, insolation is greatest during summer, as this is the period during which that part of the planet is closest to the sun. Insolation is decreased by cloud cover and increasing sea state. The choppier the water's surface, the greater the percentage of insolation reflected due to the increased surface area.

The heat content of the atmosphere and ocean remains constant over time despite short-term, local variations in insolation. This steady state is maintained by the net transport of heat from low to high latitudes. Western boundary currents carry relatively warm water from low latitudes to the polar regions. At these latitudes, net heat loss is achieved through radiation to the atmosphere. The cooled water is returned to low latitudes by the eastern boundary currents.

Latitudinal variations in insolation give rise to convection currents in the atmosphere and ocean. In the atmosphere, these air currents form the major wind bands that are the driving forces behind geostrophic circulation. Convection currents in the ocean cause thermohaline circulation. Thus insolation affects temperature throughout the entire water column. But as shown in Figure 4.4, this influence is latitudinally variable.

In the mid and low latitudes, these profiles are characterized by three temperature regimes. At the sea surface lies a homogeneous zone called the **mixed layer**. The temperature and depth of this zone are controlled by local insolation and wind mixing. Vertical mixing induced by winds transports heat to an average maximum depth of 300 m. Below the mixed layer, temperature decreases with increasing depth. This zone is called the **thermocline**. Below approximately 1000 m, water temperatures are constant with depth. This thermally homogeneous region is called the **deep zone**.

The difference in temperature between the deep zone and the mixed layer decreases with increasing latitude. In the most extreme case, (i.e., polar regions), the water column is nearly isothermal. This is due to intense cooling at the sea surface, accompanied by strong wind mixing. If cooling increases

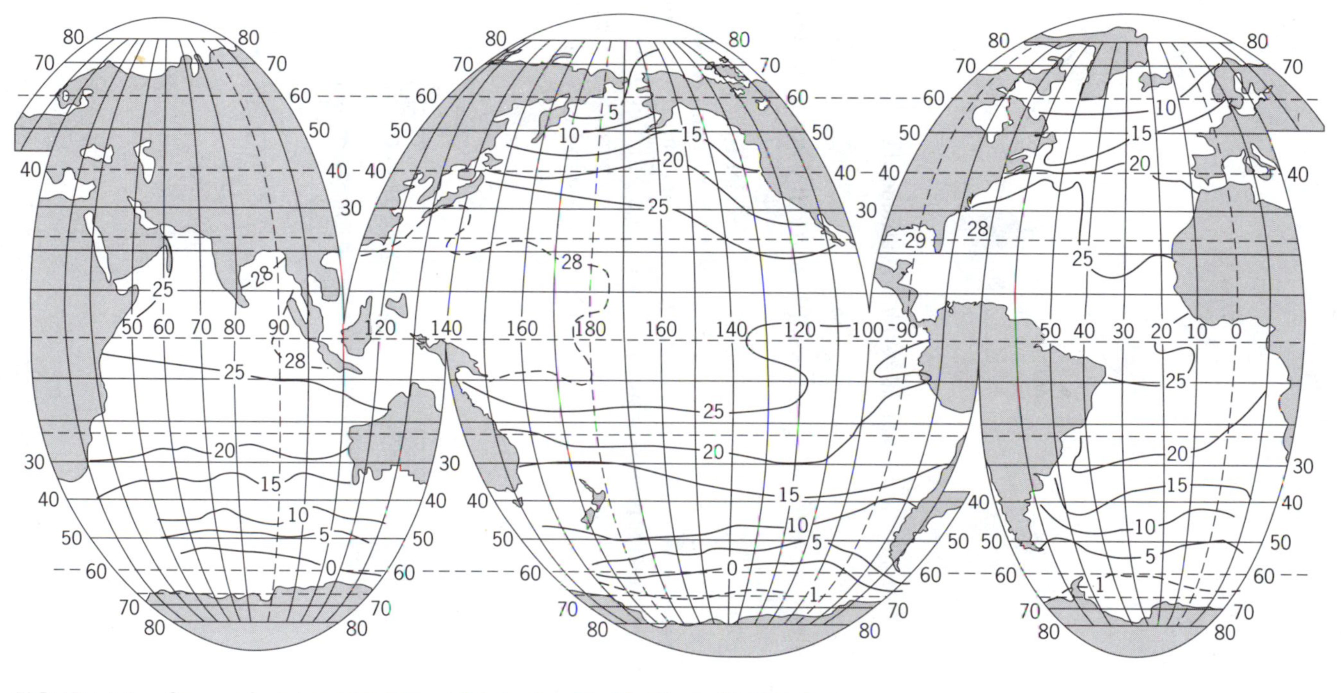

FIGURE 4.2. Sea-surface temperatures (°C) during the Northern Hemisphere summer. *Source:* From *Introductory Oceanography,* 5th ed., H. V. Thurman, copyright © 1988 by Merrill Publishing Company, Columbus, OH, p. 200. Based on data from *The Oceans,* H. U. Sverdrup, M. W. Johnson, and R. H. Fleming, copyright © 1941 by Prentice Hall, Inc., Englewood Cliffs, NJ, p. 230. Reprinted by permission.

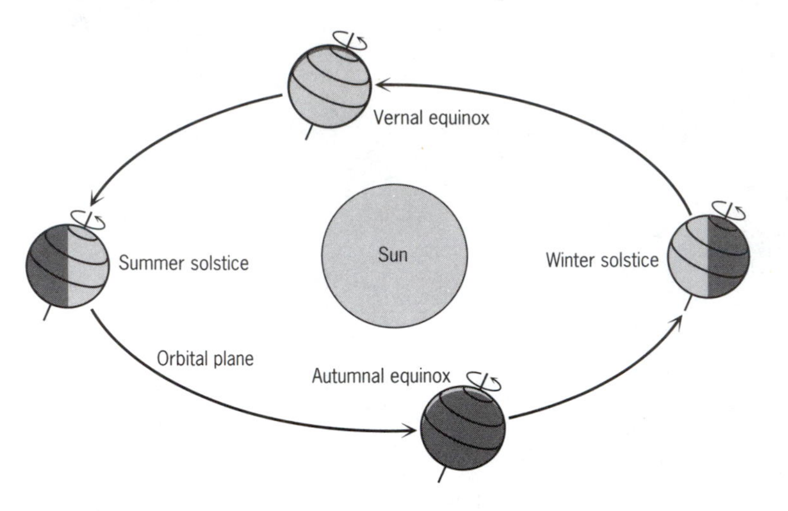

FIGURE 4.3. Earth's annual revolution around the sun.

the density of the water mass sufficiently, it will sink and engage in thermohaline circulation.

The thermocline is steepest at low latitudes due to high surface-water temperatures. Seasonal variations in the depth and steepness of the thermocline occur at mid latitudes. During the summer, increased insolation elevates the surface-water temperatures producing the summer thermocline shown in Figure 4.4. Winter storms destroy the summer thermocline by cooling the surface water and deepening the mixed layer via wind mixing.

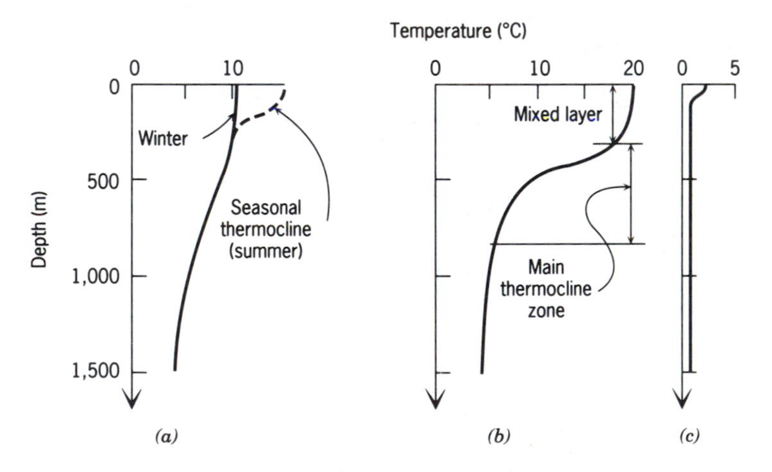

FIGURE 4.4. Average temperature profiles for the open ocean at (*a*) mid latitudes, (*b*) low latitudes, and (*c*) high latitudes. *Source:* From *Oceanography: An Introduction*, 4th ed., D. E. Ingmanson and W. J. Wallace, copyright © 1989 by Wadsworth, Inc., Belmont, CA, p. 106. Reprinted by permission.

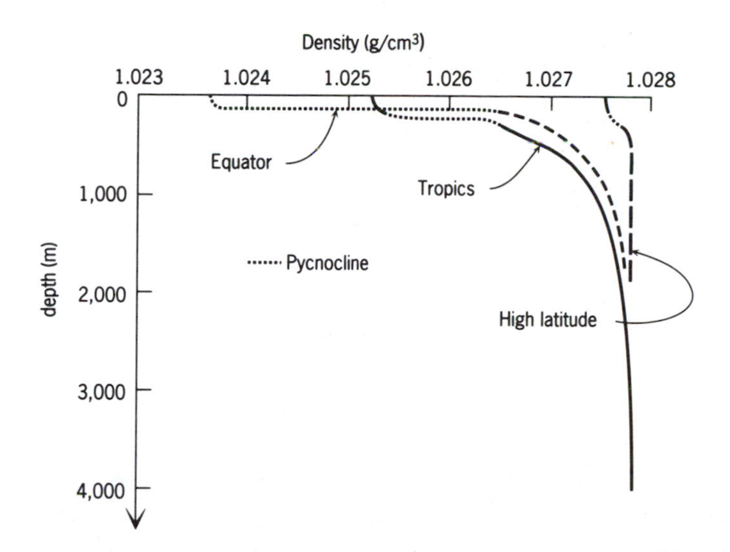

FIGURE 4.5. Typical density profiles of ocean water. From *Oceanography: An Introduction,* 4th ed., D. E. Ingmanson and W. J. Wallace, copyright © 1989 by Wadsworth, Inc., Belmont, CA, p. 106. Reprinted by permission.

Due to the force of gravity, the density of seawater increases with increasing depth. As shown in Figure 4.5, the vertical profiles of density are mirror images of the temperature profiles. This is due to the strong influence of temperature on the vertical variations in density. At low and mid latitudes, the thermocline produces a strong density gradient called the **pycnocline**. This **density stratification** inhibits vertical mixing. In comparison, little energy is required to induce vertical mixing at high latitudes due to the absence of a strong pycnocline.

TEMPORAL AND SPATIAL VARIATIONS IN SALINITY

Salinity is most variable in coastal regions due to the influence of river and ground-water runoff. In the open ocean, surface-water salinities are largely controlled by the relative balance between water loss through evaporation and water gain through precipitation. Latitudinal variations in insolation give rise to variations in the rates of these processes. As shown in Figure 4.6, low and high latitudes are characterized by excess precipitation. Mid latitudes experience net evaporative loss of water. As a result, the salinity of surface seawater is highest at mid latitudes, as shown in Figure 4.7.

As with temperature, the vertical profiles of salinity vary with latitude. Unlike the temperature profiles, salinity does not follow uniform depth trends. As shown in Figure 4.8, the deep waters tend to have lower salinities than the surface waters at mid and low latitudes. This reflects the relatively low salinity of surface water at polar regions, which is the ultimate source of these deep-water masses.

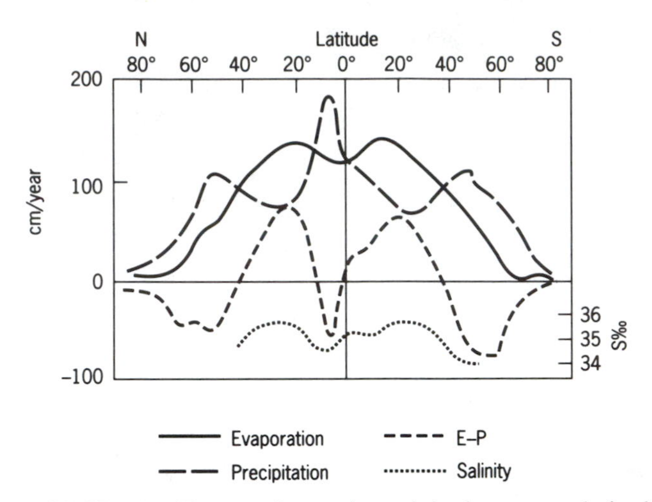

FIGURE 4.6. Evaporation and precipitation versus latitude. *Source:* From *Oceanography: An Introduction,* 4th ed., D. E. Ingmanson and W. J. Wallace, copyright © 1989 by Wadsworth, Inc., Belmont, CA, p. 94. Reprinted by permission.

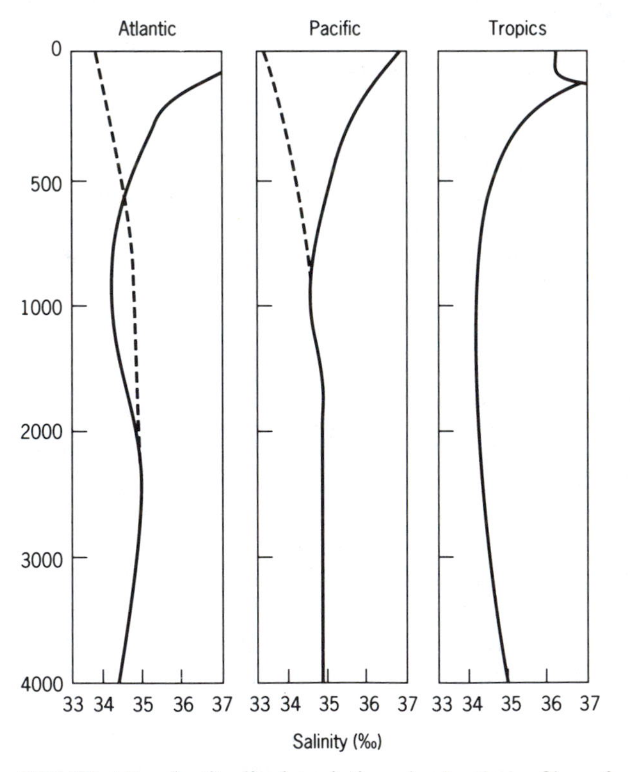

FIGURE 4.7. Latitudinal variations in depth profiles of salinity. High latitude salinities are given by the dashed lines. *Source:* From *Oceanography: An Introduction,* 4th ed., D. E. Ingmanson and W. J. Wallace, copyright © 1989 by Wadsworth, Inc., Belmont, CA, p. 94. Reprinted by permission.

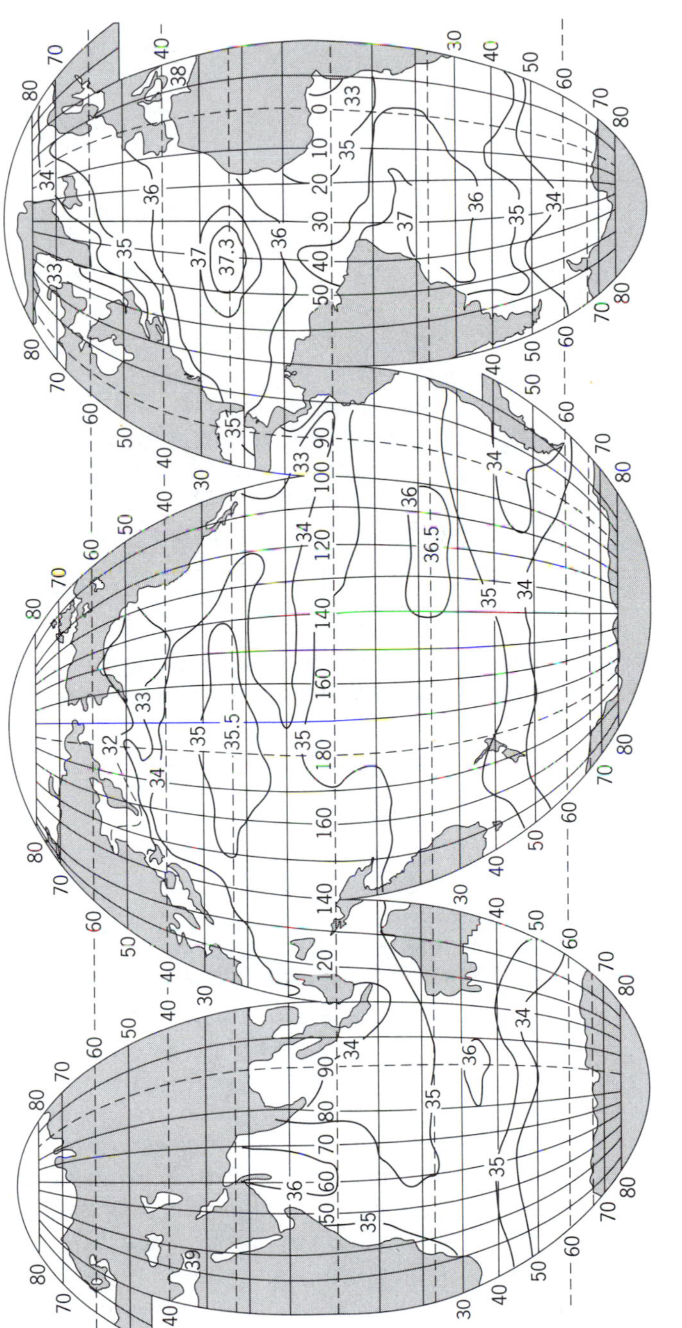

FIGURE 4.8. Salinity of surface seawater in the world's oceans. *Source:* From *Introductory Oceanography*, 5th ed., H. V. Thurman, copyright © 1988 by Merrill Publishing Company, Columbus, OH, p. 201. Based on data from *The Oceans*, H. U. Sverdrup, M. W. Johnson, and R. H. Fleming, copyright © 1941 by Prentice Hall, Inc., Englewood Cliffs, NJ, p. 230. Reprinted by permission.

TRANSPORT OF HEAT AND SALT VIA WATER MOVEMENT

How Water Moves

Water motion in the ocean is the result of two processes, advection and turbulence. Advection causes water to experience large-scale net displacement, whereas turbulence involves the random motion of water molecules. Most advective transport in the ocean is the result of currents. The geostrophic currents move surface water at velocities of 1 to 10 cm/s. In comparison, thermohaline currents transport deep water at velocities ranging from 0.1 to 0.01 cm/s.

Advection can move water both horizontally and vertically, such as during thermohaline circulation. This cyclic flow of water is initiated by the cooling of surface seawater in polar regions. The decrease in temperature increases the density of the water. The resultant water mass sinks until it reaches a level of matching density. Such convection currents are an example of the vertical advection of water. Continued sinking causes the deep water to be pushed horizontally along the depth of matching density. The resultant deep-water current is an example of horizontal advection.

Water masses of lower density are produced by sinking at subpolar latitudes. These are termed intermediate waters as they do not sink as deeply as the deep and bottom water masses. The temperature and salinity signatures of the major water masses are given in Table 4.1.

Advection is much faster than molecular diffusion and turbulence, so water masses retain the temperature and salinity acquired when they were last at the sea surface. Thus the pathway of thermohaline circulation can be traced by following the unique temperature and salinity signatures of the deep-water masses. This is called the core technique of water mass tracing, and an example is given in Figure 4.9. Eventually, turbulent mixing between adjacent water masses causes the loss of these unique temperature and salinity signatures.

Turbulent mixing is inhibited by density gradients, such as the pycnocline. Since vertical density stratification is stronger, horizontal turbulence can operate over much longer space scales than vertical turbulence. The Gulf Stream Rings are an example of lateral turbulence. For most horizontal ocean currents, it is not technically possible to distinguish water motion due to lateral turbulence from that caused by horizontal advection.

Vertical turbulence is also referred to as eddy diffusion. As the name suggests, this process is similar to molecular diffusion. In the absence of a concentration gradient, both involve the random motion of particles. In some situations, eddy diffusion is strong enough to break down vertical density stratification by homogenizing temperature and salinity gradients. Energy to fuel this process is supplied in the form of winds at the sea surface, current shear between adjacent water masses, and bottom friction at the sediment–water interface. Nevertheless, turbulent mixing is usually 10^3 to 10^5 times slower than advective transport.

TABLE 4.1
Major Water Masses of the World Ocean

Water mass	*Temperature (°C)*	*Salinity (‰)*
Central water masses		
N. Atlantic water (NAC)	8–19	35.1–36.5
S. Atlantic water (SAC)	6–17	34.7–36.0
W. North Pacific water (NPC)	6–18	34.0–34.9
W. South Pacific water (SPC)	10–17	34.5–35.6
Indian water (IC)	7–16	34.5–35.6
High-latitude surface water masses		
Atlantic subarctic water	4–5	34.6–34.7
Pacific subarctic water	3–6	33.5–34.4
Subantartic water	3–10	33.9–34.7
Antarctic circumpolar water	0–2	34.6–34.7
Intermediate water masses		
Arctic intermediate water (NAI)	3–5	34.7–34.9
N. Pacific intermediate water (NPI)	4–10	34.0–34.5
Antarctic intermediate water (AI)	3–7	33.8–34.7
Mediterranean intermediate water (MI)	6–12	35.3–36.5
Red Sea intermediate water (RSI)	8–12	35.1–35.7
Deep and bottom water masses		
N. Atlantic deep and bottom water (NAD and B)	2–4	34.8–35.1
Antarctic bottom water (AB)	−0.4	34.7

Source: From *The World Ocean: An Introduction to Oceanography*, W. A. Anikouchine and R. W. Sternberg, copyright © 1981 by Prentice Hall, Inc., Englewood Cliffs, NJ, p. 219. Reprinted by permission. After *The Oceans*, H. U. Sverdrup, M. W. Johnson, and R. H. Fleming, copyright © 1941 by Prentice Hall, Inc., Englewood Cliffs, NJ, p. 741. Reprinted by permission.

Since advection and turbulence move seawater, both can alter salinity without affecting the ratios of the major ions. For example, the advection of a saline water mass into a body of low-salinity water will displace the original water mass, thereby increasing the salinity of the water at that location. Turbulence can also alter salinity if it induces mixing between adjacent water masses of different salinity. As shown in Figure 4.10, the complete mixture of equal volumes of two such water masses produces a new water mass of intermediate salinity. If this turbulent mixing occurs below the sea surface, temperature will also be conserved.

Such mixing occurs at the interface between adjacent water masses as they move along the path of thermohaline circulation. The mixing creates temperature and salinity gradients at the boundary between the water masses. Since both temperature and salinity are conserved during this mixing process,

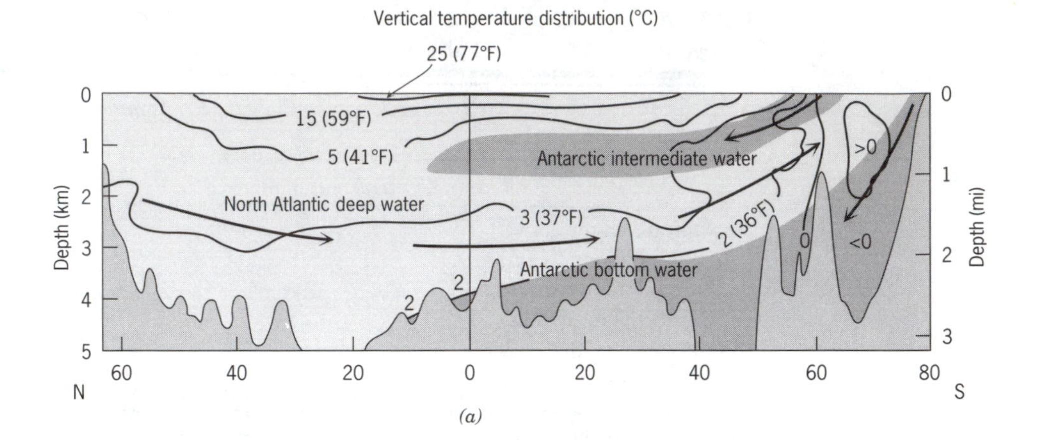

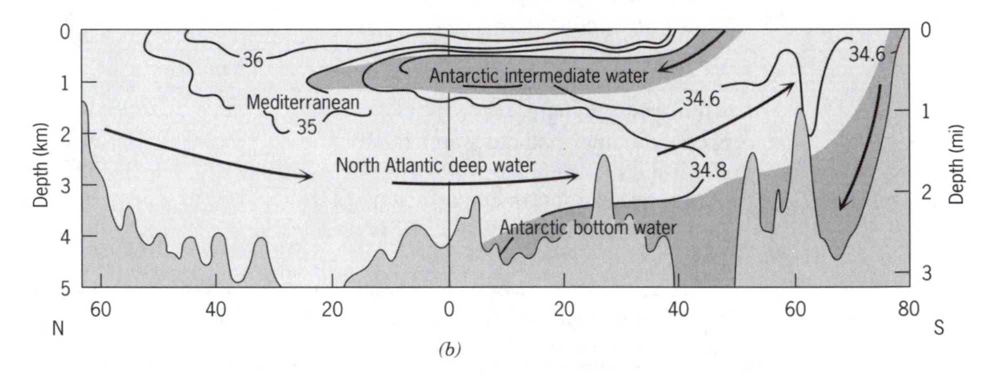

FIGURE 4.9. Atlantic ocean subsurface circulation as illustrated by (*a*) temperature and (*b*) salinity gradients. The cores of the water masses are labeled. *Source:* From *Introductory Oceanography,* 5th ed., H. V. Thurman, copyright © 1988 by Merrill Publishing Company, Columbus, OH, p. 208. Based on data from *The Oceans,* H. U. Sverdrup, M. W. Johnson, and R. H. Fleming, copyright © 1941 by Prentice Hall, Inc., Englewood Cliffs, NJ, p. 230. Reprinted by permission.

they vary linearly, as shown in Figure 4.11*a*. Nonlinear T–S diagrams (Figure 4.11*b*) are produced by the admixture of more than two water masses or by nonconservative behavior. As discussed in Chapter 3, deviations from conservative behavior are relatively small and largely restricted to the sea surface and coastal regions. Thus in the absence of nonconservative behavior, T–S diagrams can be used to identify the sources of most water masses.

Advection–Diffusion Equation

Beneath the sea surface, the concentrations of the major ions are controlled by turbulence and advection over time scales on the order of the mixing time of the ocean (1000 y). A mathematical model has been developed to describe

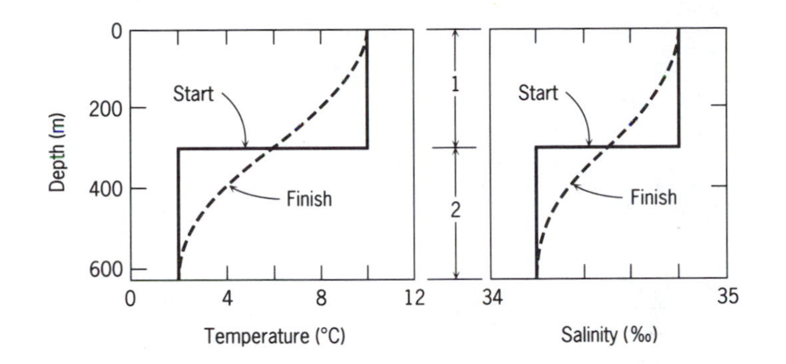

FIGURE 4.10. Conservative mixing of water masses. *Source:* From *Oceanography: A View of the Earth,* 4th ed., M. G. Gross, copyright © 1987 by Prentice Hall, Inc., Englewood Cliffs, NJ, p. 169. Reprinted by permission.

the distribution of the conservative ions as a function of these processes. This model describes how the concentration of a conservative solute in an infinitesimally small volume of the ocean can be altered by the transport of seawater through that volume. This transport occurs via advection and turbulence, which moves water through the x, y, and z dimensions as shown in Figure 4.12.

The rate of change in concentration of a conservative solute, C, at some fixed point, x, which is caused by turbulent mixing is given by Fick's Second Law:

$$\frac{\partial[C]}{\partial t} = D_x\left[\frac{\partial}{\partial x}\left(\frac{\partial[C]}{\partial x}\right)\right] = D_x\left[\frac{\partial^2[C]}{\partial x^2}\right] \tag{4.1}$$

where D_x is the turbulent mixing coefficient for water motion in the x direction.

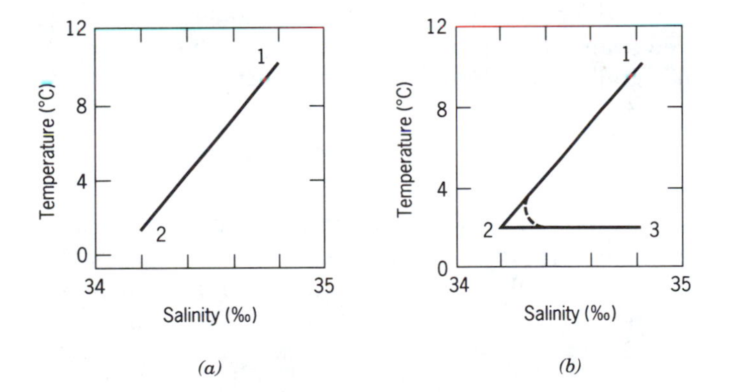

FIGURE 4.11. T–S Diagram indicating the presence of (*a*) two water masses and (*b*) multiple water masses. From *Oceanography: A View of the Earth,* 4th ed., M. G. Gross, copyright © 1987 by Prentice Hall, Inc., Englewood Cliffs, NJ, p. 169. Reprinted by permission.

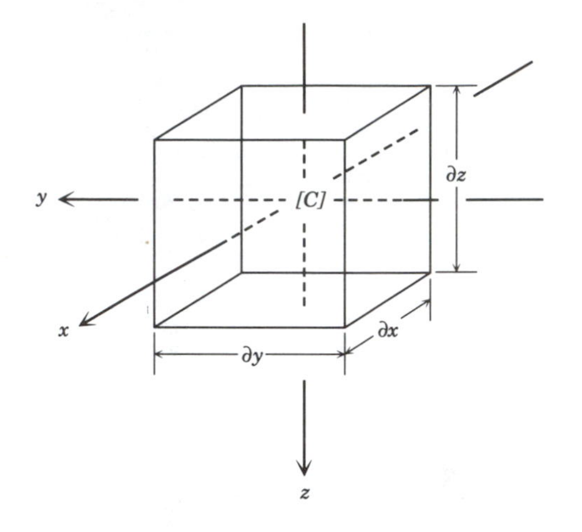

FIGURE 4.12. Directions of seawater transport through an infinitesimally small volume of the ocean. The spatial frame of reference is fixed in place, while advection and turbulence move seawater through the box. This can cause the concentration of a solute, C, to change over time.

For D_x to be a constant, the rate of turbulent mixing in the x direction must be constant over time. If not, a more complicated form of Fick's Second Law must be used.

Equations similar to 4.1 can be written for turbulent mixing in the y and z dimensions. Thus, the complete contribution of turbulent transport to the rate at which the solute concentration changes is given by

$$D_x\left[\frac{\partial^2[C]}{\partial x^2}\right] + D_y\left[\frac{\partial^2[C]}{\partial y^2}\right] + D_z\left[\frac{\partial^2[C]}{\partial z^2}\right] \tag{4.2}$$

As shown in Figure 4.13, turbulent mixing coefficients are much larger than those of molecular diffusion. As a result, the effects of turbulence on the distribution of a solute in solution almost always swamp those of molecular diffusion. The only exceptions occur in locations where water flow is restricted, such as in the pore waters of marine sediments.

The change in $[C]$ caused by the advective transport of seawater along the x dimension is given by

$$\frac{\partial[C]}{\partial t} = -v_x\left[\frac{\partial[C]}{\partial x}\right] \tag{4.3}$$

where v_x is the rate of water advection in the x direction, which is assumed to be constant over time. By analogy, the rate of concentration change in the infinitesimally small volume due to advection through the x, y, and z dimensions is given by

$$-v_x\left[\frac{\partial[C]}{\partial x}\right] - v_y\left[\frac{\partial[C]}{\partial y}\right] - v_z\left[\frac{\partial[C]}{\partial z}\right] \tag{4.4}$$

Summing the effects of three-dimensional turbulent and advective transport yields

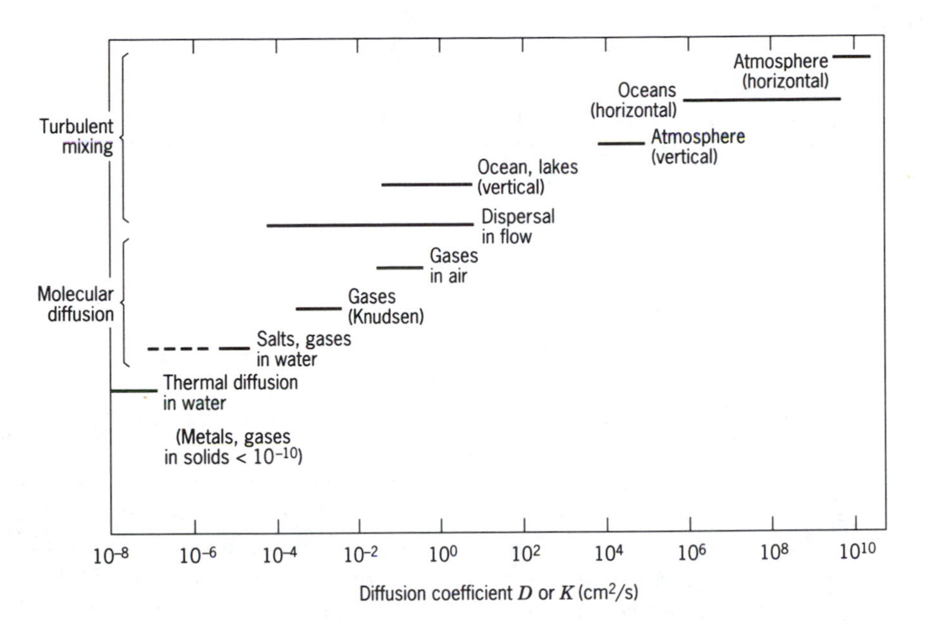

FIGURE 4.13. Range of molecular diffusivity and turbulent mixing coefficients in natural environments. *Source:* From *Geochemical Processes: Water and Sediment Environments,* A. Lerman, copyright © 1979 by John Wiley & Sons, Inc., New York, p. 57. Reprinted by permission.

$$\frac{\partial[C]}{\partial t} = D_x\left[\frac{\partial^2[C]}{\partial x^2}\right] + D_y\left[\frac{\partial^2[C]}{\partial y^2}\right] + D_z\left[\frac{\partial^2[C]}{\partial z^2}\right] - v_x\left[\frac{\partial[C]}{\partial x}\right] - v_y\left[\frac{\partial[C]}{\partial y}\right] - v_z\left[\frac{\partial[C]}{\partial z}\right] \tag{4.5}$$

For the steady-state condition, the concentration of the solute does not change over time. In this case, $\partial[C]/\partial t = 0$, so Eq. 4.5 becomes

$$0 = D_x\left[\frac{\partial^2[C]}{\partial x^2}\right] + D_y\left[\frac{\partial^2[C]}{\partial y^2}\right] + D_z\left[\frac{\partial^2[C]}{\partial z^2}\right] - v_x\left[\frac{\partial[C]}{\partial x}\right] - v_y\left[\frac{\partial[C]}{\partial y}\right] - v_z\left[\frac{\partial[C]}{\partial z}\right] \tag{4.6}$$

In regions where horizontal gradients are small, $\partial[C]/\partial x \approx \partial[C]/\partial y \approx \partial^2[C]/\partial x^2 \approx \partial^2[C]/\partial y^2 \approx 0$, so Eq. 4.6 can be reduced to one dimension:

$$0 = D_z\left[\frac{\partial^2[C]}{\partial z^2}\right] - v_z\left[\frac{\partial[C]}{\partial z}\right] \tag{4.7}$$

The concentration gradients and density can be inferred from vertical salinity profiles, leaving only two unknowns, the eddy diffusion constant and the rate of vertical advection. Though unique solutions for these constants cannot be obtained from the salinity data, the ratio, D_z/v_z, can be determined. This ratio can be used to calculate chemical reaction rates from the vertical concentration profiles of nonconservative solutes.

To describe the rate of concentration change of a nonconservative solute, S, a chemical reaction term, J, must be incorporated into Eq. 4.7, that is,

$$0 = D_z\left[\frac{\partial^2[C]}{\partial z^2}\right] - v_z\left[\frac{\partial[C]}{\partial z}\right] + J \tag{4.8}$$

With the value of D_z/v_z determined from the salinity data and the vertical concentration gradient of S, a unique solution for J can be obtained. The validity of such a model is demonstrated by how well the equation, with all of its constants, fits the data. An example is given in Figure 4.14 using a vertical profile of dissolved oxygen concentrations.

The dashed line represents the curve generated by Eq. 4.8 if O_2 is assumed to behave conservatively, (i.e., $J = 0$). The actual concentrations are less than the values predicted by this solution due to biological removal, which occurs as a result of the respiration of organic matter. To fit Eq. 4.8 to these data, the net removal rate of O_2 (i.e., J), must equal 10^{-3} ml O_2 L^{-1} y^{-1}.

The model fits the oxygen data only over a limited depth range. This is because D_z/v_z can only be determined through depths where significant vertical

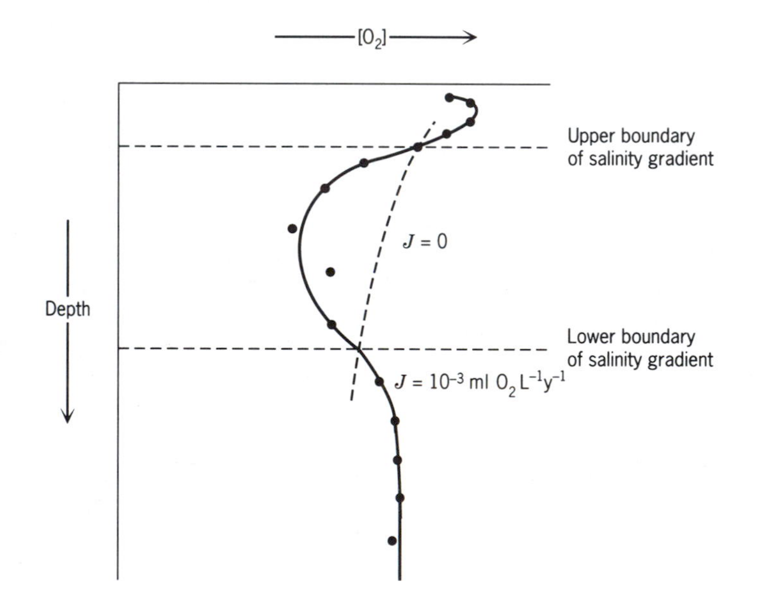

FIGURE 4.14. Schematic application of the one-dimensional advection–diffusion model to a vertical dissolved O_2 concentration profile. J is the O_2 uptake rate. *Source:* After H. Craig, reprinted with permission from the *Journal of Geophysical Research,* vol. 76, p. 5080, copyright © 1971 by the American Geophysical Union, Washington, DC.

salinity gradients exist. Therefore the model can only be applied to depth zones over which adjacent water masses are being vertically mixed.

In some cases, a better fit is obtained by using a chemical removal term that is concentration dependent. If the chemical reaction proceeds at a rate governed by first-order kinetics, J can be replaced by $-k[S]$, where k is the chemical reaction rate constant.

$$0 = D_z\left[\frac{\partial^2[C]}{\partial z^2}\right] - v_z\left[\frac{\partial[C]}{\partial z}\right] - k[S] \tag{4.9}$$

In some cases, the data is best fit with a J term that depends in a nonlinear way on the concentration of the nonconservative solute. No mechanistic understanding of the chemical reaction is implied by the mathematical form of J. It is no more than a function that provides the best empirical fit to the data. Despite these limits, such one-dimensional advection–diffusion models have provided important information on the rates at which physical and chemical processes occur in the ocean and the pore waters of marine sediments.

SUMMARY

The salinity of seawater is controlled by local variations in the rate of water transport through the hydrological cycle. These variations are largely due to temporal and spatial variations in **insolation**. Insolation varies over time due to changes in sea state, cloud cover, and distance from the sun. Insolation is greatest at low latitudes, where the angle of the sun is highest. This causes the temperature of the surface seawater to increase with decreasing latitude. Variations in insolation also drive geostrophic and thermohaline circulation. Thus insolation also influences the vertical profiles of temperature. At low and mid latitudes, three temperature regimes are present, the **mixed layer**, **thermocline**, and **deep zone**. Due to low atmospheric temperatures and strong winds, the water column is isothermal at high latitudes.

The larger the difference between surface and deep-water temperatures, the steeper the density gradient, or **pycnocline**. **Density stratification** inhibits vertical mixing. As a result, deep-water masses tend to retain the temperature and salinity they acquired when they were last at the sea surface. The path of advective flow of deep-water masses can be traced from these unique temperature and salinity signatures until turbulent mixing with adjacent water masses homogenizes the waters. Since this mixing is conservative, the source waters of the deep-water masses can be identified from T–S diagrams.

The salinity of surface seawater varies with latitude as a result of shifts in the relative rates of water loss through evaporation and water gain through precipitation. Since net evaporative water loss is greatest at mid latitudes, this is where surface-water salinities are highest. Salinity does not increase uniformly with depth. This is due to the relatively low salinity of the high-latitude surface waters that are the source waters for most deep-water masses.

Temporal variations in salinity are largely restricted to the sea surface and coastal regions due to the influence of rainfall and river runoff.

Despite short-term local variations, the heat content of the atmosphere and ocean remains constant over time scales of years. This steady state is maintained by the net transport of heat from low to high latitudes by the western and eastern boundary currents.

The relative rates of vertical advection and eddy diffusion in the deep ocean can be calculated from temperature and salinity gradients using a mathematical model. This information has been used to calculate rates at which nonconservative substances are added or removed from the deep sea.

CHAPTER 5

ION SPECIATION IN SEAWATER

INTRODUCTION

Elements are subject to chemical change as they pass through the ocean-sediment portion of the crustal-ocean factory. For example, some minerals dissolve and others precipitate, acids react with bases, and dissolved metals bind with electron donors called ligands. In the absence of external forces, chemical reactions proceed in a direction that ultimately leaves the reactants and products in their lowest energy states. This stable condition is termed **thermodynamic equilibrium**.

If the reaction stoichiometry and equilibrium constants are known, the equilibrium concentrations of the reactants and products can be calculated. Such calculations have demonstrated that calcareous shells should spontaneously precipitate in surface seawater but dissolve in deep waters. Thus surface-dwelling organisms need spend less metabolic energy in forming their calcareous shells as compared to deep-dwelling organisms. Equilibrium calculations have also been used to predict the chemical fate of pollutants introduced into the ocean.

Many reactions that occur in seawater do not attain thermodynamic equilibrium. But even in these cases, theoretical equilibrium calculations are still useful as they predict the spontaneous direction of reactions. They can also provide some insight into the relative rates of chemical reactions.

This chapter contains a description of the standard methods used to calculate the equilibrium concentrations of the various forms, or "species," of the elements present in seawater. Because of the high concentration of ionic solutes, seawater does not behave as a thermodynamically ideal solution. Thus special techniques must be used to correct for this nonideal behavior. The computational approaches covered in this chapter are applied throughout the text to aid in examining the chemical reactions that occur within the biogeochemical cycles of various elements.

NONIDEAL SOLUTION BEHAVIOR

Consider the following general reaction, where reactants A and B combine to yield products C and D. The molar reaction stoichiometry is given by the coefficients, a, b, c, and d, respectively.

$$aA + bB \rightleftharpoons cC + dD \qquad (5.1)$$

At equilibrium, the rate of the forward reaction is equal to that of the reverse. Since the concentrations of all the species are constant, the following **equilibrium constant** can be written for this reaction:

$$K_{eq} = \frac{[C]^c[D]^d}{[A]^a[B]^b} \qquad (5.2)$$

In aqueous solutions with high concentrations of ions, interactions between solutes are common. These interactions can have significant effects on solution behavior. For example, consider the dissolution of NaCl in water.

$$\text{NaCl(s)} \rightleftharpoons \text{Na}^+\text{(aq)} + \text{Cl}^-\text{(aq)} \qquad (5.3)$$

At equilibrium, the products are so strongly favored over the reactants that this reaction can be considered to go to completion. Thus if 1 mole of salt is dissolved in 1 kg of water, the concentrations of Na^+ and Cl^- should both equal 1 mol/kg or 1 molal (*m*).

The dissolution of salt lowers the freezing point of water. The extent of this lowering is given by

$$\Delta t = -nk_f m \qquad (5.4)$$

where Δt is the freezing point depression in °C, k_f is the freezing point depression constant for the solvent, m is the molality of the undissolved solute, and n is the number of moles of ions (or uncharged species) produced by the dissolution of 1 mole of solute.

Since NaCl generates 2 ions for each mole of NaCl(s) dissolved, $n = 2$. For water, $k_f = 1.86°\text{C kg } H_2O^{-1} \text{ mol solute}^{-1}$. Thus $\Delta t = -(2)(1.86)(1) = -3.72°\text{C}$. But through experimental observations, chemists have observed that the freezing point depression for a 1 m NaCl solution is $-3.01°\text{C}$. Why isn't the freezing point lowered all the way to $-3.72°\text{C}$?

The answer lies in Eq. 5.4. Since k_f is a constant and m is set by the conditions of the experiment, the only way to produce a lower freezing point depression is for n to be somewhat less than 2. In other words, the effective concentrations of the ions must be somewhat less than their total concentrations. If this nonideal behavior is not taken into account, calculations of solution properties, such as freezing point depression, will be in error.

The effective concentration, or **activity**, of an ion can be thought of as that part of the solute pool which exerts an influence on a particular chemical reaction. Thus equilibrium constants for solutions with high ion concentrations should be written in terms of the effective concentrations of the ions, that is,

$$K_{eq} = \frac{\{C\}^c\{D\}^d}{\{A\}^a\{B\}^b} \qquad (5.5)$$

where { } is used to represent ion activities.

The activity of an ion, i, is related to its total concentration by

$$a_i = \gamma_i m_i \qquad (5.6)$$

TABLE 5.1
Method for Calculating Ionic Strengths of Solutions

Solute (*i*)	*Molality* (m_i)	*Ionic Charge* (z_i)	z_i^2	$m_i z_i^2$
Na^+	0.5	+1	1	0.5
Cl^-	0.5	−1	1	0.5
K^+	0.2	+1	1	0.2
SO_4^{2-}	0.1	−2	2	0.4

$$\sum_i m_i z_i^2 = 1.6\,m$$
$$I = \tfrac{1}{2}(1.6)$$
$$= 0.8\,m$$

where a_i is the activity of the ion, γ_i is its activity coefficient, and m_i is the total molality of the solute. As γ is dimensionless, a_i has the units of molality. In dilute solutions, the activity coefficient of an ion can be calculated from the Debye–Hückel equation:

$$\log \gamma_i = -Az_i^2\sqrt{I} \tag{5.7}$$

where A is a constant that is a characteristic of the ion, z_i is its charge, and I is the **ionic strength** of the solution.

Ionic strength is a measure of the total concentration of charge contributed by the ions present in a solution. It is calculated as

$$I = \tfrac{1}{2}\sum_i m_i z_i^2 \tag{5.8}$$

where m_i is the molality of each ion and z_i is the ion's charge. An example of an ionic strength calculation is given in Table 5.1.

Equation 5.7 is valid only for $I < 0.01\,m$. Most natural waters, such as river ($0.0021\,m$) and seawater ($0.7\,m$), have much higher ionic strengths. Suitable activity coefficients are given by more complex equations, such as those listed in Table 5.2. Alternative methods must also be used to calculate their ionic strengths.

Despite the additional complexity, all the equations are functionally equivalent. That is, the activity coefficients approach a value of 1 as the ionic strength of the solution is decreased to $0\,m$. Thus in dilute solutions, $a_i \approx m_i$. In other words, the effective concentration of an ion decreases with increasing ionic strength. On the other hand, activity coefficients of uncharged solutes can be greater than 1. As described below, this is due to the relatively small degree of electrostatic attraction that can exist between uncharged solutes and the solvent or with other solutes present in the solution.

The activities of uncharged solutes are not as greatly affected by increasing ionic strength because electrostatic interactions are the source of most of the

TABLE 5.2
Various Expressions for the Calculation of Single Ion Activity Coefficients

Approximation	*Equation*[a]	*Approximate Applicability [ionic strength (M)]*
Debye–Hückel	$\log\gamma = -Az^2\sqrt{I}$	$<10^{-2}$
Extended Debye–Hückel	$= -Az^2\dfrac{\sqrt{I}}{1 + Ba\sqrt{I}}$	$<10^{-1}$
Güntelberg	$= -Az^2\dfrac{\sqrt{I}}{1 + \sqrt{I}}$	$<10^{-1}$ useful in solutions of several electrolytes
Davies	$= -Az^2\left(\dfrac{\sqrt{I}}{1 + \sqrt{I}} - 0.2I\right)$	<0.5
Brönsted–Guggenheim	$\ell n\gamma_S = \ell n\gamma_{DH_S} + \sum_j A_{S_j}(C_j) + \sum_j\sum_k B_{S_{jk}}(C_j)(C_k) + \cdots$	≤ 4

Source: From *Aquatic Chemistry*, W. Stumm and J. J. Morgan, copyright © 1981 by John Wiley & Sons, Inc., New York, p. 135. Reprinted by permission.
[a]Values for the constants can be found in Stumm and Morgan (1981).

nonideal solution behavior of ions. These interactions can be classified somewhat arbitrarily into two categories: **nonspecific** and **specific**.

The nonspecific ones include long-range interactions between solute ions, as well as those between the ions and the solvent. For example, ionic solutes in an aqueous solution exert electrostatic forces that impose some degree of order in the arrangement of the water molecules. These forces increase with decreasing distance of separation between the ion and the water molecules. The region in which the interaction between solute and solvent is greatest is termed the primary solvation shell and is illustrated in Figure 5.1.

In the primary solvation shell, the force of attraction is large enough to cause clustering of water molecules and hence a slightly higher density. This phenomenon is referred to as electrostriction (see Figure 3.1). The water molecules in the secondary solvation shell are still close enough to the ions to experience some electrostatic attraction, but the degree of ordering is much less than in the primary shell. At greater distances, the water molecules are too far from the ions to be significantly influenced by electrostatic attraction. This region is termed the bulk solution.

Strong electrolytes, such as $Na^+(aq)$ and $Cl^-(aq)$, are so well hydrated that they are too far apart to interact directly with each other, even in solutions of great ionic strength. Nevertheless, they can still exert some relatively weak indirect forces on each other. The intensity of these nonspecific interactions depends on the ionic strength and chemical composition of the solution.

Ions that are so well hydrated that they experience only nonspecific interactions are termed "free ions" (Figure 5.2*a*). Ions that are not as well hydrated can come into closer contact. In cases where ions are close enough to share some of their primary solvation shells, the resulting strong electrostatic

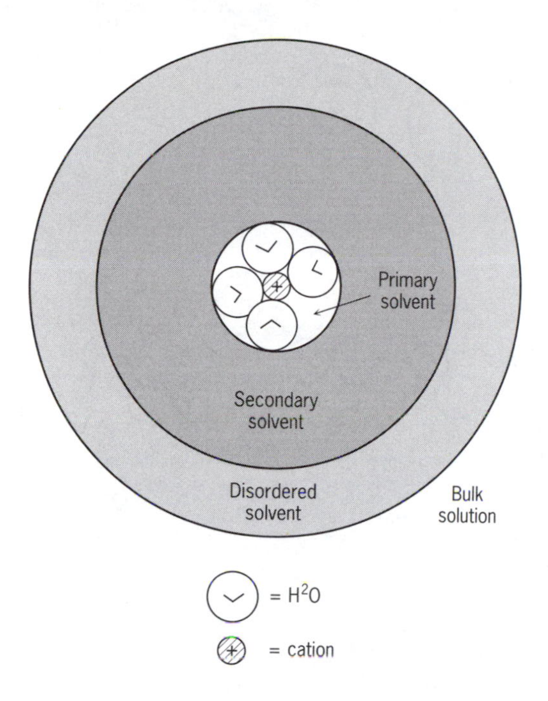

FIGURE 5.1. The various solvent regions around a metal ion. *Source:* From *Principles of Aquatic Chemistry*, F. M. M. Morel, copyright © 1983 by John Wiley & Sons, Inc., New York, p. 239. Reprinted by permission. After *Metal Ions in Solution*, J. Burgess, copyright © 1978 by John Wiley & Sons, Inc., New York, p. 20. Reprinted by permission.

attraction is referred to as **ion pairing** (Figure 5.2*b*). Ion pairing can occur among groups of ions to create ternary (n=3), quaternary (n=4), and other higher-order associations. These short-range associations are examples of specific interactions.

COMPETITIVE COMPLEXING

Ions and other solutes that come into direct contact (Figure 5.2*c*) are close enough to share electrons. When electrons are shared, atoms are held together more strongly than in ion pairs. These strong specific interactions have fixed geometries, such as those illustrated in Figure 5.3. They are termed **coordination complexes**. Most involve a metal cation (M^+) and an electron donor, referred to as a **ligand** (L). Ligands need not be anions. For example, water is an effective ligand because it is able to share its nonbonded electron pair with a cation. Thus hydrated ions are technically a type of coordination complex.

As shown in Figure 5.3, most metals are able to simultaneously share electrons with several ligands. The number of ligands a metal ion can accommodate depends on how many coordination sites it possesses. This is determined by its electronic configuration.

The formation of coordination complexes can be viewed as a competition among ligands for one or more of the coordination sites on the metal ion. To occupy most of the coordination sites, a ligand must form the strongest associations with the metal ions. The strongest of these are covalent bonds.

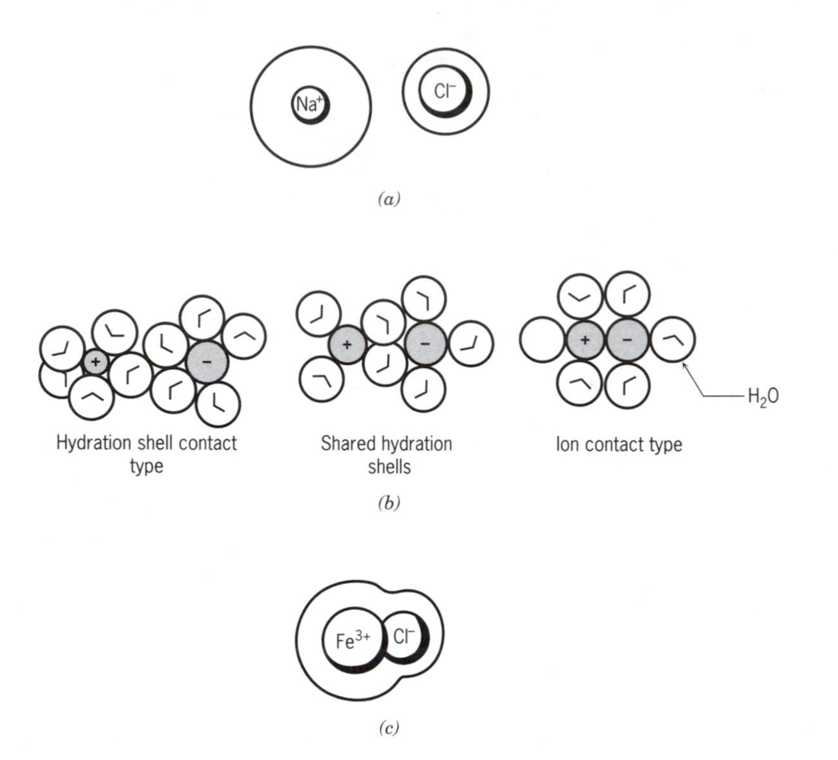

FIGURE 5.2. (*a*) Free ions, (*b*) ion pairs, and (*c*) a complex ion. *Source:* After *Principles of Aquatic Chemistry,* F. M. M. Morel, copyright © 1983 by John Wiley & Sons, Inc., New York, p. 240. Reprinted by permission.

Specific interactions are best thought of as a continuum that range in strength from weak associations, as in ion pairs, to strong ones, as in coordination complexes. Thus the formation of both ions pairs and coordination complexes can be represented by the following equilibrium expression:

$$aM^{+}(\text{aq}) + bL^{-}(\text{aq}) \rightleftharpoons M_aL_b(\text{aq}) \quad (5.9)$$

where

$$K_{\text{eq}} = \frac{\{M_aL_b(\text{aq})\}}{\{M^{+}(\text{aq})\}^{a}\{L^{-}(\text{aq})\}^{b}} \quad (5.10)$$

In solutions containing several metals and ligands, many ion pairs and complexes can be present. The equilibrium concentrations of each species can

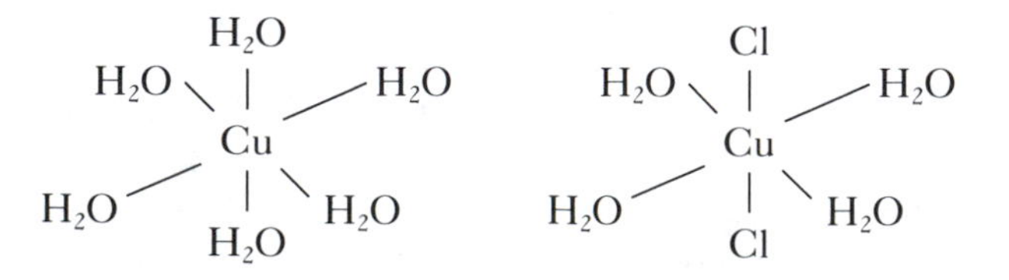

FIGURE 5.3. Various types of aqueous complexes.

be calculated if the equilibrium constants for all the ion pairs and complexes are known. In this type of calculation, the equilibrium constants are defined in terms of ion activities. These activities are the products of the concentrations of the species and their ionic activity coefficients. Ionic activity coefficients do not include the specific effects of ion pairing or complexing. They do include the effects of nonspecific interactions. Thus they are dependent on the temperature, pressure, and chemical composition of the solution. Values for the ionic activity coefficients of the major ions and their ion pairs at 35‰, 25°C, and 1 atm pressure are given in Table 5.3.

Most marine chemistry books tabulate equilibrium constants as K^0_{eq}, which are the values of thermodynamic equilibrium constants for $I = 0\ m$ and 25°C. These values can be extrapolated to other solution conditions using Eq. 5.11. This equation has been derived from thermodynamic considerations.

$$K_c = \frac{[M_aL_b]}{[M^+]^a[L^-]^b} = \frac{(\gamma_{M^+})^a(\gamma_{L^-})^b}{\gamma_{M_aL_b}} K^0_{eq} \tag{5.11}$$

The resultant equilibrium "constant", K_c, is convenient for speciation calculations because it is defined in terms of concentrations. On the other hand, it is

TABLE 5.3
Ionic Activity Coefficients for the Major Ions and Their Ion Pairs at 35‰, 25°C, and 1 atm

Dissolved Species	*Activity Coefficient*
$NaHCO_3^0$	1.13
$MgCO_3^0$	1.13
$CaCO_3^0$	1.13
$MgSO_4^0$	1.13
$CaSO_4^0$	1.13
Na^+	0.76
HCO_3^-	0.68
$NaCO_3^-$	0.68
$NaSO_4^-$	0.68
KSO_4^-	0.68
$MgHCO_3^+$	0.68
$CaHCO_3^+$	0.68
K^+	0.64
Mg^{2+}	0.36
Ca^{2+}	0.28
SO_4^{2-}	0.12
CO_3^{2-}	0.20

Source: From R. M. Garrels and M. E. Thompson, reprinted with permission from the *American Journal of Science*, vol. 260, p. 61, copyright © 1962 by the American Journal of Science, New Haven, CT.

not truly a constant as its value is determined by the solution conditions under which the activity coefficients were determined. If ionic activity coefficients are used to compute K_c, this equilibrium "constant" will not include the effects of ion pairing or complexing. Thus, K_c can be used to compute ion speciation as described below.

Calculating Speciation in a Multi-Ion Solution

The equilibrium concentrations of the species that are present in a multi-ion solution can be calculated by solving a set of simultaneous equations, as shown in the following example. Consider an aqueous solution produced by the dissolution of 1.00 mole of $MgSO_4(s)$ and 1.00 mole of $CaF_2(s)$ in 1.00 kg of water. Thus the total concentration of each element is 1.00 m, except for F, which is 2.00 m.

Upon dissolution, the following species are present: $Mg^{2+}(aq)$, $Ca^{2+}(aq)$, $SO_4^{2-}(aq)$, $F^-(aq)$, $MgSO_4^0(aq)$, $MgF^+(aq)$, $CaSO_4^0(aq)$, and $CaF^+(aq)$. A **mass balance equation** can be written to express the total concentration of each element in terms of its species's concentrations. The mass balance equation for Mg is

$$[Mg] = [Mg^{2+}] + [Mg^{2+}]_{MgSO_4^0} + [Mg^{2+}]_{MgF^+} \tag{5.12}$$

Since each mole of $MgSO_4^0$ and MgF^+ contains 1 mole of Mg, Eq. 5.12 can be simplified to

$$[Mg] = [Mg^{2+}] + [MgSO_4^0] + [MgF^+] = 1.00\,m \tag{5.13}$$

The other mass balance equations are then

$$[Ca] = [Ca^{2+}] + [CaSO_4^0] + [CaF^+] = 1.00\,m \tag{5.14}$$

$$[S] = [SO_4^{2-}] + [MgSO_4^0] + [CaSO_4^0] = 1.00\,m \tag{5.15}$$

$$[F] = [F^-] + [MgF^+] + [CaF^+] = 2.00\,m \tag{5.16}$$

The large number of unknowns can be reduced by using equilibrium constants to rewrite the ion pair concentrations in terms of the concentrations of the free ions. To do this, K_c must be calculated from Eq. 5.11. For the purposes of this calculation, the appropriate ionic activity coefficients are approximated by the values given in Table 5.3 and the K_{eq}^0's from the data in Appendix X.* The calculation of K_c is illustrated below for $MgSO_4^0$:

$$K_{c_{MgSO_4^0}} = \frac{\gamma_{MG^{2+}}\,\gamma_{SO_4^{2-}}}{\gamma_{MgSO_4^0}} K^0_{eq_{MgSO_4^0}} \tag{5.17}$$

$$= \frac{(0.29)(0.17)}{(1.13)} 10^{2.36} = 10.0$$

*Note that the treatment of significant figures is not rigorous in this calculation. At least one extra significant figure is maintained throughout each step of the calculation to prevent round-off errors. The final results as listed in Table 5.4 are all reported with two significant figures to demonstrate that mass balance was achieved.

Thus:

$$K_{c_{MgSO_4^0}} = \frac{[MgSO_4^0]}{[Mg^{2+}][SO_4^{2-}]} \quad (5.18)$$

can be rearranged to

$$[MgSO_4^0] = 10.0\,[Mg^{2+}]\,[SO_4^{2-}] \quad (5.19)$$

Doing the same for MgF^+ yields

$$[MgF^+] = 18.3\,[Mg^{2+}]\,[F^-] \quad (5.20)$$

To initiate the solution of Eqs. 5.13 through 5.16, it must be temporarily assumed that the anions are unpaired, that is, [S] = $[SO_4^{2-}]$ = 1.00 *m* and [F] = $[F^-]$ = 2.00 *m*. Thus Eqs. 5.19 and 5.20 become

$$[MgSO_4^0] = 10.0\,[Mg^{2+}] \quad (5.21)$$

$$[MgF^+] = 36.6\,[Mg^{2+}] \quad (5.22)$$

Substituting Eqs. 5.21 and 5.22 into the mass balance equation for Mg (Eq. 5.13) yields

$$1.00 = [Mg^{2+}] + 10.0\,[Mg^{2+}] + 36.6\,[Mg^{2+}] \quad (5.23)$$

Solving for $[Mg^{2+}]$,

$$[Mg^{2+}] = 0.021\ m \quad (5.24)$$

The analogous equations for calcium are

$$1.00 = [Ca^{2+}] + 7.99\,[Ca^{2+}] + 6.55\,[Ca^{2+}] \quad (5.25)$$

and

$$[Ca^{2+}] = 0.064\ m \quad (5.26)$$

Corrected values for $[SO_4^{2-}]$ and $[F^-]$ can now be obtained by substituting the calculated values of $[Mg^{2+}]$ and $[Ca^{2+}]$ into Eqs. 5.15 and 5.16 as follows:

$$1.00 = [SO_4^{2-}] + \{(10.0)(0.0210)[SO_4^{2-}]\} + \{(7.99)(0.0644)[SO_4^{2-}]\} \quad (5.27)$$

$$[SO_4^{2-}] = 0.58\ m \quad (5.28)$$

$$2.00 = [F^-] + \{(18.3)(0.0210)[F^-]\} + \{(3.27)(0.0644)[F^-]\} \quad (5.29)$$

$$[F^-] = 1.3\ m \quad (5.30)$$

With these values for $[F^-]$ and $[SO_4^{2-}]$, the cation concentrations can be recalculated, yielding $[Mg^{2+}]$ = 0.034 *m* and $[Ca^{2+}]$ = 0.10 *m*.

At this point, one **iteration**, or calculation cycle, of the simultaneous equations presented in Eqs. 5.13 through 5.16 has been completed. The entire calculation is then reiterated until little change in the calculated concentrations of the ions is observed from one cycle to the next. As shown in Figure 5.4, this usually takes several iterations.

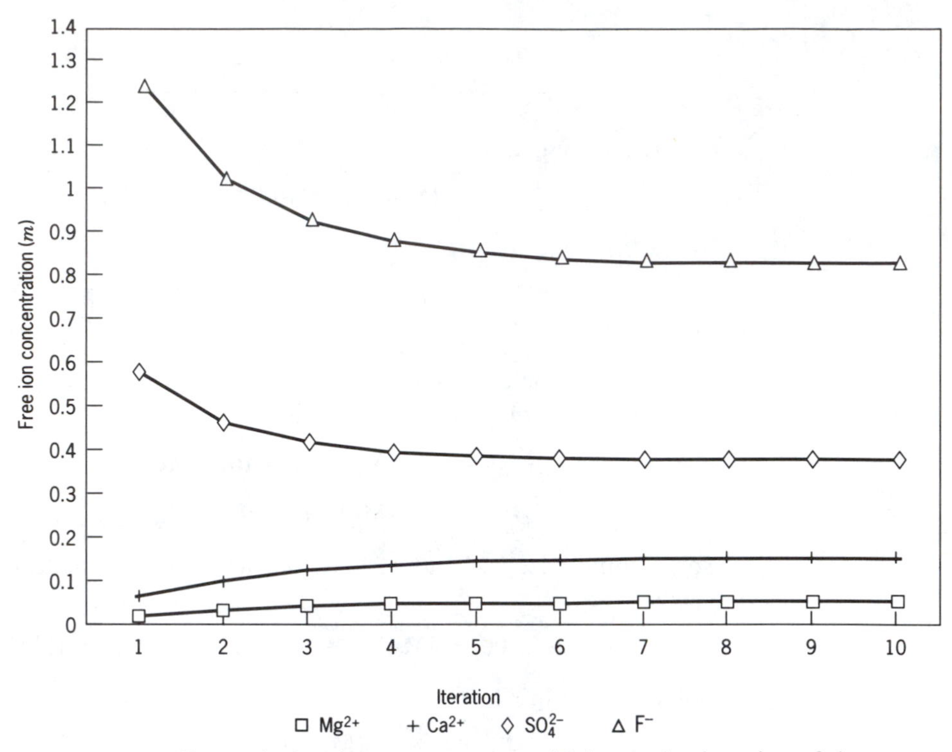

FIGURE 5.4. Change in free ion concentration with increasing iteration of the sample speciation calculation.

At the end of the calculation, the final concentrations of the ions ($[Mg^{2+}] = 0.050\ m$, $[Ca^{2+}] = 0.15\ m$, $[SO_4^{2-}] = 0.37\ m$ and $[F^-] = 0.83\ m$) are used to calculate the ion pair concentrations. For $MgSO_4^0$, Eq. 5.19 becomes

$$[MgSO_4^0] = (10.0)(0.050)(0.37) = 0.19\ m \tag{5.31}$$

Similarly,

$$[MgF^+] = (18.3)(0.050)(0.83) = 0.76\ m \tag{5.32}$$

$$[CaSO_4^0] = (7.99)(0.15)(0.37) = 0.44\ m \tag{5.33}$$

$$[CaF^+] = (3.27)(0.15)(0.37) = 0.41\ m \tag{5.34}$$

The results of such speciation calculations are usually reported as percentages of the element present in a particular form, as shown in Table 5.4. For example, the speciation calculation predicts that 5.0 percent of the Mg should be unpaired, 19 percent present as $MgSO_4^0$, and 76 percent as MgF^+. Note that mass balance is achieved (i.e., the sum of each column is 100%).

SPECIATION IN SEAWATER

Because of the great variety and abundance of solutes in seawater, speciation calculations require the solution of large sets of simultaneous equations. This

TABLE 5.4
Results of Sample Calculation after 10 Iterations

	% [Ca]	% [Mg]	% [S]	% [F]
$[Ca^{2+}]$	15	—	—	—
$[Mg^{2+}]$	—	5.0	—	—
$[SO_4^{2-}]$	—	—	37	—
$[F^-]$	—	—	—	42
$[CaSO_4^0]$	44	—	44	—
$[CaF^+]$	41	—	—	20
$[MgSO_4^0]$	—	19	19	—
$[MgF^+]$	—	76	—	38

is done by computer, using programs such as MINEQL and REDEQL. Due to the great number of species involved, these calculations also require a substantial amount of thermodynamic data. In addition, the values used must be relevant for the solution of interest, in terms of its chemical composition, ionic strength, temperature and pressure. Because of the technical difficulties associated with direct detection of most ion species, the thermodynamic data are limited in terms of quality and quantity. For most species, data are available for only a very limited set of environmental conditions. Marine chemists are working to improve the precision and accuracy of the data, thus the speciation results presented below are subject to change.

Major Ions

The major and minor ion speciation in seawater was first calculated by Garrels and Thompson in 1962. Their results for seawater at 25°C, 1 atm pressure, and 35‰ salinity are presented in Table 5.5, along with two other sets based on improved thermodynamic data. While these improvements have lead to some changes in the results of the speciation calculations, all demonstrate that the major cations are present predominantly as the free ions. In comparison, a significant amount of sulfate is ion paired. In these calculations, chloride is assumed to be wholly unpaired. This is supported by direct observations.

Calcium and magnesium are paired to a much larger extent than sodium and potassium, with approximately 10 percent of these divalent ions present as the neutral sulfate pairs. These pairs contain between 20 and 40 percent of the sulfate. Significant amounts of the minor anions (bicarbonate, carbonate, and fluoride) are also paired, predominantly with sodium and magnesium. Since the cation concentrations are much larger than the anion concentrations (with the exception of Cl^-), ion pairing influences the relative abundance of the anions much more than that of the cations.

The results of these calculations are dependent on salinity, temperature, and pressure. As shown in Table 5.6, these effects can be significant and demonstrate the need to use the appropriate thermodynamic data.

TABLE 5.5
Major and Minor Ion Speciation in Seawater at 25°C, 1 atm, and 35‰[a]

	Na^+			Mg^{2+}			Ca^{2+}			K^+		
	G	*H*	*P + H*	*G*	*H*	*P + H*	*G*	*H*	*P + H*	*G*	H	P + H
Species	*0.4752 m*[b]	*0.4823 m*	*0.4822 m*	*0.0540 m*	*0.05485 m*	*0.05489 m*	*0.0104 m*	*0.01062 m*	*0.01063 m*	*0.0100 m*	0.01020 *m*	0.01062 *m*
Free ion	99	99.0	97.7	87	89.9	89.2	91	91.5	88.5	99	98.5	98.9
MSO_4	1.2	1.0	2.2	11	9.2	10.3	8	7.6	10.8	1	1.5	1.1
$MHCO_3$	0.01	0.0	0.1	1	0.6	0.1	1	0.7	0.3	—	0.0	—
MCO_3	—	0.0	0.0	0.3	0.3	0.1	0.2	0.2	0.3	—	0.0	—
Mg_2CO_3	—	—	—	—	—	0.0	—	—	—	—	—	—
$MgCaCO_3$	—	—	—	—	—	0.0	—	—	0.1	—	—	—

	SO_4^{2-}			HCO_3^-			CO_3^{2-}			F^-		
	G	*H*	*P + H*	*G*	*H*	*P + H*	*G*	*H*	*P + H*			*P + H*
Species	*0.0284 m*	*0.02909 m*	*0.02906 m*	*0.00238 m*	*0.00186 m*	*0.00213 m*	*0.000269 m*	*0.00011 m*	*0.000171 m*			*0.000080m*
Free ion	54	62.9	39.0	69	74.1	81.3	9	10.2	8.0			51.0
NaX	21	16.4	37.1	8	8.3	10.7	17	19.4	16.0			—
MgX	21.5	17.4	19.5	19	14.4	6.5	67	63.2	43.9			47.0
CaX	3	2.8	4.0	4	3.2	4.0	7	7.1	21.0			2.0
KX	1	0.5	0.4	—	0.0	0.4	—	0.0	—			—
Mg_2CO_3	—	—	—	—	—	—	—	—	7.4			—
$MgCaCO_3$	—	—	—	—	—	—	—	—	3.8			—

Source: From *Aquatic Chemistry*, W. Stumm and J. J. Morgan, copyright © 1981 by John Wiley & Sons, Inc., New York, p. 366. Reprinted by permission. Data from (G) R. M. Garrels and M. E. Thompson, reprinted with permission from the *American Journal of Science*, vol. 260, p. 63, copyright © 1962 by the *American Journal of Science*, New Haven, CT; (H) J. S. Hanor, reprinted with permission from *Geochimica et Cosmochimica Acta*, vol. 33, p. 896, copyright © 1969 by Pergamon Press, Elmsford, NY; and (P+H) R. M. Pytkowicz and J. E. Hawley, reprinted with permission from *Limnology and Oceangraphy*, vol. 19, p. 230, copyright © 1974 by the American Society of Limnology and Oceanography, Seattle, WA.

[a]Values are percentages.

[b]The total concentration of ion used in the speciation calculation.

TABLE 5.6
Effect of Temperature and Pressure on Sulfate Species Speciation in Seawater

T(°C)	P (atm)	Free SO_4^{2-} (%)	$NaSO_4^-$ (%)	$MgSO_4^0$ (%)	$CaSO_4^0$ (%)
25	1	39.0	38.0	19.0	4.0
2	1	28.0	47.0	21.0	4.0
2	1000	39.0	32.0		
		42.0	35.0	19.0	4.0

Source: From *Aquatic Chemistry*, W. Stumm and J. J. Morgan, copyright © 1981 by John Wiley & Sons, Inc., New York, p. 367. Reprinted by permission. Data from (1) F. J. Millero, reprinted with permission from *Geochimica et Cosmochimica Acta*, vol. 35, p. 1097, copyright © 1971 by Pergamon Press, Elmsford, NY; and (2) D. R. Kester and R. M. Pytkowicz, reprinted with permission from *Geochim. Cosmochim. Acta*, vol. 34, p. 1047, copyright © 1970 by Pergamon Press, Elmsford, NY.

Trace Metals

In comparison to the major and minor ions, a much larger fraction of the dissolved trace metals are present as coordination complexes and ion pairs. As shown in Table 5.7, the ion associations that are probably most abundant in aerobic waters are those with hydroxide, chloride, sulfate, and carbonate. In anoxic waters, sulfide complexes are important. Associations with **dissolved organic matter (DOM)** are significant in coastal waters where high concentrations of this ligand are present.

In most cases, the ligands are present at far greater concentrations than the trace metals, (i.e., $[L(aq)] \gg [M(aq)]$). Thus ion associations between the trace metals and ligands influence the speciation of the former to a far greater extent than that of the latter. Since the metals can be considered as the limiting reactants, the effects of metal-metal competition for the ligands can be neglected. Thus speciation calculations for each trace metal can be done independently.

As with all speciation calculations, the results for the trace metals depend on the values used for the equilibrium constants and activity coefficients. Direct detection of trace metal species is complicated by their low concentrations and analytical interferences caused by high concentrations of the major ions. For many species, either no thermodynamic data exist or a wide range in values has been reported by various authors. Thus the calculated speciation of the trace metals is much more tentative than for the major ions. Examples of some results of trace metal speciation calculations are given in Table 5.8 for copper, lead, zinc, and cadmium. These metals are of great interest as they are pollutants, and some of their ion species are toxic at relatively low concentrations.

TABLE 5.7
Speciations, Concentrations, and Distribution Types of Trace Elements in Ocean Water

Element	*Probable Main Species in Oxygenated Seawater*	*Range and Average Concentration at 35‰ Salinity*[a]	*Type of Distribution*[b]
Li	Li^+	25 μmol kg^{-1}	Conservative
Be	$BeOH^+$, $Be(OH)_2^0$	4–30 pmol kg^{-1}; 20 pmol kg^{-1}	Nutrient-type and scavenging
B	H_3BO_3	0.416 mmol kg^{-1}	Conservative
C	HCO_3^-, CO_3^{2-}	2.0–2.5 mmol kg^{-1}; 2.3 mmol kg^{-1}	Nutrient-type
N	NO_3^- (also as N_2)	<0.1–45 μmol kg^{-1}; 30 μmol kg^{-1}	Nutrient-type
O	O_2 (also as H_2O)	0–300 μmol kg^{-1}	Mirror image of nutrient-type
Al	$Al(OH)_4^-$, $Al(OH)_3^0$	(5–40 nmol kg^{-1}; 20 nmol kg^{-1})	Mid-depth minima
Si	H_4SiO_4	<1–180 μmol kg^{-1}; 100 μmol kg^{-1}	Nutrient-type
P	HPO_4^{2-}, $NaHPO_4^-$, $MgHPO_4^0$	<1–3.5 μmol kg^{-1}; 2.3 μmol kg^{-1}	Nutrient-type
Sc	$Sc(OH)_3^0$	8–20 pmol kg^{-1}; 15 pmol kg^{-1}	Surface depletion
Ti	$Ti(OH)_4^0$	(<20 nmol kg^{-1})	?
V	HVO_4^{2-}, $H_2VO_4^-$, $NaHVO_4^-$	20–35 nmol kg^{-1}; 30 nmol kg^{-1}	Slight surface depletion
Cr	CrO_4^{2-}, $NaCrO_4^-$	2–5 nmol kg^{-1}; 4 nmol kg^{-1}	Nutrient-type
Mn	Mn^{2+}, $MnCl^+$	0.2–3 nmol kg^{-1}; 0.5 nmol kg^{-1}	Depletion at depth
Fe	$Fe(OH)_3^0$	0.1–2.5 nmol kg^{-1}; 1 nmol kg^{-1}	Surface depletion, depletion at depth
Co	Co^{2+}, $CoCO_3^0$, $CoCl^+$	(0.01–0.1 nmol kg^{-1}; 0.02 nmol kg^{-1})	Surface depletion, depletion at depth
Ni	Ni^{2+}, $NiCO_3^0$, $NiCl^+$	2–12 nmol kg^{-1}; 8 nmol kg^{-1}	Nutrient-type
Cu	$CuCO_3^0$, $CuOH^+$, Cu^{2+}	0.5–6 nmol kg^{-1}; 4 nmol kg^{-1}	Nutrient-type and scavenging
Zn	Zn^{2+}, $ZnOH^+$, $ZnCO_3^0$, $ZnCl^+$	0.05–9 nmol kg^{-1}; 6 nmol kg^{-1}	Nutrient-type
Ga	$Ga(OH)_4^-$	(0.3 nmol kg^{-1})	?
Ge	H_4GeO_4, $H_3GeO_4^-$	≤7–115 pmol kg^{-1}; 70 pmol kg^{-1}	Nutrient-type
As	$HAsO_4^{2-}$	15–25 nmol kg^{-1}; 23 nmol kg^{-1}	Nutrient-type
Se	SeO_4^{2-}, SeO_3^{2-}, $HSeO_3^-$	0.5–2.3 nmol kg^{-1}; 1.7 nmol kg^{-1}	Nutrient-type
Br	Br^-	0.84 mmol kg^{-1}	Conservative
Rb	Rb^+	1.4 μmol kg^{-1}	Conservative
Sr	Sr^{2+}	90 μmol kg^{-1}	Slight surface depletion
Y	YCO_3^+, YOH^{2+}, Y^{3+}	(0.15 nmol kg^{-1})	?
Zr	$Zr(OH)_4^0$, $Zr(OH)_5^-$	(0.3 nmol kg^{-1})	?
Nb	$Nb(OH)_6^-$, $Nb(OH)_5^0$	(≤50 pmol kg^{-1})	?
Mo	MoO_4^{2-}	0.11 μmol kg^{-1}	Conservative
(Tc)	TcO_4^-	No stable isotope	—
Ru	?	?	?
Rh	?	?	?
Pd	?	?	?

Ag	$AgCl_2^-$	(0.5–35 pmol kg^{-1}; 25 pmol kg^{-1})	Nutrient-type
Cd	$CdCl_2^0$	0.001–1.1 nmol kg^{-1}; 0.7 nmol kg^{-1}	Nutrient-type
In	$In(OH)_3^0$	(1 pmol kg^{-1})	?
Sn	$SnO(OH)_3^-$	(1–12 pmol kg^{-1}, ~ 4 pmol kg^{-1})	High in surface waters
Sb	$Sb(OH)_6^-$	(1.2 nmol kg^{-1})	?
Te	TeO_3^{2-}, $HTeO_3^-$	?	?
I	IO_3^-	0.2–0.5 μmol kg^{-1}; 0.4 μmol kg^{-1}	Nutrient-type
Cs	Cs^+	2.2 nmol kg^{-1}	Conservative
Ba	Ba^{2+}	32–150 nmol kg^{-1}; 100 nmol kg^{-1}	Nutrient-type
La	La^{3+}, $LaCO_3^+$, $LaCl^{2+}$	13–37 pmol kg^{-1}; 30 pmol kg^{-1}	Surface depletion
Ce	$CeCO_3^+$, Ce^{3+}, $CeCl^{2+}$	16–26 pmol kg^{-1}; 20 pmol kg^{-1}	Surface depletion
Pr	$PrCO_3^+$, Pr^{3+}, $PrSO_4^+$	(4 pmol kg^{-1})	Surface depletion
Nd	$NdCO_3^+$, Nd^{3+}, $NdSO_4^+$	12–25 pmol kg^{-1}; 20 pmol kg^{-1}	Surface depletion
Sm	$SmCO_3^+$, Sm^{3+}, $SmSO_4^+$	2.7–4.8 pmol kg^{-1}, 4 pmol kg^{-1}	Surface depletion
Eu	$EuCO_3^+$, Eu^{3+}, $EuOH^{2+}$	0.6–1.0 pmol kg^{-1}; 0.9 pmol kg^{-1}	Surface depletion
Gd	$GdCO_3^+$, Gd^{3+}	3.4–7.2 pmol kg^{-1}; 6 pmol kg^{-1}	Surface depletion
Tb	$TbCO_3^+$, Tb^{3+}, $TbOH^{2+}$	(0.9 pmol kg^{-1})	Surface depletion
Dy	$DyCO_3^+$, Dy^{3+}, $DyOH^{2+}$	(4.8–6.1 pmol kg^{-1}; 6 pmol kg^{-1})	Surface depletion
Ho	$HoCO_3^+$, Ho^{3+}, $HoOH^{2+}$	(1.9 pmol kg^{-1})	Surface depletion
Er	$ErCO_3^+$, $ErOH^{2+}$, Er^{3+}	4.1–5.8 pmol kg^{-1}; 5 pmol kg^{-1}	Surface depletion
Tm	$TmCO_3^+$, $TmOH^{2+}$, Tm^{3+}	(0.8 pmol kg^{-1})	Surface depletion
Yb	$YbCO_3^+$, $YbOH^{2+}$	3.5–5.4 pmol kg^{-1}; 5 pmol kg^{-1}	Surface depletion
Lu	$LuCO_3^+$, $LuOH^{2+}$	(0.9 pmol kg^{-1})	Surface depletion
Hf	$Hf(OH)_4^0$, $Hf(OH)_5^-$	(<40 pmol kg^{-1})	?
Ta	$Ta(OH)_5^0$	(< 14 pmol kg^{-1})	?
W	WO_4^{2-}	0.5 nmol kg^{-1}	?
Re	ReO_4^-	(14–30 pmol kg^{-1}; 20 pmol kg^{-1})	?
Os	?	?	?
Ir	?	?	?
Pt	?	?	?
Au	$AuCl_2^-$	(25 pmol kg^{-1})	?
Hg	$HgCl_4^{2-}$	(2–10 pmol kg^{-1}; 5 pmol kg^{-1})	?
Tl	Tl^+, $TlCl^0$; or $Tl(OH)_3^0$	60 pmol kg^{-1}	Conservative
Pb	$PbCO_3^0$, $Pb(CO_3)_2^{2-}$, $PbCl^+$	5–175 pmol kg^{-1}; 10 pmol kg^{-1}	High in surface waters, depleted at depth
Bi	BiO^+, $Bi(OH)_2^+$	$\leqslant 0.015$–0.24 pmol kg^{-1}	Depletion at depth

Source: From *Chemical Oceanography*, vol. 8, K. W. Bruland (eds.: J. P. Riley and R. Chester), copyright © 1983 by Academic Press, Orlando, FL, pp. 172–173. Reprinted by permission.

[a]Parentheses indicate uncertainty about the accuracy or range of concentration given.

[b]See Chapter 11 for a discussion of the distribution types.

TABLE 5.8
Species Distribution in Seawater for Copper, Lead, Zinc, and Cadmium[a]

Metal Species	*Copper*	*Lead*	*Zinc*	*Cadmium*
Me^{2+}	3	1.8	39.0	1.9
$MeCl^{+}$	1	8.6	15.8	29.1
$MeCl_2$/$Me(OH)Cl$	0.6	12.6	10.0	37.2
$MeCl_3^-$/$MeCl_4^{2-}$	—	—	3.8	31.0
$MeSO_4$	0.4	—	15.8	—
$MeOH^{+}$	2.8	30	0.6	—
$Me(OH)_2$	1	0.5	0.1	0.8
$MeCO_3$	80	43.0	13.0	—
$Me(CO_3)_2^{2-}$	—	3.7	1.8	—
Me-humate	11	—	—	—

Source: From *The Role of the Oceans as a Waste Disposal Option,* U. Förstner, W. Ahlf, W. Calmano, and M. Kersten (ed.: G. Kullenberg), copyright © 1986 by Riedel Publishing Co., Dordrecht, Holland, p. 605. Reprinted with permission from Kluwer Academic Publishers. Data from (1) F. J. Millero, p. 131 in *River Inputs to Ocean Systems* (eds.: J. M. Martin, J. D. Burton, and D. Eisma), reprinted with permission from UNIPUB, copyright © 1980 by UNEP/SCOR, Rome; (2) H. W. Nürnberg, p. 216 in *Trace Element Speciation in Surface Waters and Its Ecological Implications* (ed.: G. G. Leppard), reprinted with permission from Plenum Press, copyright © 1983 by Plenum, Publishing Corporation, New York; and (3) M. Bernhard, E. D. Goldberg, and A. Piro, p. 57 in *The Nature of Seawater* (ed.: E. D. Goldberg), reprinted with permission from Abakon Verlagsgesellscaft, copyright © 1975 by Abakon Verlagsgesellscaft, Berlin, Germany.

[a]Values are percentages.

Organic Matter

Dissolved organic matter contains negatively charged functional groups that can form ion pairs and complexes with dissolved metals. In estuarine waters, where DOM concentrations exceed 10 mg/L, a significant amount of the trace metals are present as organic complexes. By affecting the speciation of trace metals, organic matter can play an important role in controlling the **bioavailability** of these elements. This has important biological consequences as some metals are micronutrients and others are toxins. Our understanding of the role of organic matter in trace metal speciation is greatly limited. This is mostly due to our lack of knowledge concerning the chemical nature and behavior of most of the dissolved organic matter in the ocean. This subject is addressed at length in Chapter 23.

Though a great variety of dissolved organic compounds are present in seawater, the functional groups responsible for metal complexation are primarily R—COOH, R—OH, R_2—NH, R—NH_2, and R—SH, where R represents any organic structure. Many organic compounds contain more than one of these groups and thus are able to form multiple bonds or associations with metals. This is called chelation. Some examples are given in Figure 5.5.

Most dissolved organic compounds have relatively high affinities for metals. Nevertheless, they have limited influence on the cation speciation in seawater due to their low concentrations. The major cations are typically present at 1000-fold higher concentrations than the dissolved organic compounds or trace metals. Thus most of the binding sites on DOM are occupied by the major cations (primarily calcium and magnesium). While this has little impact on the speciation of the major ions, it greatly limits the relative amount of trace metals that can be complexed by DOM in seawater.

The organic compounds most effective at binding with trace metals are the ones with very high and specific affinities for these ions. The most extreme examples are the biomolecules present in marine organisms. These compounds

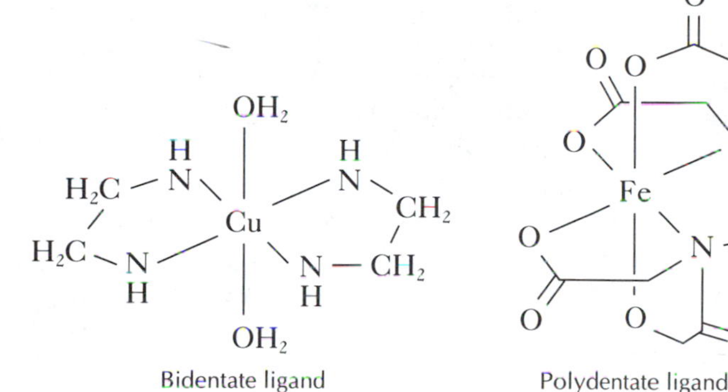

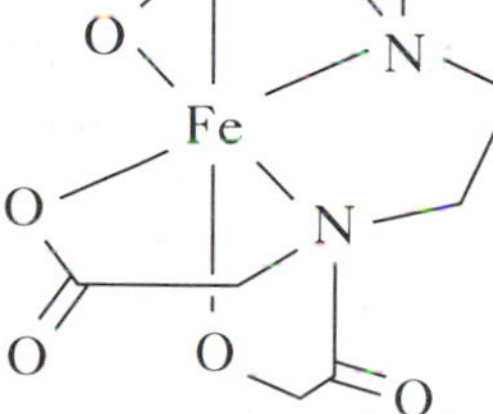

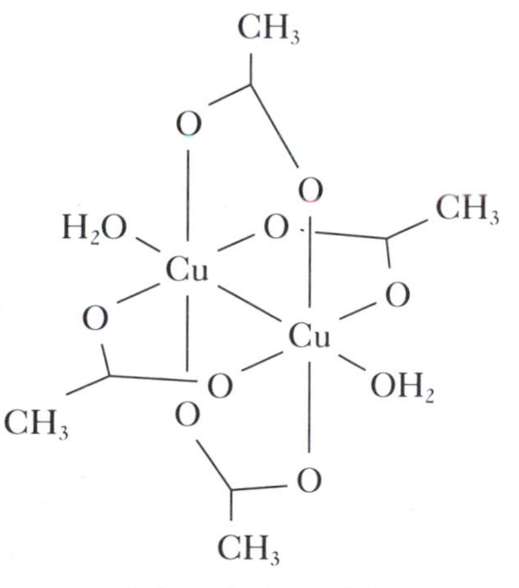

FIGURE 5.5. Examples of chelation. *Source:* From *Principles of Aquatic Chemistry*, F. M. M. Morel, copyright © 1983 by John Wiley & Sons, Inc., New York, p. 240. Reprinted by permission.

are so effective that they enable organisms to selectively concentrate certain trace metals in their tissues up to 1 million-fold over the dissolved metal concentrations in seawater. This great specificity is probably related to the three-dimensional geometry of the biomolecules in a manner analogous to that which enables enzymes to bind only certain substrates.

Some of these biomolecules are released by marine organisms into seawater. The ones with the highest trace metal affinities are large molecules that contain nitrogen, oxygen, and sulfur atoms in aromatic rings. An example, that of a ferrichrome, is given in Figure 5.6. Because of their extraordinarily high affinities for Fe^{3+} ($K \approx 10^{20}$ to 10^{60}), these biomolecules are thought to play an important role in the intracellular transport of iron. They are ubiquitous in marine bacteria, blue-green algae, and fungi. If these biomolecules have similarly high affinities for other trace metals, they could bind significant amounts of trace metals in situations where their concentrations are high, such as in dense algal blooms.

Most of the DOM in seawater is composed of compounds collectively referred to as **humic substances**. Though variable in composition, they tend to be large molecules with many negatively charged functional groups. Humics are thought to be produced by the condensation and cross-linking of organic compounds released during microbial decomposition of particulate organic matter. Because humics are so variable in structure and relatively inert, they are chemically difficult to characterize. Their structure and chemistry is discussed at length in Chapter 23. Due to their great number of negatively charged functional groups, humic substances have high affinities for metals. Lab experiments suggest that they can bind a significant amount of the trace metals in estuarine waters where DOM concentrations exceed 10 mg/L.

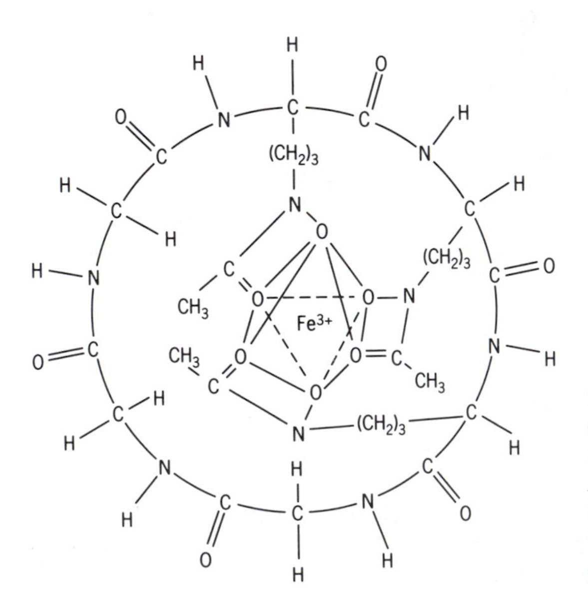

FIGURE 5.6. Molecular structure of hydroxamate ferrichrome. *Source:* From *Principles of Aquatic Chemistry,* F. M. M. Morel, copyright © 1983 by John Wiley & Sons, Inc., New York, p. 275. Reprinted by permission.

The results obtained from speciation calculations have demonstrated the importance of monitoring species activities when investigating the uptake, nutrition, and toxicity of trace metals in marine microorganisms, zooplankton, and some fish. These studies are usually done by observing the response of marine organisms to various amounts of trace metals. The organisms are typically cultured in laboratory vessels so that metal addition can be controlled. The results of two such experiments are given in Figure 5.7.

These experiments demonstrate that even at low concentrations, copper is toxic to plankton. This toxicity is decreased by addition of a chelator such as ethylenediaminetetraacetic acid (EDTA) or Tris. As shown in Figure 5.7*a*, EDTA appears to be more effective at decreasing toxicity, since at any given total copper concentration, cell motility is higher in its presence as compared to Tris. As indicated in Figure 5.7*b*, cell motility is inversely related to the activity of the free cupric ion, regardless of which chelator is present.

These results suggest that copper uptake, and hence toxicity, is lowered by the presence of an organic chelator. This is probably due to the formation of an organo-copper complex that phytoplankton cannot assimilate. If governed by equilibrium thermodynamics, the formation of these complexes causes a decrease in concentration of the other copper species, some of which are evidently toxic. These experiments demonstrate that measuring total copper concentration is not a good way to monitor water quality, since toxic responses can be independently affected as a result of complexation with chelators. Instead, the activity of one of the ion species, such as the free ion, should be measured.

In most cases, reaction with organic compounds reduces trace metal toxicity. An exception is Hg, as its methylated form is taken up more readily than Hg^0. Not all trace metals are toxic; some limit plankton growth when present at low concentrations. These metals are referred to as micronutrients. As shown in Figure 5.7*c*, plant growth is enhanced by the addition of dissolved zinc. Addition of a chelator reduces this enhancement. These results suggest that zinc becomes biologically unavailable to plankton when it is bound in organic complexes.

Metal uptake by phytoplankton is probably initiated by complexation with some ligand which is a part of the exterior of the plant's cell wall. Toxic metals like Cu^{2+} are chemically similar to micronutrients like Zn^{2+}. Thus the two metals would be expected to compete for complexation sites on the cell wall. In seawater, the concentrations of both the toxic and essential trace metals are uniformly low. This suggests plankton growth could be simultaneously controlled by the scarcity of essential trace metals and inhibition by toxic ones.

Acids and Bases

In addition to the formation of ion pairs and complexes, many equilibria that occur in the ocean involve the reaction of acids (proton donors) and bases (proton acceptors). The most abundant weak acids and bases present in seawater are listed in Table 5.9.

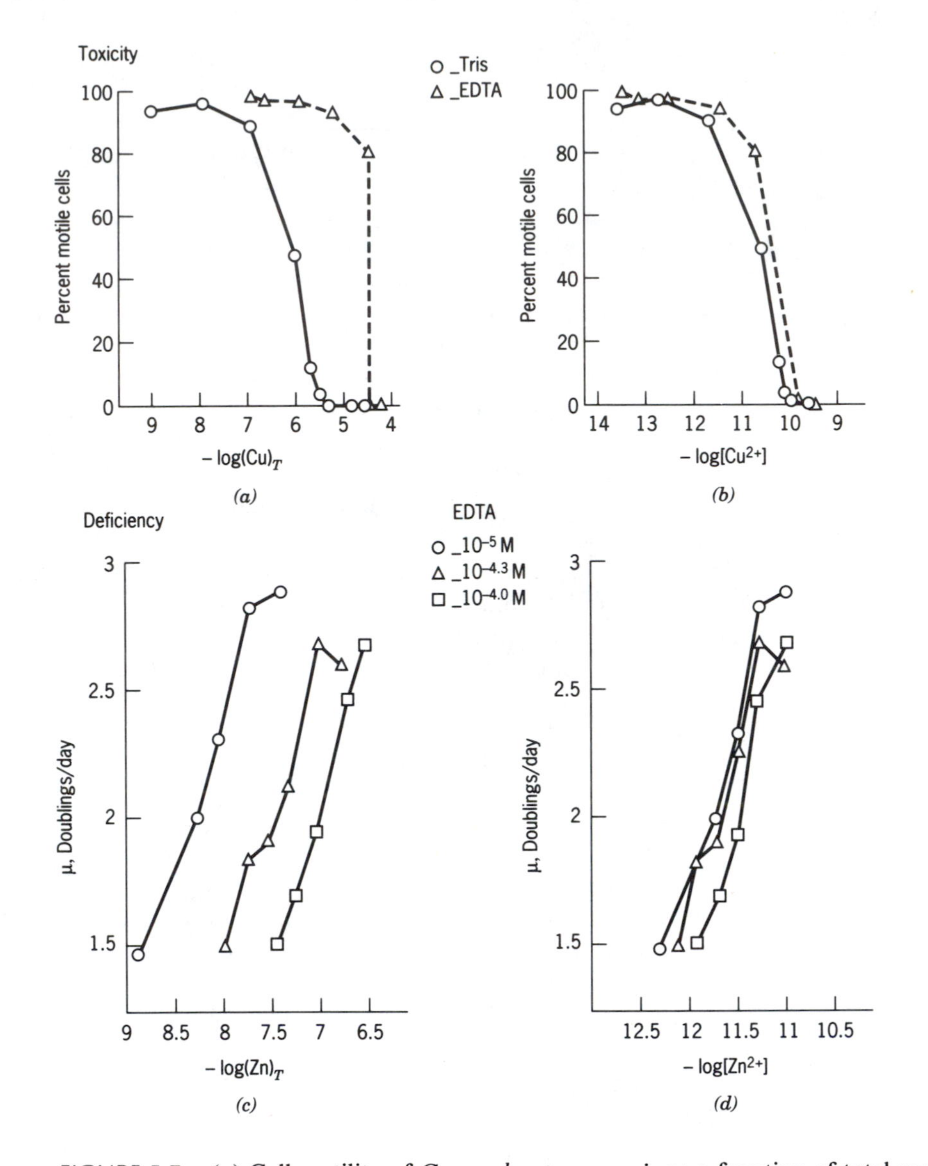

FIGURE 5.7. (*a*) Cell motility of *Gonyaulax tamarensis* as a function of total copper and (*b*) cupric ion activity for two chelators, Tris and EDTA. (*c*) Growth Rate of *Thallassiosira weissflogii* as a function of total zinc and (*d*) zinc ion activity for three concentrations of EDTA. *Source:* From *Principles of Aquatic Chemistry,* F. M. M. Morel, copyright © 1983 by John Wiley & Sons, Inc., New York, p. 302. Reprinted by permission. Copper results from D. A. Anderson and F. M. M. Morel, reprinted with permission from *Limnology and Oceanography,* vol. 23, p. 292, copyright © 1978 by the American Society of Limnology and Oceanography, Seattle, Wash. Zinc results from M. A. Anderson, F. M. M. Morel, and R. R. L. Guillard, reprinted with permission from *Nature,* vol. 276, pp. 70–71, copyright © 1978 by Macmillan Journals, Ltd., London.

TABLE 5.9
Weak Acids and Bases in Aerobic Seawater

	Seawater		
	Warm Surface	*Deep Atlantic*	*Deep Pacific*
Carbonate (mM)	2.1	2.3	2.5
Silicate (μM)	<3	30	150
Ammonia (μM)	<0.5	<0.5	<0.5
Phosphate (μM)	<0.2	1.7	2.5
Borate (mM)	0.4	0.4	0.4

Source: From *Principles of Aquatic Chemistry*, F. M. M. Morel, copyright © 1983 by John Wiley & Sons, Inc., New York, p. 128. Reprinted by permission. Data from *Chemical Oceanography*, vol. 1, J. P. Riley and G. Skirrow, copyright © 1965 by Academic Press, Orlando, FL, p. 241. Reprinted by permission.

The speciation of these compounds involves reaction with water, which is called **hydrolysis**. For acids, this reaction is given by

$$\mathrm{H}A(\mathrm{aq}) \rightleftharpoons \mathrm{H}^+(\mathrm{aq}) + A^-(\mathrm{aq}) \tag{5.35}$$

where A^- is the conjugate base of the acid HA. For bases, the hydrolysis reaction is given by

$$B(\mathrm{aq}) + \mathrm{H_2O(l)} \rightleftharpoons B\mathrm{H}^+(\mathrm{aq}) + \mathrm{OH}^-(\mathrm{aq}) \tag{5.36}$$

where B is a base. The thermodynamic equilibrium constant for the acid hydrolysis is given by

$$K_a = \frac{\{\mathrm{H}^+(\mathrm{aq})\}\{A^-(\mathrm{aq})\}}{\{\mathrm{H}A(\mathrm{aq})\}} \tag{5.37}$$

and for bases by

$$K_b = \frac{\{B\mathrm{H}^+(\mathrm{aq})\}\{\mathrm{OH}^-(\mathrm{aq})\}}{\{B(\mathrm{aq})\}} \tag{5.38}$$

This assumes $\{H_2O\} = 1$, which is nearly true even in seawater. For example, $\{H_2O\} = 0.98$ at 35‰, 25°C and 1 atm.

The larger the value of the equilibrium constant, the greater the strength of the acid or base. Very strong acids, such as HCl and HNO_3, can be considered as completely dissociated in aqueous solutions. Thus the dissolution of 1 mole of HCl in 1 L of water will produce a solution which is 1 M in H^+. Di- and triprotic strong acids, such as H_2SO_4 and H_3PO_4, generate mixtures of the conjugate bases, (i.e., HSO_4^-, SO_4^{2-}, $H_2PO_4^-$, HPO_4^{2-}, and PO_4^{3-}), whose concentrations decrease with increasing ionic charge. Dissolution of 1 mole of a weak acid will produce a solution which has a $[H^+] < 1$ M because a significant amount of HA will be present at equilibrium.

The increase in $[H^+]$ produced by the dissociation of an acid drives the following equilibrium to the left:

$$H_2O \rightleftharpoons H^+(aq) + OH^-(aq) \tag{5.39}$$

thereby causing a decrease in $[OH^-]$, since

$$K_w = \{H^+(aq)\}\{OH^-(aq)\} \tag{5.40}$$

is a constant. As with any equilibrium constant, K_w is a function of temperature and pressure. The effect of the former can be significant, as shown in Table 5.10. The effect of pressure also becomes significant at values greater than 100 atm, which is equivalent to water depths exceeding 1000 m.

H^+ activity is usually reported as **pH,** which is defined as

$$pH = -\log\{H^+\} \tag{5.41}$$

Thus at 25°C and 1 atm, a solution with a pH which is less than 7 has

TABLE 5.10
Equilibrium Constants for Dissociation of Water, 0 to 60°C

T(°C)	$K_w \times 10^{-14}$
0	0.1139
5	0.1846
10	0.2920
15	0.4505
20	0.6809
25	1.008
30	1.469
35	2.089
40	2.919
45	4.018
50	5.474
55	7.297
60	9.614

Source: From *The Physical Chemistry of Electrolytic Solutions*, H. S. Harned and B. B. Owen, copyright © 1958 by Reinhold, New York, p. 638. Reprinted by permission.

$\{H^+(aq)\} > \{OH^-(aq)\}$ and is termed **acidic**. For pH's greater than 7, $\{H^+(aq)\} < \{OH^-(aq)\}$, so the solution is termed **basic**.

The speciation of polyprotic acids is complicated by the occurrence of multiple hydrolyses. For example, the dissolution of carbonic acid involves the following equilibria:

$$CO_2(aq) + H_2O(l) \rightleftharpoons H_2CO_3(aq) \tag{5.42}$$

$$H_2CO_3(aq) \rightleftharpoons H^+(aq) + HCO_3^-(aq) \tag{5.43}$$

$$HCO_3^-(aq) \rightleftharpoons H^+(aq) + CO_3^{2-}(aq) \tag{5.44}$$

As a result of these reactions, the carbonate system can buffer against changes in pH caused by addition of acid as shown by the following:

$$H^+(aq) + HCO_3^-(aq) \rightarrow H_2CO_3(aq) \tag{5.45}$$

$$H^+(aq) + CO_3^{2-}(aq) \rightarrow HCO_3^-(aq) \tag{5.46}$$

or by addition of a base:

$$OH^-(aq) + HCO_3^-(aq) \rightarrow H_2O(l) + CO_3^{2-}(aq) \tag{5.47}$$

$$OH^-(aq) + H_2CO_3(aq) \rightarrow H_2O(l) + HCO_3^-(aq) \tag{5.48}$$

On time scales of less than a few million years, these buffering reactions keep the pH of seawater constant at a value of 8. The other weak acids, except for borate, have a negligible impact on the pH and buffering ability of seawater due to their relatively low concentrations.

Speciation calculations can be performed for the weak acids and bases in a fashion similar to that presented earlier for the major ions. Appropriate equilibrium constants can be found in Appendix X. The results of these calculations are presented as a function of pH in Figure 5.8. At the pH of seawater, the dominant species are carbonate, bicarbonate, ammonium, hydrogen phosphate, dihydrogen phosphate, and boric and silicic acid.

As noted already for carbonate and sulfate, some conjugate bases are subject to a significant amount of ion pairing. This is also seen in phosphate, whose speciation is strongly influenced by ion pairing and acid–base reactions, as shown below (Figure 5.9).

Minerals

Some solutes in seawater participate in equilibria that involve minerals. These reactions can be represented by

$$AB(s) \rightleftharpoons A^+(aq) + B^-(aq) \tag{5.49}$$

where the thermodynamic equilibrium constant is

$$K_{eq} = \frac{\{A^+(aq)\}\{B^-(aq)\}}{\{AB(s)\}} \tag{5.50}$$

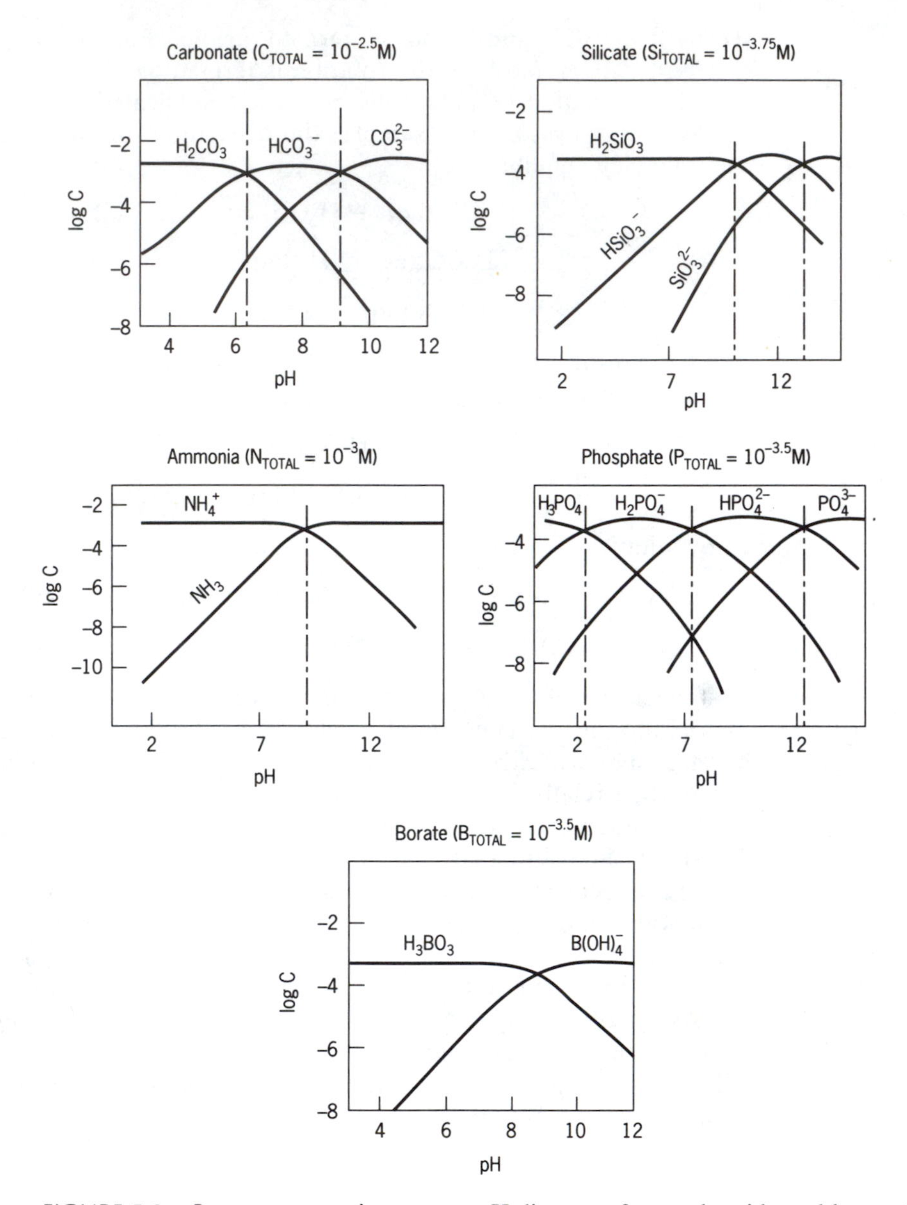

FIGURE 5.8. Log concentration versus pH diagrams for weak acids and bases that are most common in seawater. *Source:* From *Principles of Aquatic Chemistry,* F. M. M. Morel, copyright © 1983 by John Wiley & Sons, Inc., New York, p. 129. Reprinted by permission.

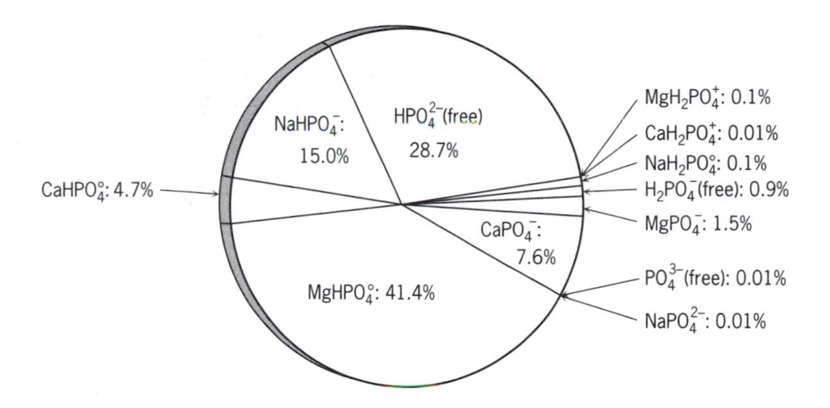

FIGURE 5.9. Ion pairing of phosphate species. *Source:* From E. Atlas, K. Johnson, and R. M. Pytkowicz, reprinted with permission from *Marine Chemistry,* vol. 4, p. 253, copyright © 1976 by Elsevier Science Publishers, Amsterdam, The Netherlands.

For pure mineral phases, $\{AB(s)\} = 1$* and Eq. 5.50 becomes

$$K_{sp} = \{A^+(aq)\}\{B^-(aq)\} \tag{5.51}$$

where K_{sp} is called the **solubility product.**

The solution behavior of a mineral can be predicted by comparing the observed **ion activity product (IAP)** to the K_{sp}. If the IAP is greater than the K_{sp}, the solution is supersaturated with respect to that mineral. In this case, precipitation should proceed spontaneously until the ion activities are decreased to levels dictated by the K_{sp}. Conversely, if the IAP is less than the K_{sp}, the solution is undersaturated with respect to that mineral, so dissolution should proceed spontaneously until the ion activities are increased to levels dictated by the K_{sp}. As with any equilibrium, mineral solubility is a function of temperature, pressure, and the ionic strength of the solution.

For some minerals, solubility is enhanced by the formation of ion pairs and complexes. This is called the salting-in effect. Thus, solubility calculations in seawater must take into consideration the effects of ion speciation.

The surface of most solids, both colloids and particulate material, possess negatively charged sites. These sites bind cations. This complexation reaction is usually referred to as **adsorption** or **ion exchange** due to its ready reversibility. In locations where the concentration of these solids are high, adsorption can have a significant impact on ion speciation.

Colloids are particles that range in diameter from 100 to 1000 Å and can be thought of as gels or sols. In seawater, their composition is greatly variable. Some are inorganic substances, such as amorphous silica and polymetallic

*This is only true in dilute solutions, because at high ionic strengths, the adsorption of ions alters the mineralogy of the solid's surface, thereby affecting its activity.

oxyhydroxides, whereas others are composed of organic compounds. Many are composites of organic and inorganic materials.

Some organic colloids are aggregations of biopolymers. Marine organisms commonly exude and excrete biopolymers such as gelatin, starch, and proteins. The high ionic strength of seawater causes the aggregation of these dissolved organic compounds. The formation of biocolloids is probably aided by bioflocculation, which is caused by the coagulation of bacteria and plankton onto surfaces and each other. The most abundant type of organic colloids are aggregations of humic substances.

All colloids have relatively high affinities for dissolved trace metals. Some remain suspended in seawater, whereas others continue to aggregate until they become dense enough to sink. Such particles continue to adsorb metals during their transit through the water column. This process represents a major sink for dissolved trace metals.

Other Equilibria

Ion speciation is also controlled by other chemical reactions, some of which attain equilibrium. Those that involve gases are the subject of the next chapter. The others are **redox reactions**, which involve a competition for electrons. Since most of these reactions are controlled by the biochemical activities of marine organisms, they rarely attain equilibrium. But even in these cases, theoretical equilibrium calculations provide useful information because they predict the spontaneous direction of the redox reactions. They can also provide some insight into the relative rates of the reactions.

SUMMARY

Many chemical changes that elements undergo as they pass through the ocean and sediments involve chemical **equilibria**. If the reaction stoichiometry and **equilibrium constants** are known, the equilibrium concentrations of the reactants and products can be calculated. To do this, seawater must be treated as a **nonideal solution**. This behavior is the result of solute–solute and solute–solvent interactions that occur in solutions of high **ionic strength**. Because of these interactions, thermodynamic equilibrium constants must be defined in terms of **activities**, or effective concentrations.

Effective concentrations are somewhat less than the total concentrations of the solutes due to **specific** and **nonspecific interactions**. The latter occur as **ion pairs** and **coordination complexes**. The formation of these species can be thought of as the result of a competition among **ligands** for coordination sites on the metal cations. The greater the affinity of the ligand for a metal, the larger the equilibrium constant for the ion complex. These equilibrium constants can be used to generate **mass balance equations** for each solute. The resulting set of equations can be solved **iteratively** to obtain unique values for the concentration of each ion species. Due to the great number of solutes, speciation calculations for seawater are best done by computer.

The results demonstrate that the major ions are mostly present as hydrated ions, except for sulfate, which is present predominantly as $NaSO_4^-$ and $MgSO_4^0$. In comparison, significant amounts of the minor ions, such as fluoride and carbonate, are ion paired. Trace metals tend to form coordination complexes with the hydroxide, chloride, sulfate, and carbonate ions.

Dissolved organic matter (DOM) possesses many negatively charged functional groups. As a result, some compounds have relatively high and specific affinities for dissolved trace metals. Examples include **humic substances** and other high-molecular weight compounds exuded by marine microbes. A significant amount of trace metals is complexed in estuarine waters, where DOM concentrations are high. This complexation is of importance to marine plankton as it limits the **bioavailability** of trace metals. The formation of organometallic complexes can be growth enhancing, if the bound metal is toxic. On the other hand, complexation can limit plant growth by reducing the bioavailability of essential metals. Toxic metals and organic matter are common components of marine pollution. Thus speciation calculations must be done to predict the impacts of these pollutants on marine life.

Some ion species present in seawater are products of the **hydrolysis** of acids and bases. Equilibria involving the most abundant of the weak acid species, carbonate and bicarbonate, maintain the **pH** of seawater at a relatively constant value of 8. Ion speciation in seawater is also influenced by equilibria that involve minerals. In this case, the species's activities can be used to predict mineral solubility. Minerals will spontaneously precipitate in solutions that are **supersaturated**. Conversely, they will spontaneously dissolve in solutions that are **undersaturated**. The degree of saturation is determined by comparing the observed **ion activity product (IAP)** to the **solubility product** (**K_{sp}**).

Most solids, including colloids and particulate matter, possess negatively charged functional groups on their surfaces. Thus they influence metal speciation through complexation on their surfaces. This process is termed **adsorption**, or **ion exchange**, since it is readily reversible. A significant amount of these particles settle to the seafloor. As a result, metal adsorption onto sinking particles is a major mechanism by which trace metals are removed from the ocean.

Another important class of equilibrium reactions are those governed by competition for electrons. These **redox reactions** are controlled by the metabolic processes of marine organisms and as a result rarely attain equilibrium. But even in this case, theoretical equilibrium calculations provide useful information as they predict the spontaneous direction of the redox reactions. They can also provide some insight into the relative rates of redox reactions. The redox chemistry of seawater is the subject of the next part of this text.

The results obtained from speciation calculations are dependent on the quality of the thermodynamic data used. Since equilibrium constants and activity coefficients are functions of temperature, pressure, and the chemical composition of the solution, appropriate values must be used. In practice, this is difficult to do as the quality and quantity of thermodynamic data are still greatly limited. Marine chemists are working to improve the precision and accuracy of the data. Thus the speciation results presented in this chapter are subject to change, particularly for the trace metals and organic ligands.

CHAPTER 6

GAS SOLUBILITY AND EXCHANGE ACROSS THE AIR–SEA INTERFACE

INTRODUCTION

Gases are involved in many biogeochemical cycles. For example, the marine biogeochemical cycle of organic matter is largely controlled by the competing processes of photosynthesis and respiration. During photosynthesis, plants consume carbon dioxide gas (CO_2) and produce oxygen gas (O_2). Animals, bacteria, and other heterotrophic microorganisms use O_2 to respire organic matter. Without O_2, these organisms die. Massive fish kills now common to the New York bight are the result of O_2 deprivation caused indirectly by pollution. Fish kills can also occur in mariculture ponds if the supply rate of O_2 is inadequate to meet the O_2 demand of the animals.

In addition to in situ production during photosynthesis, O_2 is supplied to the surface waters of the ocean by dissolution of atmospheric O_2. The ability of a surface water mass to dissolve O_2 has important consequences for life in the deep sea. Since photosynthesis is restricted to the euphotic zone, the only source of O_2 to the deep sea is through the sinking of surface water masses. Having recently been at the sea surface, these water masses are relatively enriched in O_2. The degree of this enrichment depends on several factors that are discussed in this chapter. If thermohaline circulation were to stop, deep-dwelling aerobic organisms would die.

Carbon dioxide gas is also exchanged between the ocean and atmosphere at the sea surface. The ability of seawater to absorb this gas directly affects global climate because atmospheric CO_2 is a greenhouse gas. The possible warming of Earth's atmosphere as a result of excess CO_2 production from the burning of fossil fuels is the subject of Chapter 25.

Many other gases exchanged at the **air–sea interface** have anthropogenic sources that threaten to overwhelm the natural supply. Like CO_2, some may contribute to the greenhouse effect. Other gases cause acid rain and affect the ozone layer. Thus our understanding of the natural cycling of these gases has become essential to determining the impact of pollution.

In this chapter, the fundamentals of aqueous gas chemistry are described, such as the factors that determine gas solubility and the rate of gas exchange across the air–sea interface. This background information is very important as the remainder of this text deals with biogeochemical processes that frequently involve gases.

DALTON'S LAW OF PARTIAL PRESSURES

The most abundant gases present in the ocean and atmosphere are listed in Table 6.1. In the atmosphere, each gas exerts a pressure, called a **partial pressure**, that is independent of the pressures of the other gases. The symbol used to represent the partial pressure of a gas A is P_A. The total pressure (P_T) exerted by the gases is the sum of the partial pressures of the component gases. This is termed Dalton's Law of Partial Pressures. This law implies that the behavior of each gas is independent of all others. Hence the ideal gas laws, which are defined in terms of total pressure, can be used interchangeably for partial pressures.

TABLE 6.1
Composition of the Atmosphere

Constituent	*Formula*	*Abundance by Volume*	
Nitrogen	N_2	78.084 ± 0.004%	
Oxygen	O_2	20.948 ± 0.002%	
Argon	Ar	0.934 ± 0.001%	
Water vapor	H_2O	Variable (% to	ppm)
Carbon dioxide	CO_2	348	ppm[a]
Neon	Ne	18	ppm
Helium	He	5	ppm
Krypton	Kr	1	ppm
Xenon	Xe	0.08	ppm
Methane	CH_4	2	ppm
Hydrogen	H_2	0.5	ppm
Nitrous oxide	N_2O	0.3	ppm
Carbon monoxide	CO	0.05 to 0.2	ppm
Ozone	O_3	Variable (0.02 to 10	ppm)
Ammonia	NH_3	4	ppb
Nitrogen dioxide	NO_2	1	ppb
Sulfur dioxide	SO_2	1	ppb
Hydrogen sulfide	H_2S	0.05	ppb

Source: From *The Handbook of Environmental Chemistry*, M. Schidlowski, (ed.: O. Hutzinger), copyright © 1986 by Springer-Verlag, Heidelberg, Germany, p. 152. Reprinted by permission. Data from 1) *U.S. Standard Atmosphere*, reprinted with permission from NOAA/NASA/U.S. Air Force, copyright © 1976 by NOAA/NASA/U.S. Air Force, Washington, DC, p. 33; and 2) *Evolution of the Atmosphere*, J. C. G. Walker, reprinted with permission from Macmillan, copyright © 1977 by Macmillan Publishing Co., New York, p. 20.
[a] as of 1987.

At sea level, the total gas pressure in the atmosphere is relatively constant, being approximately 1 atm. Minor variations are caused by physical phenomena, such as storms, which lower atmospheric pressure. Others are caused by chemical processes, such as seasonal variations in plant and animal activity, which cause small fluctuations in the partial pressures of CO_2 and O_2.

GAS SOLUBILITY

Gas molecules are continually passing into and out of the sea surface. When the rates of exchange are equal, the gas is said to be at equilibrium. At gaseous equilibrium, the atmospheric and aqueous concentrations remain constant over time. As such, equilibria are a type of steady state in which the transport processes are reversible. Net transfer across the air–sea interface occurs if the gas is not present at its equilibrium concentration. If present in excess in seawater, a gas will undergo net transport from the ocean to the atmosphere until its aqueous concentration is lowered to the equilibrium level. Likewise, a gas will undergo net transport into the ocean if its aqueous concentration is less than the equilibrium level.

Gaseous equilibrium can be represented by the following expression:

$$A(\mathrm{g}) \rightleftharpoons A(\mathrm{aq}) \tag{6.1}$$

where A(g) and A(aq) represent gas A in the gaseous state and aqueous solution, respectively. For a dilute solution in which concentrations are approximately equal to activities, the thermodynamic equilibrium constant for gas exchange is given by

$$K_{\mathrm{eq}} = \frac{[A(\mathrm{aq})]}{[A(\mathrm{g})]} \tag{6.2}$$

$[A(\mathrm{g})]$ is usually expressed as a partial pressure, P_A. Since $PV = nRT$, P_A is related to $[A(\mathrm{g})]$ as follows:

$$[A(\mathrm{g})] = \frac{n}{V} = \frac{P_A}{RT} \tag{6.3}$$

Substituting P_A/RT into Eq. 6.2 and solving for $[A(\mathrm{aq})]$ yields

$$[A(\mathrm{aq})] = \frac{K_{\mathrm{eq}}}{RT} P_A \tag{6.4}$$

This is known as **Henry's Law**, and is usually written as

$$[A(\mathrm{aq})] = K_H \times P_A \tag{6.5}$$

where Henry's Law Constant, K_H, is equal to K_{eq}/RT.

The partial pressure of CO_2 in our atmosphere is presently 348 ppm or 348 μatm. The K_H for CO_2 at 0°C and 35‰ is 0.065. Thus the concentration of CO_2 which would be achieved if seawater reached gaseous equilibrium with the atmosphere would be 23 μM.

Marine chemists also express gas concentrations in units of ml gas/L seawater (SW). In this unit, the amount of gas is expressed as the volume it would occupy if at standard temperature and pressure (STP). Since 1 mole of gas occupies 22,400 ml at STP,

$$\frac{\text{ml gas}}{\text{L SW}} = \frac{\text{mol gas}}{\text{L SW}} \times \frac{22{,}400 \text{ ml gas}}{\text{mol gas}} \tag{6.6}$$

Thus Henry's Law can be converted to units of ml gas/L SW by multiplying both sides of Eq. 6.5 by 22,400. This yields

$$\frac{\text{mol gas}}{\text{L SW}} \times 22{,}400 = \frac{\text{ml gas}}{\text{L SW}} = K_H \times 22{,}400 \times P_A \tag{6.7}$$

The product, $22{,}400 \times K_H$, is termed the **Bunsen Solubility Coefficient (α_A)**. It can be thought of as the milliliters of gas present in 1 L of water if equilibrium is achieved at STP. Thus the equilibrium gas concentration, in units of ml gas/L SW, is given by

$$[A(\text{aq})] = \alpha_A \times P_A \tag{6.8}$$

where α_A has the units of ml gas L SW^{-1} atm gas^{-1}.

Since K_H and α are inversely proportional to temperature, gas solubility increases with decreasing temperature. As shown in Figure 6.1, this effect is nonlinear for the higher molecular weight gases, which also tend to be more soluble than the smaller ones. Gas solubility also decreases with increasing salinity and hydrostatic pressure. Empirical equations for each gas have been developed which relate the Bunsen Coefficient to variations in temperature and salinity.

These equilibrium concentrations can be thought of as the gas concentration that a water mass would attain if it were to equilibrate with the atmosphere at its in situ temperature and salinity. This is termed the **normal atmospheric equilibrium concentration**, or **NAEC**. The NAECs for the most abundant atmospheric gases are given in Table 6.2 for seawater of 35‰ salinity and temperatures of 0° and 24°C. These NAECs were calculated using Eq. 6.8 and values of the Bunsen Solubility Coefficient as given in Figure 6.1. The NAECs of O_2 and N_2 at other temperatures and salinities are given in Appendix VIII.

The solubility relationships given in Eqs. 6.5 and 6.8 can be used to compute the partial pressure that a gas exerts in a solution. For example, the average concentration of CO_2 in cold surface water is 10 μM. Using the K_H for 0°C, this gas would exert a partial pressure of 150 μatm in the surface seawater. Since this is less than its atmospheric partial pressure (348 μatm), this seawater should experience a net influx of CO_2 from the atmosphere.

DEVIATIONS FROM NAEC

In the example given above, the in situ gas concentration is less than that of its NAEC, so the seawater is said to be **undersaturated** with respect to CO_2.

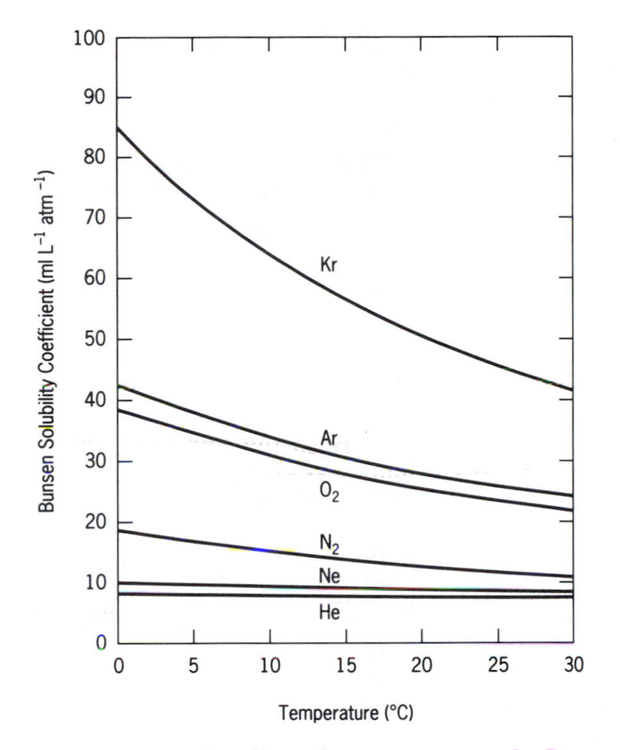

FIGURE 6.1. The effect of temperature on the Bunsen solubility coefficients of some atmospheric gases for seawater of 35‰. *Source:* From *Chemical Oceanography*, W. S. Broecker, copyright © 1974 by Harcourt, Brace and Jovanovich, Publishers, Orlando, FL, p. 119. Reprinted by permission.

If the observed in situ gas concentrations exceed the NAEC, the seawater is said to be **supersaturated** with respect to the gas. If the in situ concentration is equal to the NAEC, the seawater is said to be **saturated**. The degree of gas saturation is usually expressed as

$$\% \text{ saturation} = \frac{[A] \text{ in situ}}{\text{NAEC for } A} \times 100 \tag{6.9}$$

At equilibrium, seawater is 100 percent saturated. If the percentage of saturation is greater than 100, the seawater is supersaturated. If the percentage of saturation is less than 100, the seawater is undersaturated.

Most deviations from NAECs are the result of **nonconservative** behavior involving chemical processes that remove or supply the gas faster than the seawater can reequilibrate with the atmosphere. For example, O_2 is often supersaturated in the mixed layer due to rapid production by phytoplankton.

The gases that are relatively inert, such as Rn, Ar, He, and N_2, are usually present at 100 percent saturation since their concentrations are largely controlled by physical processes. As with the major ions, exceptions to this

TABLE 6.2
Normal Atmospheric Equilibrium Concentrations of Various Gases in Surface Ocean Water

Gas	Partial Pressure in Dry Air (atm)	Equilibrium Concentration in Surface Seawater (ml/L)[a] 0°C	24°C
H_2	5×10^{-7}		
He	5.2×10^{-6}	4.1×10^{-5}	3.4×10^{-5}
Ne	1.8×10^{-5}	1.7×10^{-4}	1.5×10^{-4}
N_2	.781	14	9
O_2	.209	8.8	5.5
Ar	9.3×10^{-3}	.36	.22
CO_2	3.2×10^{-4}	.47	.23
Kr	1.1×10^{-6}	8.1×10^{-5}	4.9×10^{-5}
Xe	8.6×10^{-8}	1.2×10^{-5}	$.6 \times 10^{-5}$

Source: From *Chemical Oceanography*, W. S. Broecker, copyright © 1974 by Harcourt, Brace and Jovanovich, Publishers, Orlando, FL, p. 118. Reprinted by permission.
[a]Salinity of the seawater is assumed to be 35‰.

conservative behavior do occur. Some of these deviations are caused by physical processes as described below.

Postequilibration Temperature Changes

If the temperature of a water mass is altered after having been isolated from the sea surface, apparent supersaturations can result. Such alterations can result from the mixing of water masses that have different temperatures. As shown in Figure 6.2, the complete mixing of equal volumes will produce a new

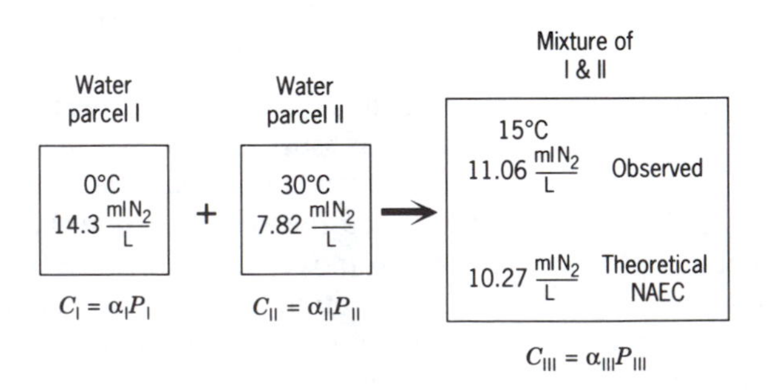

FIGURE 6.2. Changes in percent saturation caused by the mixing of two water masses.

water mass with a temperature intermediate between the two original water masses. If the water masses contain a conservative gas, such as N_2, the gas concentration in the new water mass will also be halfway between the NAECs of the original water masses. In this example, mixing produces a water mass with a gas concentration of 11.06 ml N_2/L and temperature of 15°C. But Appendix VIII lists the NAEC for N_2 at 15°C as 10.27 ml N_2/L. Therefore the mixing has caused an apparent supersaturation of 114 percent.

How can this be? No additional gas was added to the water. As shown in Figure 6.3*a*, the NAEC is always somewhat less than the actual gas concentration produced by the mixing of the water masses. The cause of this is illustrated in Figure 6.3*b* (i.e., the Bunsen Coefficient is not directly proportional to temperature). Due to the concave nature of the curve, supersaturations, rather than undersaturations, are always the result.

Such apparent supersaturations can be produced during turbulent mixing between adjacent deep-water masses of different temperatures. The greatest temperature differences occur near hydrothermal vents, which introduce high-temperature vent fluids (up to 350°C) into the cold (~0°C) deep waters. Such mixing produces the most marked supersaturations and can be used as an indicator of the presence of hydrothermal fluids.

Changes in Atmospheric Partial Pressure

Changes in the atmospheric partial pressure of a gas can also produce apparent deviations from saturation. If this change is rapid, the surface waters will not have enough time to equilibrate with the new atmospheric partial pressure. Thus the dissolved gas concentration will not be in equilibrium with this new partial pressure. As mentioned above, the atmospheric partial pressures of some gases do vary on a seasonal basis.

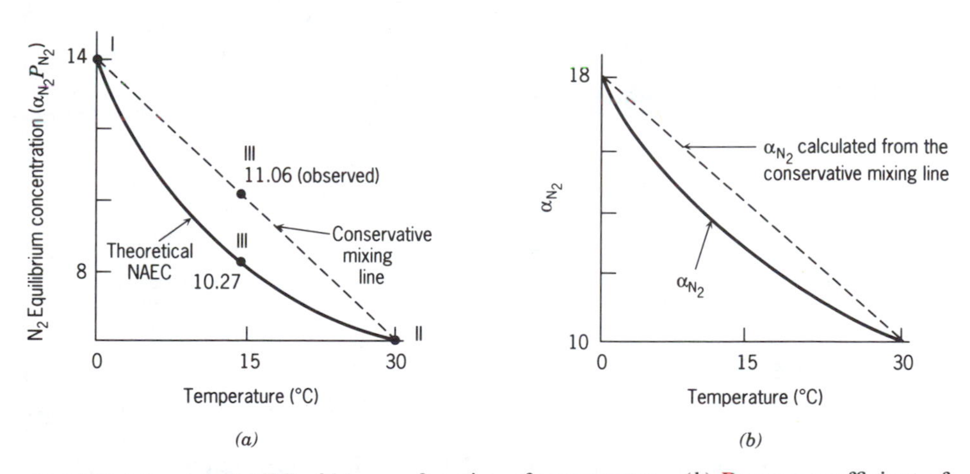

FIGURE 6.3. (*a*) NAEC of N_2 as a function of temperature. (*b*) Bunsen coefficient of N_2 as a function of temperature.

Kinetic Considerations

The establishment of equilibrium takes time. If environmental conditions change faster than equilibrium can be attained, deviations from NAECs result. For example, rapid changes in temperature can cause disequilibrium. Wind stress can cause supersaturation by increasing turbulence to the point where gas is forced into the sea surface by bubble injection. Deviations from equilibrium can also result from barriers to gas exchange, such as sea-surface slicks. For example, slicks left by oil spills can reduce the rate of influx of O_2 from the atmosphere and eventually cause fish and other animals to suffocate.

RATES OF GAS EXCHANGE

Thin-Film Model

In situations where equilibrium is not attained, the rate of gas exchange across the air–sea interface can be inferred from kinetic models of gas behavior. The most commonly used model is illustrated in Figure 6.4. In this model, a thin layer of stagnant water is assumed to act as a barrier to gas exchange across the air–sea interface. The atmosphere above the **thin film** is assumed to be

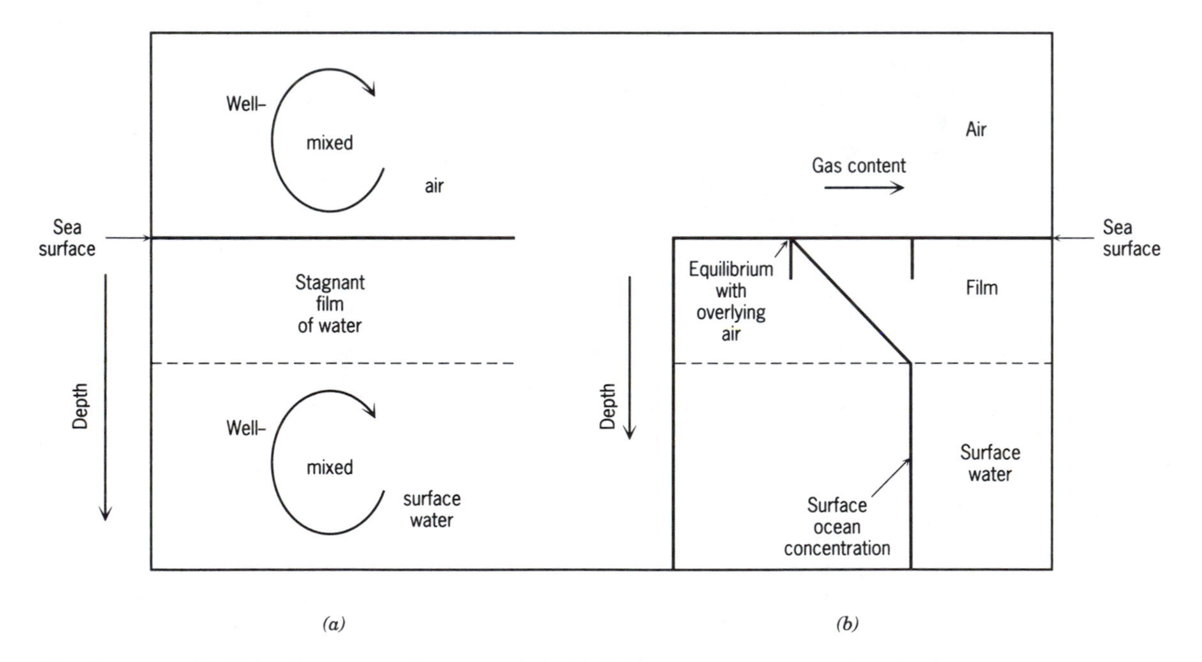

FIGURE 6.4. (*a*) Thin-film gas exchange model. (*b*) Relative gas concentrations in the atmosphere, thin film, and underlying surface water. *Source:* From *Chemical Oceanography,* W. S. Broecker, copyright © 1974 by Harcourt, Brace and Jovanovich, Publishers, Orlando, FL, p. 126. Reprinted by permission.

well mixed, as is the water below the thin film. The top of the thin film is assumed to be in gaseous equilibrium with the atmosphere. Due to contact with the mixed layer, the bottom of the thin film has the same gas concentration as in the mixed layer.

Gas moves through the stagnant layer by the process of molecular diffusion. If the gas concentration at the top of the stagnant layer is different from that at the bottom, net diffusion of gas molecules occurs. As with ionic solutes, net diffusion proceeds from the region of higher concentration to the region of lower concentration until the solution is homogeneous. When this occurs, gaseous equilibrium between the atmosphere and the mixed layer is achieved. Since the number of molecules entering the sea surface equals the number escaping per unit time, no net flux of gas occurs at equilibrium.

In the case of disequilibrium, the net flux of gas molecules (F_A) is directly proportional to the concentration gradient ($d[A]/dz$) in the stagnant layer, as given by Eq. 6.10:

$$F_A = D_A \frac{d[A]}{dz} = D_A \frac{[A(\text{aq})]_{\text{top}} - [A(\text{aq})]_{\text{bottom}}}{z} \tag{6.10}$$

where z is the film thickness and D_A is the molecular diffusivity coefficient. The concentration gradient is estimated from $[A(\text{aq})]_{\text{top}} - [A(\text{aq})]_{\text{bottom}}$, which is the difference in gas concentration between the top and bottom of the thin film. In practice, the concentration at the top of the layer is inferred from the gas's atmospheric partial pressure. The concentration at the bottom of the layer is assumed to be equal to that of the mixed layer. The net gas flux is usually reported as the number of moles that pass through a square meter of sea surface in 1 year. For example, the net helium flux out of the sea is 1.8×10^{-6} mol m^{-2} y^{-1} at 0°C.

The net diffusive flux depends on the magnitude of the concentration gradient. The larger the difference in concentration between the top and bottom of the stagnant film, the greater the flux of gas across the air–sea interface. Likewise, the greater the thickness of the film, the smaller the flux of gas across the air–sea interface. From the measurement of gas fluxes and concentration gradients, oceanographers have inferred that the stagnant-film thickness ranges from 10 to 60 μm.

The thickness of the stagnant film is enhanced by the presence of a sea-surface microlayer. These microlayers are films of DOM that range from 50 to 100 μm in thickness. They also tend to be enriched in metals, most of which are complexed with the organic matter. Bacteria and plankton are thought to be the source of this DOM, since these organisms are also present at elevated concentrations in the sea-surface microlayers. Photochemical oxidation of this DOM appears to be a large source of carbon monoxide, dimethylsulfide, and bromine, causing a net flux of these gases from the ocean. The occurrence of sea-surface microlayers is patchy as they are dispersed by breaking waves.

Molecular diffusivity coefficients range in magnitude from 1×10^{-5} to 4×10^{-5} cm^2/s. As shown in Table 6.3, D_A's increase with increasing temperature

TABLE 6.3
Molecular Diffusivity Coefficients of Various Gases in Seawater

Gas	*Molecular Weight, (g/mol)*	*Diffusion Coefficient ($\times 10^{-5}$ cm²/s)* 0°C	24°C
H_2	2	2.0	4.9
He	4	3.0	5.8
Ne	20	1.4	2.8
N_2	28	1.1	2.1
O_2	32	1.2	2.3
Ar	40	.8	1.5
CO_2	44	1.0	1.9
Rn	222	.7	1.4

Source: From *Chemical Oceanography*, W. S. Broecker, copyright © 1974 by Harcourt, Brace, and Jovanovich Publishers, Orlando, FL, p. 127. Reprinted by permission. See Broecker and Peng (1982), p. 119, for data sources.

and decreasing molecular weight. This causes gas fluxes to increase with increasing temperature and decreasing molecular weight.

In comparison, eddy diffusivity constants range in magnitude from 1 to 100 cm²/s. Thus when vertical turbulence is present, as in the water below the stagnant film, transport due to mixing overwhelms that from molecular diffusion. In other words, molecular diffusion is a significant transport mechanism only under still (stagnant) conditions. As a result, stagnant films greatly slow the rate at which gaseous equilibrium with the atmosphere is attained.

When vertical turbulence is strong, the stagnant-film thickness is reduced. This is usually the result of strong winds, as shown in Figure 6.5. The choppier the waves, the greater the area across which gas can exchange. Thus winds increase gas fluxes by increasing the surface area of the air–sea interface and by causing bubble injection.

The rate of gas exchange across the air–sea interface can be expressed as the ratio of D_A to z. D_A has units of cm²/s, and z has units of micrometers ($\mu m = 10^{-4}$ cm). Thus this ratio has units of cm/s and can be thought of as the rate at which a column of gas is pushed through the water column. This rate is called a piston velocity. For D_A's of approximately 2×10^{-5} cm²/y and z's that average 40 μm, piston velocities are approximately 5×10^{-3} cm/s or 1600 m/y. This is equivalent to pushing a 4-m-high column of gas through the sea surface every day.

Piston velocities can be used to compute the time required for gaseous equilibration with the atmosphere. If the mixed layer ranges in depth from 20

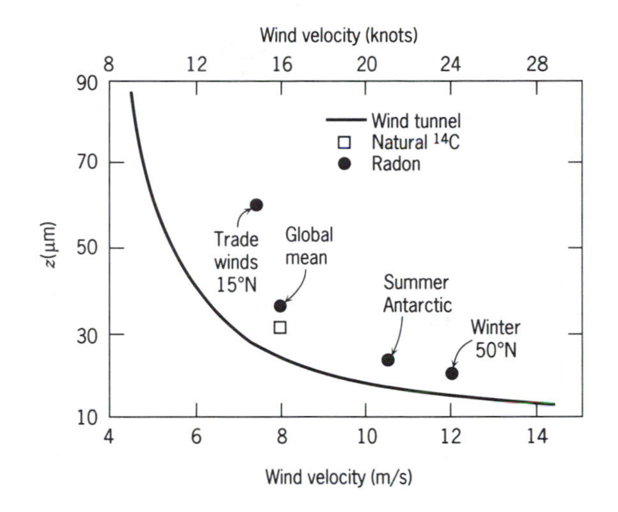

FIGURE 6.5. The effect of wind velocity on thin-film thickness. The solid line represents results obtained from measurements made in wind tunnels. In situ measurements were made from distributions of the naturally occurring radioisotopes of carbon and radon as explained in Chapter 28. From *Tracers in the Sea,* W. S. Broecker and T.-H. Peng, copyright © 1982 by the Lamont-Doherty Geological Observatory, Palisades, NY, p. 128. See Broecker and Peng (1982) for data sources.

to 100 m, the length of time required to push all of the gas through this zone is 20 to 100 m divided by 4 m/day, or 5 to 20 days.

Surface-Renewal Model

In the surface-renewal model, the sea surface is viewed as a region occupied by slabs of water. These slabs are transported to the sea surface as a result of turbulent mixing. While at the sea surface, the top of the slab reaches gaseous equilibrium with the atmosphere, while its bottom retains the chemical composition of the underlying mixed layer. Thus a gas concentration gradient is set up in the slab. Turbulent mixing eventually returns the slab to the mixed layer. Its site at the sea surface is immediately occupied by another slab. If the rate of slab exchange is fast, the mixed layer will attain gaseous equilibrium with the atmosphere.

Thus the gas flux across the air–sea interface is determined by the frequency at which the slab is replaced or "renewed." The net diffusive flux is related to the frequency of slab renewal (θ) by

$$\text{Flux} = \sqrt{\frac{D_A}{\theta}}([A(\text{aq})]_{\text{top}} - [A(\text{aq})]_{\text{bottom}}) \tag{6.11}$$

where $[A(\text{aq})]_{\text{top}}$ is the gas concentration at the top of the slab when it is at the sea surface and $[A(\text{aq})]_{\text{bottom}}$ is the concentration at the bottom of the slab. If a concentration gradient exists, the less time the slab spends at the sea surface, the greater the net flux. This replacement time is functionally equivalent to the film thickness in the thin-film model. Though the gas flux in the surface-renewal model is proportional to the square root of D_A, the results from the two models are similar because D_A is small.

NONCONSERVATIVE GASES

Deviations from NAECs are also caused by chemical reactions that occur in the atmosphere and ocean. Marine organisms are the major source or sink of nearly all nonconservative gases. The exceptions are those gases with large anthropogenic inputs, such as the higher molecular weight halogenated hydrocarbons.

For CO_2, H_2S, and NH_3, gas solubility is augmented by reaction with water as shown in Eqs. 6.12 through 6.14. These weak acids and bases partially dissociate, thereby increasing their solubility.

$$CO_2 + H_2O \rightleftharpoons H_2CO_3 \rightleftharpoons H^+ + HCO_3^- \rightleftharpoons 2H^+ + CO_3^{2-} \quad (6.12)$$

$$H_2S + H_2O \rightleftharpoons HS^- + H_3^+O \quad (6.13)$$

$$NH_3 + H_2O \rightleftharpoons NH_4^+ + OH^- \quad (6.14)$$

The extent to which a gas is nonconservative is often reflected in its degree of deviation from saturation. For example, O_2 supersaturations as high as 120 percent have been observed in the euphotic zone. Such high deviations are possible because the rate of O_2 supply from photosynthesis is much faster than the rate at which exchange across the air–sea interface can reestablish gaseous equilibrium. O_2 undersaturations as low as 0 percent have been observed in stagnant waters located in the thermocline. In these regions, the rate of O_2 uptake via aerobic respiration exceeds the rate at which water motion can resupply the gas. Not all chemical reactions proceed at rates fast enough to create significant deviations from NAECs. For example, while N_2 gas is consumed by nitrogen-fixing plankton and produced by denitrifying bacteria, these processes are so slow that water motion and gas diffusion keep the concentration of N_2 close to that of the NAEC.

As mentioned above, the activities of humans have begun to significantly alter the global fluxes of some gases. Some examples are given in Table 6.4.

The combustion of fossil fuels is responsible for increasing the rates of CO_2, NO_x, O_3, and SO_x production. Some are greenhouse gases and some contribute to acid rain. Deforestation is thought to be the cause of the elevated methane fluxes, as this gas is produced by termites during the decomposition of dead wood. Methane is also produced by microbial activity in rice paddies and in the rumen, or guts, of cows. Some of the gases are synthetic compounds invented by humans. These include the chlorinated hydrocarbons, chloroform and freons, which are used as solvents and refrigerants. Air currents have transported freons into the upper atmosphere where they undergo photochemical reactions with ozone. As a result, a significant amount of ozone has been

TABLE 6.4
Net Global Gas Fluxes Across the Air–Sea Interface

Element	Compound	Controlling Resistance	Net Global Air–Sea Flux		Present Atmospheric Level (1987)[e]	Global Man[c] % Human Impact	Marine Flux as % of Global Cycle[d]
			Direction[a]	Magnitude[b]			
H	H_2	r_w	+	10^{12}			
C	HCHO	$r_a \sim r_w$	−	10^{13}			
	CH_4	r_w	+	$\sim 10^{13}$	1.66 ppmv	~240	~2 ↑
	C_2H_6	r_w	+	10^{12}	~ppbv	?	?
	C_3H_8	r_w	+	10^{12}	~ppbv	?	?
	C_2H_4	r_w	+	10^{12}	~0.1 ppbv	?	?
	C_3H_6	r_w	+	10^{14}	~0.1 ppbv	?	?
	CO_2	r_w	−	8×10^{15}	348 ppmv	25	~50 ↓
N	N_2O	r_w	+	6×10^{12}	~310 ppb	7	10–30 ↑
	NO	r_w	+(?)		10–100 pptv	>50	↓ as NO_x?
	NO_2	?	+(?)		10–100 pptv	>50	↓ as NO_x?
	NH_3	r_a	+(?)	?	?10–20 pptv	?	↓ ?
O	O_3	r_w	−	6×10^{14}	?~30–50 ppbv	>50%?	↓ ?
S	Total volatile S		+	$34\text{–}170 \times 10^{12}$ (as S)			
	H_2S	r_w	+	15×10^{12}			
	$(CH_3)_2S$	r_w	+	$30\text{–}80 \times 10^{12}$	~100 pptv	~100	30 ↑
	CS_2	r_w	+	0.3×10^{12}			
	COS	r_w	+	0.8×10^{12}	500 pptv	?<5	Dominant natural marine ↑
	SO_2	r_a	−	5×10^{12}	?10–100 pptv	?~100	↓
Cl	CH_3Cl	r_w	+	$3\text{–}8 \times 10^{12}$	600 pptv	?<5	Major natural marine source ↑

(Table continues on following page)

TABLE 6.4 *Continued*
Net Global Gas Fluxes Across the Air–Sea Interface

Element	Compound	Controlling Resistance	Net Global Air–Sea Flux: Direction[a]	Net Global Air–Sea Flux: Magnitude[b]	Present Atmospheric Level (1987)[e]	Global Man[c] % Human Impact	Marine Flux as % of Global Cycle[d]
	CCl_4	r_w	–	$\sim 10^{10}$	~150 pptv	?large	??
	CCl_3F	r_w	–	3×10^9	~220 pptv	totally manmade	(Trace <1) ↓
	CCl_2F_2	r_w	–	2×10^9	~335 pptv	totally manmade	(Trace <1) ↓
I	CH_3I	r_w	+	$3\text{–}13 \times 10^{11}$	1–20 pptv	?	Major natural marine source

Source: From T. E. Graedel, P. G. Brewer, and F. S. Rowland, reprinted with permission from *Applied Geochemistry*, vol. 3, p. 41, 1988, copyright © 1988 by Pergamon Press, Elmsford, New York. After P. S. Liss and W. G. N. Slinn, *Proc. of the NATO Adv. Study Inst. on Air–Sea Exchange of Gases and Particles*, copyright © 1983 by D. Reidel Publishing Co., Dordrecht, pp. 278–279. Reprinted with permission from Kluwer Academic Publishers, Dordrecht, Holland.

[a] +, sea→air; –, air→sea.

[b] Units are *g* (of the *compound*)/y.

[c] $\frac{\text{Observed} - \text{pre-industrial}}{\text{Pre-industrial}} = \frac{\text{Atmospheric perturbation}}{\text{Natural level}} \times 10^2$.

[d] Oceanic source, ↑; oceanic sink, ↓.

[e] Gas concentrations are given in terms of volume, e.g., 1 ppmv = 1 L of gas in 10^6 L of air.

destroyed. As the natural ozone layer in the upper atmosphere protects the earth from damaging high-energy UV radiation, concern over its loss has lead to an international effort to phase out of the use of freon.

SUMMARY

Gases are an important component of marine biogeochemical cycles. The most abundant atmospheric gases are introduced into the ocean via molecular diffusion across the **air–sea interface**. Net diffusion ceases when the gas concentration in seawater reaches the **normal atmospheric equilibrium concentration (NAEC)**. At this level, gaseous equilibrium with the atmosphere is attained because the rate of gas diffusion into the ocean is equal to the rate of diffusion out of the ocean. The concentration at which this occurs depends on the temperature and the salinity of the seawater, as well as the hydrostatic pressure. Gas solubility is also related to the chemical composition of the gas and atmospheric pressure. The latter relationship is given by **Henry's Law**, which states that gas solubility is directly proportional to the **partial pressure** of that gas in the overlying atmosphere.

Marine chemists measure gas solubility as the number of milliliters of gas that dissolve in 1 L of seawater when the seawater is equilibrated with 1 atm of the gas. The solubility "constant" determined under these conditions is called the **Bunsen Solubility Coefficient (α_A)**. Thus the NAEC of a gas A is calculated by multiplying $\alpha_A \times P_A$, where the value of α_A is corrected for salinity and temperature effects.

Observed deviations from the NAEC are expressed as undersaturations and supersaturations. Most are caused by chemical processes that proceed at rates faster than either gas exchange or water motion can reestablish the NAEC. Gases affected by such chemical processes are termed **nonconservative**. When gaseous disequilibria exist, the net diffusive flux of gas can be calculated from Fick's first law. At the sea surface, net transport is thought to occur by molecular diffusion across a **thin film** of stagnant water located at the air–sea interface. The gas concentration at the top of the film is maintained by gaseous equilibrium with the atmosphere. The gas concentration at the bottom is maintained by contact with the underlying waters of the mixed layer.

The flux of gas through the stagnant film is directly proportional to the concentration gradient within the film. Thus the larger the concentration difference between the top and bottom of the film and the smaller the depth of the film, the greater the flux. Since winds decrease the film, they act to enhance gas fluxes. Fluxes also increase with increasing temperature, due to the effect of heat on the rate of molecular diffusion. The flux of a gas is also dependent on its chemical composition, as the rate at which gases diffuse increases with decreasing molecular weight.

Human activities are a large source of some gases. This has lead to significant impacts on the global biogeochemical cycles of such gases as CO_2, NO_x, SO_x, and O_3.

PROBLEM SET 1

1. Calculate the thickness of a salt layer left after evaporating an ocean of 35‰ salinity and 4-km depth. The density of sea salt is 2.2 g/cm³.
2. Calculate the concentrations in mol/kg of the six major ions in seawater of 38‰ salinity.
3. A sample of deep water from the Pacific Ocean has the following gas content:

$$[N_2] = 12.91 \text{ ml } N_2/\text{L SW}$$
$$[O_2] = 5.26 \text{ ml } O_2/\text{L SW}$$

The pertinent Bunsen Solubility Coefficients are

$$\alpha_{N_2} = 16.5 \text{ ml } N_2 \text{ atm } N_2^{-1} \text{ L SW}^{-1}$$
$$\alpha_{O_2} = 34.6 \text{ ml } O_2 \text{ atm } O_2^{-1} \text{ L SW}^{-1}$$

Calculate the percentage of saturation of N_2 and O_2 in this sample of deep water. Are the samples saturated, supersaturated, or undersaturated with respect to O_2 and N_2?

4. What differences would you expect in the surface-water concentrations of inert gases at high as compared to low latitudes? Explain your answer.
5. Harvey (1955) has shown that a large part of the deep water in the eastern basins of the Atlantic Ocean contains dissolved oxygen at levels of 5.25 cc per liter. To what concentration in μmol/kg does this correspond, if the salinity of the water is 36.5‰ and the temperature is 1°C?
6. On March 30, 1966, the R/V *Anton Brun* found an upwelling zone off the west coast of Peru. At a depth of 10 m the water contained 2.91 ml O_2/L. The temperature was 15.8°C and the salinity 34.889‰. What was the percent saturation of oxygen?

   After setting a drougue and following the upwelled water mass as it flowed north for 4 days, a sample taken at 10 m contained 6.05 ml O_2/L. The temperature was 16.04°C and salinity = 34.855‰. What was the percentage saturation of oxygen now? What do you think happened to the water mass?

7. The net photosynthetic production of oxygen is 10×10^{-7} ml O_2 cm^{-2} s^{-1} in a turbulently mixed region of seawater. This seawater has a steady-state O_2 concentration of 8.0 ml O_2/L SW even though its saturation concentration is 6.0 ml O_2/L SW, because it is separated from the atmosphere by a stagnant thin film. The diffusion coefficient of O_2 is 1.2×10^{-5} cm²/s. What is the thickness of the thin film?
8. A salt solution is prepared by dissolving 0.090 moles of NaCl(s) and 0.090 moles $MgCl_2$(s) in 1 kg of water. Also added are 0.010 moles of NaF(s) and Na_2SO_4(s). Assuming that the only ion pairs that form are MgF^+, $MgSO_4^0$, NaF^0, and $NaSO_4^-$, compute the ion speciation of this solution. Use the ion activity coefficients given in Table 5.3 and the log K^0_{eq}'s given in Appendix X. Assume log K^0_{eq} for MgF^+ is 1.50. Report the answers in a form similar to that of Table 5.4.

SUGGESTED FURTHER READINGS

Colloids

YARIV, S., and H. CROSS. 1979. *Geochemistry of Colloid Systems for Earth Scientists*. Springer-Verlag, New York, 450 pp.

Gases

BROECKER, W., and T.-H. PENG. 1982. *Tracers in the Sea*. Lamont-Doherty Geological Observatory, New York, 690 pp.

Ion Speciation and General Principles of Aquatic Chemistry

DREVER, J. I. 1982. *The Geochemistry of Natural Waters*. Prentice Hall, Englewood Cliffs, NJ, 388 pp.

MOREL, F. M. 1983. *Principles of Aquatic Chemistry*. John Wiley & Sons, Inc., New York, 446 pp.

STUMM, W., and J. J. MORGAN. 1981. *Aquatic Chemistry: An Introduction Emphasizing Chemical Equilibria in Natural Waters*. John Wiley & Sons, Inc., New York, 780 pp.

Seawater Composition

HOLLAND, H. D. 1978. *The Chemistry of the Atmosphere and Oceans*. John Wiley & Sons, Inc., New York, 351 pp.

PART TWO

It appears that in natural habitats, organisms capable of mediating the pertinent (thermodynamically favored) redox reactions are nearly always found.

Werner Stumm and James Morgan, *Aquatic Chemistry*, 1981

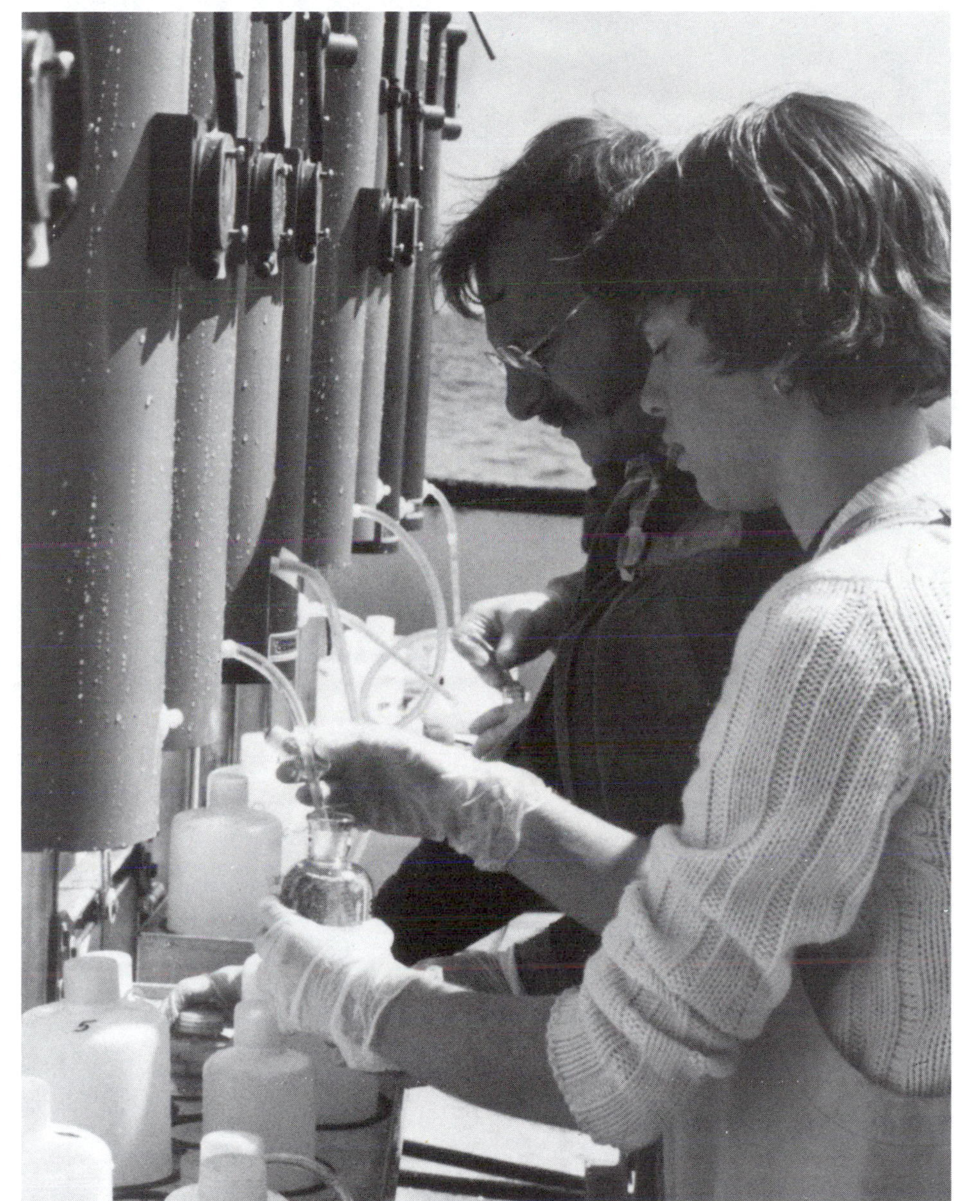

THE REDOX CHEMISTRY OF SEAWATER

CHAPTER 7

THE IMPORTANCE OF OXYGEN

INTRODUCTION

The biogeochemical cycles of the minor and trace elements are strongly influenced by redox reactions. Most of these reactions are mediated by marine organisms and fueled ultimately by energy derived from solar radiation. The latter is captured and transformed by photosynthetic organisms into the chemical forms, O_2 and organic matter. This stored energy is then extracted by heterotrophic organisms to fuel their metabolic processes. Collectively, photosynthesis and respiration control the biogeochemical cycles of most of the minor and trace elements. The elements most strongly affected are those whose concentrations in living tissues are much higher than in ambient seawater (i.e., carbon, oxygen, nitrogen, phosphorus, sulfur, and many trace metals).

These redox processes do not reach equilibrium due to the continuous supply of sunlight and hence plant activity. Nevertheless, thermodynamic principles can be used to predict the relative stability of the redox species, as well as the spontaneous direction and relative rates of redox reactions. The calculations used to make these theoretical predictions are presented in this chapter.

The results from such calculations suggest that the redox chemistry of seawater should be largely controlled by the reduction of O_2 and the oxidation of organic matter, because these are the most abundant of the strong oxidizing and reducing agents. The other chapters in Part 2 are a discussion of the effects of nonequilibrium redox processes on the marine biogeochemical cycles of the trace and minor elements.

BASIC CONCEPTS IN ELECTROCHEMISTRY

Half-Cell Reactions

Redox reactions occur when electrons are transferred between atoms or molecules. Most first-year chemistry students have performed the following redox reaction. When metallic zinc is placed in a beaker containing an aqueous solution of copper sulfate, a vigorous exothermic reaction ensues. At its conclusion, the zinc has dissolved, the solution has lost its blue tint, and an orange solid has formed. The reaction that occurred is the following:

$$Zn(s) + Cu^{2+}(aq) \rightleftharpoons Zn^{2+}(aq) + Cu(s) \tag{7.1}$$

during which electrons have been transferred from the zinc to the copper. This electron flow can be monitored by conducting the reaction in an electrochemical cell fitted with a voltmeter, as illustrated in Figure 7.1. Because the electron flow from the zinc to the copper electrode is spontaneous, this is termed a galvanic cell. In this cell, the reactants and products are kept in separate compartments, so that the electrons are forced to flow through a wire.

Since the reactants are present at nonequilibrium concentrations, the following reaction occurs at the zinc electrode:

$$Zn(s) \rightarrow Zn^{2+}(aq) + 2e^- \tag{7.2}$$

and at the copper electrode:

$$Cu^{2+}(aq) + 2e^- \rightarrow Cu(s) \tag{7.3}$$

As the zinc electrode becomes oxidized, more positive charge (i.e., $Zn^{2+}(aq)$) is introduced into this side of the cell. To maintain electroneutrality throughout the solution, sulfate ions diffuse across the porous plug from the copper to the zinc side of the cell.

Electrons spontaneously flow from the zinc to the copper electrode because copper has a greater affinity for electrons than does zinc. Equation 7.3 is an example of a **reduction** reaction, during which the reactant gains electrons. Equation 7.2 is an example of an **oxidation** reaction, during which the reactant loses electrons. In this example, $Cu^{2+}(aq)$ causes the oxidation of Zn(s) and hence is called an **oxidizing agent**. Similarly, the Zn can be thought of as a **reducing agent**. Equations 7.2 and 7.3 are termed **half-cell reactions**. Since free electrons are not found in nature, half-reactions must occur in pairs, with each oxidation being accompanied by a reduction. The overall chemical change is called a reduction–oxidation, or redox reaction.

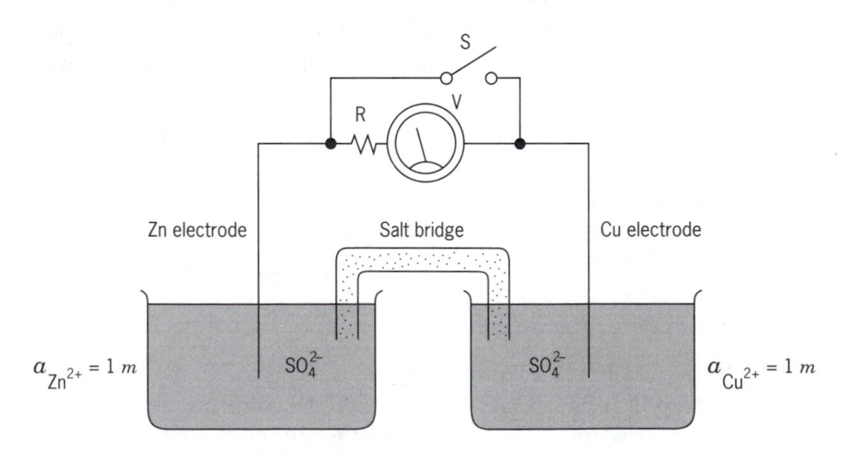

Oxidation Numbers

In the reduction half-cell reaction, the charge on the copper is reduced from +2 to 0. Likewise, in the oxidation half-cell reaction, the charge on the zinc increases from 0 to +2. Chemists have extended this concept of relative charge to chemically bound atoms. A set of rules has been developed so that the relative charge, or oxidation number, of an atom in a molecule can be calculated. The most commonly used rules are given below; others can be found in Appendix XI.

1. The oxidation number of a monoatomic ion is equivalent to its charge—e.g., the oxidation number of Cu^{2+}(aq) is II.
2. The oxidation number of an atom in its elemental state is 0—e.g., the oxidation number of atomic oxygen in O_2 is 0, as is the oxidation number of atomic copper in Cu(s).
3. The oxidation number of oxygen in most molecules is −II. Notable exceptions are peroxides (R—OOH), in which oxygen's oxidation number is −I and O_2, in which oxygen's oxidation number is 0.
4. The oxidation number of atomic hydrogen is I, except when it is in hydrides; then it is −I.

Oxidation numbers are commonly written as Roman numerals, often in parentheses—e.g., Cu(II)O. An increase in oxidation number indicates that the atom has been oxidized, whereas a decrease indicates that the atom has been reduced. With these rules, the oxidation number of an atom in a molecule can be calculated by constructing a charge-balance equation as shown below for chromium in the dichromate ion, $Cr_2O_7^{2-}$:

Total charge on $Cr_2O_7^{2-}$

$$= (\text{number of Cr atoms})(\text{oxidation number of Cr}) + (\text{number of O atoms})(\text{oxidation number of O}) \quad (7.4)$$

$$-2 = (2)(\text{Cr}) + (7)(-2) \quad (7.5)$$

$$\text{Cr} = \text{VI} \quad (7.6)$$

Energetics of Redox Reactions

Reactions proceed spontaneously in the direction that causes the energy of the products and reactants to decrease. When the energies of the reactants and products have reached an equal, minimal level, there is no longer any driving force to cause changes in the concentrations of the chemicals. The resulting stable state is referred to as thermodynamic equilibrium.

Several forms of energy can be transformed or transferred during a chemical reaction. Thus the total energy of a chemical system must be considered when predicting the direction and extent of spontaneous reactions. This total energy is referred to as the **Gibbs free energy, *G***. Chemical systems that have large free energies are unstable and will spontaneously react until the equilibrium state is achieved. In doing so, energy is often given off to the surroundings and can be used to do work. The resulting **free energy change, ΔG**, is a measure of how far the original reaction mixture was from equilibrium. By convention, ΔG is negative for a spontaneous reaction. As shown below, the ΔG of a reaction can be calculated from various types of thermodynamic data. The more negative the result, the more favored the reaction and the greater the amount of energy that can be released to the surroundings.

The galvanic cell pictured in Figure 7.1 is not at equilibrium. If switch S is closed, electrons will spontaneously flow from the zinc (anode) to the copper (cathode) electrode. This flow will continue until the reactants and products attain their equilibrium concentrations. If switch S is opened before the cell reaches equilibrium, the electron flow will be interrupted. The voltmeter would register a positive voltage, which is a measure of the degree to which the redox reaction drives electrons from the anode to the cathode. Since this voltage is a type of energy that has the potential to do work, it is referred to as a **cell potential**, or redox potential, E_{cell}.

Like ΔG, the magnitude of E_{cell} is a measure of how far a reaction mixture is from equilibrium. It is related to ΔG by

$$\Delta G = -nF(E_{cell}) \tag{7.7}$$

where F, Faraday's constant $= 23.066$ kcal V^{-1} mol electrons transferred^{-1} and $n =$ mol electrons transferred. For the Cu/Zn cell illustrated in Figure 7.1, $E_{cell} = +1.10$ V; thus $\Delta G = -(2 \text{ mol})(23.066 \text{ kcal V}^{-1} \text{ mol}^{-1})(+1.10 \text{ V}) = -50.7$ kcal. Since ΔG is negative, this reaction is spontaneous.

For redox reactions conducted at standard conditions,* such as those shown in Figure 7.1, Eq. 7.7 can be written as

$$\Delta G^0 = -nF(E^0_{cell}) \tag{7.8}$$

where ΔG^0 is the standard free energy change and E^0_{cell} is the standard redox potential.

The degree to which electrons are driven from the anode to the cathode can also be thought of as an **electron activity** $\{e^-\}$. This activity is analogous to an electron pressure, rather than to a concentration, and is usually expressed as

$$pe = -\log \{e^-\} \tag{7.9}$$

The **pe** in a galvanic cell is related to the redox potential by

$$pe_{cell} = \frac{F}{2.303\,RT} E_{cell} \tag{7.10}$$

where $R = 1.987 \times 10^{-3}$ kcal °K^{-1} mol^{-1} and T is given in °K.

---

*Standard electrochemical conditions are (1) temperature = 25°C; (2) pressure = 1 atm; and (3) all solutes maintained at unit activity (i.e., $a = 1$).

E_{cell} is the sum of the energies, or half-cell potentials, contributed by the half-cell reactions. Thus E_{cell} for any redox reaction can be calculated by summing its half-cell potentials. These half-cell potentials cannot be measured directly because oxidations and reductions always occur in pairs. Instead, half-cell potentials are determined indirectly by measuring the redox potential of a galvanic cell in which one half-cell reaction is provided by a **standard hydrogen electrode (SHE)**. Such a cell is shown in Figure 7.2 for the Zn/Zn^{2+} couple.

In this cell, the following reaction occurs spontaneously at the anode:

$$Zn(s) \rightarrow Zn^{2+}(aq) + 2e^- \tag{7.11}$$

and at the cathode:

$$2H^+(aq) + 2e^- \rightarrow H_2(g) \tag{7.12}$$

Summing the two yields the redox reaction:

$$Zn(s) + 2H^+(aq) \rightarrow Zn^{2+}(aq) + H_2(g) \tag{7.13}$$

The E_{cell} of this standard cell is +0.76 V. By international convention, the half-cell potential of the hydrogen reduction is assigned a value of exactly 0 V. Thus the half-cell potential of the zinc oxidation is equal to E^0_{cell} (i.e., +0.76 V). This voltage is called the standard half-cell potential and is represented by the symbol E^0_h, to indicate that it was determined against a standard hydrogen electrode.

As shown in Table 7.1, E^0_h values are usually listed for half-cell reactions written as reductions. Thus the standard half-cell reduction potential for the Zn^{2+}/Zn couple is −0.76 V. Half-cell reductions that are "strong" enough to spontaneously oxidize $H_2(g)$ have a positive E^0_h. Conversely, half-cell reductions that have a negative E^0_h are not "strong" enough to oxidize $H_2(g)$. Instead, they spontaneously proceed as oxidations, causing the reduction of $H^+(aq)$.

These tabulated values can be used to calculate the E^0_{cell} for any reaction, as illustrated in Table 7.2 for the Zn/Cu galvanic cell. The redox reaction is

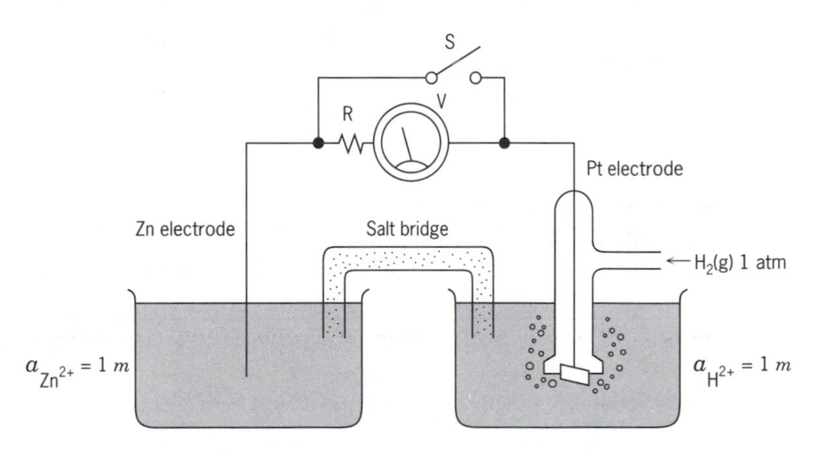

FIGURE 7.2. Galvanic cell used for measuring E°_h.

TABLE 7.1
Standard Electrode Potentials for Selected Half-Reactions

Reaction	*Log K at 25°C*	*Standard Electrode Potential (V) at 25°C*
$Na^+ + e^- = Na(s)$	−46	−2.71
$Zn^{2+} + 2e^- = Zn(s)$	−26	−0.76
$Fe^{2+} + 2e^- = Fe(s)$	−14.9	−0.44
$Co^{2+} + 2e^- = Co(s)$	−9.5	−0.28
$V^{3+} + e^- = V^{2+}$	−4.3	−0.26
$2H^+ + 2e^- = H_2(g)$	0.0	0.00
$S(s) + 2H^+ + 2e^- = H_2S$	+4.8	+0.14
$Cu^{2+} + e^- = Cu^+$	+2.7	+0.16
$AgCl(s) + e^- = Ag(s) + Cl^-$	+3.7	+0.22
$Cu^{2+} + 2e^- = Cu(s)$	+11.4	+0.34
$Cu^+ + e^- = Cu(s)$	+8.8	+0.52
$Fe^{3+} + e^- = Fe^{2+}$	+13.0	+0.77
$Ag^+ + e^- = Ag(s)$	+13.5	+0.80
$Fe(OH)_3(s) + 3H^+ + e^- = Fe^{2+} + 3H_2O$	+17.1	+1.01
$IO_3^- + 6H^+ + 5e^- = \frac{1}{2}I_2(s) + 3H_2O$	+104	+1.23
$MnO_2(s) + 4H^+ + 2e^- = Mn^{2+} + 2H_2O$	+43.6	+1.29
$Cl_2(g) + 2e^- = 2Cl^-$	+46	+1.36
$Co^{3+} + e^- = Co^{2+}$	+31	+1.82

Source: From *Aquatic Chemistry*, W. Stumm and J. J. Morgan, copyright © 1981 by John Wiley & Sons, Inc., New York, p. 437. Reprinted by permission.

spontaneous because the half-reaction (Cu^{2+}/Cu) with the larger reduction E_h^0 (+0.34V) acts as the oxidizing agent. Since the other half-reaction (Zn^{2+}/Zn) proceeds as an oxidation, the sign of its half-cell reduction potential must be reversed from −0.76 to +0.76 V. The largest E_{cell} is generated by pairing the half-cell reaction that has the largest reduction E_h^0 with the one which has the smallest. Conversely, the closer the reduction E_h^0's of the half-cell reactions, the smaller the E_{cell}.

TABLE 7.2
Computing E_{cell}^0 from E_h^0's[a]

Location	*Reaction*	E_h^0 *(V)*
Cathode	$Cu^{2+}(aq) + 2e^- \rightarrow Cu(s)$	+0.34
Anode	$Zn(s) \rightarrow Zn^{2+}(aq) + 2e^-$	+0.76
Redox reaction	$Zn(s) + Cu^{2+}(aq) \rightarrow Cu(s) + Zn^{2+}(aq)$	+1.10

[a]The redox reaction and E_{cell}^0 are calculated by summing the half-reactions and E_h^0's, respectively.

The results obtained under standard conditions can be used to predict thermodynamic behavior at other concentrations and temperatures. To derive the necessary equations, consider the general redox reaction:

$$a\mathrm{OX}_1 + b\mathrm{RED}_2 \rightleftharpoons c\mathrm{RED}_1 + d\mathrm{OX}_2 \tag{7.14}$$

which can be viewed as the sum of the reduction half-reaction:

$$a\mathrm{OX}_1 + ne^- \rightleftharpoons c\mathrm{RED}_1 \tag{7.15}$$

and oxidation half-reaction:

$$b\mathrm{OX}_2 + ne^- \rightleftharpoons c\mathrm{RED}_2 \tag{7.16}$$

The latter is written as a reduction, such as would appear in a standard table of half-cell reductions. If this redox reaction (Eq. 7.14) proceeds spontaneously to the right, the half-cell reduction potential, $E^0_{\mathrm{h}_1}$, is greater than the half-cell reduction potential, $E^0_{\mathrm{h}_2}$. In other words, species 1 is a stronger reducing agent than species 2.

From thermodynamic principles, chemists have demonstrated that the free energy change at nonstandard conditions, ΔG, is related to the free energy change under standard conditions, ΔG^0, by

$$\Delta G = \Delta G^0 + RT\ln Q \tag{7.17}$$

where Q is the reactant quotient and is similar in form to the equilibrium constant, K_{eq}, that is,

$$Q = \frac{\{\mathrm{RED}_1\}^c\{\mathrm{OX}_2\}^d}{\{\mathrm{OX}_1\}^a\{\mathrm{RED}_2\}^b} \tag{7.18}$$

Substituting Eqs. 7.7 and 7.8 into Eq. 7.17 yields the **Nernst Equation**:

$$E_{\mathrm{cell}} = E^0_{\mathrm{cell}} - \frac{RT}{n\mathrm{F}}\ln Q \tag{7.19}$$

Since 2.303 log x = ln x, the Nernst Equation applied at 25°C becomes

$$E_{\mathrm{cell}} = E^0_{\mathrm{cell}} - \frac{0.0592}{n}\log Q \tag{7.20}$$

At equilibrium, $E_{\mathrm{cell}} = 0$ and $Q = K_{\mathrm{cell}}$, which is the thermodynamic equilibrium constant for the redox reaction. Thus at equilibrium, Eq. 7.20 becomes

$$E^0_{\mathrm{cell}} = \frac{0.0592}{n}\log K_{\mathrm{cell}} \tag{7.21}$$

$E^0_{\mathrm{cell}} \neq 0$ because it is measured under conditions of disequilibrium (i.e., 1 M concentrations of all solutes).

By substituting in terms of pe_{cell} (as per the definition given in Eq. 7.10), the Nernst Equation becomes

$$pe_{\mathrm{cell}} = pe^0_{\mathrm{cell}} - \frac{\log Q}{n} \tag{7.22}$$

and at equilibrium

$$pe^0_{\text{cell}} = \frac{\log K_{\text{cell}}}{n} \tag{7.23}$$

Thus for a one-electron transfer, $pe^0_{\text{cell}} = \log K_{\text{cell}}$.

The pe and E_h of a half-cell reaction can similarly be related to the ΔG and K_{eq} of that half-reaction as shown below. For the half-reaction given in Eq. 7.15,

$$K_1 = \frac{\{\text{RED}_1\}^c}{\{\text{OX}_1\}^a\{e^-\}^n} \tag{7.24}$$

Rearranging,

$$\frac{1}{\{e^-\}^n} = K_1 \frac{\{\text{OX}_1\}^a}{\{\text{RED}_1\}^c} \tag{7.25}$$

taking the log of all the terms,

$$pe_1 = \frac{1}{n}\left(\log K_1 + \log \frac{\{\text{OX}_1\}^a}{\{\text{RED}_1\}^c}\right) \tag{7.26}$$

and substituting $pe^0_1 = (\frac{1}{n}) \log K_1$ (obtained by analogy from Eq. 7.23) yields

$$pe_1 = pe^0_1 + \frac{1}{n}\left(\log \frac{\{\text{OX}_1\}^a}{\{\text{RED}_1\}^c}\right) \tag{7.27}$$

At equilibrium $\{e^-\} = 1$. Thus, $pe_1 = 0$ and $K_1 = \{\text{OX}_1\}^a/\{\text{RED}_1\}^c$, so Eq. 7.27 becomes

$$pe^0_1 = \tfrac{1}{n} \log K_1 \tag{7.28}$$

and for the half-cell reaction given in Eq. 7.16,

$$pe^0_2 = \tfrac{1}{n} \log K_2 \tag{7.29}$$

Since $K_{\text{cell}} = K_1/K_2$,

$$pe^0_1 - pe^0_2 = \tfrac{1}{n} \log K_{\text{cell}} \tag{7.30}$$

The reaction with the greatest tendency to proceed spontaneously will be the one with the largest equilibrium constant. This is achieved by pairing the oxidizing agent (OX_1) with the largest pe to the reducing agent (RED_2) whose half-cell reduction has the most negative pe. In seawater, these chemicals are O_2 and organic matter, respectively.

Finally, since at equilibrium, $\Delta G^0 = -RT \ln K$,

$$\Delta G^0 = 2.303\, nRT\, (pe^0_2 - pe^0_1) \tag{7.31}$$

It can also be shown that

$$\Delta G = 2.303\, nRT\, (pe_2 - pe_1) \tag{7.32}$$

Thus ΔG, *pe*, E_h, and K contain virtually the same thermodynamic information. While E_h is the quantity that is analytically measured, *pe* is preferred by marine chemists as it is temperature independent and numerically easier to work with. A comparison of the two electron activity scales at 25°C is given in Figure 7.3. ΔG is often used to compare the relative stability of species because chemists think of this concept in terms of energy yields. The merits of each thermodynamic parameter will become evident in the next section of the chapter where the energetics of some marine redox processes are considered.

THE REDOX CHEMISTRY OF SEAWATER

Aquatic chemists have defined their own electrochemical standard state to facilitate the calculation of redox speciation in aqueous solution. In this standard state, all reactions are conducted at pH 7.0, 25°C, and 1 atm. The concentrations of all other solutes are 1 *M*. Values so obtained are designated with the subscript "w." The pe^0_w's for the most important redox couples in seawater are given in Table 7.3.

These values can be used to predict redox speciation in seawater. The approach used is illustrated by the following example. Consider the half-reactions

$$\tfrac{1}{4}O_2 + H^+ + e^- \rightleftharpoons \tfrac{1}{2}H_2O \qquad (7.33)$$

$$\tfrac{1}{8}SO_4^{2-} + \tfrac{9}{8}H^+ + e^- \rightleftharpoons \tfrac{1}{8}HS^- + \tfrac{1}{2}H_2O \qquad (7.34)$$

Since the pe^0_w for Eq. 7.33 is larger (+13.75V) than that of Eq. 7.34 (−3.75V), the former proceeds as the reduction and the latter as the oxidation.

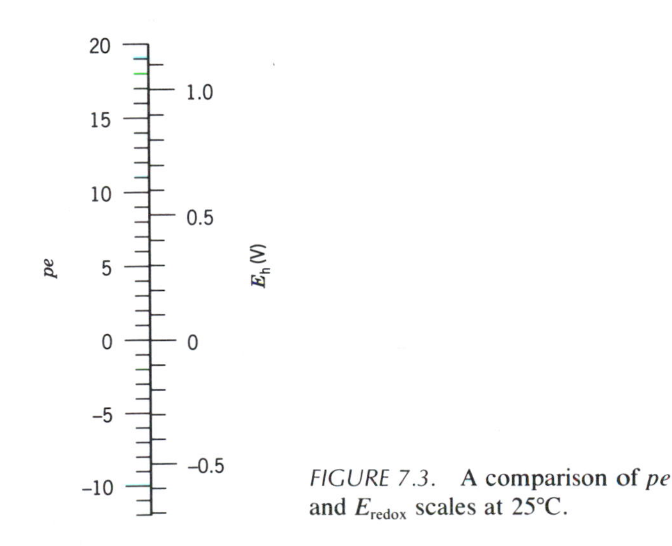

FIGURE 7.3. A comparison of *pe* and E_{redox} scales at 25°C.

TABLE 7.3
Log K, pe^0_w, and $E^0_h(w)$ of Redox Processes[a]

Reaction	pe^0 ($\equiv \log K$)	pe^{0a}_w	$E^0_h(w)$
(1) $\frac{1}{4}O_2(g) + H^+ + e^- = \frac{1}{2}H_2O$	+20.75	+13.75	+0.81
(2) $\frac{1}{5}NO_3^- + \frac{6}{5}H^+ + e^- = \frac{1}{10}N_2(g) + \frac{3}{5}H_2O$	+21.05	+12.65	+0.75
(3) $\frac{1}{2}MnO_2(s) + \frac{1}{2}HCO_3^-(10^{-3}M) + \frac{3}{2}H^+ + e^- = \frac{1}{2}MnCO_3(s) + \frac{3}{8}H_2O$	—	+3.9[b]	+0.23
(4) $\frac{1}{2}NO_3^- + H^+ + e^- = \frac{1}{2}NO_2^- + \frac{1}{2}H_2O$	+14.15	+7.15	+0.42
(5) $\frac{1}{8}NO_3^- + \frac{5}{4}H^+ + e^- = \frac{1}{8}NH_4^+ + \frac{3}{8}H_2O$	+14.90	+6.15	+0.36
(6) $\frac{1}{6}NO_2^- + \frac{4}{3}H^+ + e^- = \frac{1}{6}NH_4^+ + \frac{1}{3}H_2O$	+15.14	+5.82	+0.34
(7) $\frac{1}{2}CH_3OH + H^+ + e^- = \frac{1}{2}CH_4(g) + \frac{1}{2}H_2O$	+9.88	+2.88	+0.17
(8) $\frac{1}{4}CH_2O + H^+ + e^- = \frac{1}{4}CH_4(g) + \frac{1}{4}H_2O$	+6.94	−0.06	+0.00
(9) $FeOOH(s) + HCO_3^-(10^{-3}M) + 2H^+ + e^- = FeCO_3(s) + 2H_2O$	—	−0.8[b]	−0.05
(10) $\frac{1}{2}CH_2O + H^+ + e^- = \frac{1}{2}CH_3OH$	+3.99	− 3.01	−0.18
(11) $\frac{1}{6}SO_4^{2-} + \frac{4}{3}H^+ + e^- = \frac{1}{6}S(s) + \frac{2}{3}H_2O$	+6.03	−3.30	−0.20
(12) $\frac{1}{8}SO_4^{2-} + \frac{5}{4}H^+ + e^- = \frac{1}{8}H_2S(g) + \frac{1}{2}H_2O$	+5.25	−3.50	−0.21
(13) $\frac{1}{8}SO_4^{2-} + \frac{9}{8}H^+ + e^- = \frac{1}{8}HS^- + \frac{1}{2}H_2O$	+4.25	−3.75	−0.22
(14) $\frac{1}{2}S(s) + H^+ + e^- = \frac{1}{2}H_2S(g)$	+2.89	−4.11	−0.24
(15) $\frac{1}{8}CO_2(g) + H^+ + e^- = \frac{1}{8}CH_4(g) + \frac{1}{4}H_2O$	+2.87	−4.13	−0.24
(16) $\frac{1}{6}N_2(g) + \frac{4}{3}H^+ + e^- = \frac{1}{3}NH_4^+$	+4.68	−4.68	−0.28
(17) $\frac{1}{2}(NADP^+) + \frac{1}{2}H^+ + e^- = \frac{1}{2}(NADPH)$	−2.0	−5.5	−0.33
(18) $H^+ + e^- = \frac{1}{2}H_2(g)$	0.0	−7.00	−0.41
(19) Oxidized ferredoxin + e^- = reduced ferredoxin	−7.1	−7.1	−0.42
(20) $\frac{1}{4}CO_2(g) + H^+ + e^- = \frac{1}{24}(\text{glucose}) + \frac{1}{4}H_2O$	−0.20	−7.20	−0.43
(21) $\frac{1}{2}HCOO^- + \frac{3}{2}H^+ + e^- = \frac{1}{2}CH_2O + \frac{1}{2}H_2O$	+2.82	−7.68	−0.45
(22) $\frac{1}{4}CO_2(g) + H^+ + e^- = \frac{1}{4}CH_2O + \frac{1}{4}H_2O$	−1.20	−8.20	−0.48
(23) $\frac{1}{2}CO_2(g) + \frac{1}{2}H^+ + e^- = \frac{1}{2}HCOO^-$	−4.83	−8.33	−0.49

Source: From *Aquatic Chemistry*, W. Stumm and J. J. Morgan, copyright © 1970 by John Wiley & Sons, Inc., New York, p. 318. Reprinted by permission. See Stumm and Morgan (1970) for data sources.

[a]Values for pe^0_w apply to the electron activity for unit activities of oxidant and reductant in neutral water, that is, at pH = 7.0 for 25°C.

[b]These data correspond to $(HCO_3^-) = 10^{-3}$ M rather than unity and so are not exactly pe^0_w; they represent typical aquatic conditions more nearly than pe^0_w values do.

This yields the following redox reaction (Eq. 7.35) in which hydrogen sulfide is oxidized by O_2. This reaction is mediated by bacteria called sulfide oxidizers.

$$\frac{1}{8}HS^- + \frac{1}{4}O_2 \rightleftharpoons \frac{1}{8}SO_4^{2-} + \frac{1}{8}H^+ \tag{7.35}$$

How favored is this reaction? How much HS^- should be present at equilibrium? These are important questions as HS^- appears to be the source of chemical energy supporting the marine life that lives around hydrothermal vents, even though it is extremely toxic to other organisms.

The relative abundances of SO_4^{2-} and HS^- at equilibrium can be calculated from Eq. 7.30 using the stoichiometry given in Eq. 7.35.

$$13.75 - (-3.75) = \frac{1}{1} \log \frac{\{SO_4^{2-}\}^{1/8}\{H^+\}^{1/8}}{P_{O_2}^{1/4}\{HS^-\}^{1/8}} \tag{7.36}$$

The superscripts in the equilibrium constant become coefficients since $\log x^a = a \log x$. The $\log \{H^+\}$ term can be replaced by $-$pH to yield

$$17.50 = \frac{1}{8} \log \frac{\{SO_4^{2-}\}}{\{HS^-\}} - \frac{1}{8}\text{pH} - \frac{1}{4} \log P_{O_2} \tag{7.37}$$

Assuming the solution has a pH = 8 and is in gaseous equilibrium with the atmosphere (i.e., P_{O_2} = 0.21 atm),

$$\frac{\{SO_4^{2-}\}}{\{HS^-\}} = 10^{149} \tag{7.38}$$

In other words, virtually no HS^- should be present at equilibrium.

The equilibrium constant, which can be calculated by substituting into Eq. 7.30, is

$$17.50 = \log K \tag{7.39}$$

or $K = 10^{17.50}$. ΔG_w^0 can be calculated from Eq. 7.31 as

$$\begin{aligned} \Delta G_w^0 &= (2.303)(1\ \text{mol})(1.987 \times 10^{-3}\ \text{kcal}^{-1}\ {}^\circ\text{K mol}^{-1})(298.15^\circ\text{K}) \\ &\quad(-3.75 - 13.75) \\ &= -23.88\ \text{kcal} \end{aligned} \tag{7.40}$$

Thus bacteria should obtain large amounts of energy from oxidizing sulfide. Note that K is also large. In such cases, the reaction can be considered to proceed to completion. Usually, these redox reactions occur as a series of steps, each of which is catalyzed by a specific enzyme. Though each step is reversible, they each proceed at different rates, causing the overall redox reaction to function as a unidirectional process.

Relative Redox Intensity

The half-reaction with the highest $p\epsilon$ will force all other half-reactions to proceed as oxidations. Although seawater contains stronger oxidizing agents than O_2, these others do not exert a controlling influence on the redox chemistry of the ocean. This is due to their relatively low concentrations and slow rates of reaction. In comparison, nearly all reactions that involve O_2 proceed relatively rapidly as they are mediated by enzymes produced by a large variety of marine organisms. Because of O_2's great oxidizing power, equilibrium thermodynamics predicts that all biochemically active elements should exist primarily in their highest oxidation states.

The greater the difference in *pe* between the oxidizing and reducing agents, the greater the free energy yield. Since organisms depend on this energy to fuel their metabolic processes, the redox reaction that produces the most energy is of greatest biological benefit. In seawater, the redox reaction that yields the most energy is the aerobic oxidation of organic matter to carbon dioxide. The chemical equation which describes this redox reaction is

$$CH_2O + O_2 \rightleftharpoons CO_2 + H_2O \qquad (7.41)$$

where organic matter is represented generically by the empirical formula CH_2O. The ΔG^0_w for this reaction can be computed by substituting into Eq. 7.31 as follows:

$$\begin{aligned}\Delta G^0_w &= (2.303)(1\ \text{mol})(1.987 \times 10^{-3}\ \text{kcal}\ {}^\circ\text{K}^{-1}\ \text{mol}^{-1})(298.15^\circ\text{K}) \\ &\quad (-8.20 - 13.75) \qquad (7.42) \\ &= -29.95\ \text{kcal}\end{aligned}$$

where the values of pe^0_1 and pe^0_2 are for the half-reactions 1 and 22, as listed in Table 7.3, respectively.

When one of the reactants is depleted, the next most energetic redox reaction will occur in its stead. In the absence of organic matter, the next most energetic electron donor used by marine microbes is hydrogen sulfide. Note that this reaction produces only slightly less energy (−23.88 kcal), than does the oxidation of organic matter. The other half-cell oxidations that proceed in the presence of O_2 are listed in order of their relative redox intensity at the bottom of Figure 7.4.

The chemical equations describing these reactions are given in Table 7.4. If, during the oxidation of organic matter, O_2 is depleted first, the next-strongest oxidizing agent, nitrate, will take its place. Nitrate is reduced in a stepwise fashion to $N_2(g)$. This process is termed denitrification and is also mediated by a particular type of marine bacteria. As shown in Table 7.5, the energy yield is only slightly less than that of aerobic respiration. If a large amount of organic matter is present, the nitrate will be depleted and the next reducing agent of choice is Mn(IV). The relative strength of the other reducing agents in seawater is shown in the top half of Figure 7.4 and their reaction stoichiometry is given in Table 7.5.

As shown in this table, the chemical energy that drives these reactions forward can be either inorganic or organic in nature. Due to its relative abundance and low *pe*, organic matter is the most important reducing agent in the ocean. This organic matter is ultimately derived from biological "fixation" reactions in which oxidized inorganic carbon is converted to a reduced organic form. The most important **carbon fixation reactions** are listed in Table 7.6. Also included are the nitrogen fixation reactions, which involve the conversion of oxidized inorganic nitrogen to a reduced form readily taken up by plants. Note that this process requires an energy source (organic matter) in order to proceed spontaneously.

Photosynthesis is responsible for most of the carbon fixation and hence organic matter production on this planet. In this redox reaction, plants use

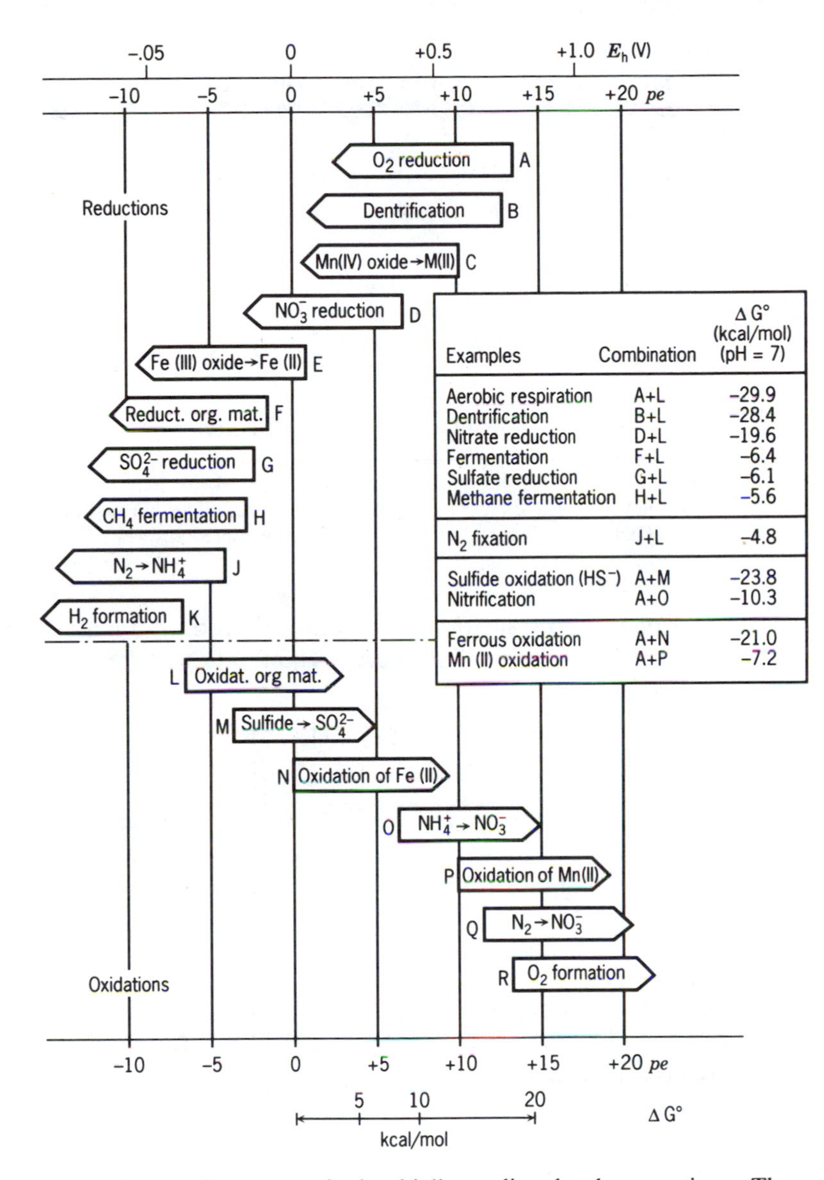

Examples	Combination	Δ G° (kcal/mol) (pH = 7)
Aerobic respiration	A+L	-29.9
Dentrification	B+L	-28.4
Nitrate reduction	D+L	-19.6
Fermentation	F+L	-6.4
Sulfate reduction	G+L	-6.1
Methane fermentation	H+L	-5.6
N_2 fixation	J+L	-4.8
Sulfide oxidation (HS^-)	A+M	-23.8
Nitrification	A+O	-10.3
Ferrous oxidation	A+N	-21.0
Mn (II) oxidation	A+P	-7.2

FIGURE 7.4. Sequence of microbially mediated redox reactions. The arrows point in the direction of the spontaneous redox reaction. The originating point of the arrows indicates the redox energy associated with each half-reaction. No information is contained in the relative length of the arrows. *Source:* From *Aquatic Chemistry*, W. S. Stumm and J. J. Morgan, copyright © 1981 by John Wiley & Sons, Inc., New York, p. 460. Reprinted by permission.

TABLE 7.4
Oxidation of Reduced Inorganic Compounds[a]

Nitrification
$\frac{1}{6}NH_4^+ + \frac{1}{4}O_2(g) = \frac{1}{6}NO_2^- + \frac{1}{3}H^+ + \frac{1}{6}H_2O;\ \Delta G_w^0 = -10.8$ kcal/mol
$\frac{1}{2}NO_2^- + \frac{1}{4}O_2(g) = \frac{1}{2}NO_3^-;\ \Delta G_w^0 = -9.0$ kcal/mol

Sulfide oxidation
$\frac{1}{8}H_2S(g) + \frac{1}{4}O_2(g) = \frac{1}{8}SO_4^{2-} + \frac{1}{4}H^+;\ \Delta G_w^0 = -23.5$ kcal/mol

Iron oxidation
$Fe^{2+} + \frac{1}{4}O_2(g) + 5H_2O = Fe(OH)_3(s) + 2H^+;\ \Delta G_w^0 = -21.0$ kcal/mol

Hydrogen oxidation
$\frac{1}{2}H_2(g) + \frac{1}{4}O_2(g) = \frac{1}{2}H_2O;\ \Delta G_w^0 = -28.4$ kcal/mol

Source: After *Principles of Aquatic Chemistry*, F. M. M. Morel, copyright © 1983 by John Wiley & Sons, Inc., New York, p. 330. Reprinted by permission.
[a]Accomplished by chemolithotrophs, mostly autotrophs.

solar energy to transform inorganic carbon into organic carbon, (i.e., $CO_2 + H_2O \rightarrow CH_2O + O_2$). The energy required to fix 0.25 mole of inorganic carbon into organic form is given by

$$\Delta G_w^0 = (2.303)(1\ \text{mol})(1.987) \times 10^{-3}\ \text{kcal}\ ^\circ\text{K}^{-1}\ \text{mol}^{-1})(298.15^\circ\text{K})(13.75 - (-8.20)) = +29.95\ \text{kcal} \quad (7.43)$$

This reaction is not spontaneous. To form 0.25 mole of organic matter requires the input of solar energy in an amount equal to the energy obtained from the aerobic oxidation of 0.25 mole of organic matter. Photosynthesis is listed in

TABLE 7.5
Oxidation of Organic Compounds Represented Generically as "CH_2O"[a]

Aerobic respiration
$\frac{1}{4}\text{"}CH_2O\text{"} + \frac{1}{4}O_2(g) = \frac{1}{4}CO_2(g) + \frac{1}{4}H_2O;\ \Delta G_w^0 = -29.9$ kcal/mol

Denitrification
$\frac{1}{4}\text{"}CH_2O\text{"} + \frac{1}{5}NO_3^- + \frac{1}{5}H^+ = \frac{1}{4}CO_2(g) + \frac{1}{10}N_2(g) + \frac{7}{20}H_2O;\ \Delta G_w^0 = -28.4$ kcal/mol

Sulfate reduction
$\frac{1}{4}\text{"}CH_2O\text{"} + \frac{1}{8}SO_4^{2-} + \frac{1}{8}H^+ = \frac{1}{4}CO_2(g) + \frac{1}{8}HS^- + \frac{1}{4}H_2O;\ \Delta G_w^0 = -6.1$ kcal/mol

Methane fermentation
$\frac{1}{4}\text{"}CH_2O\text{"} = \frac{1}{8}CO_2(g) + \frac{1}{8}CH_4(g);\ \Delta G_w^0 = -5.6$ kcal/mol

Hydrogen fermentation
$\frac{1}{4}\text{"}CH_2O\text{"} + \frac{1}{4}H_2O = \frac{1}{4}CO_2(g) + \frac{1}{2}H_2(g);\ \Delta G_w^0 = -1.6$ kcal/mol

Source: After *Principles of Aquatic Chemistry*, F. M. M. Morel, copyright © 1983 by John Wiley & Sons, Inc., New York, p. 330. Reprinted by permission.
[a]Accomplished by chemoorganotrophs, all heterotrophs.

TABLE 7.6
Carbon and Nitrogen Fixation Reactions

Carbon fixation[a]
$\frac{1}{4}CO_2(g) + \frac{1}{4}H_2O = \frac{1}{4}\text{"}CH_2O\text{"} + \frac{1}{4}O_2(g);\ \Delta G^0_w = +29.9$ kcal/mol
$\frac{1}{4}CO_2(g) + \frac{1}{2}H_2S(g) = \frac{1}{4}\text{"}CH_2O\text{"} + \frac{1}{16}S_8(col)^b + \frac{1}{4}H_2O;\ \Delta G^0_w = +4.7$ kcal/mol
$\frac{1}{4}CO_2(g) + \frac{1}{6}NH_4^+ + \frac{1}{12}H_2O = \frac{1}{4}\text{"}CH_2O\text{"} + \frac{1}{6}NO_2^- + \frac{1}{3}H^+;\ \Delta G^0_w = +17.8$ kcal/mol
Nitrogen fixation[c]
$\frac{1}{6}N_2(g) + \frac{1}{3}H^+ + \frac{1}{4}\text{"}CH_2O\text{"} + \frac{1}{4}H_2O = \frac{1}{3}NH_4^+ + \frac{1}{4}CO_2(g);\ \Delta G^0_w = -4.8$ kcal/mol

Source: After *Principles of Aquatic Chemistry*, F. M. M. Morel, copyright © 1983 by John Wiley & Sons, Inc., New York, p. 330. Reprinted by permission.
[a]Accomplished by autotrophs.
[b]col = colloidal.
[c]Accomplished by nitrogen (N_2) fixers.

Figure 7.4 as "O_2 formation." The O_2 produced by photosynthesis is available to engage in the oxidation reactions listed in the bottom half of Figure 7.4.

The second two carbon fixation reactions listed in Table 7.6 are accomplished by a few species of bacteria. To fix carbon, these microorganisms use energy obtained from the oxidation of reduced inorganic chemicals, such as hydrogen sulfide and ammonium.

Metabolic Classification of Organisms

To aid in discussions of marine redox chemistry, microorganisms are usually identified in terms of their metabolic processes. The three classification schemes most commonly used are presented in Figure 7.5. Plants are termed **photoautolithotrophs** to indicate that their energy source is sunlight, their carbon source is inorganic carbon (so they are carbon fixers), and their electron donor (H_2O) is inorganic. Animals are simply referred to as **heterotrophs**, as they use organic matter as a source of carbon, energy, and electrons. Most fungi, protozoans, and bacteria do the same. Some exceptions are the photosynthetic purple sulfur bacteria which, like the plants, are photoautolithotrophs. Bacteria that oxidize either ammonium, nitrite, or sulfide are examples of chemoautolithotrophs.

Redox Speciation

The theoretical equilibrium redox speciation of seawater can be calculated from the equations and *pe* values given in this chapter. The results are usually presented in the form of *pe*–pH diagrams, as illustrated in Figure 7.6 for the biologically important elements. These plots are used to display the relative abundances of the redox species under a wide range of environmental conditions.

The equations for the boundary lines presented in Figure 7.6*a* are calculated as follows. The redox couple that defines the upper boundary line is given by

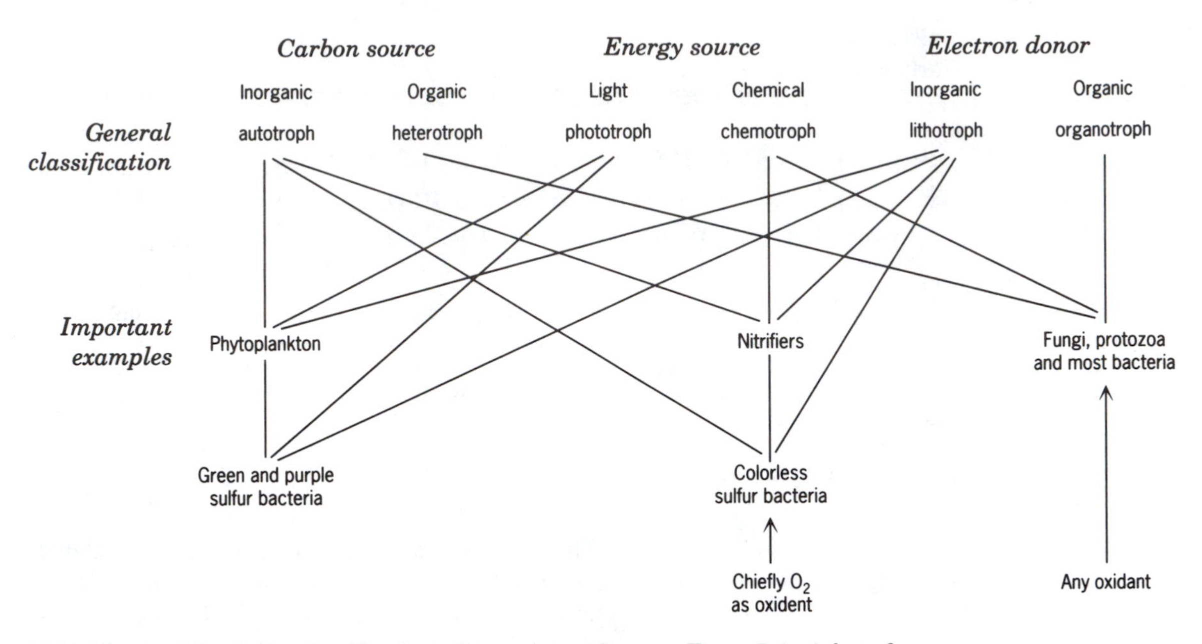

FIGURE 7.5. Metabolic classification of organisms. *Source:* From *Principles of Aquatic Chemistry,* F. M. M. Morel, copyright © 1983 by John Wiley & Sons, Inc., New York, p. 331. Reprinted by permission.

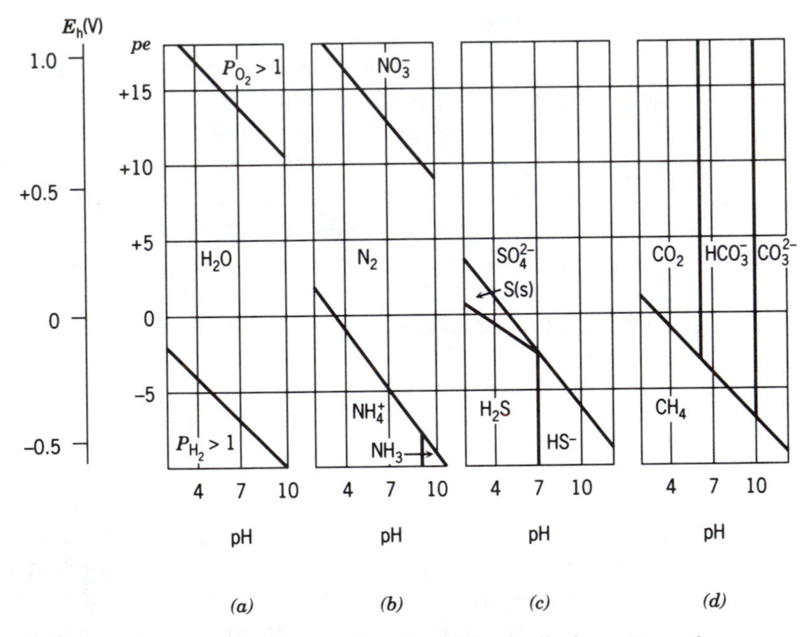

FIGURE 7.6. *pe*–pH diagrams for the biologically important elements at 25°C: (*a*) oxygen, (*b*) nitrogen, (*c*) sulfur, and (*d*) carbon. *Source:* From *Aquatic Chemistry,* W. S. Stumm and J. J. Morgan, copyright © 1981 by John Wiley & Sons, Inc., New York, p. 450. Reprinted by permission.

Eq. 7.33. Its equilibrium constant is

$$K = \frac{1}{P_{O_2}^{1/4}\{H^+\}\{e^-\}} \tag{7.44}$$

Inverting and then taking the log of both sides yields

$$-\log K = \log P_{O_2}^{1/4} + \log\{H^+\} + \log\{e^-\} \tag{7.45}$$

or,

$$\log K = -\tfrac{1}{4}\log P_{O_2} + \text{pH} + pe \tag{7.46}$$

At equilibrium, $\log K = pe^0$ for this one-electron transfer. Since pe^0 for this reaction is $+20.75$, Eq. 7.46 becomes

$$20.75 + \tfrac{1}{4}\log P_{O_2} = \text{pH} + pe \tag{7.47}$$

Thus at $P_{O_2} = 1$ atm,

$$pe = -\text{pH} + 20.75 \tag{7.48}$$

This equation can be redefined for other partial pressures of O_2, but this effect is minor as Eq. 7.47 is not significantly altered by variations in the P_{O_2} term.

The redox couple that defines the lower boundary line is

$$H^+ + e^- \rightleftharpoons \tfrac{1}{2}H_2 \tag{7.49}$$

Its equilibrium constant is

$$K = \frac{P_{H_2}^{1/2}}{\{H^+\}\{e^-\}} \tag{7.50}$$

Taking the log of both sides and rearranging:

$$\log K = \tfrac{1}{2}\log P_{H_2} - \log\{H^+\} - \log\{e^-\} \tag{7.51}$$

or,

$$\log K = \tfrac{1}{2}\log P_{H_2} + \text{pH} + pe \tag{7.52}$$

By convention, $\log K = pe^0 = 0$ for the H^+/H_2 couple. Substituting this value into Eq. 7.52 yields

$$-\tfrac{1}{2}\log P_{H_2} = \text{pH} + pe \tag{7.53}$$

Thus at $P_{H_2} = 1$ atm,

$$pe = -\text{pH} \tag{7.54}$$

As with the upper boundary line, Eq. 7.53 is not significantly influenced by variations in P_{H_2}.

The region that lies above the upper boundary (high pH and *pe*) corresponds to environmental conditions where water is thermodynamically unstable and is spontaneously oxidized to O_2. Under these conditions, the partial pressure of O_2 in equilibrium with liquid water will exceed 1 atm. In the region that lies below the lower boundary (low pH and *pe*), water is also thermodynamically unstable and will be spontaneously reduced to H_2. Under these conditions, the partial pressure of H_2 in equilibrium with liquid water will exceed 1 atm.

Seawater is not naturally subject to gas pressures that exceed 1 atm. Thus, the region of the diagram where water is stable defines the *pe*–pH conditions most likely to exist in seawater. Within this region, conditions of high *pe* are usually coincident with high pH. Conditions of low pH and low *pe* are caused by the aerobic oxidation of large amounts of organic matter. The removal of O_2 lowers the *pe* of the system, whereas the concurrent production of CO_2 lowers the pH. Waters in which most of the O_2 has been removed are termed anaerobic or anoxic.

When O_2 is present, it is the dominant oxidizing agent due to its high *pe*, abundance, and reactivity. Its pe_w^0 is so much larger than that of all other species that the equilibrium *pe* of aerobic seawater should be close to +13.75. In reality, the overall *pe* of seawater, like its pH, is the result of many competitive and interactive reactions, not all of which attain equilibrium. Nevertheless, the use of the equilibrium approach is not totally unreasonable as the redox reactions proceed somewhat independently. Furthermore, most achieve a type of steady state that approximates equilibrium.

Redox reactions tend to proceed independently for two reasons. First, reactions tend to occur in a sequence dictated by their relative energy yields. Thus a closed container of seawater will first undergo aerobic redox reactions, then anaerobic ones if sufficient organic matter is present to deplete the O_2. Similarly, sediments tend to be aerobic at the surface and anaerobic at depth. This separation is also enhanced by the poisoning effect some redox chemicals have on the enzymes of competing processes. Without the catalytic effect of enzymes, many redox reactions proceed at very slow rates.

Second, redox processes tend to be separated in time and space due to the relative sluggishness of solute transport. For example, molecular diffusion is the major mechanism by which solutes can be transported through the pore waters of sediments. In many cases this process is slower than the chemical reaction rates and thus prevents the attainment of redox equilibrium. Likewise, gas exchange across the air–sea interface can be inhibited by the presence of dense algal mats. In this case, the underlying water tends to go anaerobic.

As shown in Figure 7.6*b*, nitrate should be the dominant species at the *pe* and pH of seawater. The large amounts of N_2 that are actually present in seawater suggest that redox equilibrium is not attained. This is likely due to the kinetic problems associated with breaking the triple bond in N_2 during nitrogen fixation. In comparison, sulfate dominates at the *pe* and pH of aerobic seawater, as predicted in Figure 7.6*c*, whereas sulfide is favored under the reducing conditions commonly caused by the decomposition of large amounts of organic matter. As shown in Figure 7.6*d*, all forms of dissolved organic matter are thermodynamically unstable in aerobic seawater. The presence of substantial amounts of DOM demonstrates that the speciation of carbon is not regulated by thermodynamic equilibrium. In the next chapter, the biogeochemical cycles of carbon and nitrogen are considered from the perspective of the relative rates of their redox reactions.

Electron Transfer on the Intracellular Level

Redox energy is transferred between molecules within cells by the reactions shown in Table 7.7. **Electron carriers** such as nicotinamide adenine dinucleotide phosphate (NADP) and nicotinamide adenine dinucleotide (NAD) transfer energy via redox reactions. Others, like adenosine triphosphosphate (ATP), transport energy by transferring phosphate groups. This process is termed phosphorylation.

The Global Biogeochemical Redox Cycle

The transfer of redox energy can be represented as the global biogeochemical cycle illustrated in Figure 7.7. This cycle is driven by solar energy. Through the process of photosynthesis, solar energy is converted into thermodynamically unstable chemical species. This is initiated by the transfer of electrons from water to carbon dioxide, creating electron-rich organic matter and electron-poor O_2. This organic matter is thermodynamically unstable in the presence of O_2. The actions of nonphotosynthetic organisms tend to restore the system to

TABLE 7.7
Some Cellular Energy Transfer Reactions

Redox Half-Reactions (Reduction)	pe^0_w
$NAD^+ + 2H^+ + 2e^- = NADH^+ + H^+$	−5.4
$NADP^+ + 2H^+ + 2e^- = NADPH + H^+$	−5.5
2 Ferredoxin(Ox) + $2e^-$ = 2 ferredoxin(Red)	−7.1
Ubiquinone + $2H^+ + 2e^-$ = ubiquinol	+1.7
2 Cytochrome C(Ox) + $2e^-$ = 2 cytochrome C(Red)	+4.3
Phosphate Exchange Half-Reactions (Hydrolysis)[a]	ΔG^0_w(kcal/mol)
Phosphoenol pyruvate = pyruvate + P_i	−14.8
Phosphocreatinine = creatinine + P_i	−10.3
Acetylphosphate = acetate + P_i	−10.1
Adenosine triphosphate (ATP) = ADP + P_i	
37°C, pH = 7.0, excess Mg^{2+}	−7.3
25°C, pH = 7.4, 10^{-3} *M* Mg^{2+}	−8.8
25°C, pH = 7.4, no Mg^{2+}	−9.6
Adenosine diphosphate (ADP) = AMP + P_i	−7.3
Glucose-1-phosphate = glucose + P_i	−5.0
Glucose-6-phosphate = glucose + P_i	−3.3
Glycerol-1-phosphate = glycerol + P_i	−2.2

Source: From *Principles of Aquatic Chemistry*, F. M. M. Morel, copyright © 1983 by John Wiley & Sons, Inc., New York, p. 338. Reprinted by permission. See Morel (1983) for data sources.

[a]P_i = inorganic phosphate.

equilibrium by catalyzing the oxidation of organic matter. During these reactions, electrons supplied by the organic matter cause O_2, nitrate, sulfate, and carbon dioxide to be reduced to water, N_2, H_2S, and CH_4, respectively. The oxidation of organic matter regenerates the carbon dioxide required for photosynthesis.

The reduced compounds produced from the oxidation of organic matter are also thermodynamically unstable in the presence of O_2. Their ensuing oxidation regenerates the oxidized species (NO_3^-, SO_4^{2-}, and CO_2) needed to reinitiate the cycle. In addition, the electrons are transferred back to regenerated water, which is then available to participate in photosynthesis.

The solar energy has two possible fates: Either it becomes buried in the sediments or soil as organic matter or is dissipated as heat. The latter is one of the products of metabolism. If insolation and the concentrations of the redox species remain constant over time, this heat would have to be radiated from the planet to maintain a steady-state energy balance with respect to incoming solar radiation. These biological processes also affect the global heat budget by controlling the atmospheric levels of greenhouse gases, such as CO_2.

PHOTOCHEMICAL REDOX REACTIONS

Other photochemical reactions besides photosynthesis occur in surface seawater. Most are initiated by **oxygen free radicals** that are produced by the interaction of O_2 and sunlight. These free radicals are stronger oxidizing agents than O_2, but they are relatively unstable and occur at very low concentrations. Thus their influence is limited to the surface waters, where their concentrations are highest. Here they appear to cause the oxidation of a significant amount

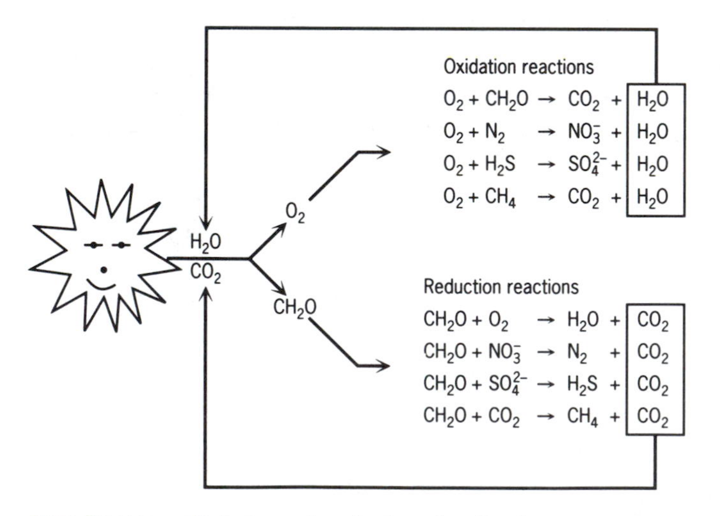

FIGURE 7.7. Global geochemical cycle of redox energy.

of dissolved metals* and organic matter. These photochemical reactions are thought to be partially responsible for the formation of humic substances.

SUMMARY

The biogeochemical cycles of the minor and trace elements are strongly influenced by **redox reactions**. These reactions involve the transfer of electrons between atoms and molecules. The flow of electrons continues until the reactants and products attain their lowest **free energy** levels. This state is known as equilibrium. The **cell potential (E_{cell})** and **free energy change (ΔG)** are both a measure of how far the reactants and products are from the equilibrium state. The tendency of electrons to flow from reduced to oxidized species can also be expressed as an **electron activity**, or **pe**.

For bookkeeping purposes, redox reactions can be thought of as combinations of **oxidation** and **reduction half-reactions**. The **oxidation number** of an atom that undergoes an oxidation increases, whereas that of an atom that undergoes a reduction decreases. The **electrode potential** of these half-cell reactions is determined by measurement in a **standard hydrogen cell** at 25°C, 1 atm, and 1 *M* concentrations of all solutes. Standard half-cell potentials can be combined to determine the redox potential of a reaction at standard state. Combinations that produce large positive potentials are most strongly favored and proceed spontaneously.

The redox potential under nonstandard conditions can be calculated from the **Nernst Equation**. Other equations can be derived to interrelate the free energy change, redox potential, *pe*, and electrode potential at standard, as well as nonstandard states, such as at equilibrium.

These equations can be used to calculate the redox speciation of seawater. The results indicate that the O_2/H_2O couple is the dominant oxidizing agent due to its relatively high *pe*, abundance, and reactivity. For similar reasons, organic matter is the most important reducing agent in seawater. This electron donor is the product of photosynthesis, which is an example of a **carbon fixation reaction** fueled by solar energy. Because photosynthetic organisms use inorganic carbon, solar energy, and water as their electron donor, they are termed **photoautolithotrophs**. The organic matter and O_2 produced by this process drive the **heterotrophic** metabolism of most animals, bacteria, protozoans, and fungi.

In the absence of O_2, other electron acceptors (**oxidizing agents**) drive the oxidation of organic matter. These processes occur in order of energy yield. In the absence of organic matter, microorganisms use other reduced compounds as electron donors (**reducing agents**), also in order of their energy yield. These redox reactions tend to be separated in time and space for two reasons. First, they occur in order of energy yield. In some cases, reactants inhibit competing

---

*There are exceptions to this. For example, Mn in MnO_2 appears to be reduced to Mn^{2+} by photochemically generated superoxide, O_2^-.

redox reactions by binding with their competitors' enzymes in a fashion that deactivates them. Without this catalytic effect, most redox reaction rates are very slow. Second, redox reactions are controlled by the relative rates of physical processes that act to transport solutes in the ocean. In locations, such as pore waters, where this rate of transport is slow, redox equilibrium is often not attained. Nevertheless, biologically mediated redox reactions tend to reach a steady state that closely approximates equilibrium conditions. Thus the equilibrium approach can be used to calculate redox speciation, though deviations, especially in the biologically active elements, are commonly observed.

Redox energy is transported within cells through the transfer of electrons between specialized molecules called **electron carriers**. These include **NADP**, **NAD**, and **ATP**. On a global level, redox energy is transferred through a biogeochemical cycle driven by solar energy. Photosynthetic organisms use this solar energy to synthesize molecules that are thermodynamically unstable. The actions of nonphotosynthetic organisms tend to restore these chemicals to stable form through a series of redox reactions. Eventually the solar energy is either buried in the sediments as organic matter or released to the environment as heat, which is then lost from the ocean via radiation and conduction. In this way, a steady-state balance of redox energy is maintained.

Though the O_2/H_2O couple is the dominant oxidizing agent in seawater, small amounts of stronger oxidizing agents are present in the surface waters. They are produced by the interaction of sunlight and O_2. The resulting **oxygen free radicals** are extremely reactive. In the surface waters where their concentrations are highest, they oxidize significant amounts of dissolved metals and organic matter and appear to be partially responsible for the formation of humic substances.

CHAPTER 8

ORGANIC MATTER: PRODUCTION AND DESTRUCTION

INTRODUCTION

In the last chapter, organic matter was identified as the most important electron donor in the marine environment. As such, it provides the energy needed to drive most of the biologically mediated redox reactions. Marine chemists use the term organic matter to refer collectively to any and all organic compounds. Though they vary greatly in molecular weight and structure, these compounds are composed primarily of the elements carbon, nitrogen, oxygen, phosphorus, and sulfur.

The oceanic reservoirs of nitrogen and phosphorus are relatively small. As a result, their distribution in seawater is controlled by the biologically mediated redox processes that also drive the biogeochemical cycle of organic matter. Hence nitrogen and phosphorus are termed **biolimiting elements**. In comparison, much greater amounts of carbon and sulfur are present in seawater. Since biological processes have less of an impact on their marine distributions, they are termed **biointermediate elements**. Most of the hydrogen and oxygen in the ocean is present as water. Thus biological processes have little impact on the bulk distribution of these elements in the sea. On the other hand, these processes do strongly affect the concentrations of hydrogen- and oxygen-containing species, such as $CO_2(g)$, $O_2(g)$, $H_2(g)$, and $H^+(aq)$.

In this chapter, the biogeochemical cycling of organic matter is discussed. To do this, a model compound is used to represent an average molecule of organic matter. Chemical equations are written to illustrate the impact of metabolic processes on this average organic matter and the biolimiting elements. The effect of these processes on the biogeochemical cycle of oxygen is also discussed. These concepts will be used in later chapters to examine the biogeochemical cycles of nitrogen and phosphorus.

THE PRODUCTION OF ORGANIC MATTER

Since phytoplankton are the most abundant marine primary producers, they are the dominant source of marine organic matter. The average atomic ratio of C to N to P in marine phytoplankton is 106 to 16 to 1. This is called the **Redfield–Richards Ratio**, after the scientists who conducted the first geographically comprehensive survey of the elemental composition of marine phytoplankton.

Based on this ratio an average molecule of phytoplanktonic organic matter can be represented by the empirical formula $C_{106}(H_2O)_{106}(NH_3)_{16}PO_4$. Thus the process of photosynthesis is given by

$$106\,CO_2 + 122\,H_2O + 16\,HNO_3 + H_3PO_4 \rightarrow (CH_2O)_{106}(NH_3)_{16}H_3PO_4 + 138\,O_2 \quad (8.1)$$

This reaction illustrates that, in addition to carbon dioxide and water, phytoplankton also require dissolved inorganic nitrogen and phosphorus. Due to their relatively low concentrations in seawater, inorganic nitrogen and phosphorus are usually the limiting reactants in photosynthesis. Since adding these chemicals to seawater stimulates plant growth, the biolimiting elements are also referred to as plant **nutrients**.

Equation 8.1 illustrates only the stoichiometry of the process of photosynthesis. The actual chemical species that are taken up, or **assimilated**, by the plant are somewhat variable. The process of assimilation occurs via active transport of ions across the cell membrane. Thus inorganic carbon is probably taken up as bicarbonate (HCO_3^-), which is in thermodynamic equilibrium with $CO_2(aq)$. In addition to nitrate (NO_3^-), nitrogen is assimilated as the species nitrite (NO_2^-) and ammonium (NH_4^+). Phosphorous is probably assimilated as phosphate (PO_4^{3-}) and sulfur as sulfate (SO_4^{2-}). Though sulfur is an important constituent of organic matter, its chemistry is not considered further here, as the production and consumption of organic matter has little impact on sulfate ion concentrations due to its relatively great abundance in seawater.

Photosynthetic organisms require visible light. Insolation is the source of this light but does not penetrate far beneath the sea surface due to absorption by water molecules and particulate matter. Thus photosynthesizing plants are limited to a relatively shallow surface layer, which is termed the **euphotic** zone. The bottom of the euphotic zone is defined as the depth at which 1 percent of the insolation remains unabsorbed. The greater the insolation and the lower the turbidity, the deeper the euphotic zone. But even in the clearest waters at the lowest latitudes, the euphotic zone is limited to the top 200 m of the ocean.

Though plant growth is restricted to the surface ocean, photosynthesis is the ultimate source of almost all the organic matter that supports heterotrophic activity in the sea. As shown in Figure 8.1, the feeding activities of animals cause organic matter to be transferred up the food chain.

Phytoplankton are the dominant primary producers in the open ocean. The only exceptions occur in the relatively isolated ecosystems that are located around hydrothermal vents and other types of underwater seeps. The geological activity at these sites causes reduced inorganic compounds to be released into the deep waters. Some bacteria at these sites are able to fix inorganic carbon using energy obtained from the oxidation of the reduced inorganic compounds. In these locations, the in situ production of organic matter exceeds that supplied

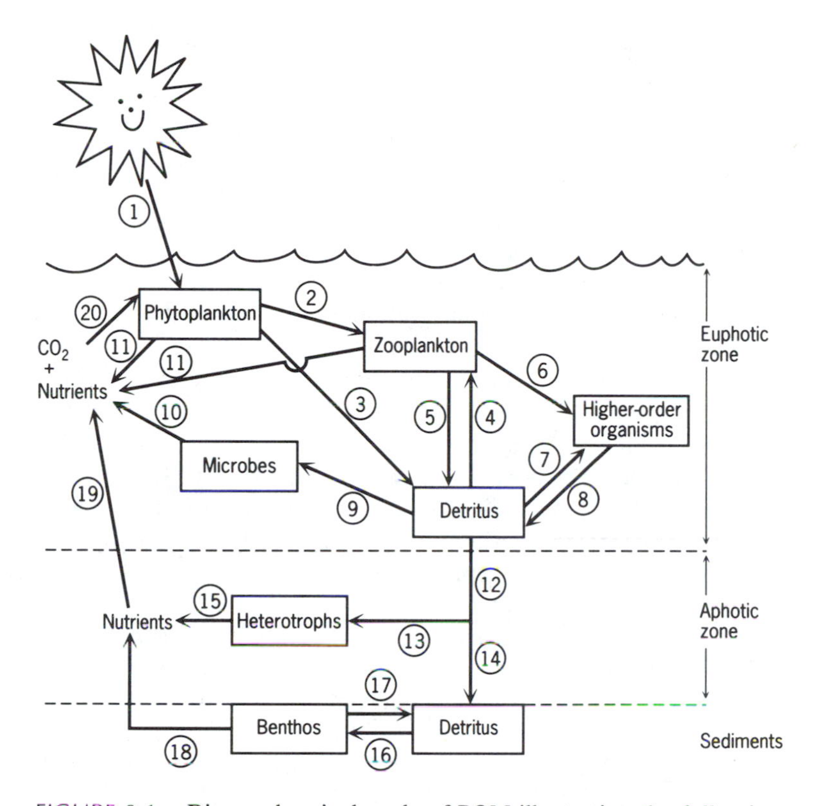

FIGURE 8.1. Biogeochemical cycle of POM illustrating the following processes: (1) Photosynthesis, (2) consumption, (3) death, (4) consumption of detritus, (5) excretion of POM and death, (6) consumption, (7) consumption of detritus, (8) excretion of POM and death, (9) bacterial degradation, (10) nutrient regeneration, (11) excretion of nutrients, (12) sinking POM, (13) consumption, (14) sedimentation, (15) nutrient regeneration, (16) consumption, (17) excretion of POM and death, (18) nutrient regeneration, (19) nutrient transport via vertical advection and eddy diffusion, (20) nutrient assimilation.

indirectly from photosynthesis through sinking **particulate organic matter (POM)**. Thus chemoautolithotrophic bacteria are the dominant primary producers in these ecosystems. Most are aerobes and hence their metabolic activities require O_2, which is ultimately derived from photosynthesis.

THE AEROBIC DESTRUCTION OF ORGANIC MATTER

The particulate organic matter in the ocean is composed of a wide variety of materials including excreta, molts, aggregated dissolved organic matter (DOM), and living, as well as dead tissues. Most of the nonliving, or **detrital**, fraction is decomposed by the respiratory activities of bacteria, fungi, protozoans, and animals. This process can be represented as

$$(CH_2O)_{106}(NH_3)_{16}H_3PO_4 + 138\,O_2 \rightarrow 106\,CO_2 + 122\,H_2O + 16\,NHO_3 + H_3PO_4 \qquad (8.2)$$

The complete decomposition of POM returns the nitrogen, phosphorous, and carbon to their soluble forms. Hence, this process is termed **nutrient regeneration** or **remineralization**.

Since organic matter is the sole source of chemical energy for heterotrophic organisms, POM is rapidly remineralized upon death or excretion. Nevertheless, a small amount sinks fast enough to reach the seafloor before becoming completely remineralized. But since respiration occurs through all depths, less than 1 percent of the POM produced by primary production survives the trip to the seafloor. Even as part of the sediments, POM is subject to respiration. Though the effects of the benthos are greatest at the sediment–water interface, remineralization is a significant sink of POM through the top 10 cm of sediment. Thus the organic carbon content of marine sediments is usually less than 1 percent by mass, except in coastal areas, where values as high as 10 percent have been observed.

Equation 8.2 illustrates only the stoichiometry of respiration. The actual metabolic processes usually occur as a series of coupled reactions. For example, during the initial stages of remineralization, nitrogen is released from organic matter in the form of ammonium. If O_2 is present, the ammonium is rapidly oxidized by certain species of bacteria to nitrite and by others to nitrate. This process is termed nitrification. In the surface waters, a significant fraction of the ammonium is assimilated by phytoplankton before it can become oxidized.

IMPACT ON O_2 CONCENTRATIONS

Since respiration consumes O_2, deep-water masses tend to be undersaturated with respect to this gas. The thermocline is the site of the largest undersaturations for several reasons: First, the strong density stratification in the thermocline inhibits vertical mixing. This isolates the thermocline from the mixed layer and hence the atmospheric pool of O_2. Second, the flux of POM to the thermocline is larger than that to the deep zone, so the O_2 demand is larger. Third, horizontal advection is relatively sluggish in this depth zone. Since molecular diffusion is too slow to transport a significant amount of O_2 into the thermocline, the respiration of POM causes a large net removal of O_2 from the thermocline. Thus the in situ O_2 concentrations are lower than the NAEC.

Below the thermocline, undersaturations are not as pronounced due to the relatively faster supply of O_2 from thermohaline circulation and smaller flux of sinking organic detritus. As noted in Chapter 4, deep-water masses are created at the sea surface in polar zones where cooling gives rise to convection currents. Gas solubility is enhanced at lower temperatures, so the sinking water masses are O_2 rich. As the water masses travel horizontally through the ocean basins, O_2 is consumed during the respiration of organic matter. This organic matter has two origins—either it was present in the water mass at the time of its sinking from the sea surface, or it was transported into the water mass sometime later by the sinking of detrital POM. Thus, as a water mass moves through the ocean basins, its in situ $[O_2]$ decreases due to the continuing decomposition of organic matter.

The amount of O_2 consumed since a water mass was last at the sea surface can be calculated if its original gas concentration is known. Assuming that the water mass was in gaseous equilibrium with the atmosphere at the time it sank from the sea surface, the original concentration is equal to the NAEC. The difference between the in situ O_2 concentration and the NAEC is called the **AOU**, or **apparent oxygen utilization**, of the water mass. In mathematical terms,

$$\mathrm{AOU} = \mathrm{NAEC} - [O_2]_{\text{in situ}} \tag{8.3}$$

AOU is usually expressed in units of μmol O_2/kg SW. This is an "apparent" term because deviations from the NAEC can also result from physical processes, such as the warming of a water mass after it has been isolated from the sea surface.

The higher the AOU, the greater the amount of O_2 removed since the water mass was last at the sea surface. Thus, AOU increases with increasing distance from the site at which the deep-water mass was formed. Since the AOU increases with the age of the water mass, the pathway of thermohaline circulation can be traced from the distribution of AOU in the deep sea. As shown in Figure 8.2, the AOU in deep water is lowest in polar regions, suggesting these areas are the sites of deep-water formation.

An estimate of the amount of organic matter respired since a deep-water mass was last at the sea surface can be inferred from its AOU and the stoichiometry given in Eq. 8.2. The respiration of 1 mole of average marine plankton detritus requires the oxidation of 106 moles of organic carbon. As shown below, this requires 106 moles of O_2:

$$106\,CH_2O + 106\,O_2 \rightarrow 106\,CO_2 + 106\,H_2O \tag{8.4}$$

Similarly, Eq. 8.2 indicates that 16 moles of organic nitrogen must be oxidized to nitrate. As shown below, this requires 32 moles of O_2:

$$16\,NH_3 + 32\,O_2 \rightarrow 16\,HNO_3 + 16\,H_2O \tag{8.5}$$

Since phosphorous is not oxidized during the respiration of organic matter, it does not contribute to the O_2 uptake. Thus, 138 moles of O_2 are consumed during the respiration of 1 mole of average plankton detritus. Stated another way, the molar ratio of organic carbon respired to O_2 consumed is 106 to 138.

The amount of POM remineralized since a deep-water mass was last at the sea surface can be calculated using this ratio. For example, the lowest AOU shown in Figure 8.2 occurs in the North Pacific. The amount of organic carbon that must be oxidized to cause this AOU is

$$190\ \mu\text{mol}\ O_2/\text{kg SW} \times \frac{106\ \text{mol C}}{138\ \text{mol}\ O_2} = 146\ \mu\text{mol C/kg SW} \tag{8.6}$$

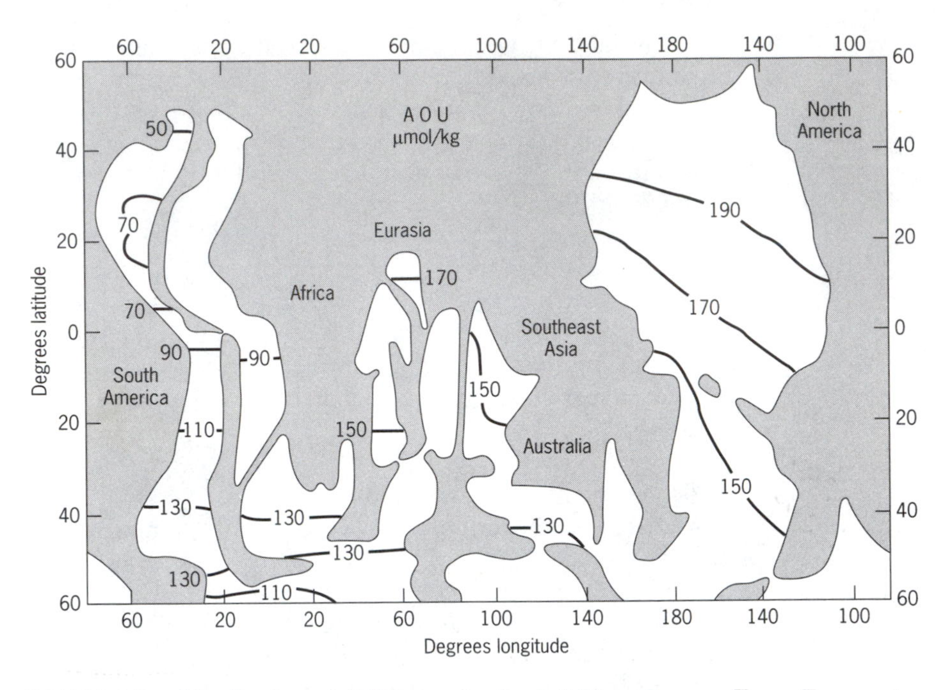

FIGURE 8.2. Distribution of AOU at a depth of 4000 m. *Source:* From *Tracers in the Sea,* W. S. Broecker and T.-H. Peng, copyright © 1982 by the Lamont-Doherty Geological Observatory, Palisades, New York, p. 133. Reprinted by permission. See Broecker and Peng (1982) for data sources.

In other words, this process causes the inorganic carbon content of the seawater to be increased by 146 µmol/kg.

Deviations from the stoichiometry given in Eq. 8.2 are common. Thus, these type of calculations only provide estimates of the effects that POM remineralization can have on the chemical composition of seawater. These deviations are the result of several phenomena, such as variability in the elemental composition of phytoplankton. In addition, not all of the POM that is respired has the elemental composition of phytoplankton. For example, the C to N ratios of zooplankton and fecal pellets are higher than 106:16. Equation 8.2 also assumes that all of the organic carbon and nitrogen are completely oxidized, which is often not the case. Furthermore, nitrogen is respired faster than carbon, so the C to N ratio of organic matter increases as it decomposes.

Despite these deviations, the N to P ratio of dissolved inorganic nitrogen and phosphorus is approximately the same as in marine phytoplankton. This concurrence suggests that either the phytoplankton determine the N to P ratio in seawater or vice versa. Evidence in support of the former is seen in the stabilizing role phytoplankton play in the biogeochemical cycles of nitrogen and phosphorus. One example of such a feedback is described below.

The growth of marine phytoplankton is nutrient limited, so plants remove virtually all of the inorganic nitrogen and phosphorus that is present in surface seawater. In the event that plants were to run out of nitrogen first, more could be obtained through the process of nitrogen fixation. Since this process is energetically expensive, nitrogen fixers would only be ecologically competitive under conditions of sustained nitrogen limitation. In such a case, this supply of "new" nitrogen would enable the phytoplankton to continue synthesizing organic matter with an N to P ratio of 16 until the phosphorus was depleted. Further growth would require some of the planktonic POM to be remineralized. Assuming that this remineralization followed the stoichiometry of Eq. 8.2, the nutrients would be regenerated in the Redfield–Richards Ratio. This hypothetical scenario suggests that marine phytoplankton are ultimately phosphorus limited.

THE ANAEROBIC DESTRUCTION OF ORGANIC MATTER

If the rate of O_2 removal at a particular location is close to or exceeds its rate of supply from thermohaline circulation, O_2 concentrations will be present at low or even nondetectable levels. Such oxygen-deficient zones occur in areas where the overlying primary productivity is high and the rate of deep-water motion is relatively slow. The former ensures a large flux of organic matter and the latter a slow supply of O_2. These conditions are commonly found in upwelling areas located in regions where thermohaline circulation is restricted, as shown in Figure 8.3.

At some of these locales, **aerobic respiration** had depleted the O_2 concentrations to levels that permit anaerobic microorganisms to metabolize any remaining organic matter. As described in Chapter 7, nitrate and sulfate can function as electron acceptors in the absence of O_2, thus enabling the oxidation of organic matter. The stoichiometry of denitrification and sulfate reduction is given in Eqs. 8.7 and 8.8.

$$(CH_2O)_{106}(NH_3)_{16}H_3PO_4 + 84.8\,HNO_3 \rightarrow 106\,CO_2 + 148.8\,H_2O + 42.4\,N_2 + 16\,NH_3 + H_3PO_4 \quad (8.7)$$

$$(CH_2O)_{106}(NH_3)_{16}H_3PO_4 + 53\,SO_4^{2-} \rightarrow 106\,CO_2 + 106\,H_2O + 16\,NH_3 + 53S^{2-} + H_3PO_4 \quad (8.8)$$

organic oxidation using SO_4^{2-} as the electron acceptor

Note that in the absence of O_2, the remineralized nitrogen and sulfur remain in reduced form (i.e., NH_3 and S^{2-}). Hydrolysis of the CO_2 produces acid ($H_2O + CO_2 \rightarrow H_2CO_3 \rightarrow H^+ + HCO_3^-$), which causes the reduced forms to be almost wholly converted to NH_4^+ and HS^-.

Due to its higher free energy yield, denitrification is thermodynamically favored over sulfate reduction and will proceed until the nitrate is depleted. If organic matter is present, sulfate reduction will then occur. These conditions are not found in open ocean waters but have been observed in coastal waters near upwelling zones and in the underlying sediments.

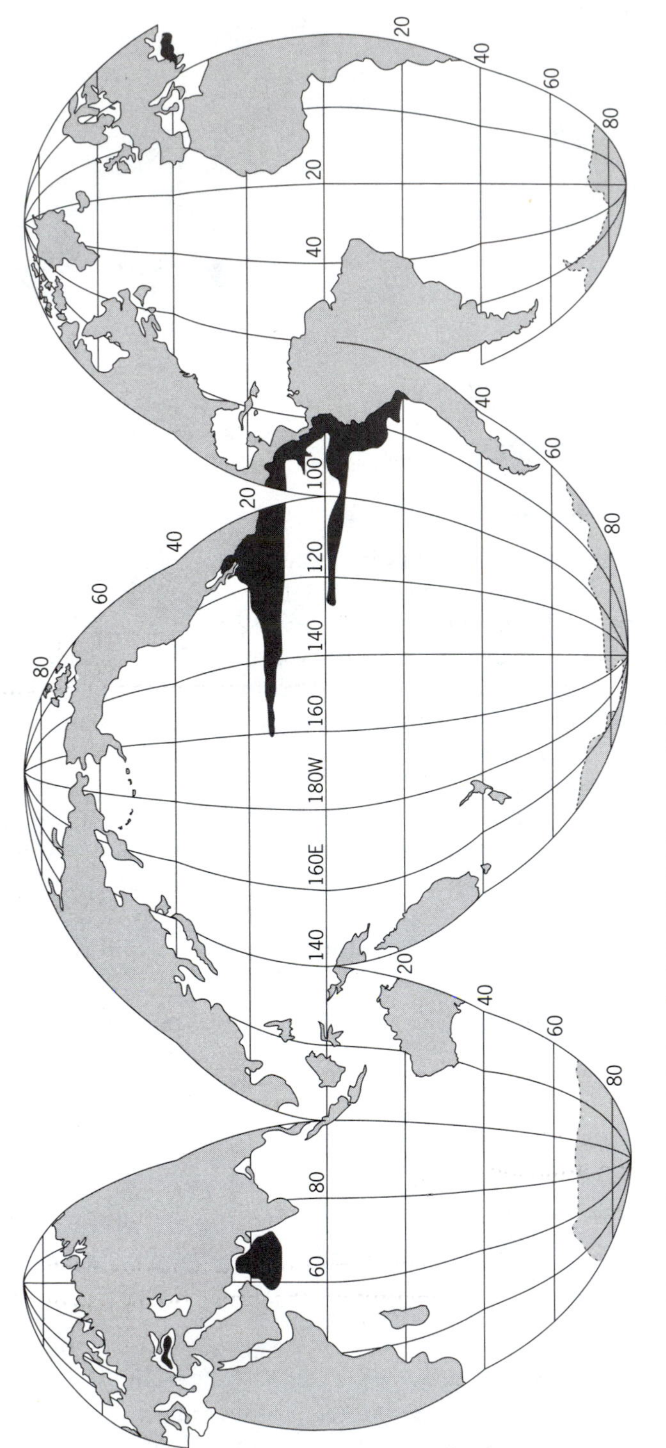

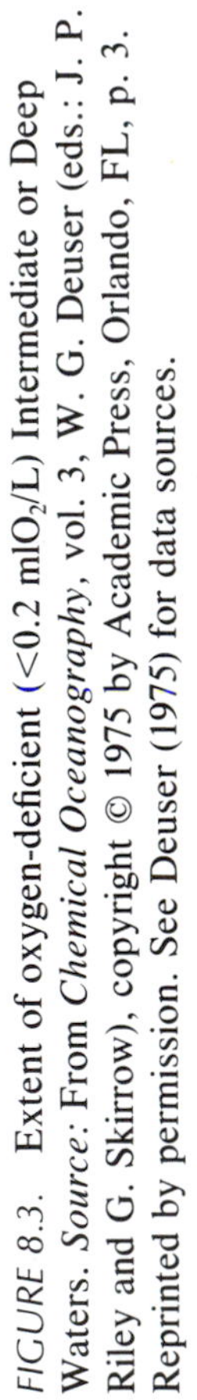

FIGURE 8.3. Extent of oxygen-deficient (<0.2 mlO_2/L) Intermediate or Deep Waters. *Source*: From *Chemical Oceanography*, vol. 3, W. G. Deuser (eds.: J. P. Riley and G. Skirrow), copyright © 1975 by Academic Press, Orlando, FL, p. 3. Reprinted by permission. See Deuser (1975) for data sources.

THE INFLUENCE OF ORGANIC MATTER ON THE GLOBAL BIOGEOCHEMICAL OXYGEN CYCLE

The presence of free oxygen (O_2) in Earth's atmosphere is a unique feature among the planets of our solar system. Another of Earth's unique features, life, is largely responsible for this accumulation of O_2. The planet's atmosphere was not always O_2-rich. During the early stages of Earth's formation, all of its atmospheric gases were derived from volcanic emissions or the degassing of volcanic rocks. These atmospheric gases were mostly reduced species, such as ammonia, sulfur dioxide, carbon monoxide, hydrogen, methane, hydrogen chloride, hydrogen sulfide, and water.

Before life evolved on the planet, O_2 was produced solely through the process of photodissociation,

$$H_2O(g) + UV \rightarrow H_2(g) + O_2(g) \tag{8.9}$$

This reaction occurs in the upper atmosphere when ultraviolet radiation from the sun strikes water molecules. The rate of this photodissociation has always been slow. On the surface of the early earth, O_2 was thermodynamically unstable due to the reduced nature of the chemicals that composed the fresh volcanic rocks and gases. Thus, most of the O_2 produced by photodissociation was rapidly consumed in oxidizing these substances.

The rate of O_2 production on Earth's surface increased considerably with the evolution of photosynthetic organisms. Eventually plants produced enough O_2 to exceed the amount required to oxidize the crustal rocks completely. Once O_2 was no longer a limiting reactant, its atmospheric levels began to rise.

The oxygen content of the atmosphere is now 21 percent vol/vol and appears to have been constant at this level for at least the past 65 million years. The present set point and its great stability are probably the result of a feedback loop that operates on time scales on the order of millions of years. This loop must be the result of interactions among some of the processes that constitute the biogeochemical cycle of O_2.

For any set of processes to function as this feedback loop, they must be collectively dependent on atmospheric O_2 levels and vice versa. As shown in Figure 8.4, respiration of all the organic matter in the biosphere would lower the atmospheric O_2 level by only 1 percent. Thus, this reservoir is not a likely component of the O_2 feedback loop as it is too small to have a significant impact. The weathering of igneous rock is also presently an unimportant O_2 sink as the production rate of new igneous rock is slow and all the original volcanic rock constituting the crust of the early planet has long since been oxidized. Indeed, most exposed sedimentary and metamorphic rock is now also completely oxidized, suggesting that weathering proceeds to completion once erosion exposes the rocks to the atmosphere. In other words, the rate of oxidation of these rocks is limited by their erosion rate, not by the abundance

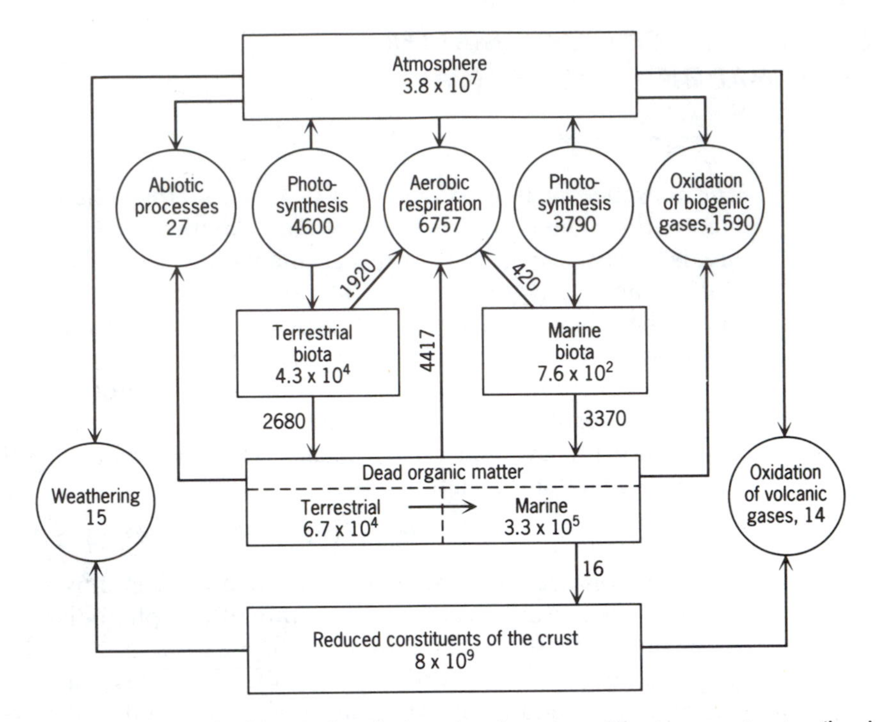

FIGURE 8.4. The biogeochemical cycle of oxygen. The transport rates (in circles) and reservoir sizes (in boxes) are given in units of 10^{12} mol O_2/y and 10^{12} mol O_2, respectively. Also shown are the potential O_2 sinks, i.e., the reservoirs of reduced materials in the biosphere, sediments, and crust. The sizes of these reservoirs are given in terms of the amount of O_2 needed to oxidize the reduced materials completely. Since this is mostly organic carbon, the complete oxidation of the entire marine biospheric reservoir (7.6×10^{14} moles of organic carbon) would consume 7.6×10^{14} moles of O_2, assuming the stoichiometry given in Equation 8.4. *Source:* From *The Natural Environment and the Biogeochemical Cycles,* J. G. C. Walker (ed.: O. Hutzinger), copyright © 1980 by Springer-Verlag, Heidelberg, Germany, p. 89. Reprinted by permission.

of atmospheric O_2. Thus the crustal weathering sink cannot be the factor that stabilizes atmospheric O_2 levels.

The mechanism that stabilizes the atmospheric O_2 level is most likely one that produces, rather than consumes, this gas. Photosynthesis itself is not a net O_2 source. Once a plant dies, all of the O_2 generated during the biosynthesis of its POM will be consumed during its oxidation. But if the detrital POM is buried prior to oxidation, the O_2 will remain in the atmosphere. Thus the greater the amount of POM buried, the higher the atmospheric O_2 levels and vice versa. In other words, the amount of POM buried should be inversely related to the atmospheric O_2 level. This process has the potential to act as a stabilizing mechanism only if the amount of POM buried is large enough to have a significant impact on the amount of O_2 in the atmosphere.

The continental margins are the only locations where the amount of buried sedimentary organic carbon is large enough to account for a substantial amount of O_2 production. These sediments have relatively high carbon contents because they are locations where primary productivity is high. This ensures a large flux of POM to the sediments. In addition, a greater percentage of POM is preserved in these sediments because 1) high sedimentation rates cause the POM to be rapidly buried; 2) the POM takes less time to reach the seafloor due to the relatively shallow water depths; and 3) the large POM flux creates oxygen-deficient conditions in the bottom waters. These conditions enhance the preservation of POM because remineralization rates are slower in the absence of O_2.

Terrestrial organic matter is not a significant source of atmospheric O_2 due to the nearly complete oxidation of soils prior to their burial on land. Little terrestrial organic matter is transported to the seafloor, so most of the buried organic carbon is produced by marine plankton.

The mechanism by which the burial of organic sediments could act to stabilize the atmospheric O_2 level is illustrated by two examples. First, consider a perturbation that causes a decrease in the partial pressure of atmospheric O_2. Such a decrease would cause dissolved O_2 concentrations in the ocean to decline. These O_2-deficient conditions would enhance the preservation of POM in the sediments. Thus the burial rate of POM would eventually increase, raising the O_2 production rate and hence the atmospheric O_2 level. Alternatively, if some perturbation were to cause the partial pressure of O_2 to increase, the rate of organic matter oxidation would increase. This accelerated rate of O_2 consumption would lower the atmospheric level and also decrease the burial rate of POM.

If either of these perturbations was permanent, a lower limit on the time required to reestablish steady state would be given by

$$\frac{\text{Total amount of atmospheric } O_2}{\begin{array}{c}\text{Burial rate of sedimentary}\\ \text{organic matter in the crust}\end{array}} = \frac{3.8 \times 10^{19}\ \text{mol } O_2}{16 \times 10^{12}\ \text{mol } O_2/\text{y}} \tag{8.10}$$

$$\approx 2.4 \text{ million years}$$

If the global biogeochemical cycle of organic matter is in steady state, 2.4 million years is also the minimum amount of time required to pass the entire atmospheric O_2 reservoir through the sediments. This assumes these sediments are completely oxidized once plate tectonics lifts them above sea level and exposes them to O_2. If oxidation has been incomplete, the organic carbon content of the sediments should have increased over time. Since no evidence for this has been observed in the sedimentary record, sedimentary organic carbon appears to be completely oxidized during each erosional cycle. Thus the proposed negative feedback loop should operate on time scales of no less than several million years.

Further support for the importance of POM in controlling atmospheric O_2 levels is seen in the following. If all the nutrients present in a parcel of newly created deep water were to be converted into POM by phytoplankton, the respiration of this amount of reduced carbon would consume all the dissolved O_2. The concentration of the latter is set by equilibration with the atmosphere. The occurrence of this redox equivalence is probably not coincidental.

If the burial of POM stabilizes the atmospheric O_2 level, the set point must be determined by factors that influence the production and destruction of POM. These include any processes that affect 1) the nutrient content of seawater, 2) the elemental composition of the plankton, or 3) the solubility of O_2. For example, a warming of the atmosphere would decrease gas solubility. The atmospheric O_2 level would have to increase to maintain the dissolved O_2 concentrations. To do this, more organic carbon would have to be buried in the sediments. Likewise, a change in the rate of plate tectonics could affect the rate of nutrient supply to the ocean by altering the rate of erosion of sedimentary rocks.

SUMMARY

Organic matter is composed of the elements carbon, hydrogen, oxygen, nitrogen, and sulfur. This material is of great importance in the marine environment as it is the source of electrons that drives most biologically mediated redox reactions. As a result, the biogeochemical cycle of organic matter has a great influence on the cycles of these biologically active elements. The effect is largest for nitrogen and phosphorus due to their relatively low abundances. These **biolimiting elements** are also classified as **nutrients** because their availability limits plant growth. The biogeochemistry of organic matter has less of an impact on the marine distributions of carbon and sulfur due to their greater abundances. Hence they are termed biointermediate elements. Most of the hydrogen and oxygen in the ocean is present as water. Thus biological processes have little impact on the bulk distribution of these elements in the sea. On the other hand, these processes do strongly affect the concentrations of hydrogen- and oxygen-containing species, such as $CO_2(g)$, $O_2(g)$, $H_2(g)$, and $H^+(aq)$.

Organic matter is synthesized from inorganic compounds by autotrophs. In the marine environment, phytoplankton produce most of the **particulate organic matter (POM)** through the process of photosynthesis. In addition to inorganic carbon, the plants must **assimilate** inorganic nitrogen, phosphorus, and sulfur. Extensive observations of the elemental composition of phytoplankton have demonstrated that the average atomic ratio of C to N to P in the tissues of these plants is 106 to 16 to 1. This is termed the **Redfield-Richards Ratio**.

Some of the POM synthesized by plants is consumed and thus passed up the food chain. This POM is eventually converted into **detrital** form through excretion or the death of the organisms. This POM is decomposed or **remineralized** by the metabolic activities of heterotrophic organisms. If O_2 is present, remineralization occurs via the process of **aerobic respiration**. When oxidation is complete, the particulate nitrogen and phosphorus are returned to soluble form, and hence this process is referred to as **nutrient regeneration**. Due to the large supply of O_2, most POM synthesized in the **euphotic** zone is rapidly remineralized. Nevertheless, a small fraction sinks fast enough to reach the seafloor. Since respiration occurs through all depths, only 1 percent of the POM synthesized by plants survives to become buried in the sediments. As a result, most marine sediments have a very low organic carbon content.

Respiration below the euphotic zone causes undersaturations with respect to O_2. These undersaturations are most pronounced in the thermocline since the waters are relatively stagnant in this zone and receive a relatively large flux of POM. The amount of O_2 consumed since a deep-water mass was last at the sea surface is given by its **apparent oxygen utilization (AOU)**, which is the difference between the in situ O_2 concentration and the NAEC. POM continually sinks into deep-water masses as they travel through the ocean basins. Thus the AOU increases as deep-water masses age.

In the absence of O_2, certain species of microorganisms use nitrate and sulfate as electron acceptors and thus are able to respire organic matter. Oxygen-deficient conditions are caused when the supply of POM is large relative to that of O_2. Such conditions are commonly found in coastal upwelling areas and in the sediments.

The stoichiometry of aerobic respiration and the AOU can be used to predict changes in water chemistry that should result from the aerobic decomposition of POM. These predictions can be invalidated by deviations from the assumed stoichiometry. Nevertheless, the N to P ratio in average phytoplankton is similar to that of the dissolved nutrients. This suggests that the seawater N to P ratio is controlled by the production and destruction of POM. These processes also have a large impact on the biogeochemical cycle of O_2. In particular, the relative stability of atmospheric O_2 levels is thought to be the result of a negative feedback loop that involves the production of O_2 via the burial of POM in coastal sediments.

CHAPTER 9

VERTICAL SEGREGATION OF THE BIOLIMITING ELEMENTS

INTRODUCTION

Marine organisms have had a large impact on the chemical evolution of the planet. Production of organic matter by photosynthetic organisms has made Earth's atmosphere O_2-rich. Marine plants have also been responsible for sequestering a large amount of carbon in marine sediments.

These organisms also have a large impact on the marine biogeochemical cycles of the biolimiting elements, such as nitrogen, phosphorus, and silicon. The distributions of these elements in seawater and the sediments are also influenced by physical factors, such as water motion, climate, and river runoff. As a result of the interaction of these physical and biological processes, the biolimiting elements have much higher concentrations in the deep than the surface waters. The reasons for this are discussed below using a box model, which also provides estimates of the oceanic residence times and recycling efficiencies of these elements.

SURFACE-WATER DEPLETIONS, BOTTOM-WATER ENRICHMENTS

At mid and low latitudes, the biolimiting elements tend to be present at low, and often undetectable, levels in the surface waters. Beneath the mixed layer, concentrations increase greatly with increasing depth. As shown in Figure 9.1, this transition zone usually coincides with the thermocline. In the deep zone, uniform concentrations are observed. In general, deep-water concentrations are somewhat lower than those at the bottom of the thermocline.

The deep-water enrichment and surface-water depletion of the biolimiting elements are a consequence of how biogenic particles are produced and destroyed in the ocean. The availability of the dissolved form of these elements

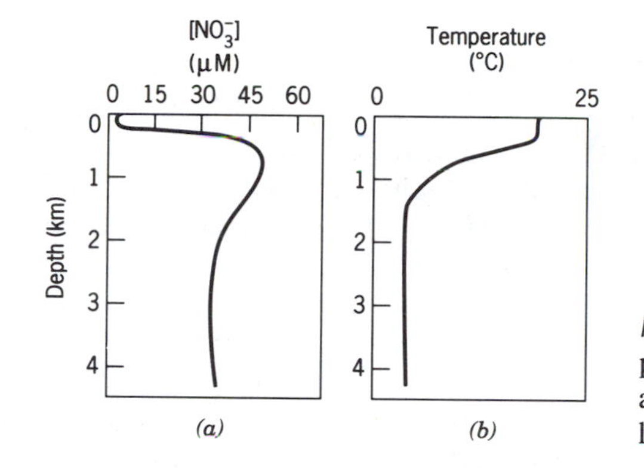

FIGURE 9.1. Idealized depth profiles of (*a*) a biolimiting element and (*b*) temperature at mid latitudes.

in surface seawater limits the growth of phytoplankton. Thus, plants tend to extract nearly all of these species from the mixed layer. Some of the elements, such as nitrogen and phosphorus, are incorporated into the tissues, or *soft parts*, of the plants. Others, such as silicon, end up in the shells, or *hard parts*, which are deposited by many species of phytoplankton.

The soft parts are eventually remineralized by bacteria. This returns the nitrogen and phosphorus to soluble form. Likewise, the dissolution of the siliceous hard parts resolubilizes silicon. When these processes occur in the surface waters, the biolimiting elements are rapidly assimilated by phytoplankton and thus returned to particulate form.

Some of the biogenic particles are carried below the mixed layer by the sinking of detritus and the feeding activities of animals. As illustrated in Figure 8.1, this particulate matter is also subject to remineralization, but plants do not grow below the mixed layer, so the regenerated nutrients remain dissolved. Since strong density stratification in the thermocline inhibits vertical mixing, most of these nutrients remain trapped below the mixed layer. The ocean has not become a biological desert, so some of these nutrients must eventually be returned to the surface ocean. This occurs when nutrient-enriched deep waters are carried back toward the sea surface as part of the return flow of thermohaline circulation and by eddy diffusion.

The nutrient enrichment in the thermocline is larger than in the deep zone due to differences in the rates of thermohaline circulation and the particle flux. Since remineralization causes the particle flux to decrease with increasing depth, the lower concentrations in the deep zone are partially due to a decreased supply of biogenic particles. As horizontal advection is relatively slow in the thermocline, these waters are in place long enough to accumulate a large amount of regenerated nutrients. The nearly homogeneous distribution of nutrients in the deep zone is due to the absence of a strong density gradient below the thermocline. Since vertical water motion is not inhibited, turbulence mixes adjacent deep-water masses. This prevents the formation of strong nutrient concentration gradients.

INTERPRETING DEPTH PROFILES

The oceanographic parameters most commonly measured include water temperature and salinity, as well as nutrient, dissolved O_2, and carbon concentrations. Depth profiles of these data, such as the set illustrated in Figure 9.2, are used to assess the nature and relative rates of biogeochemical processes. With this knowledge, marine chemists interpret the profiles of other chemical species. In Chapters 11 and 12, this technique is used to interpret depth profiles of some trace metals in seawater and the sediments.

The temperature and salinity profiles are used to determine the origin of water masses and their relative rates of motion. In the open ocean, vertical variations in water density are largely determined by water temperature. Thus the size of the temperature gradient in the thermocline can be used to assess the intensity of density stratification. This information is important because such density stratification isolates the mixed layer from the deep zone. The effects of this are evident in the concentration profiles, as all (except for dissolved silicon) have gradients which coincide with the thermocline.

The relative rates of respiration and photosynthesis can be assessed by examining the O_2, nutrient, and dissolved carbon profiles. The effects of photosynthesis are clearly seen in the low carbon and nutrient concentrations of the surface water. The O_2 concentrations are high due to contact with the sea surface and production by the plants. POM is remineralized in the surface waters, but the plants must be assimilating the regenerated nutrients fast

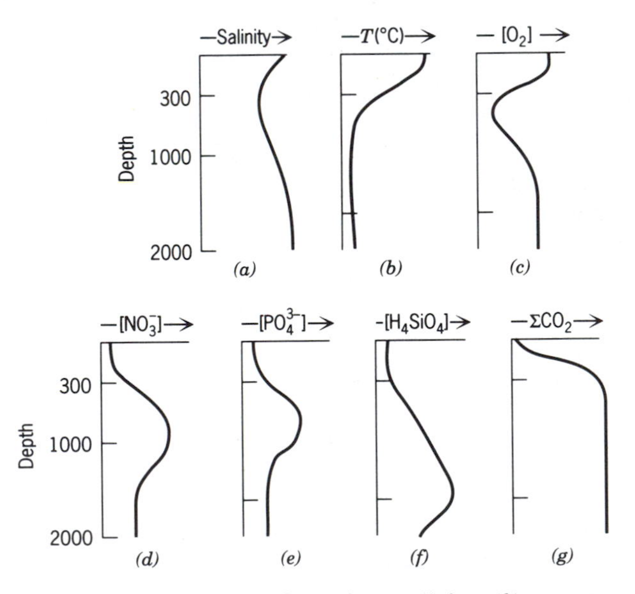

FIGURE 9.2. Depth profiles of (*a*) salinity, (*b*) temperature, (*c*) dissolved O_2, (*d*) nitrate, (*e*) phosphate, (*f*) dissolved silicon, and (*g*) total dissolved inorganic carbon concentrations (ΣCO_2) at mid latitudes.

enough to keep their concentrations very low. In comparison, only respiration occurs below the mixed layer. Because the products of POM remineralization are not removed, the nutrient and dissolved carbon concentrations increase below the thermocline. The concomitant decline in O_2 concentrations suggests that remineralization occurs via the process of aerobic respiration.

The well-mixed nature of the waters above and below the thermocline is demonstrated by their thermal and chemical homogeneity. O_2 concentrations decline at the top of the thermocline as the resupply of this gas is inhibited by the decreased amount of vertical mixing. The lowest O_2 concentrations are found in the thermocline. The minimum in O_2 concentration coincides with the nutrient concentration maxima. This suggests that the waters in the thermocline have received the greatest amount of sinking POM and are relatively stagnant. The bottom of the thermocline is marked by an abrupt change in the temperature, O_2, and nutrient gradients. This suggests the presence of two water masses, the upper one forming the bottom of the thermocline and the lower one establishing the top of the deep zone. In general, large changes in gradients suggest the presence of multiple water masses.

Not all biolimiting elements behave identically. For example, the dissolved silicon maximum occurs at a greater depth than the nitrate and phosphate maxima. This reflects the different mechanisms by which these elements are resolubilized. In contrast to the nutrients, silicon is remineralized when siliceous shells dissolve. The profiles suggest that this occurs at a greater depth than does the bulk of the POM remineralization. Geographic variations in the depth profiles of these biolimiting elements are considerable. The causes for this are the subject of the next chapter.

BROECKER BOX MODEL

The vertical distribution of biolimiting elements is characterized by deep-water enrichments and surface-water depletions. This **vertical segregation** is caused by the remineralization of biogenic particles in the deep sea. Not all particulate matter that sinks into the deep zone is remineralized. Some survives to become buried in the sediments. How much of the biogenic particle flux from the surface waters is remineralized in the deep zone? How much is lost from the ocean by burial in the sediments? What effect does this have on the concentrations of the biolimiting elements? Answers to these questions are important as the availability of these elements ultimately limits global marine productivity.

The degree to which a biolimiting element is remineralized can be thought of as its **recycling efficiency**. Using a simple mathematical model devised by Dr. W. S. Broecker, the recycling efficiencies of the biolimiting elements have been calculated from their surface, deep, and river water concentrations. In this model, the waters of the ocean are split into two reservoirs, a surface and a deep layer. The former represents the warm waters of the mixed layer and the top half of the thermocline; the latter is composed of the relatively cold waters of the bottom half of the thermocline and the deep zone.

As shown in Figure 9.3, the only communication between the two reservoirs is assumed to occur through upwelling and downwelling. The thermocline inhibits vertical circulation. Thus downwelling, which is achieved by the sinking of deep-water masses, occurs primarily in polar regions where the thermocline is absent. This deep water is upwelled by the return flow of thermohaline circulation that occurs throughout the ocean.

If the size of the two reservoirs remains constant over time, the global rate of upwelling must equal the global rate of downwelling. Since the surface areas of the two reservoirs are equal, the global rate of water transport between the boxes can be expressed as the annual exchange of a water layer (v_{mix}) that is 300 cm thick. This value of v_{mix} (300 cm/y) was determined from radiocarbon (^{14}C) distributions in seawater.

This model also assumes that materials enter the ocean either through river runoff or upwelling. All other sources, such as atmospheric fallout, hydrothermal emissions, and groundwater seepage, are neglected as they supply insignificant amounts of the biolimiting elements. Likewise, the model assumes that the biolimiting elements are transported from the surface ocean either through the sinking of biogenic particles or downwelling. These particles are considered to be composed of both hard and soft parts deposited by marine organisms. The sole pathway by which these elements leave the ocean is assumed to be through the burial of unremineralized hard and soft parts in the

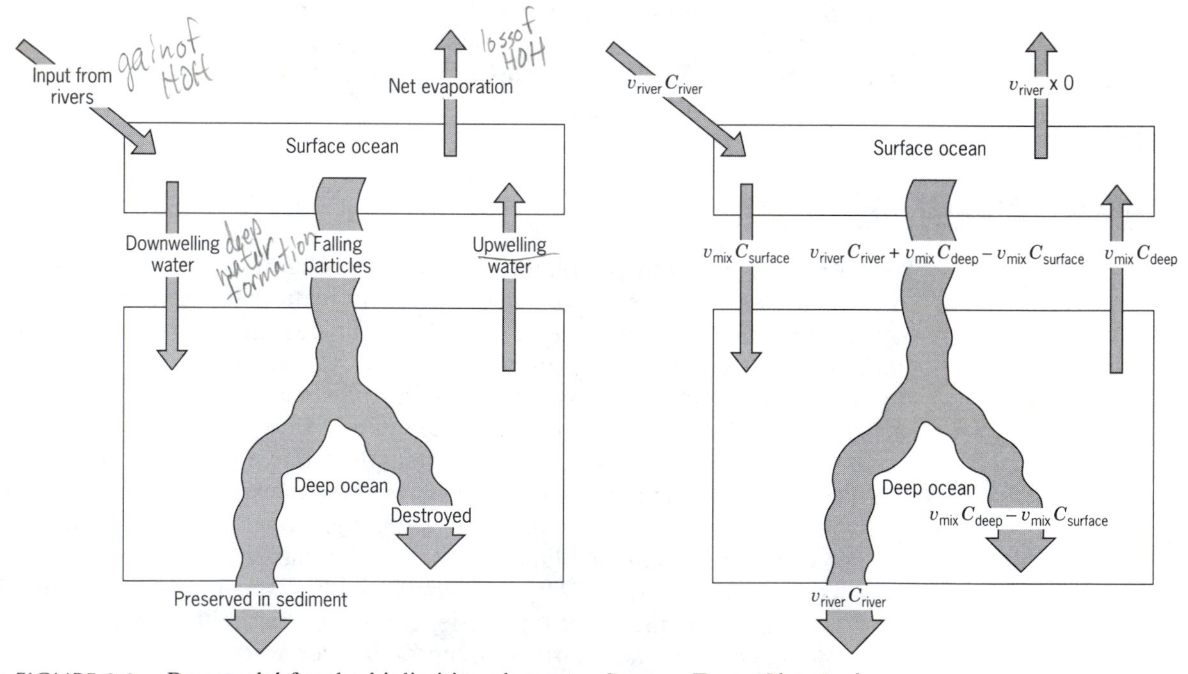

FIGURE 9.3. Box model for the biolimiting elements. *Source:* From *Chemical Oceanography*, W. S. Broecker, copyright © 1974 by Harcourt, Brace and Jovanovich, Publishers, Orlando, FL, pp. 14–15. Reprinted by permission.

sediments. This model considers only the transport of materials into and out of the two reservoirs; thus details as to what happens to the elements while they reside in the boxes are not needed. The only requirement is that their distributions within a box be homogeneous.

If a biolimiting element is in steady state, then its concentration in the two reservoirs remains constant over time. In addition, the amount of element transported into a box must equal the amount transported out. For the surface-water reservoir, this mass balance is given by

$$\begin{matrix}\text{Upwelling} \\ \text{flux}\end{matrix} + \begin{matrix}\text{River runoff} \\ \text{flux}\end{matrix} = \begin{matrix}\text{Downwelling} \\ \text{flux}\end{matrix} + \begin{matrix}\text{Particle} \\ \text{flux}\end{matrix} \tag{9.1}$$

The first three fluxes can be calculated by multiplying the biolimiting element's concentration (C) by the annual water transport (v); that is,

$$\text{Flux} = v \times C = \frac{\text{cm}}{\text{y}} \times \frac{\text{mol}}{\text{L}} \times \frac{\text{L}}{1000\ \text{cm}^3} = \frac{\text{mol}}{\text{cm}^2\ \text{y}} \tag{9.2}$$

Therefore, the upwelling flux is the product of the deep-water concentration. C_{deep}, and the annual amount of water upwelled, v_{mix}. The downwelling flux is the product of the surface-water concentration, $C_{surface}$, and the annual amount of water downwelled, which is also equal to v_{mix}. The flux of materials carried by river runoff is the product of the river-water concentration, C_{river}, and the annual amount of water transported into the oceans by rivers, v_{river}. The latter is expressed as the depth of a layer of water that would be produced by spreading the annual river-water input across the entire surface area of the ocean.

The biogenic particle flux is represented as P. Substituting these symbols into Eq. 9.1 and solving for P yields

$$P = v_{mix}\, C_{deep} + v_{river}\, C_{river} - v_{mix}\, C_{surface} \tag{9.3}$$

The surface-ocean recycling efficiency is calculated as the fraction of the element in the surface box that is removed in particulate form. This fraction, g, is given by

$$g = \frac{\begin{matrix}\text{Amount of particles} \\ \text{leaving the surface} \\ \text{ocean}\end{matrix}}{\begin{matrix}\text{Total input to the} \\ \text{surface ocean}\end{matrix}} = \frac{v_{mix}\, C_{deep} + v_{river}\, C_{river} - v_{mix}\, C_{surface}}{v_{mix}\, C_{deep} + v_{river}\, C_{river}} \tag{9.4}$$

This equation can be converted algebraically by dividing the right side through by v_{mix} and C_{river} to yield

$$g = 1 - \frac{\dfrac{C_{surface}}{C_{river}}}{\dfrac{C_{deep}}{C_{river}} + \dfrac{v_{river}}{v_{mix}}} \tag{9.5}$$

Multiplying the numerator and denominator in the last term of this equation by v_{mix}/v_{river} yields

$$g = 1 - \frac{\dfrac{v_{mix}}{v_{river}} \dfrac{C_{surface}}{C_{river}}}{\dfrac{v_{mix}}{v_{river}} \dfrac{C_{deep}}{C_{river}} + 1} \tag{9.6}$$

The annual amount of river water entering the ocean is 32×10^{15} L/y. Assuming that the average area of the ocean is equal to that at the sea surface (3.6×10^{18} cm^2), this input represents the annual addition of a layer of water approximately 10 cm deep. Since v_{mix} is 300 cm/y, $(v_{mix}/v_{river}) = 30$. So g becomes

$$g = 1 - \frac{30 \dfrac{C_{surface}}{C_{river}}}{\left(30 \dfrac{C_{deep}}{C_{river}}\right) + 1} \tag{9.7}$$

Thus to compute the surface-water recycling efficiency of an element, all that must be known are its surface, deep, and river water concentrations. For phosphorus, $C_{deep}/C_{surface} = 5$ and $C_{surface}/C_{river} = 0.25$; thus $g = 0.95$. This means that if enough time has elapsed for the complete exchange of water between the two reservoirs, then 95 percent of the phosphorus that enters the surface box is removed in particulate form. Detailed studies of nutrient dynamics in the mixed layer indicate that the average atom is recycled 10 times before escaping in particulate form to the deep sea.

If a steady state exists, then the flux of an element into the ocean must equal its flux out; that is,

$$v_{river} C_{river} = \text{Particle flux to the sediments} \tag{9.8}$$

The particle flux to the sediments can be calculated by first defining f as

$$f = \frac{\text{Particle flux to the sediments}}{\text{Total particle flux}} \tag{9.9}$$

This is the fraction of the total particle flux, P, that survives the descent through the deep-water box to become buried in the sediments.

The particle flux to the sediments can be calculated by multiplying f and P, since

$$\frac{\text{Particle flux to the sediments}}{\text{Total particle flux}} \times P = \begin{array}{c}\text{Particle flux}\\ \text{to the sediments}\end{array} \tag{9.10}$$

The particle flux to the sediments represent the sole route by which the element is lost from the ocean, so

$$\frac{\text{Particle flux to the sediments}}{\text{Total particle flux}} \times P = v_{river} C_{river} \tag{9.11}$$

or

$$f \times P = v_{\text{river}} C_{\text{river}} \tag{9.12}$$

Substituting the definition of P given in Eq. 9.3 and solving for f yields

$$f = \frac{v_{\text{river}} C_{\text{river}}}{v_{\text{river}} C_{\text{river}} + v_{\text{mix}} C_{\text{deep}} - v_{\text{mix}} C_{\text{surface}}} \tag{9.13}$$

This is simplified algebraically to

$$f = \frac{1}{1 + \dfrac{v_{\text{mix}}}{v_{\text{river}}} \dfrac{C_{\text{deep}}}{C_{\text{river}}} - \dfrac{C_{\text{surface}}}{C_{\text{river}}}} \tag{9.14}$$

Substituting in $v_{\text{mix}}/v_{\text{river}} = 30$ gives

$$f = \frac{1}{1 + \left(30 \dfrac{C_{\text{deep}}}{C_{\text{river}}}\right) - \left(\dfrac{C_{\text{surface}}}{C_{\text{river}}}\right)} \tag{9.15}$$

For phosphorus, $f \approx 0.01$. This means that only 1 percent of the particle flux that enters the deep-water box during a mixing cycle survives to become buried in the sediments. Ninety-nine percent is remineralized in the deep water.

The overall recycling efficiency of a biolimiting element is given by the fraction of the river input that is buried in the sediments during one complete mixing cycle. This is calculated as

$$f \times g = \frac{\text{Particle flux to the sediments}}{\text{Total particle flux}} \times \frac{\text{Total particle flux}}{\text{Total input to surface seawater}} = \frac{\text{Particle flux to the sediments}}{\text{Total input to surface seawater}} \tag{9.16}$$

For phosphorus, $f \times g = 0.01 \times 0.95 \approx 0.01$. This means that only 1 percent of the phosphorus introduced into the ocean by river runoff is removed to the sediments during each mixing cycle.

Oceanographers have determined that it takes 1000 years for water to be completely exchanged between the two reservoirs. Thus the residence time of a biolimiting element in the ocean is given by

$$\tau = \frac{\text{Total amount of the element in the ocean}}{\text{Removal rate}} = \frac{1}{\dfrac{f \times g}{1000 \text{ y}}} \tag{9.17}$$

where the total amount of an element in the ocean is represented as unity since $f \times g$ is the fraction of element removed. Since $f \times g$, or 1 percent of the phosphorus, is removed from the ocean in 1000 y, removal of 100 percent requires 100,000 y. So each atom of phosphorus has only a 1 percent chance of escaping from the ocean during any particular mixing cycle. On average, each atom must spend 100 mixing cycles in the ocean, getting exchanged between the surface- and deep-water reservoirs, before becoming buried in the sediments.

The surface-water volume is 10 times smaller than the deep water. Since the rates of downwelling and upwelling are equal, seawater must spend one-tenth of its time in the surface-water box. The residence time of surface water is then 1/10 × 1000, or 100y. Since the volume of surface water is small, as is $c_{surface}$ for the biolimiting elements, downwelling cannot be a significant removal mechanism. Thus the particle flux, P, must be responsible for most of the removal of the biolimiting elements from the surface box. To maintain a 20-fold deep-water enrichment for phosphorus, the particle flux must operate 20 times faster than the downwelling flux. This also requires the residence time of phosphorus in the surface-water box to be 20 times smaller than that in the deep water.

Thus, during one mixing cycle of the ocean, an atom of phosphorus spends, on average, only 5y (1/20 of 100 y) in the surface box. The atom spends the rest of the mixing cycle (995 y) in the deep water. After 100,000 y, it escapes to the sediments. In comparison to the ocean, transport rates in the rock cycle are slow. The average atom of phosphorus spends 200 million years buried in the sediments before geologic processes cause it to be uplifted, eroded, and carried back into the ocean.

BIOLIMITING VS. BIOINTERMEDIATE VS. BIOUNLIMITED

The box model results can be used to develop a quantitative definition of a biolimiting element. Those elements with $g \approx 1$ are almost completely removed from the surface seawater in particulate form. If they also have an $f << 1$, the particles are almost completely remineralized in the deep water. Thus a large deep-water enrichment is established without much loss of the element from the ocean. As shown in Table 9.1, these are the characteristics of the biolimiting elements as exemplified by phosphorus and silicon.

The greater deep-water enrichment seen in silicon ($C_{deep}/C_{surface} = 32$) as compared to phosphorus is due to its smaller recycling efficiency in both the surface and deep ocean. Thus a greater fraction of the silicon is transported to the deep sea and the sediments. This difference in behavior is due to the different mechanisms by which phosphorus and silicon are remineralized. Phosphorous is resolubilized through the aerobic respiration of POM. This occurs at shallower depths than the dissolution of siliceous shells and also causes concentration maximum of dissolved silicon to be deeper than that of phosphorus. Due to its relatively slow rate of dissolution, silicon has a shorter residence time in the ocean than phosphorus.

Elements characterized by $g << 1$ and $f \approx 1$ have little or no vertical concentration gradient because an insignificant fraction of these chemicals is incorporated into biogenic particles. Sodium and sulfur are examples of such

TABLE 9.1
Calculation Summary for the Elements Phosphorous, Silicon, Barium, Calcium, Sulfur, and Sodium

Category	*Element*	$\frac{C_{surface}}{C_{river}}$	$\frac{C_{deep}}{C_{river}}$	g	f	$f \times g$	τ (y)
Biolimiting	P	0.15	3.0	0.95	0.01	0.01	1×10^5
	Si	0.02	0.7	0.97	0.05	0.05	2×10^4
Biointermediate	Ba	0.1	0.3	0.70	0.14	0.10	1×10^4
	Ca	24.8	25	0.01	0.14	0.0013	8×10^5
Biounlimited	S	5000	5000	—	—	0.0001	2×10^7
	Na	50,000	50,000	—	—	0.00001	2×10^8

Source: From *Chemical Oceanography*, W. S. Broecker, copyright © 1974 by Harcourt, Brace, and Jovanovich Publishers, Orlando, FL, p. 21. Reprinted by permission.

biounlimited elements. Their relatively long residence times suggest that removal via the sinking of a biogenic particles is also very slow.

Elements with intermediate values of f and g are classified as **biointermediate**. In the case of barium, the fraction removed as particles from the surface water is similar to that of a biolimiting element. But a much greater amount of these particles escapes from the ocean, causing its $C_{deep}/C_{surface}$ to be approximately 3. As a result of its large particle flux out of the ocean, barium's residence time is quite short.

In contrast, relatively little of the calcium is removed from the surface water in particulate form, though a significant amount of this flux survives to become buried in the sediments, mostly as calcareous shells. Because the particle flux is small relative to the large amount of calcium dissolved in seawater, $C_{deep}/C_{surface}$ is nearly equal to 1.

The biochemical reactions associated with metabolic processes require many elements other than carbon, nitrogen, oxygen, hydrogen, sulfur, and silicon. Those essential ones that are present at low concentrations have the potential for behaving as biolimiting, or at least biointermediate, elements. As indicated in Table 9.2, many are trace metals. Their marine biogeochemistry

TABLE 9.2
The Role of the Elements in Life Processes

Elements necessary for the life processes
H, B, C, N, O, F, Na, Mg, Si, P, S, Cl, K, Ca, V, Mn, Fe, Co, Ni, Cu, Zn, Br, I

Elements probably necessary for life
Al, Ti, As, Sn, Pb

Elements probably not necessary for life
He, Li, Be, Ne, Ar, Sc, Cr, Ga, Ge, Se, Kr, Rb, Sr, Y, Zr, Nb, Tc, Ru, Rh, Pd, Ag, Cd, In, Sb, Te, Xe, Cs, Ba, La, Rare earths, Hf, Ta, W, Re, Os, Ir, Pt, Au, Hg, Tl, Bi, Po, At, Rn, Fr, Ac, Th, Pa, U

Source: From *Marine Chemistry*, R. A. Horne, copyright © 1969 by John Wiley & Sons, Inc., New York, p. 244. Reprinted by permission.

is discussed in Chapters 11 and 12. Even some nonessential elements, such as cadmium, exhibit biolimiting behavior (i.e., surface-water depletions and deep-water enrichments). These elements are chemically similar to the essential ones and hence tend to be strongly enriched in biogenic particles even though they are toxic at high enough concentrations.

SUMMARY

The vertical concentration profiles of the biolimiting elements are characterized by surface-water depletions and deep-water enrichments. This **vertical segregation** is the result of interactions between biological processes and thermohaline circulation. Concentrations are low and often undetectable in the surface waters due to nearly complete extraction by phytoplankton. Some biolimiting elements, such as nitrogen and phosphorus, are incorporated into the tissues, or *soft parts*, of the plants. Others, such as silicon, are deposited as shells, or *hard parts*. Most of these biogenic particles are remineralized in the mixed layer and rapidly reassimilated by the phytoplankton.

A small fraction sinks out of the surface waters. Most is remineralized and remains dissolved because phytoplankton can not photosynthesize at these depths. Density stratification inhibits vertical mixing, so the resolubilized elements are retained in these waters. Concentrations tend to be lower in the deep zone than in the thermocline due to differences in the supply of sinking POM and the rate of horizontal advection. As a result, the waters in the thermocline accumulate larger amounts of regenerated nutrients and other resolubilized biolimiting elements. These materials, as well as those in the deep zone, are eventually transported back to the mixed layer via the return flow of thermohaline circulation and eddy diffusion.

Some of the biogenic particles pass through the deep zone without becoming remineralized and hence become buried in the sediments. The degree to which a biolimiting element is remineralized is given by its **recycling efficiency**. This efficiency, as well as the residence time of the element, can be calculated from a simple box model. The results indicate that the biolimiting elements are efficiently recycled in the deep waters, but not in the surface waters. As a result, the biolimiting elements have fairly long residence times and larger deep-water enrichments. Because siliceous shells dissolve slower than POM is remineralized, dissolved silicon has a greater deep-water enrichment and shorter residence time than phosphorus. This also causes the deep-water dissolved silicon maximum to be deeper than the phosphorus maximum.

In comparison, elements that are not incorporated into biogenic particles in significant amounts tend not to have vertical concentration gradients. These **biounlimited** elements are characterized by very long residence times. Elements with moderate concentration gradients are termed **biointermediate**. Though these elements are incorporated in biogenic particles, either the flux is too small to have much of an impact on their seawater concentrations, or most of the sinking particles are buried in the sediment and hence cannot contribute to a deep-water enrichment. Any element which is incorporated into biogenic particles and is present in low concentrations has the potential to be biolimiting. Many of these are trace metals.

The depth profiles of the nutrients are often used in conjunction with those of salinity, temperature, and dissolved O_2 to assess the types and relative rates of biogeochemical processes that are occurring at a particular site. Marine chemists often compare these profiles to those of other elements to provide insights into the biogeochemistry of the latter.

CHAPTER 10

HORIZONTAL SEGREGATION OF THE BIOLIMITING ELEMENTS

INTRODUCTION

Surface-water concentrations of the biolimiting elements are uniformly low due to nearly complete extraction by phytoplankton. In comparison, the deep-water concentrations exhibit considerable geographic variability. These differences in the intensity of vertical segregation are the result of advective transport in the deep zone that is driven by thermohaline circulation.

The interaction between this advective transport and the biogeochemical processes that cause vertical segregation is the subject of this chapter. As a result of this interaction, the nutrients, as well as other biolimiting elements, have become enriched in the deep waters of the Pacific as compared to the Atlantic Ocean. These global nutrient distributions have a great impact on the ocean, as the growth of most phytoplankton is nutrient limited. Oceanographers hypothesize that nutrient distributions in the ocean are maintained at steady state by a feedback loop that is largely controlled by the metabolic activities of marine organisms. This control is thought to be achieved through processes, such as CO_2 uptake and release, which influence climate and hence the rate and flow patterns of thermohaline circulation.

GEOGRAPHIC DIFFERENCES IN DEPTH PROFILES

The surface-water concentrations of the biolimiting elements are uniformly low. Thus geographic variations in the intensity of vertical segregation are largely the result of differences in deep-water concentrations, as indicated in Figure 10.1. In general, the deep-water nutrient concentrations are highest in the Pacific Ocean and lowest in the Atlantic.

This pattern is reversed for the O_2 profiles, as illustrated in Figure 10.2. A detailed explanation of these trends requires a description of the global pattern of thermohaline circulation, which is given below.

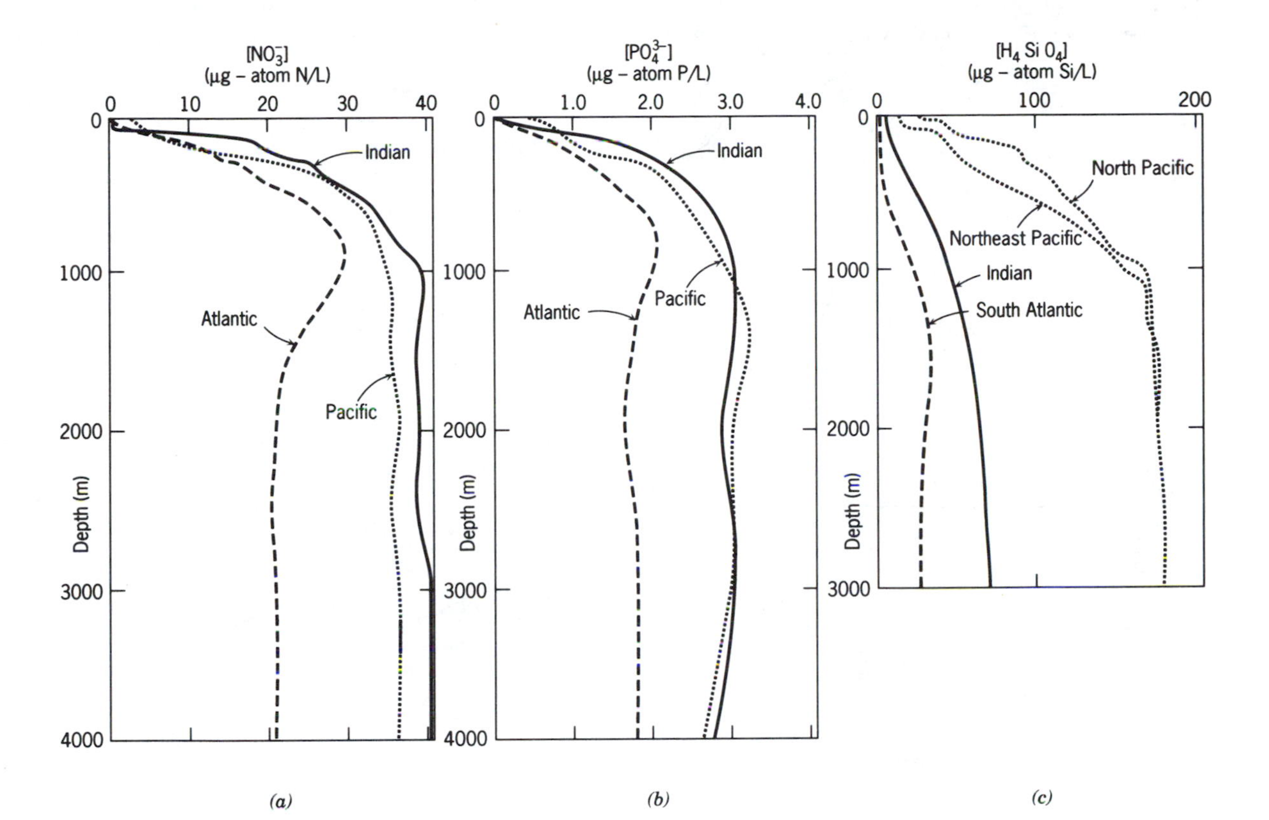

FIGURE 10.1. Vertical distribution of (*a*) nitrate, (*b*) phosphate, and (*c*) dissolved silicon in the Atlantic, Pacific, and Indian oceans. Note that 1 μg-atom/L is equivalent to 1 μ*M*. Thus 1 μg-atom NO_3-N/L is equivalent to 1μmol of dissolved nitrogen (in the form of NO_3^-) per liter of seawater. *Source:* From *The Oceans,* H. U. Sverdrup, M. W. Johnson, and R. H. Fleming, copyright © 1941 by Prentice Hall, Inc., Englewood Cliffs, New Jersey, p. 242. Reprinted by permission. See Sverdrup et al. (1942) for data sources.

THE GLOBAL PATTERN OF THERMOHALINE CIRCULATION

The global pattern of thermohaline circulation is illustrated in Figure 10.3. This circulation is initiated by the sinking of surface waters in polar regions. The formation of deep-water masses is presently limited to the subpolar North Atlantic and the South Atlantic. North Atlantic Deep Water (NADW) is created in the Norwegian Sea between Greenland and Iceland by the sinking of cold surface water. Antarctic Bottom Water (AABW) is formed by the sinking of subsurface water along the continental shelf of the Weddell Sea which is located south of the Atlantic Ocean. NADW and AABW are the densest water masses in the open ocean and thus compose most of the deep zone.

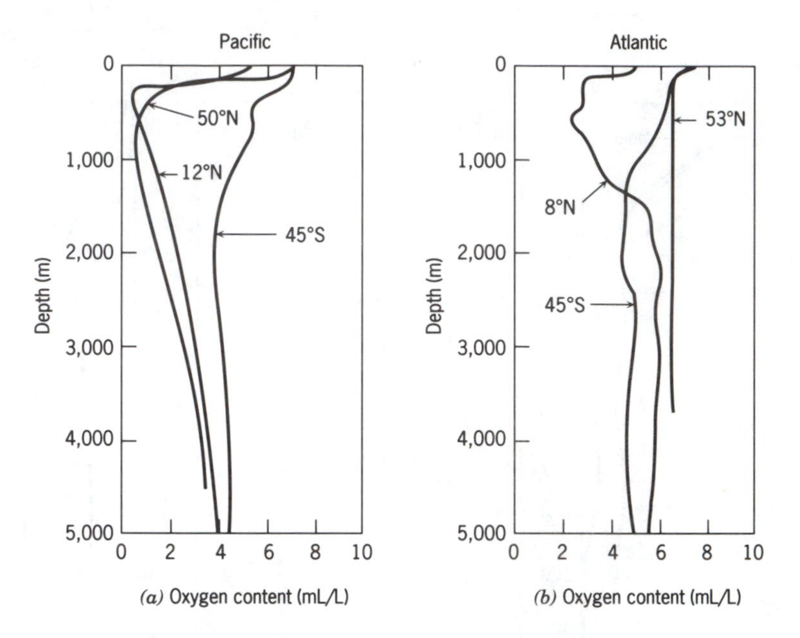

FIGURE 10.2. Depth profiles of oxygen concentrations from (*a*) the Pacific and (*b*) the Atlantic oceans. *Source:* From *Oceanography: An Introduction,* 4th ed., D. E. Ingmanson and W. J. Wallace, copyright © 1989 by Wadsworth, Inc., Belmont, CA, p. 99. Reprinted by permission.

After sinking into the deep zone, these water masses travel horizontally. Turbulent mixing with the overlying water masses alters their temperature and salinity signatures. Nevertheless, enough of their original composition is retained to determine the pathway of flow through the deep ocean using the core technique of water mass tracing. The flow patterns of these deep-water masses at the 4000-m depth interval are illustrated in Figure 10.4.

NADW flows into the South Atlantic, where it meets and mixes with AABW. The water masses then flow east, forming a circumpolar current that supplies deep water to the Indian and Pacific oceans. Since deep-water masses are not presently being formed in either of these oceans, their deep zones are composed entirely of water that has traveled from the Atlantic Ocean as NADW or AABW. As a result, the oldest deep water is located in the North Pacific Ocean.

In contrast to the restricted locations of deep-water formation, the return flow of thermohaline circulation occurs throughout the ocean basins. Deep waters are transported back to the sea surface by a type of upwelling that is driven by turbulence. If thermohaline circulation thoroughly mixes the waters of the ocean, the upwelled waters must eventually be returned to the site of deep-water formation. Oceanographers hypothesize that this is achieved by the actions of surface currents. This process is represented in Figure 10.3 by the dashed lines.

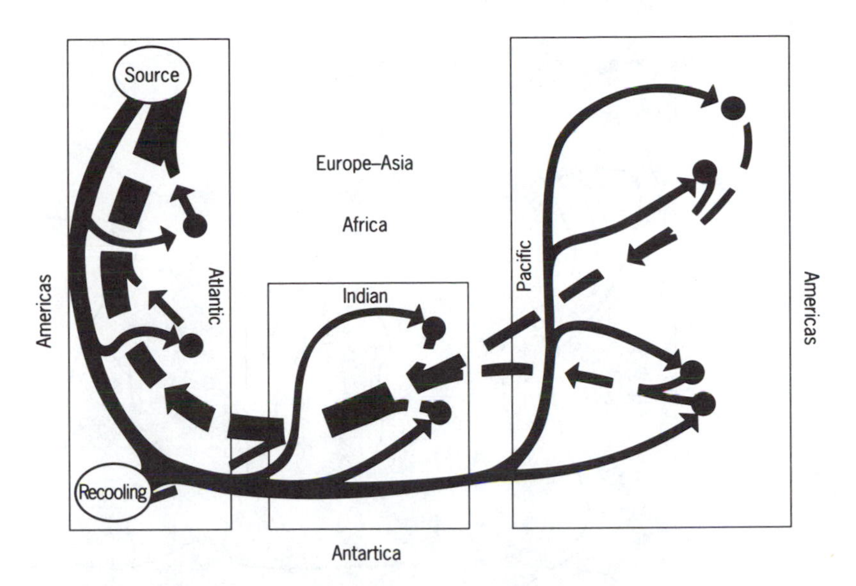

FIGURE 10.3. Idealized map of the deep-water flow (solid lines) and surface-water flow (dashed lines). The open circles designate the sinking of NADW in the Norwegian Sea and the recooling of water along the perimeter of the Antarctic Continent due to admixture with AABW. The blackened circles indicate the distributed upwelling that balances this deep-water generation. *Source:* From *Chemical Oceanography,* W. S. Broecker, copyright © 1974 by Harcourt, Brace and Jovanovich, Publishers, Orlando, FL, p. 25. Reprinted by permission.

HORIZONTAL SEGREGATION OF NUTRIENTS IN THE DEEP ZONE

Phytoplankton produce biogenic particles in the surface waters of all the ocean basins. Most of these particles sink to the deep sea and then are remineralized. The rain of biogenic particles causes the nutrient concentration of the deep-water masses to increase as they move through the ocean basins. As illustrated in Figure 10.5, the further a deep-water mass has traveled from its site of formation, the greater the amount of particles it will have accumulated. Thus the nutrient concentrations increase as the water mass "ages." Since the oldest deep waters are located in the North Pacific, the overall impact has been to push the nutrients into the deep zone and towards the Pacific Ocean. The latter effect is termed **horizontal segregation**.

Nutrients are carried back to the sea surface by the slow upwelling that constitutes the return flow of thermohaline circulation. The size of the resulting horizontal gradient is determined by the rate of water motion, the flux of biogenic particles, and the recycling efficiency of the nutrient. If a steady state exists, the deep-water concentration gradient must be the result of a balance between the rates of nutrient supply and removal as illustrated by the following analogy.

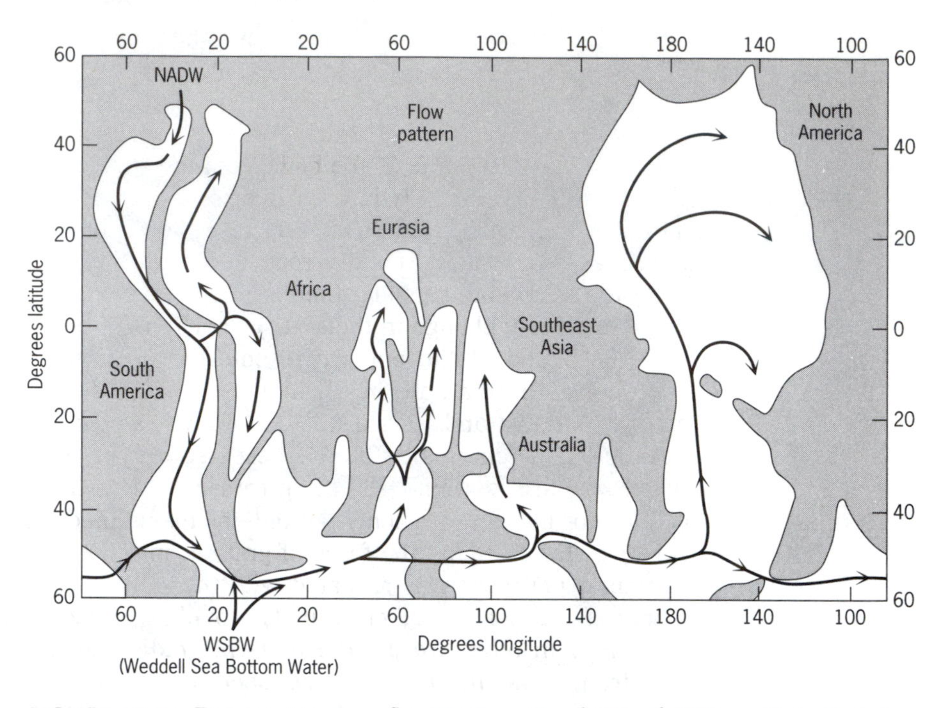

FIGURE 10.4. Deep water mass flow pattern at a depth of 4000 m. Note that the boundaries of the ocean basins are illustrated as the 4000-m bathymetric contour. The Atlantic Ocean is on the left, the Indian in the middle, and the Pacific Ocean is on the right. The long "spit" running down the Atlantic Ocean is the Mid-Atlantic Ridge. *Source:* From *Tracers in the Sea,* W. S. Broecker and T.-H. Peng, copyright © 1982 by the Lamont-Doherty Geological Observatory, Palisades, New York, p. 33. Reprinted by permission.

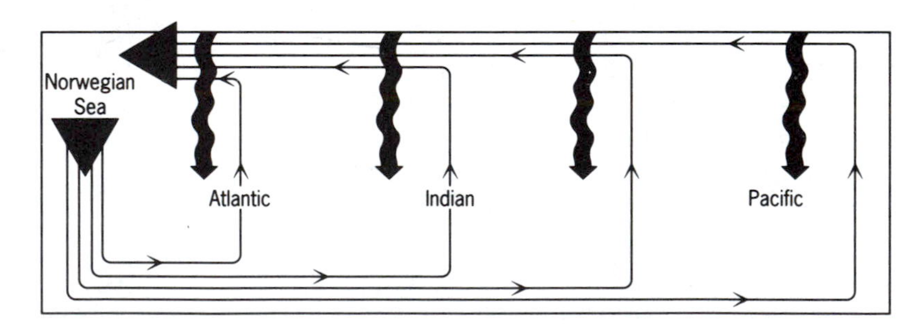

FIGURE 10.5. Idealized vertical section running from the North Atlantic to the North Pacific showing the major advective flow pattern (solid lines) and the rain of biogenic particles (wavy lines). *Source:* From *Chemical Oceanography,* W. S. Broecker, copyright © 1974 by Harcourt, Brace and Jovanovich, Publishers, Orlando, FL, p. 25. Reprinted by permission.

Imagine that you are asked to partially fill an empty swimming pool with Ping-Pong balls. Even if you throw the balls into the shallow end, gravity carries them to the deep end, where they pack together neatly. You add balls until they extend halfway across the shallow end, as shown in Figure 10.6*a*.

Now, you pick up one of the balls from the shallow end, walk around the pool and throw it vigorously into the deep end (Figure 10.6*b*). This causes one of the deep-end balls to be pushed up into the shallow end. If you continue this process, you will eventually recycle all the balls from the shallow end through the deep end, pushing them back to the shallow end again, without having caused any change in the level of balls in the pool.

The deep end of the pool represents the Pacific Ocean with its higher nutrient concentration and the shallow end, the Atlantic. The forces driving the nutrients down and towards the Pacific Ocean (the biogenic particle flux and thermohaline circulation) are represented by you and gravity. If the balls (or nutrients) are continuously being transported to the deep end, a steady-state level (or nutrient gradient) can only be maintained if the balls are pushed up (or upwelled), against the force of gravity, into the shallow end.

If this experiment was repeated with the same number of balls, but in a pool with a steeper slope, more balls would be present in the deep end at steady state. In other words, the nutrient gradient would be steeper, if the forces driving the nutrients to the deep water were increased. Chemical evidence from ancient marine sediments suggests that these forces have changed over time, particularly during ice ages. The impact of these perturbations on the horizontal segregation of nutrients is discussed in the last section of this chapter.

The present-day deep-water concentration gradients of O_2, dissolved silicon, and nitrate are illustrated in Figure 10.7. In the case of O_2, the concentrations decrease with increasing age of the deep water because O_2 is removed during the remineralization of POM.

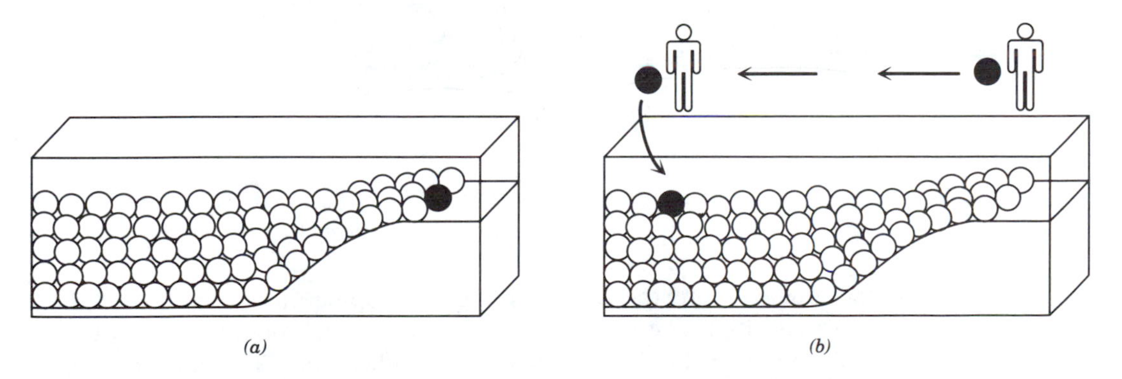

FIGURE 10.6. The maintenance of a steady-state gradient via internal cycling. Cross-sectional view of a swimming pool partially filled with Ping-Pong balls. (*a*) The initial position of the Ping-Pong balls. (*b*) The movement of one Ping-Pong ball over time. Note that the overall distribution of balls does not change.

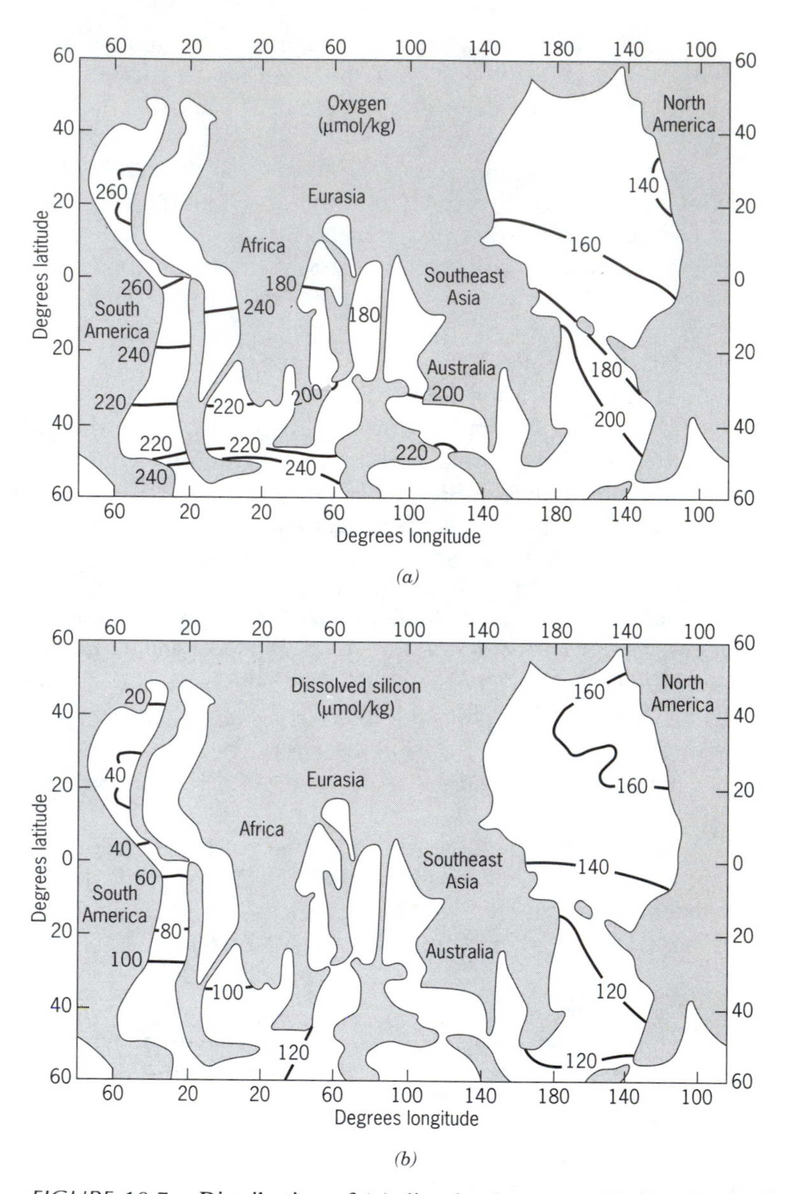

FIGURE 10.7. Distribution of (*a*) dissolved oxygen, (*b*) dissolved silicon, and

(continued on next page)

These gradients are the cause of the geographic variations depicted in Figures 10.1 and 10.2. The O_2 depth profiles also reflect the influence of water temperature on gas solubility and density stratification. The uniformly large concentrations observed at high latitudes are due to intense cooling, which increases O_2 solubility and eliminates density stratification. At these locations, vertical mixing occurs throughout the water column and transports O_2 to the deep sea.

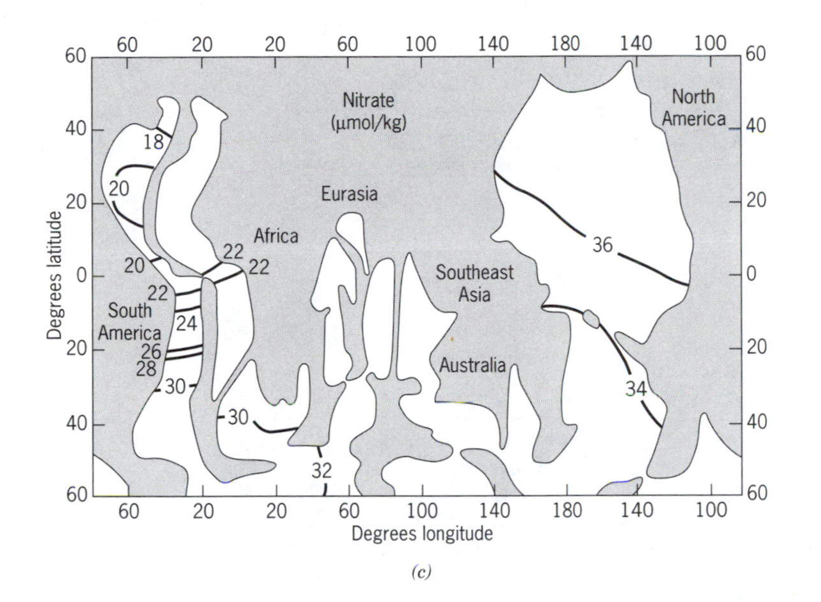

(c)

FIGURE 10.7 (Cont.) (*c*) nitrate at 4000 m depth in the world's major ocean basins. *Source:* From *Tracers in the Sea,* W. S. Broecker and T.-H. Peng, copyright © 1982 by the Lamont-Doherty Geological Observatory, Palisades, New York, pp. 30–31. Reprinted by permission. See Broecker and Peng (1982) for data sources.

The O_2 content of the surface waters is lower at mid latitudes due to higher temperatures, which lower gas solubility. The thermocline is characterized by a concentration minimum that increases in intensity from the Atlantic to the North Pacific. A much larger decrease occurs in the deep waters. Horizontal segregation is not as intense at mid depths because intermediate waters are formed in the Pacific Ocean, as well as in the Atlantic. This also causes the Atlantic Ocean to have the most pronounced mid-depth nutrient maxima, as relatively "young," nutrient-poor water is present in its deep zone. By the time the deep waters reach the Pacific Ocean, their nutrient concentrations have increased to levels that match those at the bottom of the thermocline. Note that the O_2 minimum is missing from the vertical profile from 45°S (Figure 10.2*b*) because of close proximity to the site of AABW formation.

The silicon profiles are notably different from those of nitrogen and phosphorus. Because this element is remineralized by the relatively slow dissolution of hard parts, it has a relatively low recycling efficiency. This leads to a higher degree of both vertical and horizontal segregation. In Chapter 9, $C_{deep}/C_{surface}$ was used as a measure of the intensity of vertical segregation. This can also be expressed as the concentration differences, $C_{deep} - C_{surface}$. This difference is greater in the Pacific than in the Atlantic due to horizontal segregation. Since the surface-water concentrations are close to zero, this ratio is approximately equal to $(C_{deep})_{Pacific}/(C_{deep})_{Atlantic}$.

As indicated in Table 10.1, this ratio is greater than 1 for all the biolimiting elements, as their deep-water concentrations are greater in the Pacific than in

TABLE 10.1
Horizontal Segregation of Biologically Utilized Elements Between the Deep Atlantic and the Deep Pacific

Element	$\frac{(C_{deep} - C_{surface})_{Pacific}}{(C_{deep} - C_{surface})_{Atlantic}}$
Nitrogen (as NO_3^-)	2
Phosphorus	2
Carbon	3
Silicon	5
Barium	4

Source: From *Chemical Oceanography*, W. S. Broecker, copyright © 1974 by Harcourt, Brace, and Jovanovich, Publishers, Orlando, FL, p. 23. Reprinted by permission. Data from A. C. Redfield, reprinted with permission from *American Scientist*, vol. 46, p. 207, copyright © 1958 by Sigma Xi, New Haven, CT.

the Atlantic Ocean. The higher the ratio, the greater the degree of horizontal segregation. The relatively slow remineralization of silicon is responsible for its greater enrichment in the deep waters of the Pacific. The high degree of horizontal segregation exhibited by barium suggests that this element is also transported to the deep sea as a component of hard parts.

Horizontal Segregation and the Redfield–Richards Ratio

As a deep-water mass ages, it accumulates the products of POM remineralization. The resulting changes in water chemistry can be predicted if it is assumed that (1) the degrading POM has the Redfield–Richards Ratio of C to N to P and (2) the POM is completely remineralized by the process of aerobic respiration as presented in Eq. 8.2. If so, the regeneration of 1 mole of phosphate would remove 138 moles of O_2 and produce 16 moles of nitrate. Thus, deep-water nitrogen concentrations should increase 16 times faster than those of phosphorus. Likewise, the O_2 concentrations should decline 138 times faster than the phosphorus increases. To determine whether in situ nutrient regeneration conforms to this ideal behavior, deep-water concentrations are plotted as shown in Figure 10.8. If the nutrients have been regenerated by the ideal behavior described above, the deep-water nitrate and phosphate concentrations should generate a best-fit line with a slope equal to 16. That of O_2 and phosphate should be -138.

An indicated in this figure, considerable deviations from the ideal behavior have been observed. Such behavior could be produced by the decomposition of POM, such as feces, molts, and animal tissues, which does not have the

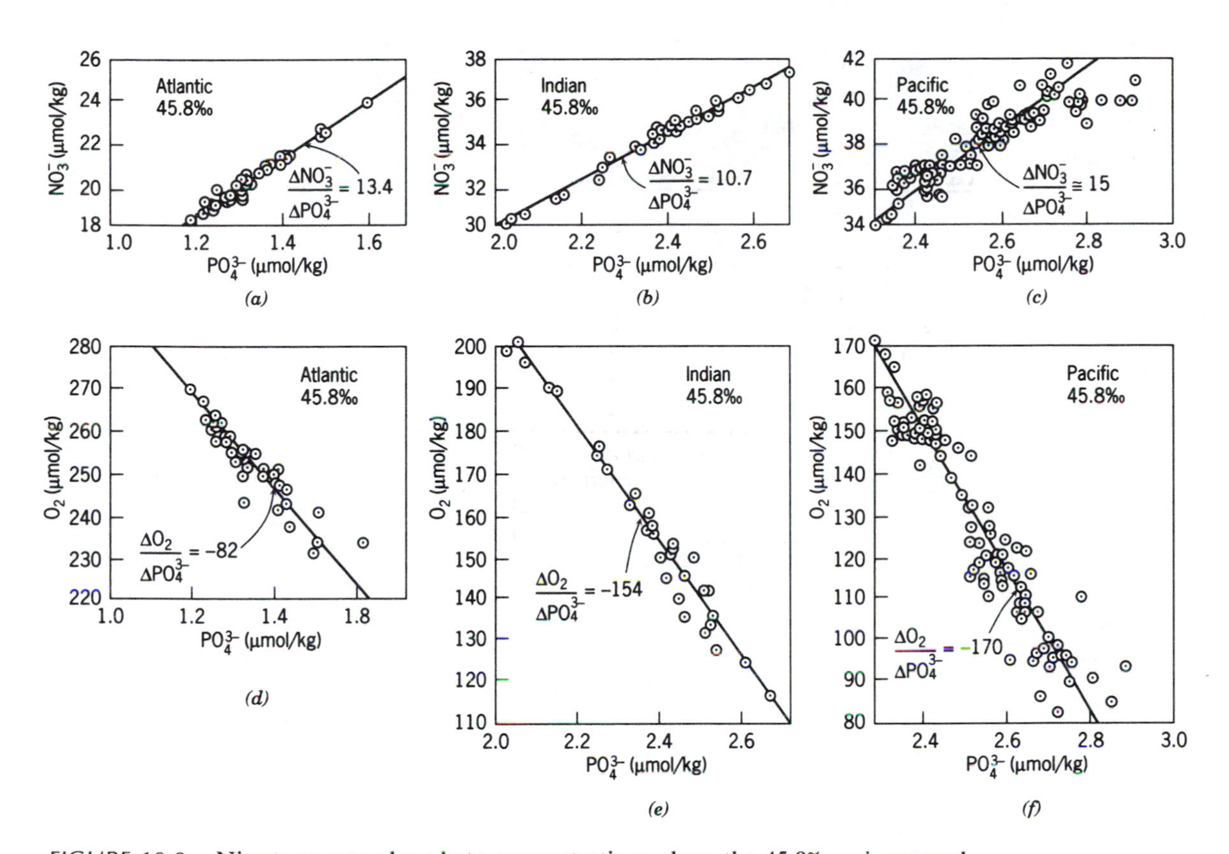

FIGURE 10.8. Nitrate versus phosphate concentrations along the 45.8‰ σ_t isopycnal surface (~2500 m depth) in the (*a*) Atlantic, (*b*) Indian, and (*c*) Pacific Oceans. Dissolved oxygen versus phosphate concentrations along the 45.8‰ σ_t isopycnal surface (~2500 m depth) in the (*d*) Atlantic, (*e*) Indian, and (*f*) Pacific Oceans. The slopes of these lines represent the proportions by which these constituent concentrations are altered by the remineralization of POM in the deep sea. *Source:* From *Tracers in the Sea,* W. S. Broecker and T.-H. Peng, copyright © 1982 by the Lamont-Doherty Geological Observatory, Palisades, New York, p. 141. Reprinted by permission. See Broecker and Peng (1982) for data sources.

Redfield–Richards Ratio. It could also be the result of the incomplete oxidation of plankton detritus, if the organic compounds that degrade do not have the Redfield–Richards Ratio. For example, the preferential decomposition of nitrogen-rich compounds, such as proteins, would regenerate more nitrate than would be predicted from the accompanying changes in O_2 or phosphate. Mixing with adjacent water masses can also alter nutrient and O_2 ratios.

Deviations can also be caused by remineralization which proceeds by processes other than aerobic respiration. As seen in Figure 10.9, water samples obtained from depths where denitrification has occurred have lower dissolved inorganic nitrogen concentrations ($[NO_3^-] + [NO_2^-]$) than would be predicted from their apparent oxygen utilization (AOU). During denitrification, nitrate is

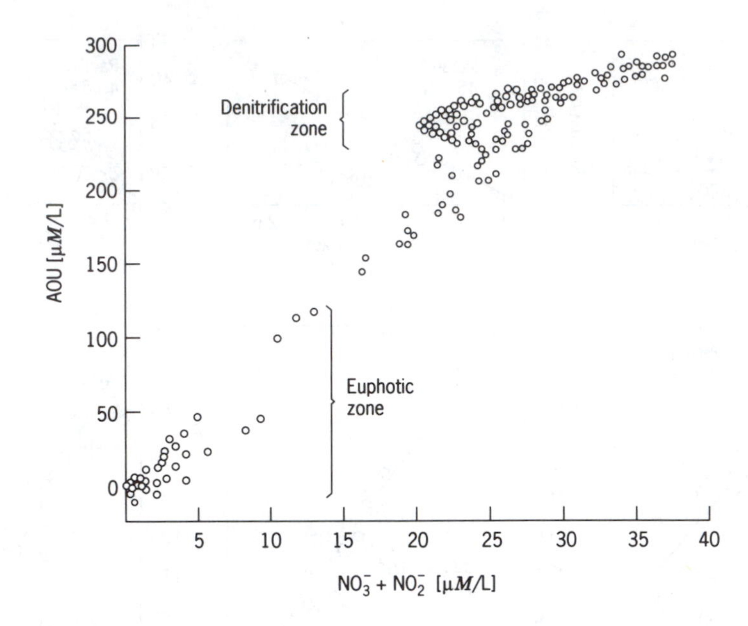

FIGURE 10.9. AOU versus $[NO_3^-] + [NO_2^-]$ in the waters of the Arabian Sea. *Source:* From W. G. Deuser, E. H. Ross, and Z. J. Mlodzinska, reprinted with permission from *Deep-Sea Research,* vol. 25, p. 435, copyright © 1978 by Pergamon Press, Elmsford, New York.

first reduced to nitrite and then to N_2, so the sum of the concentrations of both species, $[NO_3^-] + [NO_2^-]$, must be considered when assessing the regeneration of dissolved inorganic nitrogen in O_2-deficient waters. In such waters, dissolved inorganic nitrogen appears to be "missing," because denitrification converts it to N_2.

If remineralization of POM was the sole source of nutrients, a newly formed deep-water mass should have no nutrients. Such a water mass should also have an O_2 concentration equal to its NAEC. The best-fit lines in Figures 10.8*d* through *f* indicate that these deep waters have non-zero levels of phosphate at the O_2 concentrations that correspond to their NAECs. In other words, some phosphate is apparently present in newly formed deep water. The same can be demonstrated for nitrate. This suggests phytoplankton growth in the overlying surface waters is limited by some other factor, such as temperature or the availability of essential trace metals. Since all the nutrients cannot be assimilated, some sink with the water mass. These are termed preformed nutrients.

The amount of nutrients in a deep-water mass is the sum of the preformed nutrients plus the amount produced by in situ remineralization of POM. For phosphate, this is given by

$$[PO_4^{3-}] = [PO_4^{3-}]^\circ + \frac{AOU}{138} \qquad (10.1)$$

where $[PO_4^{3-}]^\circ$ is the preformed phosphate concentration. The increase in phosphate concentration due to in situ remineralization is given by AOU/138 since one phosphate molecule is formed for every O_2 molecule consumed.

The preformed nutrient concentrations can be calculated from the AOU and deep-water concentrations. If the environmental conditions at the site of water mass formation do not vary, the preformed concentration should have a constant value characteristic of a deep-water mass. As shown in Figure 10.10, the surface-water concentrations of phosphate are significantly lower in the subpolar North Atlantic as compared to the subpolar South Atlantic. Apparently plants are less effective at removing nutrients from the surface waters of the Southern Ocean. Since these profiles are from regions where deep water is formed, the preformed phosphate concentration of AABW should be lower than that of NADW. This calculated quantity can be treated as a conservative tracer of water-mass motion. In situations where neighboring water masses have large differences in preformed nutrient concentrations, the pathways of flow should be traceable over long distances. Eventually turbulent mixing will homogenize the nutrient concentrations, destroying the tracer signal.

Feedback Relations in the Marine Biogeochemical Nutrient Cycle

Physical and biogeochemical factors determine the steady-state concentrations of the biolimiting elements. Thus changes in these factors have the potential

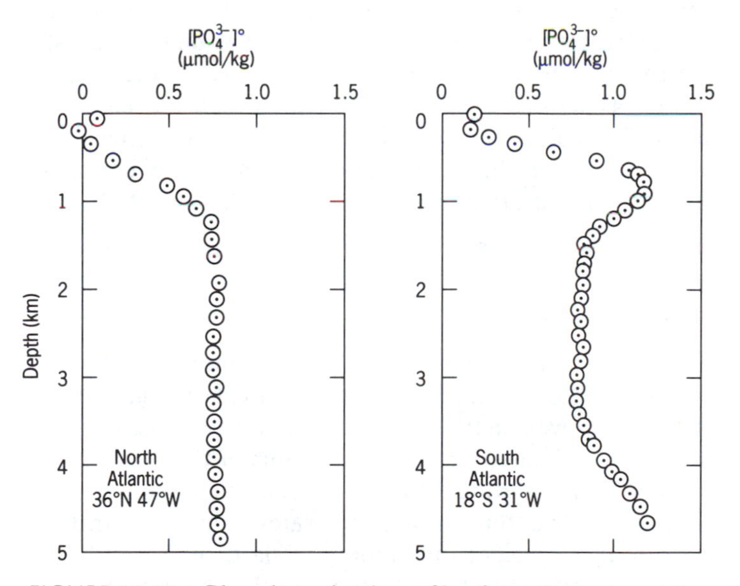

FIGURE 10.10. Phosphate depth profiles from the regions where NADW and AABW are formed. *Source:* From *Tracers in the Sea,* W. S. Broecker and T.-H. Peng, copyright © 1982 by the Lamont-Doherty Geological Observatory, Palisades, New York, p. 344. Reprinted by permission. See Broecker and Peng (1982) for data sources.

to alter nutrient concentrations throughout the ocean. Fortunately, the ocean appears to have a set of negative feedback loops that minimize the impacts of such perturbations. Some scientists liken these feedbacks to the self-regulating mechanisms that stabilize the biochemistry of organisms. Thus the current biogeochemical state of the planet is considered to be the result of, as well as stabilized by, these feedbacks. This concept is termed the Gaia hypothesis.

Likely perturbations include changes in climate, sea level, thermohaline circulation, the rate of plate tectonics, and catastrophic events, such as meteorite impacts. In turn, the ocean has the potential, through its feedback loops, of controlling global climate and thus sea level and the rate of thermohaline circulation. Much of the ongoing research in oceanography is focused on investigating how these feedbacks operate, so that the impact of human activities can be predicted.

The importance of the biogeochemical cycle of organic matter in regulating atmospheric O_2 levels was described in Chapter 8. Organic matter production and destruction, and hence nutrient cycling, is also important in stabilizing atmospheric CO_2 levels. This is discussed in Chapter 25. As noted in Chapter 8, the N to P ratio of dissolved nutrients appears to be set by the metabolic efforts of the phytoplankton. These organisms also act to stabilize nutrient concentrations. For example, a doubling of the river input of nutrients would cause a doubling in the production rate of organic matter due to the extreme nutrient limitation of phytoplankton. This will cause an increase in the flux of biogenic particles out of the surface water and in the rate of POM burial in the sediments. Thus the ocean will adjust to the increased supply of nutrients by increasing the rate at which nutrients are removed. Some time is required for the removal rates to rise to the level at which steady state is reestablished. The required time for adjustment depends on the degree to which a flux is altered, as well as the efficiency with which nutrients are removed from the ocean. This adjustment will also affect the steady-state nutrient concentrations. For example, if the perturbation given above continues, the steady-state nutrient concentrations will be permanently elevated.

SUMMARY

The deep-water enrichments of the biolimiting elements are geographically variable. In general, the intensity of vertical segregation is greatest in the North Pacific and lowest in the Atlantic. The rain of biogenic particles from the mixed layer causes the nutrient concentrations of the deep-water masses to increase as they travel through the ocean basins along the path of thermohaline circulation. The further a deep-water mass has traveled from its site of formation, the greater the amount of particles it will have accumulated. Thus the nutrient concentrations increase as the water mass "ages." Since the oldest deep waters are located in the North Pacific, the overall impact has been to push the nutrients into the deep zone and toward the Pacific Ocean. This effect is termed **horizontal segregation**.

Nutrients are carried back to the sea surface by the slow upwelling that constitutes the return flow of thermohaline circulation. The size of the deep-water concentration gradient is determined by the rates of water motion, the flux of biogenic particles, and the recycling efficiency of the nutrient. Because silicon is regenerated by the relatively slow process of shell dissolution, it exhibits a greater degree of vertical and horizontal segregation than nitrogen or phosphorus. The relatively high degree of horizontal segregation observed for barium suggests that this element is also transported to the deep sea as a component of hard parts.

As a deep-water mass ages, it accumulates the products of POM remineralization. The resulting changes in water chemistry can be predicted if the degrading POM is assumed to have the Redfield–Richards Ratio of C to N to P and is completely remineralized by the process of aerobic respiration. Thus deep-water nutrient and O_2 concentrations should be linearly related, with a slope determined by the ideal stoichiometry outlined above. Considerable deviations from these ratios can be caused by the decomposition of POM that does not have the Redfield–Richards Ratio. They can result from the incomplete oxidation of plankton detritus, if the organic compounds that degrade do not have the Redfield–Richards Ratio. Mixing with adjacent water masses can affect nutrient and O_2 ratios. Deviations are also the result of remineralization that proceeds by processes other than aerobic respiration, such as denitrification.

Some newly formed deep-water masses contain residual amounts of nutrients due to incomplete extraction by phytoplankton. If the environmental conditions at the site of water-mass formation do not vary, these preformed concentrations should be constant at a value characteristic of a deep-water mass. Thus this calculated quantity can be treated as a conservative tracer of water-mass motion. In situations where neighboring water masses have large differences in preformed nutrient concentrations, the pathways of flow should be traceable over long distances.

The ocean, like the rest of the planet, appears to have a set of negative feedback loops that stabilize the steady-state concentrations of the nutrients and many other biologically active chemicals. This is achieved through the interaction of many biogeochemical cycles. As a result, the effects of perturbations caused by changes in climate, sea level, thermohaline circulation, the rate of plate tectonics or catastrophic events, such as meteorite impacts, are minimized. In turn, the ocean has the potential, through its feedback loops, of controlling global climate and thus sea level and the rate of thermohaline circulation.

CHAPTER 11

TRACE METALS IN SEAWATER

INTRODUCTION

Virtually every element has been detected in seawater. Those which are present at concentrations less than 50 and 0.05 μmol/kg are termed the minor and **trace elements**, respectively. As shown in Figure 11.1, the trace elements are mostly metals. Despite their dissolved concentrations in seawater, iron, aluminum, and silicon compose the bulk of Earth's crust. Some trace metals are micronutrients, playing an important role in enzyme systems. The biogeochemical cycles of these trace metals are controlled by redox reactions as discussed below.

Most of our understanding of the marine chemistry of trace metals rests on research done since 1970. Prior to this, the accuracy of concentration measurements was limited by lack of instrumental sensitivity and contamination problems. The latter are still a major difficulty during sample collection due to the ubiquitous presence of metal in the hulls of research vessels, paint, hydrowires, and sampling bottles. This has proven to be an insurmountable problem in improving the accuracy of measurements of dissolved iron concentrations.

SOURCES OF TRACE METALS TO THE OCEAN

Metals are introduced into seawater by river runoff, winds, hydrothermal venting, diffusion from the sediments, and anthropogenic activities. The magnitudes of some of these fluxes are given in Table 11.1.

Rivers

Rivers are a major source of particulate and dissolved metals, both of which are mobilized during the weathering of granitic and basaltic crust. Some of the particulate trace metals are present as cations adsorbed onto the surfaces of

Trace elements <50 pmol/kg
Trace elements 0.50–50 nmol/kg
Minor elements 0.05–50 μmol/kg
Major elements 0.05–50 mmol/kg
Major elements >50 mmol/kg

Ia	IIa	IIIb	IVb	Vb	VIb	VIIb	VIIIb			Ib	IIb	IIIa	IVa	Va	VIa	VIIa
3 Li I	4 Be II											5 B III	6 C IV	7 N V	8 O 0	9 F -I
11 Na I	12 Mg II											13 Al III	14 Si IV	15 P V	16 S VI	17 Cl -I
19 K I	20 Ca II	21 Sc III	22 Ti IV (?)	23 V V	24 Cr VI	25 Mn II	26 Fe III	27 Co II	28 Ni II	29 Cu II	30 Zn II	31 Ga III	32 Ga IV	33 As V	34 Se VI	35 Br -I
37 Rb I	38 Sr II	39 Y III	40 Zr IV	41 Nb V	42 Mo VI	43 (Tc) VII	44 Ru IV ?	45 Rh III ?	46 Pd II ?	47 Ag I	48 Cd II	49 In III	50 Sn IV	51 Sb V	52 Te VI ?	53 I V
55 Cs I	56 Ba II	57 La III	72 Hf IV	73 Ta V	74 W VI	75 Re VII	76 Os (IV) ?	77 Ir III ?	78 Pt (IV, II) ?	79 Au I	80 Hg II	81 Tl (III, I)	82 Pb II	83 Bi III		

58 Ce III	59 Pr III	60 Nd III	61 (Pm)	62 Sm III	63 Eu III	64 Gd III	65 Tb III	66 Dy III	67 Ho III	68 Er III	69 Tm III	70 Yb III	71 Lu III

FIGURE 11.1. An abbreviated periodic table of the elements in seawater. The elements are placed into categories according to their average concentrations. The probable major oxidation state in aerobic seawater is also indicated. Elements in parentheses are radioactive and do not have a stable isotope. *Source:* From *Chemical Oceanography,* vol. 8, K. W. Bruland (ed.: J. P. Riley and R. Chester), copyright © 1983 by Academic Press, Orlando, FL, p. 170. Reprinted by permission.

clay minerals. The increase in ionic strength that occurs when river water meets seawater leads to the desorption of some of these trace metals. The increase in ionic strength and pH also causes these resolubilized metals to precipitate as oxyhydroxides and organometallic colloids. These removals and additions are geographically variable and poorly understood.

The dissolved metals that do reach the ocean tend to be far more reactive in seawater than the major ions. Many of these reactions cause the metals to be converted to solid form. As a result, sedimentary rocks, such as marine shales and clays, are enriched in trace metals as compared to the igneous silicates (Table 11.2). These sedimentary materials are relatively depleted in calcium and magnesium because these elements are preferentially deposited in limestone and dolomite. The relative depletion of sodium in average sedimentary materials is due to its retention as a dissolved ion in seawater.

TABLE 11.1
Oceanic Inputs of Some Trace Metals in 10^9 g/y

Element	*Mining*	*Continental and Volcanic Dust Flux*	*Industrial and Fossil Fuel Emissions*	*Atmospheric Rainout*	*Stream Load*	*Industrial and Fossil Fuel Emmissions / Atmospheric Rainout*
Cd	170	3	55	510	1,200	0.1
As	460	28	780	2,900	3,000	0.3
Hg	89	0.4	110	410	50	0.3
Se	12	7	120	200	180	0.6
Co	260	70	44	62	3,500	0.7
Ni	6,600	280	980	1,200	13,000	0.8
Zn	58,000	360	8,400	10,000	25,000	0.8
Cu	71,000	190	2,600	2,600	11,000	1.0
Sb	690	10	380	340	1,000	1.1
V	190	650	2,100	1,900	24,000	1.1
Mn	92,000	6,100	3,200	3,000	160,000	1.1
Cr	23,000	580	940	720	17,000	1.3
Mo	830	11	510	310	700	1.7
Ti	10,000	35,000	5,200	2,700	840,000	1.9
Fe	600,000	280,000	110,000	49,000	9,900,000	2.2
Al	120,000	490,000	72,000	33,000	17,000,000	2.2
Pb	35,000	59	20,000	5,700	4,700	3.5
Sm		41	12	3	900	4.0
Ag	92	0.6	50	10	130	5.0
Sn	2,400	52	430		2,900	

Source: From F. T. Mackenzie, R. J. Lahtzy, and V. Paterson, reprinted with permission from *Mathematical Sedimentology*, vol. 11, p. 101, copyright © 1979 by Plenum Press, New York.

Atmospheric Input

Some metals are deposited on the sea surface as a component of wind-borne dust and detritus. Atmospheric transport is the largest oceanic input for some metals, such as As and Pb. Unfortunately, the atmospheric particle data are limited by sampling problems caused mostly by considerable spatial and temporal variability.

Atmospheric transport is especially important in the surface waters of mid-ocean gyres due to the lack of other sources. For example, these sites are too far from land to receive an appreciable amount of riverine metals. Vertical mixing is inhibited by strong density stratification, so transport from the deep waters is restricted.

TABLE 11.2
Concentration and Relative Abundances of Metal Composition in Basalt, Granite, Average Igneous Rock, Shale, and Red Clays[a]

Element	*Granite*	*Basalt*	*Average Igneous Rock*	*Shale*	*Red Clay*
Na	26,000	18,000	22,000	10,000	15,000
K	8000	42,000	25,000	27,000	25,000
Rb	30	170	100	140	110
Cs	1	4	3	5	6
Mg	46,000	1600	24,000	15,000	21,000
Ca	76,000	5000	40,000	22,000	29,000
Sr	470	100	290	300	180
Ba	330	840	590	580	2300
Fe	86,000	14,000	50,000	47,000	65,000
Mn	1500	400	950	850	6700
Ni	130	5	70	70	225
Co	50	1	25	19	74
Cu	90	10	50	45	250
Cr	170	4	90	90	90
Th	4	18	11	12	12

Source: From *Chemical Oceanography*, W. S. Broecker, copyright © 1974 by Harcourt, Brace, and Jovanovich, Publishers, Orlando, FL, p. 97. Reprinted by permission.

[a]The average igneous rock is composed of equal parts of granite and basalt. Concentrations are given in ppm.

Diagenetic Remobilization from Nearshore Sediments

Sinking particles transport trace metals to the sediments, where chemical reactions can resolubilize a significant fraction. This process is termed **diagenetic remobilization** and is the subject of the next chapter. The resolubilized metal can diffuse across the sediment–water interface into the deep zone. Organic-rich sediments, such as those in the coastal zone, are particularly good sources of remobilized metals due to their large trace-metal enrichments and the lability of this sedimentary POM.

Hydrothermal Activity

Hydrothermal fluids associated with tectonic spreading centers tend to be enriched in trace metals. The vent fluids become enriched in these metals when seawater is heated via contact with magma several kilometers below Earth's surface. Thus the hot solution leaches metals from the basalt. Due to its increased temperature, the density of the water is lowered. The hot, metal-rich water rises through fissures until it is emitted into the deep ocean.

Most of the metals are precipitated as sulfides immediately upon entry into the deep sea. Others, such as manganese, iron, barium, lithium, and rubidium, react more slowly. Their hydrothermal fluxes represent a net addition of dissolved metals to the ocean and appear to exceed their river and atmospheric fluxes.

Anthropogenic Input

Metals are also introduced into the ocean as a result of human activities. Much enters via river and atmospheric transport. As indicated in Table 11.1, some anthropogenic fluxes exceed the natural ones. Other transport pathways include dumping, nuclear bomb explosions, and the use of metallic structures such as oil-drilling platforms. The higher atomic weight metals, called heavy metals, are particularly toxic to marine life. Even small inputs can have very large negative impacts. This has been observed for tin, which is introduced into the ocean through solubilization from organometallic paints used to prevent the biofouling of metal-hulled boats.

OCEANIC SINKS OF TRACE METALS

The trace metals are present in very low concentrations in seawater due to their rapid and efficient removal into particles. This occurs via adsorption onto surfaces, precipitation, and incorporation into biogenic particles as described below.

ADSORPTION AND PRECIPITATION UNDER OXIC CONDITIONS

The precipitation of a metal requires that its dissolved concentration exceed the level dictated by the solubility product (K_{sp}) of some mineral. But most metals are present at concentrations far below those that should exist if mineral equilibrium is established. Thus mineral equilibrium must be a relatively unimportant control on trace metal concentrations. Instead, the concentrations of dissolved metals appear to be determined by their rate of removal from seawater. This removal occurs via adsorption or precipitation onto sinking particles or by incorporation into biogenic materials and hence has been termed "The Great Particle Conspiracy."

Most particulate matter in the ocean, such as clay minerals, metal oxyhydroxides, and POM, possess a small net negative charge at the pH of seawater. Hence, metal cations are electrostatically attracted to their surfaces. As shown in Figure 11.2, dissolved metals are also adsorbed by films of organic matter that tend to coat most inorganic particles.

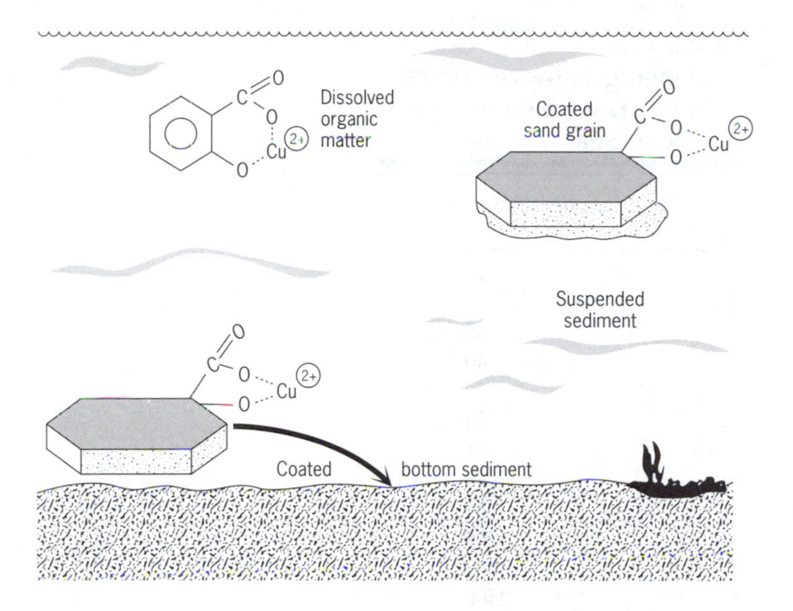

FIGURE 11.2. Complexation of metal ions by organic matter in suspended sediment and dissolved phase as exemplified by copper. *Source:* From *Organic Geochemistry of Natural Waters,* E. M. Thurman, copyright © 1985 by Martinus Nijhoff/Dr. W. Junk Publishers, Dordrecht, The Netherlands, p. 407. Reprinted by permission.

Some of these metal-enriched particles eventually sink to the seafloor, thus removing metals from the ocean. This process is termed **scavenging**. The rate and degree to which a dissolved metal is scavenged from the ocean depends on (1) its elemental nature, (2) the abundance of particulate matter, (3) the concentrations of other solutes, and (4) the water depth. Metal scavenging rates have been inferred from the concentrations of naturally occurring radionuclides, as described in Chapter 28.

The scavenging rate of a metal is usually expressed as a turnover time. A turnover time is the theoretical amount of time required to remove all of a particular chemical species from seawater by some process. Likewise, it can be expressed as the amount of time required to generate the standing stock of a particular chemical species as a result of some process that produced that chemical. As indicated in Table 11.3, the turnover time for the scavenging of many trace metals from the deep sea is close to or less than the mixing time of the ocean (1000 y). In other words, these rates of metal removal are similar to or faster than the rates of water motion. As a result, areas with high rates of scavenging can have local dissolved metal concentrations that deviate considerably from the average. For example, regions with large particle fluxes would be expected to have relatively high scavenging rates. If the scavenging rate is very rapid, metals will be removed faster than water movement can

TABLE 11.3
Scavenging Turnover Times of Some Trace Elements from the Deep Sea

Element	*Scavenging Turnover Time (y)*
Sn	10
Th	33
Fe	40
Co	40
Po	40
Ce	50
Mn	51
Pb	54
Pa	67
Sm	200
Cu	650
Sc	2500
Be	3700
Lu	4000

Source: From *Aquatic Surface Chemistry*, M. Whitfield and D. R. Turner (ed.: W. Stumm), copyright © 1987 by John Wiley & Sons, Inc., New York, p. 481. Reprinted by permission. See Whitfield and Turner (1987) for data sources.

resupply them, causing this region to have relatively low dissolved metal concentrations.

Some scavenged metals are reversibly adsorbed onto sinking particles, but others are not. These metals are probably incorporated into the crystal lattice of mineral phases, such as amorphous polymetallic oxyhydroxides. These precipitates form spontaneously in oxic seawater because most metals form very insoluble oxides.

Though most reduced metals are thermodynamically unstable in seawater, redox disequilibrium can exist as a result of relatively slow reaction rates and fast inputs. For example, manganese takes longer to oxidize than iron, so the former is carried farther into the ocean from its point of entry than the latter. Another extreme example is the co-occurrence of Se(IV) and Se(VI) at similar concentrations, although the former is thermodynamically unstable in oxic seawater. A depth profile illustrating the distribution of these two species is shown in Figure 11.3. The kinetic barrier responsible for this thermodynamic disequilibria is likely the absence of an enzyme that can catalyze the oxidation of Se(IV) to Se(VI). Similar redox disequilibria exist between Cr(III) and

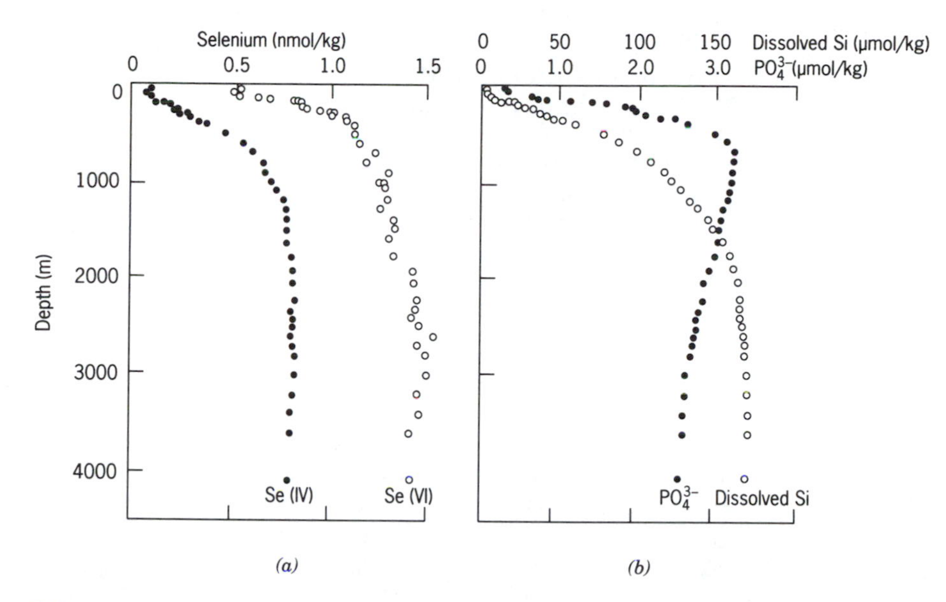

FIGURE 11.3. Vertical concentration profiles of *(a)* Se(IV) and *(b)* Se(VI) in the North Pacific. For comparison, profiles of dissolved phosphate and silicon are also shown. *Source:* From C. I. Measures and J. D. Burton, reprinted with permission from *Earth and Planetary Sciences Letters,* vol. 49, p. 105, copyright © 1980 by Elsevier Science Publishers, B.V., Amsterdam, The Netherlands.

Cr(II), Mn(IV) and Mn(II), Cu(II) and Cu(I), I(V) and I(−I), Sb(V) and Sb(III), and As(V) and As(III).

Incorporation into Biogenic Materials

As noted above, marine organisms concentrate metals in their tissues and skeletal materials. Many of these trace metals are classified as **micronutrients** because they are essential to the growth of phytoplankton. Most are components of enzyme systems, which catalyze important biochemical reactions such as glycolysis, the tricarboxylic acid cycle, photosynthesis, and protein metabolism.

The degree to which a metal is concentrated in a marine organism is given by an **enrichment factor (EF)**, which is defined as

$$\text{EF} = \frac{\text{Metal concentration in biogenic material}}{\text{Metal concentration in seawater}} \tag{11.1}$$

EFs as high as 10^5 have been observed. The major cations have the lowest values. As shown in Table 11.4, iron has the highest EF. This is likely due to its binding with biomolecules, such as ferredoxin, which have extraordinarily high affinities for iron. Table 11.4 also demonstrates that organisms differ in their degree of trace-metal enrichment. In general, lower organisms tend to have higher EFs.

TABLE 11.4
Enrichment Factors for Metals in the Tissues of Phytoplankton and Brown Algae

Element	*Plankton*	*Brown Algae*
Al	25,000	1550
Cd	910	890
Co	4600	650
Cr	17,000	6500
Cu	17,000	920
Fe	87,000	17,000
I	1200	6200
Mg	0.59	0.96
Mn	9400	6500
Mo	25	11
N	19,000	7500
Na	0.14	0.78
Ni	1700	140
P	15,000	10,000
Pb	41,000	70,000
S	1.7	3.4
Si	17,000	120
Sn	2900	92
V	620	250
Zn	65,000	3400

Source: From *The Handbook of Environmental Chemistry*, P. J. Craig (ed.: O. Hutzinger), copyright © 1980 by Springer-Verlag, Heidelberg, p. 210. After *Trace Elements in Biochemistry*, H. J. M. Bowen, copyright © 1966 by Academic Press, Orlando, FL, pp. 87–88. Reprinted by permission.

Trace metals are also enriched in shell and skeletal materials, as shown in Table 11.5. For example, the spicules of the protozoan, *Acantharia*, have large strontium enrichments. These trace metals are resolubilized much in the same way as nitrogen, phosphorus, and silicon, following the death of the organisms. Thus the depth profiles of these elements are similar to those of the nutrients.

Adsorption and Precipitation under Anoxic Conditions

Oxidation of sedimentary POM can cause anoxic conditions that permit bacteria to reduce sulfate to sulfide. The remineralization of POM also resolubilizes trace metals since these elements are strongly enriched in the tissues of phytoplankton. The resulting levels of sulfide and dissolved metals are often

TABLE 11.5
Concentration of Metals in Calcareous Shells of Foraminifera and Siliceous Shells of Radiolaria in ppm of Ash

	Ba	*Sr*	*Cu*	*Ag*	*Zn*	*Pb*	*Ti*	*V*	*Cr*	*Mn*	*Fe*	*Ni*	*B*
Foraminifera	700	400	25	3	0	10	15	trace	8	300	9	0	0
Radiolaria	5400	Major	750	0	600	105	300	0	90	0	300	57	0

Source: From *Equilibria, Nonequilibria and Natural Waters*, R. M. Pytkowicz, copyright © 1983 by John Wiley & Sons, Inc., New York, p. 300. Data from J. Greenslate, Z. Frazer, and G. Arrhenius, undated, Scripps Institute of Oceanography Report, LaJolla, CA. Reprinted by permission.

high enough to exceed the solubility product of some **metal sulfides**. As a result, metal sulfides, such as pyrite (FeS_2), are commonly found in organic-rich anoxic sediments.

Other resolubilized trace metals precipitate as replacement ions in existing solids such as fecal pellets and bone. Examples of these fossilized materials include **barite, phosphorite,** and **glauconite**. These precipitates contain small amounts of a variety of trace metals as well as other elements. As a result, their chemical composition is variable and their structure is usually amorphous, making it difficult to assign them an empirical formula.

Hydrothermal Activity

Hydrothermal activity is a major net sink for magnesium but is of unknown importance to the trace metals. In particular, low-temperature weathering of cooled basalt may be a significant sink for some trace metals.

TYPES OF METAL DISTRIBUTIONS

The horizontal and vertical distributions of dissolved trace metals are determined by their relative rates of supply and removal. The nature of this balance can be assessed from the depth profiles of the metals, which can be classified into one of the following categories: (1) a conservative type, (2) a nutrient type, (3) a surface-water enrichment, (4) a mid-depth minimum, (5) a mid-depth maximum, (6) a mid-depth minimum or maximum within a suboxic layer, or (7) a mid-depth minimum or maximum within an anoxic layer. Examples of each are given below. The profile types characteristic of each trace metal are given in Table 5.7.

Conservative-Type Distributions

A few trace metal species (Rb^+, Cs^+, and MoO_4^-) have depth distributions that are linearly related to temperature, as illustrated in Figure 11.4. This conservative behavior suggests that their concentrations are controlled by physical processes such as advection and turbulent mixing. These metals can also be classified as biounlimited as they are not significantly concentrated in biogenic materials.

Nutrient-Type Distributions

Vertical profiles with deep-water enrichments are classified as a nutrient type, if the metal concentrations exhibit a strong linear correlation with any of the nutrients. Examples are given in Figures 11.5 and 11.6. The nutrient-type profiles are characteristic of trace metals, such as Zn, which are biolimiting. Biointermediate trace metals produce profiles with a less marked deep-water enrichment, such as those seen for Cu and Ni. These metals exhibit weak correlations with the nutrients.

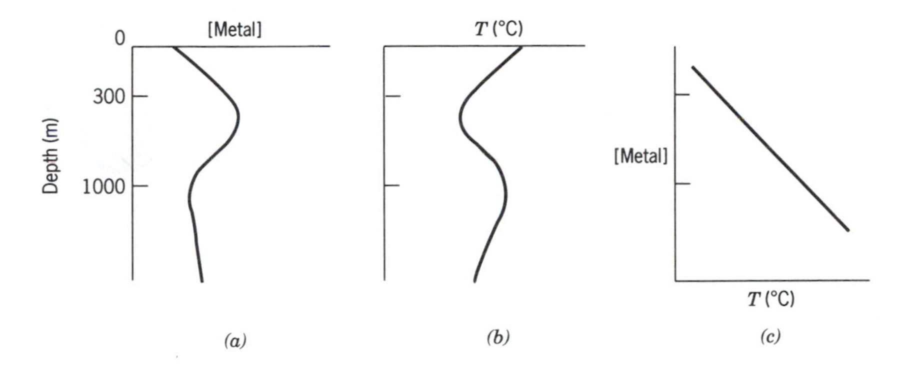

FIGURE 11.4. *(a)* Depth profile of a conservative metal. *(b)* Depth profile of water temperature. *(c)* Conservative metal concentration versus water temperature.

The nutrient-type profiles are the result of vertical segregation, which occurs much in the same way as for the nutrients. These trace metals are extracted from surface seawater by phytoplankton. Upon death, some of the biogenic material is recycled in the surface waters and some sinks below the thermocline. The degree to which the deep water becomes enriched in a trace metal depends on its EF, as well as the magnitude of the particulate metal flux and the relative rates of water motion. As shown in Table 11.6, the trace metals vary considerably in their degree of vertical segregation.

These trace metals are also subject to horizontal segregation, although advection in the Atlantic Ocean is so strong that all the horizontal concentration gradients are washed out. In contrast, relatively slower circulation has produced large gradients in the Pacific Ocean. These gradients are large enough to be used as water-mass tracers.

Three variations on these nutrient-type profiles have been observed. They include profiles characterized by 1) a mid-depth maximum, 2) a deep-water maximum, or 3) a combination of both. Mid-depth maxima are produced if the metals are regenerated at shallow depths, as with nitrate and phosphate. Cd and As(V) are examples of this. The excellent correlation of Cd with phosphate, as shown in Figure 11.6*b*, suggests that these chemicals are present in the same type of POM and that similar processes are acting to resolubilize them.

Deep-water maxima are produced by trace metals that undergo regeneration below the thermocline. These elements are thought to be carried to the deep sea as components of shell and skeletal material. Examples of this type of trace metal include zinc, barium, and germanium. As shown in Figure 11.6*a*, the close linear correlation with silicon suggests that zinc is carried to the deep sea as a component of siliceous shells and is resolubilized as the shell dissolves.

Biointermediate metals, such as Ni (Figure 11.5*a*) and Se (Figure 11.3*a*), exhibit smaller deep-water enrichments. Their concentrations do not contain enough information to deduce the phase by which these metals are transported to the deep sea. Nevertheless, the poor linear correlation between nickel and

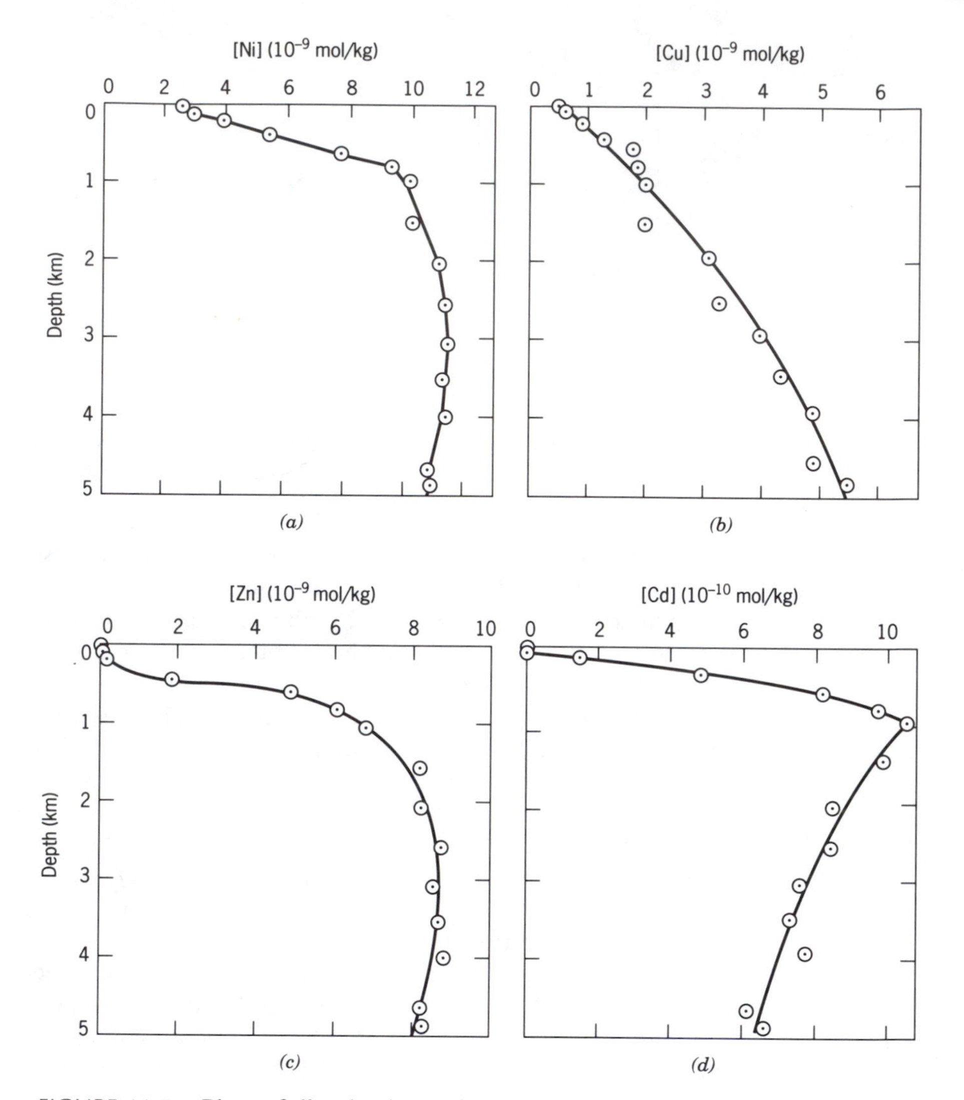

FIGURE 11.5. Plots of dissolved *(a)* Ni, *(b)* Cu, *(c)* Zn, and *(d)* Cd concentration as a function of depth in the central North Pacific. *Source:* From K. W. Bruland, reprinted with permission from *Earth and Planetary Sciences Letters,* vol. 47, pp. 189–192, copyright © 1980 by Elsevier Publishing Company, Amsterdam, The Netherlands.

barium (Figure 11.6*c*) suggests that the two elements are released into the deep sea by different mechanisms.

Surface-Water Enrichments

Surface-water enrichments are usually caused by the presence of large metal supplies to the mixed layer. Removal usually occurs through relatively rapid precipitation or adsorption reactions. As a result, the turnover time of these

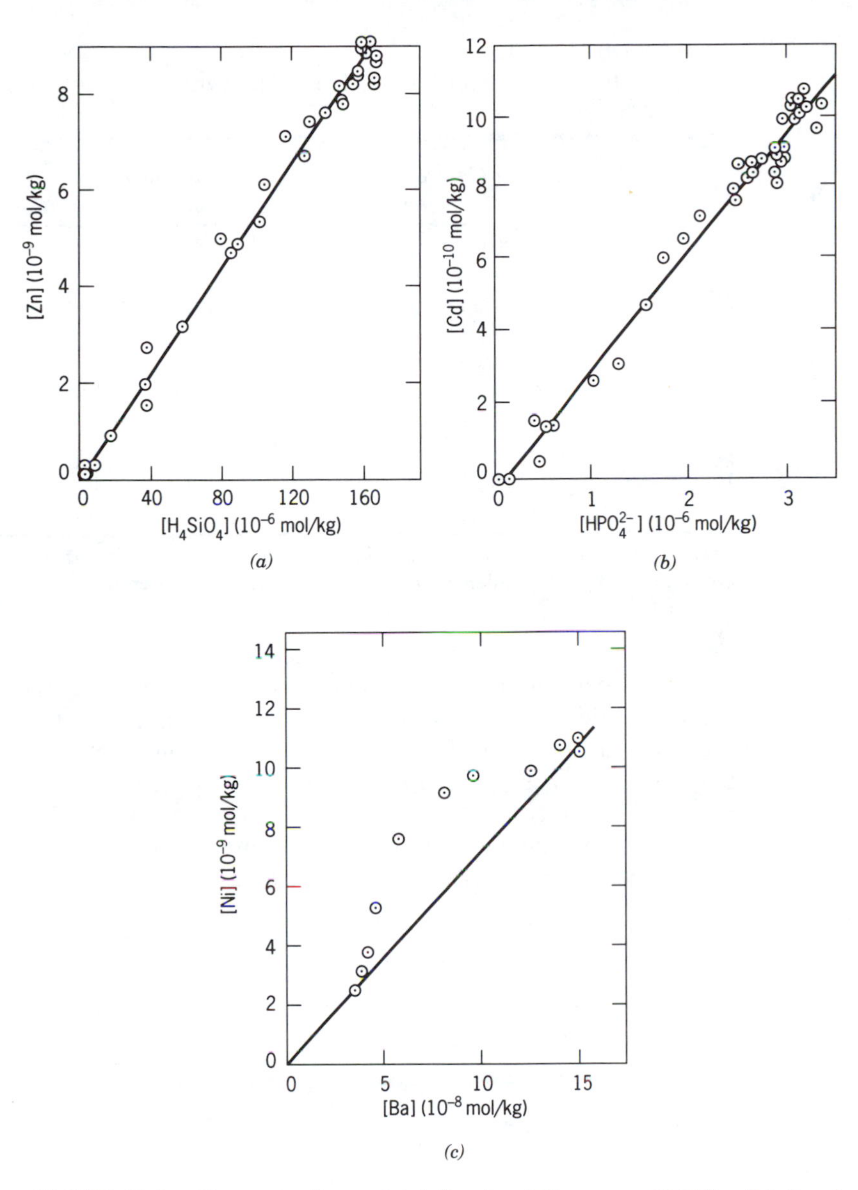

FIGURE 11.6. Concentrations of *(a)* dissolved Zn versus H_4SiO_4, *(b)* dissolved Cd versus HPO_4^{2-}, and *(c)* dissolved Ni versus dissolved Ba using the data from Figure 11.5. *Source:* From *Tracers in the Sea,* W. S. Broecker and T.-H. Peng, copyright © 1982 by the Lamont-Doherty Geological Observatory, Palisades, New York, p. 217. Reprinted by permission.

metals is much shorter than the mixing time of the water mass (i.e., the amount of time required to homogenize the water mass completely).

Several processes can supply particulate metals at large enough rates to create concentration maxima at the sea surface. For example, atmospheric transport of anthropogenic lead has created concentration maxima in the

TABLE 11.6
Fractionation of Nutrient-Type Trace Elements between Surface and Deep Water[a]

Element	*Minimum Central Gyre Surface Concentration*	*Maximum Pacific Deep-Water Concentration*	*Deep/Surface*
Cd	0.001–0.002	1.1	~1000
Zn	0.05	9	180
Ge	≤0.007	0.115	≥16
Cu	0.5	6	12
Ni	2	11	5.5
Ba	32	150	4.7
Se	0.5	2.3	4.6
Cr	2	5	2.5
I	250	450	1.8
As	1.1	1.9	1.7

Source: From *Chemical Oceanography*, vol. 8, K. W. Bruland (ed.: J. P. Riley and R. Chester), copyright © 1983 by Academic Press, Orlando, FL, p. 213. Reprinted by permission.
[a]Concentrations are given in nmol/kg.

surface waters of the North Atlantic and to a lesser extent in those of the North Pacific, as shown in Figure 11.7. Pb is injected into the atmosphere in the form of an aerosol produced by the burning of leaded gasoline. The particulate lead is washed out of the atmosphere and onto the sea surface by rainfall. Since the use of leaded gasoline has declined, surface-water concentrations have decreased. As indicated in Tables 5.7 and 5.8, dissolved lead is present primarily in the form of carbonate, hydroxide, and chloride complexes.

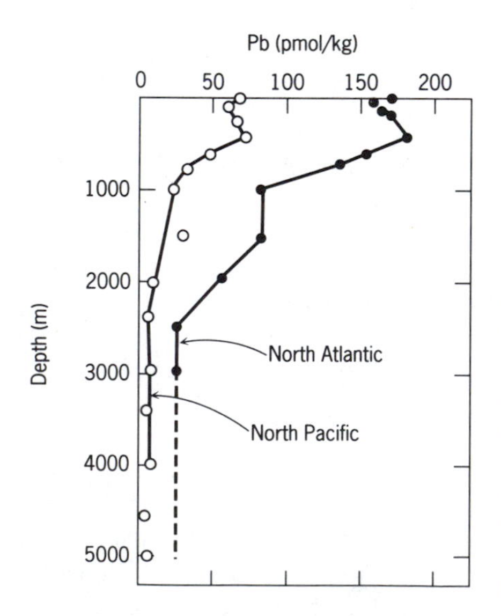

FIGURE 11.7. Vertical profiles of dissolved lead concentrations in the central North Pacific and North Atlantic. *Source:* From *Trace Metals in Seawater,* B. K. Schaule and C. C. Patterson (eds.: C. S. Wong, E. Boyle, K. W. Bruland, J. D. Burton, and E. D. Goldberg), copyright © 1981 by Plenum Press, New York, p. 491. Reprinted by permission.

Surface and subsurface maxima can also be caused by horizontal advection. For example, the riverine input of metals can be carried far out into the open ocean by surface currents, as shown for Mn in Figure 11.8. Likewise, lateral transport of trace metals released from continental shelf sediments by diagenesis can cause subsurface maxima far from their point of injection.

Biologically mediated redox reactions can also produce surface and subsurface maxima of reduced species. Reduced Cr, As, and I are present in

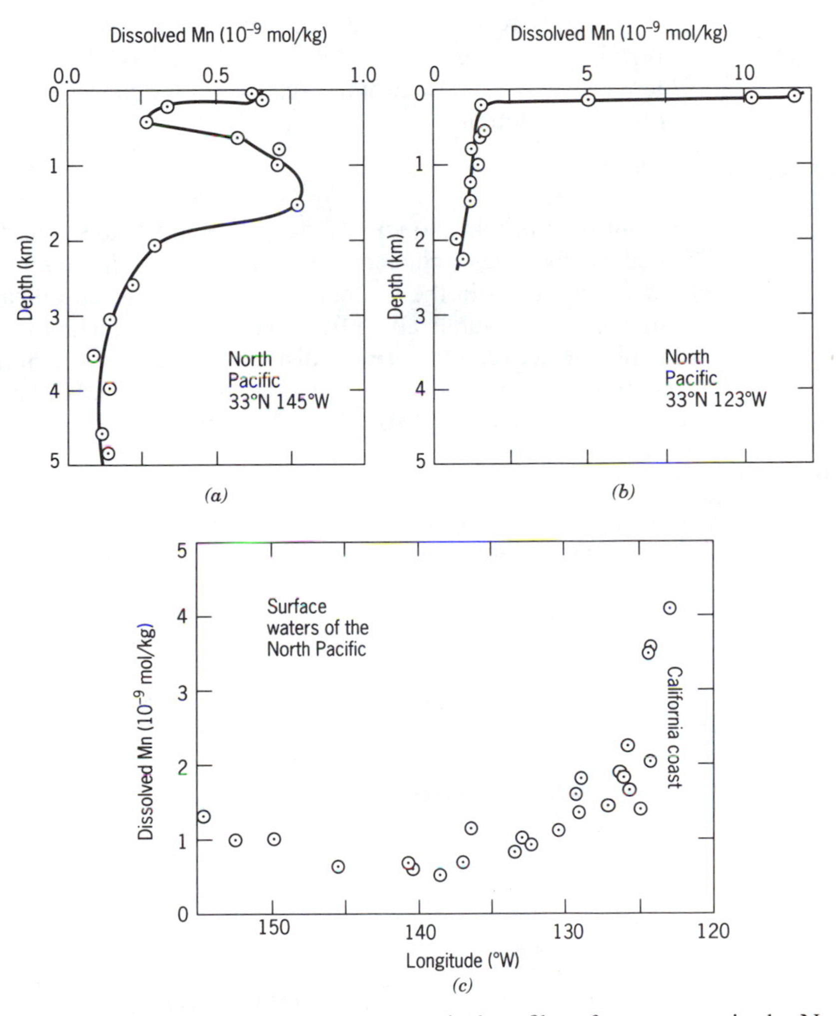

FIGURE 11.8. Upper panels show vertical profiles of manganese in the North Pacific Ocean at *(a)* an open-ocean station and *(b)* a coastal station. *(c)* The Mn content of surface water with increasing distance from the California coast. Note the tenfold scale difference in concentration between these diagrams. *Source:* From W. M. Landing and K. W. Bruland, reprinted with permission from *Earth and Planetary Sciences Letters,* vol. 49, p. 49, copyright © 1980 by Elsevier Publishing Company, Amsterdam, The Netherlands.

surface seawater as a result of such processes. Marine organisms tend to methylate toxic trace metals like As, as well as tin, antimony, and mercury. Since arsenic is a cogener of phosphorus, it interferes with the process of phosphorylation by competing with phosphorus for binding sites on the phosphorylation enzymes. Marine organisms may methylate arsenic to prevent this.

Due to the presence of O_2 in seawater, all iodine should be in the form of IO_3^- if thermodynamic equilibrium is achieved. Instead, significant amounts of iodide (I^-) are present. This reduced species is supplied to the surface ocean by river input, as well as by the reduction of iodate in anoxic waters. The iodide persists in oxic seawater because its oxidation is slow relative to the rates of water motion.

Mid-Depth Minima

The aluminum profile shown in Figure 11.9 illustrates a mid-depth minimum. The high surface-water concentrations are due to a large atmospheric flux. The high concentrations in the bottom water are due to an upward flux supported by diffusion of resolubilized Al from the sediments. The low concentrations at mid-depth reflect both the greater distance from a source of aluminum, as well as removal via adsorption onto sinking siliceous shells. Aluminum is present primarily as $Al(OH)_4^-$ and $Al(OH)_3^0$. The latter is sparingly soluble.

Mid-Depth Maxima

Mid-depth maxima are produced by mid-depth sources of metals. As shown in Figure 11.10, the mid waters in the East North Pacific contain elevated

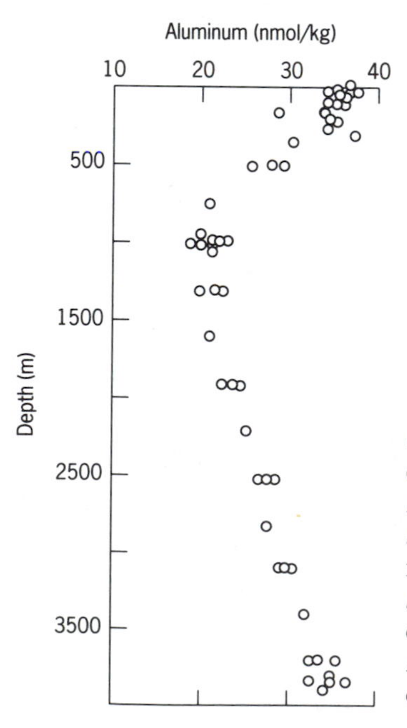

FIGURE 11.9. Vertical profile of dissolved aluminum in the North Atlantic (40°51′N, 64°10′W). *Source:* From D. J. Hydes, reprinted with permission from *Science,* vol. 205, p. 1261, copyright © 1979 by the American Association for the Advancement of Science, Washington, D.C.

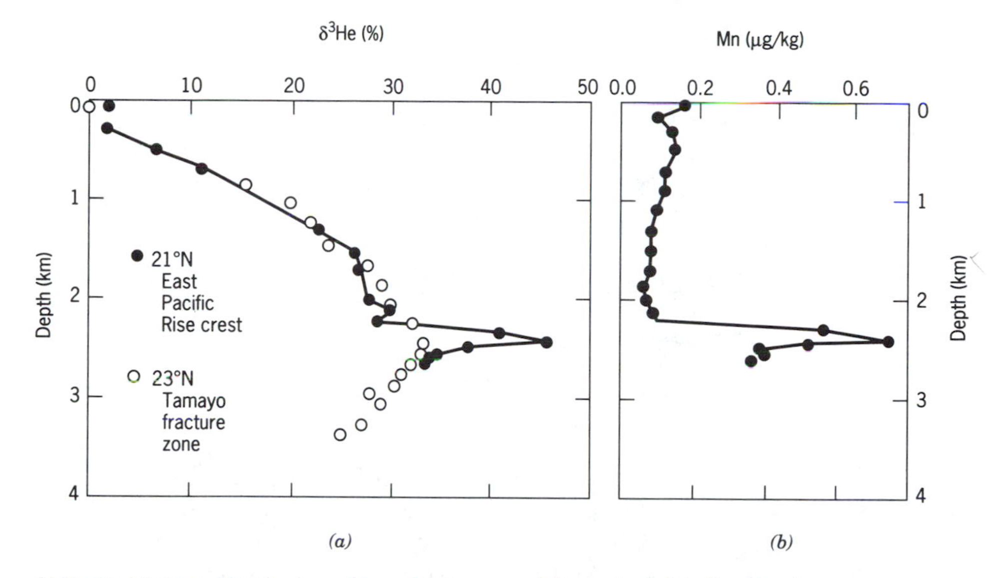

FIGURE 11.10. Vertical profiles of *(a)* excess ^3He and of *(b)* dissolved manganese at two sites in the North Pacific. The data represented by the solid circles were obtained from water located directly over the crest of the East Pacific Rise (EPR) at 21° N. *Source:* From *Tracers in the Sea,* W. S. Broecker and T.-H. Peng, copyright © 1982 by the Lamont-Doherty Geological Observatory, Palisades, New York, p. 217. Reprinted by permission. See Broecker and Peng (1982) for data sources.

levels of Mn and ^3He. Hydrothermal emissions are the source of both these chemicals. After entering the ocean, these chemicals become entrained in subsurface currents and are advected horizontally through the Pacific Ocean. Since their concentrations are ordinarily very low, this input can be used as a tracer of the pathway of hydrothermal emissions, as shown in Figure 11.8*a*.

Mid-Depth Maxima or Minima within a Suboxic Layer

Subsurface waters are often undersaturated with respect to O_2. These conditions are termed **suboxic** if some O_2 is still present. The thermocline is characterized by suboxic conditions. The most extreme O_2 depletions occur in coastal upwelling areas and in locations of restricted water flow, such as the Peru Upwelling Area and the Arabian Sea, respectively.

Under suboxic conditions, metals are reduced. Maxima result if the reduced form is more soluble than the oxidized form. For example, Mn^{2+} and Fe^{2+} maxima in suboxic waters are produced by the in situ reduction of particulate MnO_2, FeOOH (goethite), Fe_2O_3 (hematite), and Fe_3O_4 (magnatite). If the reduced species is insoluble or adsorbs strongly to particles, as with Cr^{3+}, mid-depth minima are produced. Suboxic conditions in shallow sediments can produce soluble reduced species that diffuse into the overlying bottom waters. If horizontal advection carries the chemically altered bottom water offshore, mid-depth maxima are produced.

Mid-Depth Maxima or Minima within an Anoxic Layer

Anoxic conditions are rarely found in the open ocean. Most are encountered in marginal seas, such as the Black Sea and Saanich Inlet, as well as in deep-sea trenches, such as the Cariaco Trench. Metal profiles from the Black Sea

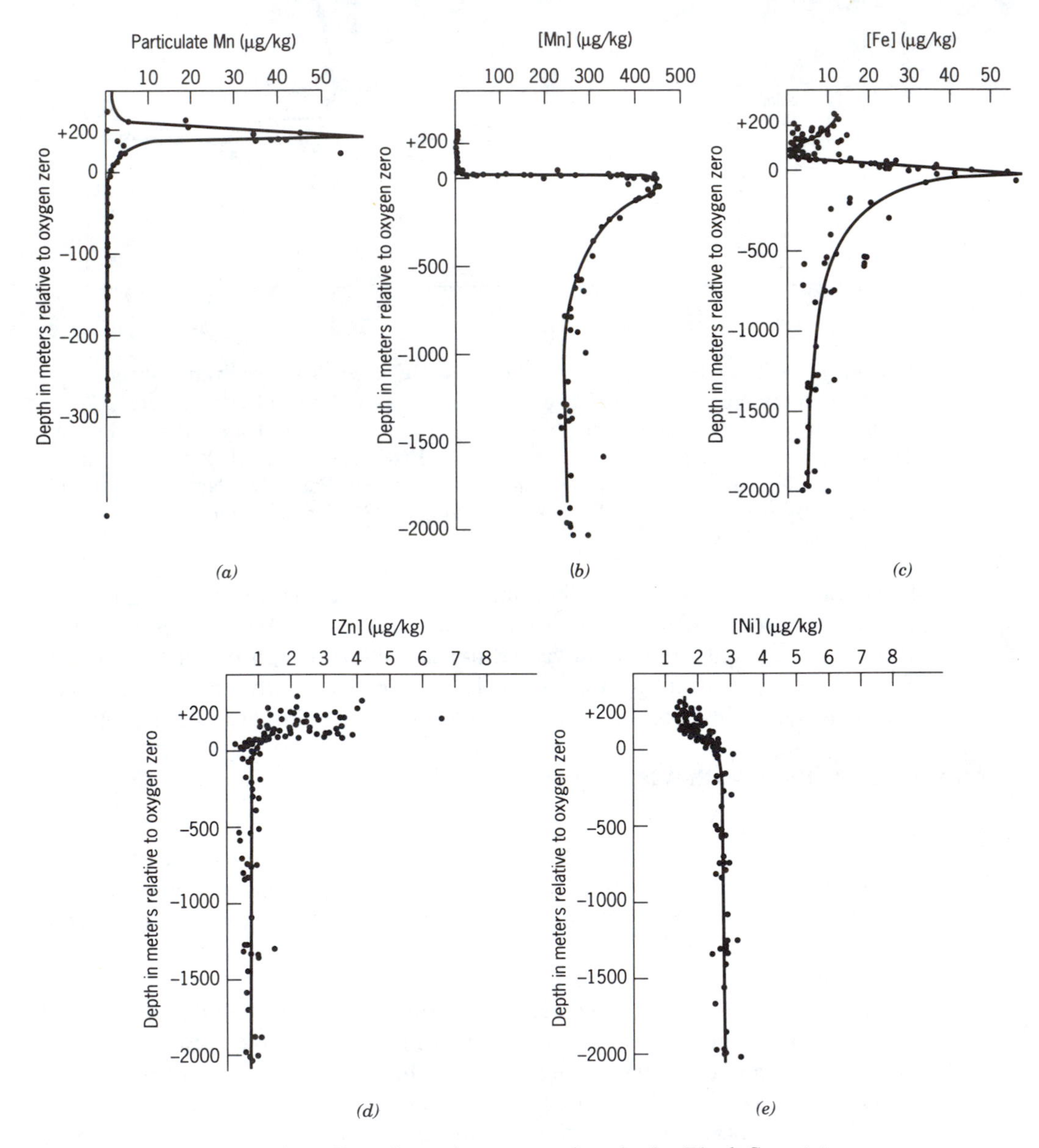

FIGURE 11.11. Vertical profiles of metal concentrations in the Black Sea. *(a)* Particulate manganese, *(b)* dissolved manganese, *(c)* dissolved iron, *(d)* dissolved zinc, *(e)* dissolved nickel,

(Continued on next page.)

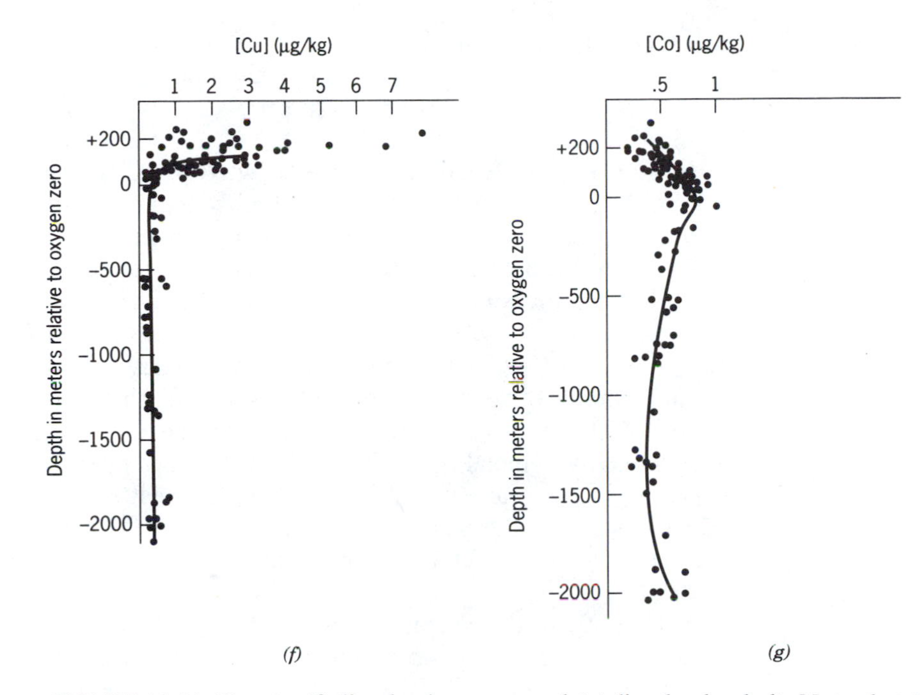

FIGURE 11.11 (Cont.) *(f)* dissolved copper, and *(g)* dissolved cobalt. Note that the depths are given relative to the level at which O_2 disappears from the water column. *Source:* From D. W. Spencer and P. G. Brewer, reprinted with permission from the *Journal for Geophysical Research,* vol. 76, pp. 5880–5888, copyright © 1971 by the American Geophysical Union, Washington, DC.

are shown in Figure 11.11. The maxima and minima are produced by the same processes that occur in suboxic waters.

Manganese is reduced to Mn^{2+} in the anoxic layer and some diffuses upward into the oxic zone. This Mn^{2+} reacts with O_2 and is converted to insoluble MnO_2. The particulate Mn sinks back into the anoxic region, where it is reduced to Mn^{2+} and hence is resolubilized. These competing reactions produce a particulate Mn (MnO_2) maximum that lies at the bottom of the oxic zone and a dissolved Mn (Mn^{2+}) maximum that lies at the top of the anoxic zone. The reduced manganese that diffuses downward is precipitated as a sulfide or carbonate mineral. This provides an upper limit to the deep-water manganese concentrations. In contrast, the nickel, zinc, and cobalt profiles do not have mid-water maxima because they are stable as the divalent cation even in the presence of O_2. Their concentrations are controlled by precipitation as sulfide or carbonate minerals.

SUMMARY

Virtually every element has been detected in seawater. Those present at concentrations less than 0.05 μmol/kg are termed trace elements. Most are metals. Though present at low levels as solutes in seawater, some of these

trace metals (iron, aluminum, and silicon) compose the bulk of the earth's crust. Others are **micronutrients**, playing an important role in enzyme systems. The biogeochemical cycles of most trace metals are controlled by redox reactions.

Metals are introduced into seawater by river runoff, winds, hydrothermal venting, **diagenetic remobilization**, and anthropogenic activities. As a whole, trace metals are more reactive in seawater than the major ions, so they are relatively enriched in the sediments. Metals are removed from seawater by nonequilibrium processes.

Under oxic conditions, dissolved metals are removed from seawater by adsorption onto particles or precipitation as oxides. Some of these metal-enriched particles eventually sink to the seafloor, thus removing metals from the ocean. This removal process is termed **scavenging**. The rate and degree to which a dissolved metal is scavenged from the ocean depends on its elemental nature, as well as the abundance of particulate matter, the concentrations of other solutes, and the water depth. Areas with high rates of scavenging can have local dissolved metal concentrations that deviate considerably from the average.

Marine organisms concentrate metals in their tissues and skeletal material. The degree to which a metal is concentrated in a marine organism is given by an **enrichment factor (EF).** EFs as high as 10^5 have been observed. In general, lower organisms tend to have higher EFs. Upon their death, these trace metals are resolubilized much in the same way as nitrogen, phosphorus, and silicon. This causes many of the trace metals to have depth profiles similar to those of the nutrients.

Though most dissolved metals are rapidly removed from seawater by adsorption or precipitation, some are returned via diagenetic remobilization from the sediments. The resolubilized metals can diffuse through the pore waters and eventually reenter the ocean. The diagenetic remobilization of trace metals under anoxic conditions causes the precipitation of **sulfide minerals**, **barite**, **phosphorite**, and **glauconite**. All contain small amounts of a variety of trace metals, as well as other elements. As a result, their chemical composition is variable and their structure is usually amorphous, making it difficult to assign them an empirical formula.

Like all other chemical species, the horizontal and vertical distributions of dissolved trace metals are determined by their relative rates of supply and removal. The nature of this balance can be assessed from the depth profiles of the metals. The concentration distribution exhibited by a trace metal can be classified as either a (1) conservative type, (2) nutrient-type, (3) surface-water enrichment, (4) mid-depth minimum, (5) mid-depth maximum, (6) mid-depth minimum or maximum within a **suboxic** layer, or as a (7) mid-depth minimum or maximum within an anoxic layer.

CHAPTER 12

DIAGENESIS

INTRODUCTION

Upon reaching the seafloor, particulate matter is subject to physical and chemical changes collectively referred to as **diagenesis**. Continued sedimentation eventually leads to deep burial and the metamorphosis of residual particulate matter. The chemical and physical changes that occur under these conditions are termed catagenesis and metagenesis, respectively. The term diagenesis refers only to those changes that occur under conditions close to those present at the time of sedimentation.

Diagenesis is an important part of the global biogeochemical cycle for several reasons: First, the sediments are a large reservoir of many elements, so global geochemical balances are dependent on the balance between burial and remobilization. In addition, the surface sediments represent a habitat for a great number and variety of marine organisms. Finally, postdepositional alteration is of importance to humans as petroleum is produced by the diagenesis and catagenesis of organic-rich sediments.

Most diagenetic chemical changes are driven by redox reactions, which involve the oxidation of organic matter. As a result, suboxic and anoxic conditions are common in marine sediments. These conditions have a profound effect on the distribution of trace metals, as discussed below. The sediment chemistry of other elements is covered in Part 3.

SOURCES OF TRACE METALS TO SEDIMENTS

Trace metals are delivered to the sediments in the form of biogenic detritus, clay minerals, and hydrogenous precipitates. Biogenic detritus is relatively enriched in trace metals as marine organisms concentrate these elements in their soft and hard parts. Upon the death of these organisms, the biogenic detritus sinks, carrying with it a significant amount of metals. Only a small fraction of this metal reaches the seafloor because most particulate matter is remineralized in the deep water by microbes. Clay minerals can contain substantial amounts of trace metals in their crystal lattices and adsorbed to their surfaces. Some of these metals are mobilized following sedimentation. Most hydrogenous precipitates, such as polymetallic oxyhydroxides, are metal-rich minerals that form by abiogenic precipitation.

PHYSICAL CHANGES AND THE ONE-DIMENSIONAL ADVECTION-DIFFUSION MODEL

Physical changes occur as particles are buried beneath the sediment–water interface. These include alterations in particle size, **porosity**, and mechanical strength. Some are caused by the burrowing and feeding actions of the benthos, as illustrated in Figure 12.1. The resulting biological stirring of the sediments is termed **bioturbation**. One particularly important consequence of bioturbation is the introduction of relatively O_2-rich bottom water into the sediments. This enhancement in O_2 supply is analogous to the aeration of soil by earthworms. Bioturbation can occur as deeply as 1 m below the sediment surface but is most intense in the top 10 cm. Thorough mixing also inhibits the formation of chemical gradients that would otherwise result from diagenesis.

The activities of some members of the benthos act to increase the size of sedimentary particles. For example, deposit feeders ingest the sediment whole and then expel any undigested particles in a consolidated mass called a fecal pellet. Other organisms alter the particle size and mechanical strength of the sediments by reinforcing the walls of their burrows with mucus-like exudates that "glue" the sedimentary grains together.

As particles settle onto the sediment surface, the addition in mass increases the pressure on the underlying materials. As a result, sediments are **compacted**, causing an increase in density, as well as the expulsion of some pore water. The resulting upward **advection** of water can transport significant amounts of

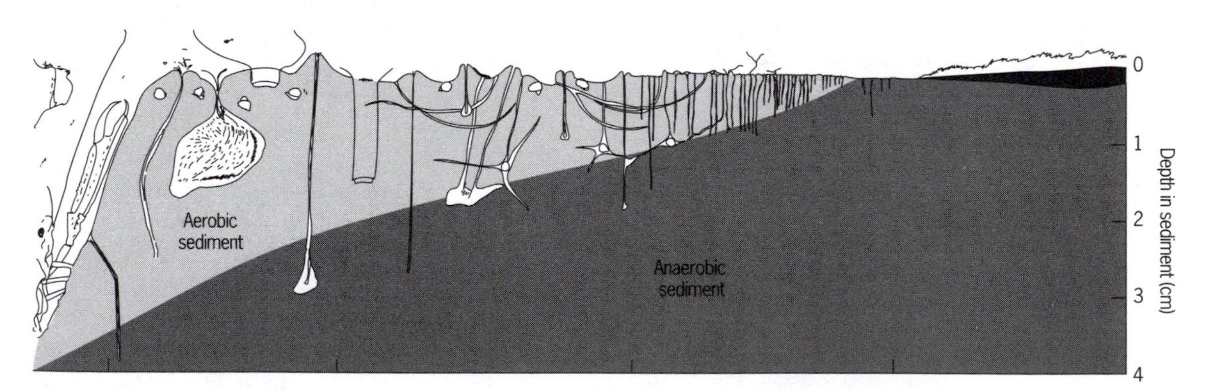

FIGURE 12.1. Bioturbating activities of Some Benthic Marine Organisms. *Source:* From T. H. Pearson and R. Rosenberg, reprinted with permission by *Ambio,* vol. 5, p. 79, copyright © 1976 by the Royal Swedish Academy of Sciences, Elmsford, New York.

solutes. Solutes can also move through the pore waters as a result of diffusion. This process causes the net transport of solutes from regions of higher to lower concentration. The vertical concentration gradients are strongest, so the effect of diffusion on the concentration of a solute C can be described by the following form of Fick's Second Law:

$$\frac{\partial[C]}{\partial t} = D_z \frac{\partial^2[C]}{\partial z^2} \qquad (12.1)$$

where $[C]$ = concentration of solute, t = time, z = depth beneath the sediment-water interface, and D_z = the vertical diffusivity coefficient for solute C. **Pore-water diffusion** is hindered by interactions with particles, such as collisions and adsorption. This causes the value of D_z for sedimentary diffusion to be somewhat less than that for simple ionic diffusion (1×10^{-5} to 2×10^{-5} cm^2/s).

The effect of pore-water advection on solute concentrations is given by

$$\frac{\partial[C]}{\partial t} = -U_z \frac{\partial[C]}{\partial z} \qquad (12.2)$$

where U is the rate of vertical pore-water advection. This is analogous to the water velocity term in Eq. 4.3. In the surface sediments, porosity does not vary much with depth. As a result, the rate of water advection is directly related to the rate of particle advection. The latter is given by the sedimentation rate (s), so Eq. 12.2 can be approximated by

$$\frac{\partial[C]}{\partial t} = -s \frac{\partial[C]}{\partial z} \qquad (12.3)$$

Combining the effects of diffusion and advection yields the following one-dimensional advection-diffusion equation for a conservative solute:

$$\frac{\partial[C]}{\partial t} = D_z \frac{\partial^2[C]}{\partial z^2} - s \frac{\partial[C]}{\partial z} \qquad (12.4)$$

If the solute undergoes any chemical changes, a reaction term must be added to Eq. 12.5. If no further information is available, the reaction rate is assumed to be first-order with respect to the solute concentration. Thus the one-dimensional advection-diffusion equation for a nonconservative solute is given by

$$\frac{\partial[C]}{\partial t} = D_z \frac{\partial^2[C]}{\partial z^2} - s \frac{\partial[C]}{\partial z} - k[C] \qquad (12.5)$$

where k = first-order rate constant for the net chemical removal of the nonconservative solute, C. In the event that the solute undergoes more than one reaction, this rate represents the net effect of all chemical changes on the solute concentration.

If the solute is in steady state ($\partial C/\partial t = 0$) and k, D_z, and s are constants, Eq. 12.5 becomes an ordinary differential equation, whose solution is

$$[C] = [C_o]e^{\frac{(s - [s^2 + 4kD_z])^{1/2}}{2D_z} z} \qquad (12.6)$$

where $[C_o]$ is the solute concentration at the sediment surface ($z = 0$) and $[C] \rightarrow 0$ as $z \rightarrow \infty$.

This mathematical model can be used to calculate the reaction rate constant (k) from the pore-water concentrations of C, if D_z and s are known. Equations derived from Fick's First Law are used to compute the solute flux across the sediment–water interface using the same data set.

More sophisticated approaches define the reaction rate as the sum of chemical removal and addition terms, whose functionalities are not necessarily first-order. Information on the reaction kinetics is empirically obtained by determining which algorithmic representation of the rate law best fits the data. This approach provides limited information on the nature of the diagenetic reaction, as the rate law is not directly related to the reaction mechanism.

The types of pore-water concentration profiles that can result from diagenetic production or removal are illustrated in Figure 12.2. A linear concentration gradient will result if no advection and reaction occurs between depths A and B, as shown in Figure 12.2*a*. Thus the solute concentration is controlled by diffusion from a source located at, or below, depth B to a sink located at, or above, depth A. A gradient that is convex (Figure 12.2*b*) will result from chemical removal or downward advection of pore water between

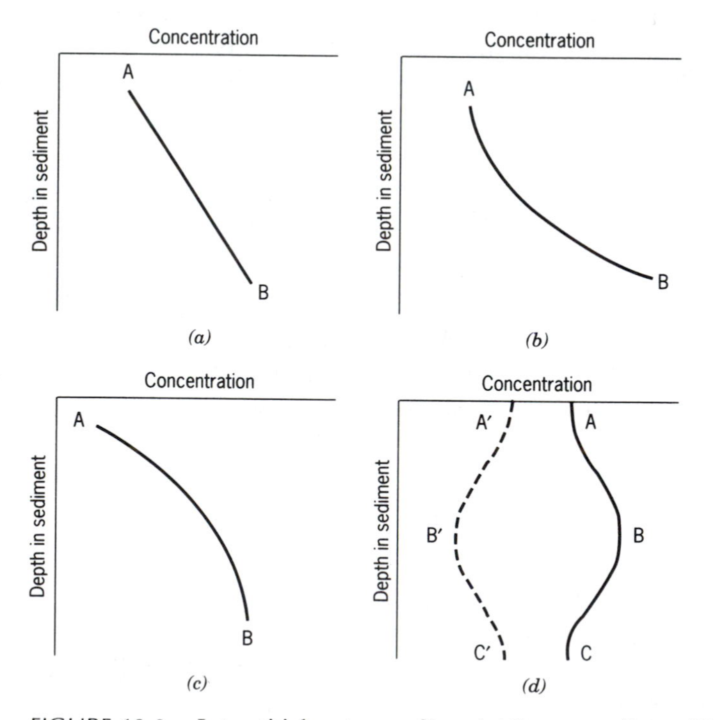

FIGURE 12.2. Interstitial water profiles: *(a)* linear gradient, *(b)* convex slope, *(c)* concave slope, and *(d)* a mid-depth maximum and minimum. *Source:* From *The Geochemistry of Natural Waters,* J. I. Drever, copyright © 1982 by Prentice Hall, Inc., Englewood Cliffs, New Jersey, p. 324. Reprinted by permission.

depths A and B. A gradient that is concave (Figure 12.2*c*) will result from chemical production or from upward advection of pore water between depths A and B. If the solute concentrations at depths A and C are equal, production at B will generate a maximum, while consumption will result in a minimum, as shown in Figure 12.2*d*.

In some cases, diagenesis produces enough solute to thermodynamically favor the precipitation of mineral phases. Sand grains act as crystallization nuclei, so mineral precipitates tend to fill the sedimentary pores. This cementation process is mostly due to the deposition of calcium carbonate and quartz. Deep burial causes the dissolution of some of these precipitates because mineral solubility increases with pressure. This effect is called pressure solution. At depths greater than several thousand meters below the sediment–water interface, pressures are high enough to crush and recrystallize these particles, thus producing sedimentary rock.

REDOX CONDITIONS IN MARINE SEDIMENTS

As in the water column, redox conditions in the sediment are controlled by the relative supplies of organic matter and electron acceptors, such as O_2, NO_3^-, and SO_4^{2-}. Since oceanic sediments are isolated from the atmosphere, O_2 is supplied only through contact with the bottom waters. As shown in Figure 12.3*a*, the top 8 to 15 cm of open-ocean sediments are oxic because the rate of organic matter deposition is slow relative to the rate of O_2 resupply by thermohaline circulation. At greater depths, benthic respiration is fast enough

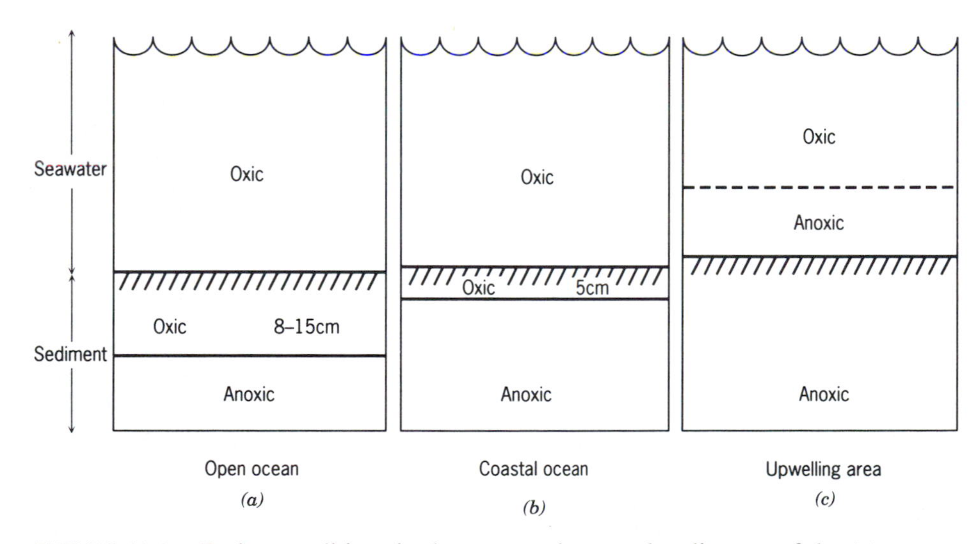

FIGURE 12.3. Redox conditions in the water column and sediments of the *(a)* open ocean, *(b)* coastal ocean, and *(c)* upwelling areas.

to maintain suboxic and anoxic conditions. In the coastal ocean, the rate of organic matter deposition is usually fast enough to cause the redox interface to lie within 5 cm of the sediment surface. In upwelling areas and salt-marsh estuaries, the anoxic zone often extends into the bottom waters due to relatively high POM fluxes.

Bioturbation affects the depth of the redox boundary by enhancing the rate of O_2 supply to the pore waters. Bioturbation is caused by animals, so little occurs below the redox boundary. This causes the oxic zone to be characterized by relatively homogeneous chemical distributions, whereas the anoxic sediments are more apt to have well-defined concentration gradients.

Ammonium, phosphate, carbon dioxide, and sulfate are produced during the oxic remineralization of organic matter. The hydrolysis of carbon dioxide produces acid, which can cause the dissolution of calcareous shells. This process elevates the pore-water concentrations of carbonate and bicarbonate. In the presence of O_2, the ammonium is oxidized to nitrate.

Below the redox boundary, bacteria oxidize organic matter using nitrate, manganese, iron, sulfate, or carbon dioxide as electron acceptors. The free energy yields of these reactions are given in Table 12.1 as the number of kilojoules produced by the oxidation of 1 mole of glucose. This molecule is used as an experimentally tractable approximation of "average" sedimentary organic matter.

Since aerobic respiration has the highest free energy yield, it is favored over all other organic matter oxidations if O_2 is present. As a result, this reaction dominates the oxic zone. At some depth below the sediment–water interface, the rate at which O_2 can be resupplied through advection and diffusion is less than the rate at which it is consumed by aerobic respiration. This causes O_2 concentrations and E_h to decrease with increasing depth as shown in Figure 12.4*a* and *b*.

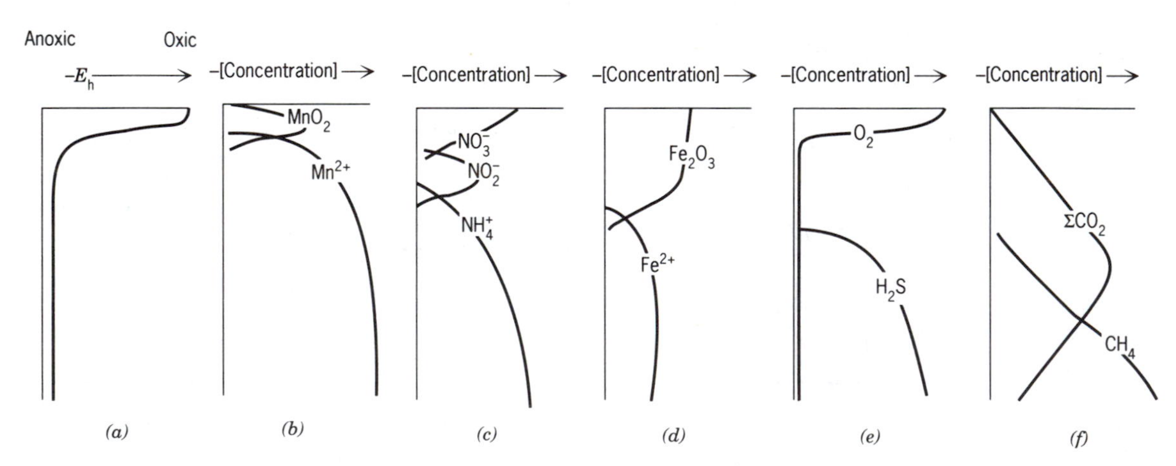

FIGURE 12.4. Idealized depth profiles of *(a)* E_h, *(b)* MnO_2 and Mn^{2+}, *(c)* NO_3^-, NO_2^-, and NH_4^+, *(d)* Fe_2O_3 and Fe^{2+}, *(e)* O_2 and H_2S, and *(f)* ΣCO_2 and CH_4 concentrations in a marine sediment.

TABLE 12.1
Oxidation Reactions of Sedimentary Organic Matter[a]

1. By Organic matter (aerobic respiration)
$(CH_2O)_{106}\,(NH_3)_{16}\,(H_3PO_4) + 138\ O_2 \rightarrow 106\ CO_2 + 16\ HNO_3 + H_3PO_4 + 122\ H_2O$ $\Delta G^0_w = -3190$ kJ/mol of glucose
2. By MnO_2
$(CH_2O)_{106}\,(NH_3)_{16}\,(H_3PO_4) + 236\ MnO_2 + 472\ H^+ \rightarrow 236\ Mn^{2+} + 106\ CO_2 + 8\ N_2 + H_3PO_4 + 366\ H_2O$ $\Delta G^0_w = -3090$ kJ/mol (Birnessite) -3050 kJ/mol (Nsutite)[b] -2920 kJ/mol (Pyrolusite)
3. By nitrate (denitrification)
$(CH_2O)_{106}\,(NH_3)_{16}\,(H_3PO_4) + 94.4\ HNO_3 \rightarrow 106\ CO_2 + 55.2\ N_2 + H_3PO_4 + 177.2\ H_2O$ $\Delta G^0_w = -3030$ kJ/mol $(CH_2O)_{106}\,(NH_3)_{16}\,(H_3PO_4) + 84.8\ HNO_3 \rightarrow 106\ CO_2 + 42.4\ N_2 + 16\ NH_3 + H_3PO_4 + 148.4\ H_2O$ $\Delta G^0_w = -2750$ kJ/mol
4. By Fe_2O_3
$(CH_2O)_{106}\,(NH_3)_{16}\,(H_3PO_4) + 212\ Fe_2O_3$ (or 424 FeOOH) $+ 848\ H^+ \rightarrow 424\ Fe^{2+} + 106\ CO_2 + 16\ NH_3 + H_3PO_4 + 530\ H_2O$ (or 742 H_2O) $\Delta G^0_w = -1410$ kJ/mol (Hematite, Fe_2O_3) -1330 kJ/mol (Limonitic goethite, FeOOH)
5. By Sulfate (Sulfate reduction)
$(CH_2O)_{106}\,(NH_3)_{16}\,(H_3PO_4) + 53\ SO_4^{2-} \rightarrow 106\ CO_2 + 16\ NH_3 + 53\ S^{2-} + H_3PO_4 + 106\ H_2O$ $\Delta G^0_w = -380$ kJ/mol
6. By fermentation
$(CH_2O)_{106}\,(NH_3)_{16}\,(H_3PO_4) \rightarrow 53\ CO_2 + 53\ CH_4 + 16\ NH_3 + H_3PO_4$ $\Delta G^0_w = -350$ kJ/mol

Source: From P. N. Froelich, G. P. Klinkhammer, M. L. Bender, N. A. Luedtke, G. R. Heath, D. Cullen, P. Dauphin, D. Hammond, B. Hartman, and V. Maynard, reprinted with permission by *Geochimica et Cosmochimica Acta*, vol. 43, p. 1076, copyright ©1979 by Pergamon Press, Elmsford, NY.

[a] CH_2O is used to represent organic matter having an average oxidation state. Nitrogen is treated as if it all starts as R—NH_2, a primary amine. Phosphorous is calculated as if it all starts as glucose-1-phosphate. The ΔG^0_w are presented as kilojoules per mole of glucose ($C_6H_{12}O_6$). Multiplication by 17.67 will convert these values to kilojoule per Redfield-molecule mole.

[b] Composition = $Mn_{1-x}M_xO_{2-2x}(OH)_{2x}$.

In this suboxic zone, bacteria reduce NO_3^- to N_2, $Mn(IV)O_2$ to Mn^{2+}, and Fe^{3+} to Fe^{2+}, causing the concentrations of these reactants to decrease and their products to increase with depth. The iron that is reduced is probably resolubilized from oxyhydroxides, which are collectively represented as Fe_2O_3 in Table 12.1. This relatively labile phase contains only a small fraction of the sedimentary iron. Most is present as a component of crystalline minerals and so is largely inaccessible to bacteria.

The energy yield from sulfate reduction is so low that sulfate cannot effectively compete for organic matter in the presence of O_2. O_2 also appears to "poison" sulfate-reducing enzymes. Thus sulfate reduction does not proceed until O_2 is depleted and the other electron acceptors have been reduced to relatively low concentrations. The depth at which this occurs is marked by the top of the hydrogen sulfide gradient. The concentration of CO_2 increases with depth in the oxic, suboxic, and sulfate-reducing portion of the anoxic zone as it is produced by the oxidation of organic matter regardless of the electron acceptor. At greater depths, the depletion of sulfate leaves organic matter to oxidize itself through a disproportionation reaction in which some of the carbon is oxidized to CO_2 and some is reduced to CH_4. This is termed methane fermentation or methanogenesis. The oxidation of organic matter can proceed via the reduction of CO_2, causing the CO_2 concentration to decrease with depth. Ammonium concentrations increase with depth below the suboxic layer as no O_2 is present to oxidize the remineralized nitrogen.

Phosphate is remineralized during the oxidation of organic matter. It is also mobilized by the dissolution of phosphate-bearing minerals such as calcium carbonate, hydroxyapatite, and fluoroapatite. Unlike the other products of remineralization, pore-water phosphate concentrations are regulated only by mineral solubility and not by redox reactions, since phosphorus exists nearly entirely in the +5 oxidation state.

Kinetic barriers usually prevent equilibrium from being attained, so the redox zones are usually not as well defined, as depicted in Figure 12.4. The causes of redox disequilibria are traceable to physical and chemical effects associated with biological processes. Indeed, disequilibrium is a requirement for life because it is the driving force that provides a continuing source of energy to fuel metabolic processes. Plants are the ultimate source of this disequilibrium, as they synthesize organic matter that is thermodynamically unstable.

Disequilibrium is also maintained by slow reaction rates caused by transition states with high activation energies. These kinetic barriers are lowered by catalysts, which in the marine environment are mostly enzymes. In the absence of a suitable enzyme, reaction rates can be very slow. For example, the concentration of Mn^{2+} in oxic seawater is higher than predicted from the solubility of MnO_2 due to the very slow rate at which divalent manganese is oxidized. Oxidation rates are faster in the sediment due to the presence of manganese-oxidizing microbes.

The introduction of O_2 into the sediments by bioturbation can cause localized and temporary redox disequilibria. If bioturbation is slow enough, microbial activity will eventually return the sediment to redox equilibrium. On the other hand, "permanent" oxic zones can be established around biological structures, such as worm burrows, due to the relatively high rate of O_2 supply. Burrow walls are sites of intense microbial activity due to the abundant supply of organic matter from mucus secretions. Bacterial growth is also enhanced by the removal of inhibitory metabolites which is facilitated by the relatively rapid rates of water movement in the burrows. As a result, resolubilized metals, such as iron and manganese, are precipitated and thus concentrated near burrow walls. Old burrows can be easily identified by these dark-colored mineral deposits.

Disequilibrium can also be caused by patchiness in the distribution of sedimentary organic matter. Microbes will flourish in those parts of the sediments composed of organic-rich particles such as fecal pellets. Intense microbial activity can remove all the O_2 within and around these organic particles. The result is an anoxic microzone that has redox conditions quite different from those of the neighboring sediments.

POSTDEPOSITIONAL MIGRATION OF METALS

As a particle is buried, its depth below the sediment–water interface increases, as shown in Figure 12.5. If the particle flux and all other environmental conditions remain constant, the positions of the redox boundaries (relative to the sediment–water interface) remain stationary over time. Thus, during burial, particles are moved downward through stationary redox boundaries.

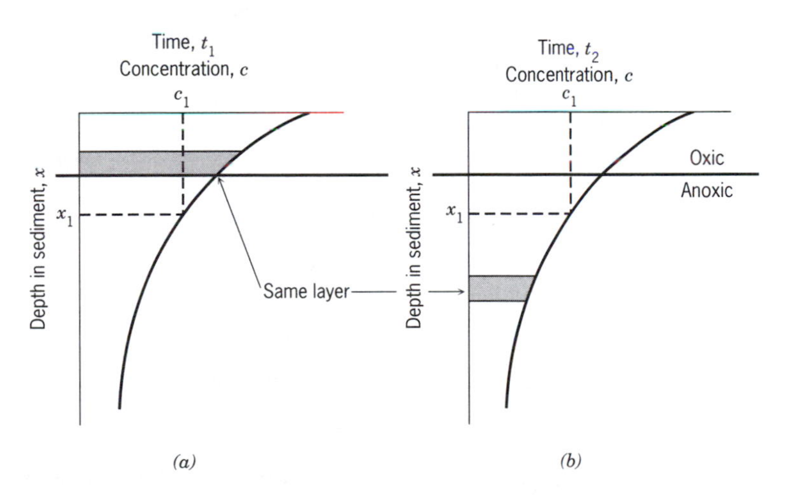

FIGURE 12.5. The movement of a layer of sediment below the sediment–water interface at *(a)* some initial time and *(b)* some later time. *Source:* From *Principles of Chemical Sedimentology,* R. A. Berner, copyright © 1971 by McGraw-Hill Book Co., New York, p. 88. Reprinted by permission.

When the particles enter the anoxic zone of the sediments, trace metals, such as manganese and iron, are reduced. This produces the dissolved manganese maximum illustrated in Figure 12.6. As ions, the reduced species diffuse through the pore water along concentration gradients. The ions that diffuse upward eventually encounter the oxic zone and are precipitated as the trace metal is oxidized. For manganese, this reaction is

$$2Mn^{2+}(aq) + O_2(g) + H_2O(l) \rightarrow 2MnO_2(s) + 4H^+(aq) \qquad (12.7)$$

As a solid, the manganese is trapped at the redox boundary. This produces the total manganese maximum seen in the oxic part of the sediment column in Figure 12.6. The position of the redox boundary is indicated by the steep E_h gradient.

If the downward flux of dissolved manganese is large enough, $MnCO_3$ will spontaneously precipitate. Thus, solubility equilibrium with this mineral limits the Mn^{2+} concentrations in the deep sediments. The slight increase in the total manganese concentration suggests that $MnCO_3$ has been deposited at depths near the bottom of the profile.

If the redox boundary is fixed relative to the position of the sediment–water interface. the total manganese maximum must also remain stationary.

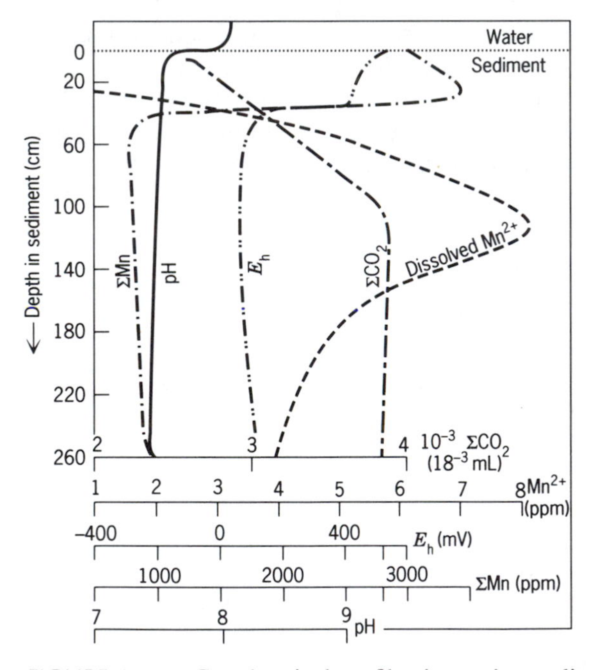

FIGURE 12.6. Geochemical profiles in marine sediments showing vertical distributions of total Mn, dissolved Mn, E_h, pH and total dissolved carbon. *Source:* From D. A. Crerar and H. L. Barnes, reprinted with permission from *Geochimica et Cosmochimica Acta,* vol. 38, p. 280, copyright © 1974 by Pergamon Press, Elmsford, New York.

This is achieved by the resolubilization of any particulate manganese that passes through the redox boundary. Once reduced, the manganese is mobilized and migrates upward until it encounters the redox boundary and is reprecipitated. Assuming that the only oxidation occurring in the sediments is that of manganese, the position of the total manganese maximum is maintained by a steady-state balance between the downward flux of O_2 (F_{O_2}) and the upward flux of Mn^{2+} ($F_{Mn^{2+}}$), which is given by the following mass balance equation:

$$F_{O_2} = -\tfrac{1}{2}F_{Mn^{2+}} \tag{12.8}$$

since 2 Mn^{2+} ions are oxidized for every molecule of O_2 reduced.

Thus trace metals are subject to a continuing cycle of dissolutions and reprecipitations that produce a large enrichment at the redox boundary. This is analogous to the industrial process of zone refining of ores. The segregation of manganese is of particular importance as this appears to be a significant source of trace metals for some manganese nodules.

The total iron maximum is not as pronounced as that of manganese because this element is mostly present as a constituent of crystalline minerals and hence is not readily mobilized. Iron is solubilized in the oxic sediments by remineralization of POM. This iron precipitates as oxyhydroxides, which are solubilized when they are buried below the redox boundary. The solubilized iron that diffuses upward reprecipitates at the redox boundary as oxyhydroxides. That which diffuses downward precipitates as carbonates.

Zn, Ni, and Co do not participate in redox reactions because they exist predominantly in the +2 oxidation state. In the oxic zone, they form soluble complexes with DOM and coprecipitate with iron and manganese as amorphous oxides. In the anoxic zone, these trace metals precipitate as sulfides due to the very low solubilities of these minerals. Thus these metals are not as strongly segregated in the sedimentary column as are iron and manganese.

SUMMARY

Upon reaching the seafloor, particulate matter is subject to physical and chemical changes that are collectively referred to as **diagenesis**. These processes are important because the sediments are a large reservoir for many elements, so global geochemical balances are dependent on the balance between their burial and remobilization. In addition, the surface sediments represent a habitat for a great number and variety of marine organisms. Finally, diagenesis is the first stage by which petroleum is produced in organic-rich sediments.

Trace metals are delivered to the sediments in the form of biogenic detritus, clay minerals, and hydrogenous precipitates. The physical changes that occur as particles are buried beneath the sediment–water interface include alterations in particle size, **porosity**, and mechanical strength. Some are the result of **bioturbation**, which acts to stir the sediments, as well as to enhance the rate of O_2 supply. As a result, the oxic zone of sediments is characterized by relatively homogeneous chemical distributions, whereas the anoxic zone is more apt to have well-defined concentration gradients.

As sediments are buried, the resulting **compaction** produces water motion, which is a type of advection. Solutes move through the sediments as a result of **pore-water advection** and **diffusion**. A one-dimensional advection-diffusion model has been developed that enables the calculation of first-order reaction rate constants (k) from the concentration gradients of nonconservative pore-water solutes. Other equations have been developed to provide an estimate of the solute flux across the sediment–water interface.

The top 8 to 15 cm of open-ocean sediments are oxic because the rate of organic matter deposition is slow relative to the rate of O_2 resupply by thermohaline circulation. At greater depths, benthic respiration is fast enough to maintain suboxic and anoxic conditions. Larger fluxes of POM cause the redox interface to lie at shallower depths in the sediments underlying coastal regions and upwelling areas.

Since aerobic respiration has the highest free energy yield, this reaction dominates the oxic zone. In the underlying suboxic zone, bacteria reduce NO_3^- to N_2, $Mn(IV)O_2$ to Mn^{2+}, and Fe^{3+} to Fe^{2+}. Due to its relatively low free energy yield, sulfate reduction does not proceed until O_2 is depleted and the other electron acceptors have been reduced to low concentrations. Under even more reducing conditions, fermentation reactions occur.

Redox equilibrium is often not attained in the sediments due to slow reaction rates. These kinetic barriers are lowered by catalysts, which in the marine environment are mostly enzymes. The introduction of O_2 into the sediments by bioturbation can cause localized and temporary redox disequilibria. Patchiness in the distribution of sedimentary organic matter can also lead to redox disequilibria as organic-rich particles are sites of intense microbial activity that can create anoxic microzones.

As sediments are buried, they pass through redox boundaries that remain stationary with respect to depth beneath the sediment-water interface. In the anoxic zone, some remineralized trace metals are reduced. Those reduced species that are soluble migrate up into the oxic zone, where they precipitate as insoluble oxides. Otherwise, they migrate down into the anoxic zone, where they precipitate as carbonates, hydroxides, or sulfides. As a result, some trace metals, such as iron and manganese, become highly concentrated at the redox boundary.

PROBLEM SET 2

1. Write the oxidation numbers for all of the atoms in the following compounds:

N_2	N ____		
NO_2^-	N ____	O ____	
NO_3^-	N ____	O ____	
Fe_2O_3	Fe ____	O ____	
FeOOH (goethite)	Fe ____	O ____	H ____
MnO	Mn ____	O ____	
MnO_2	Mn ____	O ____	
$S_2O_3^{2-}$	S ____	O ____	

2. Which reaction will proceed most strongly in favor of the products?
   a. $\Delta G_w^0 = +28.6$ kcal/mol; $E_{cell}^0 = -1.23$ V; $pe_w^0 = -13.75$.
   b. $\Delta G_w^0 = -28.6$ kcal/mol; $E_{cell}^0 = +1.23$ V; $pe_w^0 = +13.75$.
   c. $\Delta G_w^0 = 0$ kcal/mol; $E_{cell}^0 = 0$ V; $pe_w^0 = 0$.
   d. $\Delta G_w^0 = +100$ kcal/mol; $E_{cell}^0 = -100$ V; $pe_w^0 = 100$.

3. Answer the following questions for—

$$2\ Mn^{2+} + O_2 + 2H_2O \rightarrow 2\ MnO_2 + 4H^+$$

   a. The oxidation number of Mn changes from ____ to ____ .
   b. The oxidation number of O changes from ____ to ____ .
   c. The oxidizing agent is ____ .
   d. The element that gets reduced is ____ .
   e. Write the half-cell oxidation reaction. *Hint: Use Table 7.1.*
   f. Write the half-cell reduction reaction. *Hint: Use Table 7.3.*
   g. The E_{cell}^0 for the reaction is ____ .
   h. If equilibrium is achieved under the environmental conditions (pH = 8.1) found in seawater, the E_{cell}^0 for this reaction would be +0.81 V. Compute the equilibrium constant for this reaction in seawater.
   i. This reaction does not reach equilibrium in seawater. Compute the disequilibrium concentration of Mn^{2+} in seawater assuming that the pe_{cell} of this reaction is equal to the pe_w^0 of the dominant oxidizing agent in seawater, i.e. 13.75 and an $[O_2] = 6$ ml/L at 25°C.

4. Calculate the concentration of H_2S that should be present in oxic seawater if thermodynamic equilibrium is achieved in seawater of 35‰, 0°C, and pH 8.1. The concentration of H_2S in seawater is actually 2×10^{-7} M. Why doesn't the calculated value agree with the experimental data?

5. At a site being considered for the culture of clams, the seawater that is supplied to the region was found to have a total phosphate concentration of 0.5 μmol/kg, and the nitrate concentration was 20 μmol/kg. The clams involved were filter feeders, whose food source was to be phytoplankton grown in the same seawater. Would you recommend fertilizing this water? If so, why, and with what?

6. The concentration of phosphate in a local bay increases by 0.01 μM over a 4-hour period during the night. Calculate the expected change in nitrogen (over the same time period) based on the Redfield–Richards Ratio. State all assumptions used.

SUGGESTED FURTHER READINGS

Diagenesis

FROELICH, P. N., G. P. KLINKHAMMER, M. L. BENDER, N. A. LUEDTKE, G. R. HEATH, D. CULLEN, P. DAUPHIN, D. HAMMOND, B. HARTMAN, and V. MAYNARD. 1979. Early Oxidation of Organic Matter in Pelagic Sediments of the Eastern Equatorial Atlantic: Suboxic Diagenesis. *Geochimica et Cosmochimica Acta*, 43:1075–1090.

LERMAN, A. 1979. *Geochemical Processes: Water and Sediment Environments*. John Wiley & Sons, Inc., New York, 481 pp.

Redox Chemistry

BRECK, W.G. 1974. Redox Chemistry. In: *The Sea: Marine Chemistry*, vol. 5 (ed: E. Goldberg). John Wiley & Sons, Inc., New York, pp. 153–179.

DREVER, J. I. 1982. *The Geochemistry of Natural Waters*. Prentice Hall, Englewood Cliffs, NJ, 388 pp.

OLAUSSON, E., and I. CATO (eds.). 1980. *Chemistry and Biogeochemistry of Estuaries*. John Wiley & Sons, Inc., New York, 452 pp.

MOREL, F. M. 1983. *Principles of Aquatic Chemistry*. John Wiley & Sons, Inc., New York, 446 pp.

STUMM, W., and J. J. MORGAN. 1981. *Aquatic Chemistry: An Introduction Emphasizing Chemical Equilibria in Natural Waters*. John Wiley & Sons, Inc., New York, 780 pp.

Trace Element Chemistry

BROECKER, W., and T.-H. PENG. 1982. *Tracers in the Sea*. Lamont-Doherty Geological Observatory, New York, 690 pp.

BRULAND, K. W. 1983. Trace Elements in Sea Water. In: *Chemical Oceanography*, vol. 8 (eds.: J. P. Riley and R. Chester) Academic Press, New York, pp. 157–221.

Vertical and Horizontal Segregation of the Biolimiting Elements

BROECKER, W., and T.-H. PENG. 1982. *Tracers in the Sea*. Lamont-Doherty Geological Observatory, New York, 690 pp.

PART THREE

All the rivers run into the sea; yet the sea is not full.

The Bible, Ecclesiastes 1:7

THE CHEMISTRY OF MARINE SEDIMENTS

CHAPTER 13

CLASSIFICATION OF SEDIMENTS

INTRODUCTION

The chemical composition of seawater is largely regulated by biogeochemical processes that cause dissolved materials to be converted to solid forms. These solids are then deposited on the seafloor. Thus the sediments represent a very important reservoir in the crustal-ocean factory. The sediments are also important because they contain our only record of past conditions in the ocean.

Marine sediments are heterogeneous in composition and geographically variable in distribution. The causes for this are the subject of Part 3. They involve the factors that control the production of particles, their transport to the seafloor, and their preservation in the sediments. Several models are also presented that describe the sediment's role in controlling the chemical composition of seawater. The schemes most commonly used to classify marine sediments are discussed in this chapter.

SEDIMENTS: FIVE CLASSIFICATION SCHEMES

Marine sediments are composed of **unconsolidated** particles that blanket the bedrock of the seafloor. They vary greatly in chemical composition, particle size, origin, sedimentation rate, and geographic distribution. These characteristics are commonly used to classify marine sediments.

The deposits which are located on the continental margin are termed **neritic**. Those that overlie oceanic crust, such as on the abyssal plains and mid-ocean ridges, are referred to as **oceanic**. Sediments can also be categorized on the basis of grain diameter. Since grains can be irregularly shaped, the longest diameter is generally used to classify the particle as either a clay, silt, sand, granule, pebble, cobble, or boulder using the criteria given in Table 13.1.

The types of particles that fall into each of these size classes are shown in Figure 13.1. Most sedimentary particles are either sand-, **silt-**, or **clay-sized**. Note that this classification scheme conveys no information on the mineral composition of the particles. Thus beach sand, though commonly composed of quartz and feldspar grains, can also contain shells of radiolaria and foraminifera.

TABLE 13.1
Wentworth Scale of Grain Size for Sediments

Particle	*Minimum Size (mm)*
Boulder	256
Cobble	64
Pebble	4
Granule	2
Sand	
Very coarse sand	1
Coarse sand	1/2
Medium sand	1/4
Fine sand	1/8
Very fine sand	1/16
Silt	
Coarse silt	1/32
Medium silt	1/64
Fine silt	1/128
Very fine silt	1/256
Clay	
Coarse clay	1/640
Medium clay	1/1024
Fine clay	1/2360
Very fine clay	1/4096
Colloid	1/4096

Source: From C. K. Wentworth, reprinted with permission from the *Journal of Geology*, vol. 30, p. 381, copyright © 1922 by the University of Chicago Press, Chicago.

The particles that compose marine sediments have two basic origins. As shown in Figure 13.2, either they are created in situ from dissolved compounds or they are carried to the ocean in solid phase from the land, atmosphere, Earth's interior, or outer space.

The thickness of a deposit depends on (1) the rate of particle supply to the seafloor, (2) the degree to which the particles are preserved following sedimentation, and (3) the age of the underlying crust. The latter determines the length of the time over which sedimentation has taken place. The average deposit is 0.5 km thick. The thinnest layers overlie the youngest oceanic crust, which is located at the mid-ocean ridges and rises. As shown in Figure 13.3, the thickest layers have accumulated on the continental margins. Some of these deposits exceed 1 km in thickness.

The continental margin deposits are thickest because of rapid sedimentation rates, as well as the advanced age of the underlying crust. Since these deposits

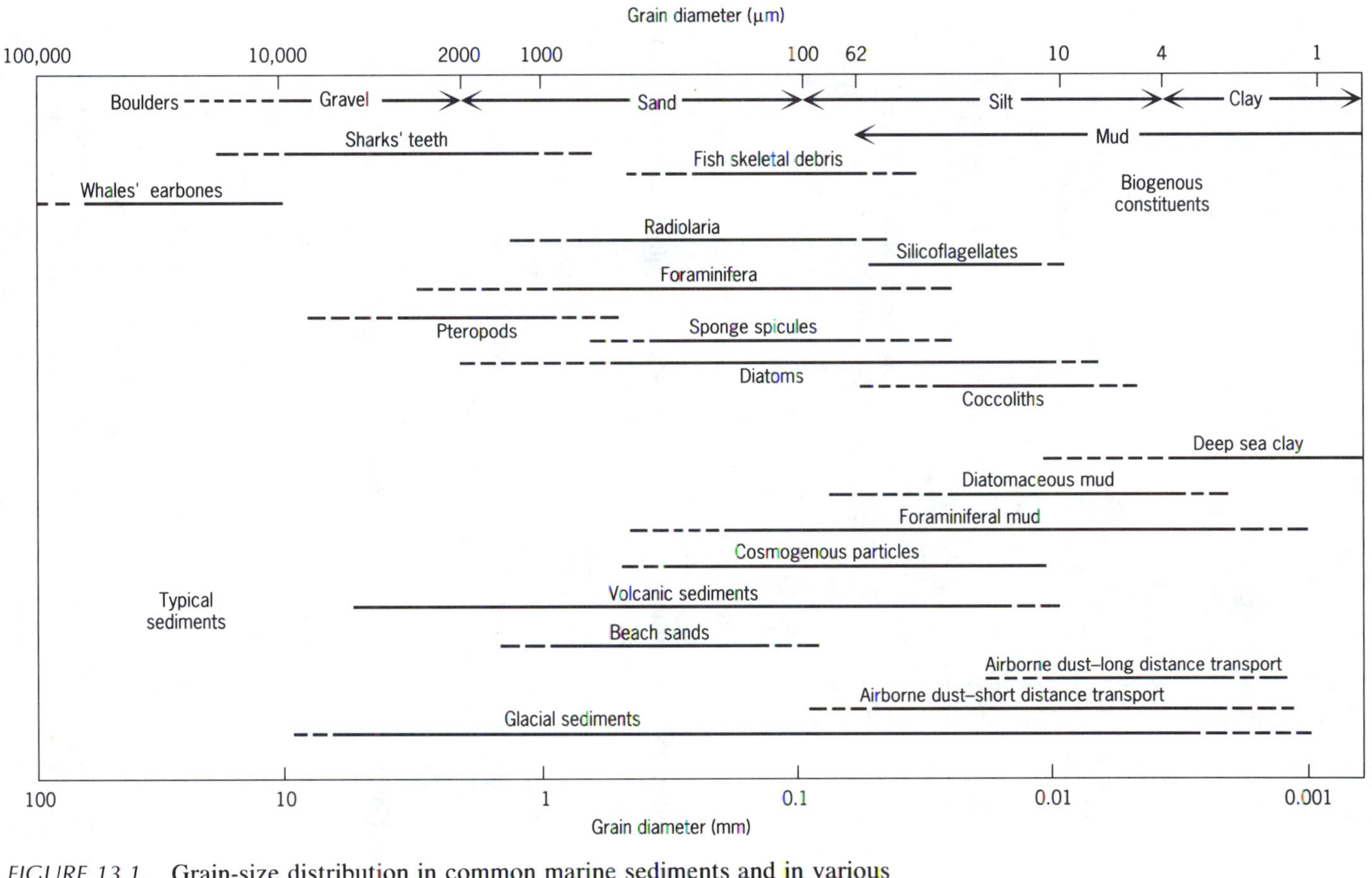

FIGURE 13.1. Grain-size distribution in common marine sediments and in various sediment sources. *Source:* From *Oceanography: A View of the Earth,* 4th ed., M. G. Gross, copyright © 1987 by Prentice Hall, Inc., Englewood Cliffs, New Jersey, p. 81. Reprinted by permission.

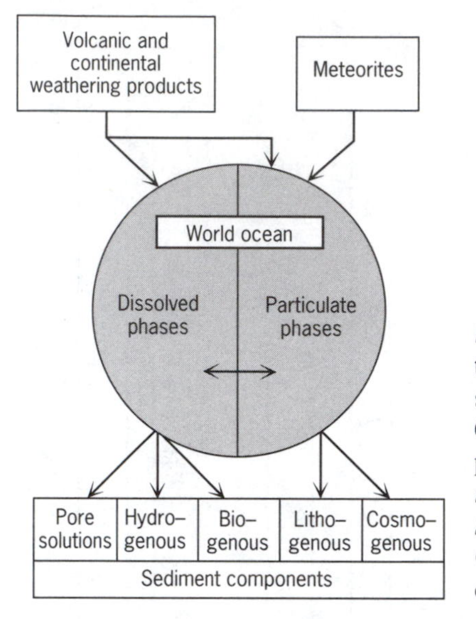

FIGURE 13.2 The classification of the components of marine sediments. *Source:* From E. D. Goldberg, reprinted with permission from *Transactions of the New York Academy of Sciences,* vol. 27, p. 8, copyright © 1964 by the New York Academy of Sciences, New York.

form at rates exceeding 1 cm/1000 y, they are termed **nonpelagic** sediments. Those that are deposited at rates slower than 1 cm/1000 y are termed **pelagic** sediments. As shown in Table 13.2, most oceanic sediments fall into this category. These slow rates reflect the great horizontal and vertical distance of these deposits from any particle source. Sedimentary deposits that are no longer in the process of forming are termed **relict**. Changes in sea level can cause a deposit to be cut off from its particle source and thus cease to accumulate. Relict sediments are subject to erosion if bottom or contour currents are strong enough to resuspend particles.

PELAGIC SEDIMENTS

Many particles reach the seafloor by sinking through the water column. This particle-by-particle accumulation is termed **pelagic sedimentation**. Sinking rates depend on particle size, shape, and density. The effect of the former is shown in Table 13.3. The faster a particle sinks, the shorter the time it is subject to decomposition or dissolution and hence the greater its chances for accumulation in the sediments. In the open ocean, the smallest particles (the clays) can take decades to reach the seafloor. During this time, the particles can experience significant horizontal transport if entrained in a thermohaline current.

As shown in Table 13.4, the sedimentation rate of pelagic deposits is geographically variable. This is due to spatial variations in the supply rate of clay minerals as these particles constitute the bulk of pelagic sediments.

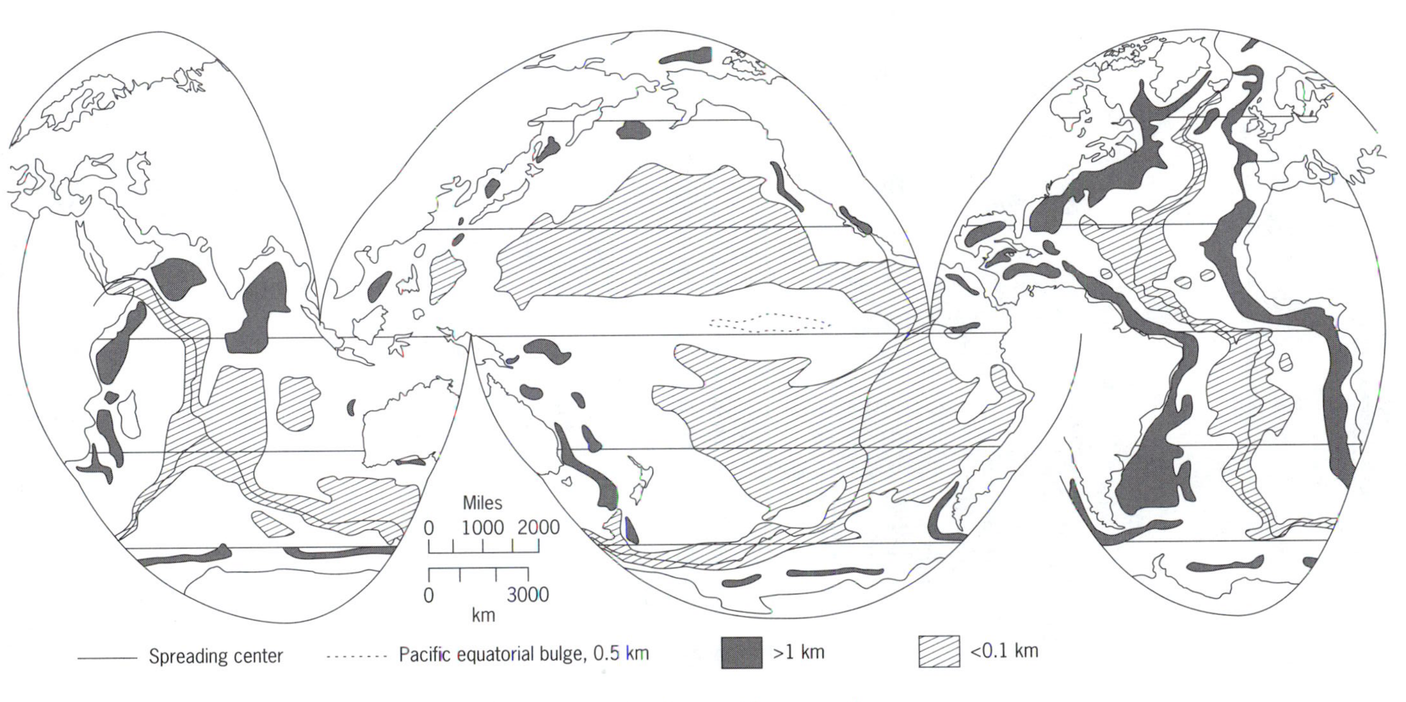

FIGURE 13.3. Sediment thickness to acoustic basement in the world ocean. *Source:* From *The Geology of Continental Margins,* C. A. Burk and C. D. Drake (ed.: W. H. Berger), copyright © 1974 by Springer-Verlag, Heidelberg, Germany, p. 217. Reprinted by permission.

TABLE 13.2
Typical Sedimentation Rates

Area	*Average Sedimentation Rate (cm/1000 y)*
Continental margin	
Continental shelf	30 (15–40)[a]
Continental slope	20
Fjord (Saanich Inlet, British Columbia)	400
Fraser River delta (British Columbia)	700,000
Upper Gulf of Thailand	400–1100
Marginal ocean basins	
Black Sea	30
Gulf of California	100
Gulf of Mexico	10
Clyde Sea	500
Deep-ocean sediments	
Coccolith muds	1 (0.2–3)
Deep-sea muds	0.1 (0.03–0.8)

Source: From *Oceanography: A View of the Earth*, 4th ed., M. G. Gross, copyright © 1987 by Prentice Hall, Inc., Englewood Cliffs, NJ, p. 89. Reprinted by permission.
[a]Observed range is given in parentheses.

Lithogenous Sediments

Clay-size particles that are less than 1 μm in diameter comprise the bulk of pelagic sediments. Most are weathered pieces of continental or oceanic crust transported to the ocean by rivers, glaciers, or winds. Due to their crustal source, these particles are classified as **lithogenous**. Winds can transport clay- and sand-sized continental debris across the ocean. This is termed **aeolian** transport. If the wind strength diminishes, the particles drop onto the sea

TABLE 13.3
Settling Velocity of Some Common Marine Particles

Particle	*Diameter (mm)*	*Rate of Settling in Sea Water[a] (cm/h)*	*Distance traveled in Sinking 1000 m Through a Current of 1 cm/s*	*Time to Settle 4 km*
Very fine sand	0.1	1472	2.4 km	11 days
Silt	0.05	31	116	1.4 y
Clay	0.001	0.147	24,500	310 y

Source: From *The World Ocean: An Introduction to Oceanography*, W. A. Anikouchine and R. W. Sternberg, copyright © 1981 by Prentice Hall, Inc., Englewood Cliffs, NJ, p. 399. Reprinted by permission.
[a]S = 34‰; T = 10°C.

TABLE 13.4
Sedimentation Rates of the Noncarbonate Fraction of Pelagic Sediments

	Rate (mm/10³ y)	
Location	*Mean*	*Range*
South Pacific	0.45	0.3–0.6
North Pacific	1.5	0.4–6.0
South Atlantic	1.9	0.2–7.5
North Atlantic	1.8	0.5–6.2

Source: From *Introduction to Marine Chemistry*, J. P. Riley and R. Chester, copyright © 1971 by Academic Press, Orlando, FL, p. 289. Reprinted by permission. See Riley and Chester (1971) for data sources.

surface and slowly settle to the seafloor. Most pelagic oceanic deposits are formed by the pelagic sedimentation of particles carried to the ocean by aeolian transport.

The **clay minerals**, quartz, illite, kaolinite, chlorite, and montmorillonite, are the most abundant lithogenous components of pelagic sediments. Most are clay-sized. If 70% or more of a deposit is composed of clay- and/or silt-sized lithogenous debris, it is termed a **clay** or **mud**. Such deposits are also described by their color (e.g., red-brown clays or gray-green muds).

Hydrogenous Sediments

Some of the particles in pelagic sediments are formed by the precipitation of solutes from seawater. If the reaction is abiogenic, the particles are termed **hydrogenous**, or **authigenic**. While marine organisms probably aid in the precipitation of most of these minerals, their actions are indirect, at least as compared to the biogenic precipitation of their hard parts. Some examples of hydrogenous precipitates include polymetallic oxyhydroxides and iron–manganese nodules. Since trace metals tend to coprecipitate, these hydrogenous precipitates do not have well-defined crystal structures and are variable in chemical composition.

Other hydrogenous minerals are formed from the precipitation of sea salt that occurs in some semi-isolated bodies of water located at low latitudes. These deposits are collectively referred to as evaporites.

Some clay minerals, such as montmorillonite and phillipsite, are also classified as hydrogenous particles even though they are not strictly the result of precipitation reactions. Instead, they are produced by the substitution of cations into an existing crystal lattice through a process called reverse weathering. Substitution of cations into biogenic detritus creates other so-called hydrogenous minerals, such as barite, phosphorite, and glauconite.

Biogenous Sediments

Biogenous sediments are composed of detrital (nonliving) particles that were originally synthesized by marine organisms. Most are structural components, such as shell, bone, and exoskeletons. Also present are tissues and excretions, such as fecal pellets and molts. Since most of the organic material is resolubilized before it can sink to the seafloor, the remains of hard parts, such as the calcareous and siliceous shells of plankton, are the most abundant biogenic particles in the sediments. Other hard parts that are present in lesser amounts include the bones of vertebrates, such as fish, and the exoskeletons of crustaceans, which are composed of apatite and chitin, respectively. Since the latter is organic, it is not well preserved in the sediments. As illustrated in Table 13.5, biogenic particles contain significant amounts of impurities, including magnesium and trace metals. The minerals celesite ($SrSO_4$) and barite ($BaSO_4$) are also deposited in the hard parts of a limited number of marine organisms.

If shells compose more than 30% by mass of the sediments, the deposit is termed an **ooze**. If the shells are predominantly composed of calcium carbonate or silica, the deposit is termed a calcareous or siliceous ooze, respectively. As shown in Table 13.6, if one type or species of marine organism has contributed most of the shell, the deposit is named after that organism (e.g., a radiolarian or globigerina ooze).

The oceanic sediments contain less than 1 percent by mass organic matter. On the continental margins, contributions range from 1 to 10 percent. Though organic detritus accounts for a relatively insignificant fraction of the sediment mass, it is an important reservoir of labile trace metals.

Cosmogenous Sediments

Fragments of extraterrestrial material are constantly entering Earth's atmosphere. Small **cosmogenous** particles are also produced by the fragmentation of meteorites as they pass through the earth's atmosphere. Due to friction generated by collision with atmospheric gases, these particles become very hot and partially melt. They tend to refreeze in spherical or teardrop shapes, as shown in Figure 13.4. Cosmogenous particles that have a diameter less than 500 μm are called **microtektites**. The ones that fall onto the sea surface are transported to the sediments by the process of pelagic sedimentation.

The unique chemical composition of this cosmogenous debris has provided a possible explanation as to why approximately 70 percent of the species of organisms on this planet were driven to extinction over a relatively short time interval approximately 66 million years ago. Evidence for this mass extinction has been observed in marine sediments throughout all the ocean basins. In a contemporaneous layer that was deposited at the end of the Cretaceous Period, the shells of many species of marine plankton abruptly vanished from the

TABLE 13.5
Percentage Composition of Skeletal Material

Substance	*Foraminifera (Orbitolites marginatis)*	*Coral (Oculina diffusa)*	*Calcareous alga (Lithophyllum antillarum)*	*Lobster (Homarus sp.)*	*Phosphatic brachiopod (Discinisca lamellosa)*	*Siliceous sponge (Euplectella speciosa)*
Ca	34.90	38.50	31.00	16.80	26.18	0.16
Mg	2.97	0.11	4.36	1.08	1.45	0.00
CO_3^{2-}	59.70	58.00	62.50	22.40	7.31	0.24
SO_4^{2-}	—	—	0.68	0.52	4.43	0.00
PO_4^{3-}	tr[a]	tr	tr	5.45	34.55	0.00
SiO_2	0.03	0.07	0.04	0.30	0.64	88.56
$(Al,Fe)_2O_3$	0.13	0.05	0.10		0.44	0.32
Organic matter, etc.	2.27	3.27	1.32	53.45	25.00	10.72

Source: From *The Oceans*, H. U. Sverdrup, M. W. Johnson, and R. H. Fleming, copyright © 1941 by Prentice Hall, Inc., Englewood Cliffs, NJ, p. 231. Reprinted by permission.

[a]tr = trace

TABLE 13.6
Marine Sediment Types

Component	*Material*	*Minimum Amount Required (% by weight)*	*Sediment Type*
Lithogenous	Clay	70	Brown clay
	Sand	70–80	Terrigenous sands
	Silt and clay mixture, some sand	70–80	Terrigenous muds
Biogenous	Globigerinids	30[a]	Globigerina ooze
	Pteropods	30[a]	Pteropod ooze
	Coccoliths	30[a]	Coccolith ooze
	Diatoms	30[b]	Diatom ooze
	Radiolarians	30[b]	Radiolarian ooze
Hydrogenous	Authigenic minerals and compounds	—	Authigenic sediments (e.g., manganese nodules)
Cosmogenous	Microtektites	—	Cosmogenous sediments

[a]Calcareous remains.
[b]Siliceous remains.

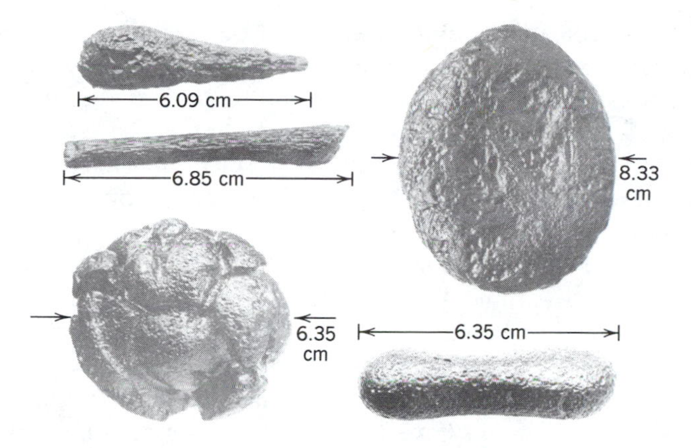

FIGURE 13.4. Microtektites. Courtesy of V. E. Barnes, University of Texas at Austin, Austin, TX.

sedimentary record. This sedimentary layer is also characterized by a large enrichment in the rare element iridium.

Since meteorites are known to have high levels of iridium, some scientists have suggested that the sediment enrichment was produced by the impact of a very large comet, which either disintegrated in Earth's atmosphere or exploded upon impact. This hypothesis assumes a huge amount of airborne debris was produced and distributed globally by the winds. If so, some would eventually have fallen onto the sea surface and settled onto the seafloor. Such a large quantity of atmospheric fallout would also have caused a temporary decrease in insolation, altering the climate and causing a decline in plant growth. Since plants form the base of the food web, all organisms would have been negatively affected.

Volcanic ejecta also contain high levels of iridium. Thus the sediment enrichment could also have been caused by an abrupt and large increase in volcanic activity. Evidence for this is suggested by high levels of volcanic ash, soot, and shocked minerals in the iridium-enriched layer. Other geochemical characteristics of this sediment layer appear to have been caused by acid rain and tsunamis, both of which are byproducts of volcanic activity. A coincident reversal in magnetic field orientations suggests that this sediment was deposited during a period of intense geological activity. This activity could have been triggered by a global-scale catastrophic event, such as the impact of a huge comet.

If Earth's crust was penetrated, great quantities of magma would have been released rapidly and explosively. Though the resulting eruptions would have obliterated all traces of any impact crater, massive flood basalts would have been produced. The geologic record indicates that many flood basalts were produced at the end of the Cretaceous Period. The ages of other flood basalts coincide with periods of lesser mass extinctions. The frequency of these occurrences is approximately 30 million years. This suggests that increases in volcanic activity are caused by periodic comet showers resulting from oscillations in the sun's motion through the galaxy.

Mass extinction accelerate evolution by opening up biological niches on a global scale. Thus, such catastrophes may act as a driving force for evolution that periodically supersedes Darwinian selection. If so, the biogeochemical development of Earth must be greatly influenced by astrophysical processes that operate on a galactic scale.

This bit of paleoceanographic detective work is of more than academic interest. Scientists have used it as an analogy for what might occur globally if a number of nuclear bombs were detonated. Since the explosion of such bombs could ignite fires, the ensuing combustion might produce enough airborne particles to cause a global change in insolation and climate. The resulting "nuclear winter" could have more far-reaching effects than blast damage or radioactive fallout.

Anthropogenic Sediments

Humans have greatly accelerated the rate of introduction of terrestrial particles into the ocean. Most of this material is soil that is being eroded as a result of deforestation and agriculture. In some locations, damming of rivers has halted the input of lithogenous particles to the coastal ocean, thereby creating relict sediments and erosional shorelines. The dumping of human sewage has greatly increased the organic matter content of some coastal sediments and waters, causing periodic episodes of anoxia.

NONPELAGIC SEDIMENTS

Since nonpelagic sediments accumulate at rates exceeding 1 mm/1000 y, they are formed by processes that move a lot of particles quickly. Included are such phenomena as **turbidity currents**, **contour currents**, **volcanic eruptions**, and **ice rafting**. Because nonpelagic sediments are produced by relatively rapid and energetic transport processes, they contain a wide range of particle sizes and hence are classified as **poorly sorted** or **unsorted** deposits. In comparison, pelagic sedimentation produces **well-sorted** deposits; that is, the sediments contain only a few size classes of particles.

Rivers carry lithogenous debris out to the continental shelf. Eventually the piles grow so large that earthquakes and other forms of energy will cause them to slide down the continental slope as a fluidized mixture of sediment and water. This underwater mudslide is termed a turbidity current. Some turbidity currents are erosive enough to gouge out submarine canyons as they flow across the continental margins. The fluidized sediments are eventually deposited on the abyssal plains at the foot of the continental slope. As shown in Figure 13.5, this produces a geomorphological feature called a deep-sea fan.

These deposits are also called turbidites. They are characterized by graded bedding, as illustrated in Figure 13.6. Since the finer grain particles remain

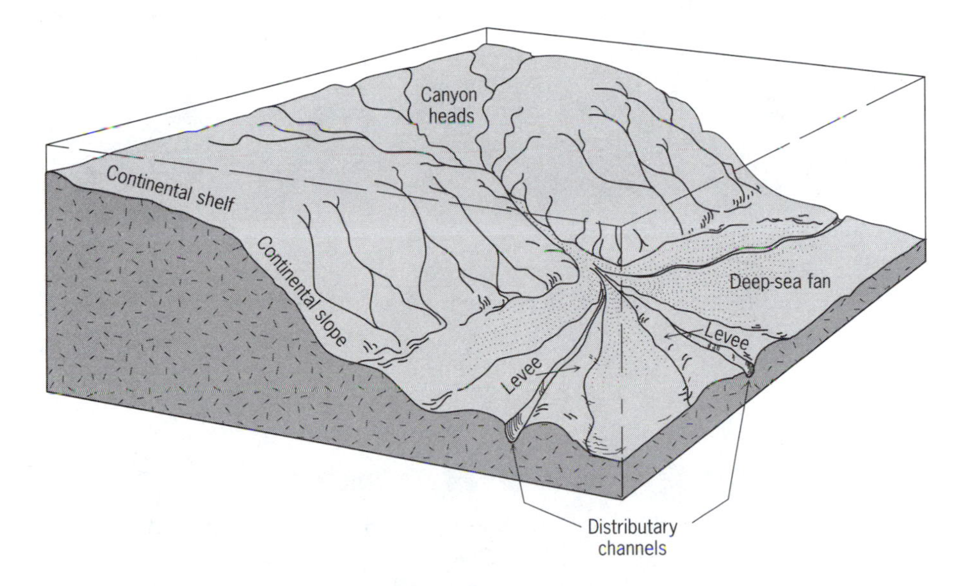

FIGURE 13.5. A submarine canyon and deep-sea fan.

suspended longest, they form the top layer of the deposit. Terrestrial debris, such as wood and leaves, are commonly found in these deposits. Most of the continental rise is composed of turbidites.

Sediments also slide down the slopes of mid-ocean ridges and rises. This phenomenon is called ponding and is responsible for smoothing out the jagged relief of the mid-ocean ridges and rises by filling in crevices and abysses as illustrated in Figure 13.7. Since these sediments contain only clay- and silt-sized particles, the resulting deposits are not graded.

Contour currents flow along the bathymetric contours of many ocean basins. Some are energetic enough to resuspend sediment, such as the one

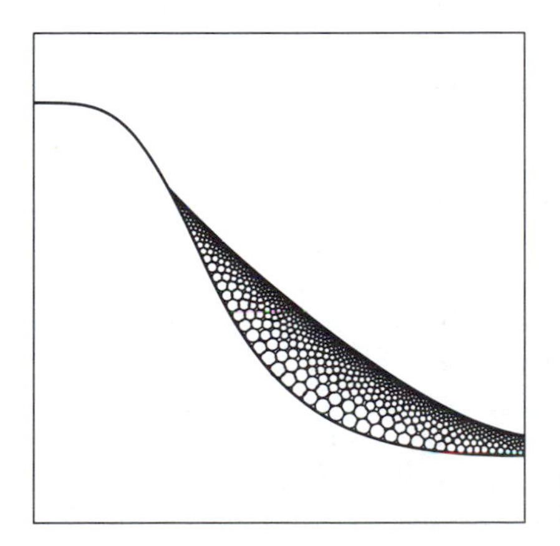

FIGURE 13.6 Graded bedding in turbidites.

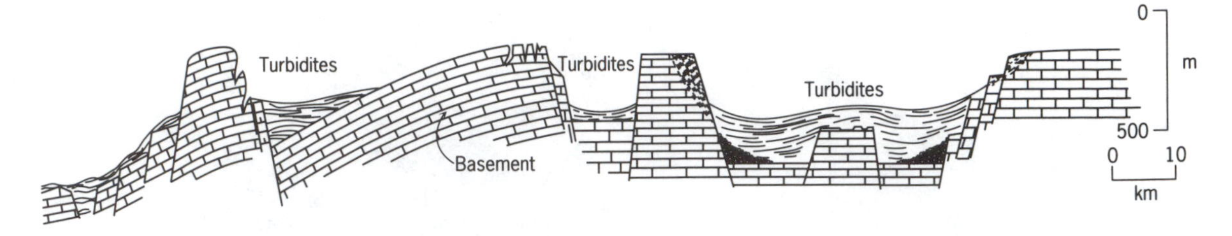

FIGURE 13.7. Sedimentary filling of depressions on mid-ocean ridges.

that flows south over the foot of the continental slope along the northwestern margin of the Atlantic Ocean. As illustrated in Figure 13.8, this contour current maintains a cloud of resuspended particles, called a nepheloid layer, in the northwestern part of the basin. While the cloud fluctuates in particle density and location, it appears to be a permanent feature in the deep zone of the North Atlantic.

Since strong contour currents can resuspend particles, they can cause some sediments to become relict deposits. Where the currents weaken, the sediment settles back down to the seafloor. This produces a patchy distribution of clay and mud along the margins of the ocean basins. Such redistributions of particles create problems for paleoceanographers because relict deposits represent gaps, or unconformities, in the sedimentary record.

Volcanic eruptions on land spew ash and glassy sand-sized fragments into the atmosphere. Ash can be transported long distances before settling onto the sea surface, but the sand-sized fragments are too heavy to be carried far. If the eruption occurs close to the coast, black sand beaches are often produced. These beaches are extremely hot, as black-colored objects absorb all wavelengths of sunlight. Because of the generally abrasive nature of sandy deposits and their high temperatures, black sand beaches are one of the most inhospitable habitats on the planet. Volcanic eruptions beneath the sea surface also eject particles into the ocean, though most basalt is extruded as pillow lavas.

Glaciers are very erosive, so as ice moves downslope to the coastline, it cuts a V-shaped channel in the crustal rock. The eroded material, which includes all particle sizes from clays to boulders, is incorporated into the glacier. When the glaciers flow into the ocean, chunks break off, forming icebergs that are then pushed by the winds and currents into the open ocean. When the icebergs melt, the glacial debris sinks to the seafloor. Since this presently occurs at latitudes greater than 60°, actively forming glacial marine deposits are restricted to high latitudes.

Some glaciers melt before reaching the coastline, leaving a pile of rubble called a terminal moraine. If the sea level rises, these deposits often become

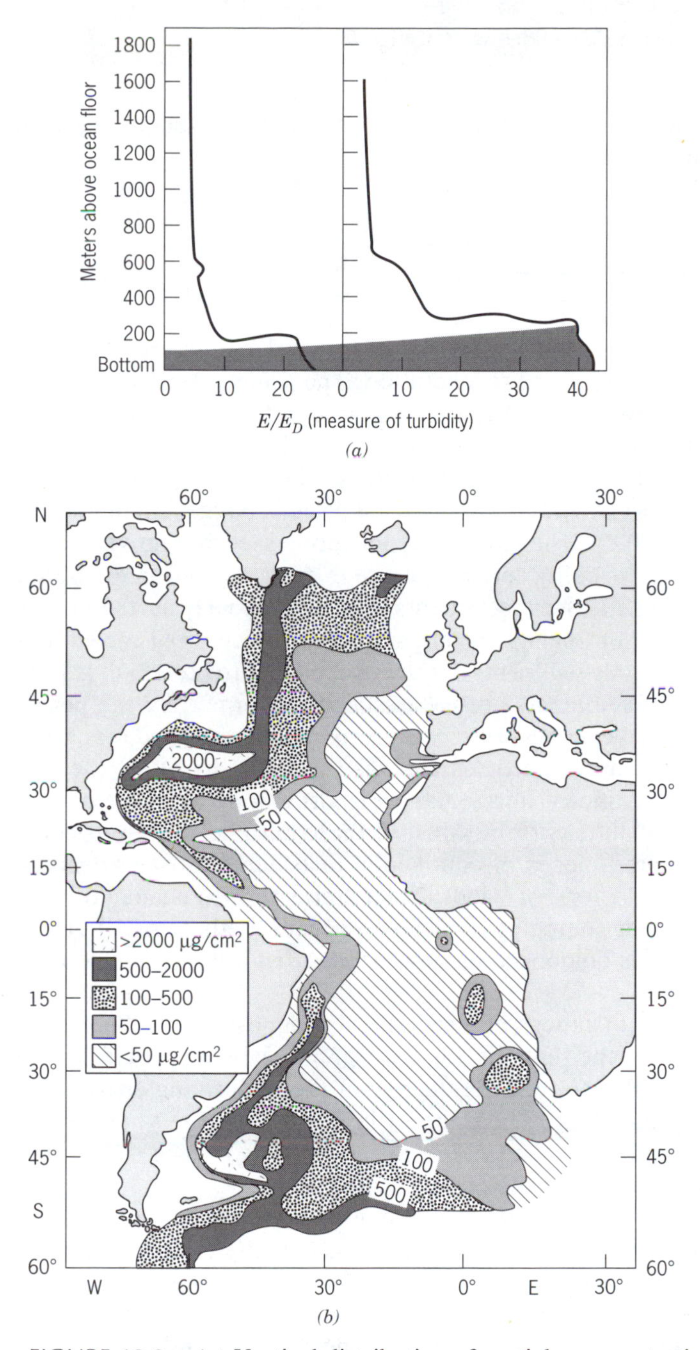

FIGURE 13.8. *(a)* Vertical distribution of particle concentrations at a site in the North Atlantic as measured by water turbidity. *(b)* Horizontal distribution of particle concentrations in the deep-water nepheloid layer of the Atlantic ocean. *Source:* From *Introductory Oceanography,* 5th ed., H. V. Thurman, copyright © 1988 by Merrill Publishing Company, Columbus, OH, p. 60. Reprinted by permission. Data from P. E. Biscaye and S. L. Eittreim, reprinted with permission from *Marine Geology,* vol. 23, pp. 161–162, copyright © 1977 by Elsevier Science Publishers, B.V., Amsterdam, The Netherlands.

islands or peninsulas. Cape Cod and Long Island are examples of terminal moraines.

SUMMARY

The sediments are a very important reservoir in the crustal-ocean factory as the seawater concentrations of many elements are controlled by their relative rates of burial and regeneration. The sediments also contain the only record of past conditions in the ocean. Marine sediments are composed of **unconsolidated** particles that vary greatly in chemical composition, particle size, origin, sedimentation rates, and location.

Most **neritic** deposits are **nonpelagic**, having accumulation rates that exceed 1 cm/1000 y. They are formed by processes that move a lot of particles quickly, such as **turbidity currents**, **contour currents**, **volcanic eruptions**, and **ice rafting**. As a result, these sediments tend to be **poorly sorted** or **unsorted** deposits. In comparison, pelagic sedimentation produces **well-sorted** deposits.

Oceanic sediments overlie oceanic bedrock, which is primarily basalt. Most oceanic sediments are **pelagic** deposits formed by the **pelagic sedimentation** of **clay-sized clay minerals**. Since sinking rates depend on particle size, shape, and density, these particles can take decades to reach the seafloor. During this time, significant horizontal transport can occur if the particle is entrained in a thermohaline current. Most of these particles are **lithogenous**; that is, they are weathered pieces of continental or oceanic crust transported to the ocean by rivers, glaciers, or winds. The latter is termed **aeolian** transport and is the most important source of particles to oceanic sediments. If 70 percent or more of a deposit is composed of clay- or **silt-sized** lithogenous debris, it is termed a **clay** or **mud**.

The thickness of a sedimentary deposit depends on (1) the rate of particle supply to the seafloor, (2) the degree to which particles are preserved following sedimentation, and (3) the age of the underlying crust. The average sediment is 0.5 km thick. The thinnest layers overlie the youngest oceanic crust which is located at the mid-ocean ridges and rises. The thickest layers have accumulated on the continental margins.

Sedimentary particles are also formed by the precipitation of solutes from seawater. If the reaction is abiogenic, the particles are termed **hydrogenous**, or **authigenic**. Some examples of hydrogenous precipitates include evaporites, polymetallic oxyhydroxides, iron-manganese nodules, clay minerals produced by reverse weathering, barite, phosphorite, and glauconite.

The most abundant **biogenous** particles in sediments are hard parts, such as shell and bone, as most organic residues are resolubilized prior to burial. These hard parts are primarily calcareous and siliceous shells deposited by plankton. They contain significant amounts of impurities, including magnesium and trace metals. If shells account for more than 30 percent by mass of the sediments, the deposit is termed an ooze.

Marine sediments contain small amounts of extraterrestrial or **cosmogenous** materials, which are mostly **microtektites** and cosmic dust. Humans have greatly accelerated the rate of introduction of terrestrial particles into the ocean. Most of this material is soil that is being eroded as a result of deforestation and agriculture. The dumping of human sewage has greatly increased the organic matter content of some coastal sediments and waters, causing periodic episodes of anoxia.

CHAPTER 14

CLAY MINERALS

INTRODUCTION

Clay minerals are a ubiquitous component of marine sediments, usually composing the bulk of a deposit. They are products of the chemical weathering of terrestrial rock and authigenic reactions that occur in seawater. These reactions result in the uptake and release of cations and thus have a large impact on the chemical composition of river and seawater. Due to their net negative surface charge, clay minerals adsorb cations and organic matter. These particles eventually settle onto the seafloor, carrying with them a significant amount of adsorbed materials. Thus the clay minerals constitute a very important reservoir in the crustal-ocean factory.

Weathering and authigenic reactions produce a variety of clay minerals. The most abundant marine forms are **illite**, **kaolinite**, **montmorillonite**, and **chlorite**. Their distributions are heterogeneous and reflect the mechanisms responsible for their transport to the ocean and burial in the sediments. The crystal structures, production, transport, and distributions of these clay minerals are discussed below.

THE STRUCTURE OF CLAY MINERALS

Clay minerals are crystalline solids that are typically less than 1 μm in diameter. They are composed primarily of the elements aluminum, oxygen, silicon, and hydrogen, which are arranged in either a tetrahedral or octahedral base unit. As shown in Figure 14.1, the corners of the tetrahedral unit are occupied by oxygen atoms, with a silicon atom in its center. A continuous sheet or layer is produced when adjacent tetrahedrons share the oxygen atoms that make up their bases. In the octahedral unit, hydroxyl groups occupy the corners and an aluminum atom is located at its center. The sharing of hydroxide groups between adjacent octahedrons also creates a continuous layer.

The clay minerals are composed of these sheets and hence are also termed layered aluminosilicates. Those that contain alternating sheets of octahedrons and tetrahedrons are called two-layer clays. The most abundant example of this group of clay minerals are the kaolinites, whose three-dimensional structure is illustrated in Figure 14.2.

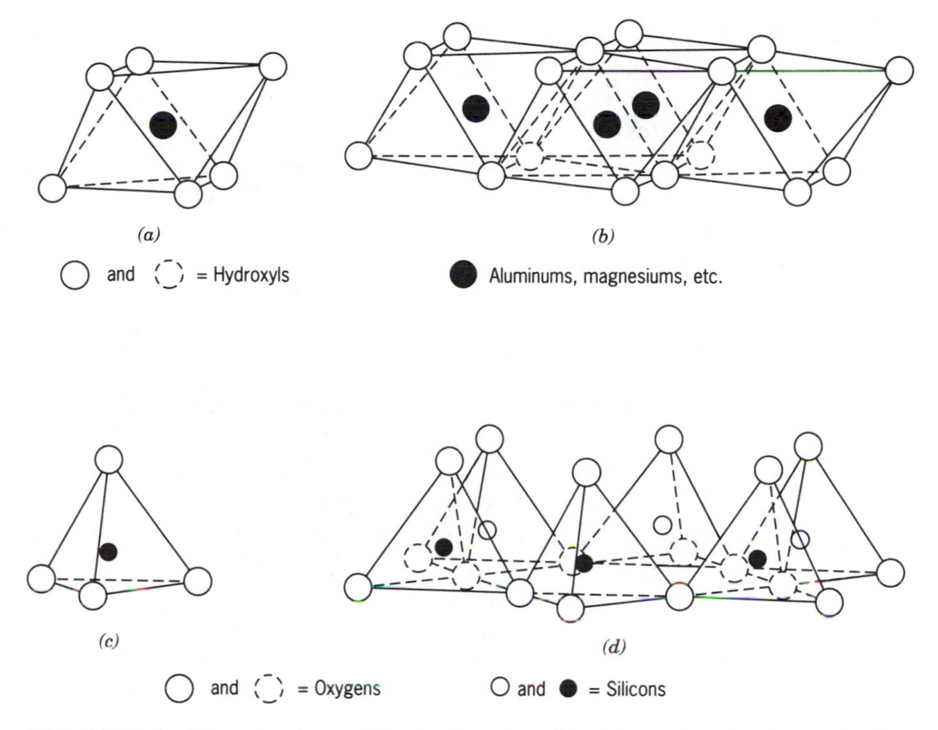

FIGURE 14.1. The structure of the basic units of the clay minerals: *(a)* octahedron, *(b)* octahedral layer, *(c)* tetrahedron, and *(d)* tetrahedral layer. *Source:* From *Clay Mineralogy*, 2nd ed., R. E. Grim, copyright © 1968 by McGraw–Hill Publishing Company, New York, p. 52. Reprinted by permission.

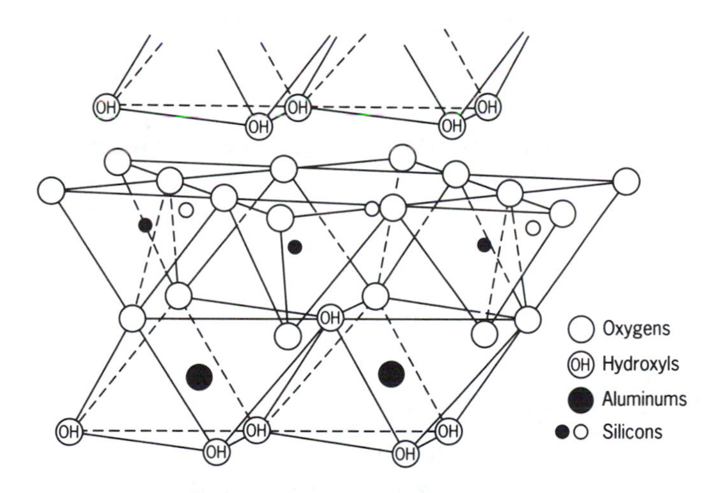

FIGURE 14.2. Crystal structure of kaolinite. *Source:* From *Clay Mineralogy*, 2nd ed., R. E. Grim, copyright © 1968 by McGraw–Hill Publishing Company, New York, p. 58. Reprinted by permission.

The repeating unit in a three-layered clay is composed of an octahedral layer sandwiched between two tetrahedral layers, as shown schematically in Figure 14.3. The most common of these are the montmorillonites, illites, and chlorites.

Due to the electronegativity of the oxygen atoms, the exterior surfaces of all the repeating units possess a small negative charge. This causes clay minerals to electrostatically attract cationic solutes. This process is termed adsorption. If the surface site is already occupied, **cation exchange** can occur as follows:

$$A^+(aq) + \text{B-clay mineral} \rightleftharpoons \text{A-clay mineral} + B^+(aq) \qquad (14.1)$$

in which cations A and B compete for the negatively charged surface sites on the clay mineral. The speciation at equilibrium is determined by the relative ion concentrations, the ability of a particular cation to compete for an adsorption site, and the **cation exchange capacity (CEC)** of the clay mineral. The latter is a measure of the amount of cations adsorbed at equilibrium. The CEC is commonly reported as the milliequivalents of positive charge which are adsorbed by 100 g of clay mineral.

The CEC of a clay mineral is directly related to its crystalline structure, as this determines the density of negative charge on its surfaces. Large charge imbalances, and hence CECs, are commonly caused by the replacement of some of the aluminum and silicon atoms by magnesium and trace metals. Some of the silicon is also replaced by aluminum. Since cations similar in ionic radius and charge are most likely to replace the aluminum and silicon, this process is termed **isomorphic substitution**.

Most isomorphic substitution is the result of chemical **weathering**. Since the conditions under which chemical weathering occurs are temporarily and spatially variable, the resulting clay minerals differ greatly in their degrees of isomorphic substitution. As shown in Figure 14.3, parentheses are used in the empirical formulae of clay minerals to indicate which elements are present in variable amounts. The CEC of the three-layer clays is higher than that of the kaolinites due to a higher degree of isomorphic substitution.

The montmorillonites are called expanding clays because their volume increases when they are placed in polar solvents such as water. This swelling is caused by the adsorption of water molecules between the repeating units. This process is reversible; the removal of the water molecules causes these clays to contract. Exchangeable cations are also adsorbed at these interlayer sites. The high density of surface charge on the illites constricts this interlayer space so that water and other large cations can not enter. Due to the strong charges, relatively small cations, such as potassium, are adsorbed but not readily exchanged. Thus the illites do not have as high a CEC as the montmorillonites.

Chlorite has a relatively low cation exchange capacity because it contains a layer of brucite, $Mg(OH)_2$, between its repeating units as illustrated in Figure 14.4. This restricts adsorption of interlayer cations.

Type of Clay	Mineral	Structure	Composition (idealized)	Cation exchange capacity (meq/100g)	Remarks
Two layer	Kaolinite	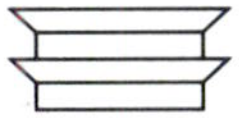	$Al_2Si_2O_5(OH)_4$	1–10	Little isomorphous substitution Small cation exchange capacities Nonexpanding
Three layer	1. Expanding (smectltes or montmorillonites)	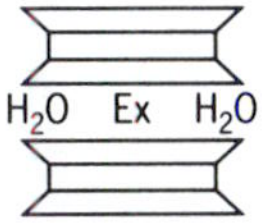 H_2O Ex H_2O	$Ex_x[Al_{2-x}Mg_x]<Si_4>O_{10}(OH)_2$	80–140	Substitution of a small amount of Al for Si in Td-sheet and of Mg, Fe, Cr, Zn, Li for Al or Mg in Oh-sheet Large CEC (Ex = Na^+, K^+, Li^+, Ca^{2+}, ...) Swell in water or polar organic compounds
	(vermiculites)	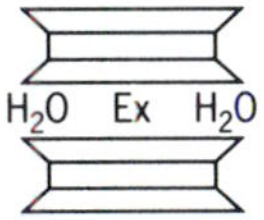 H_2O Ex H_2O	$Ex_x[Mg_3]<Al_xSi_{4-x}>O_{10}(OH)_2$	100–180	
	2. Nonexpanding (illites)	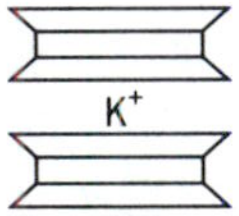 K^+	$K_{1-x}[Al_2]<Al_{1-x}Si_{3+x}>O_{10}(OH)_2$	10–140	About $\frac{1}{4}$ of Si in Td-sheet replaced by Al, similar Oh-sheet substitutions Small CEC Ex = K^+
Chlorites	Chlorite	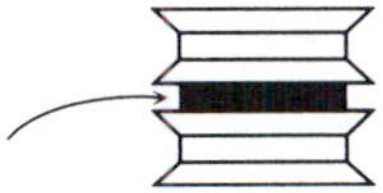 Brucite	$[Mg, Al]_3(OH)_6[Mg, Al]_3<Si, Al>_4O_{10}(OH)_2$	5–30	Three-layer alternating with brucite Brucite layer positively charged [some Al(III) replacing M(II)], partially balances negative charge on Td-Oh-Td (mica) layer Low CEC, nonswelling

Tetrahedral layer < >
Octahedral layer [ ]

Ex = exchangeable cations

FIGURE 14.3. The principal species of clay minerals. *Source:* After *Aquatic Chemistry*, W. S. Stumm and J. J. Morgan, copyright © 1981 by John Wiley & Sons, Inc., New York, p. 442; and *Clay Mineralogy*, 2nd ed., R. E. Grim, copyright © 1968 by McGraw–Hill Publishing Company, New York, p. 159. Reprinted by permission.

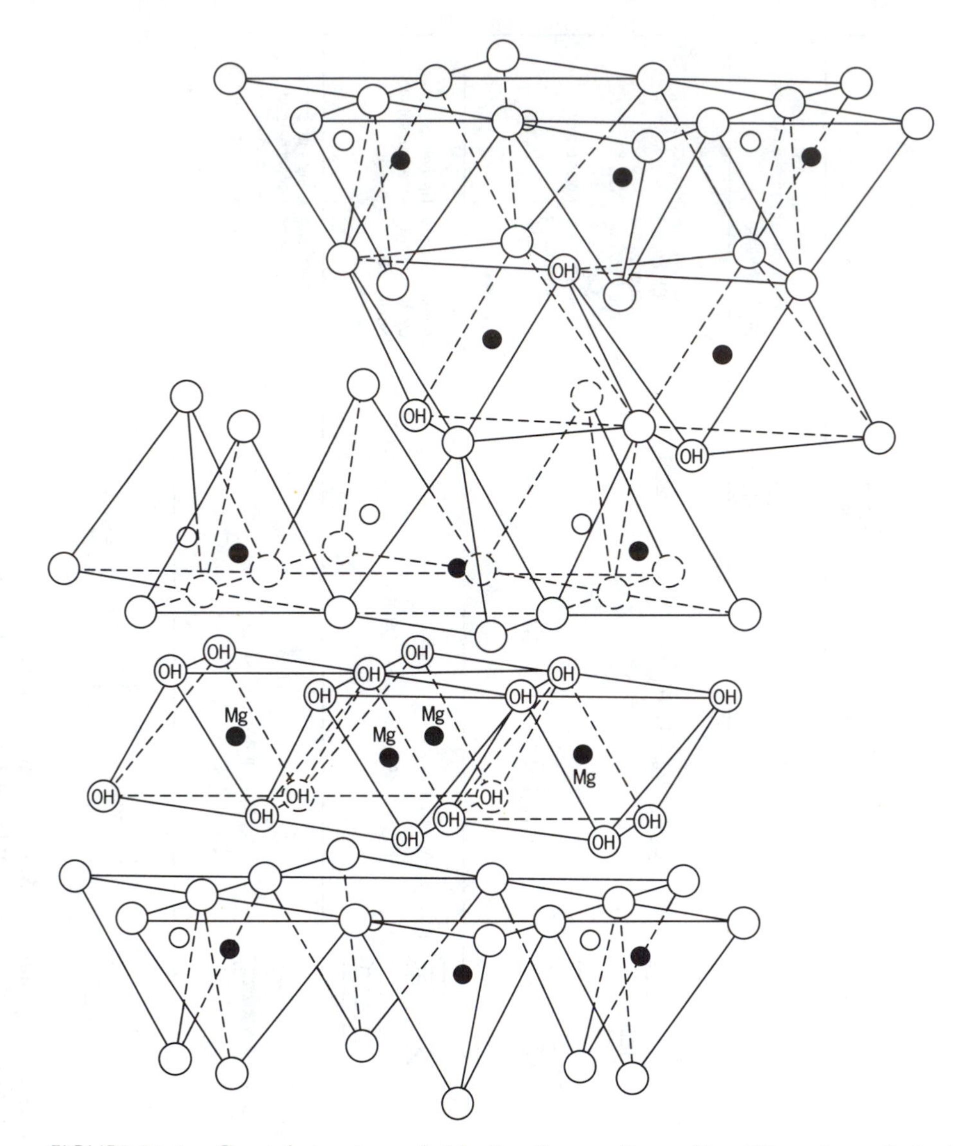

FIGURE 14.4. Crystal structure of chlorite. *Source:* From *Clay Mineralogy,* 2nd ed., R. E. Grim, copyright © 1968 by McGraw–Hill Publishing Company, New York, p. 100. Reprinted by permission.

THE PRODUCTION OF CLAY MINERALS FROM TERRESTRIAL WEATHERING

Chemical and physical processes that occur on land convert granite into a wide variety of clay minerals, most of which are sand- and clay-sized particles. These processes are collectively referred to as terrestrial weathering. As shown in Figure 14.5, weathering is initiated by physical processes that break the

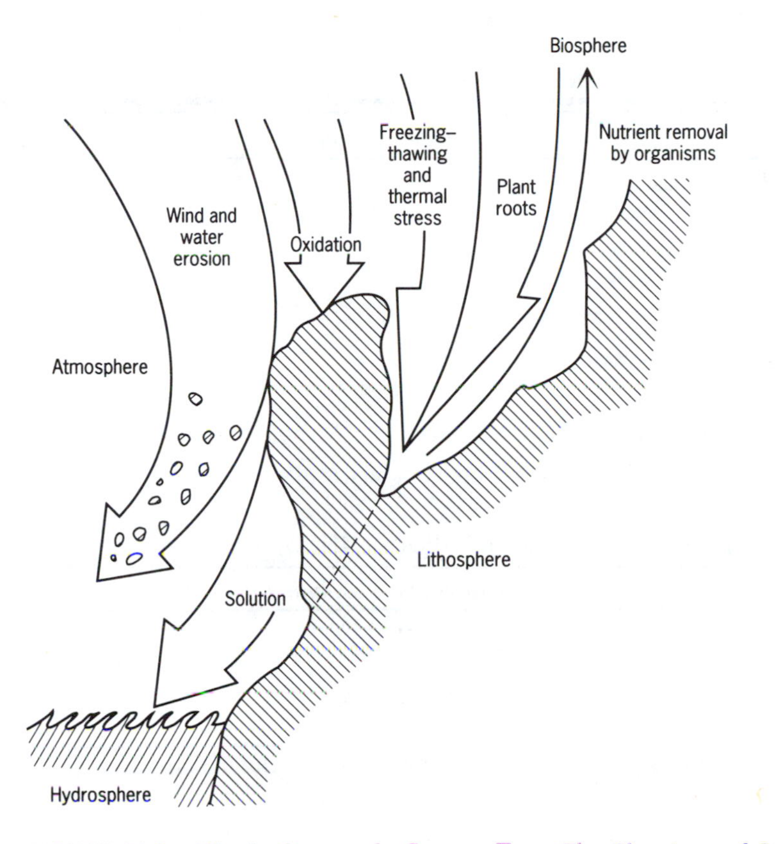

FIGURE 14.5. Weathering attack. *Source:* From *The Chemistry of Our Environment,* R. A. Horne, copyright © 1978 by John Wiley & Sons, Inc., New York, p. 534. Reprinted by permission.

parent rock into small fragments, thereby increasing the surface area that can undergo chemical weathering. Water and organisms cause much of the physical weathering of terrestrial rocks. For example, the expansion of water when it freezes in cracks and crevices fractures rock. Plants also cause rock fragmentation when their roots grow down and expand in cracks and crevices.

The newly exposed surfaces undergo chemical weathering during which bonds are broken, as well as formed. Weathering reactions can yield soluble products (congruent dissolution reactions) or a mixture of solid and soluble products (incongruent dissolution reactions). Both cations and anions are produced during chemical weathering. Solid products include clay minerals and other less-altered residues. Some examples of chemical weathering reactions are given in Table 14.1.

Chemical weathering is initiated by hydration of the parent rock's surface, as illustrated in Figure 14.6. The extent to which a rock is hydrated depends on its surface area-to-volume ratio, as well as the interatomic spacing and charge distribution in the crystal lattice.

TABLE 14.1
Examples of Typical Weathering Reactions

I. Congruent dissolution reactions

$$SiO_2(s) + 2H_2O = H_4SiO_4$$
quartz

$$CaCO_3(s) + H_2O = Ca^{2+} + HCO_3^- + OH^-$$
calcite

$$CaCO_3(s) + H_2CO_3^* = Ca^{2+} + 2HCO_3^-$$

$$Al_2O_3{\cdot}3H_2O(s) + 2H_2O = 2Al(OH)_4^- + 2H^+$$
gibbsite

$$Al_2O_3{\cdot}3H_2O(s) + 6H_2CO_3^* = 2Al^{3+} + 6HCO_3^- + 6H_2O$$

$$Ca_5(PO_4)_3(OH)(s) + 3H_2O = 5Ca^{2+} + 3HPO_4^{2-} + 4OH^-$$
apatite

$$Ca_5(PO_4)_3(OH)(s) + 4H_2CO_3^* = 5Ca^{2+} + 3HPO_4^{2-} + 4HCO_3^- + H_2O$$

II. Incongruent dissolution reactions

$$MgCO_3(s) + 2H_2O = HCO_3^- + Mg(OH)_2(s) + H^+$$
magnesite — brucite

$$Al_2Si_2O_5(OH)_4(s) + 5H_2O = 2H_4SiO_4 + Al_2O_3{\cdot}3H_2O(s)$$
kaolinite — gibbsite

$$NaAlSi_3O_8(s) + \tfrac{11}{2}H_2O = Na^+ + OH^- + 2H_4SiO_4 + \tfrac{1}{2}Al_2Si_2O_5(OH)_4(s)$$
albite — kaolinite

$$NaAlSi_3O_8(s) + H_2CO_3^* + \tfrac{9}{2}H_2O = Na^+ + HCO_3^- + 2H_4SiO_4 + \tfrac{1}{2}Al_2Si_2O_5(OH)_4(s)$$

$$CaAl_2Si_2O_8(s) + 3H_2O = Ca^{2+} + 2OH^- + Al_2Si_2O_5(OH)_4(s)$$
anorthite — kaolinite

$$CaAl_2Si_2O_8(s) + 2H_2CO_3^* + H_2O = Ca^{2+} + 2HCO_3^- + Al_2Si_2O_5(OH)_4(s)$$

$$4Na_{0.5}Ca_{0.5}Al_{1.5}Si_{2.5}O_8 + 6H_2CO_3^* + 11H_2O = 2Na^+ + 2Ca^{2+} + 4H_4SiO_4 + 6HCO_3^- + 3Al_2Si_2O_5(OH)_4(s)$$
plagioclase (andesine) — kaolinite

$$3KAlSi_3O_8(s) + 2H_2CO_3^* + 12H_2O = 2K^+ + 2HCO_3^- + 6H_4SiO_4 + KAl_3Si_3O_{10}(OH)_2(s)$$
K-feldspar (orthoclase) — mica

$$7NaAlSi_3O_8(s) + 6H^+ + 20H_2O = 6Na^+ + 10H_4SiO_4 + 3Na_{0.33}Al_{2.33}Si_{3.67}O_{10}(OH)_2(s)$$
albite — Na^+-montmorillonite

$$KMg_3AlSi_3O_{10}(OH)_2(s) + 7H_2CO_3^* + \tfrac{1}{2}H_2O = K^+ + 3Mg^{2+} + 7HCO_3^- + 2H_4SiO_4 + \tfrac{1}{2}Al_2Si_2O_5(OH)_4(s)$$
biotite — kaolinite

$$Ca_5(PO_4)_3F(s) + H_2O = Ca_5(PO_4)_3(OH)(s) + F^- + H^+$$
fluoroapatite — hydroxyapatite

$$KAlSi_3O_8(s) + Na^+ = K^+ + NaAlSi_3O_8(s)$$
orthoclase — albite

$$CaMg(CO_3)_2(s) + Ca^{2+} = Mg^{2+} + 2CaCO_3(s)$$
dolomite — calcite

(Continued on next page.)

TABLE 14.1 (Cont.)
Examples of Typical Weathering Reactions

III. Redox reactions
$MnS(s) + 4H_2O = Mn^{2+} + SO_4^{2-} + 8H^+ + 8e^-$
$3Fe_2O_3(s) + H_2O + 2e^- = 2Fe_3O_4(s) + 2OH^-$ hematite magnetite
$FeS_2(s) + 3\frac{3}{4}O_2 + 3\frac{1}{2}H_2O = Fe(OH)_3(s) + 4H^+ + 2SO_4^{2-}$ pyrite
$PbS(s) + 4Mn_3O_4(s) + 12H_2O = Pb^{2+} + SO_4^{2-} + 12Mn^{2+} + 24OH^-$ galena

Source: From *Aquatic Chemistry*, W. S. Stumm and J. J. Morgan, copyright © 1970 by John Wiley & Sons, Inc., New York, pp. 390–391. Reprinted by permission.
$^*CO_2$ (ag) + H_2CO_3 (ag)

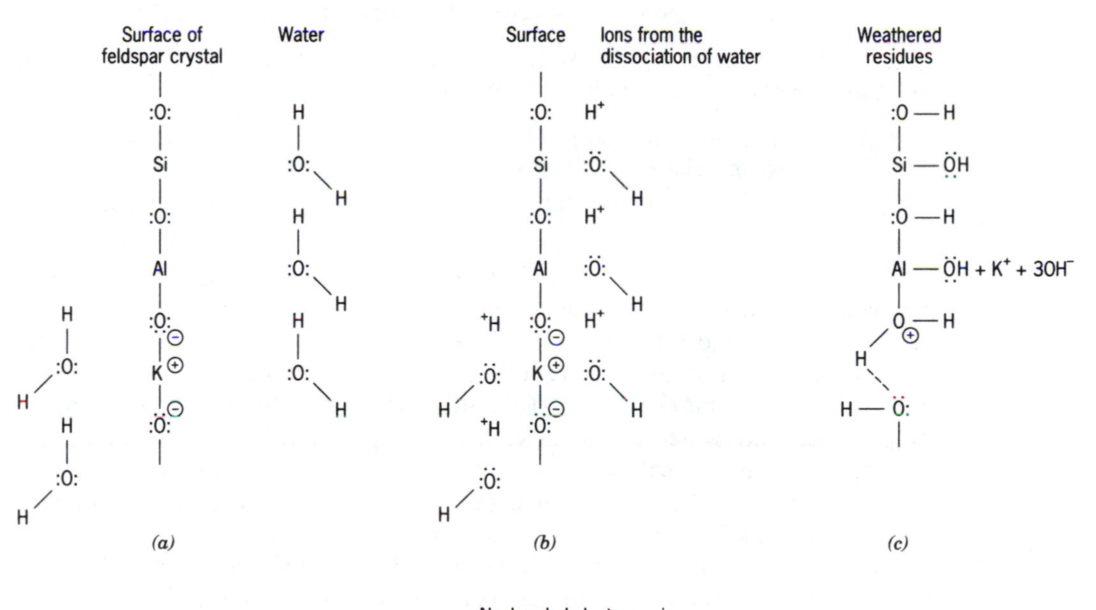

FIGURE 14.6. Decomposition of feldspar lattice by water: *(a)* hydration, *(b)* hydrolysis, and *(c)* weathered residues. Note that the Al—O and Al—Si bonds are mixtures of pure covalent and ionic forms. The bond between K and O is also a hybrid, but is much weaker than the other types of O bonds. *Source:* After *Principles of Chemical Sedimentology,* G. M. Friedman and J. E. Sanders, copyright © 1978 by John Wiley & Sons, Inc., New York, p. 138. Reprinted by permission.

The second stage of chemical weathering involves a type of hydrolysis. During this reaction, hydrogen and hydroxide ions derived from water bond with the crystal lattice. As shown in Figure 14.6*b*, this causes the breakage of some of the cation-oxygen bonds. The susceptibility of a rock to hydrolysis is determined by the degree to which it can become hydrated and the relative strength of the cation-oxygen bond. The latter is a function of the relative charge density* of the cation.

The breakage of these bonds proceeds more rapidly and to a greater extent in the presence of a stronger acid, such as carbonic acid. As shown in Eq. 5.42, H_2CO_3 is a product of the hydrolysis of $CO_2(g)$, so this acid is present in waters that are in gaseous equilibrium with the atmospheric. Pure water in equilibrium with atmospheric levels of $CO_2(g)$ has a pH of 5.6. Thus rainwater is naturally acidic, though its exact pH is determined by the dissolution of variable amounts of other acids and bases, such as H_2SO_4, HNO_3, and NH_3.

Carbonic acid reacts with the parent or primary rock to solubilize cations as follows:

$$\text{Igneous rock(s)} + H_2CO_3(aq) \rightarrow HCO_3^-(aq) + H_4SiO_4(aq) + \text{Cations(aq)} + \text{Clay minerals(s)} \quad (14.2)$$

A common example of this is the weathering of plagioclase feldspar, an abundant component of granite, to kaolinite:

$$\underset{\text{(plagioclase)}}{4Na_{0.5}Ca_{0.5}Al_{1.5}Si_{2.5}O_8(s)} + 6H_2CO_3(aq) + 11H_2O(l) \rightarrow 2Na^+(aq) + 2Ca^{2+}(aq) + 4H_4SiO_4(aq) + 6HCO_3^-(aq) + \underset{\text{(kaolinite)}}{3Al_2Si_2O_5(OH)_4(s)} \quad (14.3)$$

Due to its relative reactivity and abundance, plagioclase feldspar is the most important parent rock for all clay minerals except montmorillonite.

The amount and type of chemical weathering that occurs depends on the chemical composition of the parent rock and the weathering solution, as well as the environmental conditions. Climate is important because it determines temperature and water availability, both of which control the amount of vegetation. Plant growth increases the rate and extent of chemical and mechanical weathering because the decomposition of organic detritus produces CO_2, which enhances the dissolution and fragmentation of rocks. Topography is another important environmental control, as it determines water drainage (i.e., the length of time that water is in contact with the rock). This also affects any weathering reactions that involve redox chemistry, as the relative stagnancy of the water influences its O_2 content.

The weathering of granite under temperate conditions yields a suite of clay minerals that include, in order of decreasing abundance: illite, vermiculite, kaolinite, and chlorite. As illustrated in Figure 14.3, vermiculite is one of the

---

*The charge density of a cation is given by the ratio of its ionic charge (z) to its ionic radius (r), (i.e., z/r).

expanding three-layer clays. If weathering continues to completion, all atoms other than silicon and oxygen will be removed, leaving a solid composed entirely of silica (SiO_2). This mineral is called quartz.

Under tropical conditions, the chemical weathering of granite yields kaolinite. If the kaolinite remains on land, continued weathering will remove all of its cations, leaving gibbsite, a hydrated aluminum oxide ($Al_2O_3 \cdot 3H_2O$), as well as quartz. The intense weathering at these latitudes causes the oxidation and precipitation of iron, which is leached from the primary rock. This produces iron oxides, such as goethite (FeOOH), which are recognizable by their red–orange coloration. Since chlorite is a low-temperature weathering product, it is the dominant clay mineral only at high latitudes.

THE PRODUCTION OF CLAY MINERALS FROM AUTHIGENIC PROCESSES

After delivery to the ocean by rivers, clay minerals react with seawater. The processes that alter the chemical composition of the terrestrial clay minerals during the first few months of exposure are termed **halmyrolysis**. These include cation exchange, fixation of ions into inaccessible sites, and some isomorphic substitutions.

Most cation exchange occurs in estuaries and the coastal ocean due to the large difference in cation concentrations between river and seawater. As riverborne clay minerals enter seawater, exchangeable potassium and calcium are displaced by sodium and magnesium because the Na^+/K^+ and Mg^{2+}/Ca^{2+} ratios are higher in seawater than in river water. Trace metals are similarly displaced.

Mass balance calculations suggest that some of the major cations are removed from seawater via reactions that involve their incorporation into the crystal lattice of clay minerals. Because bicarbonate is consumed, these reactions are referred to as **reverse weathering**. An example of this process is given in Eq. 14.4:

$$2K^+(aq) + 2HCO_3^-(aq) + \underset{\text{(kaolinite)}}{3Al_2Si_2O_5(OH)_4(s)} \rightarrow \underset{\text{(illite)}}{2KAl_3Si_3O_{10}(OH)_2(s)} + 5H_2O(l) + 2CO_2(g) \quad (14.4)$$

These types of reactions do not reconstitute the original igneous silicate. Instead, a secondary clay mineral is formed. Evidence supporting the occurrence of reverse weathering has proven difficult to obtain for two reasons. First, the clay minerals produced by this process are also transported to the ocean as part of the suspended load in river runoff. Mass balance calculations suggest that only a small percentage of the sedimentary clay minerals would have to be altered by reverse weathering to account for the necessary cation removal. Because of their low concentrations and nonunique structures, the presence of any secondary clays will probably be impossible to detect. Second, the rate of reverse weathering is so slow that laboratory studies of this process are difficult to conduct.

Sediments buried for millions of years contain only illite and chlorite. These are the clay minerals predicted to result from reverse weathering. This suggests that catagenesis, rather than reverse weathering, is more likely to be the missing major cation sink. Alternatively, the chemical composition of seawater could have changed over time, such that reverse weathering reactions presently proceed at much slower rates than in the past.

Some **authigenic** clay minerals are produced by the reaction of seawater with fresh volcanic glass. This commonly occurs near mid-ocean ridges and rises or where lava from coastal volcanoes flows into the sea. Clay minerals produced by this process are primarily montmorillonite and the zeolites—phillipsite and clinoptilite.

THE TRANSPORT PATHWAYS OF CLAY MINERALS

Rivers and Oceanic Currents

River transport is responsible for most of the terrestrial input of clay minerals to the coastal ocean. The largest rivers are located primarily in the Northern Hemisphere. Many discharge into marginal seas, so much of their sediment load is deposited there. The riverine sediments deposited in the mouth of the Mississippi River form the world's largest delta.

Rivers that empty directly into the ocean deposit much of their sediment load in the zone where seawater mixes with freshwater. This is caused by a decrease in water velocity, as well as by chemical reactions that promote the precipitation and sedimentation of many materials. As a result, most terrigenous clay minerals are deposited on the continental shelf, especially the larger size fractions. Since the clay-sized fraction can remain suspended for years, some of these particles are advected out into the open ocean, where they eventually settle to the seafloor by the process of pelagic sedimentation.

Turbidity and contour currents resuspend and transport the sediments that lie on the continental margin. These sediments are redeposited when the currents weaken. In the case of the turbidity currents, this redistribution usually occurs along the foot of the continental slope and is largely responsible for the accumulation of sediments in the continental rise. The resuspension of particles by contour currents can also maintain nepheloid layers.

Winds

Winds are responsible for the transport of 25 to 75 percent of the clay minerals deposited in pelagic oceanic sediments. Most aeolian transport is caused by the major wind bands whose directions and locations are illustrated in Figure 14.7.

The winds pick up the clay minerals and carry them until the air currents weaken. The particles that fall onto the sea surface are transported to the seafloor by pelagic sedimentation. Winds that pass over deserts transport the largest amounts of clay minerals. For example, the trade winds that blow across the Sahara Desert are responsible for the high concentrations of kaolinite in the sediments of the subtropical South Atlantic.

Ice

The transport of clay minerals by ice is important only in high latitudes and thus has a major impact on the distribution of chlorite. Icebergs raft glacial debris from the land out into the sea. Since icebergs rarely travel to latitudes lower than 60°, chlorite is most abundant in the sediments located at high latitudes.

Organisms

Marine organisms affect the chemistry of clay minerals in two ways. First, filter feeders incorporate clay minerals into their fecal pellets. Since these pellets sink faster, they accelerate the rate at which clay minerals are transported to the seafloor. Second, clay minerals, like all particles in seawater, are coated with a layer of organic matter, as illustrated in Figure 11.2. This adsorbed material is DOM excreted and exuded by marine organisms, as well as produced by the partial remineralization of POM. Since trace metals are adsorbed to the organic matter and to the exchange sites on the clay minerals, pelagic sedimentation represents an important mechanism by which DOM and cations are removed from seawater.

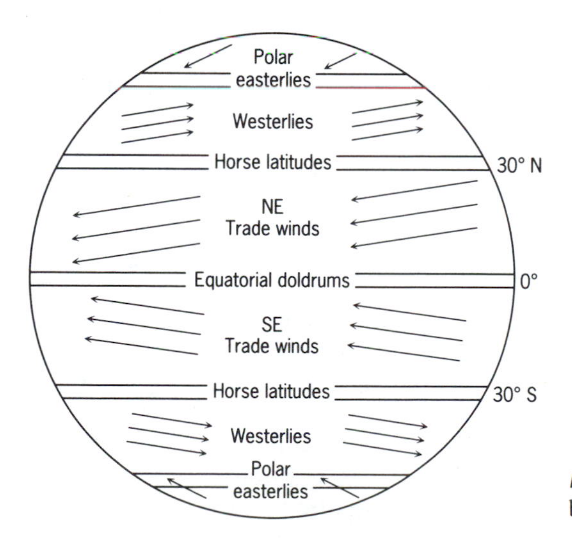

FIGURE 14.7. The major wind bands.

GLOBAL PATTERNS OF CLAY MINERAL DISTRIBUTIONS

The sediments with the highest clay mineral contents are located on the continental shelves and abyssal plains of the North Pacific Ocean. The global distribution pattern of the individual clay minerals reflects the location of their sources, as well as the mechanisms responsible for their transport. As shown in Figure 14.8, kaolinite concentrations are highest in tropical and equatorial latitudes, particularly off the western coasts of North Africa and Australia and the northeastern coast of South America. The first two are the result of aeolian transport by the Trade Winds from the Saharan and Australian deserts, respectively. The South American deposit is the result of river transport from the Amazon.

As shown in Figure 14.9, the largest concentrations of chlorite are located at high latitudes, reflecting the site of its production, as well as the limit of its transport by icebergs.

Montmorillonite and zeolite concentrations in marine sediments reflect proximity to subaerial volcanism, as these aluminosilicates are produced by the chemical alteration of basalt. As shown in Figure 14.10, montmorillonite concentrations are highest in the South Pacific due to the influence of the East Pacific Rise and in the Indian Ocean due to the 90° East Ridge. Since the Atlantic Ocean has a smaller surface area and no trenches, the rate of sedimentation of terrigenous clay minerals is very high and dilutes the contribution of all other sediment types. As a result, the sediments around the Mid-Atlantic Ridge do not contain high percentages of montmorillonite.

Illite is a result of weathering at all latitudes. As shown in Figure 14.11, it is the dominant component of the sediments except where diluted by local enrichments of other particles. For example, the low proportions in the equatorial and South Pacific are caused by the relatively large contribution of montmorillonite produced by hydrothermal activity at the East Pacific Rise.

SUMMARY

Clay minerals are crystalline, layered aluminosilicates. These clay- and sand-sized grains are usually the most abundant particle type in marine sediments. The majority are produced during the **weathering** of granite on land. This process is promoted by acid hydrolysis, the acid being supplied by the dissolution of atmospheric CO_2 in rainwater. The amount and type of products generated by terrestrial weathering are determined by local environmental conditions, as well as by the chemical composition of the parent rock and the weathering solution.

In general, **chlorite**, a three-layered clay, is the dominant clay mineral produced by low-temperature weathering at high latitudes. **Kaolinite**, a two-layered clay, dominates at low latitudes, whereas **illite**, another three-layered

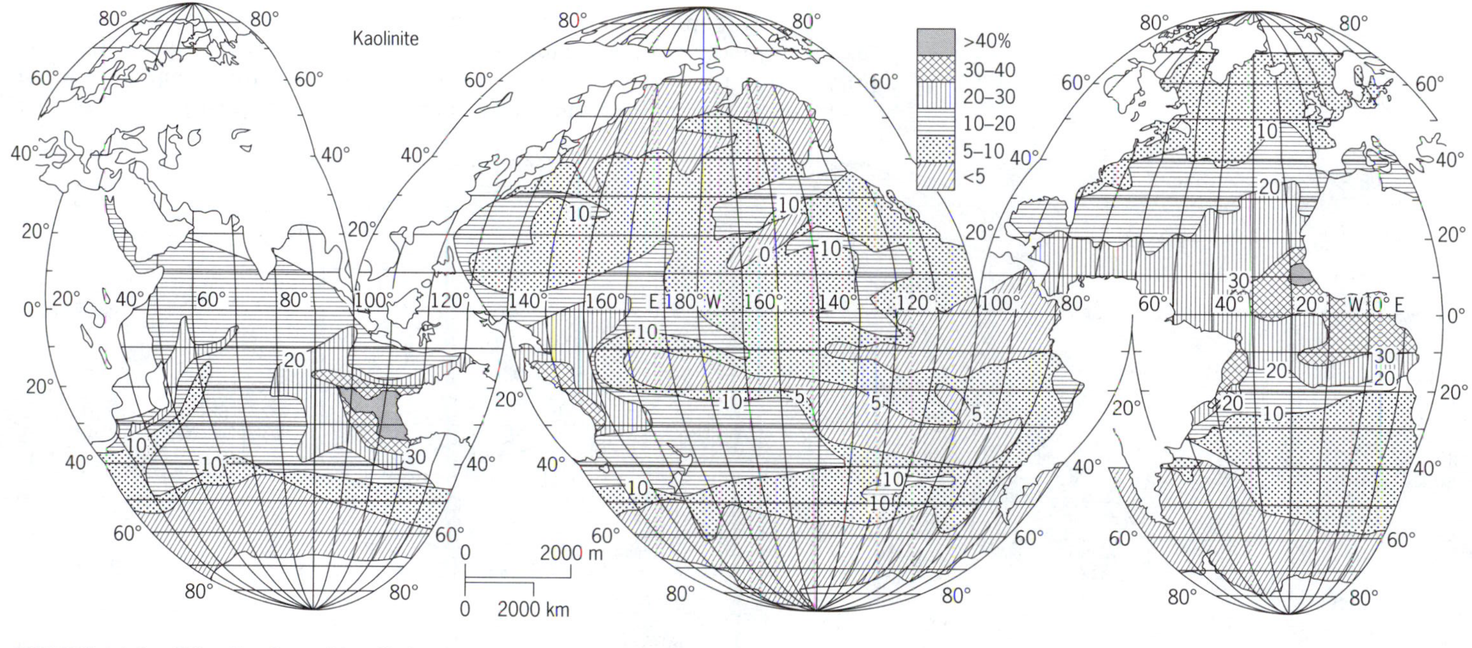

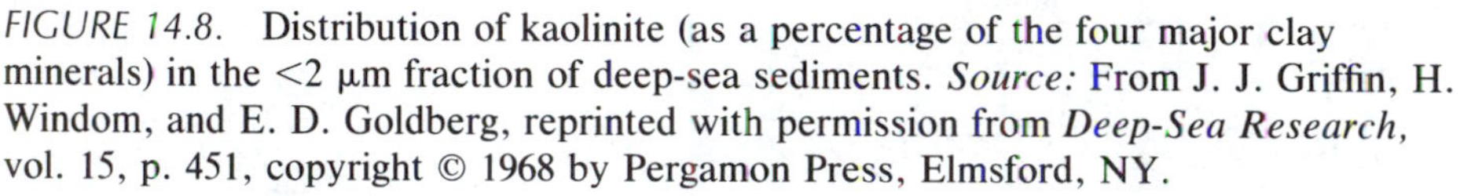
FIGURE 14.8. Distribution of kaolinite (as a percentage of the four major clay minerals) in the <2 μm fraction of deep-sea sediments. *Source:* From J. J. Griffin, H. Windom, and E. D. Goldberg, reprinted with permission from *Deep-Sea Research,* vol. 15, p. 451, copyright © 1968 by Pergamon Press, Elmsford, NY.

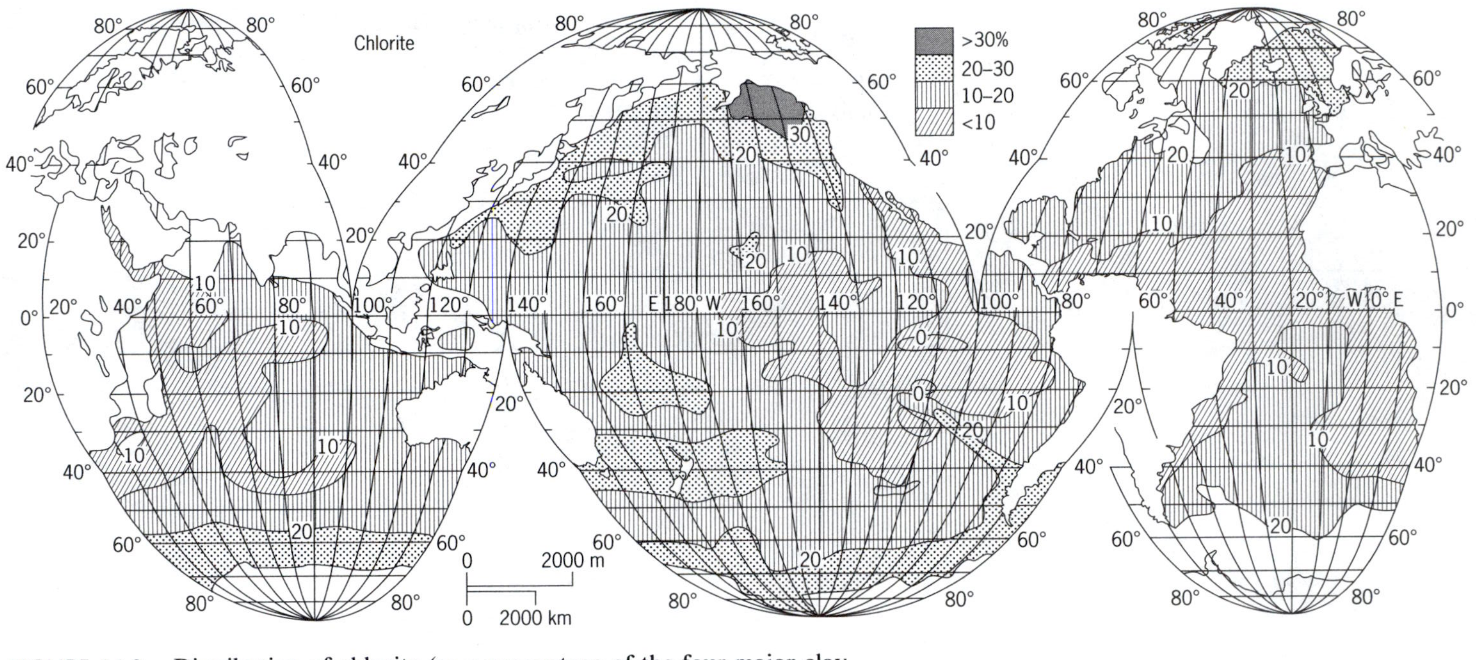

FIGURE 14.9. Distribution of chlorite (as a percentage of the four major clay minerals) in the <2 μm fraction of deep-sea sediments. *Source:* From J. J. Griffin, H. Windom, and E. D. Goldberg, reprinted with permission from *Deep-Sea Research,* vol. 15, p. 436, copyright © 1968 by Pergamon Press, Elmsford, NY.

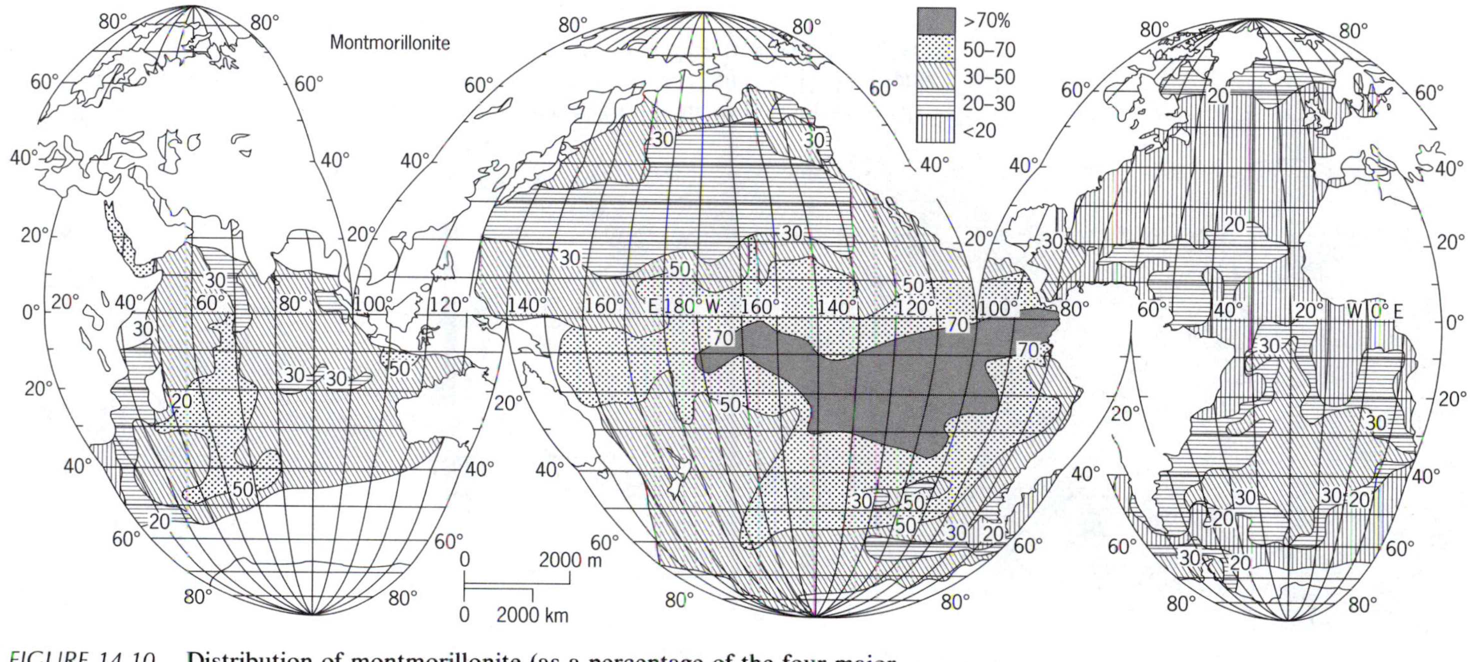

FIGURE 14.10. Distribution of montmorillonite (as a percentage of the four major clay minerals) in the <2 μm fraction of deep-sea sediments. *Source:* From J. J. Griffin, H. Windom, and E. D. Goldberg, reprinted with permission from *Deep-Sea Research*, vol. 15, p. 447, copyright © 1968 by Pergamon Press, Elmsford, NY.

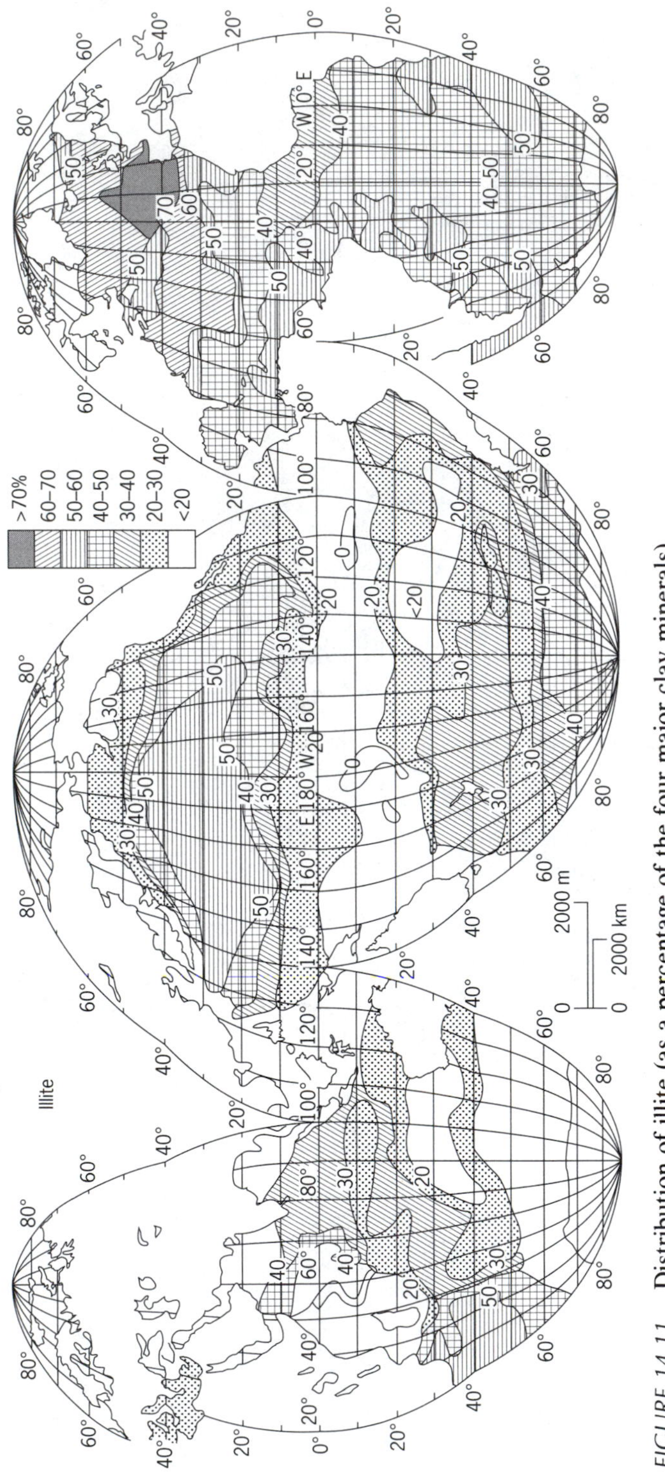

FIGURE 14.11. Distribution of illite (as a percentage of the four major clay minerals) in the <2 μm fraction of deep-sea sediments. *Source:* From J. J. Griffin, H. Windom, and E. D. Goldberg, reprinted with permission from *Deep-Sea Research*, vol. 15, p. 442, copyright © 1968 by Pergamon Press, Elmsford, NY.

clay, is produced in abundance at all latitudes. Dissolved cations and anions are also the result of chemical weathering. Both the solutes and solids are carried by rivers to the ocean. Terrestrial clay minerals are also transported to the sea by winds, ice, and organisms. The former process is responsible for most of the pelagic oceanic sediments, as the majority of grains introduced by rivers are deposited either in estuaries or on the continental margin. Some of these deposits are redistributed by ocean currents.

The reactions that clay minerals undergo during the first few months after introduction into seawater are termed **halmyrolysis**. One of the most important involves **cation exchange**, in which cations adsorbed to negatively charged sites on clay minerals are displaced by the cations that are much more abundant in seawater than in river water. These reactions reach equilibrium and are reversible.

The surfaces of clay minerals are negatively charged due to the relative electronegativity of their oxygen atoms and to **isomorphic substitution**. The latter is variable in nature and degree and causes clay minerals to differ considerably in chemical composition and **cation exchange capacity (CEC)**.

Authigenic clay minerals are thought to be produced by the process of **reverse weathering** and the alteration of fresh volcanic rock. The latter causes **montmorillonite** to be enriched in the sediments surrounding the East Pacific Rise and the mid-ocean ridge in the Indian Ocean.

The global distribution of the individual clay minerals reflects both the location of their production as well as the mechanisms responsible for their transport. As a result of aeolian transport by the Trades, kaolinite concentrations are highest in tropical and equatorial latitudes, particularly off the western coasts characterized by deserts. Chlorite is most abundant at high latitudes as it is transported by icebergs, which are restricted to latitudes greater than 60°. Montmorillonite and zeolite concentrations are highest near active spreading centers, except in the Atlantic, where the concentrated supply of terrestrial clay minerals dilutes their contribution. As a general product of weathering at all latitudes, illite is the dominant sedimentary component, except where it is diluted by local enrichments of other particles.

CHAPTER 15

CALCITE, ALKALINITY, AND THE pH OF SEAWATER

INTRODUCTION

Marine organisms have had a large impact on the distribution of carbon in the crustal-ocean factory. In addition to sequestering most of the crustal carbon in sedimentary carbonates, they have also caused a significant fraction to become buried as organic carbon. As described in Chapter 8, this has lead to the creation of an oxic atmosphere on Earth. The atmospheric CO_2 levels also appear to be stabilized by feedbacks that involve the sedimentary and oceanic reservoirs of organic and inorganic carbon.

The inorganic deposits are composed of biogenic particles created primarily by surface-dwelling plankton. Following death, the calcareous hard parts sink toward the seafloor. As most are dissolved in the deep waters, only a small fraction is buried in the sediments. Over long time periods, buried carbonates are converted to the sedimentary rocks, limestone and dolomite.

The precipitation and dissolution of calcium carbonate is governed by equilibrium reactions that also stabilize the pH of seawater on time scales less than a few million years. These reactions are also responsible for controlling the solubility of atmospheric carbon dioxide gas in seawater. The ocean's ability to absorb this gas is an important consideration in predicting the magnitude of the anthropogenic greenhouse effect. In this chapter, the equilibrium chemistry of calcium carbonate is discussed and used to explain the distribution of calcareous oozes on the seafloor.

THE FORMATION OF CALCIUM CARBONATE

Most biogenic calcium carbonate is precipitated by surface-dwelling plankton in the form of microscopic shells called **tests**. A significant but far lesser fraction is also deposited by protozoa that live below the euphotic zone. The seashells on the beach, numerous as they appear, compose an insignificant amount of the sedimentary carbonates.

Many species of plankton deposit calcareous tests. The shape, mineralogy, size, and density of their tests are unique. Partly as a result of these species-to-species differences, tests are not equally well preserved in the sediments. Those that are most abundant in marine sediments are illustrated in Figure 15.1. They are produced by **coccolithophorids**, which are phytoplankton; **foraminifera**, which are protzoans; and **pteropods**, which are gastropods. The disks produced from the fragmentation of the plant tests are called coccoliths. The empty tests of the protozoans are termed forams and have the appearance of a chambered snail shell. The hard part of the pteropod is a pen-shaped internal structure.

The coccoliths and forams are composed of the mineral calcite, while the hard parts of the pteropods are made up of aragonite. As illustrated in Figure 15.2, these minerals have different crystal structures. As a result, aragonite is more soluble than calcite, so pteropod "pens" are not as well preserved in the sediments as the tests.

Regions where biological productivity is highest would be expected to have the most concentrated deposits of sedimentary carbonates. But, as shown in Figure 15.3, the distribution of calcareous oozes seems to bear a greater relationship to the topography of the seafloor than to the geographic distribution of primary production. In particular, calcareous oozes appear most common on topographic highs, such as the mid-ocean ridges and rises. As discussed below, the preservation of calcium carbonate is a more important control on the distribution of sedimentary carbonate than is its production.

CALCIUM CARBONATE SOLUBILITY

The preservation of calcium carbonate in the sediments is controlled by thermodynamic and kinetic processes that determine the degree to which tests dissolve in the deep sea. The thermodynamic controls are better understood and are discussed first.

The amount of calcium carbonate that will spontaneously dissolve in water if thermodynamic equilibrium is attained is governed by the following reaction:

$$CaCO_3(s) \rightleftharpoons Ca^{2+}(aq) + CO_3^{2-}(aq) \quad (15.1)$$

At equilibrium, the rate of calcium carbonate dissolution is equal to the rate of its precipitation. The concentrations of the chemicals remain constant over time, so no further net dissolution occurs. Since the solution can dissolve no more calcium carbonate, it is said to be saturated. The K_{sp} for this reaction is given by

$$K_{sp} = \{Ca^{2+}\}_{saturated} \times \{CO_3^{2-}\}_{saturated} \quad (15.2)$$

where $\{Ca^{2+}\}_{saturated}$ and $\{CO_3^{2-}\}_{saturated}$ are the activities of Ca^{2+} and CO_3^{2-} at equilibrium.

Marine chemists measure ion concentrations rather than activities, so the

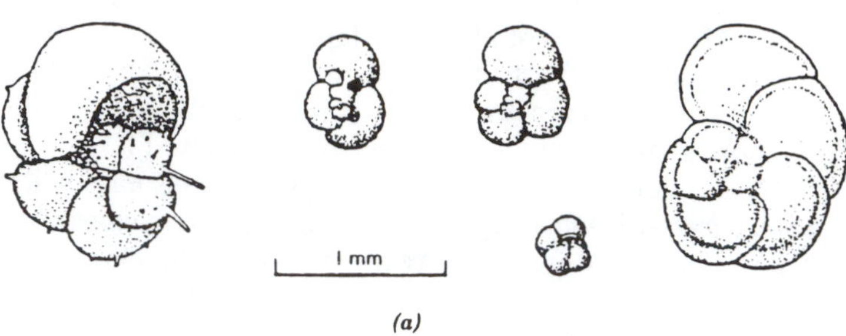

(a)

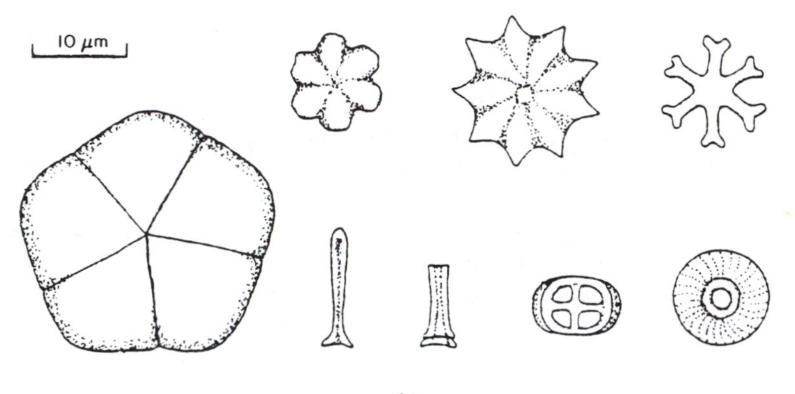

(b)

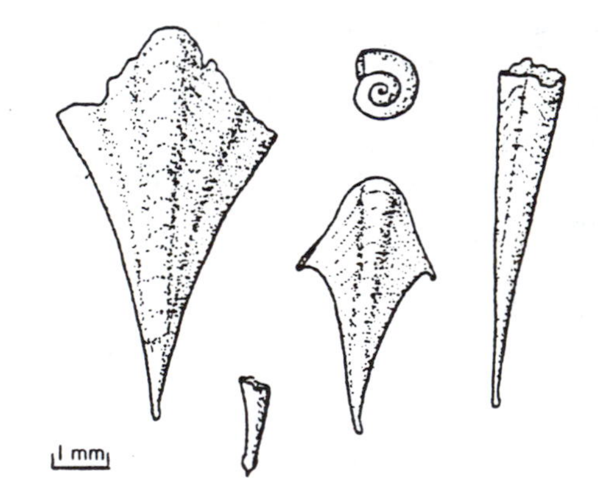

(c)

FIGURE 15.1. Calcareous remains found in deep-sea sediments: *(a)* forams, *(b)* coccoliths, *(c)* pteropod pens. *Source:* From *The Sea,* vol. 3, W. R. Reidel (ed.: M. N. Hill), copyright © 1963 by John Wiley & Sons, Inc., New York, pp. 869, 872–873. Reprinted by permission.

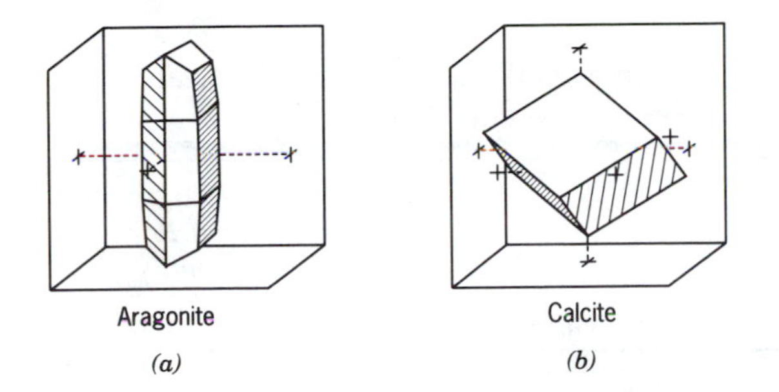

FIGURE 15.2. Crystalline forms of (a) aragonite and (b) calcite. *Source:* From *Marine Chemistry,* R. A. Horne, copyright © 1969 by John Wiley & Sons, Inc., New York, p. 214. Reprinted by permission. After *Mineralogy,* 2nd ed., L. G. Berry, B. Mason, and R. V. Dietrich, copyright © 1983 by W. H. Freeman and Co., New York, pp. 330, 340. Reprinted by permission.

solubility product for calcium carbonate is usually defined as

$$K_{sp}{}^* = [Ca^{2+}]_{saturated} \times [CO_3^{2-}]_{saturated} \tag{15.3}$$

The apparent solubility product, or $K_{sp}{}^*$, is salinity dependent due to specific and nonspecific solute-solute interactions.

Calcium carbonate solubility is also temperature and pressure dependent. As shown in Table 15.1, solubility, and hence the saturation concentrations

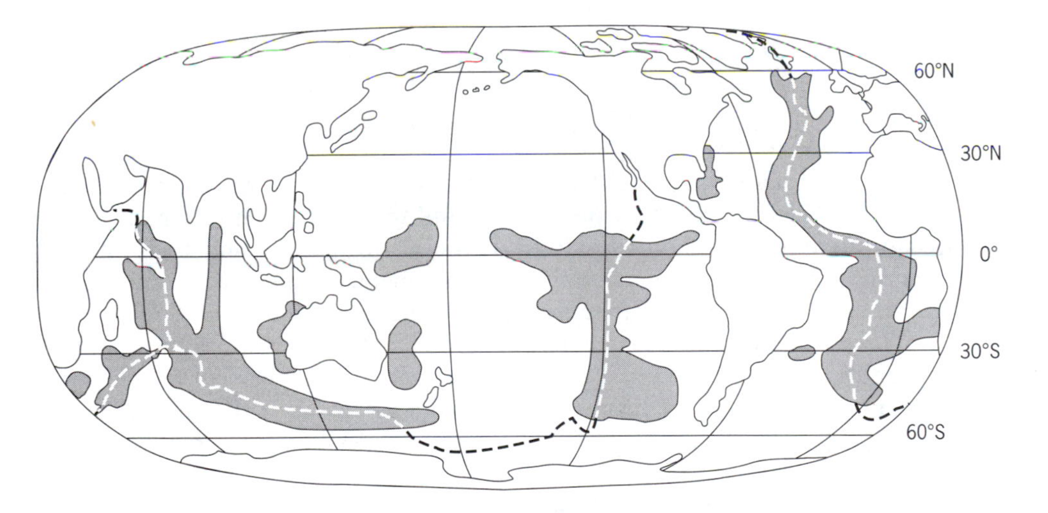

FIGURE 15.3. The global distribution of calcareous oozes on the seafloor. Sediments containing more than 75% calcium carbonate are shown by the darkened area. The axes of the mid-ocean ridges and rises are indicated by the dashed line. *Source:* from *Tracers in the Sea,* W. S. Broecker and T.-H. Peng, copyright © 1982 by the Lamont-Doherty Geological Observatory, Palisades, NY, p. 59. Reprinted by permission. See Broecker and Peng (1982) for data sources.

TABLE 15.1
Saturation Concentrations of Carbonate in Seawater as a Function of Temperature and Pressure

Temperature (°C)	*Pressure (atm)*	*Saturation Carbonate Ion Concentration, (10^{-6} mol/L)* Calcite	Aragonite
24	1	53	90
2	1	72	110
2	250	97	144
2	500	130	190

Source: From *Chemical Oceanography*, W. S. Broecker, copyright © 1974 by Harcourt, Brace and Jovanovich, Publishers, Orlando, FL, p. 38. Reprinted by permission.

of $[CO_3^{2-}]$, increases with decreasing temperature and increasing pressure. Thus calcium carbonate should be more soluble in the deep sea than in the surface waters. Because aragonite is more soluble its saturation $[CO_3^{2-}]$ is always higher than that of calcite.

Most water masses are not saturated with respect to either calcite or aragonite. The degree to which a water mass deviates from the equilibrium concentrations can be expressed as

$$D = \frac{[Ca^{2+}]_{seawater} \times [CO_3^{2-}]_{seawater}}{[Ca^{2+}]_{saturated} \times [CO_3^{2-}]_{saturated}} = \frac{\text{Ion product}}{K_{sp}^*} \quad (15.4)$$

where $[Ca^{2+}]_{seawater}$ and $[CO_3^{2-}]_{seawater}$ are the in situ concentrations in the water mass of interest.

If D is greater than 1, the water mass is supersaturated and calcium carbonate will spontaneously precipitate until the ion concentrations decrease to saturation levels. If D is less than 1, the water mass is undersaturated. If calcium carbonate is present, it will spontaneously dissolve until the ion product rises to the appropriate saturation value.

Though calcium is a biointermediate element, it is present at very high concentrations in seawater. Thus shell formation and dissolution cause its concentration to vary by less than 1 percent. Since K_{sp}^*'s are determined in seawater, $[Ca^{2+}]_{seawater} \approx [Ca^{2+}]_{saturated}$. Thus Eq. 15.4 can be simplified to

$$D = \frac{[CO_3^{2-}]_{seawater}}{[CO_3^{2-}]_{saturated}} \quad (15.5)$$

D can be used to predict the geographic distribution of sedimentary carbonates. For example, sediments lying below waters that have $D<1$ should be devoid of calcareous oozes and vice versa. Since direct measurement of

$[CO_3^{2-}]_{seawater}$ is difficult, it is usually inferred from two easily measurable parameters, alkalinity and the concentration of total dissolved inorganic carbon as described below.

Calculation of Carbonate Ion Concentrations

Dissolved inorganic carbon (DIC) is present in seawater as the species carbon dioxide, carbonic acid, carbonate, and bicarbonate. Since each of these molecules contains one atom of carbon, the **total DIC concentration (ΣCO_2)** is given by the following mass balance equation:

$$\Sigma CO_2 = [CO_2] + [H_2CO_3] + [HCO_3^-] + [CO_3^{2-}] \quad (15.6)$$

In most seawater, $[CO_2] + [H_2CO_3] <<< [HCO_3^-] + [CO_3^{2-}]$, so Eq. 15.6 can be simplified to

$$\Sigma CO_2 = [HCO_3^-] + [CO_3^{2-}] \quad (15.7)$$

The remineralization of POM produces CO_2, which is rapidly hydrolyzed to carbonic acid, bicarbonate, and carbonate via the reactions given in Eqs. 5.42 through 5.44. Since H^+ is also produced, the addition of CO_2 supplies acid to seawater. The dissolution of calcium carbonate also produces DIC. Most of the resolubilized carbonate ion is rapidly converted to bicarbonate. Thus as biogenic detritus is remineralized, ΣCO_2 increases.

As illustrated in Figure 15.4, ΣCO_2 exhibits a vertical segregation similar to that of the nutrients. Concentrations are low in surface waters due to the rapid uptake of DIC by phytoplankton. The biogenic detritus that sinks out of the surface waters is remineralized in the deep zone. Due to density stratification, the resolubilized DIC is trapped below the thermocline. As with the nutrients, ΣCO_2 increases as a water mass ages.

Carbonate and bicarbonate can buffer seawater against changes in pH as illustrated in Eqs. 5.45 through 5.48. Conjugate bases of the weak acids listed in Table 5.9 also contribute to the buffering ability of seawater. In the pH range of 4 to 8, the most important of these buffering reactions are

$$H^+ + B(OH)_4^- \rightleftharpoons B(OH)_3 + H_2O \quad (15.8)$$

$$H^+ + HPO_4^{2-} \rightleftharpoons H_2PO_4^- \quad (15.9)$$

$$H^+ + H_3SiO_4^- \rightleftharpoons H_2SiO_4 \quad (15.10)$$

$$H^+ + NH_3 \rightleftharpoons NH_4^+ \quad (15.11)$$

$$H^+ + OH^- \rightleftharpoons H_2O \quad (15.12)$$

Other buffering reactions involve organic bases and, in anoxic waters, S^{2-} and HS^-.

The total buffering ability of seawater can be thought of as the concentration of negative charge which will react with added H^+. This concentration is called

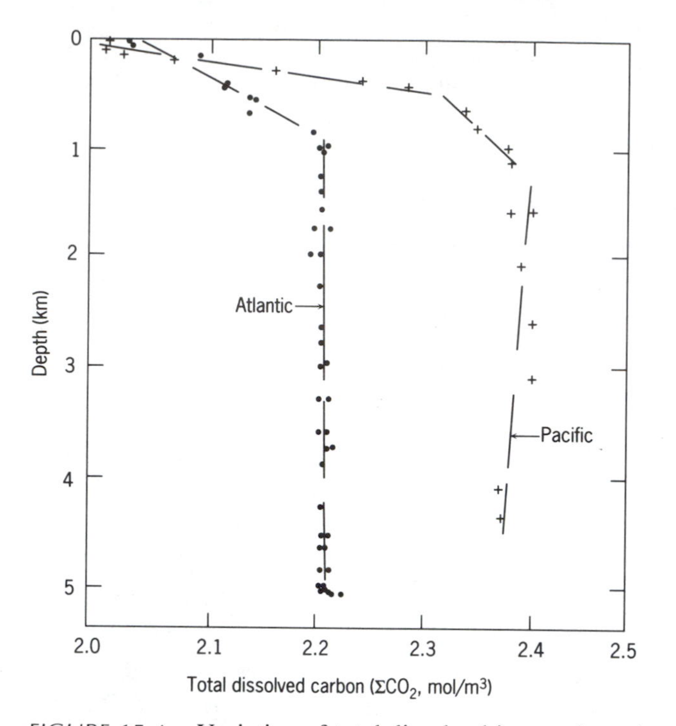

FIGURE 15.4. Variation of total dissolved inorganic carbon concentrations with depth in the Atlantic (36°N 68°W) and in the Pacific (28°N 122°W). *Source:* From *Chemical Oceanography,* W. S. Broecker, copyright © 1974 by Harcourt, Brace and Jovanovich, Publishers, Orlando, FL, p. 39. Data from Dr. R. Weiss, Scripps Institute of Oceanography, La Jolla, CA. Reprinted by permission.

the total alkalinity of seawater and is given by the following mass balance equation:

$$\begin{aligned}\text{Total alkalinity} &= 2[CO_3^{2-}] + [HCO_3^-] + [OH^-] - [H^+] \\ &+ [B(OH)_3] + [H_3SiO_4^-] + [HPO_4^{2-}] + [NH_3] \\ &+ [\text{other conjugate bases of weak acids}]\end{aligned} \tag{15.13}$$

The carbonate concentration is multiplied by 2 because each molecule has two units of negative charge. Since any H^+ already present in seawater represents an in situ sink of negative charge, its concentration is subtracted. Alkalinity is usually reported as the milliequivalents of charge present in 1 L of solution (meq/L).

The concentrations of carbonate and bicarbonate are at least 1000-fold higher than those of the other species, so the total alkalinity of seawater can be approximated by

$$\textbf{Carbonate alkalinity} = CA = 2[CO_3^{2-}] + [HCO_3^-] \tag{15.14}$$

Total alkalinity is determined by titrating a seawater sample with a strong acid, such as HCl. This acid addition affects the pH of seawater in a fashion that is largely determined by shifts in the carbonate buffering equilibria. These shifts are illustrated in Figure 15.5 using a 5 *mM* solution of Na_2CO_3, which has an initial pH of 11. A seawater titration would, of course, start at a pH ≈ 8. The resulting titration curve would also be somewhat different due to the effects of the other conjugate bases, such as borate.

Approximately 5 mL of acid is required to decrease the pH of the Na_2CO_3 solution from 11 to 9 because H^+ is consumed by carbonate. At pH 9, most of the carbonate has been converted to bicarbonate. Thus further acid addition causes the pH to decline rapidly to 7. At this and lower pH's, the production of H_2CO_3 from the reaction of H^+ and bicarbonate is strongly favored. As a result, a considerable amount of acid (approximately 5 more milliliters) is required to decrease the pH of the solution to 5. At this pH, most of the carbonate has been converted to carbonic acid, so further acid addition causes a rapid decline in pH. The carbonate alkalinity of this solution can be calculated from the amount of acid required to titrate to the bicarbonate end point.

With some minor deviations, the carbonate equilibria stabilize the pH of seawater at approximately 8. Deviations are largely the result of kinetic problems that inhibit the attainment of equilibrium. For example, the pH of the euphotic zone increases during the day and decreases at night. Because photosynthesis requires sunlight, net uptake of CO_2 occurs during the day. A net release of CO_2 occurs at night due to dark respiration. In stagnant waters subject to large POM fluxes, such as exist in some upwelling areas and estuaries, the remineralization of POM can produce acid faster than water

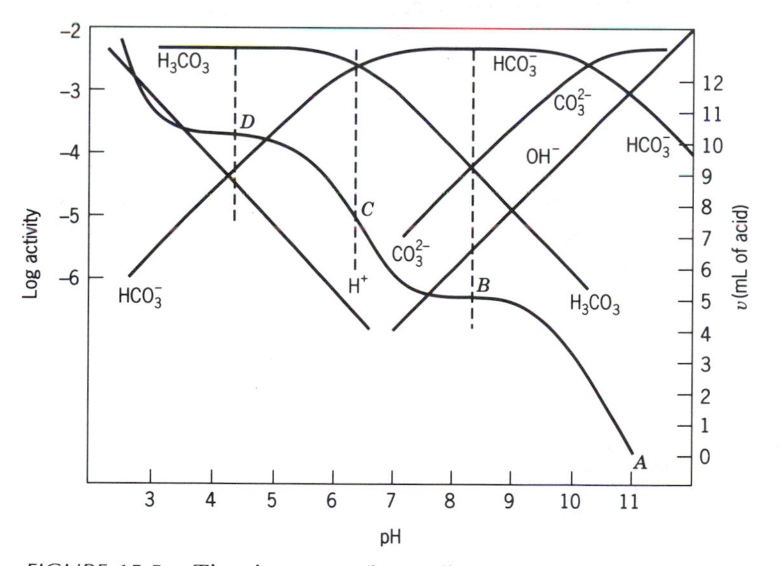

FIGURE 15.5. Titration curve (heavy line *ABCD*) for 5 mM Na_2CO_3 with acid. Also shown is the concurrent DIC speciation, with concentrations given as the log of their ion activities. *B* is the carbonate end point, *A* and *C* are regions of strong buffering, and *D* is the bicarbonate end point.

motion can resupply the carbonate buffers. The waters in these areas are characterized by acidic pH's.

The concentration of carbonate ion in seawater can be calculated from the CA and ΣCO_2 by combining Eqs. 15.7 and 15.14 to obtain

$$[CO_3^{2-}] = CA - \Sigma CO_2 \tag{15.15}$$

In practice, ΣCO_2 is inferred from the water sample's CA and in situ pH.

As shown in Table 15.2, the conjugate bases represent only a small fraction of the total amount of negatively charged species in seawater. Most is supplied by the major anions. Since seawater is electrically neutral, this negative charge must be balanced by an equivalent amount of positive charge, most of which is supplied by the major cations. The following charge balance equation describes this condition of electroneutrality:

$$[Na^+] + [K^+] + 2[Mg^{2+}] + 2[Ca^{2+}] = [Cl^-] + 2[SO_4^{2-}] + [Br^-] + 2[CO_3^{2-}] + [HCO_3^-] \tag{15.16}$$

The major cations and Br^- exhibit nearly conservative behavior; thus Eq. 15.16 can be rewritten as

$$\Sigma[\text{conservative cations}] - \Sigma[\text{conservative anions}] = 2[CO_3^{2-}] + [HCO_3^-] \tag{15.17}$$

where the ion concentrations are given in equivalents of charge per kg of seawater. This equation implies that carbonate alkalinity is a conservative property of seawater. Thus, the ratio of carbonate alkalinity to salinity should be constant. Deviations from this conservative behavior do occur. Most are caused by the precipitation and dissolution of calcium carbonate. These

TABLE 15.2
Charge Balance in Seawater

Positive			*Negative*		
Cation	*Mass* *(mol/m³)*	*Charge* *(mol/m³)*	*Anion*	*Mass* *(mol/m³)*	*Charge* *(mol/m³)*
Na^+	470	470	Cl^-	547	547
K^+	10	10	SO_4^{2-}	28	56
Mg^{2+}	53	106	Br^-	1	1
Ca^{2+}	10	20	Σ	—	604
Σ	—	606	$HCO_3^- + CO_3^{2-}$	—	2
			Σ'	—	606

Source: From *Chemical Oceanography*, W. S. Broecker, copyright © 1974 by Harcourt, Brace and Jovanovich, Publishers, Orlando, FL, p. 41. Reprinted by permission.

processes also cause total alkalinity to exhibit nonconservative behavior. Of lesser effect are processes that cause the uptake and release of the other weak acids and bases, such as NH_4^+ and NO_3^-. These particular species are consumed and released during photosynthesis and the remineralization of POM, respectively.

THE EFFECT OF SHELL FORMATION AND DISSOLUTION ON THE CARBONATE ALKALINITY AND ΣCO_2 OF SEAWATER

The dissolution of calcareous tests increases the ΣCO_2 and alkalinity of a water mass. Shell formation causes the reverse. In comparison, the remineralization of POM increases the ΣCO_2, but has no effect on the alkalinity as the hydrolysis of CO_2 produces one unit of positive charge (H^+) for each unit of negative charge (HCO_3^-). Likewise, the synthesis of POM decreases ΣCO_2, but has no effect on alkalinity.

If enough acid is added to seawater to decrease $[CO_3^{2-}]$ below saturation levels, the deficit will be replaced by the spontaneous dissolution of calcium carbonate. Hence, CA and ΣCO_2 will rise. The large size of the calcium carbonate reservoir provides the ocean with a potentially enormous reserve buffering capability. Despite this, some short-term and local variations in pH do occur as the result of kinetic problems encountered in the attainment of equilibrium. In addition to the examples given above, such disequilibria can be caused by slow shell dissolution. Furthermore, the burial of calcium carbonate in the sediments protects it from dissolution.

VERTICAL SEGREGATION OF ΣCO_2 AND ALKALINITY

The concomitant remineralization of sinking calcareous tests and POM causes the vertical segregation of pH and the DIC species, as illustrated in Figure 15.6. The high pH and low ΣCO_2 of the surface waters are the result of shell formation and the synthesis of POM. Beneath the mixed layer, the net remineralization of shell and POM causes the ΣCO_2 to increase with depth through the thermocline.

The increase in ΣCO_2 within the thermocline is accompanied by a decrease in pH caused by the hydrolysis of CO_2 produced by the remineralization of POM. This supply is so large that significant amounts of CO_2 and H_2CO_3 are present at these depths. The acid produced from the hydrolysis of CO_2 would be expected to cause the dissolution of calcium carbonate and hence result in a high carbonate ion concentration in the thermocline. The opposite is observed,

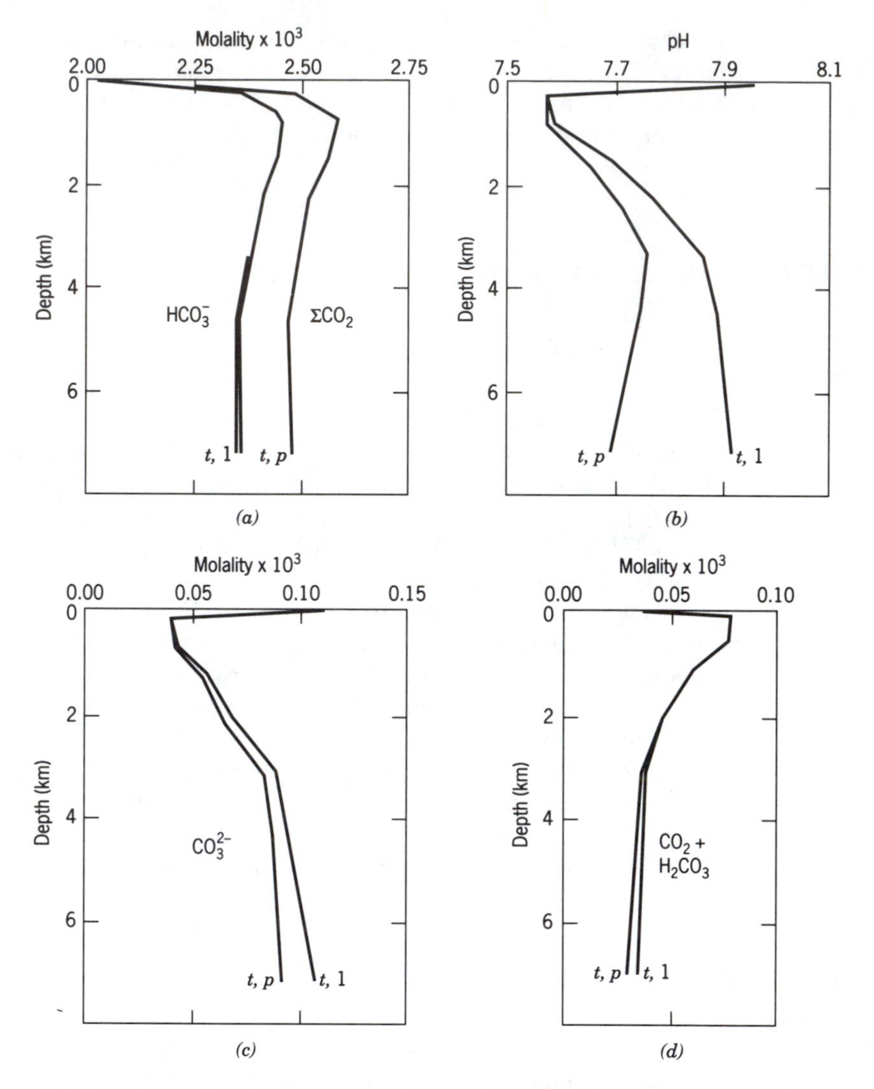

FIGURE 15.6. Vertical concentration profiles of *(a)* bicarbonate and ΣCO_2, *(b)* pH, *(c)* carbonate, and *(d)* carbon dioxide + carbonic acid concentrations. Curves labeled *t,p* have been corrected for the effects of in situ temperature and pressure on equilibrium speciation. Curves labeled *t*,1 have been corrected for the in situ temperature effect, but not for that caused by pressure. *Source:* From C. Culberson and R. M. Pytkowicz, reprinted with permission from *Limnology and Oceanography,* vol. 13, p. 414, copyright © 1968 by the American Society of Limnology and Oceanography, Seattle, WA.

as the large supply of CO_2 drives the following equilibrium strongly to the right.

$$CO_2(aq) + CaCO_3(s) + H_2O(l) \rightleftharpoons Ca^{2+}(aq) + 2HCO_3^-(aq) \qquad (15.18)$$

Thus $[HCO_3^-]$, rather than $[CO_3^{2-}]$, increases.

Since the ΣCO_2 is lower in the deep zone, these waters have evidently accumulated a smaller amount of the products of biogenic particle decomposition than has the thermocline. This reflects the relative "youth" of the deep waters, as well as the smaller flux of biogenic particles that reach these depths. Since less remineralized CO_2 has been added to these waters, the ratio of $[CO_3^{2-}]$ to $[HCO_3^-]$ is higher than in the thermocline.

HORIZONTAL SEGREGATION OF ΣCO_2 AND ALKALINITY

As illustrated in Figure 15.7, both ΣCO_2 and CA exhibit horizontal segregation similar to that of the nutrients. As a water mass moves along the pathway of thermohaline circulation, it accumulates biogenic particles. In general, for each mole of $CaCO_3$ that dissolves, 1 to 4 moles of POM is remineralized. This dissolution and remineralization causes the ΣCO_2 and CA, as well as the nutrient concentrations, to increase, as shown in Figure 15.8. As a water mass ages, its bicarbonate concentration increases and its carbonate ion concentration decreases because the continuing generation of CO_2 from the remineralization of POM pushes the equilibrium, $CO_2(aq) + CaCO_3(s) + H_2O(l) \rightleftharpoons Ca^{2+}(aq) + 2HCO_3^-(aq)$, further to the right. Thus the very highest CA and ΣCO_2 are present in the oldest deep waters (i.e., those in the North Pacific).

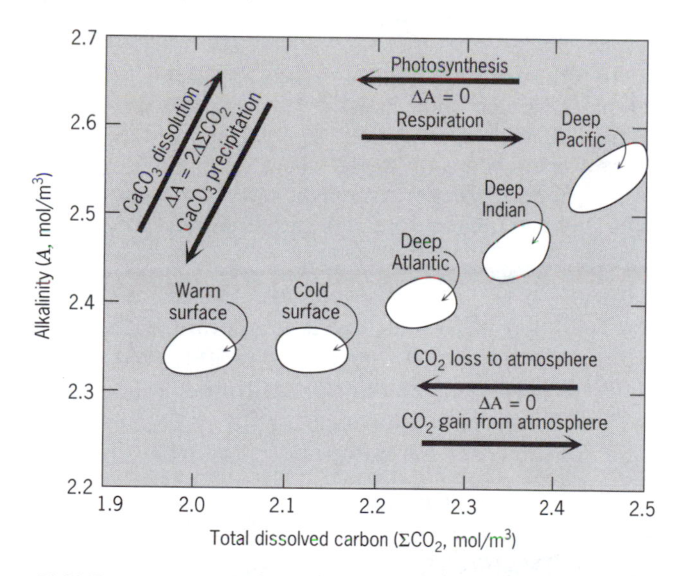

FIGURE 15.7. Relationship between the total dissolved inorganic carbon content and the alkalinity of waters from various parts of the ocean. The arrows indicate the effects of various processes occurring within the sea. *Source:* From *Chemical Oceanography,* W. S. Broecker, copyright © 1974 by Harcourt, Brace and Jovanovich Publishers, Orlando, FL, pp. 14–15. Reprinted by permission.

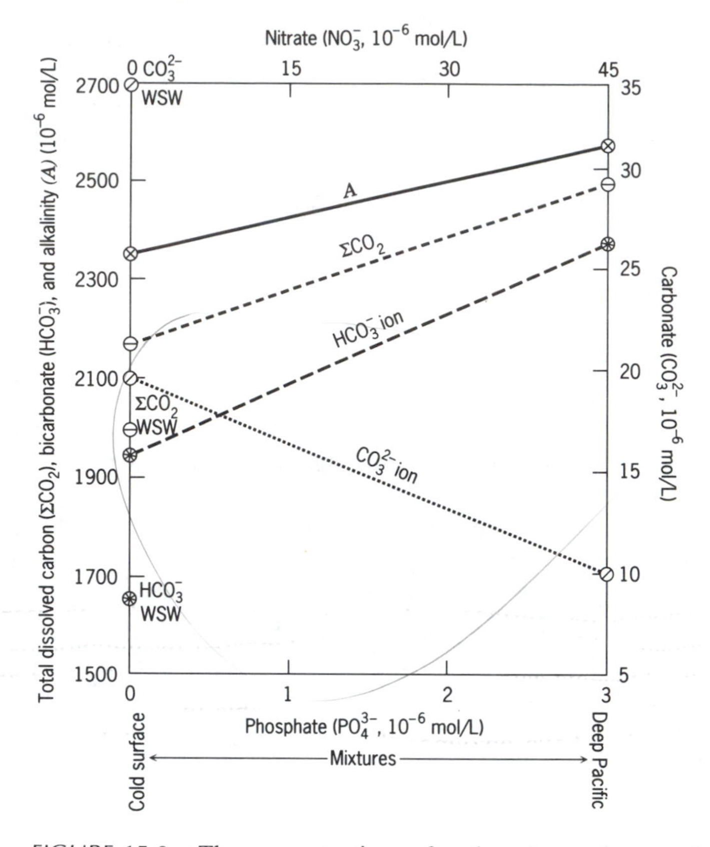

FIGURE 15.8. The concentrations of carbonate species as a function of nutrient concentrations in various water masses. The lines join the two end members, cold surface water and Pacific Deep Water. The values for warm surface waters (WSW) are shown as discrete points on the left-hand axis. *Source:* From *Chemical Oceanography,* W. S. Broecker, copyright © 1974 by Harcourt, Brace and Jovanovich, Publishers, Orlando, FL, p. 45. Reprinted by permission.

The lowest ΣCO_2 concentrations are present in warm surface waters due to the low solubility of CO_2 at higher temperatures. Since the addition of CO_2 does not affect the charge balance of seawater, cold and warm surface water masses that are in gaseous equilibrium with the atmosphere will have the same alkalinity. The relative amounts of carbonate and bicarbonate differ because the equilibrium constants for the buffering reactions are functions of temperature.

THE PRESERVATION OF CALCIUM CARBONATE IN MARINE SEDIMENTS

Thermodynamic Considerations

The preceding discussion has demonstrated that deep waters are more corrosive to calcium carbonate than the surface waters due to the production of CO_2

from the remineralization of POM. Since the amount of CO_2 added to a water mass increases as it ages, the older water masses are more corrosive to calcium carbonate than the newer ones. In other words, the carbonate ion concentration used in the definition of D, the degree of saturation, is controlled by the equilibrium reaction involving CO_2—$CO_2(aq) + CaCO_3(s) + H_2O(l) \rightleftharpoons Ca^{2+}(aq) + 2HCO_3^-(aq)$—rather than $CaCO_3(s) \rightleftharpoons Ca^{2+}(aq) + CO_3^{2-}(aq)$.

The effect of POM remineralization on calcium carbonate solubility can be observed by comparing the depths at which deep waters become undersaturated with respect to calcite and aragonite. The depth range over which $D = 1$ is called the **saturation horizon**. Shells that sink below this horizon should spontaneously dissolve. As illustrated in Figure 15.9*a*, the saturation horizon for calcite is deeper in the Atlantic than in the Pacific Ocean. If equilibrium is established, no calcareous shells should be present at depths greater than 4500 m in the Atlantic and 3500 m in the Pacific. The horizon is shallower in the Pacific due to its greater accumulation of remineralization biogenic CO_2. Because aragonite is more soluble than calcite, its saturation horizon is shallower. If equilibrium is achieved, aragonite should not be present below depths of 1 km in either ocean.

Though the surface waters are supersaturated with respect to calcium carbonate, abiogenic precipitation is hindered by unfavorable kinetics. As a result, abiogenic calcium carbonate is spontaneously formed only at very high supersaturations. Marine organisms are able to overcome this kinetic barrier because they have enzymes that catalyze the precipitation reaction. Due to favorable thermodynamics, surface-water organisms need expend less energy to deposit calcareous shells than those that live in the deep water. As described below, kinetic considerations also influence the dissolution of calcareous shells.

Kinetic Considerations

If the seafloor lies below the saturation horizon, thermodynamic considerations predict that calcium carbonate should not be preserved in the sediments. Nevertheless, calcareous shells are found in such sediments. This thermodynamic disequilibrium is caused by dissolution rates that are slow relative to the rates of sinking and burial.

The flux of calcareous shells that reaches the seafloor is called the **rain rate** of calcium carbonate. The likelihood that a shell will dissolve in transit is determined by factors that control its sinking and dissolution rates. Both are influenced by shell density. The dissolution rate is also affected by the relative size and shape of the shell, as these factors determine its surface area and hence the amount of calcium carbonate that is in contact with corrosive seawater. Thus, the less dense, thin-walled shells dissolve fastest. The denser and "rounder" shells have faster sinking rates and slower dissolution rates. Hence they reach the seafloor in a more intact state.

Dissolution rates are also affected by water chemistry. The greater the degree of undersaturation, the larger the driving force to reestablish equilibrium. Since the degree of saturation of seawater decreases with depth, so do dissolution rates. The depth at which shell dissolution starts to have a detectable

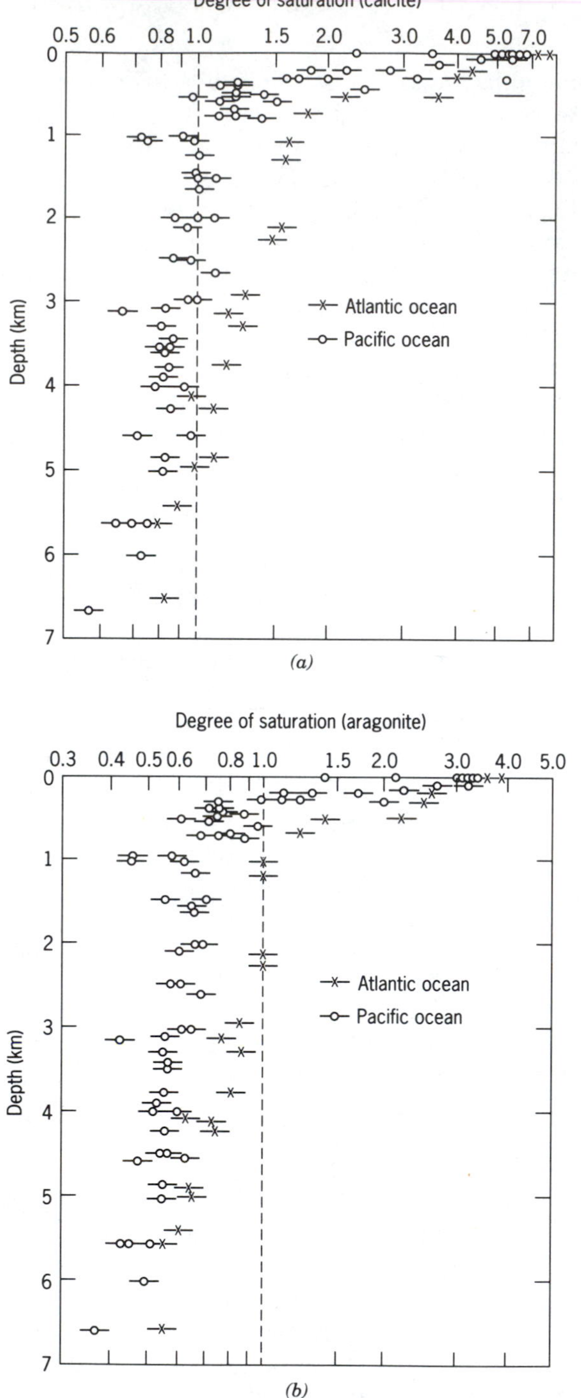

FIGURE 15.9. The Degree of supersaturation as a function of depth in the Atlantic and Pacific oceans for *(a)* calcite and *(b)* aragonite. *Source:* From Y.-H. Li, T. Takahashi, and W. S. Broecker, reprinted with permission from the *Journal of Geophysical Research,* vol. 74, p. 5521, copyright © 1969 by the American Geophysical Union, Washington, DC.

impact on the calcium carbonate content of the surface sediments is termed the **lysocline**.

As shown in Figure 15.10, the lysocline lies quite far below the saturation horizon. This reflects the combined effects of relatively rapid sinking rates and slow dissolution rates. The latter is thought to be partly due to the inhibitory effect of some dissolved substances, such as phosphate, which have mid-depth concentration maxima. Shells are also protected from dissolution by the organic matter coat that covers all marine particles. This film acts as a barrier that prevents corrosive seawater from coming into contact with the calcium carbonate.

Below the lysocline, the calcium carbonate content of the surface sediments decreases rapidly. This is due to a decline in the rain rate of calcium carbonate, as well as to an abrupt increase in the rate at which tests dissolve. The decline in the rain rate reflects the cumulative effects of longer dissolution times. The

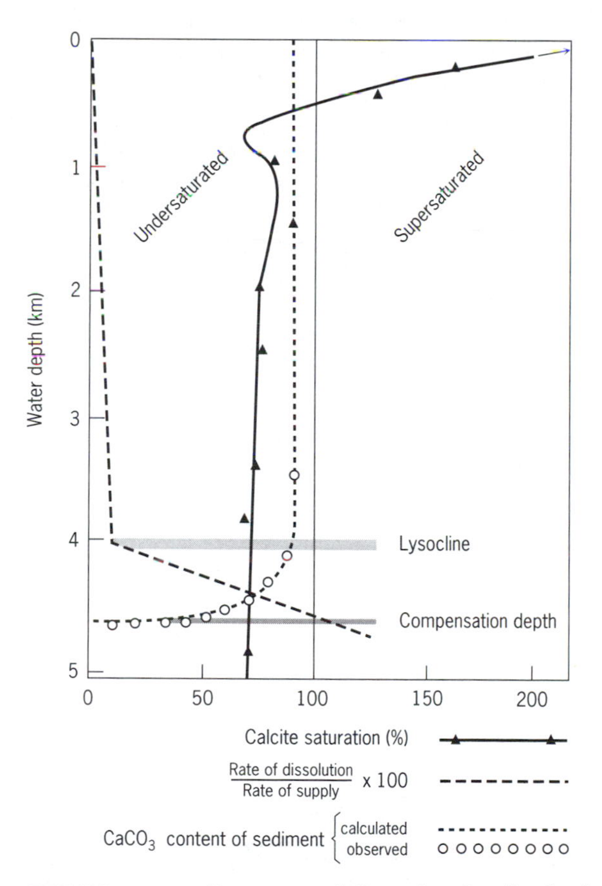

FIGURE 15.10. Parameters influencing the distribution of calcium carbonate with increasing water depth in equatorial Pacific sediment. *Source:* From *Cenozoic History and Paleoceanography of the Central Equatorial Pacific Ocean*, Tj. H. van Andel, G. R. Heath, and T. C. Moore, Jr., copyright © 1975 by the Geological Society of America, Boulder, CO, p. 40. Reprinted by permission.

abrupt increase in dissolution rate appears to be related to the manner in which the thermodynamic driving force increases with depth. In the sediments, dissolution rates are also influenced by the relative rates of molecular diffusion, the remineralization of POM, and bioturbation.

No calcium carbonate is preserved in surface sediments located at depths where the dissolution rate equals or exceeds the rain rate. The depth at which the two rates are equal is termed the **calcite compensation depth (CCD)**, or snow-line. When considered three-dimensionally, the CCD can be thought of as the bottom edge of a giant bathtub ring of calcium carbonate that encircles the ocean basins.

Due to species-specific differences in dissolution rates, the depth of the lysocline and CCD varies among shell types. As shown in Figure 15.11, these differences complicate interpretation of the calcareous fossil record. At any given site, some shells lie above, and some below, their respective lysoclines. This leads to the selective preservation of some species' tests. Thus, the absence of a certain shell type from the sedimentary record does not necessarily prove the organism was not present in the ocean at the time that the preserved shells were formed and deposited.

Rapid burial enhances the preservation of calcium carbonate deposited below the saturation horizon by isolating the shells from corrosive seawater. If burial is achieved by the accumulation of noncalcareous particles, the

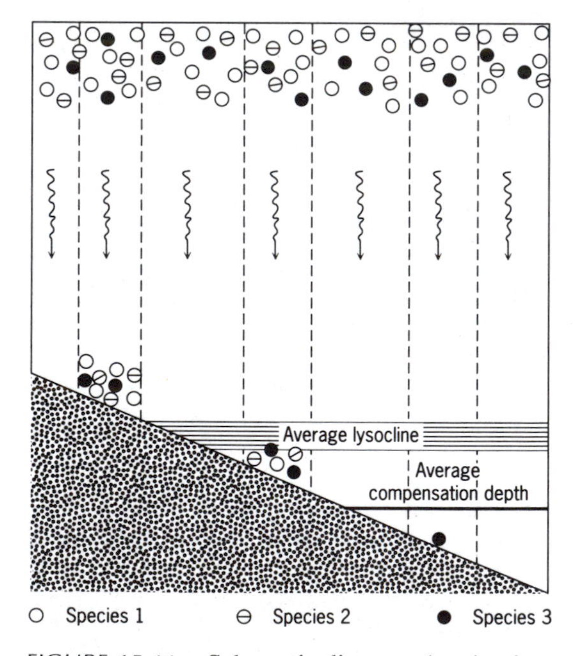

FIGURE 15.11. Schematic diagram showing increasingly selective dissolution of planktonic foraminiferal species with increasing water depth. *Source:* From *Oceanic Micropaleontology,* vol. 1, A. W. H. Bé (ed.: A. T. S. Ramsay), copyright © 1977 by Academic Press, Orlando, FL, p. 49. Reprinted by permission.

sediments will have a relatively low calcium carbonate content due to dilution. Thus, sediments with the highest calcium carbonate content are located in areas that lie above the lysocline and have small inputs of noncalcareous particles. As a result, these deposits tend to accumulate at slow rates. For example, mid-ocean ridge crests are characterized by calcareous oozes that are greater than 80% calcite by mass but accumulate at rates somewhat less than a few millimeters per thousand years. On the continental margins, the shallow depths and rapid burial enhance the preservation of calcium carbonate, but its concentration is low due to dilution by the riverborne lithogenous debris.

Paleoceanographic Interpretation of Sedimentary Carbonate

Down-core variations in the calcium carbonate content of marine sediments are partially the result of temporal changes in the environment under which the tests were deposited. For example, sediments that accumulate on the mid-ocean ridge will eventually be carried into deeper waters. This is caused by the subsidence of oceanic crust that occurs as old basalt is pushed away from active spreading centers. In the example illustrated in Figure 15.12, the sediment surface is eventually transported below the CCD. Once this occurs calcium carbonate ceases to be preserved in the accumulating sediments. If

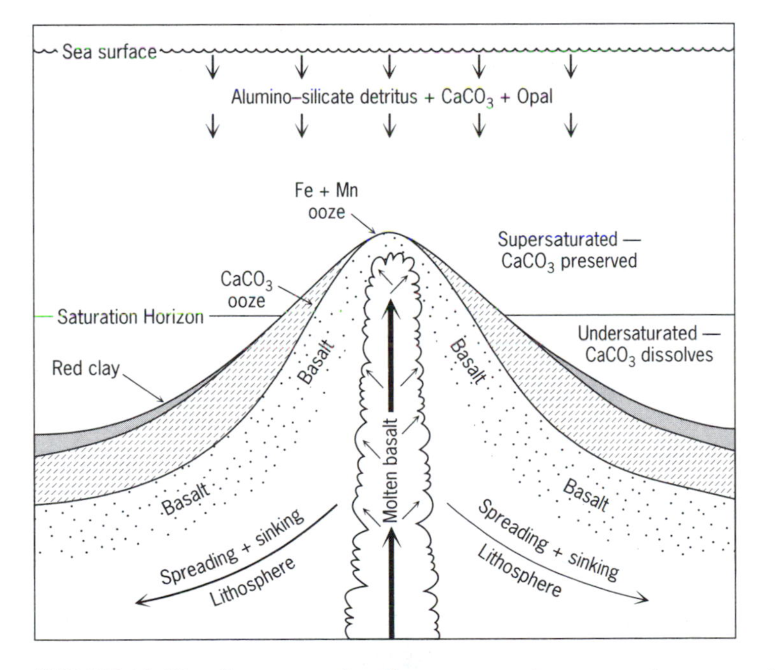

FIGURE 15.12. Sequence of sediment types that accumulate as seafloor spreading causes the oceanic crust to subside over time. *Source:* From *Chemical Oceanography,* W. S. Broecker, copyright © 1974 by Harcourt, Brace and Jovanovich, Publishers, Orlando, FL, p. 52. Reprinted by permission.

the position of the CCD has remained constant over time and the sedimentation rate is known, the depth at which calcium carbonate disappears from the sediments can be used to reconstruct the configuration of the ancient ocean basins.

Unfortunately, climatic changes have caused significant shifts in the depth of the CCD. For example, a climate change that increases the rate of deep-water production would cause an increase in thermohaline circulation. This would enhance the horizontal segregation of alkalinity and ΣCO_2 and cause the CCD in the Pacific Ocean to decrease (become shallower). As a result, the degree of calcium carbonate preservation in the sediments would also decrease, producing a layer of low carbonate content. Such interpretations of down-core variations in calcium carbonate content are complicated but appear to have potential as interpretable records of climate change.

SUMMARY

Sedimentary carbonates are the largest carbon reservoir in the crustal-ocean factory. These deposits are composed of calcareous **tests** created primarily by **coccolithophorids**, **foraminifera**, and **pteropods**. After death, the shells sink towards the seafloor. Most are remineralized in the deep zone, though a small fraction survives to be buried on the seafloor. The sediments with the highest calcium carbonate contents are found, not where overlying productivity is highest, but in locations where accumulation rates are slow and water depths shallow (i.e., mid-ocean ridge crests).

This geographic distribution is the result of thermodynamic and kinetic controls that determine the degree to which calcium carbonate is preserved in the sediments. The thermodynamic factors control the equilibrium solubility of calcium carbonate. For example, solubility increases with decreasing temperature and increasing pressure. Since these are the environmental characteristics of the deep zone, solubility increases with depth. This effect is augmented by an increase in acidity with depth caused by the hydrolysis of CO_2 generated during POM remineralization. The corrosiveness of a water mass toward calcium carbonate increases as it moves along the path of thermohaline circulation. This is caused by horizontal segregation of the products of POM remineralization.

Calcium carbonate should dissolve spontaneously in water masses that are undersaturated. Since calcium concentrations are largely invariant due to the conservative nature of this element, the degree of saturation of a water mass, *D*, is given by the ratio of its carbonate ion concentration to that which would be present at equilibrium. The carbonate ion concentration is inferred from the **carbonate alkalinity (CA)** and **total dissolved inorganic carbon concentration (ΣCO_2)** of the water sample. The CA is a measure of the total amount of negative charge titratable by a strong acid and is a conservative property. The ΣCO_2 represents the sum of the concentrations of the inorganic carbon species CO_2, CO_3^{2-}, HCO_3^-, and CO_3^{2-}.

The depth range over which seawater is saturated with respect to calcium carbonate is called the **saturation horizon**. This horizon is shallower in the Pacific than the Atlantic Ocean due to the increased corrosiveness of older deep waters. Despite undersaturated conditions in bottom waters, calcium carbonate is found in underlying surface sediments.

The presence of calcium carbonate in the sediments reflects the relative balance between its **rain rate** and dissolution rate. The depth at which the two are equal is called the **calcite compensation depth (CCD)**. If the seafloor lies above the depth, calcium carbonate accumulates. The CCD is species specific and thus leads to the selective preservation of some tests.

The depth at which dissolution makes a detectable impact on the amount of calcium carbonate preserved in the surface sediments is termed the **lysocline**. The lysocline lies considerably below the saturation horizon due to rapid sinking rates and slow dissolution rates. Sinking rates are related to shell shape and density. These factors also influence shell dissolution rates. Shell dissolution rates are inhibited by the presence of some solutes, such as phosphate, and by the presence of organic matter coats. They tend to increase with depth as a result of increasing degree of undersaturation.

Calcareous oozes form in areas that lie above the lysocline and that do not receive large inputs of noncalcareous materials (e.g., mid-ocean ridges). Though rapid accumulation rates enhance preservation, the presence of noncalcareous particles can dilute the contribution of calcium carbonate. Hence calcareous oozes are not common on continental shelves.

If the depth of the CCD has remained constant over time, down-core variations in calcium carbonate content can be used to reconstruct the shape of the ocean basins. Otherwise these variations can be used as a record of shifts in the CCD caused by changes in climate and deep-water circulation.

CHAPTER 16

BIOGENIC SILICA

INTRODUCTION

Rivers and hydrothermal emissions are a net source of dissolved silicon to the ocean. Most of this silicon is removed by the deposition of siliceous tests formed primarily by surface-dwelling plankton. A significant fraction is transported to the seafloor by pelagic sedimentation. In some locations, this **biogenic silica** constitutes more than one-third of the sediment mass. The geographic distribution of siliceous oozes reflects the biogeochemical processes that control the production rate of biogenic silica and the degree to which these particles are preserved prior to burial in the sediments.

Most silicon in the ocean is present as a constituent of mineral silicates and hence is relatively inert. Thus most of the dissolved silicon in seawater is supplied by the dissolution of biogenic silica. The growth of siliceous plankton, such as diatoms, tends to be silicon limited. Since diatoms are the most abundant phytoplankton, the biogeochemical cycling of silicon is a very important control on marine productivity. These subjects are discussed below.

THE PRODUCTION OF BIOGENIC SILICA

At the pH and ionic strength of seawater, the dominant dissolved species of silicon is **silicic acid** (H_4SiO_4). As shown in Table 16.1, the formation of siliceous hard parts is the most important mechanism by which dissolved silicon is removed from seawater. Because of vertical segregation, dissolved silicon concentrations tend to increase with depth and range from 1 to 100 μM.

Abiogenic precipitation is important only in locations, such as pore waters and estuaries, where dissolved silicon concentrations are high. Both biogenic and abiogenic precipitation produce an amorphous solid, called **opaline silica**, or **opal**, through the polymerization of silicic acid molecules. Silica also occurs in crystalline form, as exemplified by quartz (Figure 16.1). Silica differs from the mineral silicates in that its oxygen atoms are bonded to each other through three dimensions, rather than in a two-dimensional plane. As a result, the empirical formula of silica is SiO_2.

Biogenic silica is produced by **diatoms** and silico–flagellates, which are phytoplankton, and by **radiolaria**, which are protozoans. Some sponges form

TABLE 16.1
Geochemical Balance of Dissolved Silicon in the Modern Ocean Shown in Units of 10^{14} g SiO_2/y

Input		*Removal*	
Process	*Amount*	*Process*	*Amount*
Dissolved in rivers	4.3	Burial of opaline tests	10.4
Submarine weathering	0.9		
Diffusion out of the seafloor	5.7	Inorganic adsorption at river mouths	0.4
	10.9		10.8

Source: From *Marine Geology*, J. Kennett, copyright © 1982 by Prentice Hall, Englewood Cliffs, NJ, p. 479 and D. J. DeMaster, 1979, unpublished doctoral dissertation, Yale University, New Haven, CT, p. 165. After *Studies in Paleoceanography*, G. R. Heath (ed.: W. W. Hay), copyright © 1974 by the Society for Economic Petroleum Mineralogists, Tulsa, OK, p. 88. Reprinted by permission.

siliceous spires, but these constitute a small fraction of the biogenic silica. As illustrated in Figure 16.2, the siliceous tests buried in marine sediments differ greatly in shape but are usually less than 100 μm in diameter. Many were formed by organisms that are now extinct.

Diatom productivity is limited by the availability of dissolved silicon. In locations where dissolved silicon is abundant, diatoms are the dominant phytoplankton as they assimilate nutrients faster than any other type of pelagic algae. Dissolved silicon concentrations are highest in regions subject to wind-driven upwelling because nutrient-rich deep water is transported to the sea surface.

As illustrated in Figure 16.3, these areas are located at equatorial and subpolar latitudes, as well as along the western continental margins. Upwelling

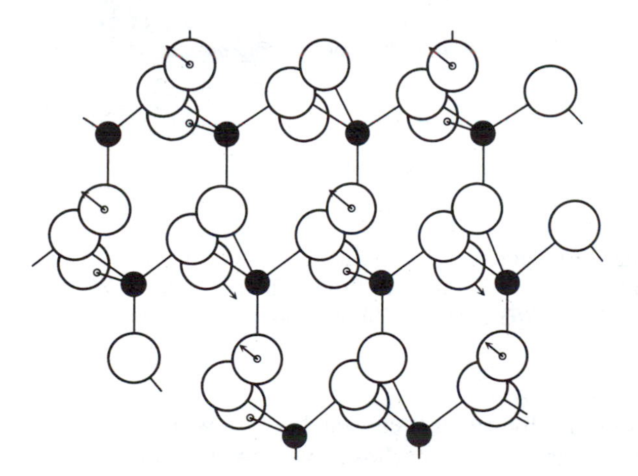

FIGURE 16.1. Three-dimensional crystal structure of quartz.

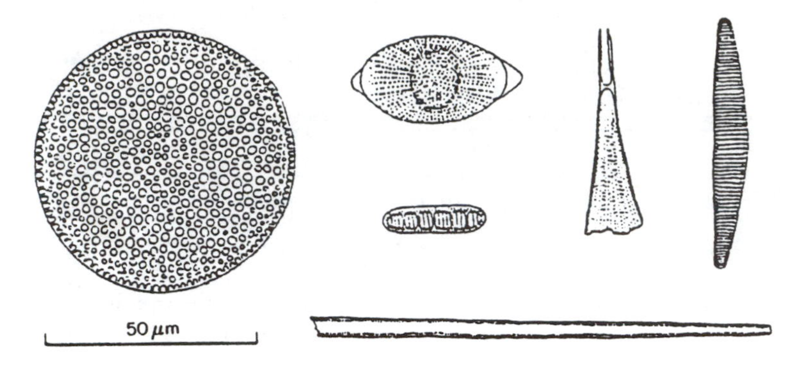

(a)

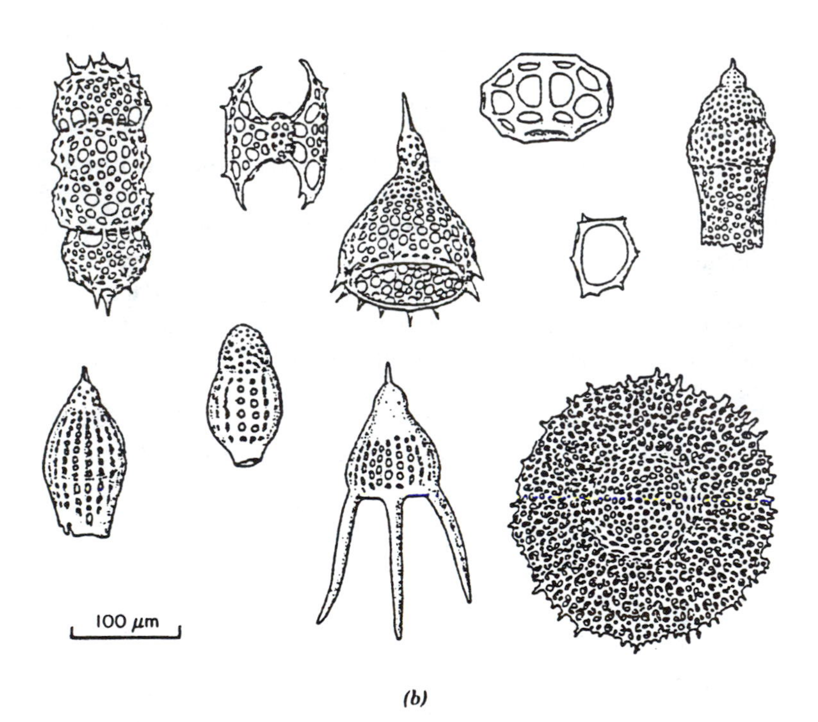

(b)

FIGURE 16.2. Biogenic siliceous remains found in deep-sea sediments from *(a)* diatoms and *(b)* radiolaria. *Source:* From *The Sea,* vol. 3, W. R. Reidel (ed.: M. N. Hill), copyright © 1963 by John Wiley & Sons, Inc., New York, pp. 870–871, 874. Reprinted by permission.

is the result of divergence caused by the wind-driven Ekman transport of water. The Trades and Westerlies drive upwelling at the equatorial and subpolar latitudes, respectively. In comparison, coastal upwelling is episodic due to fluctuations in local patterns of surface circulation.

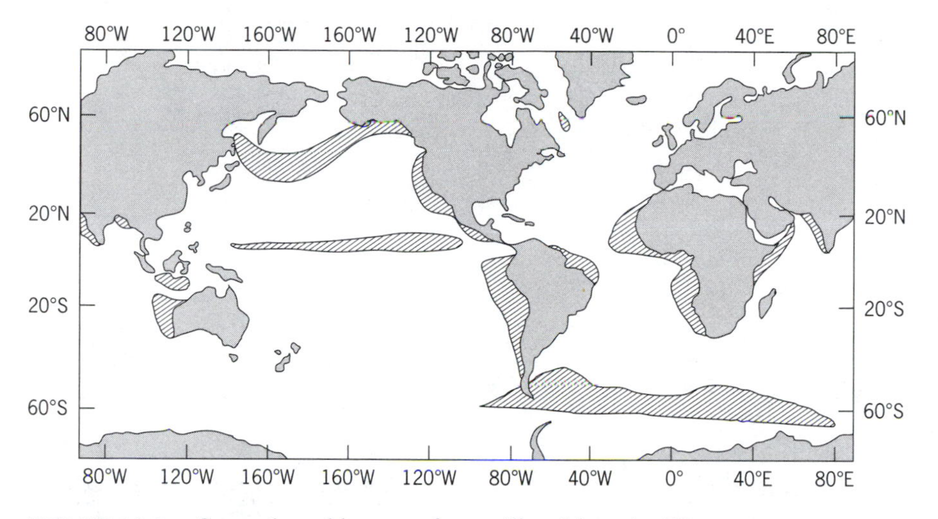

FIGURE 16.3. General world areas of upwelling driven by Ekman transport.

As illustrated in Figure 16.4*a*, diatoms are so effective in removing silicon that surface concentrations are virtually zero except in upwelling regions. The presence of nitrate at latitudes greater than 40°S indicates that dissolved silicon, rather than nitrogen, limits phytoplankton growth. The concentrations of silicon and nitrate are high at latitudes greater than 55°S as a result of supply by subpolar upwelling and relatively low uptake rates. Phytoplankton growth at these latitudes is thought to be limited by the availability of light, low temperatures, or by a micronutrient, such as dissolved iron.

After death, the siliceous tests sink toward the seafloor. As with the other biogenic particles, most are remineralized in the deep zone. The lack of density stratification in upwelling areas causes silicon and the nutrients to be recycled, as shown in Figure 16.5. Upwelling transports the regenerated nutrients and silicon back to the euphotic zone, where they are rapidly reconverted into biogenic particles, primarily by diatoms. Thus the nutrients and silicon are kept within the upwelling zone, unless they sink fast enough to become buried in the sediments. This nutrient trap is most effective when the uptake rate matches the rate of supply from upwelling. If the supply rate is larger, the nutrients will eventually be transported out of the upwelling zone by water motion. The nutrient trapping effect in upwelling areas is most pronounced for silicon as this biolimiting element exhibits the strongest degree of vertical segregation.

PRESERVATION VERSUS DISSOLUTION OF SINKING DETRITAL BIOGENIC SILICA

In contrast to calcium carbonate, all seawater is undersaturated with respect to opaline silica. Thus all siliceous tests are subject to dissolution as they sink

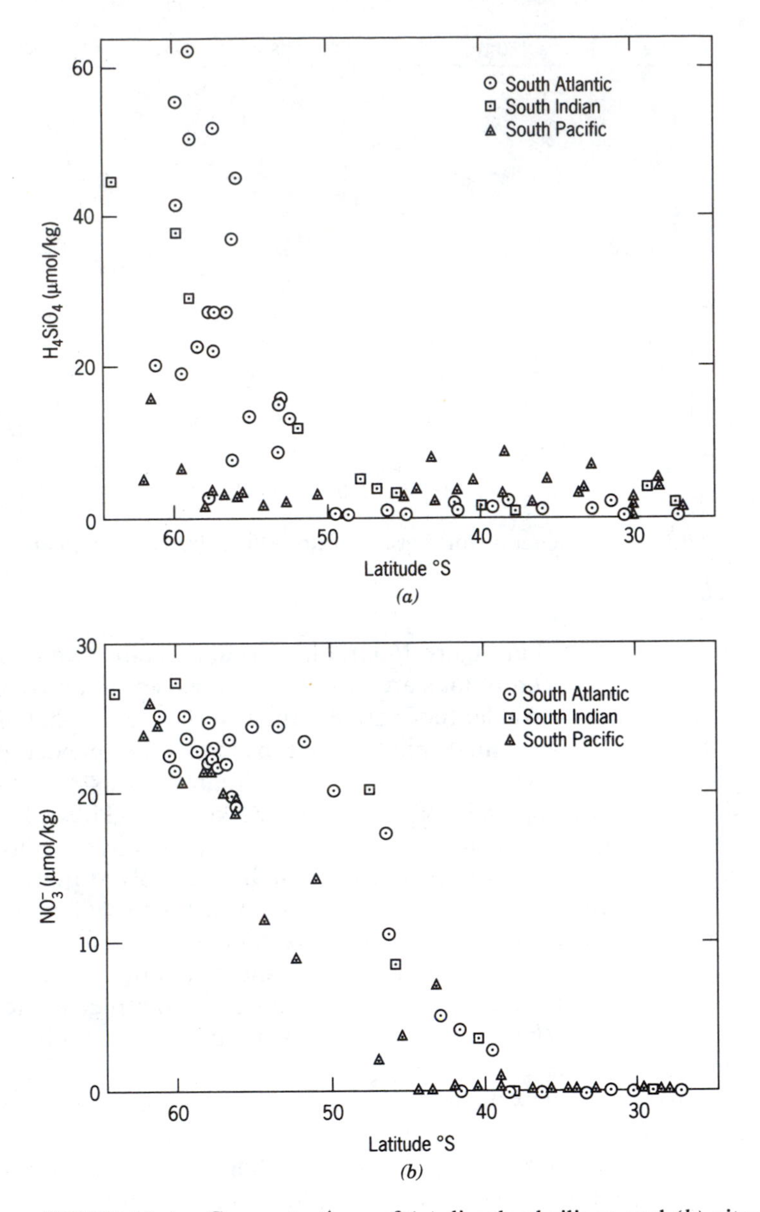

FIGURE 16.4. Concentrations of *(a)* dissolved silicon and *(b)* nitrate in surface seawater as a function of latitude in the Southern Ocean. *Source:* From *Tracers in the Sea,* W. S. Broecker and T.-H. Peng, copyright © 1982 by the Lamont-Doherty Geological Observatory, Palisades, NY, p. 39. Reprinted by permission. See Broecker and Peng (1982) for data sources.

toward the seafloor. Nevertheless, a substantial fraction is buried in the sediments, mostly as a result of slow dissolution rates. Direct observations of this have been obtained through the use of sediment traps.

Sediment traps are sampling devices that intercept sinking particles. They

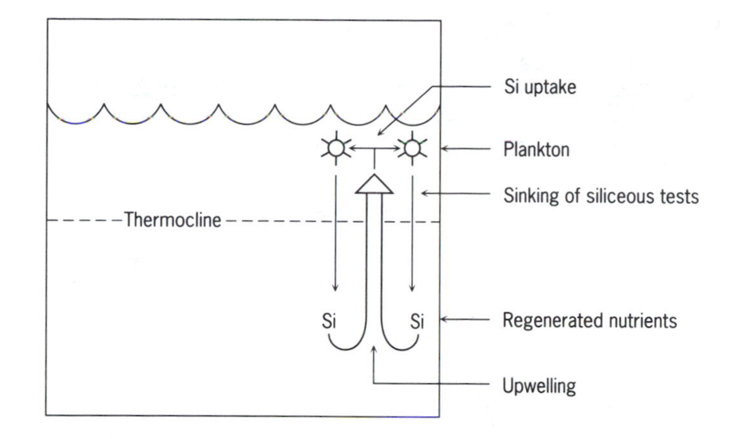

FIGURE 16.5. Nutrient trap in a strong upwelling area.

are deployed at specific depths for controlled lengths of time. As shown in Figure 16.6, the traps are usually attached to a mooring array that is recovered by activating an acoustic release.

Data obtained from two such sets of sediment traps are presented in Table 16.2 and demonstrate that the chemical composition of particulate matter is altered as it sinks toward the seafloor. At both sites, the relative proportions of organic matter and calcium carbonate decrease due to remineralization and dissolution. Since the total particle flux declines with depth, the clays constitute an increasingly larger fraction as they are relatively inert. The small or nonexistent changes in percent opal and total opal flux with depth indicate that little or no dissolution of biogenic silica occurred as these particles sank toward the seafloor.

Because opal does not dissolve rapidly, silicon is regenerated at greater depths than are nitrogen or phosphorus. This poorer recycling efficiency causes a greater degree of vertical and horizontal segregation. Slow dissolution rates also prevent seawater from attaining equilibrium concentrations of dissolved silicon.

ACCUMULATION AND PRESERVATION IN THE SEDIMENTS

As shown in Figure 16.7, the sediments with the highest concentration of opaline silica are located beneath the equatorial and subpolar upwelling areas. The factors that control this geographic distribution are (1) the rain rate of biogenic silica, (2) the degree of preservation in the sediments, and (3) the relative rate of accumulation of other particles.

The rain rate of biogenic silica is dependent on (1) the rate of its production by marine organisms, (2) the time required for a shell to reach the seafloor, and (3) shell dissolution rates. High rates of production by siliceous plankton

TABLE 16.2
Sediment Trap Data from the Atlantic and Pacific Oceans[a]

	2°S				30°N	
Atlantic	*0.4 km*	*1.0 km*	*3.8 km*	*5.1 km*	*1.0 km*	*4.0 km*
Mass Percent (%)						
Clay	7	16	25	31	7	12
$CaCO_3$	63	55	57	49	59	68
Opal	10	12	9	9	7	9
Organic matter	20	17	10	11	27	11
Flux (mol m^{-2} y^{-1})						
Organic C	0.21	0.12	0.05	0.05	0.074	0.027
$CaCO_3$	0.15	0.09	0.09	0.08	0.040	0.043
Opal	0.035	0.031	0.023	0.023	0.008	0.010
Flux (g cm^{-2} 10^3 y^{-1})						
Clay	0.2	0.3	0.4	0.5	0.005	0.08
$CaCO_3$	1.5	0.9	0.9	0.8	0.40	0.43
Opal	0.2	0.2	0.1	0.1	0.05	0.06
Atom Ratio (mol/mol)						
C/N	7.6	8.5	10.9	8.4	5.1	9.3
N/P	23.7	31.8	16.5	23.7	17.7	11.9
C/P	180	270	180	199	90	111

(Continued on next page.)

ensure a large supply of opal to the water column. The fraction reaching the seafloor is largest when transit times are shortest. A greater fraction of the particulate silica flux reaches sediments that lie in shallow waters. Thus shells that sink fastest will be preferentially preserved.

As with the calcareous tests, shell dissolution rates depend on (1) the susceptibility of a particular shell type to dissolution and (2) the degree to which a water mass is undersaturated with respect to opaline silica. The former is determined by a shell's relative density and shape. The denser tests are preferentially preserved in the sediments due to their faster sinking rates. Those with thicker walls and lower surface area-to-volume ratios are also better preserved as their dissolution rates are slower. This is why the thin, fragile spires of the radiolaria pictured in Figure 16.2*b* are the most etched and eroded.

Opal, like all marine minerals, is covered by a coating of adsorbed organic matter. This isolates the shell from contact with seawater, slowing its rate of dissolution. Shells dissolve faster in oxic as compared to anoxic conditions due to the faster remineralization of this protective organic coating. Dissolution rates are also slowed by the presence of metal ions, as they lower the solubility

TABLE 16.2 (Cont.)
Sediment Trap Data from the Atlantic and Pacific Oceans[a]

	15°N				
Pacific	*0.4 km*	*1.0 km*	*2.8 km*	*4.3 km*	*5.6 km*
Clay	2	3	2	3	4
Opal	3	9	15	15	21
$CaCO_3$	35	72	68	72	61
Organic matter	60	16	14	11	14
Flux (mol m^{-2} y^{-1})					
Organic C	0.108	0.017	0.033	0.027	0.020
$CaCO_3$	0.014	0.019	0.041	0.042	0.024
Opal	0.003	0.016	0.016	0.015	0.014
Flux (g cm^{-2} 10^3 y^{-1})					
Clay	0.01	0.01	0.01	0.02	0.02
$CaCO_3$	0.14	0.19	0.41	0.42	0.24
Opal	0.01	0.02	0.09	0.09	0.08
Atom ratio (mol/mol)					
C/N	8.4	9.6	10.6	10.2	10.0
N/P	—	—	29.1	26.1	—
C/P	—	—	309	266	—

Source: From *Tracers in the Sea*, W. S. Broecker and T.-H. Peng, copyright © 1982 by the Lamont-Doherty Geological Observatory, Palisades, NY, p. 14. Reprinted by permission. Data from S. Honjo, S. J. Manganini, and J. J. Cole, reprinted with permission from *Deep-Sea Research*, vol. 29, pp. 618–620, copyright © 1982 by Pergamon Press, Elmsford, NY.
[a]The depths at which the traps were deployed are given at the top of each column.

of silica. Metals are naturally incorporated by organisms into siliceous tests and are also adsorbed onto the shells because silica has a net negative surface charge at the pH of oxic seawater.

As with calcium carbonate, the thermodynamic solubility of opal is a function of water temperature, pressure, and the presence of other solutes. Thus, water masses differ in their degree of undersaturation. The greater the degree of undersaturation, the greater the thermodynamic driving force for dissolution and hence the faster the dissolution rate.

Most differences in the degree of undersaturation of deep waters are due to geographic variations in dissolved silicon concentrations. Thus the preservation of opal is favored in sediments that underlie deep waters with high dissolved silicon concentrations, such as in the Pacific Ocean. Preservation is also promoted by rapid burial as this isolates opal from seawater. If the opal is buried by other particle types, the relative contribution of opal to the sediment is diluted. This is why high silica contents are not observed under coastal upwelling areas or on the continental margins, despite large rain rates.

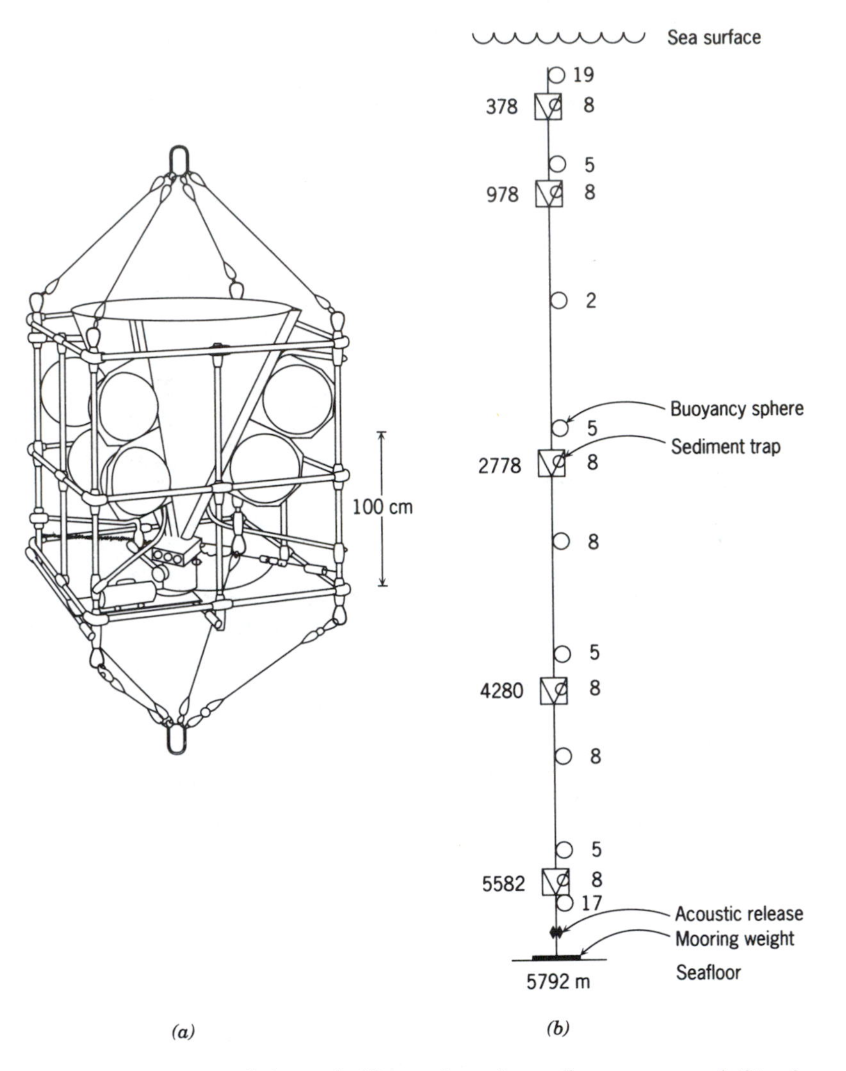

FIGURE 16.6. *(a)* Schematic illustration of a sediment trap and *(b)* schematic illustration of sediment trap mooring array. Numbers along the left side of the mooring lines are the depths of the traps. The numbers on the right side indicate the number of buoyancy spheres. A sphere contains 50 lb of positive buoyancy. The traps are retrieved by triggering the acoustic release which is located at the bottom of the mooring array. *Source:* From S. Honjo, reprinted with permission from the *Journal of Marine Research,* vol. 38, pp. 58–59, copyright © 1980 by the *Journal of Marine Research,* New Haven, CT.

Thus highly siliceous sediments result when rapid burial is caused by a large rain rate of opal and the accumulation rate of other particles is slow. These conditions cause the open-ocean upwelling areas to have silica contents in excess of 80 percent by mass.

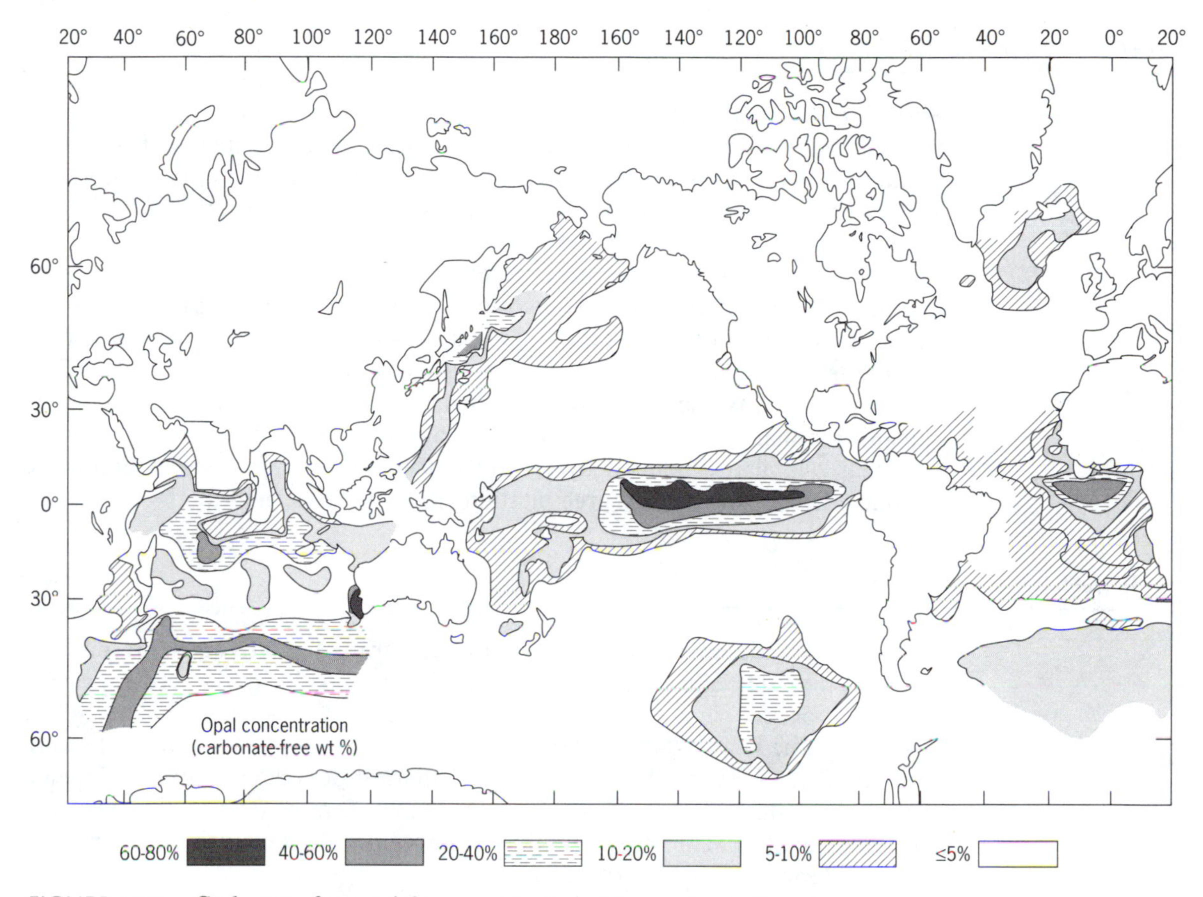

FIGURE 16.7. Carbonate-free weight percent of opal in marine sediments.
Source: From M. Leinen, D. Cwienk, G. R. Heath, P. E. Biscaye, V. Kolla, Jørn Thiede, and J. P. Dauphin, reprinted with permission from *Geology,* vol. 14, p. 200, copyright © 1980 by the Geological Society of America, Boulder, CO.

Diatoms are the dominant species at low temperatures when dissolved silicon is available. Thus diatomaceous oozes are common to the sediments of the subpolar Southern Ocean and Bering Sea. The siliceous ooze in the equatorial Pacific is composed mostly of radiolaria tests. Due to its great width, the Pacific Ocean has the largest and most intense zone of equatorial upwelling and thus has produced the most extensive equatorial deposit of siliceous ooze.

Overall, the sediments of the Pacific Ocean have the highest opal content as horizontal segregation has caused its bottom waters to have the highest dissolved silicon concentrations. In addition, the rain rate of other particles is relatively slow. The shallow depth of the CCD in this ocean causes most calcium carbonate to dissolve before reaching the seafloor. The rain rate of lithogenous detritus is also limited as a result of the trapping effect of oceanic trenches and the relatively large surface area of this ocean.

THE FORMATION OF CHERT

Though quartz is thermodynamically more stable than amorphous silica, slow crystallization rates inhibit its deposition by marine organisms. Long-term burial eventually converts biogenic silica into chert and quartz, both of which are crystalline. The alteration in structure is achieved through diagenetic and catagenetic reactions that involve the partial dissolution and reprecipitation of silica. Dissolution causes the pore-water concentrations of silicon to increase over time. Once supersaturated conditions are attained, reprecipitation usually occurs as an overgrowth on detrital grains of quartz, granite, or clay minerals, as well as on biogenic hard parts. Silicon also precipitates by absorption into biogenic hard parts and clay minerals. The latter process is a type of isomorphic substitution that occurs during reverse-weathering reactions.

Quartz is also produced by the diagenesis and catagenesis of abiogenic opaline silica, as well as by precipitation from hydrothermal fluids.

SUMMARY

Though most silicon in the ocean is in the form of relatively inert mineral silicates, **biogenic silica** is a very important reservoir in the crustal-ocean factory. Biogenic silica is amorphous and usually referred to as **opaline silica**, or **opal**. Quartz is an example of crystalline silica. Silica is distinguishable from mineral silicates by its three-dimensional sharing of oxygen atoms. Biogenic silica is deposited on the seafloor as siliceous tests formed primarily by surface-dwelling plankton. The most abundant of these are **diatoms**, which are plants, and **radiolaria**, which are protozoans.

The growth of these organisms is limited by the availability of dissolved silicon, which occurs primarily as **silicic acid** (H_4SiO_4). Where dissolved silicon is present, such as in upwelling areas, diatoms are the dominant phytoplankton. After death, the siliceous tests sink toward the seafloor. Due to less efficient recycling in the surface waters, the vertical and horizontal segregation of silicon is more marked than for the nutrients. This also causes significant enrichments of silicon in upwelling areas. Because of slow dissolution rates, a larger fraction of the silica flux reaches the seafloor as compared to POM or calcium carbonate.

The geographic distribution of sedimentary silicon reflects the relative rain rate of biogenic silica to the seafloor, its degree of preservation in the sediments, and the relative rate of accumulation of other particles. The rain rate of biogenic silica is dependent on (1) the rate of its production by marine organisms, (2) the time required for a shell to reach the seafloor, and (3) shell dissolution rates. The latter is dependent on (1) the susceptibility of a particular shell type to dissolution and (2) the degree to which a water mass is undersaturated with respect to opaline silica.

In contrast to calcium carbonate, all seawater is undersaturated with respect to biogenic silica. The degree of undersaturation varies with the temperature, pressure, and chemical composition of the water mass. Dissolution rates are slower in water masses with higher dissolved silicon concentrations (i.e., the Pacific Ocean). Preservation is also enhanced by rapid burial. If this is achieved by a rapid rain rate of biogenic silica in the absence of other particles, the resulting sediments have high silicon contents. This is observed beneath open ocean upwelling areas, such as the equatorial Pacific, Bering Sea, and subpolar Southern Ocean. Though the production rate of biogenic silica is high in coastal upwelling areas and over the continental shelves, the underlying sediments do not have high opal contents due to dilution by other sedimenting particles.

Deep burial of biogenic silica eventually converts its amorphous structure to that of crystalline chert and quartz. This occurs through diagenetic reactions that involve the dissolution and reprecipitation of sedimentary silica. Both minerals are also produced in marine sediments by the maturation of abiogenic silica and precipitation from hydrothermal emissions.

CHAPTER 17

EVAPORITES

INTRODUCTION

On the early Earth, ions were mobilized from volcanic rocks by chemical weathering. Rivers and hydrothermal emissions transported these chemicals into the ocean, making seawater salty. Since that time, the original salts have been recycled through the crustal-ocean factory by becoming buried in the sediments, metamorphosed into rock, uplifted, and reweathered. Now most salts entering the ocean are derived from the chemical weathering of sedimentary rocks, such as shale, limestone, chert, and evaporites. For a few of the major ions, volcanism continues to be either a significant source or sink.

Evaporites contain a large fraction of the crustal sodium, calcium, chlorine, sulfate, potassium, and magnesium. Thus their formation and dissolution can affect the salinity of the ocean. Both can affect global climate, since the heat capacity of seawater is a function of its salinity.

Humans have mined evaporites for at least the past 6000 y. For example, the evaporite mineral trona ($NaHCO_3 \cdot Na_2CO_3 \cdot 2H_2O$) was used by the ancient Egyptians to preserve mummies. Evaporite salts continue to be used for food preservation and in industrial processes. Petroleum geochemists use sedimentary evaporites as indicators of the likely presence of petroleum deposits because they form an impermeable barrier behind which migrating petroleum tends to pool.

Evaporites are usually layered deposits composed of minerals that precipitated in stages. The mineral type, texture, and sequence of deposition can provide information on the hydrogeochemical conditions under which the evaporites formed. This sedimentary record has been used to reconstruct paleoceanographic conditions as far back as 3.45 billion years ago. Evaporite composition also provides the most convincing evidence that the chemical composition of seawater has not varied much over the past 500 million years. The mineralogy, distribution, and formation of evaporites are the subjects of this chapter. The role of evaporite formation and dissolution in determining the salinity of seawater is discussed in Chapter 21.

FORMATION OF EVAPORITES BY EVAPORATION OF A FIXED VOLUME OF SEAWATER

The evaporation of a fixed volume of seawater causes the precipitation of minerals in a sequence determined by their relative solubility. As shown in

Figure 17.1, removal of half the water causes calcite and **aragonite** to precipitate. Continued evaporation leads to the formation of **gypsum** ($CaSO_4 \cdot 2H_2O$), followed by **anhydrite** ($CaSO_4$). Once 90% of the water has been removed, **halite** (NaCl) precipitates, along with some magnesium salts. Potassium salts deposit last. This sequence is a function of temperature and the extent to which equilibrium is maintained between the phases. Each of these precipitations alters the ion ratios in the remaining seawater. Since the Rule of Constant Proportions is violated, density, rather than salinity, must be used to monitor the increasing saltiness of the brine.

In nature, deviations from equilibrium are common. Thus depositional sequences vary considerably from the ideal presented in Figure 17.1. Deviations are caused by rapid changes in environmental conditions, such as fluctuations in water supply rates. The sequence of evaporite mineral deposition is also

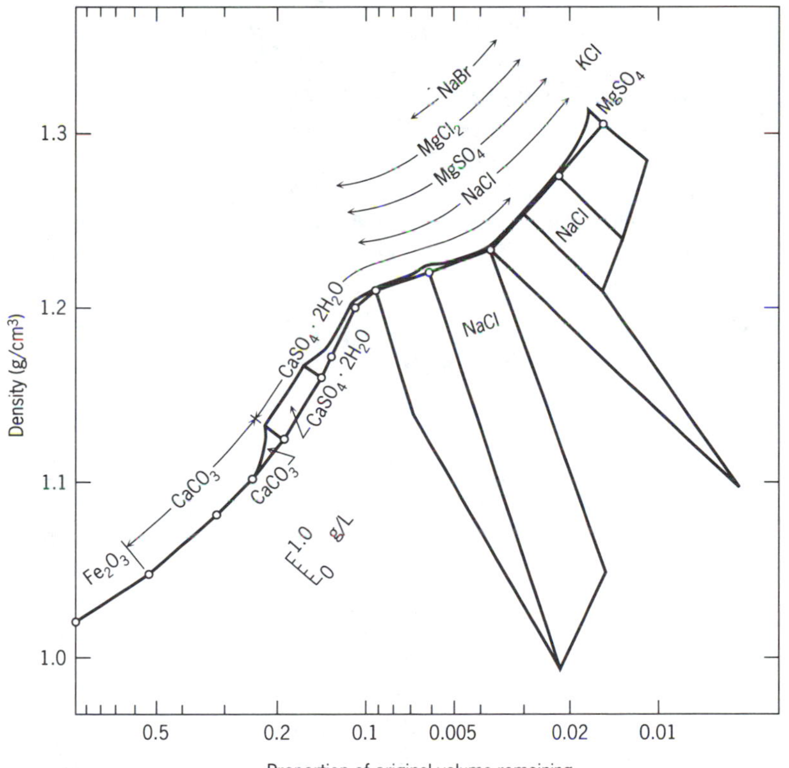

FIGURE 17.1. Brine density versus proportion of original volume remaining in the closed-system evaporation of mediterranean seawater (initial salinity = 35‰). The concentration of minerals produced (in grams per liter of water) is indicated by the lengths of the lines drawn perpendicular to the curve. *Source:* From *Principles of Sedimentology,* G. M. Friedman and J. E. Sanders, copyright © 1978 by John Wiley & Sons, Inc., New York, p. 527. Reprinted by permission. See Friedman and Sanders (1978) for data sources.

affected by changes in sea level, the rate of solute input, and the sedimentation of other particles.

Evaporite minerals are not well preserved in the sedimentary record because of diagenesis and catagenesis. The high temperatures, pressures, and pore-water salinities characteristic of deep burial cause mineral alterations such as the conversion of gypsum into anhydrite. Thus, the sedimentary record reflects not only the environmental conditions under which the evaporite was formed, but also those under which diagenesis and catagenesis occurred. Because of variability in postdepositional alteration, it is often difficult to assess the hydrogeochemical conditions under which an evaporite was deposited.

Though most evaporites are marine in origin, some are produced by the concentration of terrestrial waters. The ions present in these waters are supplied by the chemical weathering of continental rock. Terrestrial waters are more variable in composition than seawater, causing greater variability in the mineralogy of the nonmarine evaporites. Due to relatively high carbonate concentrations, nonmarine evaporites are composed primarily of carbonate minerals. In contrast, the marine forms are mostly chloride and sulfate salts.

The absence of carbonate minerals in most ancient marine evaporites has been interpreted as evidence for little variation in the chemical composition of seawater over the past 500 million years. Beyond this period, the evaporite record is difficult to interpret because diagenesis and catagenesis have caused substantial and variable alteration of the original minerals. Nevertheless, the predominance of dolomite ($CaCO_3 \cdot MgCO_3$) in evaporites precipitated during the late Precambrian (600 million years before present) suggests the magnesium-to-calcium ion ratios were higher in the ancient ocean.

METEOROLOGICAL AND GEOLOGICAL SETTINGS

Evaporites form in areas where the rate of water loss exceeds the rate of water gain. Evaporation is the sole mechanism for the former, while rainfall, surface flow off the land, and groundwater seeps can act to supply water and ions to the ocean. As illustrated in Figure 4.6, net evaporation in the open ocean presently occurs between 15° and 40°N and between 0° and 35°S. As a result, these are also the locations of the saltiest surface seawater (Figure 4.8) and the locations of most modern evaporites (Figure 17.2).

The formation of modern evaporites is also enhanced by coastal upwelling, which produces arid conditions along coastlines. Cold water rising to the sea surface cools the overlying air mass, causing any water vapor to condense. Since the air mass is dry, it is a poor source of rainwater for these coastal land areas. As a result, many coastal areas located near upwelling zones, such as those in Peru, Chile, northwestern Africa, and the Baja Peninsula, have large deserts and evaporitic sediments.

Evaporite formation is episodic because of fluctuations in weather, climate, water circulation, and sea level. For example, variations in wind velocity and

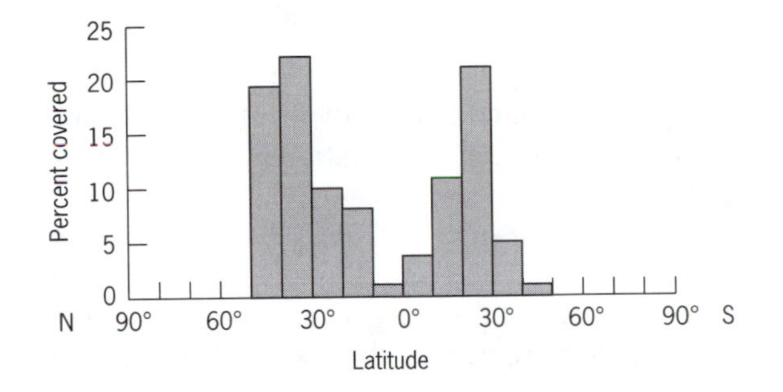

FIGURE 17.2. Latitudinal distribution of the world's modern evaporitic sediments as a percentage of total sediment. *Source:* From *Evaporite Sedimentology,* J. K. Warren, copyright © 1989 by Prentice Hall, Inc., Englewood Cliffs, NJ, p. 15. Reprinted by permission.

humidity can alter net evaporation rates. Changes in temperature can also alter mineral solubility. In deposits with high accumulation rates, seasonal variations produce small-scale layering. Large-scale depositional changes are the result of long-term shifts in climate, water circulation, and sea level.

Net water loss leading to evaporite formation is most commonly achieved by the semi-isolation of a body of seawater from the ocean. A decline in sea level can isolate a marginal sea if the water surface drops below the depth of the sill that connects the basin to the ocean. Similarly, a decrease in sea level would create shallow water depths over the continental shelf, greatly restricting water circulation. Sea level is controlled by climate, tectonic activity, and isostacy.

The growth of large evaporites takes long periods of time. Thus extended periods of stability in sea level are necessary. Under such conditions, evaporite deposition is controlled by plate tectonics and water circulation. The former determines the geological stability of a coastal land mass and its relative elevation with respect to sea level. Slow circulation enhances the rate of evaporite deposition.

The rate of sea level change has been relatively fast over the past few million years due to the great frequency of ice ages. As a result, "modern" evaporites are relatively rare and small in comparison to the ancient salt giants, some of which are 2 km thick and extend for hundreds of square kilometers. Evaporite formation is currently restricted to tidal mud flats (**sabkhat**) and coastal salt lakes (**salinas**). In contrast, the ancient evaporites were deposited in geological settings that no longer exist.

Much of the geological interpretation of the sedimentary record is based on the concept of *uniformitarianism*, which considers the present as a key to the past. This approach can not be applied to the study of ancient evaporites as the geologic conditions that led to their formation do not presently exist. Scientists accustomed to modern geological processes have difficulty in comprehending the relatively bizarre phenomena that must be invoked to explain the formation of the ancient salt giants.

MODERN EVAPORITES

The geologic settings in which marine evaporites are currently being deposited are illustrated in Figure 17.3. The locations of specific examples are given in Table 17.1.

A marine sabkha is an intertidal mud flat common in highly arid regions such as the Arabian Gulf and the Baja Peninsula. Sabkhat are commonly located on broad, flat coastal plains where the subtidal zone is populated by coral reefs. This barrier creates shallow-water lagoons that restrict water flow in the intertidal zone. Evaporation causes the salinity of this trapped water to rise. In the Arabian Sea, which is the world's warmest (20° to 34°C), the salinity of the lagoon seawater ranges from 54‰ to 67‰. Due to high biological activity, the sediments are dominated by algal mats, fecal pellets, shells, and other biogenic materials.

Evaporites form throughout the supratidal zone. The lower supratidal receives seawater only during spring high tides and is characterized by gypsum deposits. In some parts of the Arabian Gulf, this region extends 2.5 km inland. In this location, the middle supratidal, which is flooded by seawater less than once a month, is 1.5 km wide. Here gypsum and aragonite are deposited and diagenetically altered to anhydrite and dolomite, respectively. This is one of the few locations where active dolomite formation has been observed.

Periodic drying causes the surface of the tidal mudflats to break into leathery polygonal chips that are tinged with whitish halite crusts. The upper

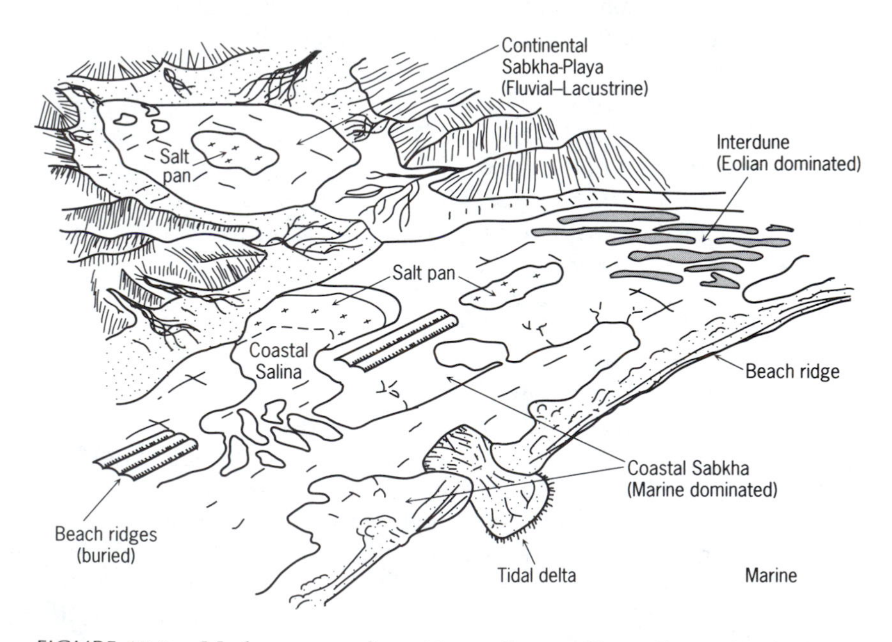

FIGURE 17.3. Modern evaporite settings. *Source:* From *Evaporite Sedimentology*, J. K. Warren, copyright © 1989 by Prentice Hall, Inc., Englewood Cliffs, NJ, p. 20. Reprinted by permission.

TABLE 17.1
Documented Modern Sea-Margin Evaporites

Sabkha
Abu Dhabi, Arabian Gulf
Sabkha Matti, Arabian Gulf
Bardawil Lagoon, North Sinai coast
Western Nile Delta
Gulf of Suez coast
Tunisian coast
Northwest Australian coast
Spencer Gulf, Australia
Baja California, Mexico
Laguna Madre, Texas
Salina
South Coast, Australia
Hutt & Leeman Lagoons, Western Australia
Lake MacLeod, Western Australia
Solar Lake, Gulf of Elat
Ras Muhammad, South Sinai coast
Tunisian coast
Pleistocene coast, Sicily
Pekelmeer, Netherland Antilles

Source: From *Evaporite Sedimentology*, J. K. Warren, copyright © 1989 by Prentice Hall, Englewood Cliffs, NJ, p. 41. Reprinted by permission. See Warren (1989) for data sources.

supratidal receives seawater only from aeolian transport and rare catastrophic floods. This zone is about 4.8 km wide and is covered mostly by anhydrite and halite. Drying is so intense that the winds remove the upper surface. Aeolian transport of this material can produce offshore deposits of evaporites if the relocated materials are rapidly buried.

The deposits in each zone vary in texture, as well as mineralogy, as shown in Figure 17.4. Some of these unique characteristics are preserved and can be used to determine the conditions under which an evaporite was deposited, but usually diagenesis and relocation by winds, waves, and turbidity currents make such interpretations difficult.

In the event of a decline in sea level, the supratidal zone will move seaward, causing evaporites to deposit on top of old lagoonal sediments. Rising sea level "drowns" these evaporites. Low-amplitude fluctuations build up laminated sediments in which layers of biogenic oozes and organic-rich muds alternate with evaporites.

Sabkhat also form inland, where river input and saline groundwater seeps contribute salt and water, forming an evaporitic pan. As illustrated in Figure 17.4, these continental sabkhat are far more isolated from the ocean than a marine sabkha. They also contain far less biogenic detritus.

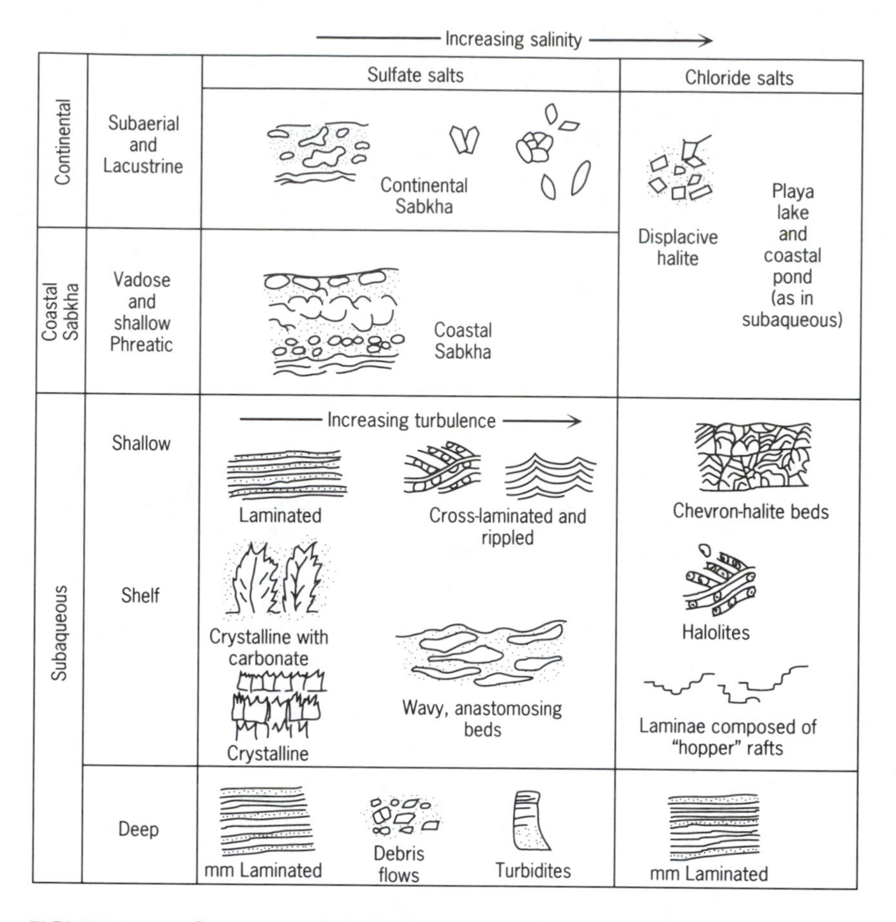

FIGURE 17.4. Summary of the various evaporitic textures indicative of particular physical environments. *Source:* From *Evaporite Sedimentology,* J. K. Warren, copyright © 1989 by Prentice Hall, Inc., Englewood Cliffs, NJ., p. 15. Reprinted by permission. See Warren (1989) for data sources.

Saline lakes, or salinas, are another type of continental evaporite. Most have formed in depressions behind calcareous dunes along the margins of the Mediterranean Sea, the Arabian Gulf, and the western coast of Australia. Seeps of saline groundwater supply salt to these brine pools. In the salinas of southwestern Australia, the evaporites deposit in a bull's-eye pattern. As illustrated in Figure 17.5, stromatolites and carbonates form along the rim and gypsum is deposited at the lake's center. These are the only locations where gypsum is presently depositing subaqueously. In some of the salinas, as much as 10 m have been deposited over the past 6000 y.

Stromatolites are domal, pillar-like structures composed of fossilized layers of algal mats. Although the organic detritus has been largely replaced by calcium carbonate, enough of the original texture remains to be identifiable as a biogenic deposit. The oldest known fossils are stromatolites, which were

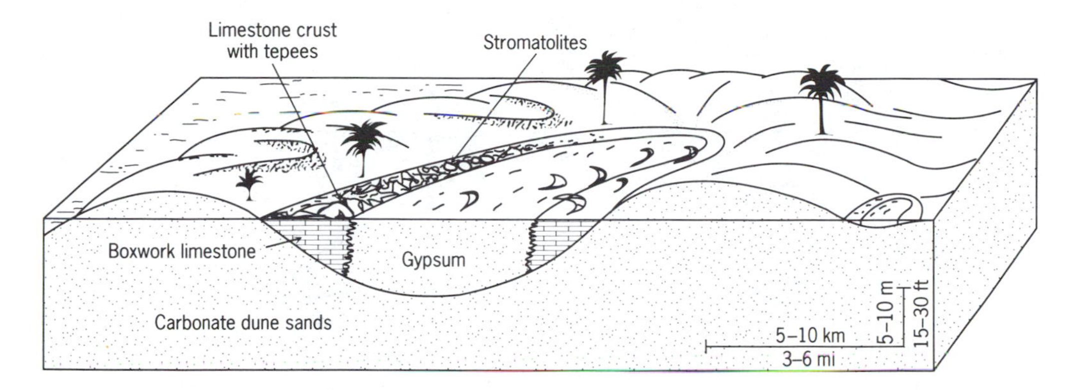

FIGURE 17.5. Depositional facies in the salinas of Southwestern Australia. *Source:* From J. K. Warren and G. C. St. C. Kendall, reprinted with permission from the *American Association of Petroleum Geologists Bulletin,* vol. 69, p. 1017, copyright © 1985 by the American Association of Petroleum Geologists, Tulsa, OK.

deposited over 3 billion years ago, probably by blue-green algae or cyanobacteria.

ANCIENT EVAPORITES

Geologic Variations in the Rate of Evaporite Deposition

The oldest known marine evaporites were deposited 3.45 billion years ago in what is now western Australia. They appear to have precipitated as shallow-water carbonates, suggesting that the chemical composition of seawater was significantly different. The presence of sulfate salts indicates the atmosphere was oxic; otherwise sulfide minerals would have been deposited.

As shown in Figure 17.6, the sedimentary record contains a greater volume of evaporites from the Proterozoic Era (0.6–2.6 billion years before present) than from the Archean Era (more than 2.6 billion years ago). Though this could be the result of better preservation, most geochemists think a change in hydrogeochemical conditions occurred that increased the rate of evaporite formation. Evaporites from the Proterozoic also show evidence of having been deposited in shallow marine, coastal, and riverine settings under oxic conditions. The prevalence of dolomite suggests a relatively high magnesium-to-calcium ion ratio was present in these waters.

The Phanerozic Era (2–600 million years before present) was characterized by the deposition of massive evaporites, the so-called salt giants, particularly during the Upper Cambrian, Permian, Jurassic and mid-Tertiary Periods. The Messinian evaporites of the Mediterranean Sea were deposited during the Miocene Epoch. As shown in Figure 17.7, the rate of evaporite formation was

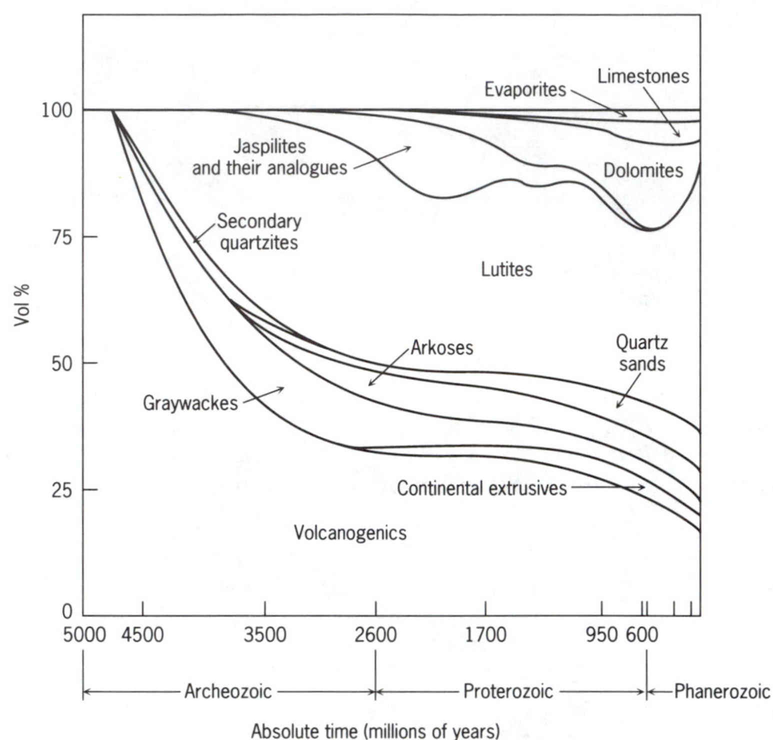

FIGURE 17.6. Volume percent of sedimentary rock as a function of age. *Source:* From A. B. Ronov, reprinted with permission from *Geochemistry,* vol. 8, p. 714, copyright © 1964 by Scripta Technica, Inc., Silver Springs, MD.

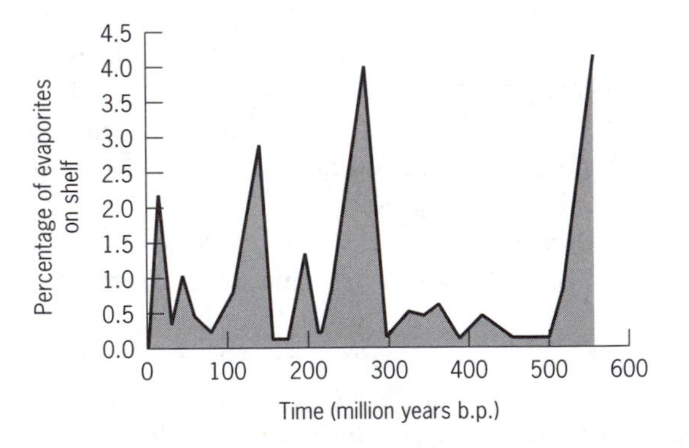

FIGURE 17.7. Distribution of evaporites as a percentage of the sediments on the continental shelves from the Cambrian to the present. (b.p., before the present.) *Source:* From *Evaporite Sedimentology,* J. K. Warren, copyright © 1989 by Prentice Hall, Inc., Englewood Cliffs, NJ, p. 2. Reprinted by permission. After A. B. Ronov, K. E. Khain, A. N. Balukhovsky, and K. B. Seslavinsky, reprinted with permission from *Sedimentary Geology,* vol. 25, pp. 319–320, copyright © 1980 by Elsevier Science Publishers, B.V., Amsterdam, The Netherlands.

very variable during the Phanerozoic, but always exceeded that of the Quaternary. The mineralogy of these massive evaporites suggests seawater had attained its present chemical composition by the beginning of this era.

The evaporites formed during the Quaternary Era (the past 2 million years) compose a relatively small part of the marine sediments. Evaporites presently account for less than 0.1% of the actively forming continental shelf sediments. This scarcity is the result of rapid, high amplitude changes in sea level caused by the frequency of ice ages, particularly during the Pleistocene Epoch. During this time interval, rising sea level has submerged shallow-water evaporites, sometimes leading to their erosion and redeposition in deeper waters. Declining sea level has exposed some deep-water evaporites and caused them to switch to a sabkha mode of formation.

Modes of Formation

The ancient massive salt deposits were laid down under geological and climatological settings that do not presently exist. Climate was warmer, so arid zones were more intense and widespread. The continental margins were broader and shallower, forming a large shallow area with restricted circulation. Sea level was relatively stable and tectonic collisions between lithospheric plates had created several large marine basins with restricted access to the open ocean.

During the Orodovician, Permian, and Cretaceous periods, the continental shelves were covered by only a few centimeters of water. This was too shallow for currents or tidal exchange to circulate the water. Evaporation increased salinities enough to cause halite and gypsum to precipitate, forming evaporites that extend over thousands of square kilometers and range in thickness from 5 to 10 km.

Evaporite deposition also occurred in basins whose shallow sills restricted water exchange with the open ocean. As illustrated in Figure 17.8, several basinal settings were possible. First, if sea level dropped below the depth of the sill, the basin was completely isolated from the ocean. As the trapped seawater evaporated, basin-wide evaporites were deposited. Most of the Mediterranean evaporites were formed in this shallow-water, deep-basin setting.

Evaporites also formed when the sill was below sea level, as illustrated in Figure 17.8*b*. In this setting, net evaporation caused dense waters to collect below the depth of the sill. Salts that precipitated in the surface waters were carried to the basin floor by pelagic sedimentation. The stagnancy of the bottom waters prevented flushing of the particles, thereby insuring their sedimentation. This type of setting produced the depositional pattern illustrated in Figure 17.9. Since the salt content of the basin water increased with increasing distance from the sill, deposits grade diagonally, as well as vertically, with the most soluble mineral (halite) having been precipitated last and furthest landward.

Some basinal evaporites appear to have been deposited by turbidity currents. As shown in Figure 17.8*c*, this caused the basin to fill with reworked

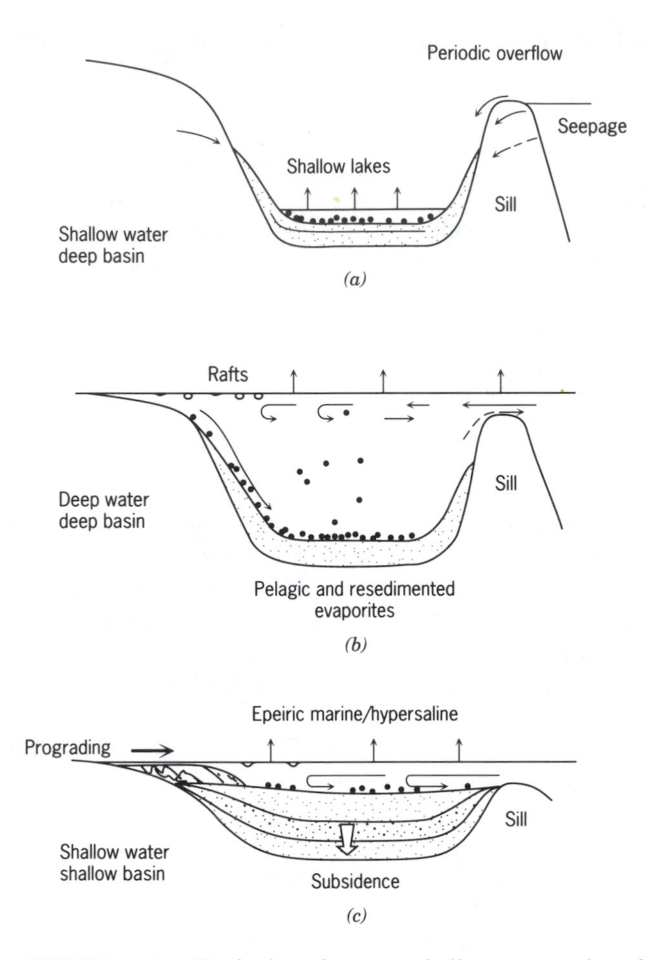

FIGURE 17.8. Basinal settings: *(a)* shallow-water, deep basin; *(b)* deep-water, deep basin; *(c)* shallow-water, shallow basin. *Source:* From *Facies Models,* 2nd ed., A. C. Kendall (ed.: R. G. Walker), copyright © 1984 by Geological Survey of Canada, Ottawa, Canada, p. 290. Reprinted by permission.

salts that had been originally deposited in shallow waters. The resulting decline in water depth promoted the in situ precipitation of evaporites. Subsidence would have kept the basin floor below the sill.

Dolomites

Dolomites compose a significant portion of ancient evaporites. Having no modern examples of dolomite precipitation beyond those of a few salinas and sabkhat, geochemists have been unable to determine conclusively how the ancient deposits must have formed. Though the dolomites of Australian salinas are produced by direct precipitation, this mechanism is thought to have been

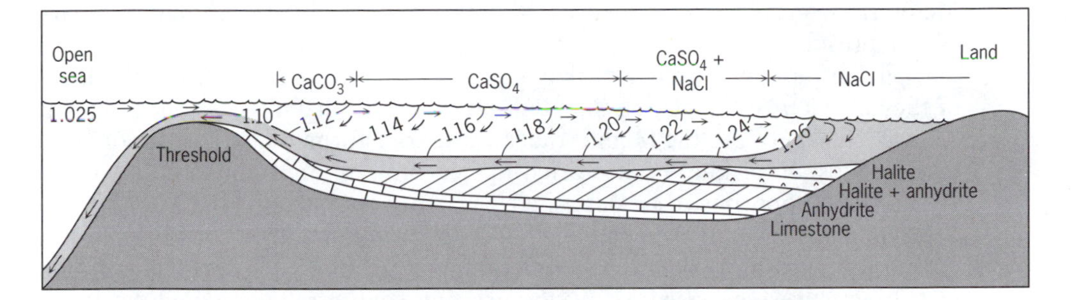

FIGURE 17.9. Schematic longitudinal profile through a semi-isolated basin located in a hot, arid climate and separated from the open sea by a narrow portal. The sill depth, although shallow, is still great enough to permit some two-way flow of surface water. The lines show inferred seawater density (g/cm^3) and the arrows show current directions. The pattern of evaporite deposition is based on the relationships between brine density and precipitate composition as shown in Figure 17.1, assuming that salt particles accumulate on the seafloor through the process of pelagic sedimentation. *Source:* From P. C. Scruton, reprinted with permission from the *American Association of Petroleum Geologists Bulletin,* vol. 37, p. 2505, copyright © 1953 by the American Association of Petroleum Geologists, Tulsa, OK.

relatively rare. Ancient dolomites were probably produced by diagenesis that involved the circulation of hypersaline brines through evaporites prior to deep burial. The ultimate source of this dolomitizing solution was probably seawater, concentrated into a dense, Mg-rich solution. Due to its density, this brine probably sank into the sediments, displaced the pore waters, and seeped through the carbonate deposits.

To convert calcium carbonate to dolomite, some of the calcium must be replaced by magnesium, requiring the partial dissolution of the carbonate. This process is promoted by contact with acidic pore water, such as occurs in organic-rich sediments where remineralization has produced carbon dioxide. This is probably why dolomites are presently forming in detrital algal mats buried beneath sabkhat. The restricted extent of these modern dolomites reflects a kinetic hindrance to precipitation. Apparently dolomite precipitation is too slow to form substantial deposits during times of rapid fluctuations in sea level.

THE GREAT SALINITY CRISIS OF THE MEDITERRANEAN SEA

The ancient evaporites of the Phanerozoic Era were deposited at rates as fast as 100 m per 1000 y. These rapid rates are thought to have been caused by a lowering of sea level resulting from tectonic activity and glaciation. Some of the largest of the salt giants are the Messinian evaporites formed in the

Mediterranean Sea during the late Miocene Epoch, 5.0 to 6.5 million years before present.

During this time, global cooling caused sea level to drop below the level of the Gibraltar Sill, isolating the Mediterranean Sea from the Atlantic Ocean. As water evaporated, the Mediterranean was transformed into a series of large inland lakes. Even these lakes dried up, causing an enormous volume (1×10^6 km^3) of evaporites to be deposited.

The modern Mediterranean Sea has a water volume of 3.7×10^6 km^3. Net evaporation presently occurs at a rate of 3.3×10^3 km^3 per year. If the Straits of Gibraltar were exposed today, the present Mediterranean would dry up in about 1000 years. Assuming a salinity of 35‰, a seasalt density of 1.35 g/cm^3, and an average basin depth of 2700 m, complete evaporation of the Mediterranean Sea would leave a layer of salt 70 m deep. Since some of the Messinian deposits are 3000 m thick, they must have been formed by the evaporation of more than one basin of seawater. In fact, to account for the volume of salt in these evaporites, the Mediterranean would have had to been refilled and evaporated at least 40 times during the later ages of the Miocene.

The presence of halite in the Messinian evaporites suggests that Mediterranean seawater must have been very salty. The absence of fossils indicates that marine organisms could not survive under these conditions. The underlying cause for this "salinity crisis" appears to have been the compounded effects of tectonic activity and a glacioeustatic change in sea level. The collision of Africa with Europe closed all connections to the ocean except for shallow sills. These were exposed by the lowering of sea level caused by an intense ice age. The presence of erosional surfaces in now deeply buried deposits in the center of the Mediterranean basin suggests evaporite deposition occurred under shallow-water conditions. These surfaces are probably ancient river valleys cut during a time of dramatically lower sea level.

Thus the basins must have been refilled in an episodic fashion such that conditions favoring shallow-water evaporite deposition were maintained. Some geologists have proposed that changes in sea level caused periodic flows of water from the Atlantic Ocean into the nearly dry Mediterranean Sea basin. This must have taken the form of a waterfall, hundreds of meters in height, which flowed over the exposed Gibraltar Sill! The episodic nature of this process is reflected in the repeating evaporite sequences found throughout the Messinian deposits.

The amount of salt sequestered in the Messinian evaporites represents about 6 percent of all the salt presently dissolved in the ocean. Its removal must have caused an average reduction of 2‰ in the salinity of the remaining seawater. This decrease could have had a significant effect on ancient climate as the colligative properties of seawater would have been altered. For example, an increase in the freezing point would have increased the amount of ice formation, thereby intensifying global cooling because ice reflects insolation. In fact, some scientists hypothesize that the formation of evaporites, as a result of the initial isolation of the Mediterranean Sea, caused the global cooling that occurred during the late Miocene.

SUMMARY

Evaporites are sedimentary rocks that contain a large fraction of the crustal sodium, calcium, chlorine, sulfate, potassium, and magnesium. Evaporites form in areas where meterological and geological conditions cause the rate of water loss to exceed the rate of water gain. As a result, evaporites are most abundant where net evaporation rates are highest (i.e., between 15° and 40°N and between 0° and 35°S).

Most evaporites are layered deposits composed of minerals that precipitated in a sequence dictated by their relative solubilities. The evaporation of seawater causes the precipitation of **aragonite**, **calcite**, **gypsum**, **anhydrite**, and **halite**, which make up the bulk of evaporitic rocks. Large amounts of dolomite are present in ancient deposits and appear to have been formed by the diagenetic and catagenetic alteration of calcite and aragonite. If diagenesis has not completely altered the rock, its mineralogy, texture, and sequence of deposition can provide information on the hydrogeochemical conditions under which the evaporites formed.

The Phanerozic Era (2–600 million years before present) was characterized by the deposition of massive evaporites called salt giants. These ancient evaporites were deposited in geological settings that do not presently exist (i.e., very shallow continental shelves and marginal seas with greatly restricted access to the ocean). Large deposits were formed due to the relative stability of sea level and higher aridity. The mineralogy of these massive evaporites suggests seawater had attained its present chemical composition by the beginning of this era.

Some of the largest salt giants are the Messinian evaporites of the Mediterranean Sea that were deposited during the Miocene. The amount of salt sequestered in these evaporites represents about 6 percent of all the salt presently dissolved in the ocean. Its removal must have caused an average reduction of 2‰ in the salinity of the remaining seawater. This decrease could have had important effects on ancient climate as the colligative properties of seawater would have been altered. For example, an increase in the freezing point would have increased the amount of ice formation, thereby intensifying global cooling because ice reflects insolation. In fact, some scientists hypothesize that the formation of evaporites, as a result of the initial isolation of the Mediterranean Sea, caused the global cooling that occurred during the late Miocene.

The evaporites that have formed during the Quaternary Era (the past 2 million years) compose a relatively small part of the marine sediments. Evaporite formation is currently restricted to tidal mud flats (**sabkhat**) and coastal salt lakes (**salinas**). As a result, evaporites presently account for less than 0.1 percent of the continental shelf sediments that are now forming. This scarcity of modern evaporites is the result of rapid, high-amplitude changes in sea level due to the frequency of ice ages, particularly during the Pleistocene Epoch.

CHAPTER 18

IRON–MANGANESE NODULES AND OTHER HYDROGENOUS MINERALS

INTRODUCTION

Much of human technology relies on a ready source of relatively pure metals. As we have begun to deplete the terrestrial reservoirs, attention has turned to metal deposits buried in marine sediments. In some locations, highly concentrated ores are present as massive **polymetallic sulfides** or **iron–manganese oxides**. The latter occur as **nodules**, **metalliferous sediments**, and crusts. These deposits are all hydrogenous precipitates containing metals derived mostly from hydrothermal activity.

Other sedimentary minerals are the result of precipitation reactions that involve the partial dissolution and reprecipitation of various types of biogenic detritus. These include phosphorite, barite, oolite, and glauconite. Some, such as phosphorite, are also being considered for commercial exploitation. The mechanisms by which these hydrogenous minerals form is discussed in this and the next chapter.

IRON–MANGANESE OXIDES

Trace metals are introduced to the ocean by atmospheric fallout, river runoff, and hydrothermal activity. The latter two are sources of soluble metals, which are primarily reduced species. Upon introduction into seawater, these metals react with O_2 and are converted to insoluble oxides. Some of these precipitates settle to the seafloor to become part of the sediments; others adsorb onto surfaces to form crusts, nodules, and thin coatings on rocks. Since reaction rates are slow, the metals can be transported considerable distances before becoming part of the sediments. Thus a significant fraction of the riverine input is deposited on the continental shelves. Most of the metals that precipitate in the open ocean are derived from hydrothermal emissions.

Trace metals tend to coprecipitate; thus the oxides are geographically variable in composition, heterogeneous, and often contain occluded seawater. This makes them difficult to define stoichiometrically. The most abundant metals in these oxides are iron and manganese, with others present in much

TABLE 18.1
Average Composition of Manganese Nodules from the Different Oceans

					Favorable North Pacific Area[c]	
Minerals	*South Pacific*[a]	*North Pacific*[a]	*West Indian*[a]	*Atlantic*[b]	*Red Clay*	*Siliceous Oozes*
Manganese	16.61	12.29	13.56	16.1	17.43	22.36
Iron	13.92	12.00	15.75	21.82	11.45	8.15
Nickel	0.433	0.422	0.322	0.297	0.76	1.16
Copper	0.185	0.294	0.102	0.109	0.50	1.02
Cobalt	0.595	0.144	0.358	0.309	0.28	0.25

Source: From *Introduction to Oceanography*, D. A. Ross, copyright © 1982 by Prentice Hall, Englewood Cliffs, NJ, p. 411. Reprinted by permission. Data from [a] D. S. Cronan, 1967, doctoral dissertation, Imperial College, University of London, London, England, and D. S. Cronan and J. S. Tooms, reprinted with permission from *Deep-Sea Research*, vol. 16, p. 344, copyright © 1969 by Pergamon Press, Elmsford, NY; [b] D. S. Cronan, reprinted with permission from *Nature, Physical Science*, vol. 235, p. 172, copyright © 1972 by Macmillan Journals, Ltd., London, England; and [c] *Ferromanganese Deposits of the North Pacific*, D. R. Horn, B. M. Horn, and M. N. Delach, copyright © 1972 by the National Science Foundation, Washington, DC, pp. 40–54. Reprinted by permission.

lower and variable amounts, as illustrated for the iron–manganese nodules in Table 18.1.

Traces of two dozen other metals have also been observed. Some polymetallic oxides are present in the crystalline phase. In the nodules, the most commonly observed crystalline structures are birnessite and todorokite. As indicated by the empirical formulae given in Table 18.2, the two differ in their degree of oxidation and metal enrichment. This is thought to be caused by differences in the redox conditions under which the nodules form.

The nodules range in diameter from 20 μm to 15 cm, with most between 1 and 10 cm. Though iron–manganese nodules are found throughout the ocean basins, their densities are highest in regions of low sedimentation rate (i.e., beneath the Southern Ocean and the mid-ocean gyres in the Pacific Ocean), as shown in Figure 18.1.

TABLE 18.2
The Most Common Minerals Found in Iron–Manganese Nodules

Mineral Type	*Empirical Formula*
Birnessite	$(Na_{0.7}Ca_{0.3})Mn_7O_{14} \cdot 2.8H_2O$ $Na_4Mn_{14}O_{27} \cdot 9H_2O$, $Mn_7O_{13} \cdot 5H_2O$ $Mn_7O_{12} \cdot 6H_2O$
Todorokite	$(Ca, Na, Mn^{2+}, K)(Mn^{4+}, Mn^{2+}, Mg)_6O_{12} \cdot 3H_2O(Na, Ca, Mn^{2+})_2Mn_5^{4+}O_{12} \cdot 3H_2O$

Source: From D. A. Crerar and H. L. Barnes, reprinted with permission from *Geochimica et Cosmochimica Acta*, vol. 38, p. 280, copyright © 1974 by Pergamon Press, Elmsford, NY.

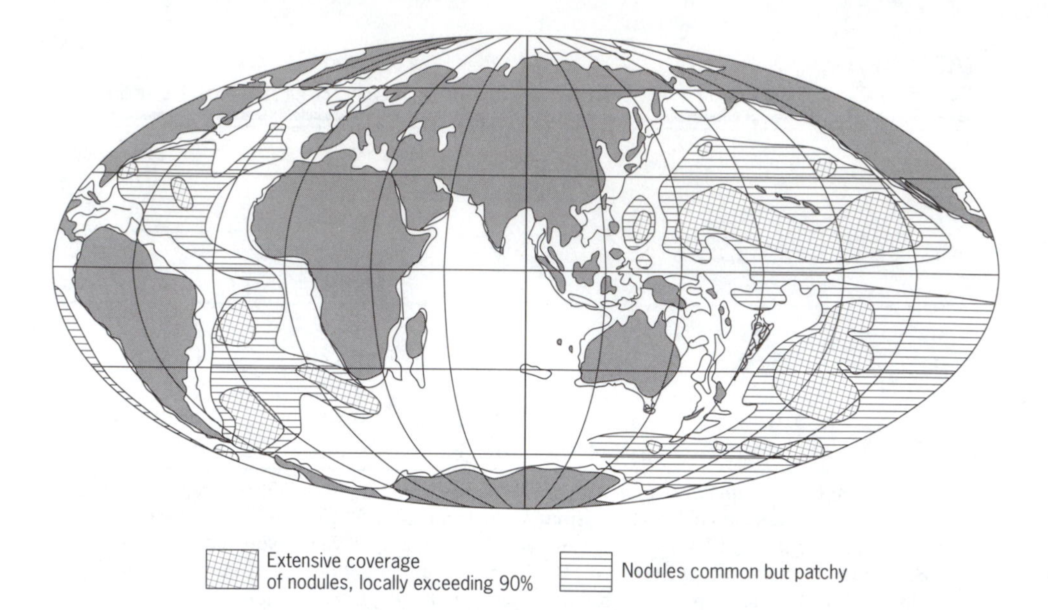

FIGURE 18.1. General distribution of Iron–Manganese nodules in the Pacific and Atlantic oceans. *Source:* From *Marine Geology,* J. Kennett, copyright © 1982 by Prentice Hall, Inc., Englewood Cliffs, NJ, p. 500. Reprinted by permission. After *Marine Manganese Deposits,* D. S. Cronan (ed.: G. P. Glasby), copyright © 1977 by Elsevier Science Publishers, B.V., Amsterdam, The Netherlands, pp. 16–21. Reprinted by permission.

In some locations, the nodules are so dense that they form a pavement, as illustrated in Figure 18.2. These surfaces inhibit erosion by shielding the underlying sediment from bottom currents.

The nodule interiors are characterized by concentric banding, as shown in Figure 18.3. These bands generally conform to the shape of a centrally located particle. These particles are thought to act as a nucleus, initiating and promoting the deposition of metals, in a fashion similar to the enhancement of raindrop formation by atmospheric dust. These precipitation nuclei are typically small rocks, bone fragments, or even shark's teeth. The banding is thought to reflect changes in growth rate and chemical composition caused by changes in the depositional environment.

The Origin of Iron–Manganese Nodules

One of the most curious aspects of the nodules is their very slow growth rate. Radiochemical analyses indicate that the nodules grow, or accrete, at rates ranging from 1 to 200 mm per million years. (The methods used to "date" nodules are described in Chapter 28.) The average rate of sedimentation on the abyssal plains, the location of most nodules, is approximately 1 mm per thousand years. Since the sediments are accumulating much faster than the nodules, the nodules should be buried. Instead, 75 percent of the nodules lie

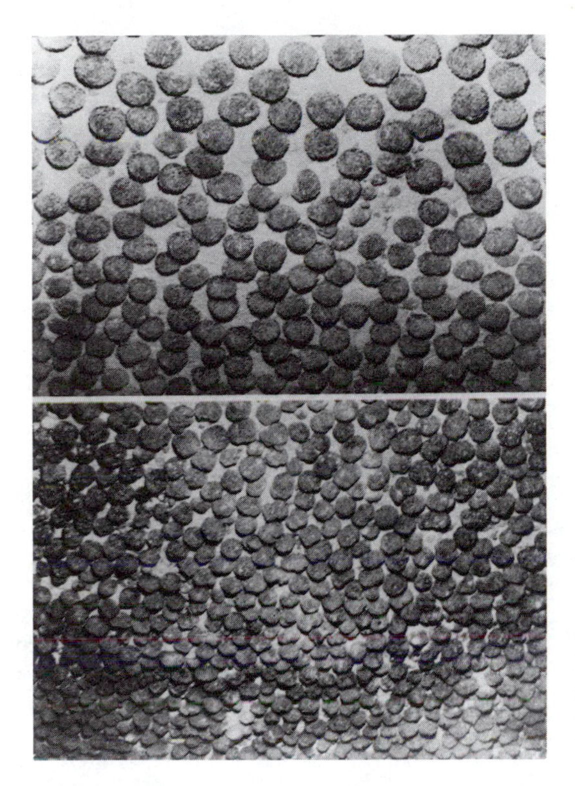

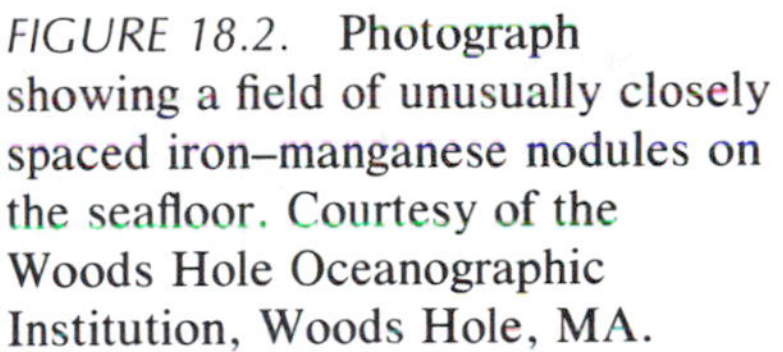

FIGURE 18.2. Photograph showing a field of unusually closely spaced iron–manganese nodules on the seafloor. Courtesy of the Woods Hole Oceanographic Institution, Woods Hole, MA.

on the sediment surface. The rest are located within the top 4 m of the sedimentary column.

Three theories have been proposed to explain this apparent paradox, as well as the great variability in chemical composition. Data were collected during the Manganese Nodule Project (MANOP) in the 1970s and 1980s to test these theories. The results are presented below.

Formation by Hydrogenous Precipitation

A true hydrogenous precipitate forms as a result of the direct precipitation of solutes from seawater. Some iron–manganese nodules appear to form in this fashion via the following precipitation reactions:

$$2Fe^{2+}(aq) + 2O_2(aq) + 2OH^-(aq) \rightarrow Fe_2O_3(s) + H_2O(l) \qquad (18.1)$$

$$2Mn^{2+}(aq) + O_2(aq) + 4OH^-(aq) \rightarrow 2MnO_2(s) + 2H_2O(l) \qquad (18.2)$$

The products are thermodynamically favored under the oxic alkaline conditions that are characteristic of most of the ocean. Reaction rates are slow, so metal oxides tend to precipitate onto detritus or preexisting nodules due to the catalytic effect of the surfaces. Metals compose 15 percent of the mass of these precipitates. Since the dissolved metal concentrations are on the order of parts per billion, hydrogenous precipitation greatly concentrates the metals.

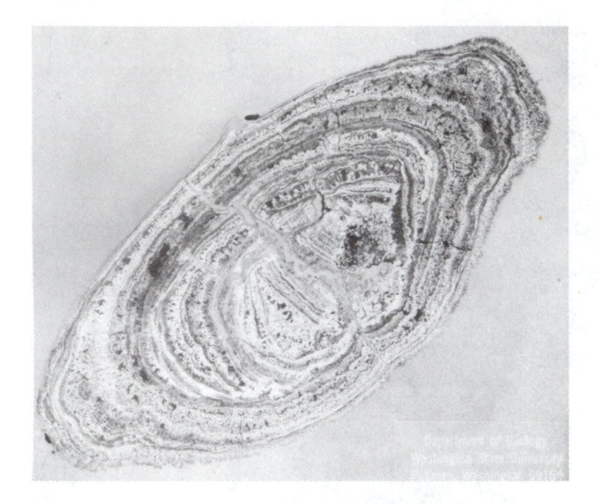

FIGURE 18.3. A polished cross-section (5.3 cm) of a manganese nodule photographed in vertical illumination. The nodule is from the North Pacific and illustrates concentric banding as well as the presence of a distinct central nucleus. Bright areas are opaque oxides; dark areas are mostly clay-sized silicate particles. The core, which nucleated the nodule's growth, consists of broken older nodule fragments. The thin smooth white layer along the left surface marks the top of the nodule as it rested on the seabed. It is enriched in Fe and depleted in Mn, Ni, and Cu compared to the bottom of the nodule. Other layers of different reflectivity, grading from bright to dark, indicate that the nature and rate of mineral deposition varied over time. As nodules increase in size, transverse as well as concentric fractures develop and enlarge, so that the nodules eventually fall apart. *Source:* From *Manganese Nodules: Research Data and Methods of Investigation,* R. K. Sorem and R. H. Fewkes, copyright © 1979 by Plenum Press, New York, p. 521. Reprinted by permission.

Since hydrogenous nodules form from solutes in seawater, their chemical composition should be independent of the underlying sediment. The ultimate source of their metals is thought to be hydrothermal effluents that are relatively enriched in the reduced species. Most of these metals are deposited close to their point of entry, forming an iron–manganese crust over the sediments near the hydrothermal vents. But due to slow reaction rates, a significant amount is transported horizontally by bottom currents in dissolved or colloidal form. Eventually these metals precipitate, often onto the surface of an iron–manganese nodule.

Formation by Suboxic Diagenesis

The growth of some nodules appears to be the result of postdepositional remobilization of metals from the underlying sediments. In areas of high biological productivity, the sedimentation rate of POM is also large. Remineralization of this POM lowers the dissolved oxygen concentrations, causing reducing conditions in the sediments. In pelagic sediments, this reducing zone is located between 1 mm and 1 m below the sediment–water interface, depending on the relative supply rates of POM and O_2. The latter is controlled by the rates of thermohaline circulation and bioturbation.

As sedimentary layers are buried, they eventually pass through the redox boundary and enter the reducing zone. Any manganese oxides in the layer will then be solubilized by the second reaction listed in Table 12.1, which describes the reduction of manganese via the oxidation of POM. Manganese is also present in the sediments as a component of POM, as this metal is concentrated in the tissues of marine organisms. This manganese is also resolubilized during the oxidation of POM.

The resolubilized metal that diffuses upward precipitates at the redox boundary as the insoluble oxide. If the redox boundary lies close to the sediment surface, precipitation can occur on detritus lying on top of, or partially submerged in, the sediment–water interface. This mechanism has been used to explain the formation of iron–manganese nodules in areas subject to moderately large POM fluxes. Nodules formed under these conditions are notable for their high manganese contents.

The observation that nodules are rarely found in estuaries and upwelling areas indicates that there is a limit to which organic matter can enhance their growth. Biogenic grains tend to be incorporated in these nodules, diluting the metal concentrations and hence decreasing their economic worth. At very high fluxes, the nodules will be buried.

Formation by Oxic Diagenesis

Most iron–manganese nodules are located in areas where the POM flux is not high enough to keep the redox boundary close to the sediment surface. The oxic conditions in these sediments inhibit postdepositional remobilization. Two mechanisms have been proposed to explain how particulate metals could be transported within such sediments so as to support the growth of iron–manganese nodules.

Nodules collected from oxic siliceous oozes in the tropical North Pacific have chemically distinct tops and bottoms. Their bottoms are relatively enriched in trace metals, suggesting that sedimentary iron and manganese are somehow mobilized under oxic conditions. At this location, most of these metals are transported to the seafloor as components of volcanic ash and biogenic hard parts. Marine chemists hypothesize that these metals are solubilized in reducing microzones maintained by rapid rates of microbial respiration. Metals mobilized within these microzones then diffuse through the sediments for substantial distances due to slow rates of oxidation. These metals would be most likely to deposit on surfaces, such as those of nodules.

Bioturbation could also play an important role in metal transport by ensuring that iron- and manganese-rich particles are brought into close contact with the nodules. Once this is achieved, only a short-term mobilization is required to transfer the metals to the nodules. While exactly the right combination of biological activities might seem unlikely, the metal content of nodules produced in oxic settings requires the precipitation of only 1 percent of the manganese, nickel, and cobalt present within a 5 cm radius.

Accretion Rates

The observation that nodules grow at widely varying rates provides further support for the existence of multiple formation mechanisms. The nodules that accrete slowest (1 mm per million years) appear to have formed by the process of hydrogenous precipitation. This accretion rate is equivalent to the annual deposition of a layer that is only one atom deep. These slow rates cause a significant amount of metal-rich seawater to become occluded between the iron–manganese oxide layers.

The nodules that form at rates on the order of tens of millimeters per million years appear to have been produced by postdepositional remobilization under oxic conditions. These nodules have relatively high copper and nickel contents. The nodules that accrete at the fastest rates (200 mm per million years) appear to have formed by postdepositional remobilization under suboxic conditions. Despite these rapid accretion rates, the ''suboxic''-type nodules account for only half of those found in areas where biological productivity is high. The other half appear to have been formed by ''oxic'' remobilization. Hydrogenous precipitation appears to produce only a small percentage of the nodules.

"Suboxic"-type nodules have been found in surface sediments where the redox barrier is located several centimeters below the sediment–water interface. Obviously, the suboxic mechanism for nodule growth can not operate across such large distances. To explain the presence of these nodules far above the redox boundary, marine chemists hypothesize that the position of the boundary has varied over time. In other words, these nodules grow only when the redox barrier is shallow. Shifts in the depth of the redox boundary are probably related to changes in the rates of supply of POM and/or O_2.

This suggests that climatological events, such as ice ages and even seasons, could alter nodule growth rates. Support for this comes from radiochemical dating of some of these ''stranded'' nodules. At depths in the nodules corresponding to 40,000 years before present, accretion rates increase from 50 to 200 mm per million years. Sediment deposited during this time period is enriched in organic matter, suggesting the occurrence of reducing conditions and an increased supply of trace metals. The concentric bands and color variations are also probably the result of changes in paleoceanographic conditions.

Even the nodules growing at the fastest rates ought to be buried but aren't. Several theories have been advanced to explain how the nodules maintain their position on the sediment surface. For example, earthquakes could rock the nodules frequently enough to keep their tops clear of sediment. Bottom currents could be strong enough to sweep the nodules free of sediment. This explanation is supported by the presence of iron–manganese nodule pavements in areas subject to fast bottom currents.

The effects of wind winnowing by bottom currents could also explain why

25 percent of the nodules are buried in the sediments. Variations in local rates of sediment deposition and erosion could cause the periodic burial and exposure of nodules. Growth would be expected only during periods of exposure. This would also explain why buried nodules appear unaffected by diagenesis (i.e., they are not buried long enough to undergo extensive alteration). Another likely explanation for the concentration of nodules at the sediment surface is linked to bioturbation. If benthic animals are active enough, their crawling and burrowing could jostle the nodules, knocking off enough sediment to prevent burial.

Geomicrobiology of Iron–Manganese Nodules

The presence of bacteria and protozoans on nodules was first reported in the 1870s as part of the published records of the Challenger Expedition. These observations are not particularly surprising given that, upon submergence in seawater, all surfaces are rapidly covered with a coating of adsorbed organic matter. This organic matter represents a concentrated food source to microbes. Hence surfaces are rapidly colonized by bacteria, followed by protozoa that feed on the bacteria.

The nodule bacteria synthesize negatively charged, extracellular polymers that act as ion exchangers adsorbing cations such as Mn^{2+}. These polymers are exuded as microfibrils, which significantly extend the surface area available for metal adsorption. The adsorbed ions are eventually converted to MnO_2 as a result of extended contact with oxic seawater.

These bacteria are consumed by protozoans such as foraminifera that tend to grow in clusters referred to as agglutinations. As protozoa feed on the bacteria, they "browse" across the nodules. This activity could help keep the nodules free of sediment. The manganese sequestered by the bacteria is passed through the digestive tract of the foraminifera and deposited in their fecal material. This manganese-rich detritus is called a stercome. The stercomata become part of the nodule, as do the calcareous tests of the foraminifera after their death. Eventually, these tests dissolve, leaving the iron–manganese oxides as a type of fossilized remain.

Rhizopod protozoans have also been observed living on nodules. These organisms incorporate manganese into their shells and protoplasm. Their stercomata are also enriched in manganese. Though deposition of manganese oxides by all of these fauna is a possible mechanism by which nodules could form under oxic conditions, no direct observations of this have been reported. In fact, microbes are just as likely to be removing as depositing manganese. Since their fecal deposits are rich in organic matter, they probably harbor reducing microzones in which manganese could be resolubilized.

Current Progress in the Mining of Iron–Manganese Nodules

The known deposits of nodules contain 10^9 tons of metals. Of greatest economic interest are nickel, copper, and cobalt. The possibility of exploiting this resource has been hampered by the legal question as to who owns the deep-sea floor. The international community has attempted to resolve this problem by formulating a joint agreement called the Law of the Sea Treaty. As of this writing, the United States is one of several major powers that has not ratified the treaty.

Several technological problems must also be solved prior to initiating a profitable mining industry. First, the sites where metal-rich nodules are most abundant must be determined. As illustrated in Figures 18.1 and 18.4, the equatorial Pacific Ocean appears, in general, to contain the densest collections of economically attractive nodules. The oxic mechanism of formation is thought to be responsible for their relatively elevated copper, nickel, and cobalt contents.

Second, a cost-efficient way must be found to remove the nodules from the seabed. One proposed technique is illustrated in Figure 18.5. Third, a cost-effective method for separating and purifying the metals must be developed. Finally, the environmental consequences of disturbing the seabed must be evaluated.

OTHER HYDROGENOUS MINERALS

In regions of high productivity, the biogenic detritus that reaches the sediment is often transformed into new mineral phases. This process can be thought of as a fossilization, where total or partial dissolution is followed by reprecipitation or isomorphic substitution. **Phosphorites**, **barites**, **glauconites**, and **oolites** are the most common examples. They are formed only in shallow waters, particularly on continental shelves in upwelling areas, as this is where the rates of sedimentation of biogenic detritus are highest. These minerals usually occur commingled as noncrystalline conglomerates.

Phosphorites are sedimentary rocks composed primarily of fluoroapatite—$Ca_5(PO_4)_3F$—and other calcium phosphate minerals. These hydrogenous precipitates form from the partial dissolution and recrystallization of skeletal and shell material. In the case of the latter, carbonate is slowly replaced by phosphate. Phosphorites occur as nodules, pebbles, slabs and conglomerates. The nodules grow slowly, at rates of 1 to 10 mm per thousand years. The conglomerates are mixtures of phosphatized carbonate grains and large fossils. On land, exposed ancient marine phosphorites are renowned for their fossils.

FIGURE 18.4. Concentration of (*a*) copper, (*b*) nickel, and (*c*) cobalt in iron–manganese nodules found on the seafloor. *Source:* From *Introduction to Oceanography*, 3rd ed., D. A. Ross, copyright © 1982 by Prentice Hall, Inc., Englewood Cliffs, NJ, pp. 411–412. Data from *Ferromanganese Deposits of the North Pacific,* D. R. Horn, B. M. Horn, and M. N. Delach, copyright © 1972 by the National Science Foundation, Washington, DC, pp. 40–54. Reprinted by permission.

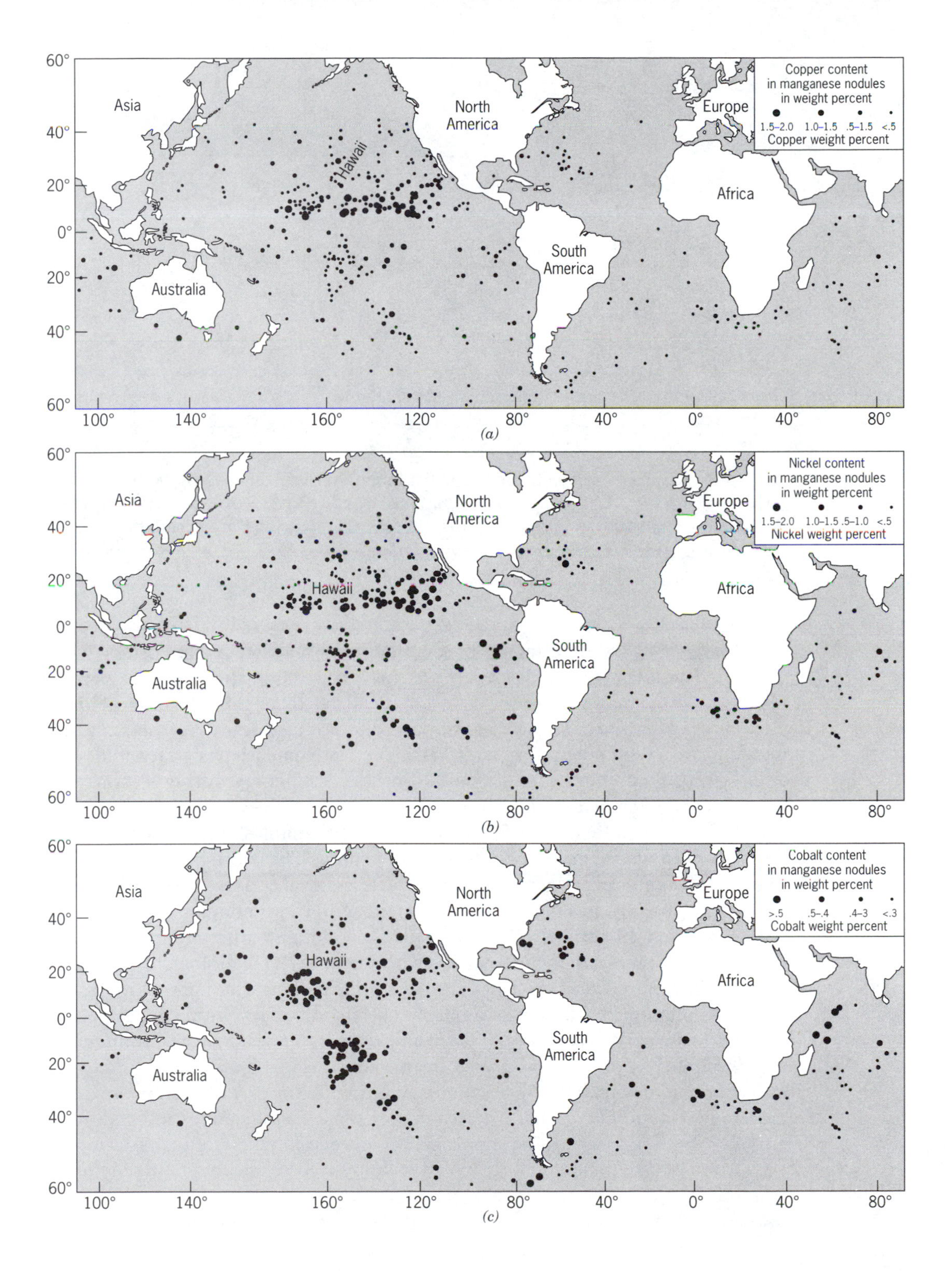
60°
40°
20°
0°
20°
40°
60°
Asia
North America
Europe
Africa
South America
Australia
Hawaii
Copper content in manganese nodules in weight percent
1.5–2.0 1.0–1.5 .5–1.5 <.5
Copper weight percent
100° 140° 160° 120° 80° 40° 0° 40° 80°
Nickel content in manganese nodules in weight percent
1.5–2.0 1.0–1.5 .5–1.0 <.5
Nickel weight percent
Cobalt content in manganese nodules in weight percent
>.5 .5–.4 .4–3 <.3
Cobalt weight percent

(a)

(b)

(c)

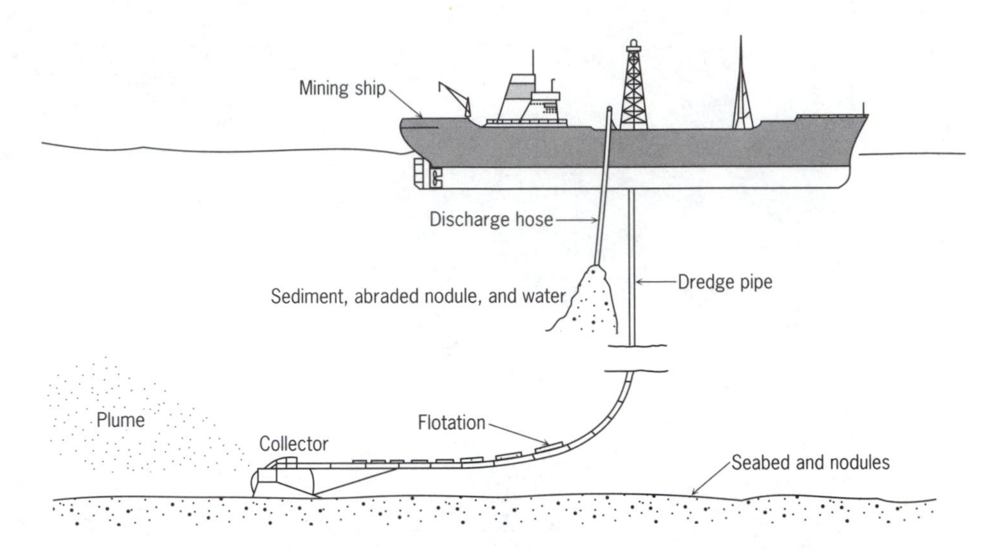

FIGURE 18.5. Proposed technology for the mining of iron–manganese nodules. *Source:* From *Introduction to Oceanography*, 3rd ed., D. A. Ross, copyright © 1982 by Prentice Hall, Inc., Englewood Cliffs, NJ, p. 417. Reprinted by permission. After J. E. Flipse, *Texas A&M Sea Grant Publication 80-205,* 1980, Texas A&M Sea Grant, Galveston, TX.

Changing rates of primary production have caused phosphorites to be deposited in layers, or beds, some of which are 300 m thick and 20 percent phosphorus by mass. Plate tectonics has caused the uplift of some of these ancient phosphorites. These deposits are our primary source of phosphorus, which is used mostly as a plant fertilizer. A small deposit, located near Charleston, South Carolina, supported the local economy during Reconstruction. Due to intensive mining, it was depleted by the 1930s.

As shown in Figure 18.6, the largest continental shelf phosphorites lie beneath current and ancient centers of upwelling. These offshore deposits are likely to be exploited in the future, once the terrestrial beds have been depleted.

Barite, $BaSO_4$, is a common component of marine sediments, constituting between 1 and 10 percent of the mass of the calcium-carbonate-free fraction. This mineral is most abundant in areas with high overlying biological productivity, as sedimenting POM appears to be the source of the barium. Resolubilization and precipitation of barium as amorphous barite occurs within fecal pellets, leading to their "fossilization." Crystalline barite appears to be produced by the precipitation of barium derived from hydrothermal emissions.

Glauconite and oolite are sand-sized fragments of nonskeletal material commonly found in carbonate sediments of continental shelves located in the tropics. Both appear to be hydrogenous precipitates. Oolites are mostly composed of calcium carbonate, which is thought to be an abiogenic precipitate formed from warm seawater supersaturated with respect to calcite and aragonite. Glauconite is an iron-rich, greenish, hydrous silicate that contains

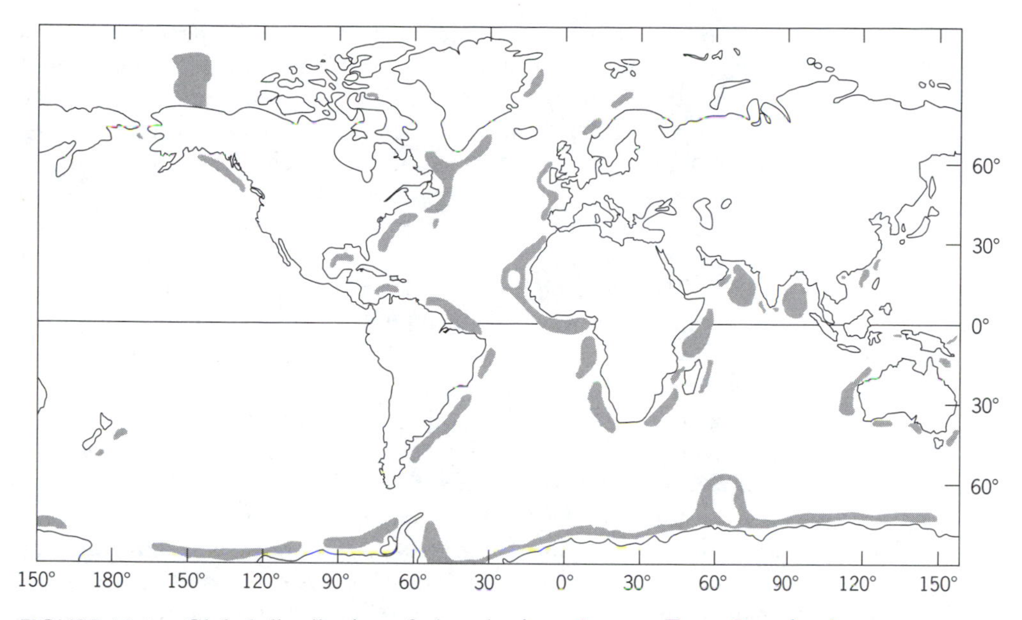

FIGURE 18.6. Global distribution of phosphorites. *Source:* From *Introduction to Oceanography,* 3rd ed., D. A. Ross, copyright © 1982 by Prentice Hall, Inc., Englewood Cliffs, NJ, p. 397. Reprinted by permission. After V. E. McKelvey and F. H. Wang, *U.S. Geological Investigation Map I-632,* 1970, U.S. Government Printing Office, Washington, DC.

fossilized biogenic detritus, such as fecal pellets and siliceous tests. It is found as nodules and encrustations. In some locations, such as the Blake Plateau, concentrations are high enough to give the sediments a greenish cast and hence are referred to as green muds.

SUMMARY

Hydrogenous precipitates form via the direct precipitation of solutes from seawater. Many are metal-rich, such as the massive **polymetallic sulfides** and **iron–manganese oxides**. The latter occur as **nodules**, **metalliferous sediments**, and crusts. Most of the metals composing these deposits are derived from hydrothermal activity and are introduced into seawater as reduced species.

These metals react with O_2 and are converted to insoluble oxides. Some settle to the seafloor to become part of the sediments; others adsorb onto surfaces to form crusts, nodules, and thin coatings on rocks. Trace metals tend to coprecipitate; thus the oxides are geographically variable in composition, heterogeneous, and often contain occluded seawater. This makes them difficult to define stoichiometrically.

Iron and manganese account for most of the metals in these oxides, though traces of at least two dozen other metals are present. Iron–manganese nodules are found in all of the ocean basins and are most abundant in the south central Pacific. They range in diameter from 20 μm to 15 cm, with most between 1 and 10 cm. As with their chemical composition, the accretion rates of the nodules are variable. Even though their accretion rates are typically 1000 times slower than the surrounding sediment, most of the nodules lie on the sediment surface.

The interior of the nodules is characterized by concentric bands that generally conform to the shape of a centrally located particle. This particle is usually a small rock, bone fragment, or shark's tooth and is thought to act as a nucleus, initiating and promoting the deposition of metals. The variability in nodule abundance, chemical composition, and accretion rate is thought to be the result of different mechanisms of deposition. Nodules appear to form as a result of (1) hydrogenous precipitation of dissolved metals from seawater, (2) diagenetic remobilization of metals from reducing sediments, or (3) diagenetic remobilization of metals from oxic sediments.

The nodules that form at very slow rates and whose chemical composition is independent of the underlying sediments are thought to be formed by hydrogenous precipitation. The nodules with the fastest accretion rates are found in sediments subject to moderate fluxes of POM. Here the redox boundary is close enough to the sediment surface so that resolubilized metals can diffuse to the nodules and then precipitate. In other sediments, the redox boundary is too deep for this to occur. Nevertheless, nodules appear to grow, albeit at slower rates, at these locations. The mobilization of metals under oxic conditions has been ascribed to the presence of reducing microzones. Only a small fraction of the metals in the sediment surrounding the nodules would have to be resolubilized to support the observed accretion rates of these nodules. Alternatively, these nodules could be located in sediments where the depth of the redox boundary shifts over time, so that accretion is episodic.

Several theories have also been proposed to explain how the nodules keep from getting buried. These include periodic rocking by earthquakes or bioturbating organisms. The browsing activities of protozoa may also aid in this process. These organisms, along with bacteria, are commonly observed on nodules and may also assist in the deposition of metals. Since the nodules contain a vast amount of precious metals, they are being considered for economic exploitation. Before this can occur, legal and technical problems will have to be solved.

Other sedimentary minerals are the result of precipitation reactions that involve the partial dissolution and reprecipitation of various types of biogenic detritus. These include **phosphorite**, **barite**, **oolite**, and **glauconite**. These hydrogenous precipitates are most common in sediments that lie beneath coastal upwelling areas due to the large fluxes of sedimenting biogenic detritus. The phosphorites are also being considered for economic exploitation.

CHAPTER 19

METALLIFEROUS SEDIMENTS AND OTHER HYDROTHERMAL DEPOSITS

INTRODUCTION

Metalliferous sediments commonly blanket the central rift valley and flanks of mid-ocean ridges and rises. The metals in these hydrogenous deposits are derived from hydrothermal fluids that are escaping through cracks in the new oceanic crust. Oceanographers had long surmised the occurrence of this activity from the geologic record. Direct observations were first made in 1977 as a result of improvements in bottom-water navigation, photography, and deep-sea submersibles. Since then, hydrothermal activity has been found to be quite extensive and a very important process in the crustal-ocean factory.

The chemical reactions that occur in hydrothermal systems are largely the result of interactions between seawater and molten basalt. As the basalt cools and freezes into a crystalline rock, some elements are solubilized and others deposited. For some elements, the resulting elemental fluxes appear to rival those from the rivers and thus could be important in controlling the steady-state composition of the ocean. Hydrothermal activity is also of biogeochemical interest as the emissions contain reduced chemicals that support a chemoautolithotrophic food chain. Some biologists hypothesize that life first evolved around such hydrothermal vents.

In addition to metalliferous sediments, massive polymetallic sulfides have been observed in various stages of formation around hydrothermal vents. Apparently, this mechanism is responsible for the formation of the economically valuable ore bodies that are now found on land. Geochemists are studying active hydrothermal systems to gain insight into the probable locations and distribution of metals in these ore deposits. Hydrothermal activity is also of interest to physical oceanographers as the emissions are enriched in rare chemicals, such as ^{3}He and Mn^{2+}, that can be used as tracers of the rate and pathways of water-mass motion.

These subjects are discussed below. Due to the limited number of observations that have been made, projections of global impacts are as yet very tentative.

THE PHYSICAL SETTING

Seafloor spreading at mid-ocean ridges and rises causes rifting of the rock that constitutes the oceanic basement. This motion is driven by volcanic eruptions during which lava is extruded onto the seafloor. As illustrated in Figure 19.1, these flows produce piles of **pillow basalts**. As this new oceanic crust cools, it crystallizes and contracts, creating fractures and fissures. Some of these cracks are large enough to be referred to as faults.

Cooling is achieved primarily through the conductive transfer of heat from the magma to the adjacent seawater. However, approximately one-third of the total heat loss occurs by convective means. This process takes place when seawater percolates down into the new crust through the fissures. The density of the seawater is lowered by conductive heating, causing it to rise spontaneously as shown in Figure 19.2.

The existing waters have temperatures as high as 400°C. The water is not converted to steam due to the boiling point elevation caused by the high pressures that exist in the deep sea. These fluids are replaced by cold seawater, which is drawn down into the hydrothermal system. The resulting convective flow transfers heat from the interior of the spreading centers to the bottom

FIGURE 19.1. Photograph of pillow basalts. Courtesy of the Woods Hole Oceanographic Institution, Woods Hole, MA.

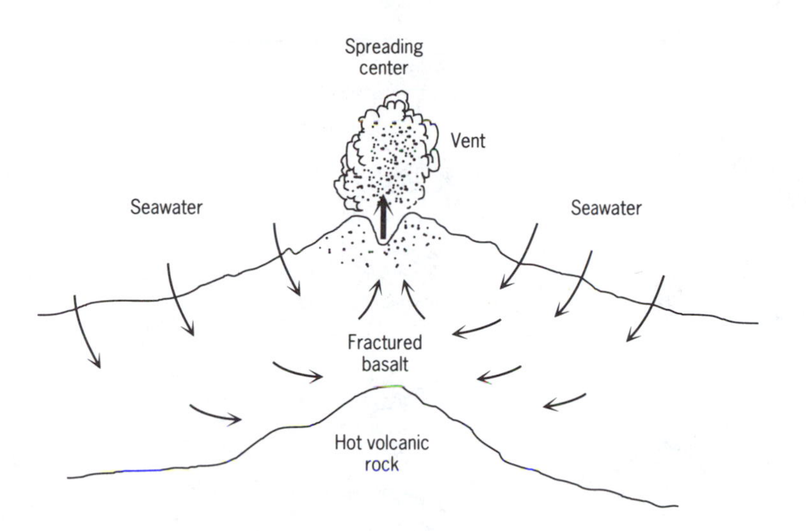

FIGURE 19.2. Hydrothermal convection cells.

waters and is termed active hydrothermal circulation. Geochemists hypothesize that water percolates as deep as 5 km below the seafloor and as far as 200 km from the ridge axis, probably through a series of these convective cells. Seawater also flows through older crust that has subsided and moved away from the ridge crest. This is termed passive hydrothermal circulation and produces exiting fluids with temperatures less than 100°C.

In both passive and active hydrothermal circulation, the rising seawater can reenter the deep ocean in several ways. Some emanates as a dissipated flow through mounds of sediment that have accumulated on top of the basalt. Some debouches onto the seafloor as a strong flow from cracks, or chimneys, as shown in Figure 19.3.

Direct observations of ongoing hydrothermal activity are limited to the few sites shown in Figure 19.4, all of which are subject to relatively rapid seafloor spreading rates (i.e., $\geq$ 3 cm/y). Since spreading centers are common to all mid-ocean ridges and rises, hydrothermal activity is probably ubiquitous.

CHEMICAL REACTIONS THAT OCCUR IN HYDROTHERMAL SYSTEMS

Seawater circulating through the oceanic ridge crest undergoes chemical reactions with the magma and basalt. The result are solutions enriched in some chemicals and depleted in others. The rock left behind is a weathered residue. The exact nature of the chemical changes depends on the water temperature and circulation rate, as well as the porosity of the rock and its chemical composition. Despite this variability and our limited number of observations,

FIGURE 19.3. A typical hydrothermal vent system. Included are basal mounds of mineral precipitates, pillow basalts, a black smoker, and various vent organisms. *Source:* From R. M. Haymon and K. C. Macdonald, reprinted with permission from *American Scientist,* vol. 73, p. 445, copyright © 1985 by Sigma Xi, Raleigh, NC.

some generalizations concerning the nature of hydrothermal chemistry have been made.

These generalizations are based on field observations and laboratory simulations. Geochemists have attempted to reproduce some of the environmental conditions that exist in hydrothermal systems by reacting basalt with seawater at pressures up to 1000 bars and temperatures from 70° to 500°C at rock-to-water ratios from 1 to 62 for periods as long as 20 months. The result from these laboratory reactions support the conclusions made from field observations quite well.

High-Temperature Reactions

High-temperature reactions occur several kilometers below the seafloor, where seawater comes into contact with magma and is heated to temperatures up to 400°C. They are thought to proceed for 60 to 80 million years following the solidification of the rock. The reactions probably cease because pelagic sedimentation and mineral precipitates fill in the fissures. In addition, the

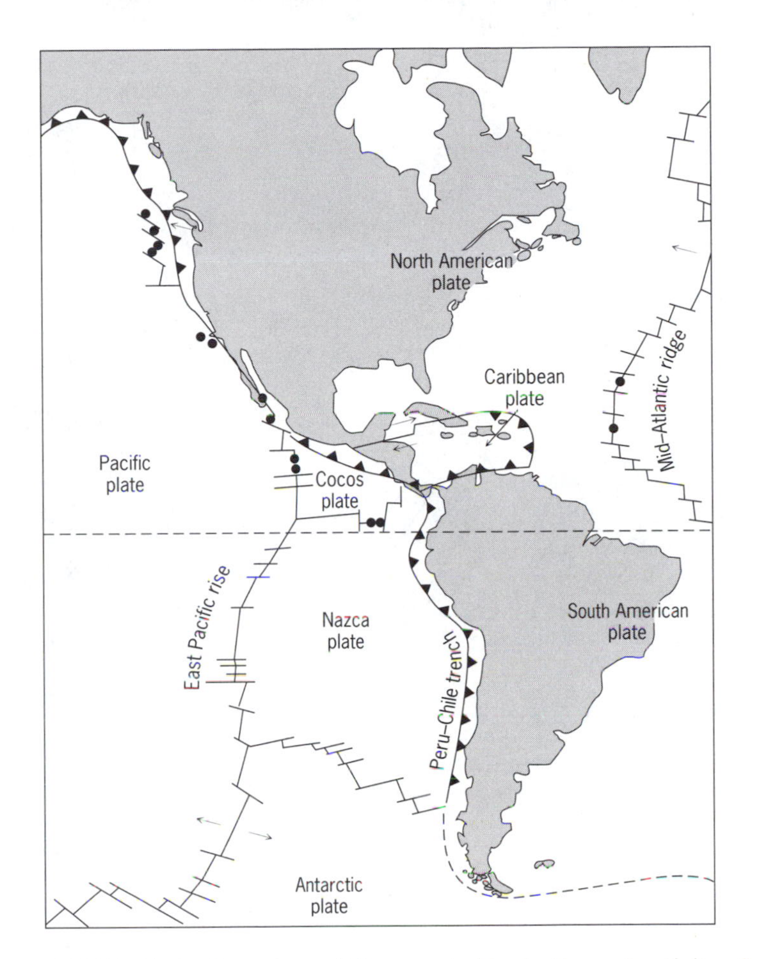

FIGURE 19.4. Locations of documented hydrothermal activity. *Source:* From R. M. Haymon and K. C. Macdonald, reprinted with permission from *American Scientist,* vol. 73, p. 442, copyright © 1985 by Sigma Xi, Raleigh, NC.

movement of the rock away from the ridge crests causes the temperature at the reaction zone to decrease.

The high-temperature reactions metamorphose the magma and rock into a metabasalt composed of the silicate minerals: albite, epidote, actinolite, chlorite, smectite (montmorillonite), mixed layer smectite–chlorite, quartz, and the serpentinites, which are magnesium-rich. Also deposited are metal-rich minerals such as pyrite (FeS_2), chalcopyrite ($CuFeS_2$), pyrrhotite ($Fe_{1-x}S$), magnetite (Fe_3O_4), and hematite (Fe_2O_3). The seawater is transformed into a hot, metal-rich, acidic solution. The reactions responsible for these chemical changes are illustrated in Figure 19.5.

As the waters begin their descent into the hydrothermal system, anhydrite precipitates because its solubility decreases with increasing temperature. Magnesium is incorporated into basalt, producing magnesium silicates and

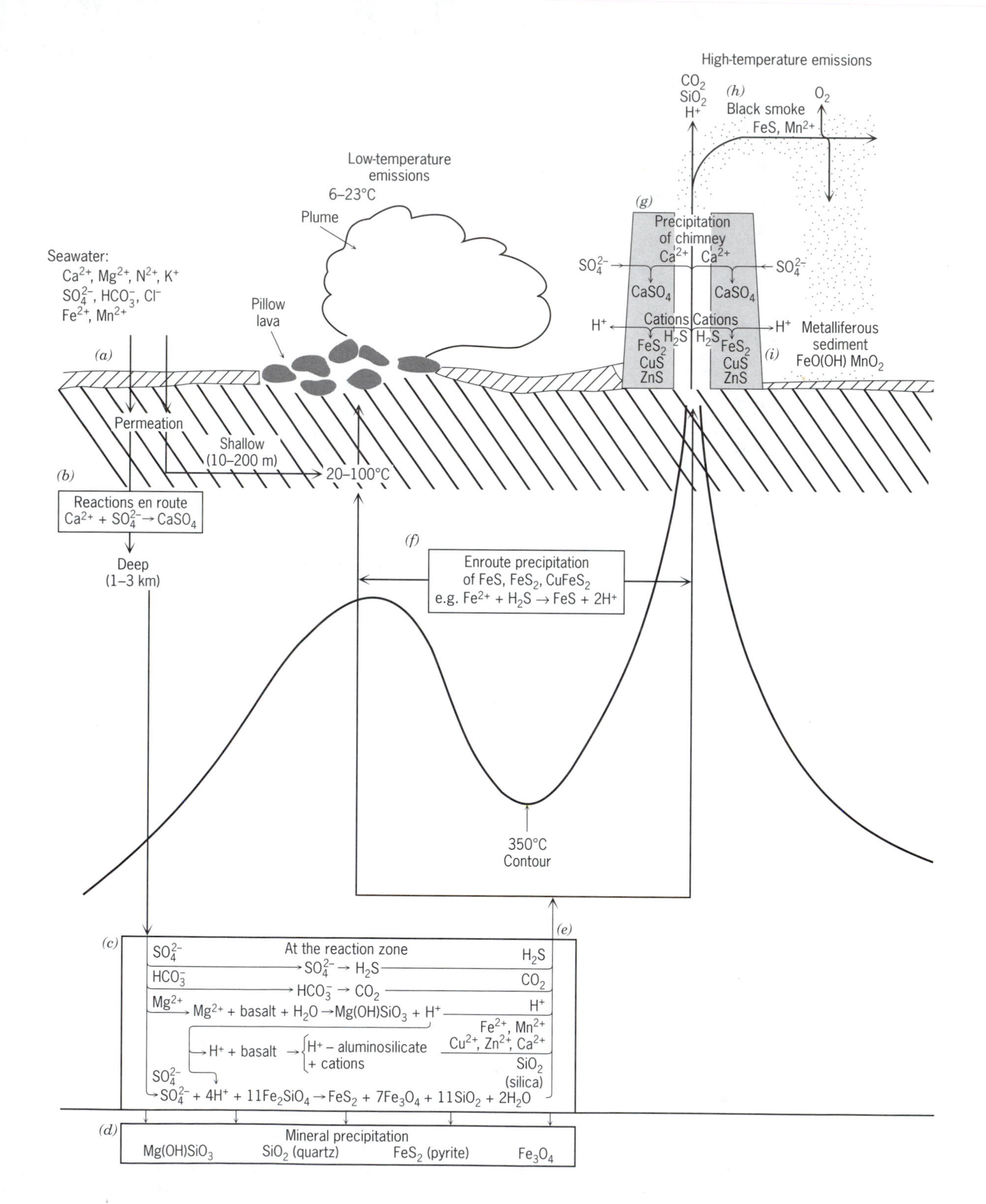

High-temperature emissions
CO2
SiO2
H+
(h)
Black smoke
FeS, Mn2+
O2
Low-temperature emissions
6–23°C
Plume
(g)
Precipitation of chimney
Ca2+ Ca2+
SO4 2−
CaSO4
CaSO4
Cations Cations
H+
H2S H2S
FeS2
CuS
ZnS
(i)
Metalliferous sediment
FeO(OH) MnO2
Seawater:
Ca2+, Mg2+, N2+, K+
SO4 2−, HCO3−, Cl−
Fe2+, Mn2+
Pillow lava
(a)
Permeation
Shallow (10–200 m)
20–100°C
(b)
Reactions en route
Ca2+ + SO4 2− → CaSO4
Deep (1–3 km)
(f)
Enroute precipitation of FeS, FeS2, CuFeS2
e.g. Fe2+ + H2S → FeS + 2H+
350°C Contour
(e)
(c)
At the reaction zone
SO4 2− → H2S
HCO3− → CO2
Mg2+ + basalt + H2O → Mg(OH)SiO3 + H+
H+ + basalt → H+ – aluminosilicate + cations
Fe2+, Mn2+
Cu2+, Zn2+, Ca2+
SiO2 (silica)
SO4 2− + 4H+ + 11Fe2SiO4 → FeS2 + 7Fe3O4 + 11SiO2 + 2H2O
(d)
Mineral precipitation
Mg(OH)SiO3
SiO2 (quartz)
FeS2 (pyrite)
Fe3O4

hydrogen ions. Some of this acid reacts with bicarbonate, converting it to CO_2. These uptakes and releases are the reverse of chemical weathering and appear to have a major role in controlling the steady-state levels of $Mg^{2+}(aq)$ and alkalinity in seawater.

Some of the resulting hydrogen ions also cause the weathering of basalt during which cations are solubilized and aluminosilicates are produced. The cations that are leached include calcium, potassium, and the reduced forms of manganese, iron, copper, and zinc. The acid also reacts with sulfate and basalt forming pyrite (FeS_2), iron oxides, quartz, and dissolved silicon. As a result, the reduced metal concentrations in the hydrothermal fluids are many orders of magnitude higher than in seawater. Hydrogen sulfide and dissolved silicon concentrations are also very high. In fact, dissolved silicon is supersaturated with respect to silica at these temperatures and pressures and precipitates as quartz veins that fill the hydrothermal conduits.

In locations where hydrothermal fluid exits directly into the bottom ocean with little subsurface dilution, large chimneys are deposited by the rapid precipitation of anhydrite and polymetallic sulfides. The precipitation reactions begin when the hot (>300°C), acidic, metal-rich fluid mixes with the cold, oxic, alkaline seawater. The calcium from the hydrothermal fluid reacts with seawater sulfate, depositing anhydrite, which forms the leading edge of a "chimney," as shown in Figure 19.6.

The formation of these anhydrite walls prevents the hydrothermal fluids flowing through the chimney from mixing with seawater. Thus, when the fluids exit from the chimneys, contact with cold, alkaline ambient seawater causes reduced metals to react rapidly with hydrogen sulfide. Because most of the sulfide precipitates are black, the hot, clear fluid is almost instantaneously converted into a black smoke. Black smokers were first discovered in 1979 on the East Pacific Rise at 21°N. A photograph of one is shown in Figure 19.7.

Initially, the sulfide particles settle onto the chimney's exterior, where they are buried by the outward growth of anhydrite. Within the chimneys, metal sulfides deposit and partially replace the anhydrite. Chimneys can build to several meters in height and their orifices range in diameter from 1 to 30 cm.

As the interior fills with sulfides, the flow is reduced and the walls cool, causing anhydrite to dissolve. Marine organisms colonize the cooling chimneys

FIGURE 19.5. Hydrothermal reactions at the mid-ocean ridges. Seawater (*a*) percolates into the oceanic crust. As it descends (*b*) calcium sulfate precipitates. At deeper levels, (*c*) the seawater reacts with hot basaltic rock to produce (*d*) mineral precipitates. (*e*) A hot, acidic, metal-rich fluid rises back toward the ocean floor. (*f*) If the rising fluid encounters cold seawater, metal sulfides are precipitated. Some precipitate in (*g*) chimneys and others (*h*) escape into the deep ocean as black smoke, which is eventually deposited, along with metal oxides, in the (*i*) sediments. *Source:* After J. M. Edmond and K. Von Damm, reprinted with permission by *Scientific American,* vol. 248, p. 84, copyright © 1983 by Scientific American, New York; and H. W. Jannasch and M. J. Mottl, reprinted with permission from *Annual Review of Microbiology,* vol. 38, p. 504, copyright © 1984 by Annual Reviews, Palo Alto, CA.

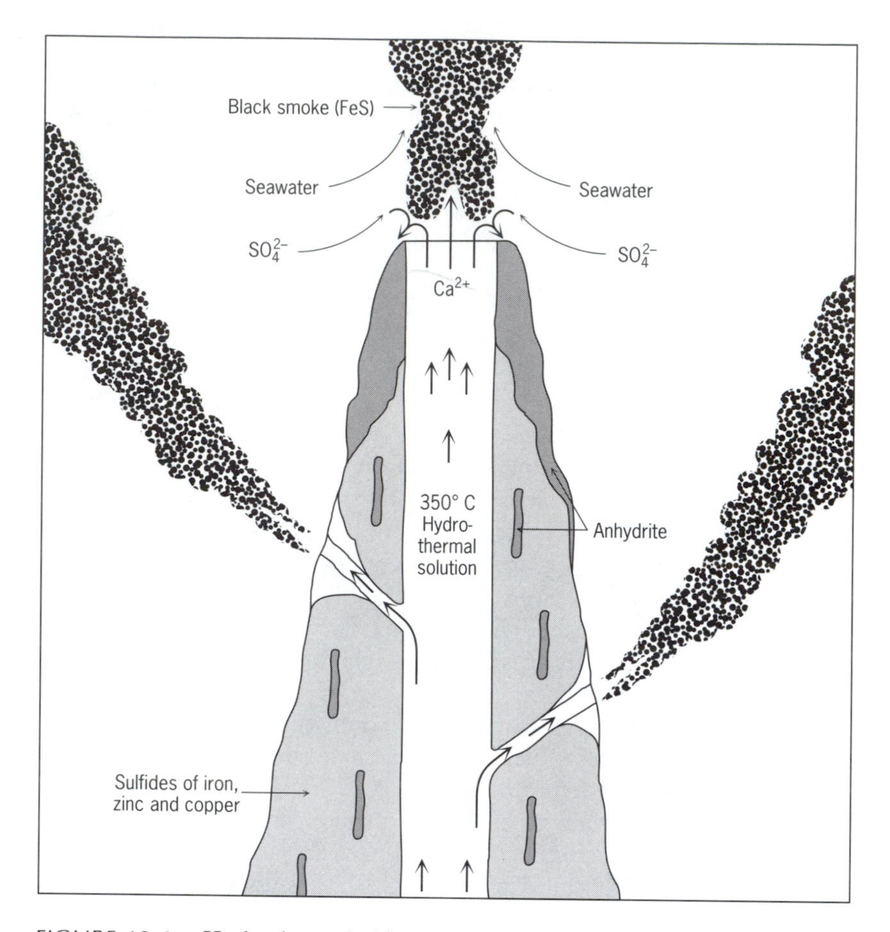

FIGURE 19.6. Hydrothermal chimney formation. *Source:* From J. M. Edmond, reprinted with permission from *Oceanus,* vol. 27, p. 16, copyright © 1984 by Oceanus, Woods Hole, MA.

and the sulfides begin to oxidize. This destabilizes the chimneys and eventually they collapse, forming basal mounds that can reach heights of 2 m or more. The relatively small size of these mounds suggests that black smokers have short lifetimes, perhaps on the order of a decade. The continuing build-up and collapse of chimneys eventually creates a massive sulfide deposit.

Low-Temperature Reactions

In some locations, subsurface mixing of relatively "fresh" seawater with high-temperature hydrothermal fluids produces solutions with temperatures less than 300°C. As shown in Figure 19.5, this is thought to occur at depths ranging between 10 and 200 m below the seafloor. The precipitates that result when these fluids enter the deep sea are milky white because they are primarily zinc sulfide. These white smokers also build chimneys, some of which are as much

FIGURE 19.7. Photograph of a black smoker. Courtesy of the Woods Hole Oceanographic Institution, Woods Hole, MA.

as 13 m high. Due to their lower temperatures, white smokers are typically encrusted with worm tubes and other organisms. The co-occurrence of white and black smokers suggests both are produced during the evolution of hydrothermal vents, as illustrated in Figure 19.8.

Warm water (≤25°C) also emanates as a relatively dissipated flow (2–10 L/s) from the tops of mounds created by the crumbling of chimneys. These flows were first observed at the Galapagos vents (86°W) and are associated with a unique biotic community. Mounds at this location reach heights of 30 m. The low temperatures of these emissions are caused by extensive subsurface mixing of the high-temperature hydrothermal fluid with "fresh" seawater. Subsurface cooling causes the sulfides to precipitate within the basalt conduits and sediments, so the existing fluid produces no "smoke."

As noted above, the dissolved silicon concentrations are so high that the hot hydrothermal fluids are supersaturated with respect to silica. As this fluid mixes with "fresh" seawater, its dissolved silicon concentration and temperature decrease. As shown in Figure 19.9, the decrease in concentration is linearly related to the temperature decrease. Thus dissolved silicon appears to behave conservatively during this low-temperature mixing process.

Since quartz precipitates in the high-temperature reaction zone, a lower limit for the temperature of the undiluted hydrothermal fluid can be obtained by extrapolating the line of linear regression shown in Figure 19.9 to the

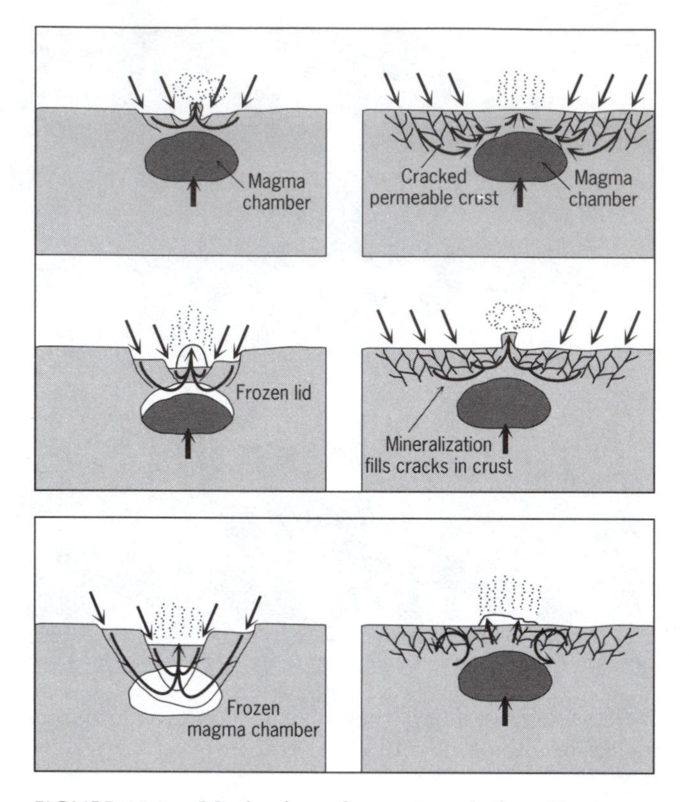

FIGURE 19.8. Mechanisms for vent evolution. Vents can evolve from (*a*) high to (*b*) low temperatures (sequence on left) or from (*c*) low to (*d*) high temperatures (sequence on right). In the case of the former, cooling of the magma chamber is accompanied by an increase in depth of hydrothermal circulation. In the case of the latter, as crustal fractures are sealed off by hydrothermal mineral precipitates, the entrainment of fresh seawater is curtailed, so relatively undiluted hydrothermal fluids are discharged. In either case, final cooling results from either (*e*) total depletion of the heat source or (*f*) effective sealing of subsurface conduits or by a combination of the two processes. *Source:* From R. M. Haymon and K. C. Macdonald, reprinted with permission from *American Scientist,* vol. 73, p. 446, copyright © 1985 by Sigma Xi, Raleigh, NC.

temperature and dissolved silicon concentrations at which the solution becomes saturated with respect to quartz. The saturation concentration is a function of temperature and pressure as illustrated in Figure 19.10. The regression line intersects the solubility curve of quartz at 350°C and 1 kilobar. This is attained at a depth 2.5 km below the sea surface.

Other chemicals behave conservatively during the low-temperature mixing process and thus can also be used as geothermometers. As shown in Figure 19.11, the hydrothermal system is a sink for magnesium. Extrapolation of this trend to zero magnesium ion concentration also yields a temperature of

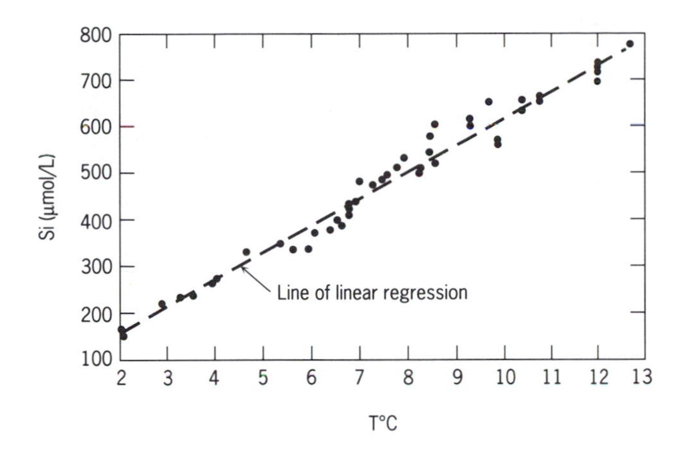

FIGURE 19.9. Dissolved silicon concentrations versus temperature in hydrothermal fluids collected from the Galapagos vents. *Source:* From J. M. Edmond, C. Measures, R. E. McDuff, L. H. Chan, R. Collier, B. Grant, L. I. Gordon, and B. Corliss, reprinted with permission from *Earth and Planetary Sciences Letters,* vol. 46, p. 11, copyright © 1979 by Elsevier Science Publishers, B.V., Amsterdam, The Netherlands.

approximately 350°C for the undiluted high-temperature hydrothermal fluid. The same has been observed for sulfate.

With this technique, the chemical composition of the undiluted high-temperature hydrothermal fluid can be reconstructed. As shown in Table 19.1, the results agree well with direct observations of the chemical composition of high-temperature fluids (350°C) collected from the black smokers at the 21°N site on the East Pacific Rise. This suggests that little subsurface mixing occurs at the 21°N vents and that the chemical composition of the undiluted hydrothermal fluid does not exhibit extreme geographic variability.

The vents also appear to remove oxygen, nitrate, copper, nickel, cadmium, chromium, uranium, and selenium. But for these elements, the concentrations extrapolate to zero at temperatures between 30° and 40°C. This suggests their removal occurs at fairly low temperatures from both the hydrothermal fluid and "fresh" seawater, probably close to the surface of the crust. In other words, they exhibit nonconservative behavior during the subsurface mixing process.

Metalliferous Sediments

Most of the particulate sulfides are deposited close to their point of entry into the deep sea. As a result, the mid-ocean ridges and adjacent abyssal plains are blanketed with metalliferous sediments. Some of these sediments have very high concentrations of zinc, iron, copper, lead, silver, and cadmium, as shown in Table 19.2.

As illustrated in Figure 19.12, the Pacific Ocean has the greatest abundance of metalliferous sediments, with the largest metal enrichments associated with

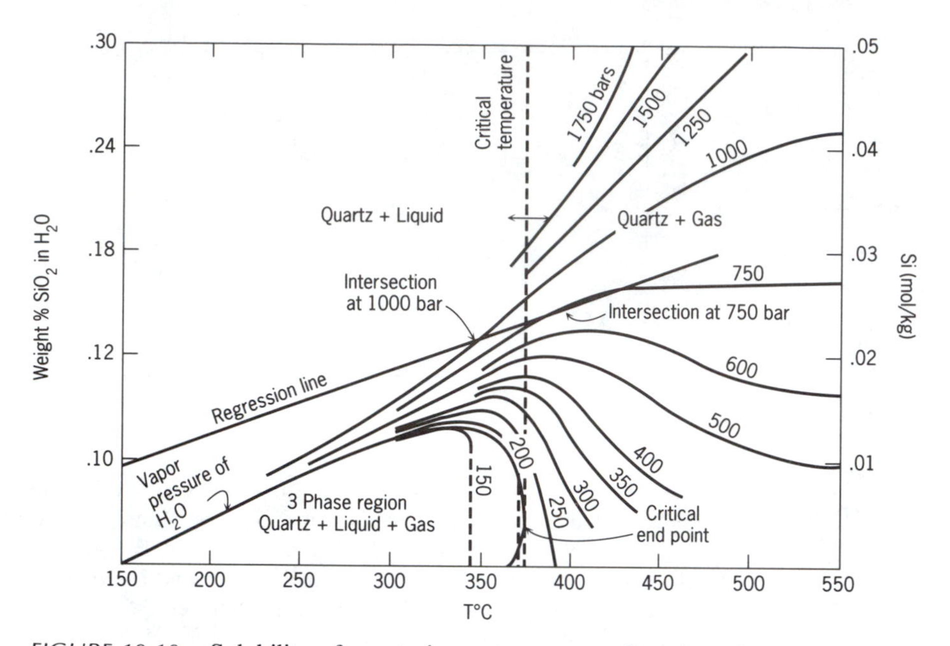

FIGURE 19.10. Solubility of quartz in vent water as a function of temperature (°C) and pressure (bars). The regression line from Figure 19.9 is also plotted and extrapolated to conditions where the solution becomes saturated with respect to quartz, e.g., 345°C at 1000 bar and 375°C at 750 bar. Under the latter conditions, the equilibrium phase of water is gas. The critical temperature is the temperature above which a gas cannot be liquified by an increase in pressure. *Source:* From J. M. Edmond, C. Measures, R. E. McDuff, L. H. Chan, R. Collier, B. Grant, L. I. Gordon, and B. Corliss, reprinted with permission from *Earth and Planetary Sciences Letters,* vol. 46, p. 13, copyright © 1979 by Elsevier Science Publishers, B.V., Amsterdam, The Netherlands.

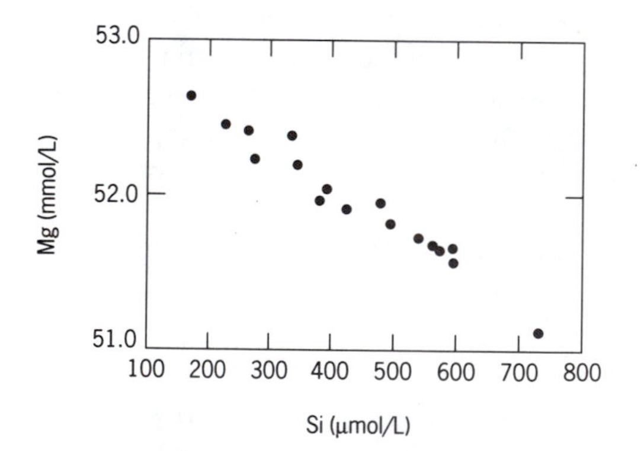

FIGURE 19.11. Magnesium versus dissolved silicon concentrations in hydrothermal fluids collected at the Galapagos vents. *Source:* From J. M. Edmond, C. Measures, R. E. McDuff, L. H. Chan, R. Collier, B. Grant, L. I. Gordon, and B. Corliss, reprinted with permission from *Earth and Planetary Sciences Letters,* vol. 46, p. 6, copyright © 1979 by Elsevier Science Publishers, B.V., Amsterdam, The Netherlands.

TABLE 19.1
Comparison of the Estimated Composition of a 350°C Hydrothermal End Member Based on Extrapolation of the Galapagos Data with That Observed at 21°N

Element	*Concentration Units*	*Galapagos*	*21°N*	*Seawater*
Li	μmol/kg	1,142–689[a]	820	28
K	mmol/kg	18.8	25.0	10.1
Rb	μmol/kg	20.3–13.4	26.0	1.32
Mg	mmol/kg	0	0	52.7
Ca	mmol/kg	40.2–24.6	21.5	10.3
Sr	μmol/kg	87	90	87
Ba	μmol/kg	42.6–17.2	95–35	0.145
Mn	μmol/kg	1,140–360	610	0.002
Fe	μmol/kg	+[b]	1,800	—[c]
Si	mmol/kg	21.9	21.5	0.160
SO_4^{2-}	mmol/kg	0	0	28.6
H_2S	mmol/kg	+	6.5	0

Source: From J. M. Edmond, K. L. Von Damm, R. E. McDuff, and C. I. Measures, reprinted with permission from *Nature*, vol. 297, p. 188, copyright © 1982 by MacMillan Journals, Ltd, London, England.

[a]Ranges in concentration reflect differences in the chemical behavior of individual vents.

[b]Nonconservative due to sub-surface mixing.

[c]Seawater concentration not accurately known.

the East Pacific Rise. These sediments become buried by lithogenous and biogenous particles as the underlying crust moves away from the active spreading center.

Fluids emitted from diffuse vents, such as those found at the Galapagos Ridge, are depleted in metallic sulfides, but are rich in reduced manganese and iron. Upon contact with seawater, the iron and manganese react with O_2 and are converted to insoluble oxides that settle to the seafloor. This produces crusts of iron–manganese oxides that cover the mounds. Since the oxidation of iron and manganese proceeds relatively slowly, considerable horizontal transport can occur before the particles settle to the seafloor. As a result, the distribution of metalliferous surface sediments is strongly controlled by the pattern of deep-water circulation. The southeastern lobe of metalliferous sediment in Figure 19.12 is produced by such circulation. This horizontal transport is the major source of metals supporting the growth of iron–manganese nodules throughout the ocean basins.

Low-temperature reactions also occur when seawater interacts with the cooling pillow basalts. This is a type of chemical weathering that occurs at temperatures less than 100°C and proceeds in stages. This process is thought to be complete within 3 million years after solidification of the basalt. First, the glassy basalt undergoes palagonitization, during which the rock is hydrated.

TABLE 19.2
Metal Enrichments in Hydrothermal Polymetallic Sulfides

	Concentrations of Selected Elements from Marine Polymetallic Sulfide Sites						
Area	*Zn (%)*	*S (%)*	*Fe (%)*	*Cu (%)*	*Pb (%)*	*Ag (ppm)*	*Cd (ppm)*
Juan de Fuca	0.63–54.0	n.a.[a]	1.8–50.5	0.0003–0.32	0.06–0.25	3–290	0–490
Guaymas Basin	30	n.a.	n.a.	1.0	0.10	300	n.a.
21°N EPR	0.12–41.8	0.74–39.7	0.61–26.2	0.13–1.3	0.04–0.61	1.6–241	20–890
Galapagos Rift	0.14	52.2	44.1	4.98	0.07	10	31

Source: From J. L. Bischoff, R. J. Rosenbauer, P. J. Aruscavage, P. A. Baedecker, and J. G. Crock, *U.S. Geological Survey Open-File Report 83–324*, 1983, U.S. Department of the Interior Geological Survey, Menlo Park, CA.
[a]n.a. = not available.

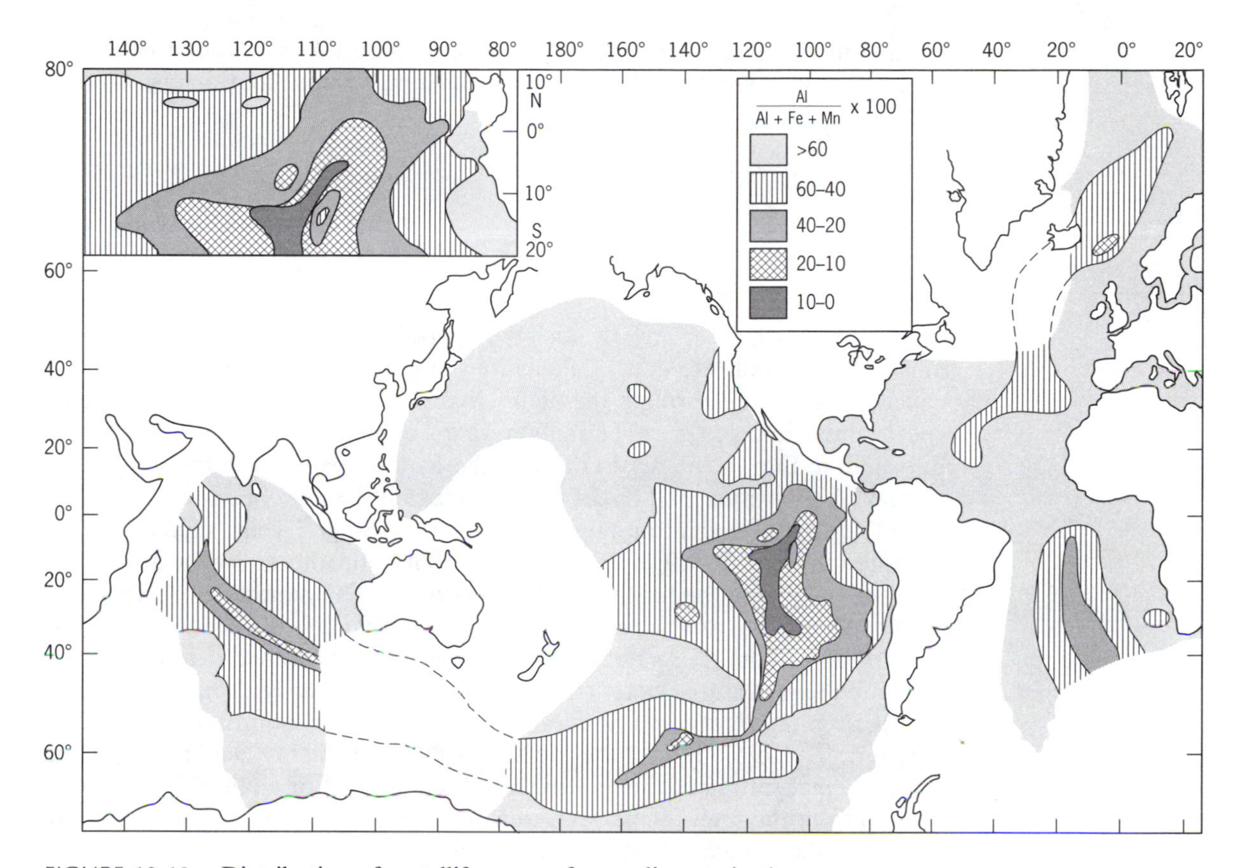

FIGURE 19.12. Distribution of metalliferous surface sediments in the ocean as indicated by the ratio Al/(Al + Fe + Mn). As these deposits are poor in aluminum relative to continental detritus, their abundance is inversely proportional to the value of the ratio. *Source:* From K. Bostrum, M. N. A. Peterson, O. Joensuu, and D. E. Fisher, reprinted with permission from the *Journal of Geophysical Research,* vol. 74, p. 3262, copyright © 1969 by the American Geophysical Union, Washington, DC.

As a result, approximately 1 percent of the water circulating through the hydrothermal system is incorporated into the weathered basalt. Next, smectites (montmorillonite) and zeolites, such as phillipsite, form as iron is oxidized. In the third stage, carbonates are precipitated. In the fourth stage, the weathered rock is compacted and dehyrdated. Because gaps exist between the pillow basalts, the reaction zone for this low-temperature weathering is thought to extend 500 m below the seafloor.

The mineral alterations proceed from the outer margin of the pillow basalts to their interior. As a result, some chemicals are leached and others deposited during the various stages of weathering. The magnitude and direction of the elemental fluxes are difficult to generalize because they depend on such variables as water temperature and circulation rate as well as the rock porosity and chemical composition. Nevertheless, potassium, manganese, sodium, cesium, rubidium, lithium, and boron appear to be taken up during the low-temperature weathering of basalt, while calcium, silicon, and magnesium are released.

ESTIMATES OF GLOBAL HYDROTHERMAL FLUXES

The global rate of water flow through the hydrothermal systems can be inferred from heat flow and ^3He data. Calculations suggest that 1.3 to 2.7 $\times$ 10^{17} g of seawater are passed through the active hydrothermal systems every year. In other words, it takes 5 to 11 million years to cycle a volume of seawater equivalent to that in the ocean through the hydrothermal systems.

Global hydrothermal fluxes of the elements have been estimated by multiplying their concentration gradient in an "average" vent, such as the one illustrated in Figure 19.9, by the annual global hydrothermal heat flux (4.9 $\pm$ 1.2 $\times$ 10^1 cal/y) and the heat capacity of seawater. The number of observations is still too limited to assess geographic and temporal variability in the chemistry of hydrothermal fluids. As noted above, variations are likely due to differences in reaction temperature, water circulation rates, water-to-rock ratio (porosity), as well as in the depth and horizontal extent of the reaction zone. Thus, the magnitude and direction of the elemental fluxes is highly dependent on the environmental setting of the hydrothermal reactions. For example, an element released from the high-temperature reaction zone can also be removed in one or more of the adjacent low-temperature reaction zones.

As a result, our present estimates of global elemental fluxes are highly uncertain and provide, at best, only an indication of the direction and order of magnitude of the chemical flows. As shown in Table 19.3, hydrothermal activity appears to be a major sink of magnesium and sulfur (as sulfate), rivaling their riverine sources. Hydrothermal emissions also appear to be a major source of lithium and rubidium and supply to a lesser but significant extent some calcium, potassium, silicon, and barium to seawater. The estimated manganese flux from the ridge crest is large enough to account for the presence of this element in all the metalliferous sediments and nodules found on the seafloor.

These chemical additions and removals cause an annual net decrease in the alkalinity of seawater equivalent to 15 percent of the total annual river input or 50 percent of that contributed by the weathering of noncarbonate continental rocks. Part of this loss results from the titration of bicarbonate by hydrogen ions released from basalt during high-temperature hydrothermal reactions.

Since the altered basalt is ultimately subducted, the mantle represents a large sink for magnesium and sulfate. The oceanic residence time of magnesium (13 million years) is close to the turnover time of water through the hydrothermal system. This suggests uptake at hydrothermal systems is the most important removal mechanism of this element from the ocean and thus acts to maintain

TABLE 19.3
Comparison of Hydrothermal Fluxes to Riverine Fluxes[a]

	Gains		*Losses*	
Constituent	*River Input*	*Hydrothermal Input*	*Sediment Loss*	*Hydrothermal Loss*
Mg^{2+}	5.3×10^{12}	0	?[b]	8×10^{12}
Ca^{2+}	12×10^{12}	3.5×10^{12}		0
Ba^{2+}	10×10^{9}	2.4×10^{9}	?	0
Li^{+}	1.4×10^{10}	16×10^{10}	?	0
K^{+}	1.9×10^{12}	1.2×10^{12}	?	0
Rb^{+}	0.4×10^{9}	2.4×10^{9}	?	0
SO_4^{2-}	4×10^{12}	0	?	4×10^{12}
F^{-}	16×10^{10}	0	?	1.1×10^{10}
Si	6×10^{12}	3×10^{12}	?	0
P	$\sim 3 \times 10^{12}$	$< 0.1 \times 10^{10}$	3×10^{10}	$< 0.1 \times 10^{10}$
ΣCO_2	$\sim 2 \times 10^{13}$	$\sim 0.1 \times 10^{13}$	2×10^{13}	0

Source: From *Tracers in the Sea*, W. S. Broecker and T.-H. Peng, copyright © 1982 by the Lamont-Doherty Geological Observatory, Palisades, NY, p. 296. Reprinted by permission. Data from J. M. Edmond, C. I. Measures, R. E. McDuff, L. H. Chan, R. Collier, B. Grant, L. I. Gordon, and J. B. Corliss, reprinted with permission from *Earth and Planetary Sciences Letters*, vol. 46, pp. 4–8, copyright © 1979 by Elsevier Science Publishers, B. V., Amsterdam, The Netherlands.
[a]All fluxes are in mol/y.
[b]Magnitude of the flux is unknown.

its steady-state balance in the ocean. The same is likely true for other elements that have large global hydrothermal fluxes. Thus hydrothermal activity appears to be of great importance in stabilizing and determining some aspects of the chemical composition of seawater. Improvements in global flux estimates require a much greater knowledge of spatial and temporal variability in the chemical composition of the hydrothermal fluids. To assess this, much more information will have to be obtained concerning vent evolution (i.e., the length of time over which a vent is active, how many vents are active at any given time, and how the chemistry of a vent changes over time).

BIOLOGY OF DEEP-SEA VENTS

Hydrothermal activity was first observed in 1977 at a series of vents that lie 2500 m below the sea surface, 280 km northeast of the Galapagos Islands. Geochemists were attempting to locate active vents by searching for temperature anomalies in the bottom waters of the East Pacific Rise. Instead, the actual discovery was made as a result of the unexpected and startling visual observations of large assemblages of clams, mussels, worms, and crabs clustered around fountains of shimmering water. The appearance of a typical vent community is shown in Figure 19.3.

This type of community appears to be a ubiquitous feature of hydrothermal vents and even cold-water seeps, though the species composition is somewhat variable. The locations of known vent and seep communities are given in Figure 19.13.

The abundance and diversity of marine life at these vents is extraordinary. Prior to the discovery of these ecosystems, marine biologists had assumed that all life in the ocean ultimately depended on the photosynthetic production of POM by phytoplankton. Since this POM is inefficiently transferred to the deep-sea, the abundance of marine organisms decreases with increasing depth. The vent ecosystems are the sole exception to this because their food chain is based on nonphotosynthetic primary producers.

Most of these primary producers are chemoautolithotrophic bacteria that obtain the chemical energy to fix carbon from the oxidation of hydrogen sulfide obtained from hydrothermal emissions. As shown in Eq. 19.1, the oxidizing agent is O_2, which is transported to the vents by thermohaline circulation.

$$CO_2 + H_2S + O_2 + H_2O \xrightarrow{\text{bacteria}} CH_2O + H_2SO_4 \tag{19.1}$$

Since O_2 is required, this chemosynthesis is not entirely independent of photosynthesis. As shown in Figure 19.14, hydrogen sulfide is analogous to the solar energy that fuels the photosynthetic fixation of carbon.

Other microbial chemoautolithotrophs, such as methanotrophs and hydrogen bacteria, have been found in and near the vents. The methane and hydrogen

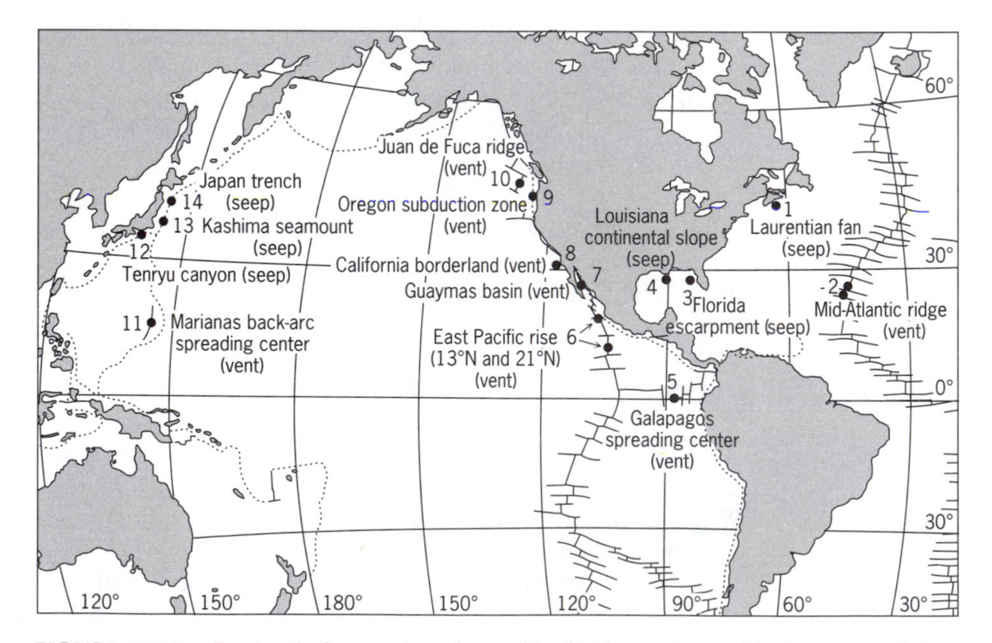

FIGURE 19.13. Presently known locations of hydrothermal vent biotic communities. *Source:* From J. F. Grassle, reprinted with permission from *Oceanus,* vol. 31, p. 44, copyright © 1989 by Oceanus, Woods Hole, MA.

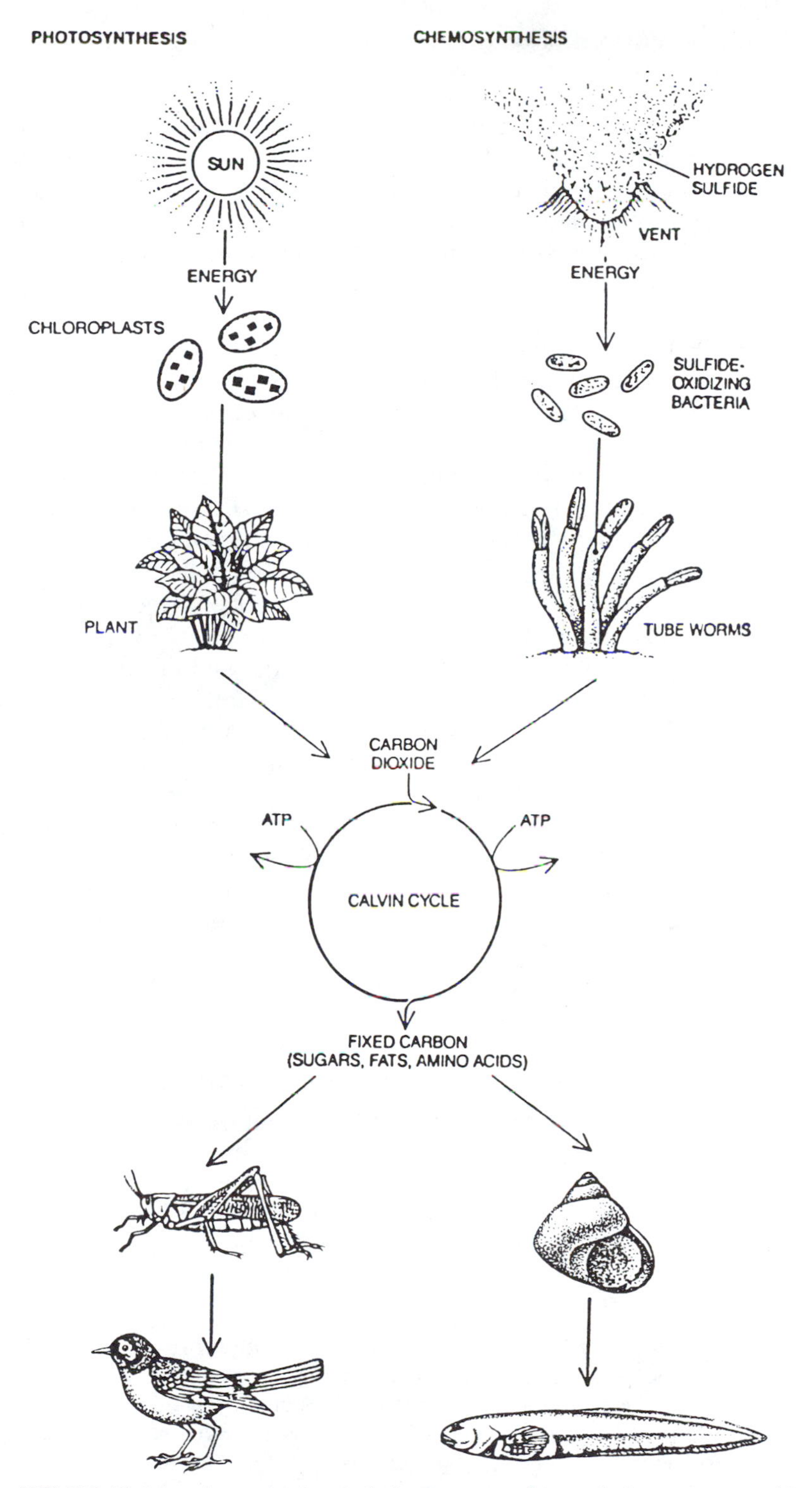

FIGURE 19.14. A comparison of photoautotrophic and chemoautotrophic food chains. *Source:* From J. J. Childress, H. Felbeck, and G. N. Somero, reprinted with permission from *Scientific American,* vol. 256, p. 117, copyright © 1987 by Scientific American, New York.

gas that issues from the vents constitute their electron energy sources. As a result of the availability of reduced compounds, the concentration of microbes in the vent waters is four times greater than that in productive surface waters and thick mats of bacteria cover the surrounding metalliferous sediments.

Many of the animals living around the hydrothermal vents do not occur elsewhere; some represent new species and one, a new genus. The most notable are the large clams (*Calyptogena magnifica*) and the giant tube worms (*Riftia pachyptila*), which reach lengths of 30 cm and 1 m, respectively. As shown in Figure 19.15, the tube worms are essentially a closed sac, having no mouth, digestive system, or other means of processing particulate food.

The worm absorbs O_2, sulfide, and CO_2 through a red respiratory plume that extends from its anterior tip. The red color is due to the presence of hemoglobin in the blood of these organisms. The chemicals are transported by the circulatory system to the trophosome that fills most of the worm's body cavity. The trophosome is composed of cells that contain dense colonies of endosymbiotic sulfur bacteria. These organisms convert the chemicals into organic matter that is then passed back into the blood, where it is aerobically respired by the worm to provide energy and nutrients.

Scientists were initially amazed by this as hydrogen sulfide had heretofore been observed to be toxic to animals at very low concentrations. Tube worms are able to tolerate sulfide because it binds to the hemoglobin at a different site than does O_2, thus preventing any toxic interference. In this way, sulfide is transported to the trophosome without becoming oxidized.

The large clams and a mussel (*Bathymodiolus thermophilus*) also have symbiotic sulfur bacteria that live on their gills. Many of the smaller vent animals do not have symbionts. They obtain their reduced carbon and nutrients by filtering POM from seawater or by consuming other animals. The nitrogen needs of these organisms also appear to be met by the chemoautotrophic fixation of N_2 by an as yet unidentified microbe.

Similar sulfur-based endosymbioses have since been observed in sulfide-rich environments, such as mangrove swamps, petroleum seeps, sewage outfall zones, and marshes. The relationship between sulfur-metabolizing bacteria and the animal host tends to be species-specific; that is, each host species harbors a unique strain of bacteria. This suggests endosymbioses with sulfur bacteria originated independently and repeatedly in diverse animal groups. The great abundance and diversity of the vent animals, as well as their relatively rapid

→

FIGURE 19.15. Internal structure of the tube worm, *Riftia pachyptila*. (*a*) Oxygen, sulfide, and carbon dioxide are absorbed through the plume filaments and transported in the blood to the cells of the trophosome. (*b*) The chemicals are absorbed into these cells, which contain dense colonies of sulfur bacteria and are converted to organic compounds that (*c*) are passed back into the circulatory system to act as an energy source for the worms. *Source:* From J. J. Childress, H. Felbeck, and G. N. Somero, reprinted with permission from *Scientific American*, vol. 256, p. 118, copyright © 1987 by Scientific American, New York.

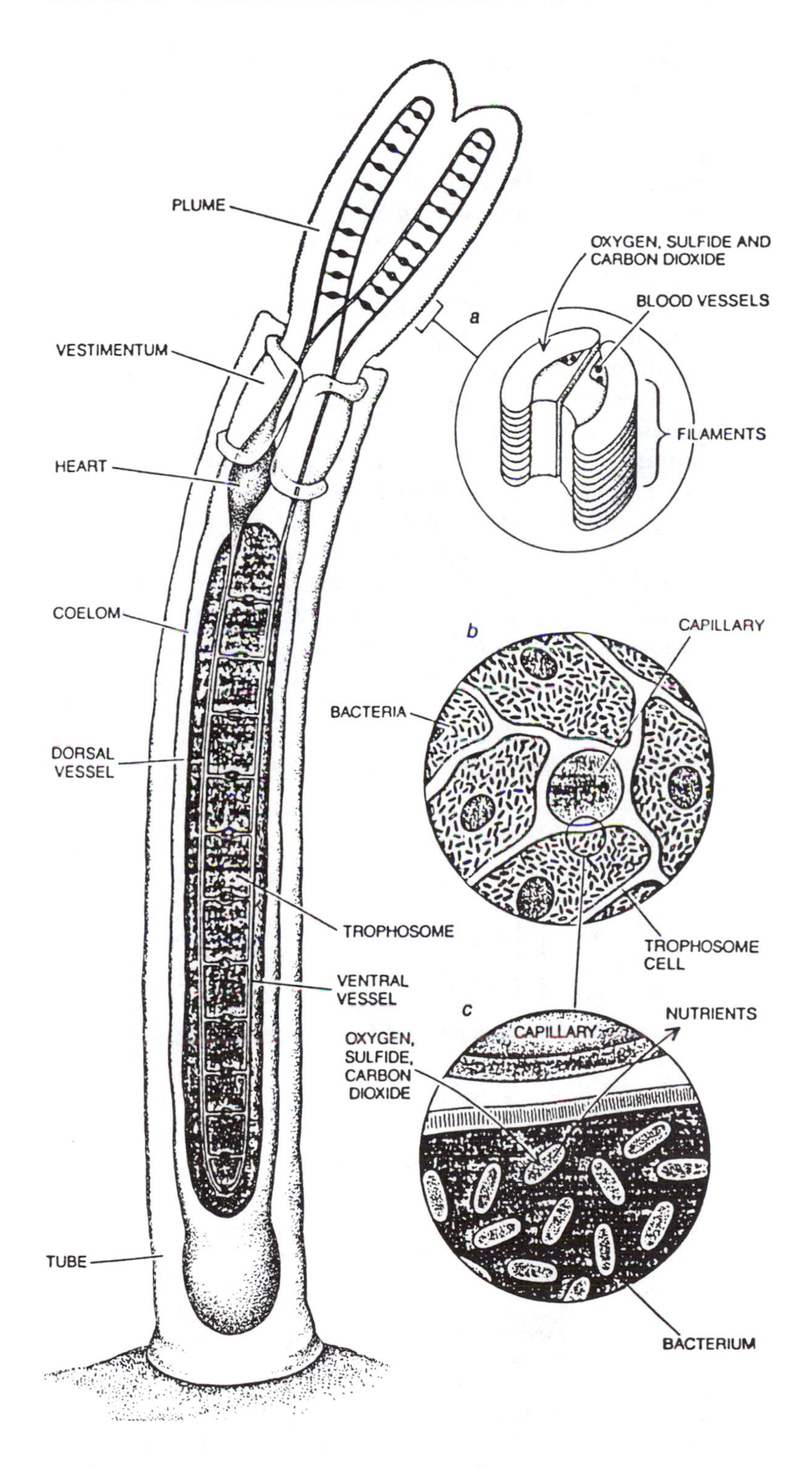
PLUME
VESTIMENTUM
HEART
COELOM
DORSAL
VESSEL
TROPHOSOME
VENTRAL
VESSEL
TUBE
a
OXYGEN, SULFIDE AND
CARBON DIOXIDE
BLOOD VESSELS
FILAMENTS
b
CAPILLARY
BACTERIA
TROPHOSOME
CELL
c
NUTRIENTS
CAPILLARY
OXYGEN,
SULFIDE,
CARBON
DIOXIDE
BACTERIUM

growth rates and large sizes, indicate that these endosymbioses are a very successful adaptation.

In 1984, another vent-type community was discovered living on sulfide emitted by cold-water seeps. These seeps are located along the continental margins off Texas and western Florida in several hundred meters of water. Petroleum appears to be an additional source of energy for the biota found at the Taxas seeps. Vent-type communities have also been discovered on the landward side of some deep-sea trenches, where the decomposition of sedimentary organic matter produces methane supporting chemosynthesis by methanotrophs. This widespread distribution of vent animals suggests these creatures have evolved efficient means for dispersal, which are as yet unknown.

Life may have evolved at hydrothermal vents in a fashion similar to that which has been suggested to have occurred at the sea surface. In the scenario presented in Figure 19.16, it is suggested that the first cells were composed of organic molecules synthesized abiotically from the reduced gases, CH_4, H_2, NH_3, and H_2S, that were supplied by high-temperature hydrothermal activity. These compounds are thought to have adsorbed onto saponite, a magnesium-rich clay mineral formed by the alteration of basalt. This adsorption positions organic compounds in orientations that promote chemical reactions between molecules. Clustering of the nonpolar ends of the resulting macromolecules could have generated the first protocells, which would then have been introduced into the deep sea as part of the hydrothermal fluids.

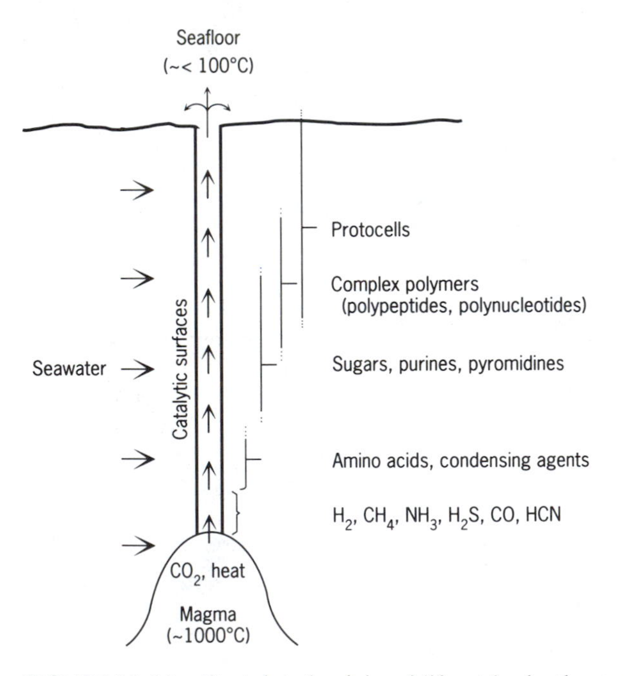

FIGURE 19.16. Postulated origin of life at hydrothermal vents. *Source:* From J. B. Corliss, J. A. Baross, and S. E. Hoffman, reprinted with permission from *Oceanologica Acta*, No. SP, p. 62, copyright © 1981 by Gauthiers-Villars, Montrouge, France.

Rocks formed by hydrothermal processes 3.5 to 3.8 billion years ago contain structures that are thought to be the remains of protocells. If this interpretation is correct, the first organisms on Earth were probably anaerobic heterotrophs. Chemoautolithotrophs and photoautotrophs would have evolved later as a result of increased competition for food resulting from increased numbers of organisms.

ORE DEPOSITS

Large ore deposits are currently thought to be products of hydrothermal activity. If so, the age and distribution pattern of these ore bodies indicate that hydrothermal activity has been widespread for billions of years. The most readily identifiable hydrothermal ores are found in **ophiolites**. These rocks are fragments of oceanic crust that escaped subduction by being thrust up onto the edge of a land mass. Only 0.001 percent of the oceanic crust has been preserved in this way. Some ophiolites contain up to 25 million tons of ore, though deposits of 2 to 5 million tons are most common. Deposits of this size are mined commercially for copper, zinc, and other metals, including gold and silver.

The locations of these ancient chunks of oceanic crust are shown in Figure 19.17. The deposits contain massive sulfides rich in copper and zinc sandwiched between layers of basaltic lava. One of the most well-studied is the Troodoos ophiolite in Cyprus, which was formed at an active spreading center during the late Cretaceous. A diagrammatic section through a typical sulfide deposit in this ophiolite is shown in Figure 19.18. The mineral veins are remnants of hydrothermal vents filled in by precipitation of sulfides.

The formation of large ore deposits requires hydrothermal venting of long duration. Sizable bathymetric lows are also needed to trap sulfide precipitates within a limited area. In some locations, deposition is enhanced by the pooling of hydrothermal fluids due to their relatively high density. This occurs in the Red Sea, which contains an active spreading center. As shown in Figure 19.19, hot-water brines produced by hydrothermal activity collect in the deepest parts of the basin. When brought to the sea surface, metallic sulfides spontaneously precipitate due to their decreased solubility at lower temperatures. Despite the sizable accumulation of metals in the deep waters of the Red Sea, concentrations are not high enough for profitable mining.

Besshi deposits are a type of massive polymetallic sulfide that lies in the midst of sedimentary rock. The latter is usually a shale that originated as a fine-grained clay. As shown in Figure 19.20, these ore bodies take the form of vertical and horizontal intrusions that lie several hundred meters above the oceanic basement. They are deposited near ridge axes located close to land masses that supply large amounts of erosional debris.

Such a Besshi deposit is currently forming in the Guaymas Basin, which is located in the middle of the Gulf of California. The hydrothermal activity at this location is associated with the northernmost extension of the East Pacific Rise. The ridge axis is buried beneath silt eroded from Mexico. The erupting

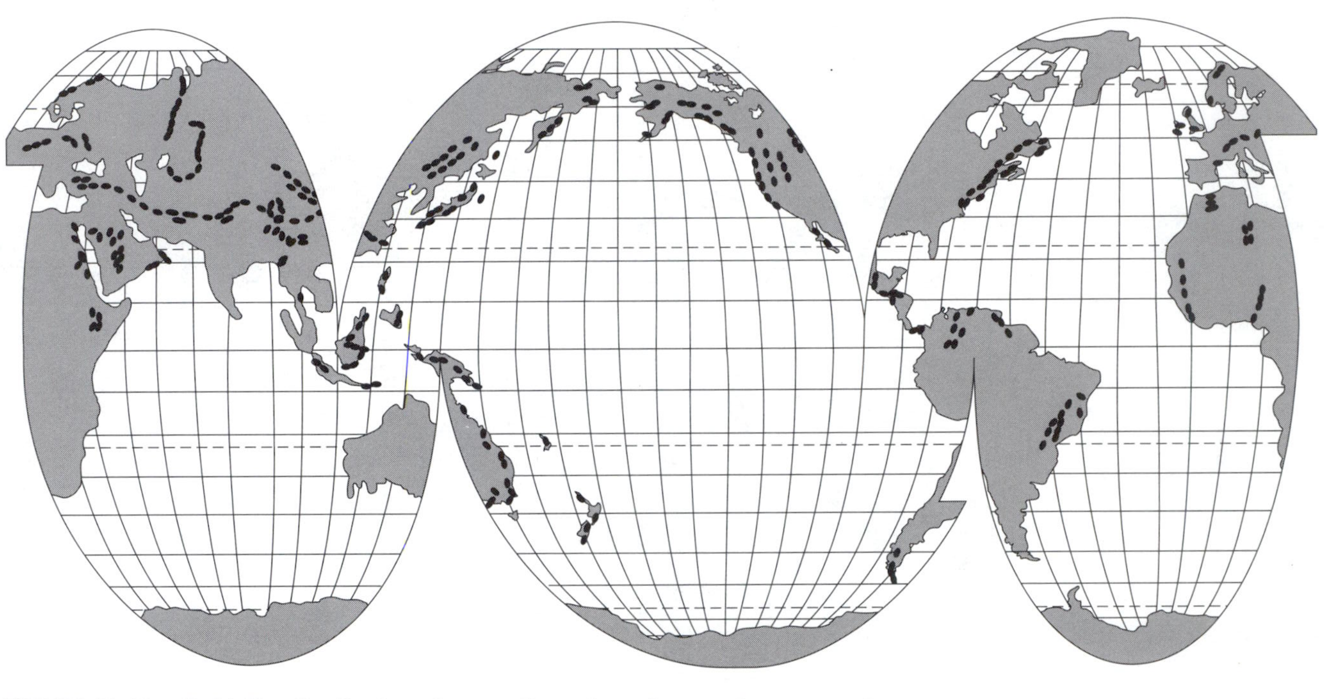

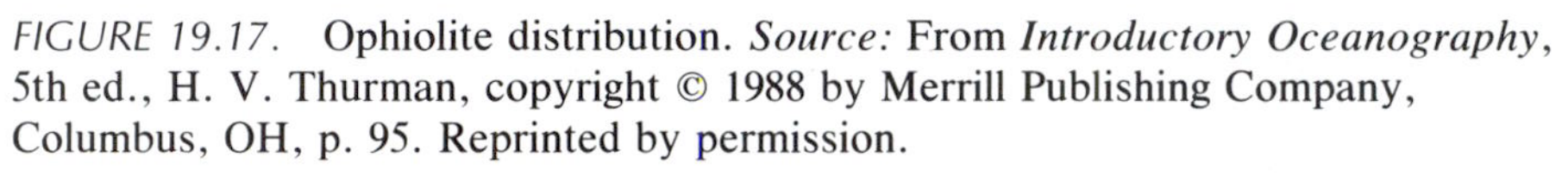

FIGURE 19.17. Ophiolite distribution. *Source:* From *Introductory Oceanography*, 5th ed., H. V. Thurman, copyright © 1988 by Merrill Publishing Company, Columbus, OH, p. 95. Reprinted by permission.

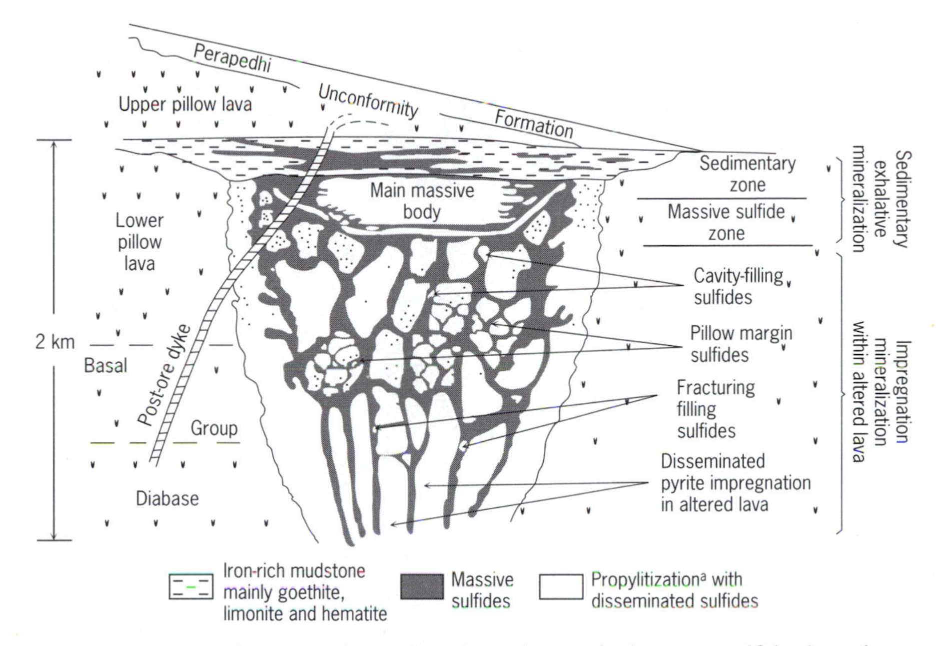

FIGURE 19.18. Diagrammatic section through a typical cyprus sulfide deposit. [a]Propylitization is the formation via hydrothermal alteration of an andestic rock consisting of such minerals as quartz, pyrite, iron oxides, calcite, chlorite, epidote, and serpentine. *Source:* From A. J. Andrews and W. S. Fyfe, reprinted with permission from *Geosciences in Canada,* vol. 3, p. 88, copyright © 1976 by the Geological Survey of Canada, Ottawa, Canada. After R. W. Hutchinson and D. L. Searle, reprinted with permission from *Soc. Mining Geol. Japan*, Special Issue 3, p. 202, copyright © 1971 by the Mineralogical Society of Japan, Tokyo.

basalt has intruded into the silt. Hot-water venting is extensive due to the permeability of the silt. Biological productivity is high in the surface waters, so the underlying sediment contains large amounts of biogenic calcium carbonate. The initially acidic hydrothermal fluids cause the shells to dissolve, raising the alkalinity and pH of the vent waters. This causes reduced metals to precipitate as sulfides within the sediments.

The Besshi deposits in the Gulf of California are also notable for the presence of petroleum hydrocarbons. These compounds appear to be produced by high-temperature cracking reactions involving the organic detritus of phytoplankton. The vent fields contain large mats of bacteria that are presumably feeding on these hydrocarbons.

SUMMARY

The fluids that are emitted from hydrothermal vents are hot, acidic, and enriched in reduced trace metals and sulfide. They are also depleted in alkalinity, sulfate, and magnesium in comparison to seawater. These chemical alterations are the result of interactions between seawater and hot basalt that occur at active spreading centers when seawater percolates down through the

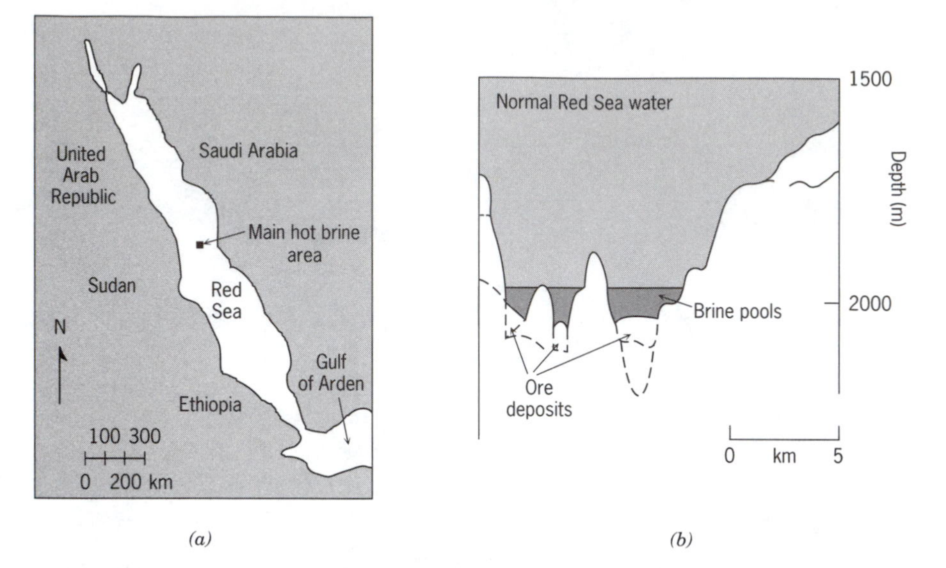

FIGURE 19.19. (*a*) Location of Red Sea hydrothermal activity and metal accumulation and (*b*) a cross-sectional view of some brine pools and ore deposits. *Source*: From J. M. Edmond, reprinted with permission from *Oceanus*, vol. 27, p. 19, copyright © 1984 by Oceanus, Woods Hole, MA.

fractures and fissures in the oceanic basement. Some of the chemical alterations take place at temperatures as high as 350°C, which exist several kilometers below the surface of the sea floor. Others occur at lower temperatures as the heated seawater spontaneously rises due to its decreased density.

The low-temperature reactions continue after the fluid has escaped into the deep ocean. If little mixing with "fresh" seawater occurs, the exiting fluids produce a black smoke immediately upon entry into the deep ocean due to the precipitation of metal sulfides. Some of these particles deposit as chimneys

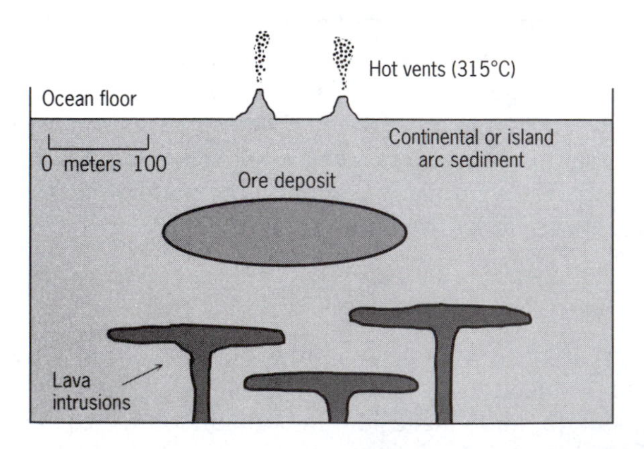

FIGURE 19.20. Besshi ores. *Source*: From J. M. Edmond, reprinted with permission from *Oceanus*, vol. 27, p. 19, copyright © 1984 by Oceanus, Woods Hole, MA.

whose leading edge is composed of anhydrite that precipitates from the reaction of hydrothermal calcium with seawater sulfate. White smokers are the result of subsurface mixing. As sulfides deposit within the chimneys, the hydrothermal flow eventually ceases. The sulfides oxidize, causing the chimneys to collapse into basal mounds. These deposits also contain metalliferous sediments derived from the pelagic sedimentation of the ''smoke'' that escaped from the chimneys and iron–manganese oxides that precipitate at much slower rates than the sulfides. As a result, the mid-ocean ridges and rises are blanketed by a layer of metalliferous sediments.

Hydrothermal circulation transports a significant amount of water and chemicals through the upper mantle. The water movement represents an important mechanism by which heat is convectively removed from spreading centers. The amount of heat thus transferred to the deep sea has been used to compute the global hydrothermal fluxes of elements. These calculations suggest that the ridge crests are a major sink of magnesium and sulfate, probably controlling the steady-state concentration of the former. They are also major sources of lithium and rubidium, as well as a significant source of calcium, potassium, silicon, and barium. These chemical flows have the net effect of lowering the alkalinity of seawater, partly because hydrothermal reactions are a source of acid.

Low-temperature reactions also occur as the cooling **pillow basalts** react with seawater. This causes the rocks to undergo a type of chemical weathering that proceeds from their margins to their interiors. The directions and magnitudes of the elemental fluxes that result from each type of hydrothermal activity are greatly variable. They are also geographically variable due to differences in water temperature and circulation rates, as well as rock porosity and chemical composition. These factors also change over time as vents evolve and die. Due to the small number of direct observations of hydrothermal activity, our ability to assess the global elemental fluxes is still greatly limited.

Hydrothermal fluids contain other reduced compounds, including hydrogen sulfide, methane, and hydrogen gas. The oxidation of these compounds by chemoautolithotrophic bacteria provides enough energy to support an abundant and diverse ecosystem around the hydrothermal vents. Many of these organisms are unique to the vent communities and are also notable for their reliance on bacterial endosymbionts. For example, tube worms obtain their reduced carbon and nutrients from sulfide-oxidizing bacteria that live within a specialized organ. Vent communities have also been observed at cold-water seeps. It is possible that life first evolved around active vents due to the large supply of chemical energy provided by the reduced compounds in the hydrothermal fluids.

When hydrothermal fluids are emitted into bathymetric lows for long periods of time, massive polymetallic sulfides are deposited. Many of these large ore bodies have been uplifted onto land by plate tectonics and are now our primary source of precious metals. Only about 0.001 percent of the oceanic crust has been preserved through such uplift. These fragments are called **ophiolites**. Their global distribution and abundance indicates that hydrothermal activity has been widespread for billions of years.

CHAPTER 20

GLOBAL PATTERN OF SEDIMENT DISTRIBUTION

INTRODUCTION

Marine sediments are classified on the basis of the most abundant particle type found in a given deposit. The relative abundance of a particle type is controlled by three factors: (1) the rate of its supply to the sediments, (2) the degree to which it is preserved in the sediments, and (3) the rate of sedimentation of other particles. Each of these has been discussed for the major sediment types, clay minerals, carbonates, biogenic silica, evaporites, and metalliferous precipitates, in Chapters 14 through 19.

Generalizing from this information, sedimentation rates are highest in areas where particles are introduced into seawater at fast rates. These particles are either transported into the ocean or produced in situ. Chemical, biological, and physical processes can destroy a large fraction before they reach the seafloor. Even those that settle onto the sediment–water interface are subject to dissolution and decomposition. The likelihood of a particle's becoming preserved in the sediments is greatest in areas where its sinking and burial rates are fastest. The relative contribution of a particle type can be diminished by rapid sedimentation of other detritus. This effect is geographically variable as it depends on the relative supply and preservation rates of all the particle types. In this chapter, the relative rates of sedimentation, preservation, and dilution of each particle type are used to explain global patterns of sediment distribution and ocean-to-ocean differences in sedimentation patterns.

TRANSPORT MECHANISMS

Rivers, wind, ice, volcanoes, and organisms are the most important agents of particulate transport. The effects of these processes can be geographically generalized as shown in Figure 20.1, which depicts the distribution of each sediment type in an "average" ocean basin.

River Transport

Most riverborne particles are clay minerals. Because of high supply rates, lithogenous particles compose the bulk of the continental margin sediments.

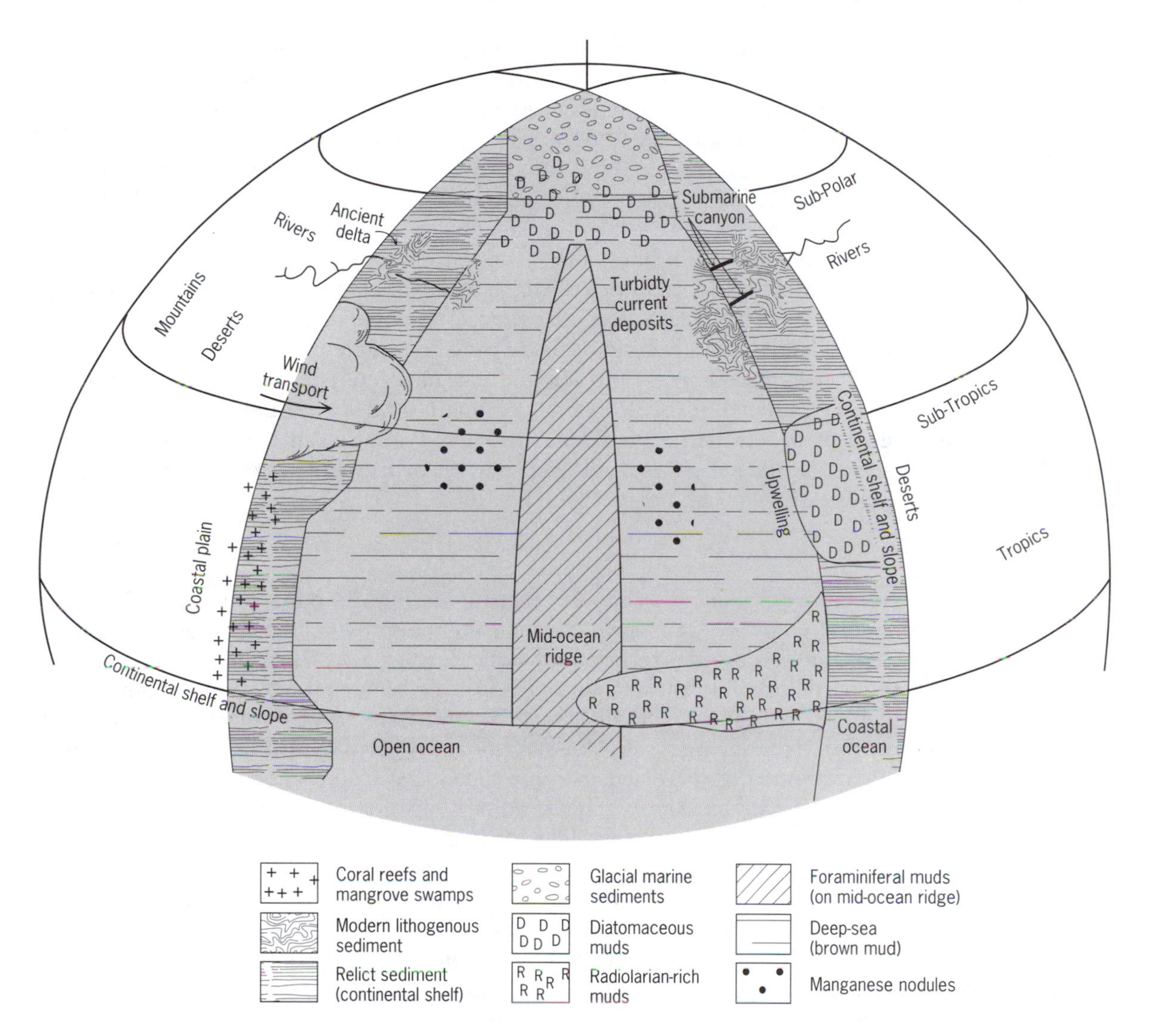

FIGURE 20.1. Schematic representation of the general distribution of sediment in a hypothetical "average" northern ocean basin. *Source*: From *Oceanography: A View of the Earth*, 4th ed., M. G. Gross, copyright © 1988 by Prentice Hall, Inc., Englewood Cliffs, NJ, p. 94. Reprinted by permission.

Relict deposits presently occupy most of the continental shelf as rising sea level has slowed, or eliminated, the supply of particles to the seaward edge of the shelf. Thus, actively forming lithogenous shelf sediments are presently restricted to areas near river mouths. Turbidity currents transport sediments from the shelf to the continental slope and rise. As a result, these locations are characterized by turbidites, composed primarily of lithogenous particles.

A small fraction of the clay-sized particles remain suspended long enough to be carried beyond the seaward limits of the continental margin. They eventually reach the seafloor as a result of pelagic sedimentation. Despite their slow sedimentation rates, they are the dominant particle type in deposits located on abyssal plains that lie below the CCD. In areas of high biological productivity, sinking rates are accelerated by incorporation into fecal pellets.

Though longshore and contour currents can resuspend and transport lithogenous sediment far from its point of entry, most clay minerals are deposited close to the latitude at which they were formed by terrestrial chemical weathering. Thus, the type of clay mineral that characterizes a sediment reflects the prevailing climatic conditions, as they determine the nature of the chemical weathering reactions. Thus kaolinite is characteristic of tropical latitudes; chlorite is found in the sediments of high latitudes; and illite, being a general product of weathering, is found throughout the ocean basins. As described below, latitudinal patterns in clay mineral distributions are most pronounced in abyssal plain sediments.

Wind Transport

As with rivers, most windborne detritus is lithogenous. Clay-sized grains of clay minerals are transported around the globe by the horizontal wind bands (i.e., the Trades, Westerlies, and Polar Easterlies). When the winds weaken, the particles fall to the sea surface and eventually settle to the seafloor. If the resulting pelagic oceanic deposit is greater than 70 percent clay minerals by mass, a deep-sea mud, abyssal clay, or red–brown mud results. The red color is contributed by oxidized iron produced during the chemical weathering of terrestrial rock and the precipitation of hydrothermal manganese.

As illustrated in Figures 14.8 through 14.11, transport by the major wind bands has produced latitudinal patterns in the distributions of the different clay minerals. For example, kaolinite concentrations are highest at low latitudes, where the Trade winds have blown across the deserts that are characteristic of this region. Chlorite, a product of chemical weathering at high latitudes, is most abundant in the sediments beneath the polar easterlies. The effect of the westerlies is most clearly seen in the distribution of detrital quartz in the abyssal plain sediments (Figure 20.2). This mineral is a product of weathering under temperate and subtropical conditions.

Ice Transport

Ice is an important transport agent only at high latitudes, where glaciers erode the land as they flow downslope to the coastline. The fragments which break off and float out to sea contain detrital particles that range in size from clay to boulders. This glacial till settles to the seafloor when the icebergs melt. Since icebergs are restricted to latitudes greater than 60°, glacial marine sediments are deposited only at high latitudes, such as in the Arctic Ocean. Most of these deposits formed during ice ages when glaciers were more abundant and widespread.

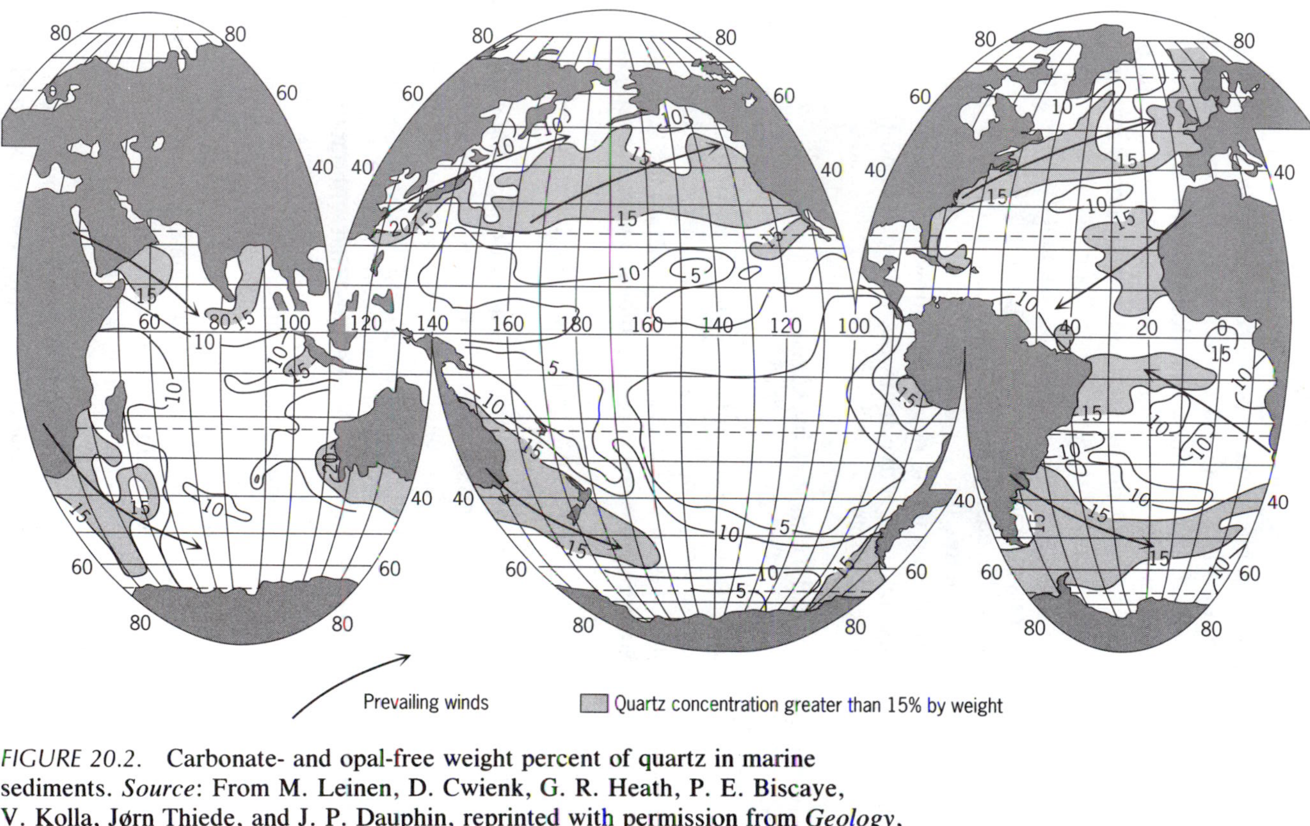

FIGURE 20.2. Carbonate- and opal-free weight percent of quartz in marine sediments. *Source*: From M. Leinen, D. Cwienk, G. R. Heath, P. E. Biscaye, V. Kolla, Jørn Thiede, and J. P. Dauphin, reprinted with permission from *Geology*, vol. 14, p. 202, copyright © 1980 by the Geological Society of America, Boulder, CO.

Particle Transport Resulting from Volcanism and Hydrothermal Activity

Underwater volcanic eruptions occur at lithospheric plate boundaries and isolated mantle hot spots. Above sea level, coastal volcanism is associated with translational and convergent plate boundaries, such as island arc volcanoes. These volcanic eruptions contribute particles to the ocean, including fragments of basalt, pillow lavas, volcanic ash, and hydrothermal precipitates. Volcanic ash can be carried great distances by winds before falling to the sea surface and settling to the seafloor.

Hydrothermal emanations contain dissolved trace elements that precipitate at varying rates once exposed to the cold, oxic conditions of the deep ocean. Metalliferous muds and manganese nodules are examples of hydrothermal precipitates that form some distance from the site at which the metals were introduced to the ocean. As shown in Figure 19.12, their distribution reflects proximity to active hydrothermal vents and the prevailing direction of deep-water currents.

Particle Production and Transport by Marine Organisms

Pelagic marine organisms create particles that have the potential to settle onto the seafloor. The degree to which these particles are preserved in the sediment depends on their chemical composition, the water depth, and chemistry of the surrounding seawater. The biogenous hard parts, such as shell and bone, are more likely to be preserved than the soft parts, such as tissues, fecal pellets, and molts. This difference reflects the ability of bacteria to decompose most POM prior to burial.

The biogenic particles that settle onto the seafloor are mostly the remains of dead organisms. Some come from creatures that lived in the surface waters, but a significant amount is derived from bottom-dwelling plankton, such as benthic foraminifera. Living organisms also sink to the seafloor. Most are bacteria and phytoplankton incorporated in rapidly sinking fecal pellets. Some marine biologists have suggested that without this supply of microbes, the deep sea would have no viable populations because the low temperatures and high pressures inhibit reproduction.

Benthic animals also produce biogenous sediments. The largest such deposits are the calcareous coral reefs created by a community of organisms that include coral animals, coralline algae, sponges, and other organisms. Coral reefs are most common in shallow waters surrounding tropical volcanic islands because the coral community requires warm water and abundant sunlight. The latter fuels photosynthesis by zooxanthallae, a mutualistic symbiont of the coral animals. Coral reefs are not common on continental shelves because they are subject to burial by the rapid sedimentation of lithogenous particles.

Only about 1 percent of the POM produced in the euphotic zone survives the trip to the seafloor to become buried in the sediments. The majority is decomposed by bacteria and protozoans. As a result, organic detritus accounts for less than 1 percent by mass of oceanic sediments. The organic content is highest in shallow-water neritic sediments, ranging from 1 to 10 percent. The most organic-rich sediments are located in salt-marsh estuaries and beneath coastal upwelling zones. As a result, most of the sedimentary organic carbon is buried on the continental margins. Since these deposits are the source beds for petroleum, most large accumulations of oil and gas are located on the continental margins.

The high organic content of neritic sediments is the result of three interdependent factors. First, the close proximity of the continents ensures a large nutrient flux that supports high levels of primary production. This is also enhanced by the shallow water depths that prevent vertical segregation of the nutrients. Although bacterial numbers are high, the particles reach the seafloor fast enough to arrive in a relatively intact state. The large POM flux creates reducing conditions in the sediment that also enhance the degree of organic matter preservation.

As shown in Table 20.1, biogenous oozes are the most common type of pelagic oceanic sediment. Calcareous oozes are more abundant than siliceous oozes, as all seawater is undersaturated with respect to opal, whereas all surface waters and 20 percent of the deep waters are saturated with respect to calcite. In addition, more calcareous shells are formed than siliceous ones.

Though productivity is moderately high along most of the continental margins, oozes are rare in neritic sediments due to dilution by lithogenous particles from the nearby continents. In contrast, the shell content of oozes on the abyssal plains averages 77 percent by mass. Values as high as 90 percent occur in deposits located beneath oceanic upwelling zones and on mid-ocean ridges and rises.

Foram oozes dominate the sediments of the mid-ocean ridges and rises. Calcareous oozes of lower carbonate contents are found on abyssal plains that

TABLE 20.1
Percentage of Pelagic Sediment Coverage in the World Oceans

	Atlantic	*Pacific*	*Indian*	*Total*
Foraminiferal ooze	65	36	54	47
Pteropod ooze	2	0.1	—	0.5
Diatomaceous ooze	7	10	20	12
Radiolarian ooze	—	5	0.5	3
Brown clay	26	49	25	38
Relative size of ocean (%)	23	53	24	100

Source: From *Marine Geology*, J. Kennett, copyright © 1982 by Prentice Hall, Englewood Cliffs, NJ, p. 457. Reprinted by permission. And *Treatise on Chemical Oceanography*, vol. 5, W. H. Berger (eds.: J. P. Riley and R. Chester), copyright © 1976 by Academic Press, Orlando, FL, p. 305. Reprinted by permission. Data from *The Oceans*, H. U. Sverdrup, M. W. Johnson, and R. H. Fleming, copyright © 1941 by Prentice Hall, Englewood Cliffs, NJ; and *The Pacific Ocean*, vol. 6, P. L. Bezrukov, copyright © 1970 by Izdat, Nauk., Moscow, USSR. Reprinted by permission.

lie above the CCD. Oceanic upwelling areas are centered at 0° and 60° latitude in the Northern and Southern hemispheres as a result of divergence caused by the major wind bands. Cold, nutrient-rich water from the base of the mixed layer is transported to the sea surface, fueling high rates of primary productivity. In the upwelling zones that occur at high latitudes, diatoms are the dominant phytoplankton, so diatomaceous oozes occur at 60°N and 60°S. The radiolarian ooze in the equatorial Pacific is the result of high overlying productivity and the scarcity of other sedimenting particles as discussed below.

DISTRIBUTION PATTERN OF SEDIMENTS IN THE WORLD OCEAN

As indicated in Figure 20.3, the distribution of sediment types within each ocean differs from the general description given in Figure 20.1. This is the result of water circulation patterns and geological features unique to each ocean basin.

Atlantic Ocean Sediments

Foram oozes are the most abundant sediment type in the Atlantic Ocean, covering 65 percent of the seafloor. The Mid-Atlantic Ridge is blanketed by oozes that are greater than 80 percent carbonate by mass. Preservation of carbonate is favored in the Atlantic because most of its seafloor lies above the CCD. This is the result of relatively shallow water depths. In addition, the CCD is relatively deep in the Atlantic because its deep water is young and has not accumulated as much remineralized CO_2 as compared to the Indian and Pacific Oceans.

Abyssal clays are found in greater abundance on the western margin of the Atlantic Ocean than the eastern. This is due to bottom topography that restricts the flow of North Atlantic Deep Water and Antarctic Bottom Water to the western side of the basin. The lower temperature of the western waters causes the CCD to be shallower. In the absence of biogenic calcite, abyssal clay deposits dominate, with most being of aeolian origin.

Significant amounts of manganese and phosphorite nodules are present on the top of the Blake Plateau, which lies at the foot of the continental margin of the southeastern United States. The Gulf Stream has eroded most of the unconsolidated sediments, leaving only a carbonate platform, which has become phosphatized. In the presence of slow sedimentation rates, manganese oxides have precipitated as nodules and pavements covering an area of 5000 km^2.

On the eastern margin, a small deposit of siliceous ooze is located slightly south of the equator. This deposit is associated with the coastal upwelling area near Walvis Bay, which is analogous to the Peru upwelling area. The geographic spread of this deposit is limited because the seafloor in this area lies above the CCD, so calcite dilutes the silica. This effect increases with increasing distance from the upwelling area.

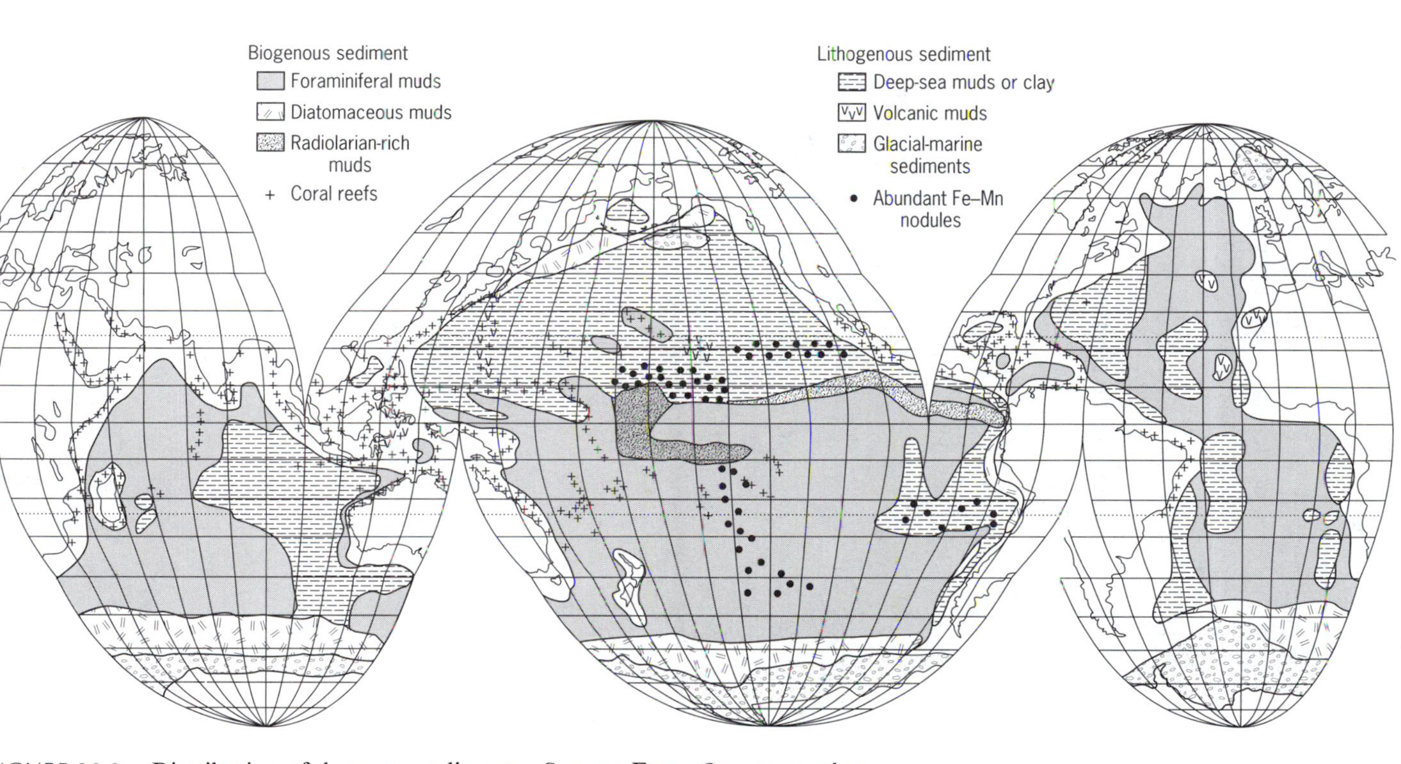

FIGURE 20.3. Distribution of deep-sea sediments. *Source*: From *Oceanography: A View of the Earth*, 3rd ed., M. G. Gross, copyright © 1982 by Prentice Hall, Inc., Englewood Cliffs, NJ, p. 99. Reprinted by permission.

Pacific Ocean Sediments

The distribution of sediment types in the Pacific Ocean is much different from that of the Atlantic. Except for the coastline of the northwest United States, the Pacific is ringed by deep-sea trenches. These trenches effectively trap all the lithogenous particles carried to the sea by river runoff. Thus other particles are relatively important in determining the composition of the sediments in this ocean.

Due to the relative scarcity of lithogenous particles, metalliferous sediments are common around the East Pacific Rise and very high densities of manganese nodules are present on the abyssal plains in the Southern Hemisphere. In these locations, the weathering products of volcanic detritus, such as montmorillonite, phillipsite, and clinoptilite, are also found in great abundance.

Upwelling in the Bering Sea, off Peru, and along the Equator supports enough primary production to produce biogenous oozes in these locations. Radiolarian rather than foram oozes are found along the Equator because the sediments lie below the CCD and are too far from land to have a significant input of lithogenous particles. The deposit in the Bering Sea is a diatomaceous ooze because at low temperatures, diatoms are the dominant phytoplankton. Abyssal clays dominate the North Pacific as this is the location where the CCD is shallowest. Aeolian transport is the source of the clay minerals that make up these deposits.

In the South Pacific, the CCD is deep enough to permit the preservation of calcareous oozes except in the center of the basin, which as a result, is covered by abyssal clays. The relatively rapid supply of hydrogenous sediments prevents the accumulation of calcareous oozes on the East Pacific Rise.

Indian Ocean Sediments

Like the Atlantic, the Indian Ocean is characterized by foram oozes along its mid-ocean ridge and most of its abyssal plains. Only in basins where the water depths exceed 5000 m are abyssal clays abundant.

Arctic Ocean Sediments

The continental shelves cover most of the seafloor in the Arctic Ocean, making this the shallowest ocean. Thus most of the sediments are neritic. Because of light limitation, primary production is inhibited, so river runoff and ice rafting supply most of the particles to this ocean. As a result, lithogenous and glacial marine sediments are most common.

Southern Ocean Sediments

The neritic sediments surrounding the continent of Antarctica are dominated by glacial marine deposits. Divergence at 60°S causes oceanic upwelling, which transports cold, nutrient-rich waters to the euphotic zone. Since diatoms are the dominant phytoplankton, diatomaceous oozes encircle Antarctica between the latitudes 50° and 70°S.

SUMMARY

The global distribution of sediment types is controlled by three factors: (1) the rate at which a particle type is supplied to the sediments, (2) the degree to which these particles are preserved in the sediments, and (3) the rate of sedimentation of other particles.

The most important mechanisms supplying particles to the ocean are rivers, wind, ice, volcanoes, and organisms. Chemical, biological, and physical processes can destroy a large fraction of these particles before they reach the seafloor. Even those that settle onto the sediment–water interface are subject to dissolution and decomposition. The likelihood of a particle becoming preserved in the sediments is greatest in areas where the sinking and burial rates are fastest. The relative contribution of a particle type can be diminished by rapid sedimentation of other detritus. This effect is geographically variable as it depends on the relative supply and preservation rates of all the particle types.

Due to the large supply of lithogenous particles, all of the coastal neritic sediments are dominated by clay minerals. The Atlantic Ocean has the greatest abundance of calcareous oozes because most of its seafloor lies above the CCD. The highest carbonate contents occur in the oozes that lie on top of the mid-ocean ridges and rises due to the relative scarcity of other particle types. The same sediment distribution is found in the Indian Ocean. Since most of the Arctic Ocean is composed of continental margins, its sediments are dominated by lithogenous clays and glacial marine deposits. The continental margins in the Southern Ocean are also characterized by glacial marine deposits. Upwelling at 60°S is responsible for the diatomaceous ooze that encircles the Antarctic continent between 50° and 70°S.

The Pacific Ocean is characterized by abyssal clays in the north because the CCD lies above the seafloor in this basin. The sole exception to this is the deposit of diatomaceous ooze in the Bering Sea produced by upwelling. The abyssal clays are mostly supplied by aeolian transport. Since the waters of the southern Pacific have not accumulated as much remineralized POM, the CCD is deeper in this basin. Hence calcareous oozes are preserved in all but the deepest part of the southern Pacific Ocean basin, which lies below the CCD. These abyssal clays contain a great abundance of manganese nodules, which precipitate from metals supplied by hydrothermal emissions from the East Pacific Rise. These metalliferous sediments dominate the East Pacific Rise. The radiolarian ooze located in the equatorial region is caused by high overlying productivity, the dissolution of calcareous shells, and the relative scarcity of lithogenous detritus.

CHAPTER 21

WHY SEAWATER IS SALTY

INTRODUCTION

In the crustal-ocean factory, the sea is viewed as a stirred reactor that segregates elements into various sedimentary reservoirs. In this chapter, several models are used to quantitatively describe this elemental segregation. These models also provide insight into why (1) the major ion ratios are constant throughout the ocean, (2) the average salinity of the ocean is 35‰, and (3) the salinity of the ocean has been relatively constant over the past 600 million years.

These subjects are of practical concern, as human activity has significantly altered the oceanic supply rates of some elements. Some of these perturbations appear to be having undesirable impacts, such as global warming, on other parts of the crustal-ocean factory. To predict the consequences of our actions, the complex set of interactions that determine and stabilize the chemistry of seawater must be understood.

THE CRUSTAL-OCEAN FACTORY

The global geochemical cycle was presented in Figure 1.1 as a crustal-ocean factory in which igneous rocks and reduced gases are transported from Earth's interior to its crust and upper mantle by volcanic activity. These substances are termed **juvenile** or **primary** materials, since this is their first trip to Earth's surface.

Once on Earth's surface, the juvenile rocks and gases are cycled through various reservoirs located within the atmosphere, hydrosphere, biosphere, crust, and upper mantle. On the early Earth, this cycling involved chemical reactions solely among the primary materials. The resulting soluble and particulate weathering products were transported to the ocean. Though some particles simply settled to the seafloor, others were chemically altered prior to burial. Most solutes were incorporated into solids that were also eventually buried in the sediments.

Assuming that these riverborne and hydrogenous solids had an average density of 1.8 g/cm^3 and were introduced into the ocean at the present rate of 2.0 to 0.65 $\times$ 10^{16} g/y, this input would have filled the ocean basins (volume

$= 1.37 \times 10^{24}$ cm^3) with sediment within 120 to 270 million years. This did not happen because the sediments are now continuously recycled through the crust and the upper mantle. During this process, the buried sediments are converted into sedimentary and metamorphic rocks that are eventually uplifted by crustal motions associated with plate tectonics. Once above the sea surface, these rocks become part of continents or oceanic islands and hence are subject to erosion. The soluble and particulate weathering products are then transported back to the ocean where they are once again recycled through the sedimentary reservoirs.

Some sediment is also recycled through the mantle during the subduction of old oceanic crust at deep-sea trenches. At these locations, the sedimentary blanket is carried down into the mantle along with the oceanic basalt. Both the crust and sediment melt and mix with the rest of the magma. Re-eruption at spreading centers creates new seafloor that is subject to high- and low-temperature weathering as a result of contact with seawater. As with terrestrial weathering, solutes and solid residues are introduced into the ocean.

At present, most mass flows on this planet involve the transport of recycled, or **secondary**, materials. This process occurs in a fashion that also stabilizes the chemical composition of seawater. As indicated by evaporite mineralogy, the concentration of the major ions and their relative contributions to salinity have remained constant over the past 600 million years. The mechanisms responsible for this are discussed below.

SOURCES AND TRANSPORT PROCESSES

As shown in Figure 21.1, the ions transported by rivers into the ocean are soluble products generated during the weathering of continental rock. The quantity and nature of the weathering products depend on climate and topography, as well as the chemical composition of the parent rock. As a result, the chemical composition of river water is spatially and temporally variable. The former is illustrated in Table 21.1.

Half the exposed rocks on the continents are shales produced by the metamorphosis of clay minerals. Second in abundance are the calcium-bearing minerals: limestone, gypsum, and anhydrite. When exposed to rainwater, these minerals dissolve, yielding $Ca^{2+}(aq)$, $CO_3^{2-}(aq)$, and $SO_4^{2-}(aq)$. The addition of carbonate increases the alkalinity of the river water. Most limestone is of biogenic origin, while gypsum and anhydrite are primarily evaporite minerals. As shown in Table 21.2, dissolution of the other abundant evaporite mineral, halite, supplies virtually all of the chloride in river water and approximately 60 percent of the sodium.

Continental rocks also contain a significant amount of granite, which is composed of igneous and metamorphic silicates. These minerals undergo chemical weathering when exposed to rainwater, which is naturally acidic due to the dissolution of atmospheric carbon dioxide. As shown in Eq. 14.2, the weathering products are silicic acid, bicarbonate, clay minerals, and cations.

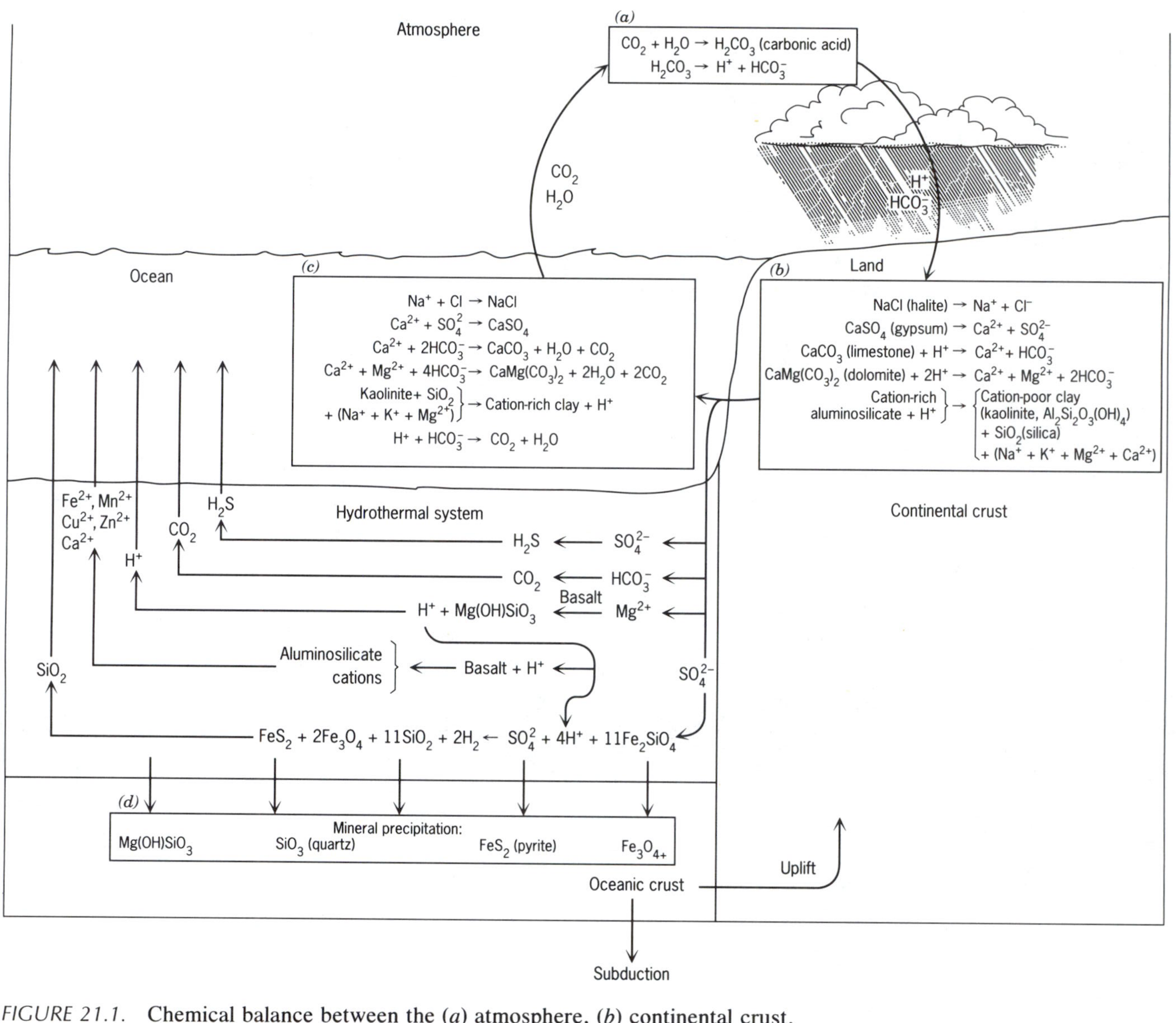

FIGURE 21.1. Chemical balance between the (*a*) atmosphere, (*b*) continental crust, (*c*) ocean, and (*d*) oceanic crust. *Source*: After J. M. Edmond and K. Von Damm, reprinted with permission from *Scientific American*, vol. 248, p. 83, copyright © 1983 by Scientific American, New York.

TABLE 21.1

Average Composition of River Water for the Different Continents[a]

Continent	HCO_3^-	SO_4^{2-}	Cl^-	NO_3^-	Ca^{2+}	Mg^{2+}	Na^+	K^+	Fe	SiO_2	Sum
North America	68	20	8	1	21	5	9	1.4	0.16	9	142
South America	31	4.8	4.9	0.7	7.2	1.5	4	2	1.4	11.9	69
Europe	95	24	6.9	3.7	31.1	5.6	5.4	1.7	0.8	7.5	182
Asia	79	8.4	8.7	0.7	18.4	5.6	9.3		0.01	11.7	142
Africa	43	13.5	12.1	0.8	12.5	3.8	11		1.3	23.2	121
Australia	31.6	2.6	10	0.05	3.9	2.7	2.9	1.4	0.3	3.9	59
World	58.4	11.2	7.8	1	15	4.1	6.3	2.3	0.67	13.1	120
[Seawater]/[river water]	25	240	2500	—	28	300	1800	150	—	0.45	—
[Anions]	0.958	0.233	0.220	0.017							1.428
[Cations]					0.750	0.342	0.274	0.059			1.425

Source: After *The Chemistry of the Atmosphere and Oceans*, H. D. Holland, copyright © 1979 by John Wiley & Sons, Inc., New York, p. 93. Reprinted by permission. Data from D. A. Livingston, *U.S. Geological Survey Professional Paper 440G*, 1963, U.S. Geological Survey, Menlo Park, CA. Reprinted by permission.

[a]Individual ion concentrations are in ppm. Total cation and anion concentrations are in milliequivalents of charge per liter. Also given is the ratio of the seawater ion concentration to that in average river water.

TABLE 21.2
Mineral Sources of the Solutes Present in Average River Water

	Anions (meq/kg)			*Cations (meq/kg)*				*Neutral Species (mmol/kg)*
Source	HCO_3^-	SO_4^{2-}	Cl^-	Ca^{2+}	Mg^{2+}	Na^+	K^+	SiO_2
Atmosphere[a]	0.58[b]	0.09[c]	0.06	0.01	≤0.01	0.05	≤0.01	≤0.01
Weathering or solution of								
Silicates	0	0	0	0.14	0.20	0.10	0.05	0.21
Carbonates	0.31	0	0	0.50	0.13	0	0	0
Sulfates	0	0.07	0	0.07	0	0	0	0
Sulfides	0	0.07	0	0	0	0	0	0
Chlorides	0	0	0.16	0.03	≤0.01	0.11	0.01	0
Organic carbon	0.07	0	0	0	0	0	0	0
Sum	0.96	0.23	0.22	0.75	0.35	0.26	0.07	0.22

Source: From *The Chemistry of the Atmosphere and Oceans*, H. D. Holland, copyright © 1979 by John Wiley & Sons, Inc., New York, p. 140. Reprinted by permission.
[a]These figures do not include soil-derived material.
[b]Largely as atmospheric CO_2.
[c]Much of this is apparently balanced by H^+.

The most important of these reactions involve the weathering of feldspar minerals and are given in Table 14.1. The weathering of orthoclase, albite, and anorthite, produce $K^+(aq)$, $Na^+(aq)$, and $Ca^{2+}(aq)$, respectively. The clay minerals that result are primarily illite, kaolinite, and montmorillonite. The weathering of kaolinite to gibbsite and the partial dissolution of quartz and chert also produce some silicic acid.

The metal hydroxides, oxides, and sulfides also tend to dissolve in naturally acidic rainwater. These minerals are mostly hydrothermal and hydrogenous in origin. Their dissolution resolubilizes the trace metals, as shown in Eqs. 21.1 through 21.3 for Fe(III) and Fe(II):

$$Fe(OH)_3(s) + 3H^+(aq) \rightarrow Fe^{3+}(aq) + 3H_2O(l) \quad (21.1)$$

$$Fe_2O_3(s) + H^+(aq) \rightarrow Fe^{3+}(aq) + 3H_2O(l) \quad (21.2)$$

$$FeS(s) + H^+(aq) \rightarrow Fe^{2+}(aq) + HS^-(aq) \quad (21.3)$$

The concentration of the divalent ion is kept low due to its rapid oxidation by O_2.

The overall effect of these weathering reactions is the addition of cations, dissolved silicon, and alkalinity to river water and the removal of O_2 and CO_2 from the atmosphere. Because the major ions are present in high concentrations in crustal rocks and are relatively soluble, they have become the most abundant solutes in seawater.

As shown in Table 21.2, about 60 percent of the sulfur in seawater is derived from the weathering of sulfides and the solution of sulfate. The rest is supplied from atmospheric sources, which are increasing due to anthropogenic emission of sulfur oxides (SO_x). Most of the calcium is derived from the dissolution of carbonates, which also supplies some of the magnesium, though the majority of the latter comes from the weathering of silicates. This is also the source of 40 percent of the riverborne sodium and most of the potassium and dissolved silicon. The dissolution of carbonates supplies 80 percent of the alkalinity in river water, with the rest derived from the weathering of igneous silicates and shales.

Though the riverine flux is the largest single source of major ions, its magnitude is not well known because of our limited knowledge of the chemical composition of the world's rivers. This is largely due to sampling difficulties caused by spatial and temporal variability in their chemical composition. As shown in Figure 21.2, most of the world's largest rivers lie in areas that are relatively inaccessible to western scientists. This has greatly limited our ability to chemically characterize "average" river water. The most recent estimates of the river-water concentrations of the minor and trace elements are given in Appendix XIII.

The available data are still too sparse to overcome sampling problems associated with spatial and temporal heterogeneity. Thus great uncertainty still exists in our estimates of riverine fluxes, particularly in the suspended load. The data suggest that the amount of particles transported by rivers is approximately equal to the mass of solutes and is much larger than the aeolian flux. The input of riverine particles has a significant impact on seawater chemistry. Most of these particles are clay minerals and have significant amounts of cations adsorbed to their surfaces. Some are desorbed when the clays enter the ocean. For calcium, this release constitutes 10 percent of the total river input. As discussed below, other ions are adsorbed from seawater and for them the clay minerals represent a sink.

Most anions in seawater are derived from volcanic gases released during the cooling of magma. In addition to Cl (as HCl) and S (as H_2S), significant amounts of carbon, nitrogen, hydrogen, and oxygen are discharged in the form of CO_2, N_2, and H_2O, respectively. The amount of primary magmatic volatiles which have been degassed thus far are given in Table 21.3. Most of the chlorine has remained in the ocean. In comparison, virtually all of the carbon and most of the sulfur has become sequestered in the sediments.

Subsurface volcanic activity is also a source of ions, which are solubilized from magma and basalt during hydrothermal reactions. Although the fluxes are poorly known, hydrothermal activity appears to be a significant source of dissolved calcium and possibly potassium and dissolved silicon (Table 19.3).

STORAGE RESERVOIRS AND REMOVAL MECHANISMS

The annual river flux of potassium into the ocean is the product of its average river water concentration (34 μmol K/kg RW) and the annual river water flux

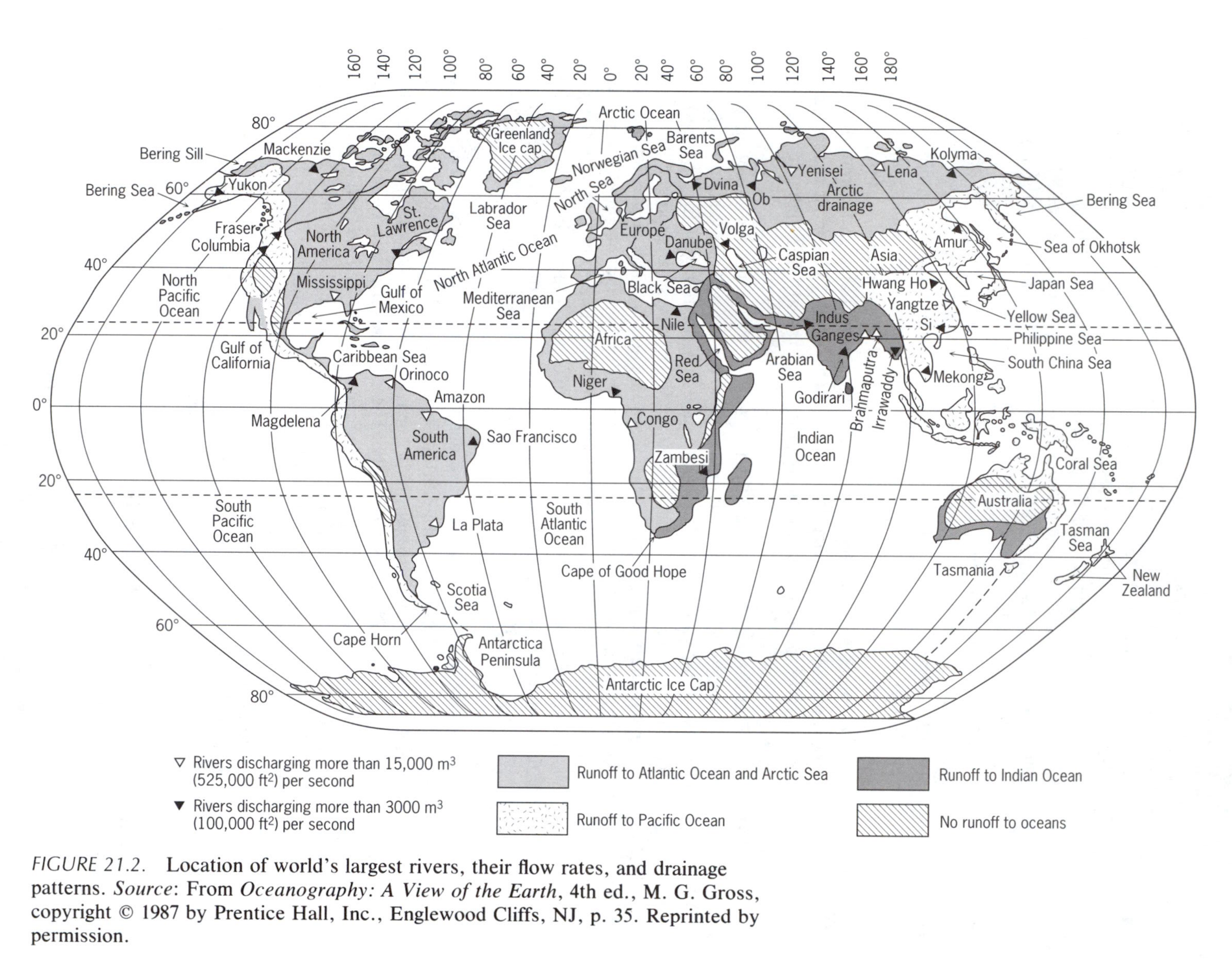

FIGURE 21.2. **Location of world's largest rivers, their flow rates, and drainage patterns.** *Source*: From *Oceanography: A View of the Earth*, 4th ed., M. G. Gross, copyright © 1987 by Prentice Hall, Inc., Englewood Cliffs, NJ, p. 35. Reprinted by permission.

TABLE 21.3
Total Magmatic Volatiles Released over Geologic Time[a]

Volatile	*Mass Released (10^{20} g)*	*Mineral Form in Mole % of Total Released*	*Soluble Form in Mole % of Total Released*
HCl	11.9	33 NaCl	67 in oceanic Cl^-
CO_2	46.4	80 in carbonate	Minor
		20 in organic carbon	Minor
H_2S	4.2	55 in FeS_2	10 in SO_4^{2-}
		35 in $CaSO_4$	
N	2.8	~100 in the atmosphere	Minor
H_2O	953	2 in clay minerals	98 in the hydrosphere

Source: From *Equilibria, Nonequilibria, and Natural Waters*, vol. 1, R. M. Pytkowicz, copyright © 1983 by John Wiley & Sons, Inc., New York, p. 28. Reprinted by permission. Data from Y.-H. Li, reprinted with permission from the *American Journal of Science*, vol. 272, p. 133, copyright © 1972 by the American Journal of Science, New Haven, CT.
[a]Also listed are their present locations and forms.

(3.6×10^{16} kg RW/y) (i.e., 1.2×10^{18} μmol K/y). The total amount of potassium presently residing in the ocean is the product of its seawater concentration (10,200 μmol K/kg SW) and the total mass of the ocean (1.4×10^{21} kg SW) (i.e., 1.4×10^{25} μmol K). Since this flux is the major source of potassium to the ocean, the present-day concentrations of potassium would have been attained after only 12 million years ([1.4×10^{25} μmol K]/[1.2×10^{18} μmol K/y]) of river-water input.

Evidently, most of the potassium that enters the ocean is removed. Since the potassium concentration in seawater has been constant for the past 500 or so million years, its removal rate must also equal its supply rate. In other words, the ocean is in steady state with respect to potassium. This is also true for the other major ions. Because some of the river water is removed through evaporation, all of the major ions are present at higher concentrations in seawater than in river water. But, as shown in Table 21.1, the degree of this enrichment varies among the ions and clearly demonstrates that seawater is not created by the simple evaporation of river water. Rather, some of each ion is also removed from the ocean, but to varying extents, with chloride being the least reactive.

Burial in the sediments and incorporation into the oceanic crust removes the major ions from the ocean for long periods of time. While some are removed primarily into one sediment or mineral type, others have multiple sinks. The chemical reactions that cause the removal of the major ions are controlled by (1) the actions of marine organisms, (2) the formation of evaporites, (3) the weathering of oceanic crust via hydrothermal processes, (4) cation exchange involving clay minerals, (5) burial of interstitial water, and (6) the reverse weathering of clay minerals. The quantitative contributions of the most important of these are given in Table 21.4 and discussed in the following sections.

The data presented in Table 21.4 suggest that the seawater concentrations of magnesium, sulfate, and sodium are close to, but not exactly in, steady state. These imbalances could reflect the poor precision of our present flux estimates or they could be the result of unrecognized sources or sinks.

Deposition of Biogenic Materials

Marine organisms control the production and destruction of calcite, pyrite, and biogenic silica, which are major elemental reservoirs of calcium, sulfur, and silicon, respectively. Calcium carbonate is the single largest sedimentary sink of major ions. Most of the calcium and carbon that have entered the ocean are now sequestered in the crust as limestone or dolomite. These minerals are primarily biogenic in origin, though evaporitic deposition has been of greater importance in the past.

Small amounts of hydrogenous calcium carbonate have also been deposited as oolites and aragonitic needle muds, as well as the iron carbonates: siderite and ankerite. The former pair are common beneath upwelling zones, where cold, deep water, rich in dissolved CO_2 and calcium, rises to the sea surface and is warmed. This warming lowers the solubility of calcium carbonate and CO_2. The latter escapes into the atmosphere, pushing the equilibrium in Eq. 15.20 to the left in favor of calcium carbonate precipitation.

The deposition of calcium carbonate is also the largest sink for alkalinity and magmatic CO_2. Calcite is the major buffer in seawater which prevents short-term fluctuations in pH. As discussed in Chapter 25, atmospheric levels of carbon dioxide are also stabilized by feedbacks that involve carbonate equilibria and hence calcite.

The burial of biogenic silica is the only significant sink for dissolved silicon. Though deep waters are undersaturated with respect to calcium carbonate and biogenic silica, some of each are preserved in the sediments due to rapid accumulation rates.

Approximately half of the sulfur that enters the ocean is removed as the iron sulfides, FeS and FeS_2. Anoxic conditions are required to produce such reduced minerals, as seawater sulfate must be reduced to sulfide. As with the evaporites, sulfide deposition rates are highest when plate tectonics isolates ocean basins, restricting circulation. Such conditions have been widespread during short periods of Earth's past. Thus deposition of iron sulfides has been episodic.

Deposition of Evaporites

Since seawater is undersaturated with respect to the sulfate and chloride salts, precipitation of evaporite minerals only occurs in isolated embayments where water flow is restricted. Thus evaporite deposition is episodic. Actively forming evaporites are presently rare, but only 0.004 percent to 0.02 percent of the

TABLE 21.4
Input–Output Balance for the Major Ions and Alkalinity[a]

	Concentration (mmol/kg)	Ocean Inventory (10^{18} mol)	River Input	Atmospheric/ Evaporite Cycling	Ion Exchange	Hydrothermal Activity	Carbonate Deposition	Net
			10^{12} mol/y					
Cl^-	545	710	10.0	(−10.0)		?		.0
Na^+	468	608	11.8	−9.3	−1.9	?		0.6
Mg^{2+}	53	69	8.0	−0.5	−1.2	−7.8		−1.5
SO_4^{2-}	28	37	3.7	−0.5		−3.8		−0.6
K^+	10	13	3.2	−0.1	−0.4	(−2.7)		.0
Ca^{2+}	10	13	17.1	−0.1	2.6	3.0	(−22.6)	.0
Alk	2.4	3.1	47.1		−0.8	−0.4	−45.2	1.3

Source: From *Principles of Aquatic Chemistry*, F. M. M. Morel, copyright © 1983 by John Wiley & Sons, Inc., New York, p. 221. Reprinted by permission. Data from R. E. McDuff and F. M. M. Morel, reprinted with permission from *Environmental Science and Technology*, vol. 14, p. 1183, copyright © 1980 by the American Chemical Society, Washington, DC.

[a]Due to lack of data, fluxes in parentheses were calculated to provide an exact mass balance. Net fluxes are the result of uncertainties in the flux data, including possibly unidentified sources and sinks.

ocean would have to be evaporitic to remove the annual river flux of sulfate and chloride.

The rate at which evaporites form is directly tied to plate tectonics. This connection is random, as crustal motions must isolate parts of the ocean for evaporitic conditions to occur. Likewise, random motions will uplift and expose evaporites to water, causing them to dissolve. This precludes the establishment of equilibrium between the evaporite minerals and the ocean. Nevertheless, the relative constancy in evaporite mineralogy over geologic time puts severe limits on the extent to which the major ion composition of seawater could have varied. As a result, changes in the concentration of any of the major ions is unlikely to have exceeded a factor of 3.

The average amount of time that evaporites spend buried in the sediments before redissolving is 100 to 200 million years. An enormous quantity of salts have been deposited through this process. The amount of halite could be as much as two to four times greater than the amount of sodium chloride that is now in the ocean. Since evaporites are the sole repository of chloride and remove 50 percent of the sulfur input, their formation is the major anion sink. They are also a significant reservoir of sodium and magnesium.

The magnesium in evaporites is in the form of dolomite, which is an authigenic carbonate. Despite its relatively large abundance in the geologic record and supersaturated state in most seawater, dolomites are presently forming at very slow rates. As a result, their mode of formation is not well understood. Most appear to be secondary evaporite minerals formed when Mg^{2+} replaces some of the Ca^{2+} in calcite and aragonite. Precipitation appears to be inhibited by kinetic barriers.

Plate Tectonics and Hydrothermal Activity

Plate tectonics also influences the major ion composition of seawater by regulating hydrothermal activity and the rate at which buried sedimentary and metamorphic rock are uplifted. The global geochemical balances of magnesium and sulfate appear to be most affected by hydrothermal processes. Magnesium is removed from seawater by high-temperature weathering at hydrothermal vents. If an increased rate of hydrothermal activity has lowered the oceanic concentration of magnesium, this could explain why modern-day dolomites are relatively rare.

Low-temperature reactions at hydrothermal vents cause the precipitation of anhydrite which removes calcium and sulfate from seawater. Sulfate is also removed as a polymetallic sulfide after first having been reduced at the high-temperature reaction zone. Alkalinity is also lost during reactions in the high-temperature zone during a type of reverse weathering reaction involving basalt.

Cation Exchange

The clay minerals carried by rivers into the ocean represent a net annual addition of 5.2×10^{15} meq of cation exchange capacity. Most of these exchange sites are occupied by calcium. Within a few weeks to months following introduction into seawater, sodium, potassium, and magnesium displace most of the calcium. As shown in Table 21.5, this uptake removes a significant fraction of the river input of magnesium and potassium.

The degree to which a cation is exchanged depends on (1) the cation exchange capacity of the clay minerals, (2) the type of exchange sites on the clay, (3) the type of cation being adsorbed, (4) the concentration of the cation, (5) the ionic strength of the solution, and (6) the temperature of the solution. Ion exchange is governed by thermodynamics, as equilibrium is rapidly attained. Thus an alteration in the cation concentration of seawater will lead to shifts in composition of adsorbed ions. Because most of the clay minerals are deposited on the continental shelf, this cation sink is restricted to nearshore waters.

Output of Interstitial Waters

At the sediment–water interface, the seafloor is approximately a 50/50 mixture of sediment and water. During burial, compaction causes the upward vertical advection of pore water. Nevertheless, a substantial amount of seawater is retained in the sediments and represents a net loss of salts from the ocean. Estimating the importance of this sink is difficult because the pore-water salinity increases with increasing depth in the sediments in a highly variable fashion. As suggested by the estimates given in Table 21.6, the output of interstitial waters is probably a significant sink only for sodium and chlorine.

TABLE 21.5
Change in Exchangeable Cations When Riverborne Clays Enter Seawater.[a]

	Average Equivalent Fraction[b]		*Change in Equivalent Fraction*[c]	*Net Removal from Ocean (g/y)*[d]	*Percentage of River Input*
	In river water	*In seawater*			
Na^+	0.04	0.47	+0.43	0.45×10^{14}	20.5 (30)[e]
K^+	0.01	0.06	+0.05	0.09×10^{14}	13
Ca^{2+}	0.60	0.16	−0.44	-0.4×10^{14}	−8
Mg^{2+}	0.25	0.32	+0.07	0.04×10^{14}	3
H^+	0.10	0	−0.10		

Source: From *The Geochemistry of Natural Waters*, J. I. Drever, copyright © 1982 by Prentice Hall, Inc., Englewood Cliffs, NJ, p. 239. Reprinted by permission. Data from F. L. Sayles and P. C. Mangelsdorf, reprinted with permission from *Geochimica et Cosmochimica Acta*, vol. 41, p. 956, copyright © 1977 by Pergamon Press, Elmsford, NY.

[a]Due to the limited precision of the data, these estimates are slightly different from those presented in Table 21.4.

[b]Equivalent fraction is the fraction of the total exchange sites occupied by that cation.

[c]A + sign indicates uptake by the clay.

[d]Assuming a suspended sediment input of 183×10^{14} g/y and a CEC of 25 meq/100 g.

[e]Value in parentheses is corrected for cyclic salts.

TABLE 21.6
Possible Removal of Ions by Burial of Interstitial Water

	Amount Removed (10^{14} g/y)	*Percentage of River Input*
Na^+	0.414	19(29)[a]
K^+	0.015	2.3
Ca^{2+}	0.016	0.3
Mg^{2+}	0.050	4.0
Cl^-	0.744	29 (46)[a]
SO_4^{2-}	0.104	3.1
HCO_3^-	0.006	0.03
SiO_2	0.0002	0.006

Source: From *The Geochemistry of Natural Waters*, J. I. Drever, copyright © 1982 by Prentice Hall, Inc., Englewood Cliffs, NJ, p. 242. Reprinted by permission.
[a]Values in parentheses are for river input corrected for cyclic salts.

Reverse Weathering

The first attempt at a model that explained the chemical composition of seawater was presented by Sillén in 1961. In this model, the major ion concentrations are assumed to be controlled by reactions that reach equilibrium. This control is suggested by the relatively homogeneous composition of seawater and the long reaction times afforded by the relatively slow turnover of water and sediments in the ocean. In this model, the cations are assumed to be supplied by igneous rocks as SiO_2, Al_2O_3, NaOH, KOH, MgO, and CaO. The anions are assumed to be supplied as the magmatic volatiles: HCl, H_2O, and CO_2. In attaining equilibrium with seawater and the sediments, these substances are assumed to have distributed themselves among several gaseous, aqueous, and solid species. The latter include quartz, kaolinite, chlorite, mica-illite, montmorillonite + calcite, and phillipsite, which represent the major sinks of Si, Al, Mg, K, Ca, and Na, respectively. Unique solutions for the equilibrium concentrations of each ion are obtained by fixing the temperature and chloride concentration. The resulting atmospheric level of CO_2 can also be calculated.

In this model, the equilibrium reactions that involve the clay minerals have the following general form:

$$\text{Cation-poor aluminosilicates} + \text{silicic acid} + \text{bicarbonate} + \text{cations} \rightarrow \text{cation-rich aluminosilicates} + \text{carbon dioxide} \quad (21.4)$$

which is that of reverse weathering. During this process, solutes mobilized by terrestrial weathering, (i.e., bicarbonate, dissolved silicon, and cations), are returned to the aluminosilicate phase. The stoichiometry of the specific reverse-weathering reactions used in the Sillén model are given with their equilibrium constants at the end of the table listed in Appendix XII.

The results from a refined version of this calculation are shown in Table 21.7. They are close to the observed major ion concentrations. Nevertheless, this model is not widely accepted as little evidence has been found for the establishment of such equilibria. In fact, concentration gradients in the ocean and pore waters suggest equilibrium is not attained. This is supported by the results of laboratory equilibration experiments that have demonstrated that these reactions are extremely slow and do not stabilize the ion concentrations at the levels observed in seawater.

The major ion concentrations are more likely controlled by a dynamic balance between their relative rates of supply to and removal from the ocean. Since these processes are irreversible, they cannot reach thermodynamic equilibrium. The quantitative importance of the various ion supply and removal mechanisms in maintaining the steady-state composition of seawater can be deduced from a mass balance calculation. The results of such a calculation were first presented by Mackenzie and Garrels in 1966 and are reproduced in Table 21.8.

In this calculation, the major ion concentrations are assumed to remain constant over time as a result of processes that remove all of their annual input. At the top of Table 21.8 is the amount of each ion supplied over one mixing cycle, (i.e., 1000 y). These ions are assumed to be removed by the deposition of calcite, halite, dolomite, pyrite, and anhydrite. For bookkeeping purposes, the effect of each type of mineral deposition is recorded as a

TABLE 21.7
Results and Components in Kramer's Ocean Model[a]

Observed Concentration (M)	*Equilibrium Concentration (M)*	*Liquid Phase*	*Solids Controlling Concentrations*
0.48	0.45	Na^+	Na-mont (E site)[b]
1.0×10^{-2}	9.7×10^{-3}	K^+	K-illite (E site)[b]
0.56	0.55 (defined)	Cl^-	0.55 (assumed)
2.9×10^{-2}	3.4×10^{-2}	SO_4^{2-}	$SrCO_3$, $SrSO_4$
1.1×10^{-2}	6.1×10^{-3}	Ca^{2+}	Phillipsite
5.4×10^{-2}	6.7×10^{-2}	Mg^{2+}	Chlorite
Variable	2.7×10^{-6}	PO_4^{3-}	OH-apatite
—	(1.7×10^{-3} atm)	CO_2	Calcite
7.0×10^{-5}	2.4×10^{-5}	F^-	F-CO_3-apatite
7.9×10^{-9}	4.7×10^{-9}	H^+	Given by electroneutrality

Source: From *Aquatic Chemistry*, W. Stumm and J. J. Morgan, copyright © 1970 by John Wiley & Sons, Inc., New York, p. 415. Reprinted by permission. Data from J. R. Kramer, reprinted with permission from *Geochimica et Cosmochimica Acta*, vol. 29, p. 935, copyright © 1965 by Pergamon Press, Elmsford, NY.

[a]Also given are the observed concentrations of the ions in seawater at 25°C and 35‰.

[b]In interpreting mass-action equilibrium, for some clays two discrete exchange sites, "C" sites and "E" sites, perhaps corresponding to interlayer and edge sites, were considered.

TABLE 21.8
Mass Balance Calculation for the Removal of River-Derived Constituents from the Ocean

Step No.	Reaction (balanced in terms of mmol of constituents used)	Constituent balance ($\times 10^{21}$ mmol) SO_4^{2-}	Ca^{2+}	Cl^-	Na^+	Mg^{2+}	K^+	SiO_2	HCO_3^-	HCO_3^- Consumed (−) Evolved (+)	CO_2 Consumed (−) Evolved (+)	Products	($\times 10^{21}$ mmol)	Total products formed (mole basis) (%)
	Amount of material to be removed from ocean in 10^8 years ($\times 10^{-21}$ mmol)	382	1220	715	900	554	189	710	3118					
1.	95.5 $FeAl_6Si_6O_{20}(OH)_4$ + 191 SO_4^{2-} + 47.8 CO_2 + 55.7 $C_6H_{12}O_6$ + 238.8 H_2O = 286.5 $Al_2Si_2O_5(OH)_4$ + 95.5 FeS_2 + 382 HCO_3^-	191	1220	715	900	554	189	710	3500	+382	−48	Pyrite Kaolinite	96 287	3 8
2.	191 Ca^{2+} + 191 SO_4^{2-} = 191 $CaSO_4$	0	1029	715	900	554	189	710	3500			$CaSO_4$	191	5
3.	52 Mg^{2+} + 104 HCO_3^- = 52 $MgCO_3$ + 52 CO_2 + 52 H_2O	0	1029	715	900	502	189	710	3396	−104	+52	$MgCO_3$ in magnesium calcite	52	2
4.	1029 Ca^{2+} + 2058 HCO_3^- = 1029 $CaCO_3$ + 1029 CO_2 + 1029 H_2O	0	0	715	900	502	189	710	1338	−2058	+1029	Calcite and/or aragonite	1029	29

5.	$715\ Na^+ + 715\ Cl^- = 715\ NaCl$	0	0	0	185	502	189	710	1338			NaCl	715	20
6.	$71\ H_4SiO_4 = 71\ SiO_{2(s)} + 142\ H_2O$	0	0	0	185	502	189	639	1338			Free silica	71	2
7.	$138\ Ca_{0.17}Al_{2.33}Si_{3.67}O_{10}(OH)_2 + 46\ Na^+ =$ $138\ Na_{0.33}Al_{2.33}Si_{3.67}O_{10}(OH)_2 + 23.5\ Ca^{2+}$	0	24	0	139	502	189	639	1338			Sodic montmorillonite	138	4
8.	$24\ Ca^{2+} + 48\ HCO_3^- = 24\ CaCO_3 + 24\ CO_2 + 24\ H_2O$	0	0	0	139	502	189	639	1200	−48	+24	Calcite and/or aragonite	24	1
9.	$486.5\ Al_2Si_{2.4}O_{5.8}(OH)_4 + 139\ Na^+ + 361.4\ SiO_2 +$ $139\ HCO_3^- = 417\ Na_{0.33}Al_{2.33}Si_{3.67}O_{10}(OH)_2 +$ $139\ CO_2 + 625.5\ H_2O$	0	0	0	0	502	189	278	1151	−139	+139	Sodic montmorillonite	417	12
10.	$100.4\ Al_2Si_{2.4}O_{5.8}(OH)_4 +$ $502\ Mg^{2+} + 60.2\ SiO_2 + 1004\ HCO_3^- =$ $100.4\ Mg_5Al_2Si_3O_{10}(OH)_8 + 1004\ CO_2 + 301.2\ H_2O$	0	0	0	0	0	189	218	147	−1004	+1004	Chlorite	100	3
11.	$472.5\ Al_2Si_{2.4}O_{5.8}(OH)_4 + 189\ K^+ + 189\ SiO_2 +$ $189\ HCO_3^- = 378\ K_{0.5}Al_{2.5}Si_{3.5}O_{10}(OH)_2 + 189\ CO_2 +$ $661.5\ H_2O$	0	0	0	0	0	0	29	−42	−189	+189	Illite	378	11

Source: From F. T. Mackenzie and R. M. Garrels, reprinted with permission from the *American Journal of Science*, vol. 764, p. 514, copyright © 1966 by the American Journal of Science, New Haven, CT.

deduction from the total amount of ions which must be removed. Since excess Na^+, Mg^{2+}, and K^+ remain, another sink must be invoked to attain steady state. The missing sink was assumed to be reverse weathering.

To remove the excess Na^+, Mg^{2+}, and K^+ via reverse weathering requires the in situ production of montmorillonite, chlorite, and illite at rates of 15×10^{14}, 5×10^{14}, and 1.4×10^{15} g/y, respectively. Since the total riverine flux of detrital clay minerals is thought to range between 0.65×10^{16} and 2.0×10^{16} g/y, the relative contribution of these secondary clays to the amount entering the ocean should range from 17 to 52 percent. Since these clays would be created from preexisting detritus, little change in the total mass of the clay minerals should occur. Since their contribution could be minor, the presence of these secondary clays is likely to be difficult to detect. Indeed, much evidence suggests that significant amounts of these clays are not present in marine sediments.

It is possible that a type of reverse weathering is occurring in unrecognized mineral phases or in other locations. For example, the incorporation of potassium into detrital clays could be producing glauconite rather than illite. Glauconite is an amorphous iron-rich illite that contains a great variety of cations, as shown by its empirical formula:

$$(K,Na,Ca)_{1.2-2.0}(Fe^{3+},Al,Fe^{2+},Mg)_{4.0}(Si_{7-7.6}Al_{1-0.4}O_{20})_4 \cdot nH_2O \quad (21.5)$$

Although some is detrital, authigenic glauconite appears to form from fecal pellets that contain grains of detrital clays. If most glauconite is authigenic, it

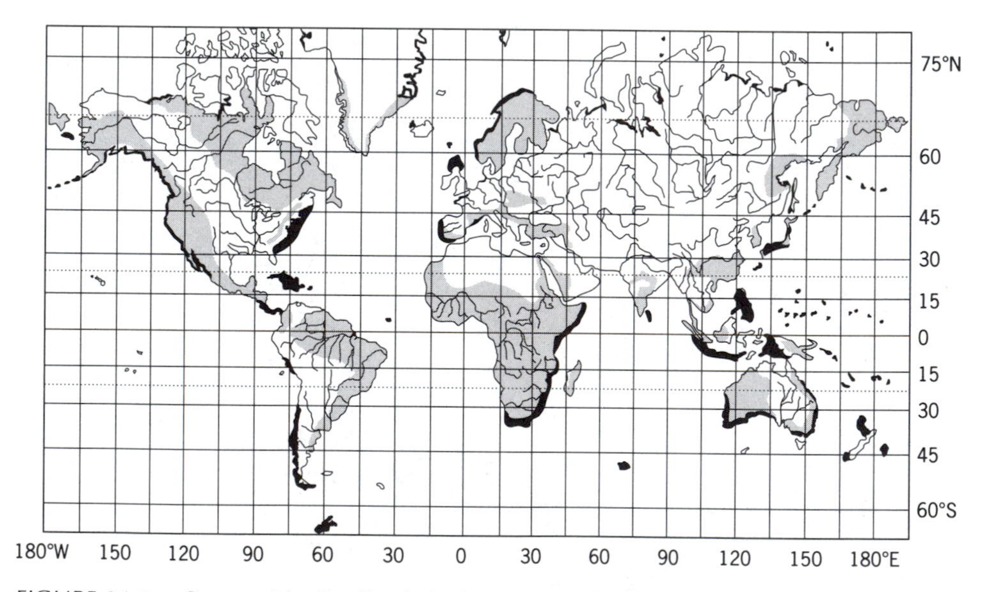

FIGURE 21.3. Geographic distribution of glauconite in marine sediments. *Source*: From E. W. Galliher, reprinted with permission from the *Bulletin of the American Association of Petroleum Geologists*, vol. 19, p. 1594, copyright © 1935 by the American Association of Petroleum Geologists, Tulsa, OK.

could represent a sink for as much as 20 percent of the riverine flux of potassium, as well as 3 percent of the riverine flux of magnesium. As shown in Figure 21.3, glauconites are most abundant in nearshore sediments due to the relatively high fluxes of sedimenting POM at these locations.

Another possible location where reverse weathering could be occurring is deep within the sediments. For example, late-stage diagenesis or catagenesis, associated with metamorphism, appears to remove magnesium from pore waters. This is seen in the decrease in pore-water [Mg^{2+}] with increasing depth, as shown in Figure 21.4.

Sulfate reduction is the likely cause of the concurrent decrease in sulfate concentrations. The resulting sulfide precipitates pyrite by reacting with iron supplied by the dissolution of iron-rich silicates and oxyhydroxides. This solubilization is favored at the low pH's generated by sulfate reduction. The decrease in magnesium appears to be caused by its isomorphic substitution for iron in the silicates and oxyhydroxides. Complete replacement could remove

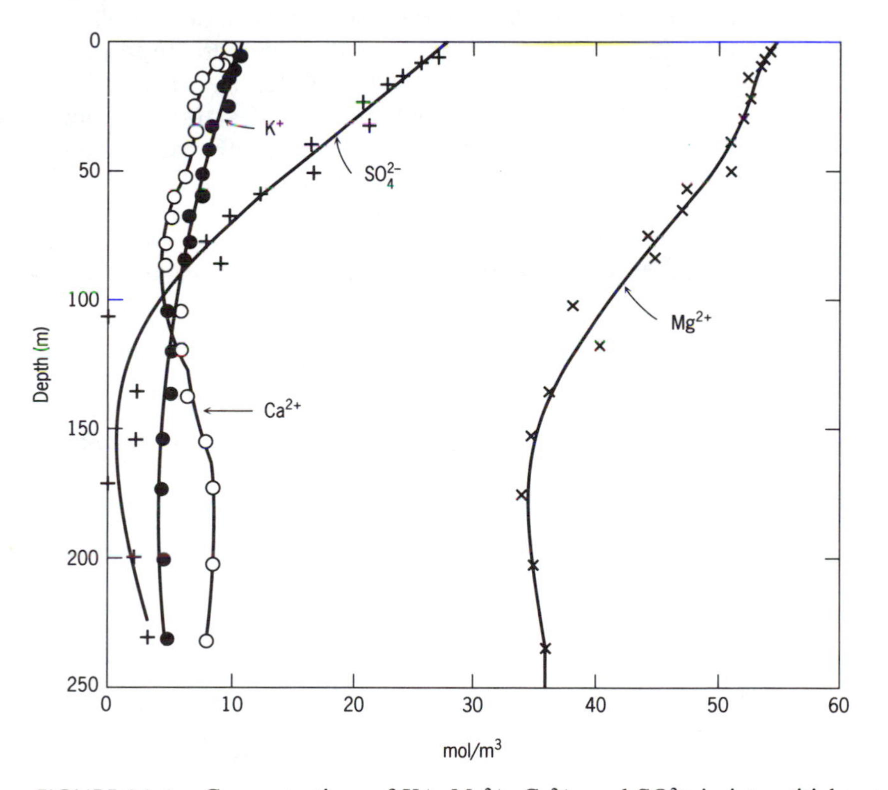

FIGURE 21.4. Concentrations of K^+, Mg^{2+}, Ca^{2+}, and SO_4^{2-} in interstitial waters from a core taken in the Caribbean Sea. *Source*: From *Chemical Oceanography*, W. S. Broecker, copyright © 1974 by Harcourt, Brace and Jovanovich Publishers, Orlando, FL, p. 188. Reprinted by permission. Data from the *Initial Reports of the Deep Sea Drilling Project*, vol. 20, F. L. Sayles, F. T. Manheim, and L. S. Waterman, (eds.: B. C. Heezen and I. D. MacGregor), 1973, U.S. Government Printing Office, Washington, DC, pp. 788–791.

20 to 30 percent of the riverine magnesium flux. This is an upper estimate as other major cations probably occupy some of the sites vacated by the iron.

With the discovery of hydrothermal activity in 1977, marine chemists realized that some of the reverse-weathering reactions required to achieve steady state were evidently occurring during high- and low-temperature reactions between seawater and basalt at active spreading centers. For example, the estimated global removals of Mg and S within the hydrothermal systems are large enough to provide the required sinks. This eliminates the relatively high dolomite and pyrite deposition rates required by the Mackenzie and Garrels model.

Reverse weathering appears to play an important role in controlling the pH and alkalinity of seawater over time scales on the order of millions of years. As shown in Figure 21.5, this stabilization is provided by a set of feedback loops that also involve the carbonate system on short time scales. Thus reverse weathering ultimately affects atmospheric CO_2 levels. Though the carbonate buffer system responds far more rapidly to changes in pH and CO_2, its buffering ability is limited in comparison to that provided by the clay minerals and is poised at a slightly different pH.

The difference between these two pH-stabilizing mechanisms is illustrated in the following example. If the CO_2 content of the atmosphere were to be doubled, much of the gas would dissolve into the ocean. This addition of CO_2 represents a source of acid, most of which would be absorbed by the carbonate-buffering system. Once equilibrium was reestablished, the pH of seawater would be only slightly lower than its initial value. The remaining acid would

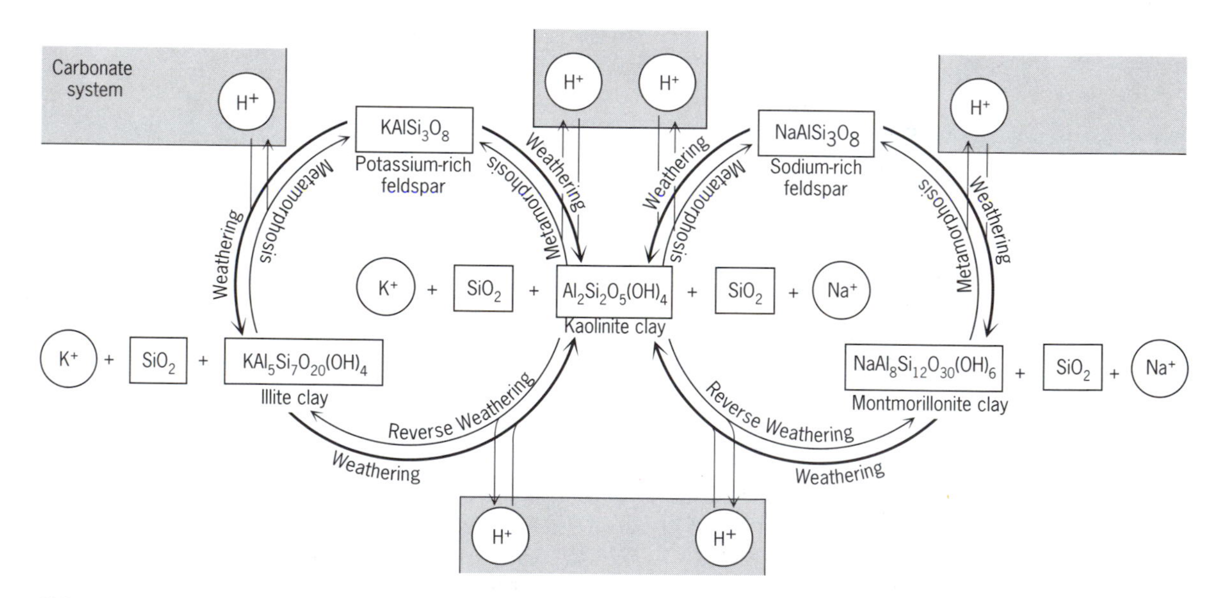

FIGURE 21.5. The role of hydrogen ions in the reverse-weathering cycle. *Source*: From R. Siever, reprinted with permission from *Scientific American*, vol. 230, p. 74, copyright © 1974 by Scientific American, New York.

be removed after the passage of several million years by reaction with the clay minerals. Since cations are released, this secondary equilibration would also cause an increase in alkalinity, as seawater must remain electrically neutral.

RESIDENCE TIMES AND STEADY STATE

The assumption that the major ions are in steady state is supported by the nearly homogenous chemical composition of their sedimentary reservoirs. The persistence of similar life forms over billions of years also suggests that the chemical composition of seawater has been fairly stable for long periods. Given a steady state, the residence time of the major ions can be calculated from Eq. 1.1, where the input rate is, in most cases, dominated by the dissolved river flux. These residence times with respect to river input ($\tau_{R,O}$) are presented in Table 21.9 along with those of the minor and trace elements.

For elements that are supplied to the ocean by routes other than river input, these tabulated values are somewhat higher than the "true" residence time. Nevertheless, they are useful for order-of-magnitude comparisons as discussed below.

TABLE 21.9

Logarithms of the Elemental Oceanic Residences Times with Respect to Their River Input (log $\tau_{R,O}$)

H 4.5																	He
Li 6.3	Be (2)											B 7.0	C 4.9	N 6.3	O 4.5	F 5.7	Ne
Na 7.7	Mg 7.0											Al 2	Si 3.8	P 4	S 6.9	Cl 7.9	Ar
K 6.7	Ca 5.9	Sc 4.6	Ti 4	V 5	Cr 3	Mn 4	Fe 2	Co 4.5	Ni 4	Cu 4	Zn 4	Ga 4	Ge	As 5	Se 4	Br 8	Kr
Rb 6.4	Sr 6.6	Y	Zr 5	Nb	Mo 5	Tc	Ru	Rh	Pd	Ag 5	Cd 4.7	In	Sn	Sb 4	Te	I 6	Xe
Cs 5.8	Ba 4.5	La 6.3	Hf	Ta	W	Re	Os	Ir	Pt	Au 5	Hg 5	Tl	Pb (2.6)	Bi	Po	At	Rn
Fr	Ra 6.6	Ac															

Ce	Pr	Nd	Pm	Sm	Eu	Gd	Tb	Dy	Ho	Er	Tm	Yb	Lu
Th (2)	Pa	U 6.4											

Source: From *The Chemistry of the Atmosphere and Oceans*, H. D. Holland, copyright © 1979 by John Wiley & Sons, Inc., New York, p. 159. Reprinted by permission. See Holland (1979) for data sources.

If J_{in} and J_{out} represent the total input and removal rates of an element, respectively, then

$$\frac{dC}{dt} = J_{in} - J_{out} \tag{21.6}$$

where C equals the total amount of the element in the ocean. If the removal rate is assumed to be proportional to the total amount of the element in the ocean, then $J_{out} = kC$, where k is the net rate constant for removal. At steady state,

$$0 = \frac{dC}{dt} = J_{in} - kC \tag{21.7}$$

or

$$\frac{1}{k} = \frac{C}{J_{in}} = \tau_c \tag{21.8}$$

which states that the residence time is inversely related to the removal rate constant. In other words, the more reactive an element, the shorter its residence time. As indicated in Table 21.9, the shortest residence times are characteristic of elements in the transition, lanthanide, and actinide series. This reflects a greater chemical reactivity caused by unfilled *d* orbitals.

The relative reactivity of an element can also be judged from its charge density, or ionic potential. This is defined as the ratio of its ionic charge, z, to its ionic radius, r. As shown in Table 21.10, elements with low ionic potentials are relatively unreactive. Since they tend to remain in solution, their residence times are relatively long. The major ions fit into this category. Elements with high ionic potentials are reactive but tend to form soluble complexes. Thus they also tend to stay in solution but are heterogeneously distributed in the ocean due to the chemical reactions they undergo in seawater. The biolimiting elements fall into this category. The elements with intermediate ionic potentials are reactive but tend to form insoluble precipitates, primarily hydroxides and oxides. As a result they are rapidly removed from the ocean and their residence times are short. The transition metals are examples of this.

TABLE 21.10
Influence of Ionic Potential on the Chemical Behavior and Residence Time of Some Elements in the Ocean

z/r	*Elements*	*Chemical Behavior*	*Residence Time*
Low	Na, K, Mg, Ca	Stay in solution as $M(H_2O)$	Long
Mid	Fe, Al, Ti	Precipitate as $M(OH)_x$	Short
High	B, C, N, P, S	Stay in solution as complex ions, eg., SO_4^{2-}, CO_3^{2-}	Medium

At the present rate of river input, enough sediment is supplied to the oceans to fill its basins every 120 to 270 million years. If the sedimentary blanket remains constant in size over time, then the residence time of an average particle must be somewhere between 120 and 270 million years. This recycling rate is constrained by plate motion. In other words, no sedimentary particle can be older than the oldest piece of oceanic crust, which is approximately 220 million years.

In comparison, the residence time of seawater in the mixed layer is 100 y and in the deep zone, approximately 900 y. Elements with residence times less than 1000 y (the mixing time of the ocean) have the potential to be heterogeneously distributed in the ocean. These include the transition elements. Elements with residence times greater than 1000 y are more likely to be homogeneous. As noted above, exceptions to this are exhibited by those elements that are rapidly recycled within the ocean.

For elements that have multiple sources or sinks, a fractional residence time (τ_i), or turnover time, can be calculated for each supply or removal process. The mean residence time of the element is then given by

$$\frac{1}{\tau_c} = \sum_{i=1}^{i=n} \frac{1}{\tau_i} = \frac{1}{\tau_1} + \frac{1}{\tau_2} + \ldots \frac{1}{\tau_n} \qquad (21.9)$$

This equation shows that the mean residence time for an element is largely controlled by the processes that have large fluxes. In other words, small or slow fluxes have an insignificant impact on the mean residence time. For example, the mean residence times for sodium and chlorine are much shorter than their $\tau_{R,O}$ because of rapid atmospheric cycling. This process involves their ejection from the sea surface via bursting bubbles. Some of the salts are redeposited on the sea surface. The rest are transported by winds over the land where they are deposited. Rivers carry these salts back to the ocean. The fractional residence time of chlorine with respect to atmospheric cycling is 73 $\times$ 10^6 years. Thus its mean residence time is 53 $\times$ 10^6 y, which is much shorter than its riverine turnover time ($\tau_{R,O}$ = 200 $\times$ 10^6 y). As shown in Table 21.11, this rapid atmospheric cycling also affects the other major ions, but to a much lesser extent.

TABLE 21.11

Percentage of Cyclic Seawater Ions by Weight Relative to Their Total Weight in Rivers

Constituent	Na^+	K^+	Mg^{2+}	Ca^{2+}	Cl^-	SO_4^{2-}	HCO_3^-
Percentage of cyclic constituent by weight in rivers	35	15	7	0.7	55	6	0.2

Source: From *Evolution of Sedimentary Rocks*, R. M. Garrels and F. T. Mackenzie, copyright © 1971 by W. W. Norton Co., New York, p. 108. Reprinted by permission.

WHAT CONTROLS THE STEADY-STATE SET POINT OF SEAWATER?

Small-scale, short-term perturbations can cause local, short-lived fluctuations in ion concentrations. Over long time periods, these concentrations average out to the steady-state level because of feedback loops that ensure a return to the original steady-state conditions. If the perturbation persists and is not too large, a new steady state can be established, with a new seawater composition. Large-scale, long-term perturbations can overwhelm the feedback loops and cause the chemical composition to vary greatly over time.

The factors that cause modern seawater to have a steady-state salinity of 35‰ involve plate tectonics and climate, as well as biogeochemical processes. Because seawater must remain electrically neutral, the anion content of the ocean automatically places limits on the cation content. The amount of major anions present in the ocean at any given time is the result of random tectonic factors. This control is twofold as tectonism determines the rate of release of magmatic gases, as well as the rate of formation and dissolution of the anion reservoirs (i.e., evaporites). Variations in anion supply are also caused by changes in the chemical composition of magma and volcanic gas (due to mantle heterogeneity) and to changes in the rate of hydrothermal activity.

Evaporite deposition and formation, as well as the weathering of other rocks, is highly dependent on climate. Thus variations in climate can alter ion input and removal rates. Perturbations in climate will also affect the hydrological cycle, which controls the amount of water in the ocean and hence its salinity. For example, the past 10^6 y have been characterized by periods of glaciation lasting approximately 100,000 y. These rapid fluctuations have caused short-term variations in ocean salinities.

Some perturbations, such as a change in the rate of terrestrial weathering or hydrothermal activity, could lead to charge imbalances. In such cases, any difference between the total anion and total cation charge concentrations would be balanced by shifts in the relative abundance of H^+ and OH^- ions. For example, an increase in continental erosion that caused a rise in the total cation charge of seawater would require a matching increase in the concentration of negative charge. This could be rapidly achieved through an increase in OH^- concentration. A large enough perturbation would cause the excess cations to form insoluble hydroxides or carbonates. Any further increase in the supply rate of excess cations would cause their precipitation rate to increase. This positive feedback would help the ocean reestablish a steady state.

An increase in pH would also shift the position of the following equilibrium in favor of the products:

$$CO_2(g) + 2OH^-(aq) + Ca^{2+}(aq) \rightarrow CaCO_3(s) + H_2O(l) \qquad (21.10)$$

This would lower the CO_2 content of the ocean and eventually that of the atmosphere. Since rainwater would be less acidic, terrestrial weathering rates would decrease, causing a slowdown in the rate at which cations are supplied to the ocean, thereby counteracting the effects of the original perturbation. At present, the charge concentrations of anions and cations are nearly equal. Thus large perturbations would be required to produce a significant shift in pH.

The cations are present at different concentrations because their removal mechanisms operate with different degrees of effectiveness. For example, the 25-fold increase in the Na^+/K^+ ratio between river water and seawater (Table 21.12) is due to the lower affinity of marine rocks for sodium as compared to potassium. In other words, the sodium sink is not as effective as the one for potassium. Likewise, the Ca^{2+}/Mg^{2+} ratio decreases by a factor of nine from river water to seawater because of the preferential removal of calcium in the form of biogenic calcite.

Changes in the relative removal rates of the cations can lead to variations in their ratios. For example, changes in the geometry of the seafloor caused by plate motions can affect the degree of preservation of biogenic calcite, silica, and POM. Changes in the type of organisms, as a result of evolution and changing habitats, can cause changes in the production rates of biogenic particles. This can also result from variations in population size. For example, the mass extinction that occurred at the end of the Cretaceous undoubtedly caused a dramatic decrease in the production rates of biogenic particles and O_2. Since dissolved O_2 concentrations also affect the rate of particle preservation, atmospheric CO_2 levels would have been altered as well.

Human activities are beginning to significantly perturb the global geochemical cycles of even the major elements. Our greatest impact thus far has been to accelerate greatly the rate at which terrestrial detritus is put into the ocean. This is largely the result of our patterns of land use that tend to destabilize soil. Due to our imperfect understanding of the feedbacks in the crustal-ocean factory, we are presently unable to predict the effect of such activities on the ocean's chemical composition.

SUMMARY

The cations and anions that make seawater salty are ultimately derived from crustal weathering and the degassing of the inner earth. Though ions continue

TABLE 21.12
Comparison of the Major Ion Ratios in River Water and Seawater

Ion Ratio	*River Water*	*Seawater*
Na^+/K^+	2.5	50
Na^+/Mg^{2+}	4	5
Na^+/Ca^{2+}	2	0.2
K^+/Mg^{2+}	2	0.1
K^+/Ca^{2+}	4.5	10
Ca^{2+}/Mg^{2+}	9	1

to be supplied to the ocean, their concentrations remain constant over time because they are removed at rates that match their supply. Most of the removal is accomplished by incorporation into the sediments or the oceanic crust. Sodium and chlorine are removed primarily as halite during the deposition of evaporites. The formation of biogenic calcite is the largest sink of calcium, carbon, and alkalinity. Magnesium is incorporated into basalt during high-temperature reactions at hydrothermal vents. Sulfate is also deposited, but mostly at low temperatures as anhydrite. This mineral is also precipitated during the formation of evaporites. Potassium is also thought to be removed by hydrothermal processes. Some may also be incorporated into an aluminosilicate phase, such as illite or glauconite, by a type of reverse-weathering process that removes alkalinity and dissolved silicon.

Some sediments are recycled through the mantle as a result of subduction at deep-sea trenches. Re-eruption at spreading centers and associated hydrothermal activity is a net source of calcium and perhaps dissolved silicon and potassium to the sea. The rest of the sediments are converted into sedimentary and metamorphic rock. Crustal motions associated with plate tectonics uplift these **secondary** materials onto land where chemical weathering resolubilizes the ions. These dissolved species are then transported back to the ocean by rivers.

The chemical homogeneity of seawater and the relatively long residence times of the major ions suggest that some of the removal mechanisms are controlled by chemical equilibria. Calculations of equilibrium speciation produce results close to the observed ion concentrations, but require the formation of extensive deposits of secondary clay minerals. No evidence for this has been found; instead it appears that hydrothermal activity is responsible for most of the removal formerly attributed to reverse weathering. Nevertheless, clay mineral equilibria could be important in controlling the pH of seawater over time scales in excess of millions of years.

The factors that cause the present oceanic steady state to have a seawater salinity of 35‰ involve plate tectonics and climate, as well as biogeochemical processes. The latter include (1) the relative reactivity of the ions; (2) the feedback loops that result from interactions among the biogeochemical cycles of carbon, oxygen, and the aluminosilicates; and (3) the requirement of electroneutrality.

Perturbations to the steady state could result from changes in the relative removal and supply rates of the ions. Changes in removal rates can result from variations in the degree of particle preservation, which is affected by factors such as the geometry of the sea floor and atmospheric O_2 levels. Removal rates are also dependent on the rate of sediment production, which is influenced by the type and abundance of marine organisms, as well as climate. Supply rates can vary as a result of changes in rock composition, river-water flow, and erosion rates. The latter is influenced by atmospheric CO_2 levels. Fortunately, the crustal-ocean factory is stabilized by feedback loops; and as a result, the chemical composition of seawater appears to have remained constant for at least the past 600 million years.

PROBLEM SET 3

1. A certain sample of seawater has a carbonate alkalinity of 2.4 meq/L and a ΣCO_2 of 2.2 m*M*. Calculate the new carbonate alkalinity, ΣCO_2, $[HCO_3^-]$, and $[CO_3^{2-}]$ if the following occur:
   a. 10 mg of carbonate shells are dissolved in 1 L of the sample.
   b. 355 mg of POM is completely remineralized. (Assume Richards-Redfield Ratio stoichiometry.)
   c. Calculate the combined effects of parts a and b on the carbonate alkalinity, ΣCO_2, $[HCO_3^-]$ and $[CO_3^{2-}]$. Explain why these additions caused the bicarbonate ion concentration to increase but left the carbonate ion concentration unchanged.
2. Using the data provided in Figure 8.2, determine the AOU of deep water nearest your coastline. Calculate how much the ΣCO_2 has increased since this water mass was last at the sea surface assuming that for each mole of particulate organic carbon (POC) respired, 1 mole of shell dissolves.
3. Seawater is able to serve as a repository for the combustion products of fossil fuel because of the following reaction:

$$CO_2 + H_2O + CO_3^{2-} \rightleftharpoons 2HCO_3^-$$

   a. What will be the impacts of this anthropogenic input on the alkalinity of seawater?
   b. What affect will this anthropogenic input have on the position of the calcite compensation depth (CCD)?
4. Draw a box model for magnesium showing its transport pathways and reservoirs in the crustal-ocean factory.
5. Explain why the steady-state concentration of dissolved Al (τ = 620 y) cannot possibly be controlled by hydrothermal processes.

SUGGESTED FURTHER READINGS

Sedimentary Rocks, Weathering, and River Input

HOLLAND, H. D. 1978. *The Chemistry of the Atmosphere and Oceans*. John Wiley & Sons, Inc., New York, 351 pp.

Carbonate Chemistry

BROECKER, W., and T.-H. PENG. 1982. *Tracers in the Sea*. Lamont-Doherty Geological Observatory, Palisades, NY, 690 pp.

Geology and Geochemistry

KENNETT, J. 1982. *Marine Geology*. Prentice-Hall, Inc., Englewood Cliffs, NJ, 813 pp.

Hydrothermal Chemistry

EDMOND, J. M., and K. VON DAMM. 1983. Hot Springs on the Ocean Floor. *Scientific American*, **248**:78–93.

JANNASCH, H. W., and M. J. MOTTL. 1985. Geomicrobiology of Deep-Sea Vents. *Science*, **229**:719–725.

THOMPSON, G. 1983. Hydrothermal Fluxes in the Ocean. In: *Chemical Oceanography*, vol. 8 (ed.: J. P. Riley and R. Chester). Academic Press, Orlando, FL, pp. 272–338.

Kinetic Models of Seawater Composition

BROECKER, W., and T.-H. PENG. 1982. *Tracers in the Sea*. Lamont-Doherty Geological Observatory, Palisades, NY, 690 pp.

MCINTYRE, F. 1970. Why the Sea Is Salt. *Scientific American*, **223**:104–115.

SIEVER, R. 1974. The Steady-State of the Earth's Crust, Atmosphere and Oceans. *Scientific American*, **230**:72–79.

PART FOUR

The microscopic organisms are very inferior in individual energy to lions and elephants, but in their united influences they are far more important than all of these animals.

C. G. Ehrenberg, 1862

ORGANIC BIOGEOCHEMISTRY

CHAPTER 22

MARINE BIOCHEMISTRY: AN OVERVIEW

INTRODUCTION

Thus far, a simple model compound—$C_{106}(H_2O)_{106}(NH_3)_{16}PO_4$—has been used to represent an average molecule of organic matter. In reality, marine organic matter is composed of a great variety of molecules, ranging from low molecular weight hydrocarbons, such as methane, to high molecular weight complexes, such as humic acids. The marine biogeochemistry of specific organic molecules is the subject of Part 4.

Organic compounds are molecules that contain carbon, with the exception of the simple oxides (CO, CO_2, HCO_3^-, H_2CO_3, CO_3^{2-}), elemental forms (diamonds and graphite), and carbonate minerals. They vary greatly in molecular weight and structure. Most contain oxygen and hydrogen. Nitrogen, sulfur, and phosphorus can also be present, but usually in lesser amounts. Most organic compounds are produced during metabolic reactions. As a result, their marine distributions are largely controlled by the activities of marine organisms, such as those illustrated in Figure 22.1. Since equilibrium is rarely achieved in biologically mediated reactions, the distribution of organic compounds is the result of kinetic, rather than thermodynamic, controls.

Organic compounds constitute a relatively small elemental reservoir. Nevertheless, they have great influence on important parts of the crustal-ocean factory. Some of these impacts are discussed in this and the rest of the chapters of Part 4. This chapter also includes a discussion of the different conceptual approaches used in the study of organic biogeochemistry and a description of the molecular structures of the most abundant classes of marine organic compounds.

THE IMPORTANCE OF ORGANIC COMPOUNDS IN THE MARINE ENVIRONMENT

Organic compounds are of great importance in the crustal-ocean factory because of their influence on marine life. For example, they constitute most

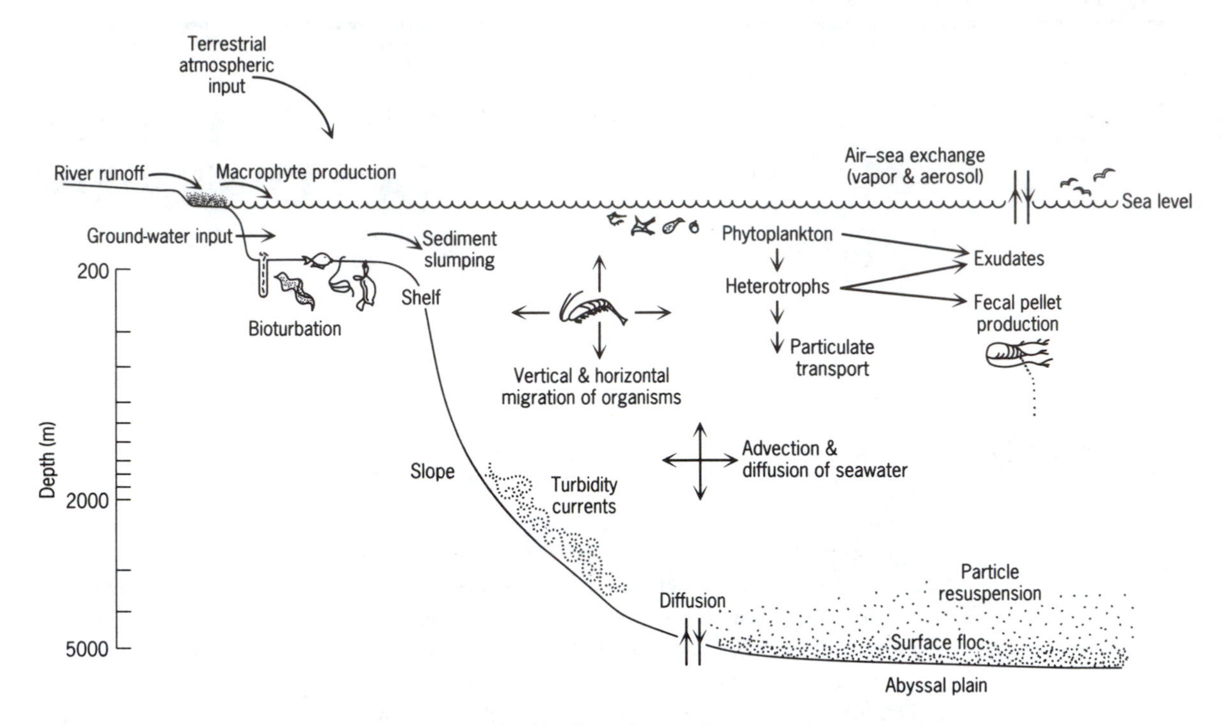

FIGURE 22.1. Factors that influence the distribution of organic matter in the marine environment.

of the tissues, as well as some of the structural parts of all organisms. Organic compounds are also the primary food source for heterotrophs. Besides consuming them, organisms use organic compounds as communicating agents. For example, many organisms excrete organic compounds, called pheromones, that attract mates. Toxic molecules are used for defense against competitors and predators. Still others are used to deactivate the toxic properties of some compounds. This is usually achieved by the formation of relatively inert complexes or by enhancing degradation rates. Due to their potent physiological activity, many of these compounds have proven useful to humans, primarily as drugs, lubricants, cosmetics, and food additives.

Organic matter also affects the mechanical properties of sediments. The largest such impact is an increase in cohesion between particles that limits resuspension. Sedimentary organic matter is the sole food source for the heterotrophic members of the benthos. These organisms also alter the physical, chemical, and geological characteristics of the sediment. For example, deposit feeders change the size distribution of particles by ''gluing'' detritus into fecal pellets. This type of bioturbation homogenizes chemical gradients and increases the supply of O_2 to the sediments. Other organisms construct burrows and feeding tubes which cause chemical and geological heterogeneity in the sediments as illustrated in Figure 12.1.

Upon introduction to seawater, all particles are rapidly covered with a coating of organic matter. Due to its net negative charge, this coating adsorbs cations. Since the particles eventually sink to the seafloor, they remove a significant amount of cations, primarily trace metals, from seawater. These coatings also represent an energy-rich food source to microorganisms. As a result, bacteria, protozoa, and fungi, are commonly found growing on particles. If the particle has a large enough surface area, it will be successively colonized by higher trophic levels, such as tunicates and sponges. This process is termed **biofouling**. To prevent this overgrowth, ship hulls are painted with toxic compounds, such as tributyltin. Unfortunately, some of these chemicals dissolve in seawater and have lethal effects on nontarget organisms.

Sedimentary organic matter is of particular interest to humans for two reasons. First, most petroleum is marine in origin, having been produced by the diagenesis and catagenesis of detrital phytoplankton tissues. Secondly, the steady-state balance of nutrients, O_2 and CO_2 in the crustal-ocean factory appears to be stabilized by feedbacks that involve the burial of organic matter in the sediments. Thus down-core variations in the organic matter content of marine sediments provide a paleoceanographic record of changes in such phenomena as primary production, water circulation, and atmospheric chemistry.

CONCEPTUAL AND ANALYTICAL APPROACHES USED TO STUDY THE MARINE BIOGEOCHEMISTRY OF ORGANIC COMPOUNDS

Due to recent improvements in analytical technology, marine chemists are now able to detect and identify a bewildering number of organic compounds. Most are present at very low concentrations. Nevertheless, many substances remain uncharacterized. Several conceptual strategies have been developed to cope with these problems and are discussed below.

Analyses of Operationally Defined Fractions

Due to analytical limitations, the concentrations of organic compounds are often determined as operationally defined fractions. For example, POM (particular organic matter) is defined as that fraction of organic matter that does not pass through a glass-fiber filter. The nominal pore size of this filter is 0.4 μm. Since the effective pore size decreases during the filtration, some of the organic matter retained on the filter is somewhat smaller than 0.4 μm. In addition, dissolved organic matter adsorbs to particles and hence constitutes part of the operationally defined POM fraction.

Despite these drawbacks, the chemical characterization of operationally defined fractions has provided insight into the biogeochemistry of organic matter. This has typically involved the measurement of such bulk properties as the %C, %N, and the C/N of POM, as well as the concentrations of particulate organic carbon (POC) and nitrogen (PON) in seawater. Similar analyses have been done on other operationally defined fractions whose acronyms are given in Table 22.1.

Spectral properties, such as the absorption of infrared light, are also used to categorize broad classes of organic matter. For example, the high molecular weight fraction of DOM (called humic substances) absorbs infrared light at many frequencies and with varying intensity, as shown in Figure 22.2. The shape of the absorbance spectrum reflects the type of bonds present in the organic compounds. Though specific structures cannot be identified, differences in peak shape and position can be used to distinguish terrestrial from marine humic substances, as shown in Figure 22.2.

Other information on complex mixtures of organic compounds can be obtained without explicitly identifying the structure of each component. For example, the source of an oil spill can be identified by using a **gas chromatograph** to "fingerprint" the petroleum. In this instrument, the complex mixture is first heated, causing the compounds to volatilize. A stream of inert gas pushes the compounds onto a long, narrow column that is packed or coated with special materials. In the former case, the column (2 to 3 m long) is packed with diatomaceous earth that has been impregnated with a high-boiling-point liquid. Alternatively, the liquid is applied as a thin film onto the inside wall of a very narrow (0.2–0.5 mm), very long (25–50 m) column. The liquid is termed the stationary phase.

The compounds dissolve in or are adsorbed by this stationary phase to a degree that is determined by their boiling points. Those compounds that are not strongly retained by the column are rapidly pushed to the other end by the stream of inert gas. The compounds that are strongly retained by the column migrate at slower rates. Since the boiling point is inversely related to molecular weight, the smallest compounds tend to elute from the column first.

The time required for elution is called the retention time. As a substance elutes from the column, its presence is signaled by a detector that transforms

TABLE 22.1
Operationally Defined Fractions of Organic Matter[a]

Operational Fraction	*Abbreviation*
Particulate organic matter	POM
Dissolved organic matter	DOM
Particulate organic carbon	POC
Dissolved organic carbon	DOC
Dissolved inorganic carbon	DIC
Particulate organic nitrogen	PON
Dissolved organic nitrogen	DON
Dissolved inorganic nitrogen	DIN

[a]Also included are some inorganic fractions.

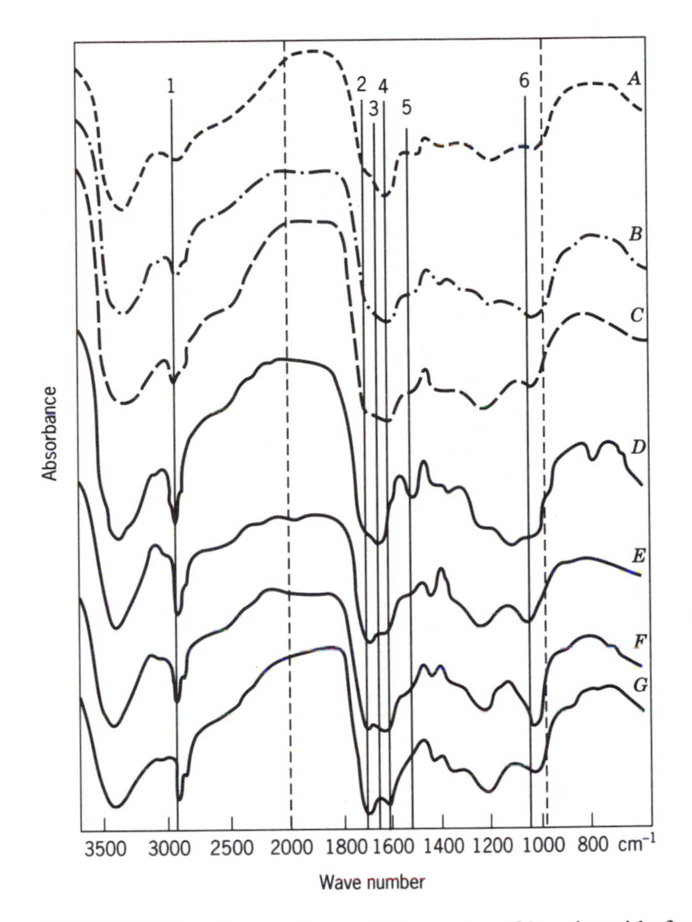

FIGURE 22.2. Comparison of IR spectra of humic acids from terrestrial soils (A–C) and marine sediments (D–G). Identification of IR bands: (1) aliphatic C—H; (2) C═O; (3 and 5) amides; (4) aromatic C═C; and (6) C—O in polysaccharides. *Source*: From A. Y. Huc, B. Durand, and F. Jacquin, reprinted with permission from the *Bulletin de L'ENSAIA De Nancy*, vol. 16, p. 71, copyright © 1974 by L'Institut National Polytechnique de Lorraine (ENSAIA), Nancy, France.

some physical or chemical property of the substance into a measurable electrical signal. This detector records the information as a peak whose area is directly proportional to the compound's abundance.

Gas chromatographs are able to completely separate the components of simple mixtures, producing chromatograms with sharp, symmetric peaks, such as those shown in Figure 22.3*a*. The components of complex mixtures are not as well separated due to similarities in molecular structures that cause their solubilities and boiling points to be nearly identical. As a result, some of these compounds co-elute and appear as large humps on the gas chromatograms, as illustrated in Figure 22.3*b*. Though the specific compounds that contribute to this hump are not identifiable, its shape is unique to the mixture and can be

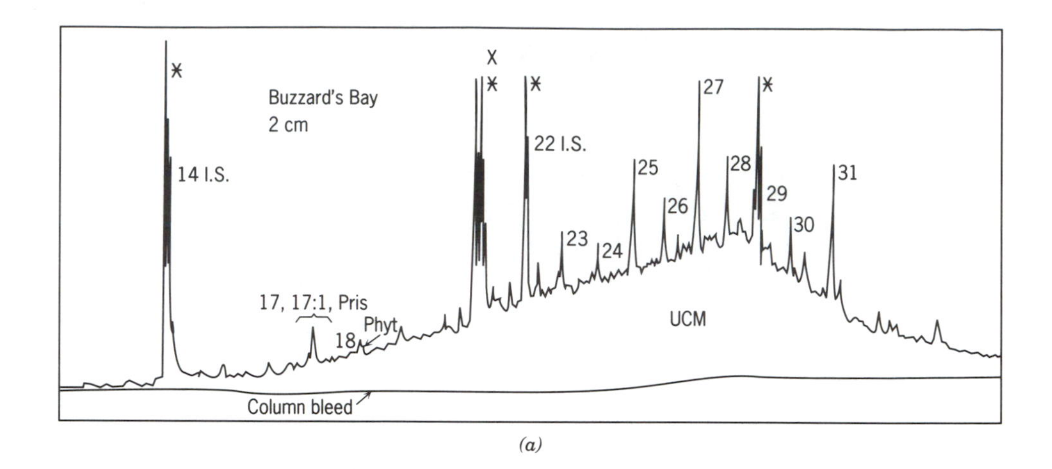

(a)

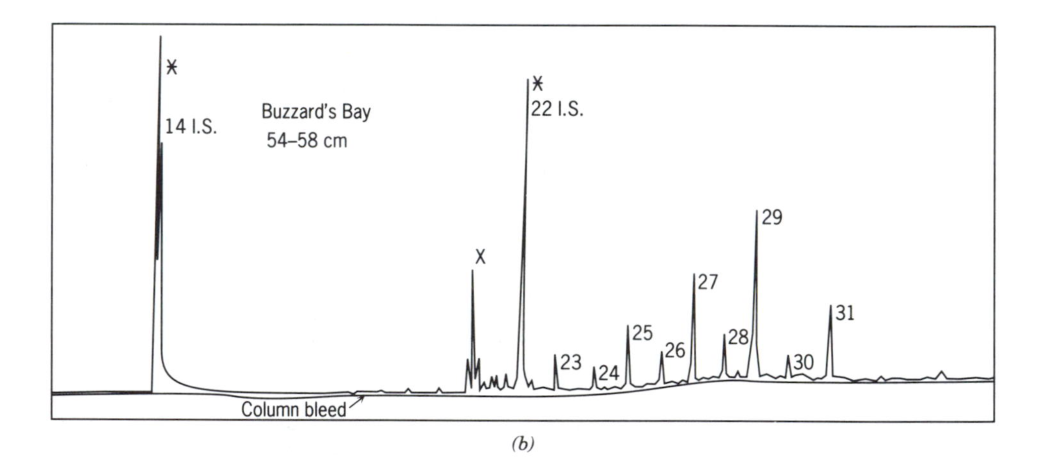

(b)

FIGURE 22.3. Gas chromatograms of anthropogenic petroleum hydrocarbons from two depths in a coastal marine sediment: (*a*) 54 to 58 cm below the sediment surface and (*b*) surface sediment. The degradation and migration of petroleum compounds is responsible for the decrease in mixture complexity with increasing depth. Some of the petroleum compounds in the surface sediments contribute to the hump which is identified only as an unresolved complex mixture (UCM). Peaks identified by an asterisk were brought on scale by use of an auto attenuator. Peaks identified as I.S. are internal standards. X is used to designate groups of unsaturated compounds of unknown structure. *Source*: From J. W. Farrington, N. M. Frew, P. M. Gschwend and B. W. Tripp, reprinted with permission from *Estuarine and Coastal Marine Science*, vol. 5, p. 797, copyright © 1977 by Kluwer Academic Publishers, Dordrecht, The Netherlands.

used as a fingerprint for matching purposes. This has been used to identify the source of petroleum present in a spill, as oil is naturally variable in chemical composition. Thus the identity of the polluter can be determined by comparing gas chromatograms of oil obtained from all likely sources, such as ships' fuel tanks. The chemical composition of petroleum is discussed further in Chapter 26 and petroleum pollution in Chapter 30.

The molecular structures of each peak in a well-resolved gas chromatogram can be identified by **mass spectrometry**. This combination of technologies is called GC-MS and is illustrated schematically in Figure 22.4. As a compound elutes off the chromatographic column it is passed into the mass spectrometer. Bombardment with a high-energy beam of electrons creates molecular ions that can spontaneously break up into fragments. The masses and relative abundances of the fragments are detected electronically and reported as a mass spectrum. An onboard computer compares this spectrum to those of known molecules to establish a best match and hence the likely identity of the compound. Once the compound has been identified, its presence in other gas chromatograms, obtained under similar operating conditions, can be determined from its retention time.

Biomarkers and Source Tracers

Certain organic molecules are synthesized by only one type of marine organism. For example, the sterol dinosterol is synthesized only by dinoflagellates. These compounds are termed **biomarkers** because their presence provides unequivocal evidence for their source (i.e., biological production by that specific organism).

In most cases, a particular compound or class of compounds is synthesized by several kinds of organisms. For example, lignins are structural polymers synthesized by all woody terrestrial plants. Since lignin is relatively inert, some is transported by rivers into the ocean and buried in nearshore sediments. Thus its relative abundance in marine sediments can be used to determine the contribution of terrestrial organic matter to the deposit.

Most organic compounds synthesized in the marine environment are very labile. They are released into seawater by either (1) exudation, (2) excretion, or (3) cell lysis following the death of the organism. Once in seawater, most are rapidly degraded by microbes. In some cases the degradation products can be used to identify the source compound. For example, phaeophytin and phytol are unique products of the decomposition of chlorophyll.

The distribution of specific compounds in the water column can also provide information on biogeochemical processes. As shown in Figure 22.5, adenosine triphosphate (ATP) concentrations are highest at the sea surface. ATP is introduced into seawater as a result of cell lysis that occurs shortly after death. Once in seawater, ATP is rapidly degraded. Thus high concentrations are the result of a rapid supply that must be supported by high rates of biological production. Since biological productivity is concentrated in the surface waters, this is the site of highest ATP concentrations.

The smaller concentration maximum located at 400 m suggests the occurrence of relatively high rates of microbially mediated decomposition of detrital POM. This maximum is probably supported by the in situ degradation of detrital POM derived from biological activity in the surface waters. This conclusion is supported by the wintertime increase in ATP concentrations at

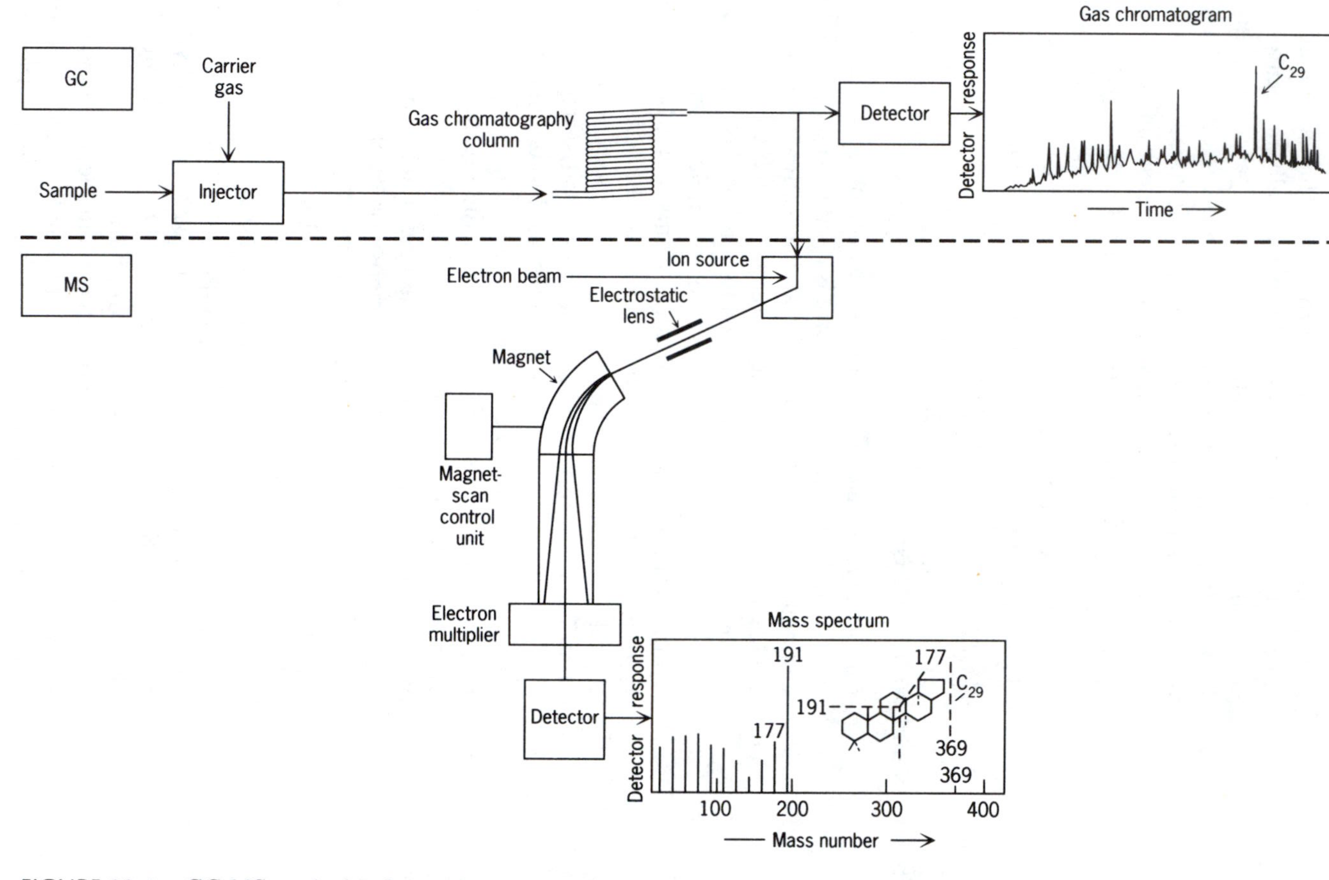

FIGURE 22.4. **GC-MS analysis of organic compounds.**

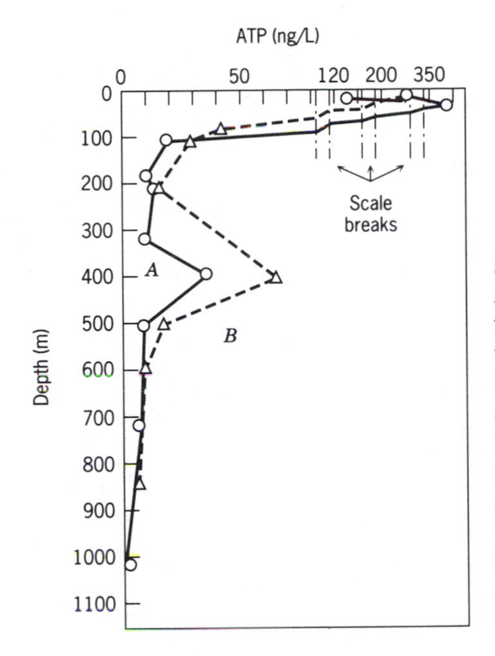

FIGURE 22.5. Depth profile of ATP concentrations at a site off the coast of California. Samples were collected in (A) May 1965 at 33°N 119°W and (B) October 1965 at 33°N 117°W. *Source*: From O. Holm-Hansen and C. H. Booth, reprinted with permission from *Limnology and Oceanography*, vol. 11, p. 515, copyright © 1966 by the American Society of Limnology and Oceanography, Seattle, WA.

mid depth. Since vertical segregation is greatest during the summer, the largest amount of POM will have sunk out of the surface waters by the beginning of winter. Apparently most of these particles are neutrally buoyant at approximately 400 m and so support increased rates of microbial activity, especially during early winter.

Isotopes as Tracers of the Source and Fate of Marine Organic Compounds

Isotopes of a particular element are atoms that have the same number of protons, but different numbers of neutrons. The naturally occurring stable isotopes of carbon, nitrogen, oxygen, and hydrogen have been used as source tracers. For example, the ratio of ^{13}C to ^{12}C is lower in terrestrial POM than in marine POM. Because these ratios are relatively constant, the contribution of terrestrial organic matter to coastal marine sediments can be inferred from its relative carbon isotope composition. These measurements are made by mass spectrometry and are discussed in Chapter 29.

Some radioactive isotopes have also been used to determine the mechanisms and rates at which some organic compounds decompose. To do this, a small amount of radioactively labelled compound is introduced into a sample of seawater or sediment. The fate of this compound is traced by periodically analyzing various fractions of the sample. The results of these studies are somewhat questionable as the seawater and sediment are not in their in situ states during the experiment. It is also assumed that the radioactively labeled compound is an appropriate analog for some fraction of ''real'' marine organic matter.

Artificial Ecosystems

As noted on page 375, laboratory experiments on seawater and sediments are not likely to produce chemical changes similar to those that occur in situ. One approach used to better mimic real conditions has been to create very large artificial ecosystems. Though designed to function as much like the natural ocean as possible, they can also be controlled and sampled efficiently. Some of the largest artificial ecosystems are the MERL tanks (Marine Ecosystems Research Laboratory) operated by the University of Rhode Island on Narragansett Bay.

As illustrated in Figure 22.6, these tanks are approximately 16 ft high and 5 ft in diameter. They contain about 1 ft of sediment and 13,000 liters of water, which is slowly trickled in from the bay. The outflow rate is set so the water volume does not change over time. With a stirring device and the top of the

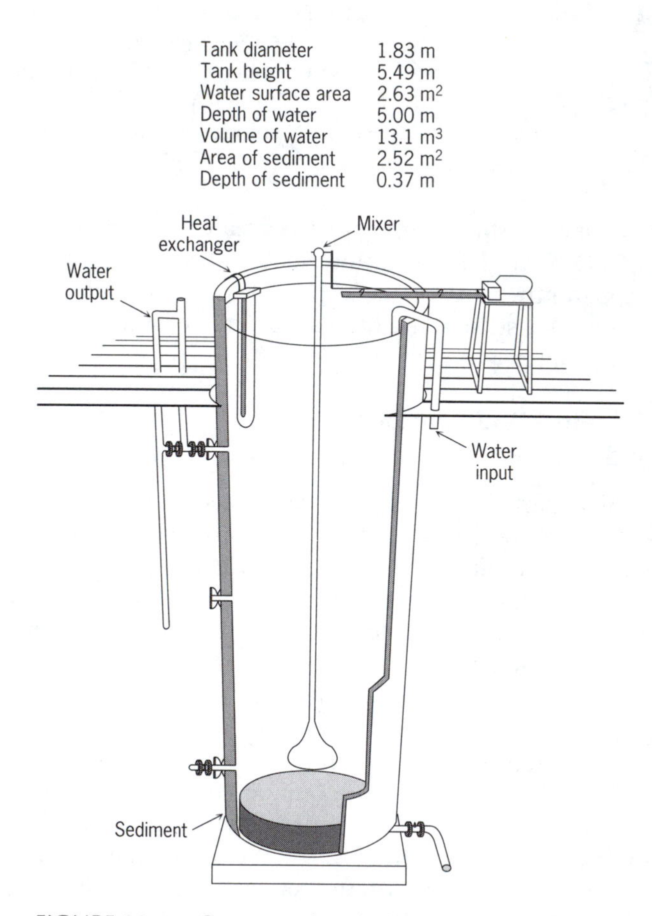

FIGURE 22.6. Cross-sectional view and dimensions of a MERL tank. *Source*: From J. B. Frithsen, A. A. Keller, and M. E. Q. Pilson, reprinted with permission from the *MERL Series, Report No. 3*, copyright © 1985 by the Marine Ecosystems Research Laboratory, University of Rhode Island, Narragansett, RI.

tank open, the tanks function much like the neighboring bay. Typical experiments involve monitoring the effects of some perturbation, like the introduction of nutrients or pesticides, on the water, sediment, and biota in the tank.

GENERAL CLASSES OF ORGANIC COMPOUNDS

Organisms obtain chemical energy by degrading organic compounds. This general process is termed **catabolism**. Some of this energy is used to build large biomolecules that provide structural support and energy storage. The general process of organic matter synthesis is termed **anabolism** and is illustrated in Figure 22.7. The reaction pathways in both anabolism and catabolism tend to be species specific. Many produce unique compounds that thus can be used as biomarkers.

A great number of reactions occur during anabolism and catabolism. As a result, organisms produce a myriad of organic molecules. These compounds are classified into groups based on similarities in structure and the presence of certain functional groups. The most common of these functional groups are listed in Table 22.2. Most organic compounds that are naturally found in the marine environment can be classified as either hydrocarbons, carbohydrates, lipids, fatty acids, amino acids, or nucleic acids. Examples of each are given below, along with some of their polymers.

Hydrocarbons

Hydrocarbons are organic compounds composed solely of carbon and hydrogen atoms. Marine organisms synthesize a great variety of these compounds. These chains of carbon atoms vary greatly in number and some are branched, as

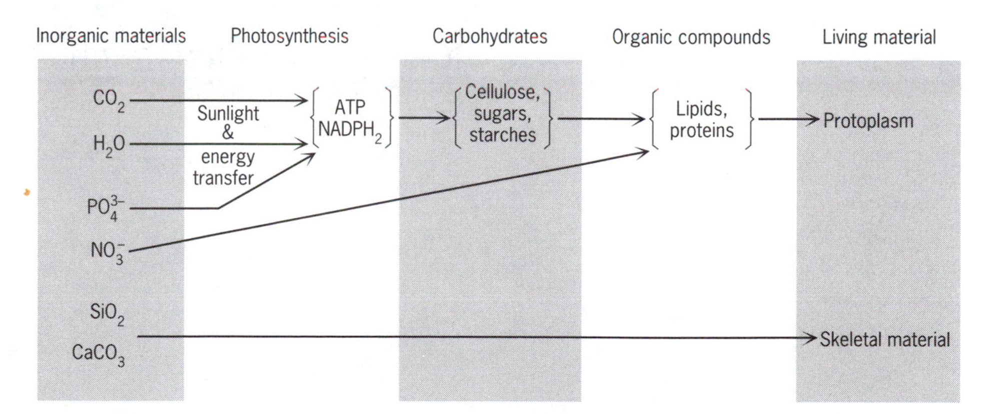

FIGURE 22.7. Various steps in the production of living material. *Source*: From *The World Ocean: An Introduction to Oceanography*, W. A. Anikouchine and R. W. Sternberg, copyright © 1981 by Prentice Hall, Inc., Englewood Cliffs, NJ, p. 379. Reprinted by permission.

TABLE 22.2
The Most Common Functional Groups Found in Naturally Occurring Organic Compounds (R = any other organic structure)

Common Name	*Molecular Formula or Structure*
Carboxylic acid group	R—COOH
Alcoholic group	R—OH
Aromatic group	R—⌬
Phenolic group	R—⌬—OH
Amine group	$R—NH_2$
	$R_2—NH$
	R_3N

illustrated in Figure 22.8. Some have double bonds. To indicate their relative hydrogen deficiency, these compounds are said to be "unsaturated."

In general, hydrocarbon production does not appear to be species specific and hence these compounds are not useful as biomarkers. Hydrocarbons are also produced by the fragmentation of large biomolecules during catagenesis and can form large deposits called petroleum.

Carbohydrates

Carbohydrates and lipids are the most important cellular reservoirs of chemical energy. The simple carbohydrates are composed of single sugar molecules, of

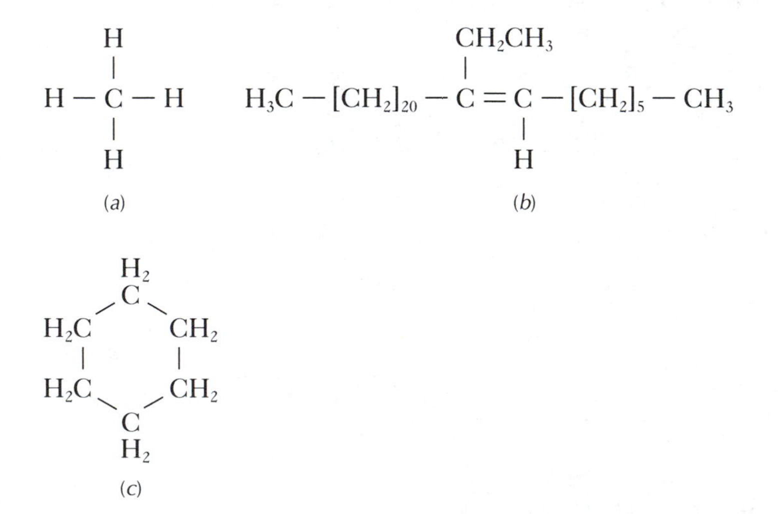

FIGURE 22.8. Structures of some hydrocarbons: (*a*) methane, (*b*) branched and unsaturated C_{31} hydrocarbon, and (*c*) cyclohexane.

which glucose is an example. Its structure, along with those of the other common simple sugars, is given in Figure 22.9.

Complex carbohydrates are formed by the linking of many simple sugars. The resulting polymers are also called polysaccharides. The most common are illustrated in Figure 22.10. Cellulose is a structural and rather chemically inert component of terrestrial plants. Alginic acid is synthesized by marine algae and has medicinal properties. The exoskeletons of crustacea are composed of chitin.

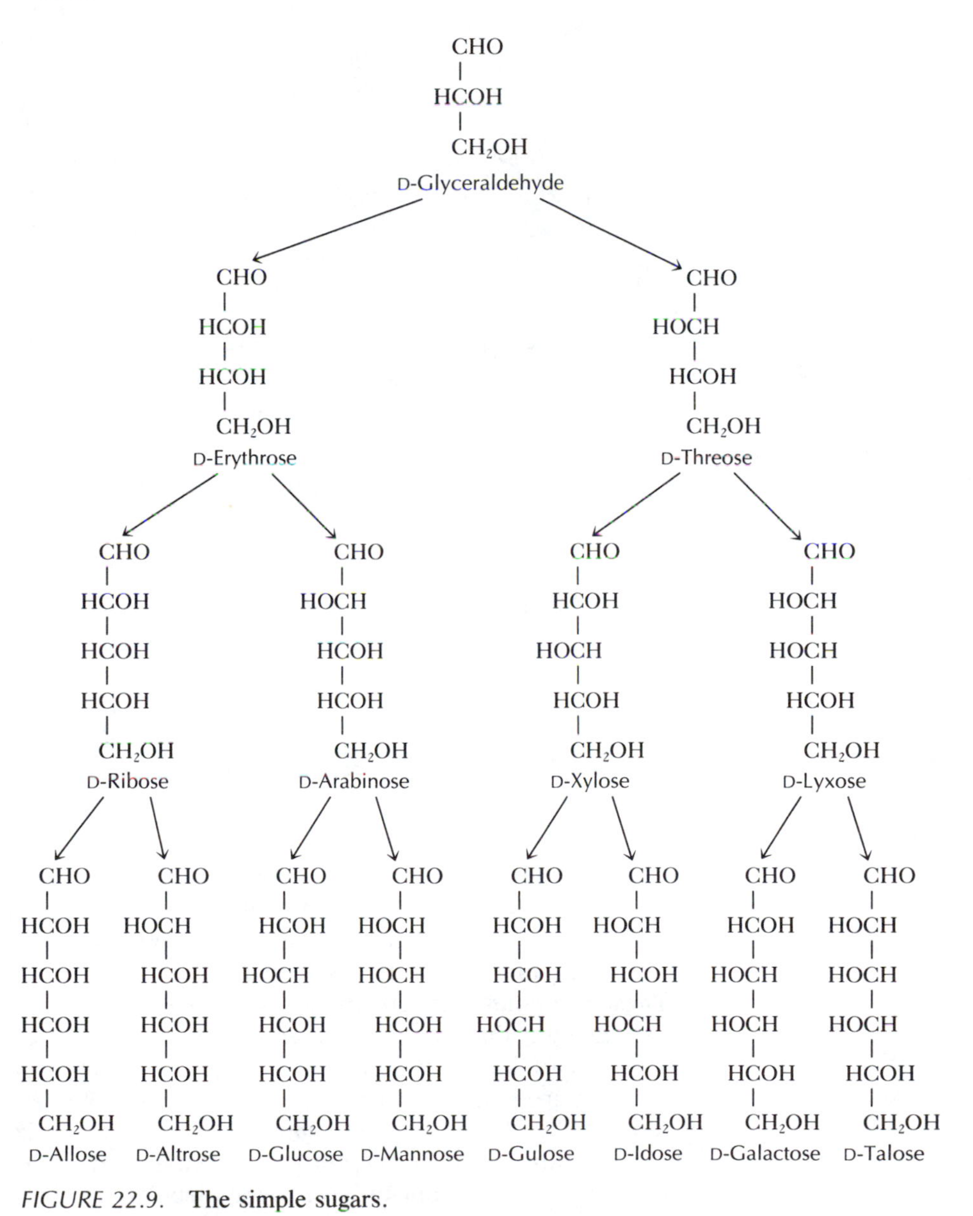

FIGURE 22.9. The simple sugars.

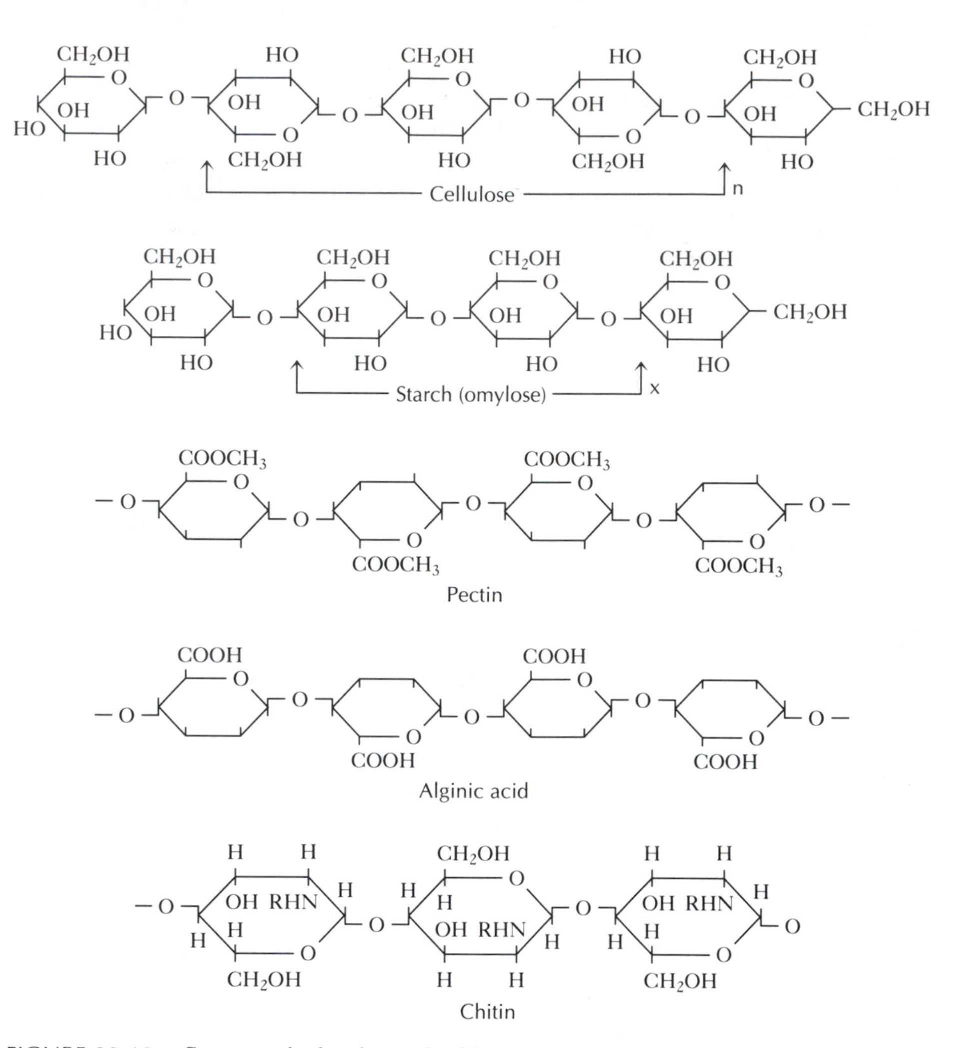

FIGURE 22.10. Some typical polysaccharides.

Lipids

Lipids are a diverse group of compounds that are soluble in nonpolar solvents and insoluble in water. As shown in Table 22.3, lipids are classified as complex or simple. The complex lipids contain various **fatty acids** that are attached to a "backbone" molecule, such as shown in Figure 22.11.

These compounds are commonly used in cells as a means of energy storage. A complex lipid that is a solid at room temperature is called a fat or wax. Those which are liquids are termed oils. The fats tend to have higher molecular weights and are synthesized primarily by animals. Waxes and oils are more common to plants. The type of complex lipids biosynthesized by an organism is an adaptation to the temperature of its environment. For example, deep-dwelling fish are more apt to have low-molecular-weight lipids than surface-

TABLE 22.3
Classification of Lipids

Lipid Type	*Backbone*
Complex	
Acylglycerols	Glycerol
Phosphoglycerides	Glycerol 3-phosphate
Sphingolipids	Sphingosine
Waxes	Nonpolar alcohols of high molecular weight
Simple	
Terpenes	
Steroids	
Prostaglandins	

dwelling fish. The relatively low freezing points of the smaller lipids keeps them from solidifying at the low temperatures encountered in the deep sea.

Fatty acids are hydrocarbons that have a terminal carboxylic acid group (R—COOH). As shown in Table 22.4, members of this group vary in chain length, degree of unsaturation, and branching. Some of these features are unique to certain organisms. For example, branched chains are produced during the microbial degradation of organic detritus.

Some phospholipids are a major component of cell walls. In these compounds, one of the alcohol groups in the backbone molecule (glycerol) is attached to an organophosphate. This group is relatively polar and hence water soluble. In comparison, the fatty acid groups are relatively insoluble. Thus the lipid molecules tend to form bilayers, as illustrated in Figure 22.12.

Simple lipids do not contain fatty acids. Pigments are examples of simple lipids. These colored organic compounds are synthesized in small amounts by most marine organisms. The most abundant are the chlorophylls, which are used by marine plants and some bacteria to convert solar energy into chemical energy. The structures of chlorophyll *a* and some other common photosynthetic pigments are illustrated in Figure 22.13. Carotene is orange and fucoxanthin is red. These compounds are characterized by a high degree of unsaturation. Note the central magnesium atom and numerous nitrogen atoms in the chlorophyll *a* molecule.

$$\begin{array}{ccccccc} H_2COH & & HOOCR & & H_2COOCR & & H_2O \\ | & & & & | & & \\ HCOH & + & HOOCR' & \longrightarrow & HCOOCR' & + & H_2O \\ | & & & & | & & \\ H_2COH & & HOOCR'' & & H_2COOCR'' & & H_2O \\ \text{Glycerol} & & \text{Three fatty acids} & & \text{Triglyceride} & & \end{array}$$

FIGURE 22.11. An example of a complex lipid.

TABLE 22.4
Some Examples of Fatty Acids

Structure	*Systematic Name*	*Common Name*
Saturated fatty acids		
$CH_3(CH_2)_{10}COOH$	*n*-Dodecanoic	Lauric
$CH_3(CH_2)_{12}COOH$	*n*-Tetradecanoic	Myristic
$CH_3(CH_2)_{14}COOH$	*n*-Hexadecanoic	Palmitic
$CH_3(CH_2)_{16}COOH$	*n*-Octadecanoic	Stearic
$CH_3(CH_2)_{18}COOH$	*n*-Eicosanoic	Arachidic
$CH_3(CH_2)_{22}COOH$	*n*-Tetracosanoic	Lignoceric
Unsaturated fatty acids		
$CH_3(CH_2)_5CH{=}CH(CH_2)_7COOH$		Palmitoleic
$CH_3(CH_2)_7CH{=}CH(CH_2)_7COOH$		Oleic
$CH_3(CH_2)_4CH{=}CHCH_2CH{=}CH(CH_2)_7COOH$		Linoleic
$CH_3CH_2CH{=}CHCH_2CH{=}CHCH_2CH{=}CH(CH_2)_7COOH$		Linolenic
$CH_3(CH_2)_4(CH{=}CHCH_2)_3CH{=}CH(CH_2)_3COOH$		Arachidonic

Carotene, fucoxanthin, and xanthophyll are examples of terpenoids. This class of compounds is constructed of multiples of the five-carbon hydrocarbon isoprene. Terpenes (Figure 22.14) may be either linear or cyclic. Some molecules contain both types of structures as exemplified by vitamin A.

Sterols are simple lipids that have a base unit composed of three six-carbon and one five-carbon ring, as shown in Figure 22.15. They are ubiquitous in marine organisms, both as components of cell membranes and as metabolic regulators. As mentioned above, some sterols reflect specific sources. Cholesterol is the dominant sterol in zooplankton, while diatoms primarily synthesize the 24-methylsterols. ß-sitosterol is synthesized primarily by higher-order land plants.

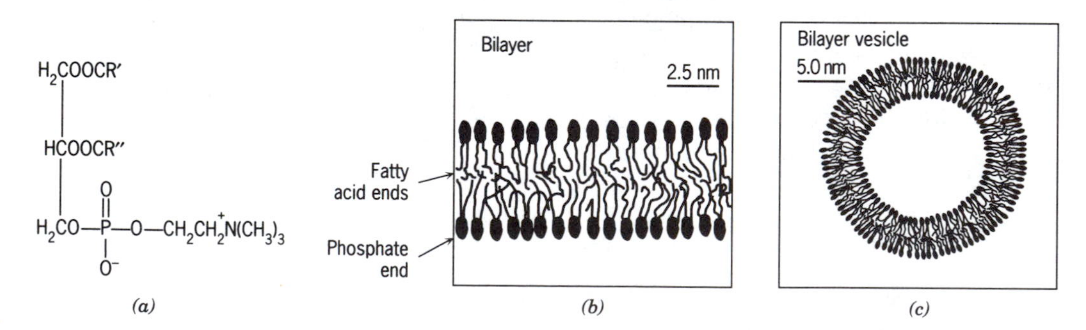

FIGURE 22.12. (*a*) A phosphoglyceride structure (e.g., phosphatidyl choline). (*b*) The orientation of these molecules in a model bilayer membrane. (*c*) The formation of a "cell" from a bilayer membrane. *Source*: From J. N. Israelachvili and N. W. Ninham, reprinted with permission from the *Journal of Colloid Interface Science*, vol. 58, p. 18, copyright © 1977 by Academic Press, Orlando, FL.

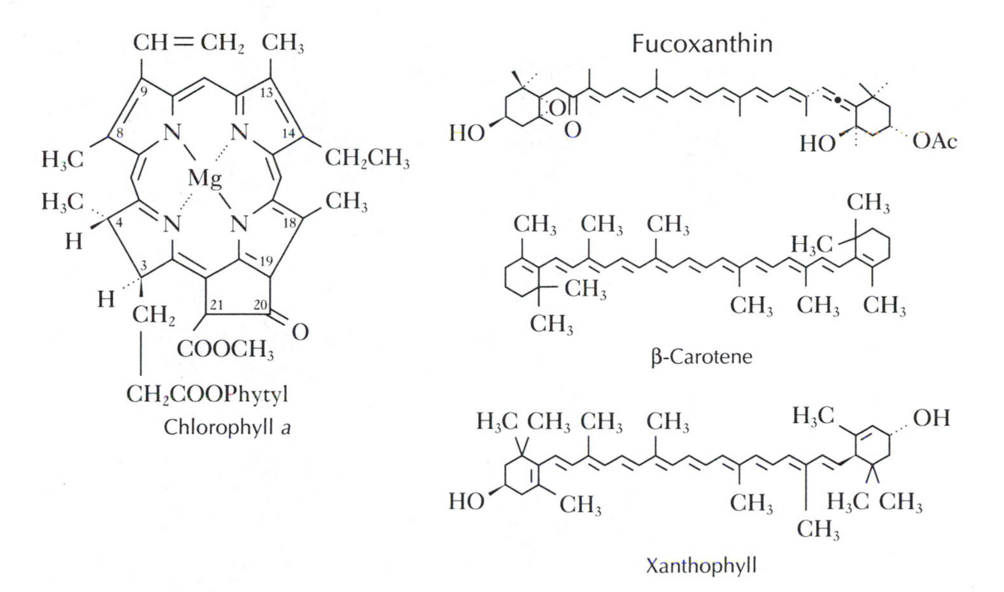

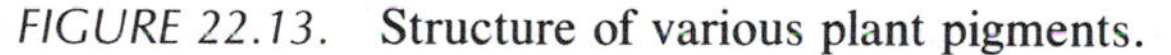

FIGURE 22.13. Structure of various plant pigments.

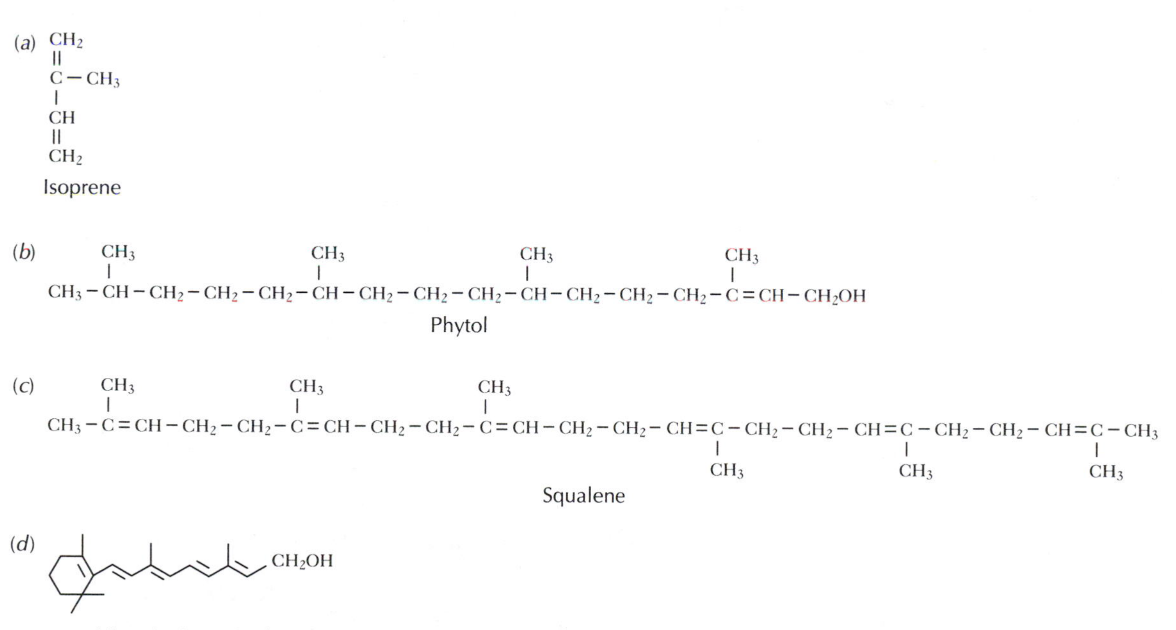

FIGURE 22.14. Examples of terpenes: (*a*) isoprene; (*b*) phytol, which is a side chain of chlorophyll *a*; (*c*) squalene; and (*d*) vitamin A.

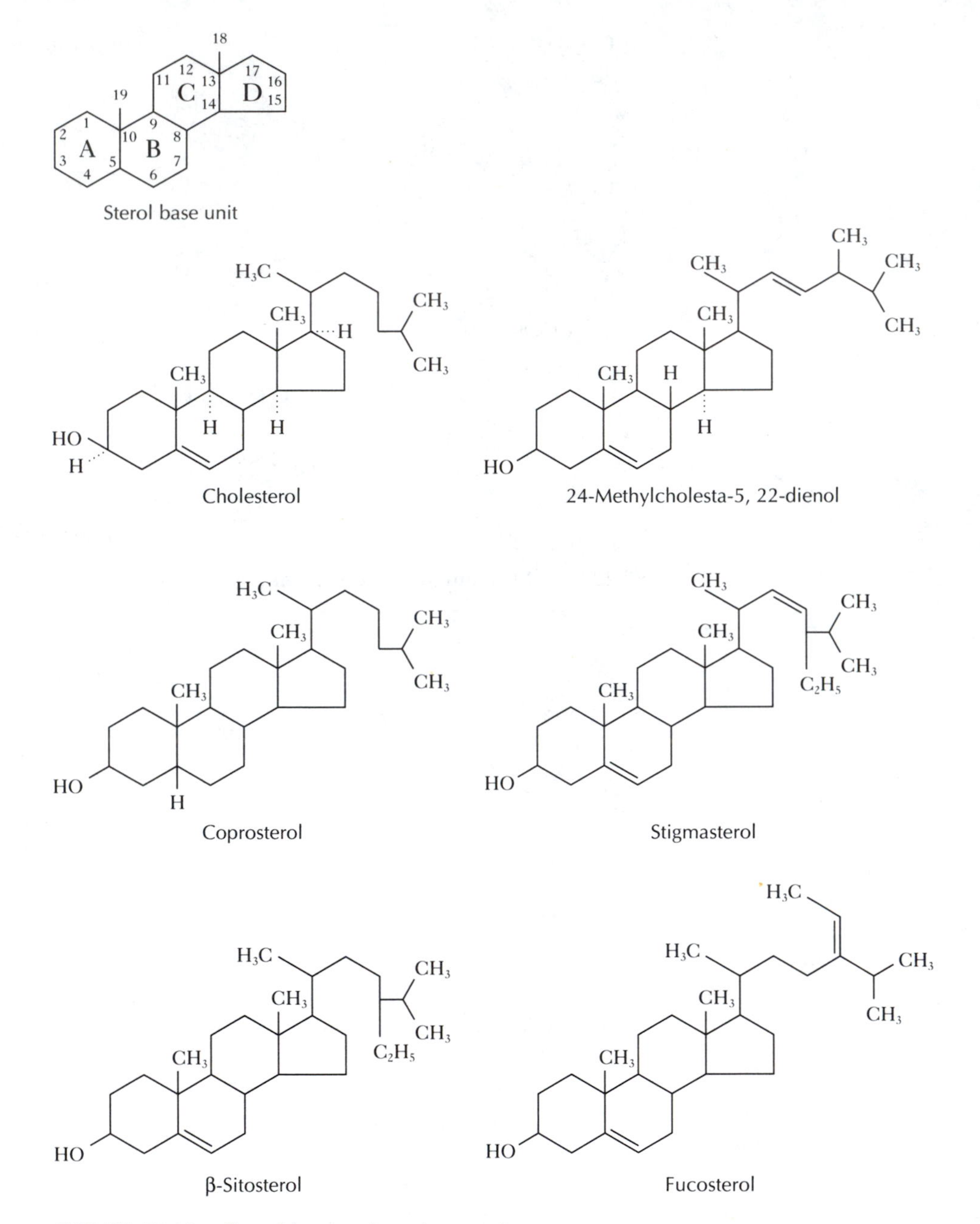

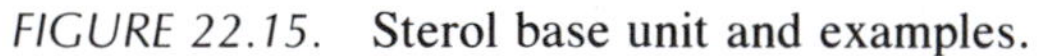

FIGURE 22.15. **Sterol base unit and examples.**

Nucleic Acids

Nucleic acids are composed of an aromatic base, a five-carbon sugar, and a phosphate group. The five aromatic bases that can be present are illustrated in Figure 22.16 and cause the nucleic acids to be relatively nitrogen rich.

An example of a nucleic acid, or mononucleotide, is given in Figure 22.17*a*. When linked together, they form polymers called polynucleotides. RNA and DNA, which store and transmit genetic information within cells, are examples of polynucleotides. Nucleic acids are also part of energy-carrying molecules such as ATP and NADP (nicotinamide adenine dinucleotide phosphate).

Amino Acids

Amino acids are the building blocks of proteins. These compounds all bear at least one amine group (R—NH_2) and a carboxylic acid group (R—COOH). The naturally occurring amino acids are listed in Table 22.5.

Amino acids become linked by reactions that occur between the amine group of one amino acid and the carboxylic acid of another. As shown in Figure 22.18, this polymerization also produces a molecule of water and hence is termed a condensation. Naturally occurring polypeptides with molecular weights in excess of 10,000 daltons are termed proteins. These biomolecules are ubiquitous in marine organisms and are not specific to particular species.

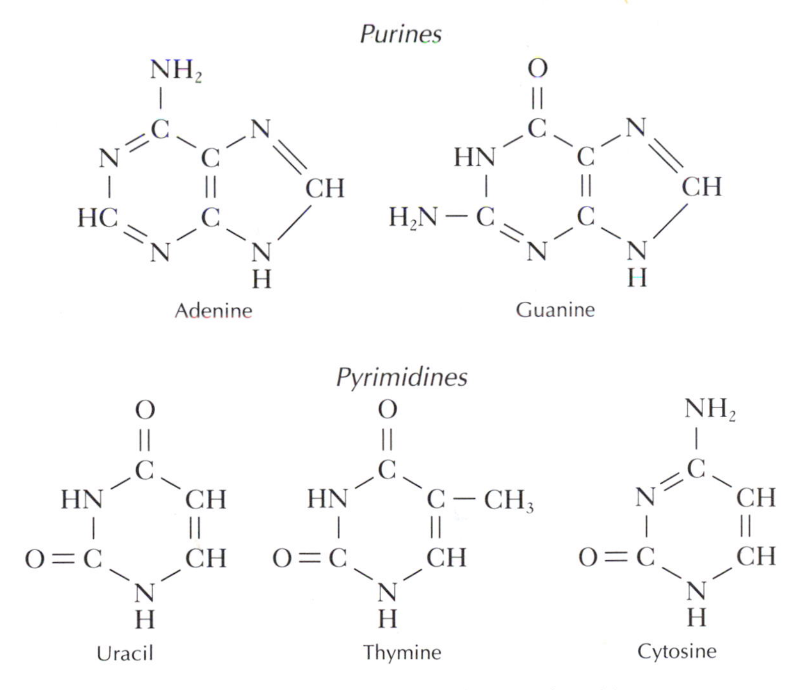

FIGURE 22.16. Aromatic bases present in nucleic acids.

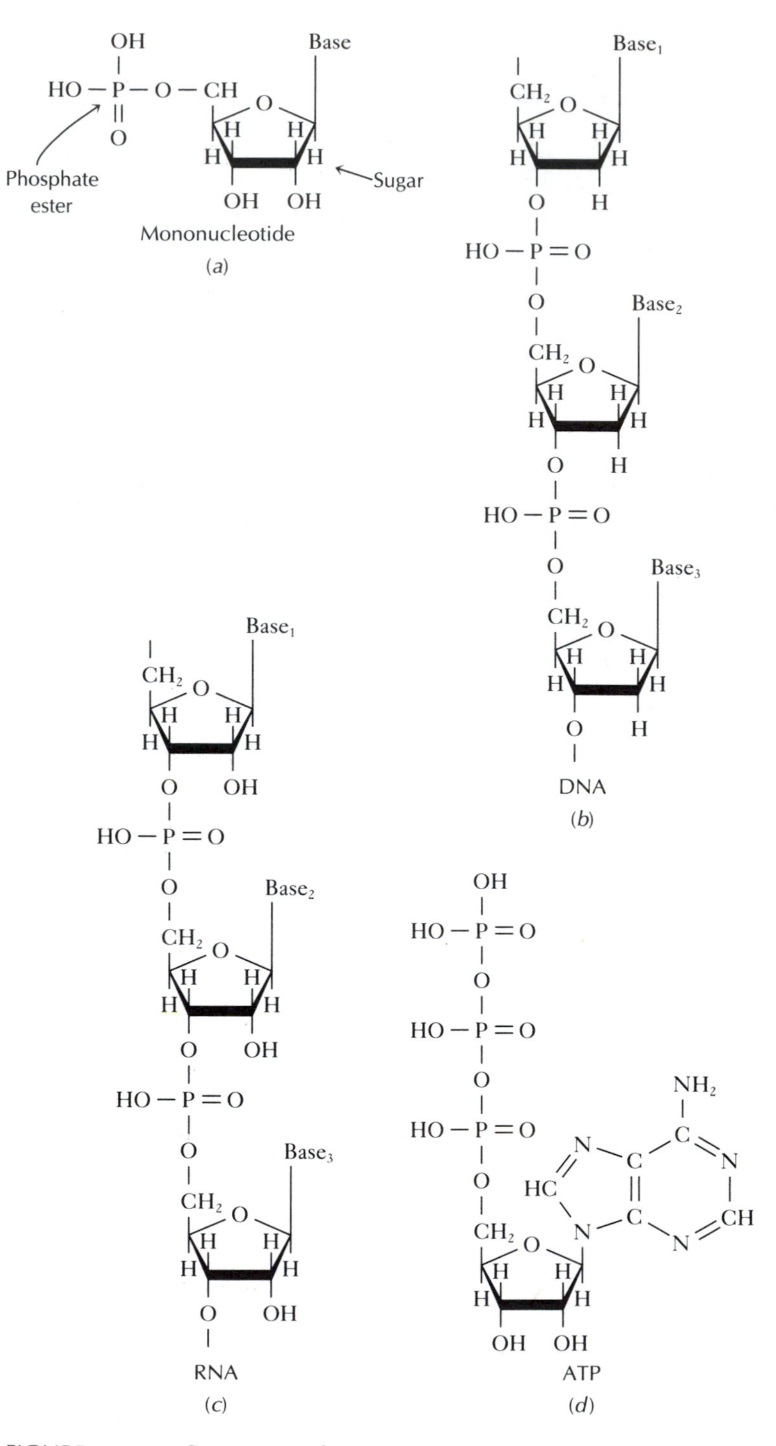

FIGURE 22.17. Structures of (*a*) a mononucleotide, (*b*) DNA, (*c*) RNA, and (*d*) ATP.

TABLE 22.5
Principal Amino Acids Derived from Proteins

I. Aliphatic amino acids

A. Monoaminiomonocarboxylic acids

Glycine

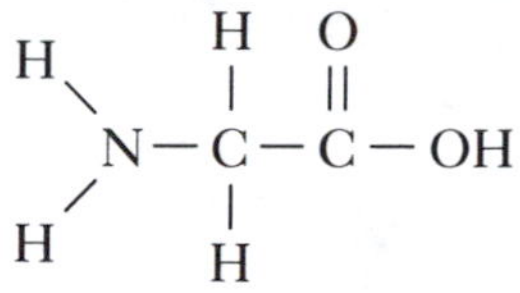

```
H      H  O
 \     |  ||
  N — C — C — OH
 /     |
H      H
```

Alanine

```
    H   H   O
    |   |   ||
H — C — C — C — OH
    |   |
    H   N
       / \
      H   H
```

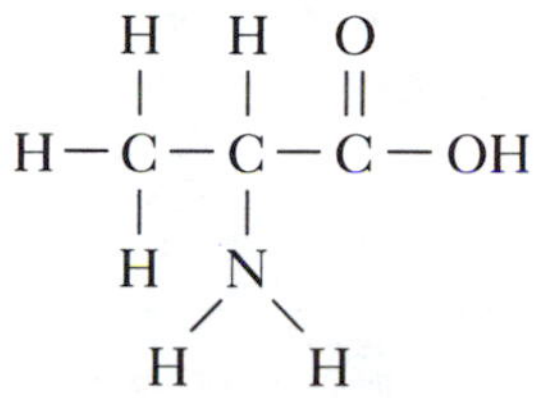

Valine	$(CH_3)_2CHCH(NH_2)COOH$
Leucine	$(CH_3)_2CHCH_2CH(NH_2)COOH$
Isoleucine	$CH_3CH_2CH(CH_3)CH(NH_2)COOH$
Serine	$HOCH_2CH(NH_2)COOH$
Threonine	$CH_3CH(OH)CH(NH_2)COOH$

B. Sulfur-containing amino acids

Cysteine	$HSCH_2CH(NH_2)COOH$
Methionine	$CH_3SCH_2CH_2CH(NH_2)COOH$

C. Monoaminodicarboxylic acids and their amides

Aspartic acid	$HOOCCH_2CH(NH_2)COOH$
Asparagine	$NH_2COCH_2CH(NH_2)COOH$
Glutamic acid	$HOOCCH_2CH_2CH(NH_2)COOH$
Glutamine	$NH_2COCH_2CH_2CH(NH_2)COOH$

D. Basic amino acids

Lysine	$NH_2CH_2CH_2CH_2CH_2CH(NH_2)COOH$
Hydroxylysine	$NH_2CHCH(OH)CH_2CH_2CH(OH_2)COOH$
Arginine	$NH_2C(NH)CH_2CH_2CH_2CH(NH_2)COOH$

Histidine

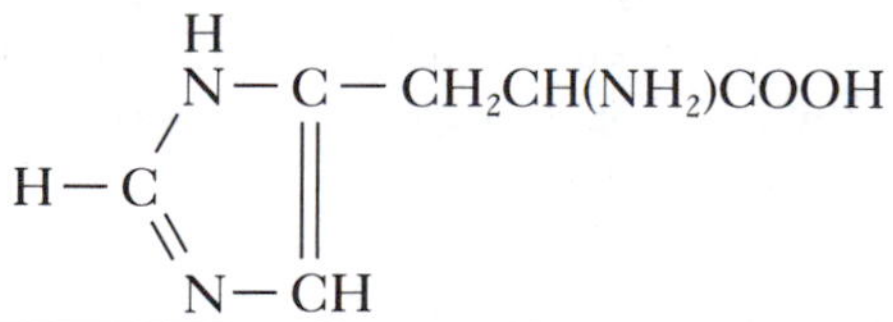

```
         H
         N — C — CH2CH(NH2)COOH
        /    ||
   H — C     ||
        \\   ||
         N — CH
```

(Continued on next page.)

TABLE 22.5 (*Continued*)
Principal Amino Acids Derived from Proteins

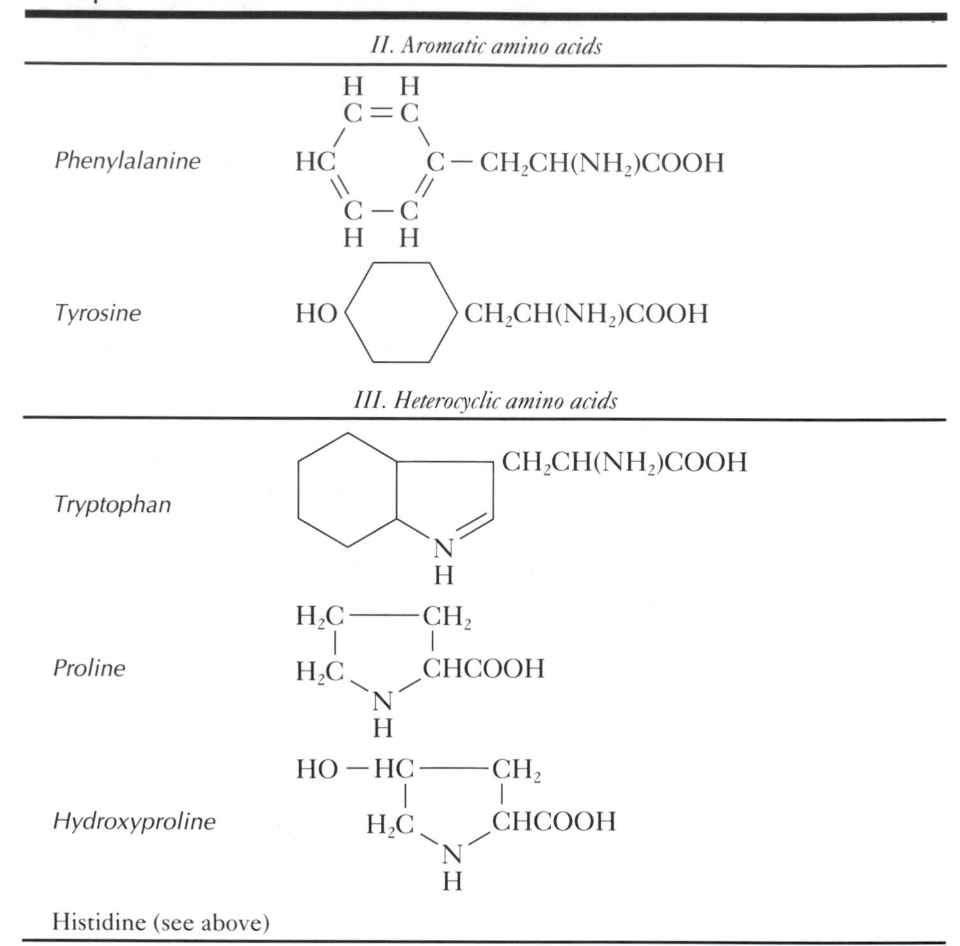

	II. Aromatic amino acids
Phenylalanine	H H / C=C / HC C — $CH_2CH(NH_2)COOH$ / C—C / H H
Tyrosine	HO ⟨ring⟩ $CH_2CH(NH_2)COOH$
	III. Heterocyclic amino acids
Tryptophan	$CH_2CH(NH_2)COOH$ / N / H
Proline	H_2C——CH_2 / H_2C CHCOOH / N / H
Hydroxyproline	HO—HC——CH_2 / H_2C CHCOOH / N / H
Histidine (see above)	

Source: From *Marine Chemistry*, R. A. Horne, copyright © 1969 by John Wiley & Sons, Inc., New York, p. 246. Reprinted by permission.

Proteins are important components of enzymes as well as of structural parts and connective tissues. For example, chitin is a polymer composed of amino acids and sugars. Amino sugars are also found in bacterial cell walls, as shown in Figure 22.19.

Humic Substances

Forty to 80 percent of marine DOM is composed of macromolecules (300–300,000 daltons) that are extremely variable in structure and elemental composition. These humic substances appear to form abiotically from fragments of biomolecules generated during the microbial degradation of organic matter.

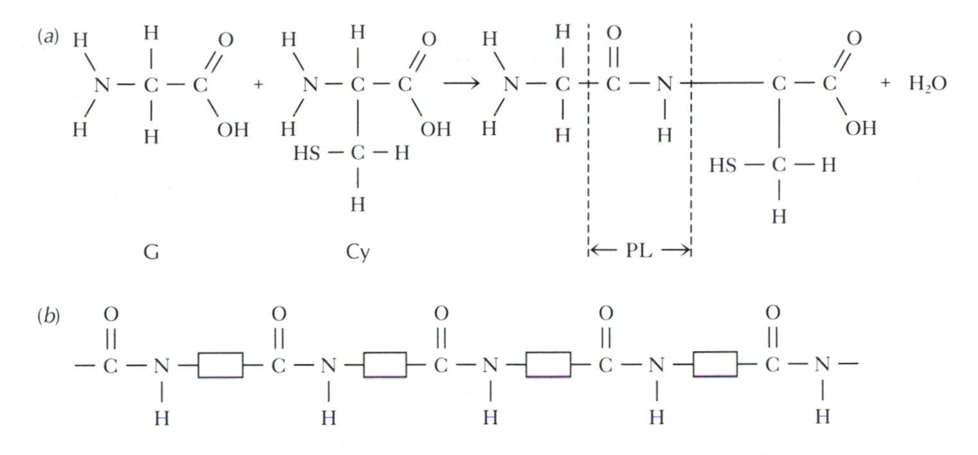

FIGURE 22.18. (*a*) The condensation of two amino acids and (*b*) diagrammatic representation of a protein, showing the peptide linkage (PL) between the amino acids.

They are also components of solids, such as sediment, soil, and peat. In addition to their compositional complexity and variability, humic substances are relatively inert. As a result, they are difficult to study, so little is known of their structure or biogeochemistry. Their marine chemistry is discussed in the next chapter.

Nitrogenous Excretion Products

Marine organisms excrete nitrogen primarily as the ammonium ion. Lesser amounts are released as the following organic species: amino acids, urea, and uric acid. The structures of the latter two are given in Figure 22.20.

Low-Molecular-Weight Carboxylic Acids

Low-molecular-weight carboxylic acids are produced and consumed during the Krebs cycle and glycolysis, so these compounds are ubiquitous in marine organisms. Phytoplankton release a significant fraction of their fixed carbon into seawater as low molecular carboxylic acids, such as those shown in Figure 22.21.

Phosphorus and Sulfur-Containing Compounds

Organic compounds that contain phosphorus include ATP, NADP, and the phospholipids. The phosphate group is often temporarily attached to other organic compounds, such as pyruvate, as part of metabolic reactions, such as glycolysis. Most organic sulfur is present as the amino acids cysteine and methionine. Microbial degradation produces low molecular weight compounds, such as dimethyl sulfide and dimethyl disulfide, which have high vapor pressures

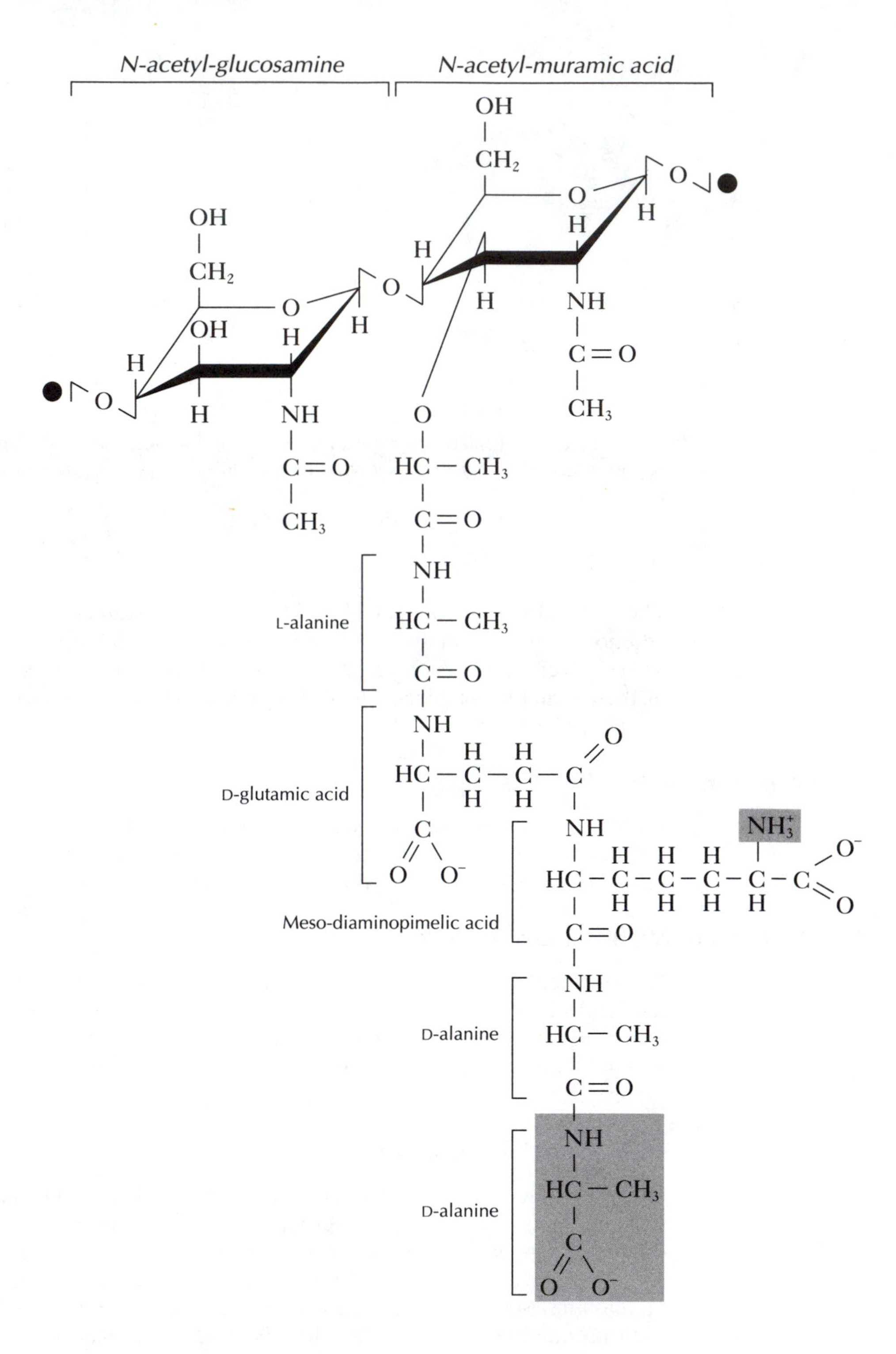
N-acetyl-glucosamine
N-acetyl-muramic acid
L-alanine
D-glutamic acid
Meso-diaminopimelic acid
D-alanine
D-alanine

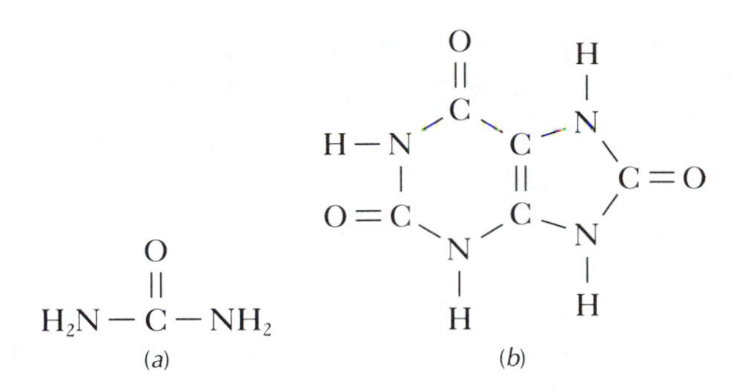

FIGURE 22.20. Structure of (*a*) urea and (*b*) uric acid.

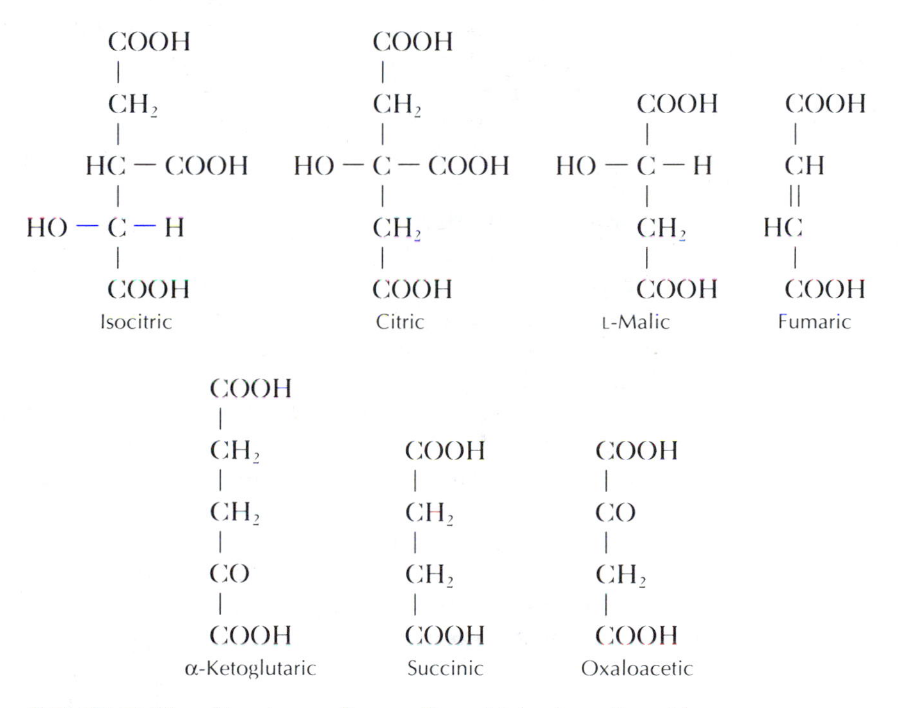

FIGURE 22.21. Structures of some di- and tricarboxylic acids.

←

FIGURE 22.19. The molecular structure of an amino sugar found in bacterial cell walls. The colored areas represent points of attachment of this macromolecule to the rest of the cell wall. The sugar units are joined end to end to form long, straight chains. The peptides form cross-links when the amino group of a meso-diaminopimelic acid in one chain replaces the terminal alanine in another chain. *Source*: From A. L. Koch, reprinted with permission from *American Scientist*, vol. 78, p. 330, copyright © 1990 by Sigma Xi, Raleigh, NC.

and are apt to diffuse across the air–sea interface. Small amounts of sulfur are also present in some polysaccharides, lipids, vitamins, and enzymes, most notably as ferrodoxin.

SUMMARY

Organic compounds constitute a relatively small elemental reservoir. Nevertheless, they have great impact on important parts of the crustal-ocean factory, largely by influencing biological activity. For example, organic compounds compose most of the tissues, as well as some of the structural parts, of marine organisms. They are also the primary food for heterotrophs, while others function as toxins that inhibit or kill competitors and predators. Still others have positive influences, such as ''neutralizing'' the toxic effects of heavy metals by forming organometallic complexes, or attracting mates. Because of their ability to provoke physiological responses, many of these compounds have proven useful to humans, primarily as drugs, lubricants, cosmetics, and food additives.

The presence of organic matter affects the mechanical properties of marine sediments and also provides food for the benthos. In turn, these organisms alter the physical, chemical, and geological characteristics of the sediment. For example, bioturbating organisms homogenize chemical gradients and increase the supply of O_2 to the sediments.

Sinking particles, like all surfaces, are covered by a film of organic matter. Due to its negative charge, this coating adsorbs metal ions and attracts heterotrophic bacteria. The former represents a significant sink for metals from the ocean and the latter promotes **biofouling**. Little of the organic matter synthesized in the euphotic zone reaches the seafloor due to efficient recycling in the water column by marine organisms. Some of the small fraction that does survive to become buried in the sediment is converted into petroleum as a result of diagenesis and catagenesis. Due to feedbacks in the crustal-ocean factory, the steady-state balance of nutrients, O_2, and CO_2 in the ocean and atmosphere is stabilized by the processes that involve the burial of organic matter. Thus down-core variations in the organic matter content of marine sediments provide a paleoceanographic record of changes in such phenomena as primary production, water circulation, and atmospheric chemistry.

Organisms obtain chemical energy by degrading organic compounds. This general process is termed **catabolism**. Some of the resulting energy is used to build large biomolecules that provide structural support and store energy. The general process of organic matter synthesis is termed **anabolism**. The reaction pathways in these processes tend to be species specific and often produce compounds that can be used as **biomarkers**. As a result of the great number of anabolic and catabolic reactions, organisms produce a myriad of organic compounds. These molecules are classified into groups based on structural similarities. In the marine environment, most organic compounds fall into one of the following general classes of biomolecules: **hydrocarbons**, **carbohydrates**, **lipids**, **fatty acids**, **amino acids**, or **nucleic acids**.

Due to analytical limitations, the concentrations of organic compounds are often determined as operationally defined fractions, such as **POM** and **DOM**. These fractions are typically characterized by bulk properties, such as their %C, C/N, and total concentrations in seawater, as well as by their unique absorbance spectra. Specific compounds and classes of compounds are identified and quantified by **mass spectrometry** following **gas chromatographic** separation. The biogeochemistry of organic matter has also been studied by tracing the fate of isotopically labelled compounds, particularly in artificial ecosystems.

CHAPTER 23

THE PRODUCTION AND DESTRUCTION OF ORGANIC COMPOUNDS IN THE SEA

INTRODUCTION

Though organic compounds constitute a relatively small reservoir in the crustal-ocean factory, they play a central role in the marine biogeochemical cycles of the biolimiting elements. Except for humic acids, all of these naturally occurring compounds are synthesized by marine organisms. Likewise, all heterotrophs require organic compounds as their source of chemical energy. The molecular structures of the most abundant of these biomolecules were given in the preceding chapter. They are present in seawater as true solutes, colloids, and solids. As shown in Figure 23.1, some of the smaller particles, such as bacteria and adsorbed colloids, are included in the operationally defined DOM fraction.

Most organic matter in the sea in present in dissolved or colloidal form. As shown in Figure 23.2, less than a quarter of the POC is alive, with most being phytoplankton and bacteria. Other marine organisms that contribute significantly to marine biomass include protozoans, fungi, and viruses. With an average concentration of 10^7 cells/ml, viruses appear to be the most numerous, but due to their small size (10^{-18} g C/cell), they account for less than 1 percent of the standing stock of marine biomass.

Because of differences in molecular structure, organic compounds vary in chemical reactivity and hence their susceptibility to transformation by marine organisms. Nevertheless, concentrations tend to be low, due to rapid uptake and release by marine organisms. The mechanisms by which organisms produce and destroy specific organic compounds are discussed below. This chapter also contains a description of the abiogenic processes that are responsible for the formation of the humic substances. This information is used to explain the marine distributions of these compounds.

THE PRODUCTION AND DESTRUCTION OF POM

The riverine transport of terrestrial POM (4.2×10^9 g C/y) is much smaller than the rate of marine primary production (4×10^{16} g C/y). The aeolian

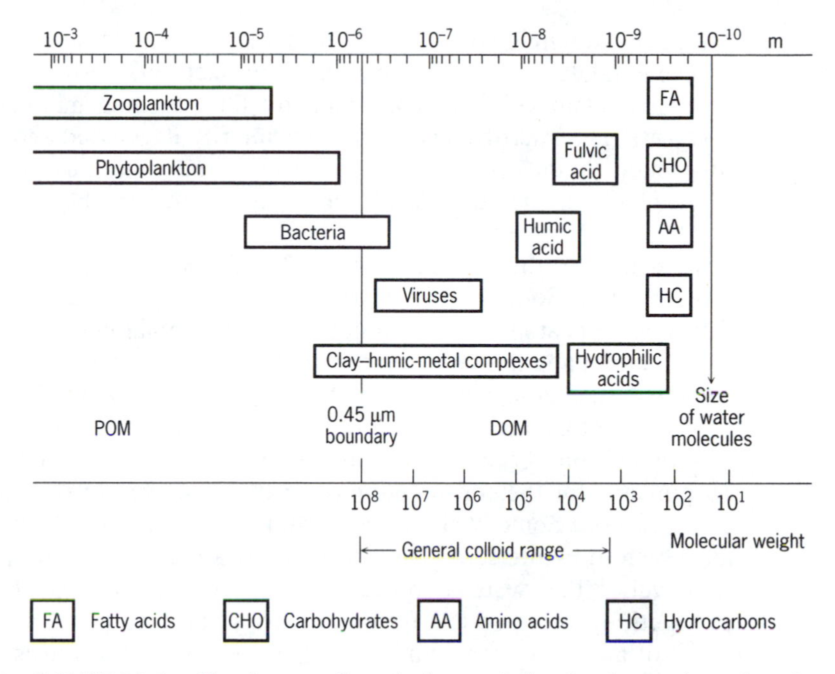

FIGURE 23.1. Continuum of particulate and dissolved organic carbon in natural water. *Source*: From *Organic Geochemistry of Natural Waters*, E. M. Thurman, copyright © 1985 by Kluwer Academic Publishers, Dordrecht, The Netherlands, p. 3. Reprinted by permission.

transport of POM is also very small. Thus most marine POM is generated in situ by primary producers, such as phytoplankton, macroalgae, and chemoautolithotrophic bacteria. Most primary production is accomplished by photoautotrophic nanoplankton (2.0–20 μm in diameter) and picoplankton (0.2–2 μm in diameter). The carbon fixed by these marine autotrophs has several possible fates. Approximately 10 percent is exuded from the plants in the form of low-

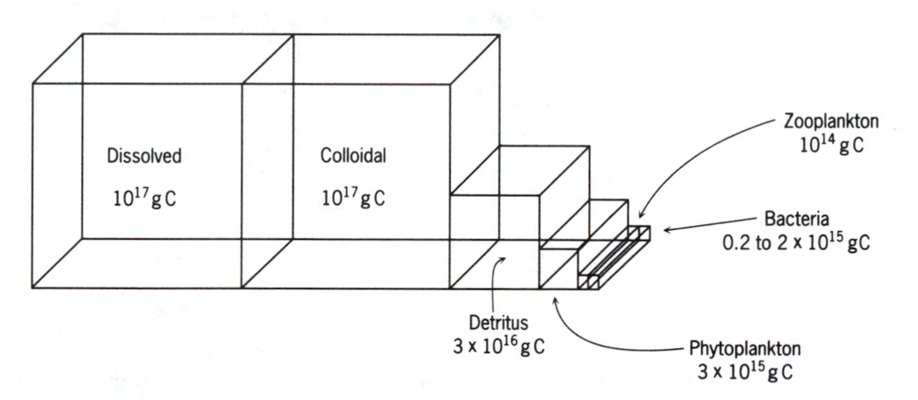

FIGURE 23.2. Distribution of organic carbon in the oceans. *Source*: After G. Cauwet, reprinted with permission from *Oceanologica Acta*, vol. 1, p. 101, copyright © 1978 by Gauthiers-Villars, Montrouge, France.

molecular-weight compounds, such as free amino acids and tricarboxylic acids. These exudates are rapidly consumed by bacteria.

Most plant cells are consumed by filter-feeding microzooplankton (20–200 μm). The microzooplankton are primarily flagellated and ciliated protozoans, such as the foraminifera, radiolaria, and tintinnids. When the uneaten phytoplankton die, their cells membranes are rapidly lysed. The resulting DOM, like the plant exudates, is rapidly taken up by bacteria. This process is responsible for the consumption of 10 to 50 percent of the photosynthetically fixed carbon. Some of this carbon is returned to seawater as DOC and DIC, but most appears to be incorporated into microbial biomass.

Bacteria and the small phytoplankton are too small to be efficiently grazed by the larger zooplankton, which are primarily copepods, euphausids, and larvae. Protozoans are able to consume bacteria, as well as the small phytoplankton, because they are also small. As a result, protozoans are responsible for the consumption of 25 to 50 percent of the photosynthetically fixed carbon. Some of this fixed carbon is passed up to higher trophic levels, as shown in Figure 23.3. Thus protozoans are an important link in the marine food web. If they were not present, most of the bacterial, and hence autotrophic production, would be lost from the marine food web.

Not all of the organic matter consumed by protozoans is passed to higher trophic levels. A substantial amount is released as DOM. This process also

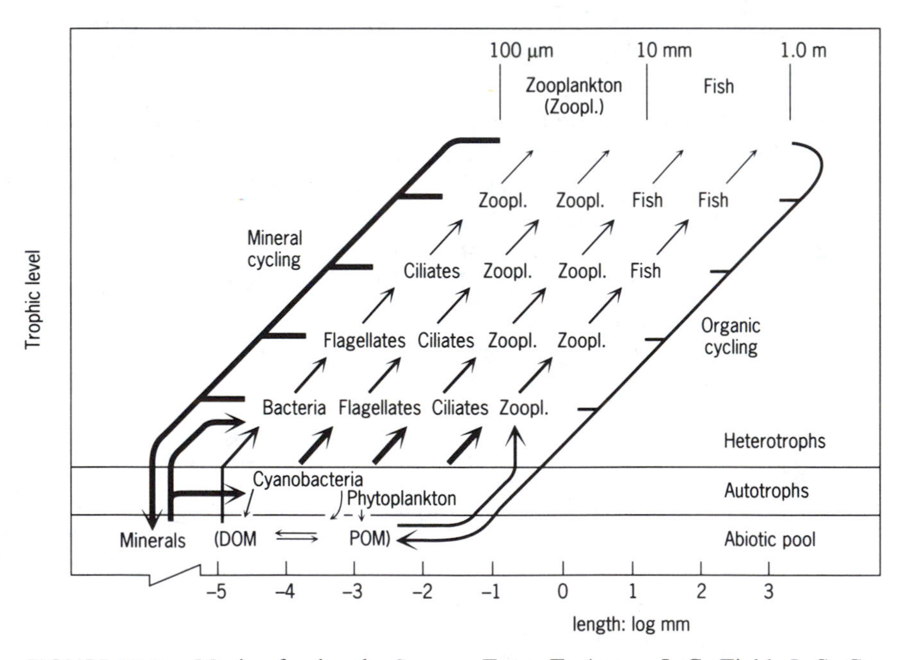

FIGURE 23.3. Marine food web. *Source*: From F. Azam, J. G. Field, J. S. Gray, L. A. Meyer-Reil, and F. Thingstad, reprinted with permission from *Marine Ecology-Progress Series*, vol. 10, p. 260, copyright © 1983 by Inter-Research, Amelinghausen, Germany.

resolubilizes nitrogen and phosphorus and hence greatly enhances nutrient recycling rates. Though larger zooplankton excrete ammonium and low molecular weight DOM, this release is significant only in ecosystems characterized by high nutrient concentrations. Under these conditions, large-celled plants, such as diatoms, are favored. Hence most of the primary production can be consumed by the larger zooplankton. Regardless, the larger zooplankton are a very important source of detrital POM, including such items as fecal pellets, egg cases, molts, mucus feeding nets, and food fragments.

The detrital pool of POM also includes dead plant and animal tissues, such as fragments of saltmarsh grass and terrestrial plants. Some of these particles aggregate to form flakes called **marine snow**. Some marine snow has been observed to contain large numbers of bacteria and protozoans whose biochemical activities cause the detrital POM to decompose as it sinks. Due to the remineralization of nitrogen and phosphorus, the seawater surrounding marine snow is often characterized by elevated nutrient concentrations and reducing conditions. The importance of these suboxic and anoxic microzones to the biogeochemical cycles of the biolimiting elements is unknown.

Large aggregates are also produced by the bioflocculation of diatoms during plankton blooms. Likewise, POM can also be created by the aggregation and precipitation of DOM from seawater. This process is accelerated by certain types of water motion, such as bubbling. Changes in environmental conditions can also cause precipitation. For example, DOM precipitates from river water in estuaries as a result of colloidal destabilization caused by increasing salinity. DOM also tends to be concentrated at boundaries, such as the sediment–water and air–sea interfaces. The latter is also characterized by enrichments of POM, such as bacteria. Under calm conditions, this organic matter forms a visible surface slick or microlayer. On windy days, the organic matter can be whipped up into an emulsion that has the appearance of a very sturdy foam.

THE CHEMICAL TRANSFORMATION OF PARTICULATE ORGANIC COMPOUNDS

Living organisms, primarily phytoplankton, make up a significant amount of the surface-water particles. Their chemical composition is somewhat species specific and is also influenced by the physiological state of the organisms. In general, 40 to 60 percent of the carbon in phytoplankton is in the form of protein. Carbohydrates and lipids compose 20 to 40 percent and 5 to 20 percent of the cellular carbon, respectively. The average elemental composition of these classes of organic compounds is given in Table 23.1.

Due to net production, the euphotic zone is characterized by high and variable POM concentrations (10–1000 μg/L). The C to N ratio of this surface-water POM ranges from 5 to 7 and is similar to that of the Redfield-Richards Ratio (C/N = 6.6). This reflects the influence of phytoplankton on the surface-water pool of POM. Below the euphotic zone, organic matter is respired faster

TABLE 23.1
Average Composition of Organic Materials

Percentage composition[a]				*Relative proportions by weight, C = 100*[b]			
Element	*Carbohydrates*	*Lipids*	*Proteins*	*Element*	*Seawater*	*Lipids*	*Proteins*
O	49.38	17.90	22.4	C	100	100	100
C	44.44	69.05	51.3	P	0.05	3.1	1.4
H	6.18	10.00	6.9	N	0.5	0.88	34.7
P		2.13	0.7	S	3150	0.45	1.6
N		0.61	17.8	Fe	0.07		0.2
S		0.31	0.8				
Fe			0.1				

Source: From *The Oceans*, H. U. Sverdrup, M. W. Johnson, and R. H. Fleming, copyright © 1941 by Prentice Hall, Inc., Englewood Cliffs, NJ, p. 230. Reprinted by permission. Data from *Textbook of Comparative Physiology*, C. G. Rogers 2nd ed., copyright © 1938 by McGraw–Hill, Inc., New York, p. 59. Reprinted by permission.
[a]Elemental Contribution by Percent Mass.
[b]Relative Elemental Contribution Normalized to Carbon.

than it is produced, so POM concentrations decline with increasing depth, as shown in Figure 23.4. In general, concentrations decrease exponentially through the thermocline and reach uniformly low (<10 μg/L) levels in the deep zone. Thus, much of the sinking POM appears to be decomposed in the thermocline. This process is partially responsible for maintaining low O_2 concentrations at mid depth. POM concentrations in the deep water exhibit little temporal or geographic variability.

Since most of the living POM is concentrated in the euphotic zone, the relative contribution of detrital POM increases with depth. The C to N ratio of this detrital POM also increases with depth because nitrogen-rich compounds (proteins) degrade faster than the nitrogen-deficient ones (carbohydrates and

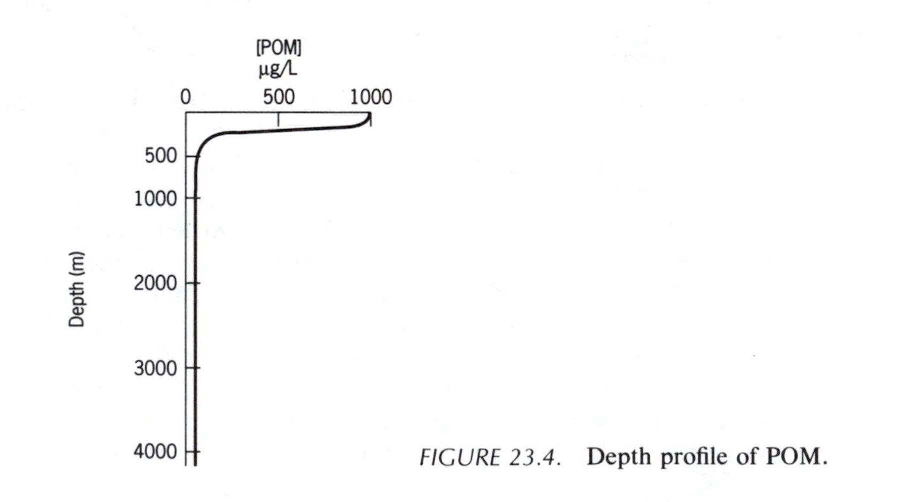

FIGURE 23.4. Depth profile of POM.

lipids). As shown in Figure 23.5, this causes the relative abundance of proteins to decrease with depth, while that of the carbohydrates increases.

In detrital POM, the large biopolymers are the first compounds to degrade. As shown in Table 23.2, this occurs by the breakage of the bonds which hold the monomeric units together. For example, proteins are decomposed into polypeptides, which are then fragmented into free amino acids. These amino acids are then converted into fatty acids and smaller molecules such as CO_2, CH_4, NH_4^+, HPO_4^{2-}, HS^-, phenols, and urea. Nucleotides, such as RNA, DNA, and ATP, are fragmented into their constituent nucleic acids, which are degraded into fragments similar to those resulting from the catabolism of amino acids. The degradation of lipids yields hydrocarbons, carbohydrates, and small carboxylic acids. These, in turn, are decomposed to CO_2 and CH_4. The polysaccharides, such as cellulose and starch, are initially broken into oligosaccharides and then into monosaccharides. Their complete degradation also yields CO_2 and CH_4. As with anabolism, the reaction mechanisms, and hence their products, are species specific.

These catabolic reactions are the result of microbial degradation, as well as consumption by higher-order organisms, such as filter-feeding zooplankton. The extent and type of transformations that detrital POM undergoes as it sinks toward the seafloor depends on (1) the types and relative abundance of deep-water heterotrophs, (2) the chemical composition of the particles, and (3) the physical packaging of the particles, as this determines their sinking rates. Most

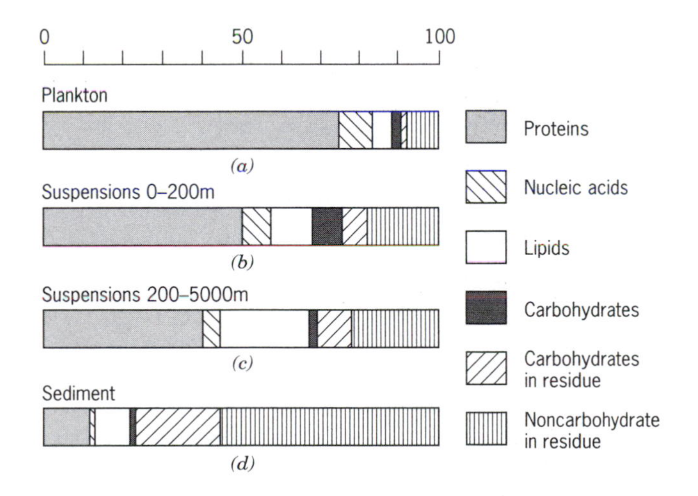

FIGURE 23.5. Biochemical composition of (*a*) plankton, (*b*) POM from the euphotic zone, (*c*) POM from the thermocline and deep waters, and (*d*) surface sediments. The scale represents the percent by mass contribution of each class of organic compound. *Source*: From G. Cauwet, reprinted with permission from *Oceanologica Acta*, vol. 1, p. 102, copyright © 1978 by Gauthiers-Villars, Montrouge, France. After A. T. Agatova and Y. A. Bogdanov, reprinted with permission from *Okeanologyia*, vol. 12, p. 234, copyright © 1972 by the Academy of Sciences of the USSR Oceanology, Moscow, USSR.

TABLE 23.2
Decomposition Products Produced from Organic Compounds Present in Detrital POM

Life Substances	*Decomposition Intermediates*	*Intermediates and Products Typically Found in Nonpolluted Natural Waters*
Proteins	Polypeptides → $RCH(NH_2)COOH$ (amino acids) → {$RCOOH$, $RCH_2OHCOOH$, RCH_2OH, RCH_3, RCH_2NH_2}	NH_4^+, CO_2, HS^-, CH_4, HPO_4^{2-}, peptides, amino acids, urea, phenols, indole, fatty acids, mercaptans
Polynucleotides	Nucleotides → purine and pyrimidine bases	
Lipids		CO_2, CH_4, aliphatic acids, acetic, lactic, citric, glycolic, malic, palmitic, stearic, and oleic acids, carbohydrates, hydrocarbons
Fats, Waxes, Oils	RCH_2CH_2COOH (fatty acids) + $CH_2OHCHOHCH_2OH$ (glycerol) → {RCH_2OH, $RCOOH$, shorter chain acids, RCH_3, RH}	
Carbohydrates		
Cellulose, Starch, Hemicellulose	$C_x(H_2O)_y$ → {monosaccharides, oligosaccharides, chitin} → {hexoses, pentoses, glucosamine}	HPO_4^{2-}, CO_2, CH_4, glucose, fructose, galactose, arabinose, ribose, xylose
Lignin	$(C_2H_2O)_x$ → unsaturated aromatic alcohols → polyhydroxy carboxylic acids	
Porphyrins and plant pigments		
Chlorophyll, Hemin, Carotenes and Xanthophylls	Chlorin → phaeophytin → hydrocarbons	Pristane, carotenoids
Complex substances formed from breakdown intermediates, e.g.,		
	Phenols + quinones + amino compounds →	Melanins, melanoidin, gelbstoffe
	Amino compounds + breakdown products of carbohydrates →	Humic acids, fulvic acids, "tannic" substances

Source: From *Aquatic Chemistry*, W. S. Stumm and J. J. Morgan, copyright © 1970 by John Wiley & Sons, Inc., New York, p. 344. Reprinted by permission.

deep-water particles are small (<20 μm), so they sink slowly (≤1 m/d) and are subject to microbial attack for extended periods of time. As a result, most small particles are composed of refractory materials, such as plankton cell walls, zooplankton exoskeletons, and precipitated humic substances. The refractory nature of this material is seen in its relatively high C to N ratio (10–13).

A relatively small number of deep-water particles are large (>20 μm) and thus sink rapidly (10–100 m/d). Most are fecal pellets produced by surface-water zooplankton. Though low in number, these large particles account for most of the mass flux of organic matter to the seafloor. Due to their rapid sinking rates, the large particles are more likely to reach the seafloor than the smaller size fraction of POM. Since they reach the seafloor relatively intact, the large particles are composed of relatively labile organic matter. Hence, they are the major food source for the benthos.

The sinking rates of small undigested particles, such as algal and microbial cells, are increased by incorporation into fecal pellets. This rapid transport, termed "the fecal pellet express," is thought to be responsible for maintaining a viable population of bacteria in the deep sea. As illustrated in Figure 23.6, it also enhances the accumulation rates of calcareous and siliceous tests in the sediments, as these particles would otherwise dissolve in transit.

Significant sampling problems are encountered in the study of POM. Because large particles are scarce in the deep sea, large volumes of water must be filtered to collect a representative number. Alternatively, a passive trap can be used to intercept sinking particles. These sediment traps (Figure 16.6) are deployed at a specific depth for times that range from days to months. The collected particles are protected from consumption by poisoning the traps. Unfortunately, these traps do not collect the entire POM flux. The fraction collected has been shown to depend on the design of the trap, particularly on its size and shape. As no standard design is used, data from different traps are not easily compared.

Though the total POM flux decreases with increasing depth, considerable variations are observed among the classes and specific compounds of organic matter, as illustrated in Figure 23.7. Rapid decreases are characteristic of chemically labile molecules. On the other hand, in situ production can produce mid-water maxima. In cases where products are species specific, their presence can be used to identify the types of marine organisms active at the sampling site. Temporal variations in these fluxes can be caused by seasonal shifts in surface productivity and diel migrations of mid-water organisms.

The sediment–water interface is the site of much biological activity. Thus POM that reaches the seafloor is subject to extensive degradation prior to burial. The sediment–water interface is a transitional zone that is best described as a loose floc populated by high concentrations of microbes. Due to high rates of organic matter decomposition, the organic carbon and nitrogen contents of these particles are decreased by an order of magnitude prior to burial. This loss continues after burial, but at much slower rates. As shown in Figure 23.8, this causes the C to N ratio in the sediments to continue to increase with

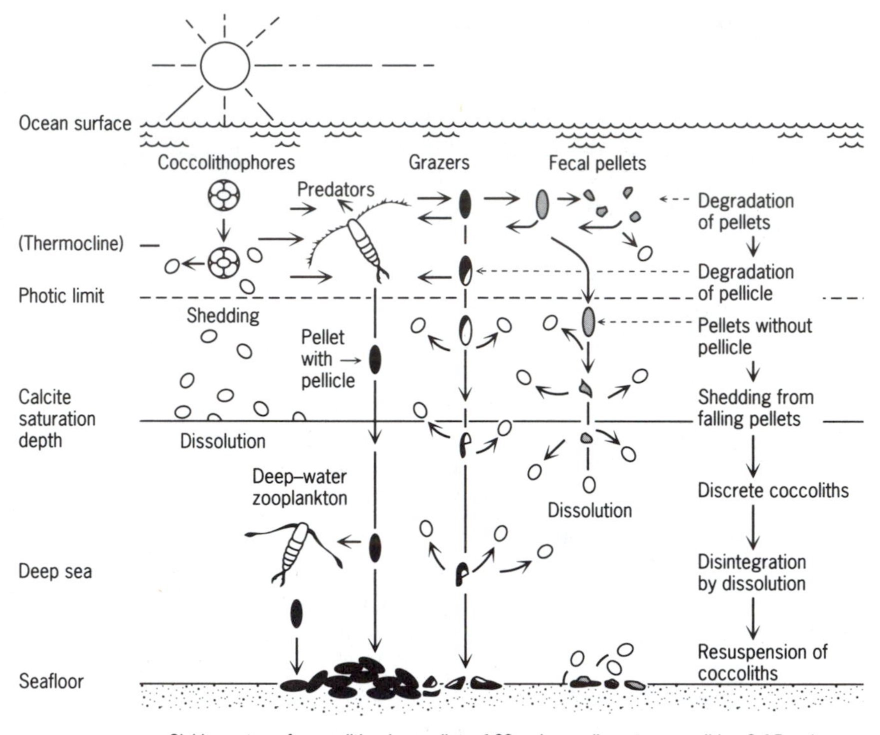

FIGURE 23.6. A schematic diagram of the relationship between the production, transportation, dissolution, and deposition of coccoliths in the ocean, demonstrating the role of fecal pellets. *Source*: From S. Honjo, reprinted with permission from *Marine Micropaleontology*, vol. 1, p. 76, copyright © 1976 by Elsevier Science Publishers, B.V., Amsterdam, The Netherlands.

depth. The chemical homogeneity of the top meter of the sediment column is likely due to bioturbation.

DISSOLVED ORGANIC COMPOUNDS

DOM is the second largest reservoir of carbon in seawater, the first being DIC. As such, DOM could represent an important source of food for heterotrophs and thus have great impact on global rates of O_2 uptake and nutrient recycling. Despite its likely importance, little is known about the quantity or chemical composition of marine DOM. This is due to the analytical difficulties associated with the identification and quantification of specific dissolved organic compounds. The low molecular weight compounds are difficult to detect because of their low concentrations and interference by salt ions. The high molecular weight compounds have proven difficult to study because of their complex and

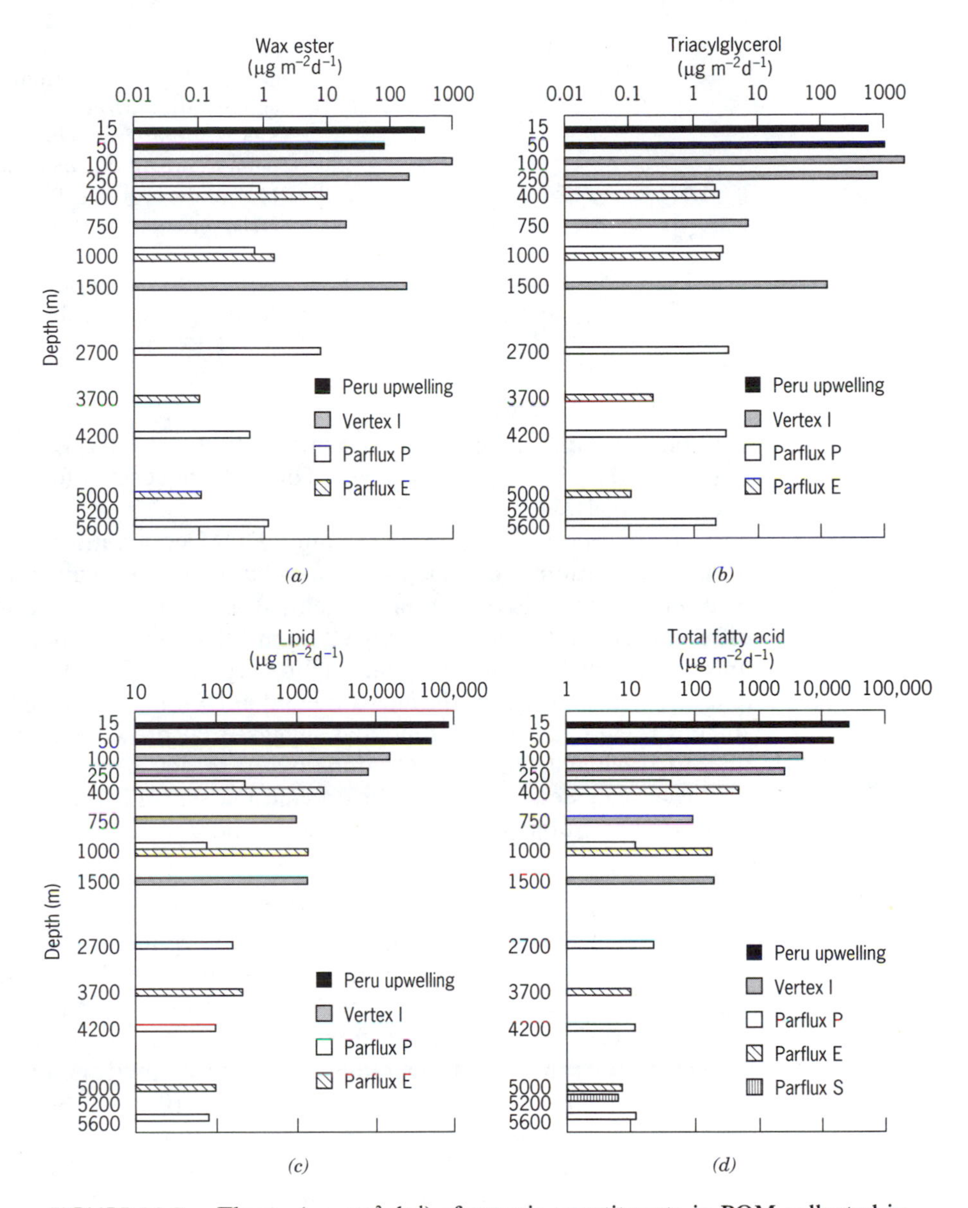

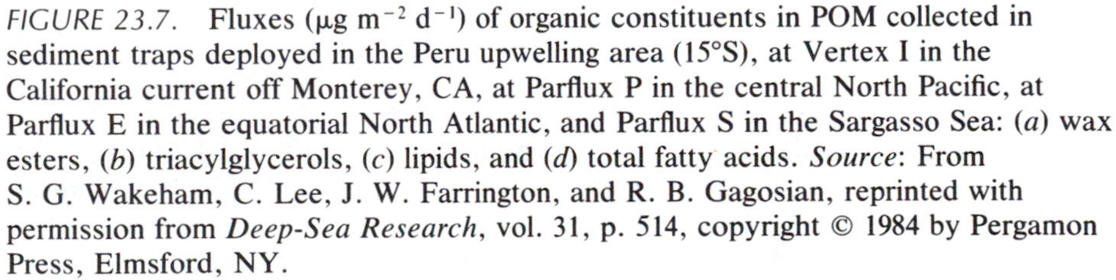

FIGURE 23.7. Fluxes ($\mu g\ m^{-2}\ d^{-1}$) of organic constituents in POM collected in sediment traps deployed in the Peru upwelling area (15°S), at Vertex I in the California current off Monterey, CA, at Parflux P in the central North Pacific, at Parflux E in the equatorial North Atlantic, and Parflux S in the Sargasso Sea: (*a*) wax esters, (*b*) triacylglycerols, (*c*) lipids, and (*d*) total fatty acids. *Source*: From S. G. Wakeham, C. Lee, J. W. Farrington, and R. B. Gagosian, reprinted with permission from *Deep-Sea Research*, vol. 31, p. 514, copyright © 1984 by Pergamon Press, Elmsford, NY.

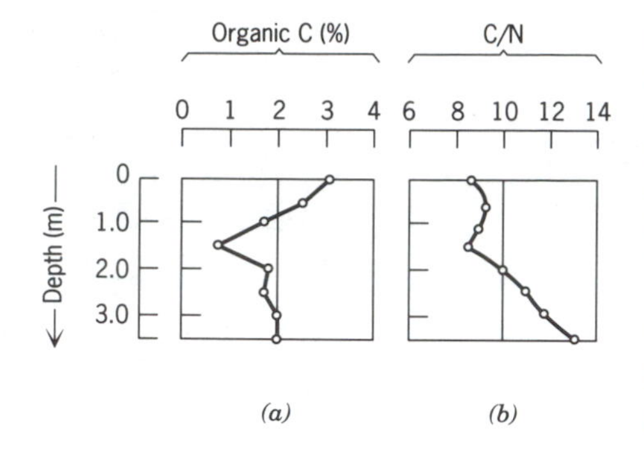

FIGURE 23.8. (*a*) % organic carbon and (*b*) C/N ratio of sedimentary organic matter in a core from the Oman Sea. *Source*: From *Humic Substances in Soil, Sediment and Water*, M. Vanderbroucke, R. Pelet, and Y. Debyser (eds.: G. R. Aiken, D. M. McKnight, R. L. Wershaw, and P. MacCarthy), copyright © 1985 by John Wiley & Sons, Inc., New York, p. 254. Reprinted by permission.

variable structures. Since these compounds are relatively inert, they do not readily undergo chemical changes. This is a necessity for any type of wet-chemical analysis.

Part of the difficulty in determining DOM concentrations also lies in the variable retention of organic colloids on filters. As a result, a variable fraction of organic matter is operationally defined as DOM. This variability is due to differences in filter porosity as well as in filtration conditions, such as in the amount of particle loading and suction applied. The development of new analytical techniques for the measurement of DOM is an active area of research. Recent results using a new method suggest that previous measurements have underestimated DOC concentrations by 50 to 400 percent. If so, our models of the biogeochemical cycling of organic matter will have to be considerably revised. This subject is addressed toward the end of this chapter.

Sources of DOM

Terrestrial DOM is carried to the ocean by winds, rivers, or groundwater runoff. Aeolian transport can occur via diffusion of volatile compounds across the air–sea interface or by adsorption onto aerosols that are deposited on the sea surface. This is a minor flux compared to the riverine input. DOC concentrations in river water range from 1 to 20 mg/L. Its annual global flux is approximately the same as for POC (0.11×10^9 g C). Some of this DOC is precipitated when river water enters estuaries and mixes with seawater. This effect is largest for terrestrial humic acids. Due to the small size of the terrestrial DOM flux, it appears that most marine DOM is produced in situ.

DOM is produced in situ as a result of excretions and exudations from animals, microorganisms, and plants. As mentioned above, phytoplankton "excrete" a significant amount of their fixed carbon in the form of low molecular weight compounds, such as those listed in Table 23.3. Though this process is variable, it is ubiquitous and appears to be associated with dark respiration.

Animals and microbes also excrete low-molecular-weight compounds. Those released by animals tend to be nitrogen-rich (e.g., uric acid, urea, amino acids, and nucleic acids). Fecal material is an indirect source of DOM as the

TABLE 23.3
Excretion Products of Algae

Products	
Glycolic acid	Polysaccharides
Amino acids	Vitamins
Volatile compounds	Polymeric organic acids
Sterols	Fatty acids
Pigments	

Source: From *Organic Geochemistry of Natural Waters*, E. M. Thurman, copyright © 1985 by Kluwer Academic Publishers, Dordrecht, The Netherlands, p. 83. Reprinted by permission. Data from G. E. Fogg, reprinted with permission from *Limnology and Oceanography*, vol. 22, p. 576, copyright © 1977 by the American Society of Limnology and Oceanography, Seattle, WA; and E. Billmire and S. Aaronson, reprinted with permission from *Limnology and Oceanography*, vol. 21, p. 138, copyright © 1976 by the American Society of Limnology and Oceanography, Seattle, WA.

solids are solubilized during microbial degradation. DOM is also released into seawater following the death of organisms due to rapid lysis of their cell membranes. Sloppy feeding breaks cell membranes, thereby releasing DOM into seawater.

The Ecological Role of DOM

Although the low-molecular-weight compounds represent a small fraction of the DOM reservoir, they play an important ecological role in the sea. These compounds have high production rates, but their concentrations tend to be very low due to rapid uptake, primarily by heterotrophic bacteria. As a result, low-molecular-weight compounds tend to have small turnover times and hence function as an important microbial food source, despite their low concentrations.

Some marine organisms release low-molecular-weight compounds in an effort to control some aspect of their environment. For example, some of these compounds act as toxins that repel or kill competitors or predators. Others act to "neutralize" toxins. Some function as attractants and are used for mating purposes. Other types of chemical communication conveyed by dissolved compounds are described in Table 23.4.

The Fate of Low-Molecular-Weight DOM in the Sea

The low-molecular-weight fraction of DOM is rapidly consumed upon introduction into seawater. As a result, the concentrations of these compounds are low

TABLE 23.4
Classes of Interorganismic Chemical Effects[a]

I. Allelochemic Effects
   A. Allomones (+ /), which give adaptive advantage to the producing organism
      1. Repellents (+ /), which provide defense against attack or infection (many secondary plant substances, chemical defenses among animals, probably some toxins of other organisms)
      2. Escape substances (+ /) that are not repellents in the usual sense (inks of cephalopods, tension-swimming substances)
      3. Suppressants (+ −), which inhibit competitors (antibiotics, possibly some allelopathics, and plankton ectocrines)
      4. Venoms (+ −), which poison prey organisms (venoms of predatory animals and myxobacteria, aggressins of parasites, and pathogens)
      5. Inductants (+ /), which modify growth of the second organism (gall, nodule, and mycorrhiza-producing agents)
      6. Counteractants (+ /), which neutralize as a defense the effect of a venom or other agent (antibodies, substances inactivating stinging cells, substances protecting parasites against digestive enzymes)
      7. Attractants (+ /)
         a. Chemical lures (+ −), which attract prey to a predator (attractants of carnivorous plants and fungi)
         b. Pollination attractants, which are without (+ 0) or with (+ +) advantage to the organism attracted (flower scents)
   B. Kairomones (/ +), which give adaptive advantage to the receiving organism
      1. Attractants as food location signals (/ +), which attract the organism to its food source, including (− +) those attracting to a food organism (use of secondary substances as signals by plant consumers, of prey scents by predators or chemical cues by parasites), (+ +) pollination attractants when the attracted organism obtains food, and (0 +) those attracting to nonliving food (response to scent by carrion feeder, chemotactic response by motile bacteria and by fungal hyphae)
      2. Inductants (/ +), which stimulate adaptive development in the receiving organism (hyphal loop factor in nematode-trapping fungi, spine-development factor in rotifers)
      3. Signals (/ +) that warn of danger or toxicity to receiver [repellent signals that have adaptive advantage to the receiver; scents and flavors that indicate unpalatability of nonliving food, predator scents]
      4. Stimulants (/ +), such as hormones, that benefit the second organism by inducing growth
   C. Depressants (0 −), wastes and so forth, that inhibit or poison the receiver without adaptive advantage to releaser from this effect (some bacterial and parasite toxins, allelopathics that give no competitive advantage, some plankton ectocrines)

II. Intraspecific Chemical Effects
   A. Autotoxins (− /), repellents, wastes, and so forth, that are toxic or inhibitory to individuals of the releasing populations, with or without selective advantage from detriment to some other species (some bacterial toxins, antibiotics, ectocrines, and accumulated wastes of animals in dense culture)

TABLE 23.4 (*Continued*)
Classes of Interorganismic Chemical Effects[a]

B. Adaptive autoinhibitors (+ /) that limit the population to numbers that do not destroy the host or produce excessive crowding (staling substance of fungi)
C. Pheromones (+ /), chemical messages between members of a species, that are signals for
   1. Reproductive behavior
   2. Social regulation and recognition
   3. Control of caste differentiation
   4. Alarm and defense
   5. Territory and trail marking
   6. Food location

Source: From R. H. Whittaker and P. P. Feeny, reprinted with permission from *Science*, vol. 171, p. 767, copyright © 1971 by the American Association for the Advancement of Science, Washington, D.C.

[a]Adaptive advantage is indicated by +, detriment by −, and adaptive indifference by 0, for the releasing organism first and the receiving organism second. The virgule (/) indicates that adaptive advantage or detriment is not specified for one side of the relationship.

(<100's μg/L). As shown in Figure 23.9, the polypeptides are most abundant. The concentration of free amino acids is much lower due to rapid microbial uptake. Most of the sugars are glucose, which is excreted into seawater by phytoplankton.

Though most of the labile fraction of DOM is consumed and degraded by microbes, some is assimilated by phytoplankton during dark respiration. The degradation products are eventually incorporated into humic substances as described below. A significant amount of DOM is also lost by adsorption onto sinking particles. The bursting of bubbles at the sea surface also causes a small amount of DOM to be volatilized or converted to POM.

The Production and Destruction of High-Molecular-Weight DOM

Forty to 80 percent of DOM is composed of high-molecular-weight compounds that are collectively referred to as **humic substances**. Their contribution is highest in deep waters due to the relative scarcity of low-molecular-weight compounds. They impart a yellow or brown cast to seawater when present at very high concentrations. Humic substances that are insoluble at acidic pH's are termed **humic acids**; those that are soluble at all pH's are termed **fulvic acids**. The latter are more abundant in seawater.

As shown in Figure 23.10, humic substances are thought to be produced by abiogenic chemical reactions that link together relatively low-molecular-weight compounds. Most of these low-molecular-weight compounds are derived from the degradation of biopolymers, such as proteins and carbohydrates.

These reactions include condensations, polymerizations, oxidations, and reductions. As shown in Figure 23.11, the nonaromatic structures tend to form

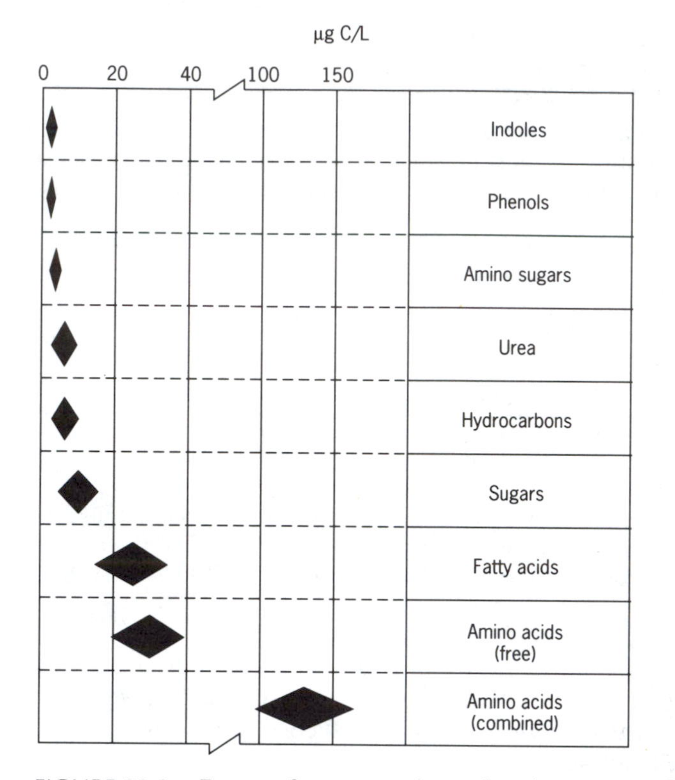

FIGURE 23.9. Range of concentrations of various types of DOC. *Source*: From *Symposium on Organic Matter in Natural Waters*, E. T. Degens (ed.: D. Hood), copyright © 1970 by the Institute of Marine Science, University of Alaska, College, Alaska, p. 79. Reprinted by permission.

highly cross-linked polymers. At the sea surface, photochemical reactions produce O_2 free radicals that accelerate the rates of some of the polymerization reactions.

The molecular structure of the resulting humic substances are very variable, partly as a result of differences in the availability of low-molecular-weight compounds. Nevertheless, they possess similar functional groups, such as aromatic structures, carboxylic acids, and amines, as shown in Figure 23.12*a*. This example is a likely structure for a humic acid. In comparison, fulvic acids tend to have a greater abundance of acidic functional groups, such as those shown in Figure 23.12*b*.

As shown in Figure 23.13, terrestrial humic substances compose a significant fraction of the riverine input of DOM to the ocean. Terrestrial humics tend to have more acidic groups and a higher degree of oxidation than marine forms. They also contain more aromatic structures and hence are more susceptible to flocculation when exposed to high concentrations of electrolytes. Other differences between terrestrial and marine humics are described in Table 23.5.

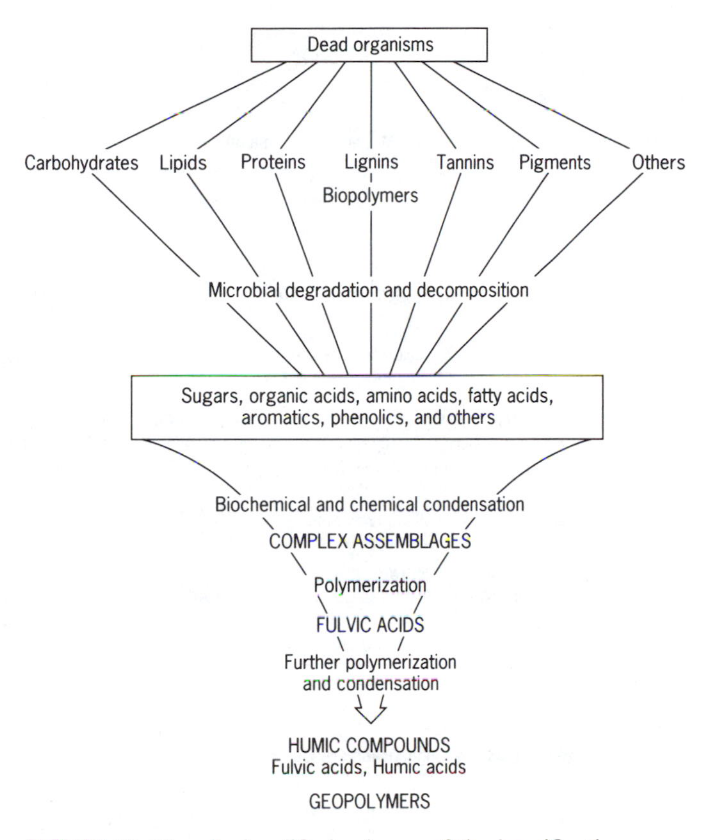

FIGURE 23.10. A simplified scheme of the humification process. *Source*: From *Geochemistry of Marine Humic Compounds*, M. A. Rashid, copyright © 1985 by Springer-Verlag, Heidelberg, Germany, p. 59. Reprinted by permission.

The presence of lignin degradation products causes the terrestrial humic substances to have a relatively high degree of acidity and aromaticity. Most of these fragments are phenolic acids (Figure 23.14*b*), which are liberated during the degradation of lignin (Figure 23.14*a*). These compounds react with other forms of degraded organic matter to generate terrestrial humic substances. Since lignin is unique to land plants, these phenolic moieties can be used as a biomarker indicating the presence of terrestrial organic matter. Their low abundances in open ocean sediments supports the hypothesis that most riverborne POM is deposited in the coastal zone.

A Proposed Cycle for Marine DOM

DOM concentrations are highest in the surface water as this is where most organic matter is synthesized. As illustrated in Figure 23.15, concentrations decrease exponentially through the thermocline. The deep waters are characterized by low and uniform concentrations. Surface-water concentrations are

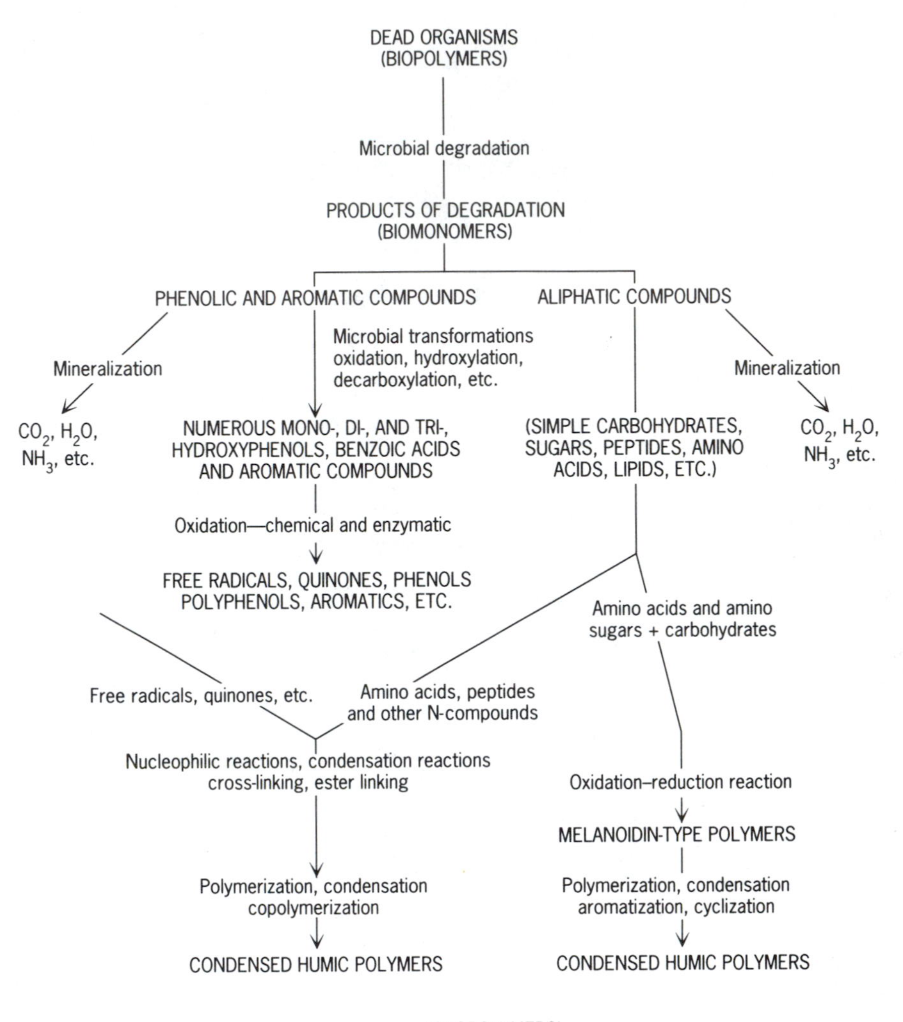

FIGURE 23.11. Major processes involved in the synthesis of humic compounds. *Source*: From *Geochemistry of Marine Humic Compounds*, M. A. Rashid, copyright © 1985 by Springer-Verlag, Heidelberg, Germany, p. 58. Reprinted by permission.

FIGURE 23.12. Model chemical structure of (*a*) a humic acid and (*b*) part of a fulvic acid. *Source*: From (*a*) D. Kleinhempel, reprinted with permission from *Albrecht Thaer Archiv*, vol. 16, p. 8, copyright © 1970 by Aus dem Institut für Acker- und Pflanzenbau Müncheberg der Deutschen Akademie der Landwirtschaftsweissen-schaften, Berlin, Germany, and (*b*) *Geochemistry of Marine Humic Compounds*, M. A. Rashid, copyright © 1985 by Springer-Verlag, Heidelberg, Germany, p. 75. Reprinted by permission.

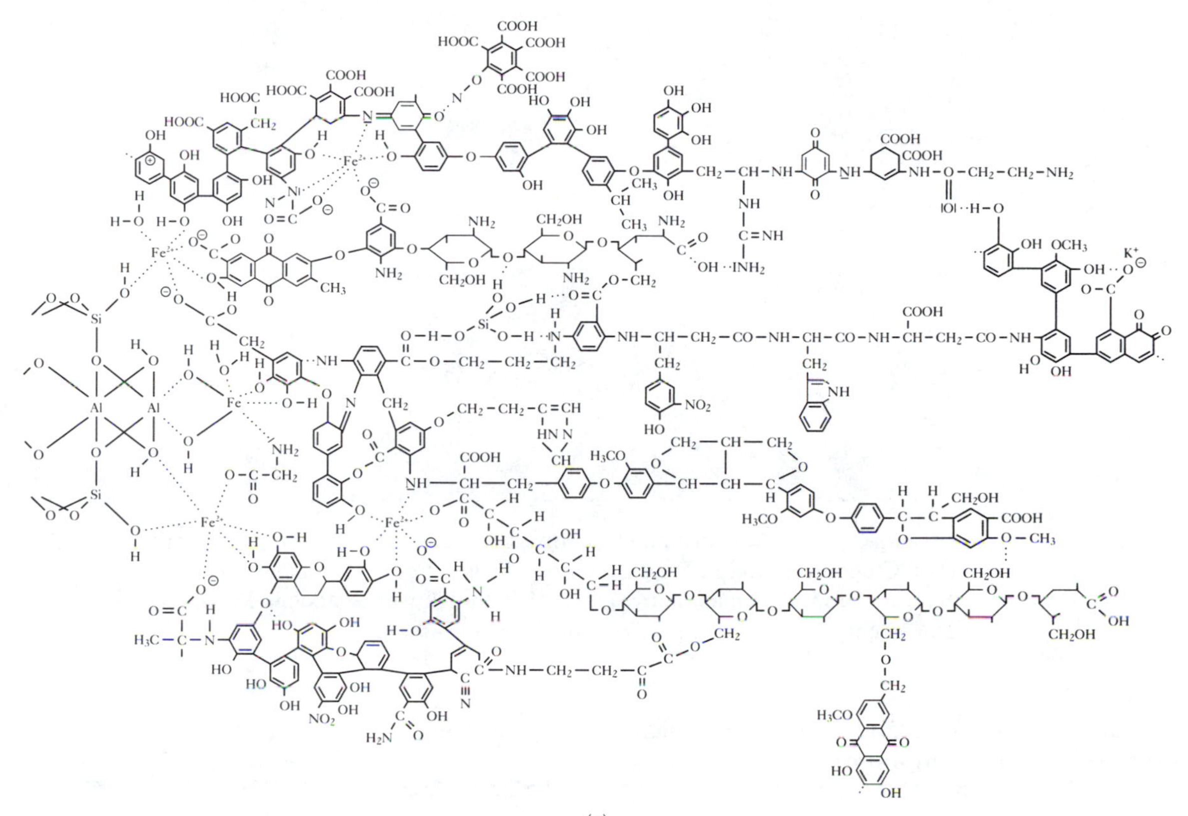

(*a*)

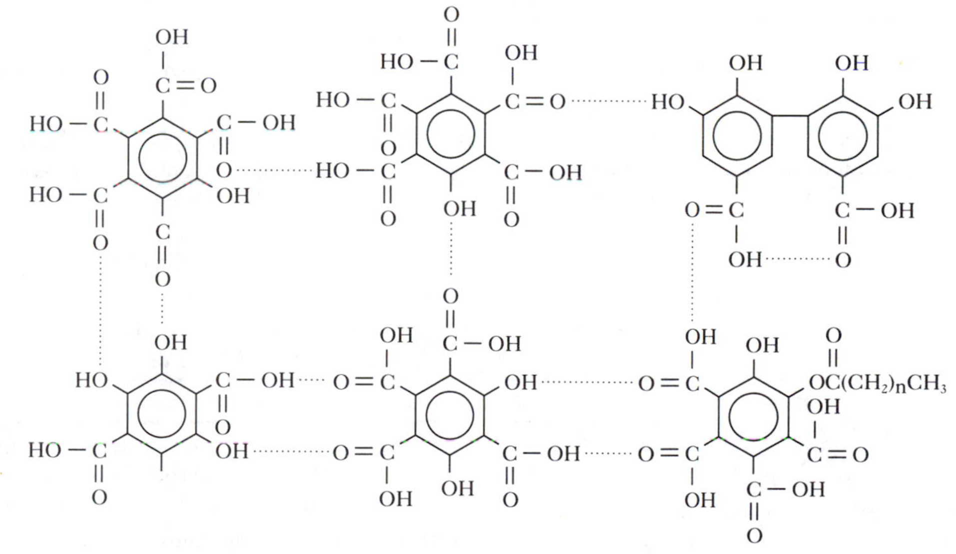

(*b*)

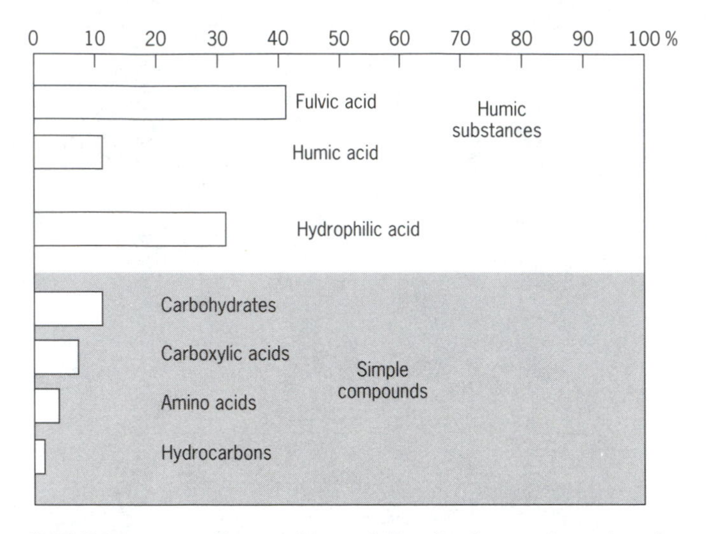

FIGURE 23.13. Composition of dissolved organic carbon in average river water with a DOC concentration of 5 mg/L. *Source*: From *Organic Geochemistry of Natural Waters*, E. M. Thurman, copyright © 1985 by Kluwer Academic Publishers, Dordrecht, The Netherlands. Reprinted by permission.

TABLE 23.5
Certain Striking Differences in the Chemical Nature and Geochemical Behavior of Marine and Soil Humic Compounds[a]

Marine Humic Compounds	*Soil Humic Compounds*
General Composition	
Marine humic compounds are rich in aliphatic substances; the concentration of aliphatic carbon is almost twice that of aromatic carbon	A strong aromatic core is generally the characteristic feature of soil humic compounds; aromatic carbon is predominant in the molecule
Aromatic compounds are present but do not constitute a predominant part of the molecule; the polyphenolic and polyaromatic components are low in concentration and exhibit low degrees of substitution	Aliphatic side chains are attached to the aromatic core, but due to the participation of lignin degradation products, aromatic moieties are more abundant
	The molecules contain more polycyclic and polyphenolic constituents
Anaerobic conditions do not favor the development of highly condensed organic molecules	A high degree of oxidation leads to a high degree of condensation in humic molecules
Elemental Composition	
The carbon contents in fulvic and humic acids are 45–50% and 50–55%, respectively	The carbon content shows broader variations; because of the strong carbon lattice, the concentration of carbon is as high as 65%
Aquatic conditions favor greater hydrogen contents	Terrestrial humic compounds have generally low concentrations of hydrogen

TABLE 23.5 (*Continued*)
Certain Striking Differences in the Chemical Nature and Geochemical Behavior of Marine and Soil Humic Compounds[a]

Marine Humic Compounds	*Soil Humic Compounds*
Elemental Composition (Continued)	
Marine organisms are rich in nitrogen and therefore the humic compounds originating from them contain high amounts of nitrogen; some sedimentary humic compounds contain up to ten times as much sulfur as the soil humic acids	The nitrogen content of soil humic compounds is generally low
Isotopic Composition	
The carbon isotopic composition of marine humic compounds is heavy; $\delta^{13}C$ values ranging from $\approx -20‰$ to $-23‰$	Lighter carbon isotopic composition ($\delta^{13}C = -25$ to $-28‰$) is characteristic of terrestrially derived humus
$\delta^{15}N$ is generally about $+9‰$ and reflects nitrate as a nitrogen source	Terrestrial plants use atmospheric nitrogen fixed by bacteria; $\delta^{15}N$ is about $+2‰$
$\delta^{2}H$ values do not show much spread; the average is about $-105‰$	The $\delta^{2}H$ of soil humic compounds show much broader values ranging from $-50‰$ to $-100‰$
Functional Groups	
The total acidity is highly variable but in general it is lower than that of soil humic matter; acidity in marine humic compounds is largely a function of carboxyl groups; the phenolic hydroxyl content is generally low; the restricted aeration of bottom sediments favors greater accumulations of carbonyl groups	Soil humic compounds generally possess higher degrees of total acidity; the acidic characteristics arise from the participation of carboxyl as well as phenolic hydroxyl groups
	Because of the high degree of aeration, soil humic material contains more carboxyl groups and fewer carbonyl groups
Alcoholic hydroxyl contents are generally high	Alcoholic hydroxyl contents are low
Functional groups constitute about 20–30% of the humic molecules	About 25% of soil humic acids and about 50–60% of a fulvic acid molecule consist of functional groups

Source: From *Geochemistry of Marine Humic Compounds*, M. A. Rashid, copyright © 1985 by Springer-Verlag, Heidelberg, Germany, p. 105. Reprinted by permission.

[a]See Chapter 29 for an explanation of δ values that are used to describe the stable isotope composition of humic substances.

highest in the coastal zone and in the surface microlayer due to the presence of large numbers of organisms.

DOM concentrations are several orders of magnitude larger than those of the POM. The high concentration of deep-water DOM suggests that this pool of organic matter is relatively refractory and hence remains in seawater. This conclusion is also supported by the advanced age of deep-water DOM as established by its ^{14}C content (6000 y at 5700 m in the north central Pacific Ocean) and the increase in its C to N ratio with increasing depth.

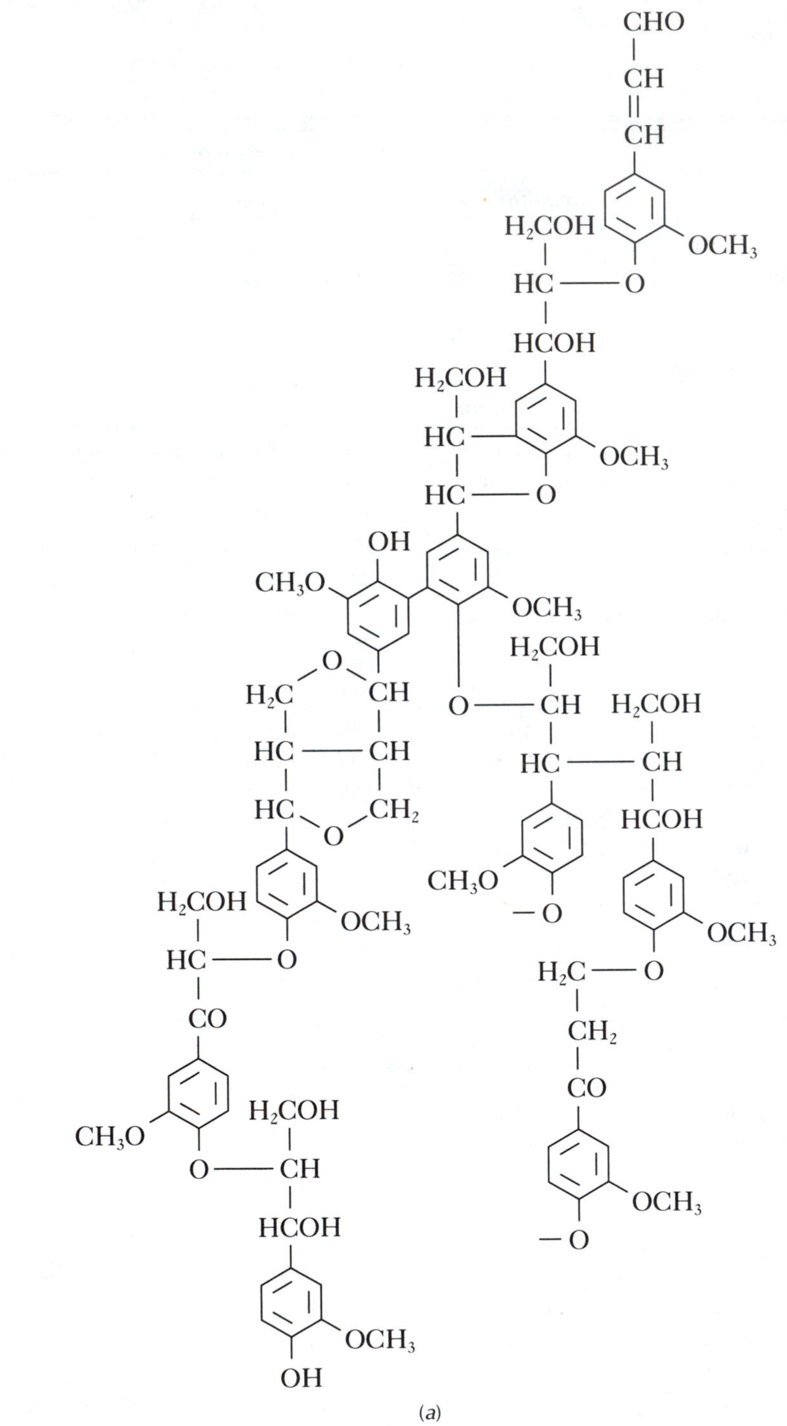
CHO
CH
CH
H$_2$COH
OCH$_3$
HC
O
HCOH
H$_2$COH
HC
OCH$_3$
HC
O
OH
CH$_3$O
OCH$_3$
H$_2$COH
O
H$_2$C
CH
O
CH
H$_2$COH
HC
CH
HC
CH
HC
CH$_2$
O
HCOH
CH$_3$O
O
H$_2$COH
OCH$_3$
OCH$_3$
HC
O
H$_2$C
O
CO
CH$_2$
CO
H$_2$COH
CH$_3$O
O
CH
OCH$_3$
HCOH
O
OCH$_3$
OH

(*a*)

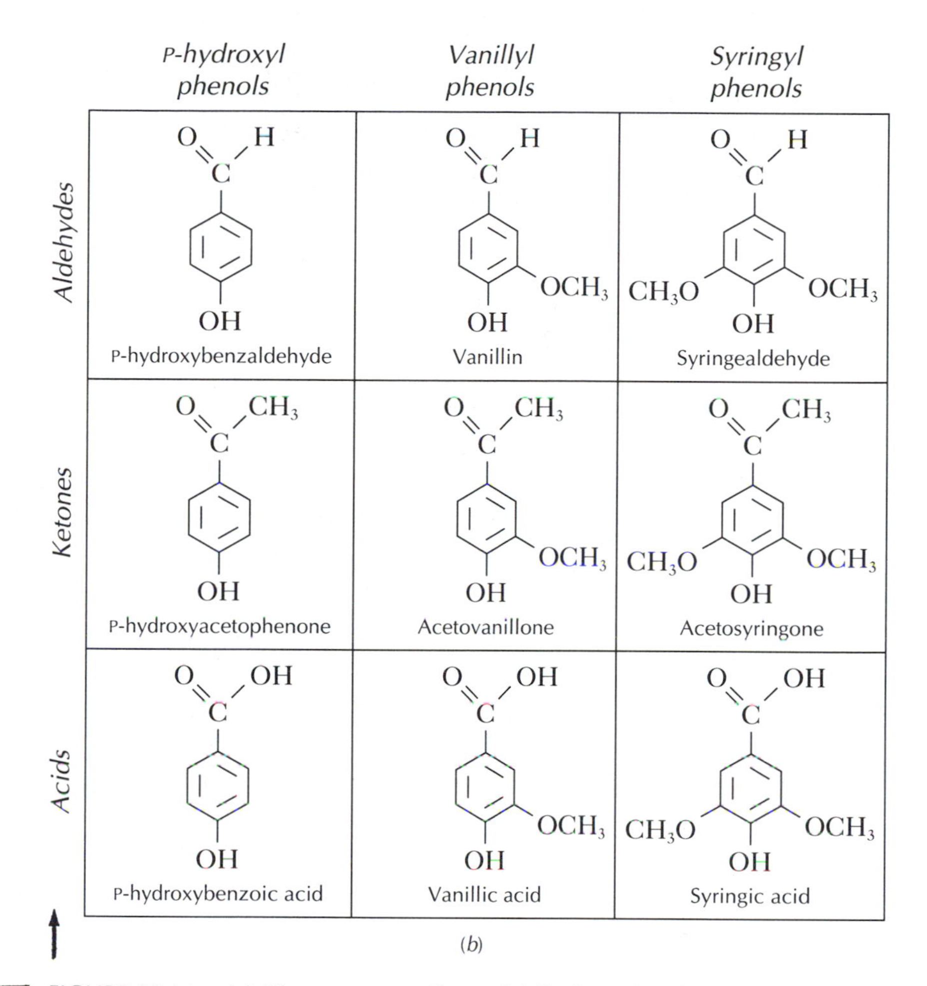

FIGURE 23.14. (*a*) The structure of a model lignin molecule. (*b*) Aromatic acids and phenols derived from lignin which are used as indicators of terrestrial plant matter. *Source*: From *Geochemistry of Marine Humic Compounds*, M. A. Rashid, copyright © 1985 by Springer-Verlag, Heidelberg, Germany, p. 47 (reprinted by permission); and J. I. Hedges and P. L. Parker, reprinted with permission from *Geochimica et Cosmochimica Acta*, vol. 40, p. 1021, copyright © 1976 by Pergamon Press, Elmsford, New York.

The radiocarbon age of surface-water DOM (1300 y) suggests that a significant fraction (approximately 50 percent) of this organic matter is also refractory. These "old" ages have also been observed in the surface-water humics. Thus it appears that labile DOM, produced by biological activity in the surface waters, is degraded into fragments that combine to form a large pool of relatively refractory DOM. The latter is then cycled somewhat conservatively through the ocean. As a result, deep-water DOM is not as vertically or horizontally segregated as is POM, O_2, or the nutrients.

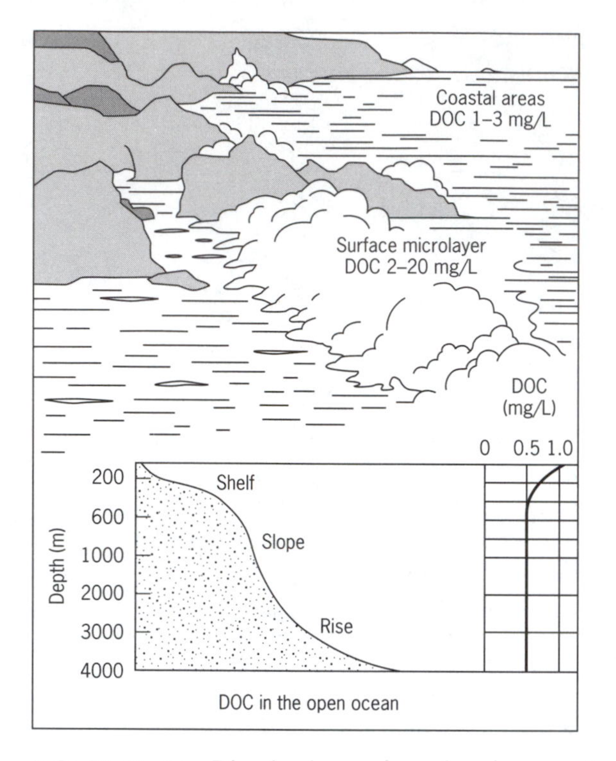

FIGURE 23.15. Dissolved organic carbon in seawater. *Source*: From *Organic Geochemistry of Natural Waters*, E. M. Thurman, copyright © 1985 by Kluwer Academic Publishers, Dordrecht, The Netherlands, p. 25. Reprinted by permission.

This characterization of the marine DOM cycle may have to be substantially revised in light of recent methodological developments in the measurement of DOM. Substantially higher concentrations of DOC and DON have been detected when filtered seawater is combusted at 680°C in the presence of a platinum catalyst. As shown in Figure 23.16, depth variations in these elevated concentrations correlate closely with those observed in such biologically controlled species as AOU and nitrate.

The strong correlations suggest that the "extra" DOM is biologically labile. This observation is somewhat perplexing as this same organic matter is unreactive in the standard lab procedures formerly used to measure DOM. Elemental analysis has shown that the "extra" DOM is relatively nitrogen enriched (C/N = 6–7) and high in molecular weight (4000–22,000 daltons). These compounds are thought to be highly cross-linked polymers derived from proteinaceous precursors by a process similar to that which produces humic substances. Diurnal variations in the concentration of this DOM suggest that the precursors are components of phytoplankton cell walls.

If this "extra" DOM truly exists, our understanding of several other major features of marine chemistry will have to be considerably revised. For example,

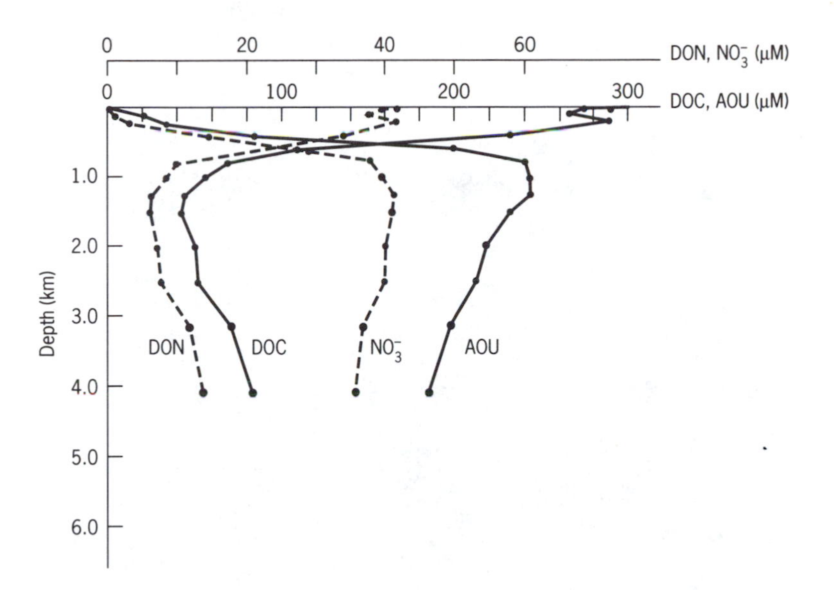

FIGURE 23.16. Depth Profiles of DOC, DON, AOU, and NO_3^- Concentrations at 20°N 137°E. *Source*: From P. M. Williams and E. R. M. Druffel, reprinted with permission from *Oceanography*, vol. 1, p. 16, copyright © 1988 by The Oceanography Society, Washington, DC. Data from Y. Sugimura and Y. Suzuki, reprinted with permission from *Marine Chemistry*, vol. 24, p. 125, copyright © 1988 by Elsevier Science Publishers, B.V., Amsterdam, The Netherlands.

if this "extra" DOM is produced from the degradation of POM that has a chemical composition close to that of detrital phytoplankton, a large pool of as-yet-undiscovered dissolved phosphorus must also be present. The vertical profiles indicate that a significant amount of this DOM is respired. Thus respiration must constitute a large sink for dissolved O_2 that has also not heretofore been detected. Due to its evidently high concentrations, this DOM could be responsible for the complexation of a large amount of dissolved metals. This has also not been considered in current models of ion speciation.

GLOBAL DISTRIBUTION OF ORGANIC CARBON

The geographic distribution of DOM, POM, and sedimentary organic carbon is closely related to that of primary production. As shown in Figure 23.17, primary production is highest in coastal waters and lowest in the open ocean. Biological activity in the coastal zone is supported by large amounts of nutrients supplied by rivers, coastal upwelling, and eddy diffusion. Because of shallow depths, coastal waters are not well stratified and hence vertical mixing is able to return regenerated nutrients to the euphotic zone. Production is lowest in the open ocean due to nutrient limitation.

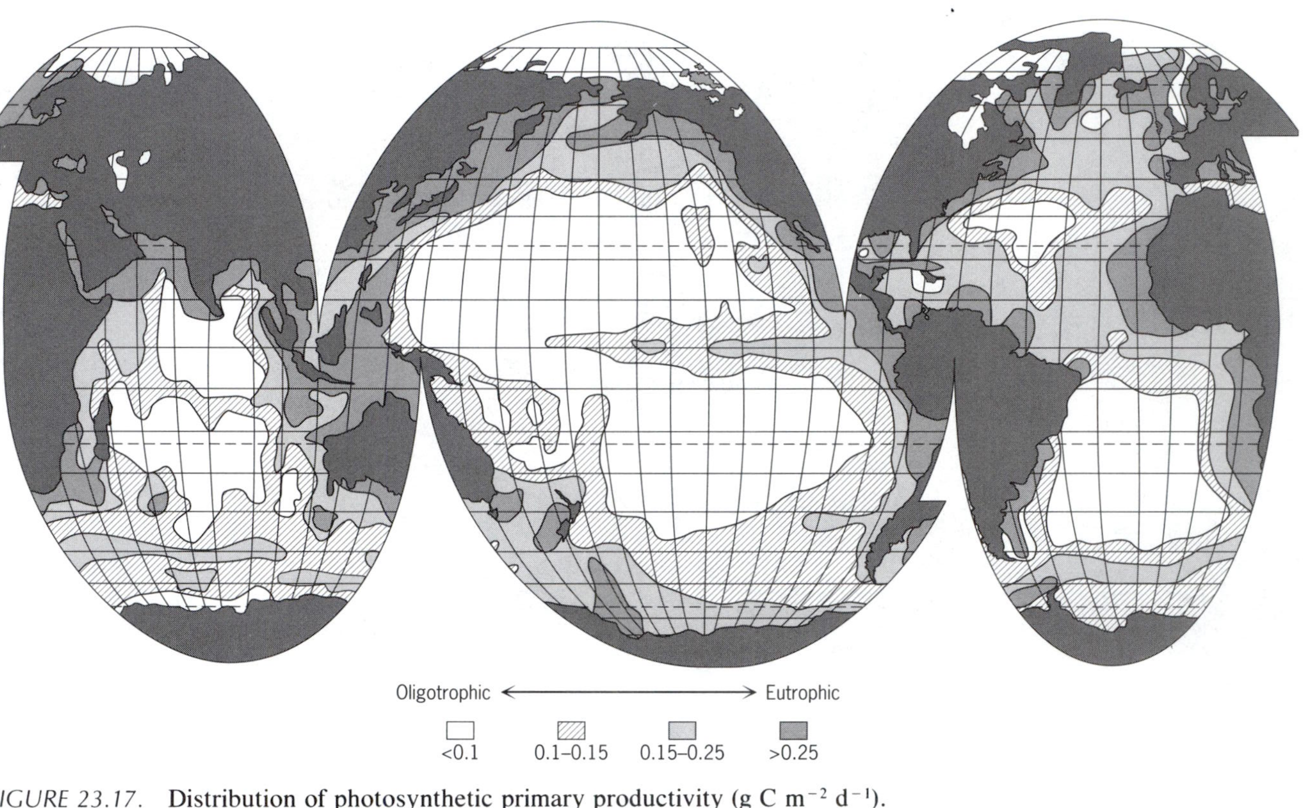

FIGURE 23.17. Distribution of photosynthetic primary productivity (g C m^{-2} d^{-1}). *Source*: From *Introductory Oceanography*, 5th ed., H. V. Thurman, copyright © 1988 by Merrill Publishing Company, Columbus, OH, p. 365. After *Scientific Exploration of the South Pacific*, O. J. Koblentz-Mishke, V. V. Volkovinsky, and J. G. Kabanova, (ed.: W. S. Wooster), copyright © 1970 by the National Academy of Sciences, Washington, DC, p. 185. Reprinted by permission.

As shown in Table 23.6, low-productivity provinces cover most of the surface area of the ocean. Despite low rates, their extensive surface areas cause the open ocean waters to support most of the organic matter production in the sea.

DOM and POM concentrations are highest in coastal upwelling areas and the lowest in the open ocean, particularly in the middle of geostrophic gyres. Because of large seasonal fluctuations in productivity, surface-water concentrations are temporally variable. Seasonal changes in the deep-water flux of large organic particles have also been observed but are relatively small compared to those in the surface waters. Due to the large concentration of refractory compounds, seasonal variations in deep-water DOM concentrations are not observed.

Because of bioturbation, benthic respiration, and slow sedimentation rates, seasonal changes in the POM flux are not well preserved in the sediments. Instead, the organic carbon composition of marine sediments reflects only long-term changes in the rates of primary production in the overlying waters. This record is somewhat altered by changes in oceanographic factors, such as water depth and redox potential, which affect the degree to which sinking POM is preserved prior to its burial in the sediments.

As shown in Figure 23.18, the organic carbon content of oceanic sediments is ≤ 1 percent by mass. In comparison, the organic carbon content in the sediments underlying coastal upwelling areas and estuaries is often greater than 10 percent. The higher organic content of these sediments reflects both higher rates of biological production in the overlying waters and the effect of a relatively short water column. The latter ensures that the sinking POM reaches the seafloor rapidly. The large supply of sinking POM also enhances preservation on the seafloor as it aids in rapid burial and creates suboxic conditions that inhibit the activities of heterotrophic organisms.

TABLE 23.6
Division of the Ocean into Provinces According to Their Level of Primary Organic Production

Province	*%Ocean*	*Area (km²)*	*Mean Productivity (g C/m²/y)*	*Total Productivity (10 tons of C/y)*
Open ocean	90.0	326.0×10^6	50	16.3
Coastal zone	9.9	36.0×10^6	100	3.6
Upwelling areas	0.1	3.6×10^5	300	0.1
Total				20.0

Source: From *The Handbook of Marine Science*, vol. II, F. G. Smith and F. A. Kalber, copyright © 1974 by CRC Press, Boca Raton, FL, p. 12. Reprinted by permission.

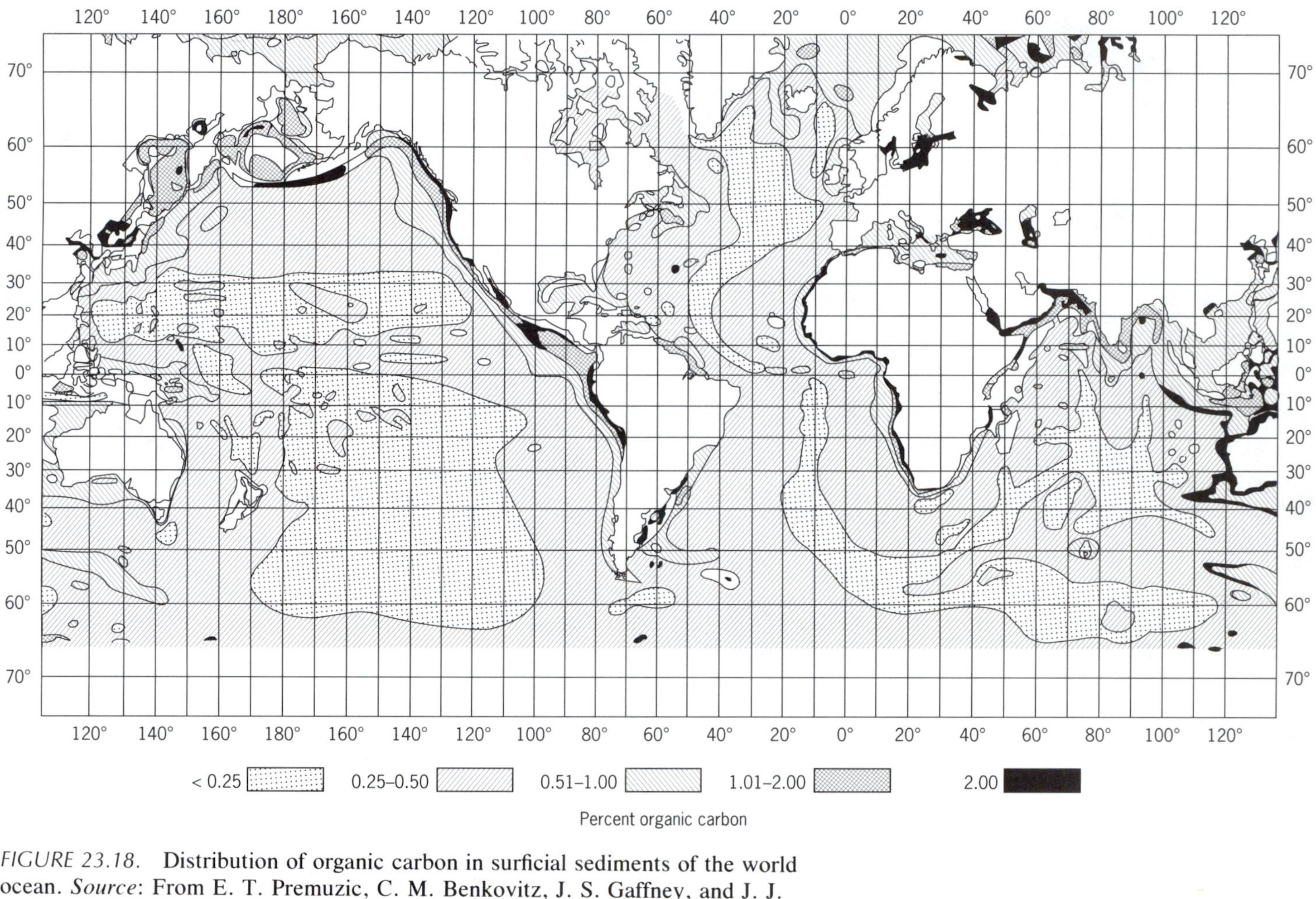

FIGURE 23.18. Distribution of organic carbon in surficial sediments of the world ocean. *Source*: From E. T. Premuzic, C. M. Benkovitz, J. S. Gaffney, and J. J. Walsh, reprinted with permission from *Organic Geochemistry*, vol. 4, p. 65, copyright © 1982 by Pergamon Press, Elmsford, NY.

SUMMARY

The atmospheric and riverine inputs of organic matter to the ocean are relatively small. Thus most marine organic matter is produced in situ as a result of biological activity. Though living organisms account for a small fraction of the POM, they are responsible for all of its production. Phytoplankton form the base of the marine web. The carbon that they fix is passed to higher trophic levels, though much is lost to excretion and exudation. Many marine organisms exude organic compounds in an effort to modify their environment (e.g., to repel or kill predators and competitors). Most POM is detrital and is composed of particles such as fecal pellets, dead tissues, molts, and aggregates, termed **marine snow**.

Because nitrogen-rich compounds are more labile, the C to N ratio of POM increases with depth, though its overall concentration declines. The steepest gradients coincide with the thermocline. The decomposition of the majority of the sinking POM in these depths is partially responsible for the presence of an O_2 minimum in the thermocline. Some POM is consumed and transformed into other compounds, causing mid-depth concentration maxima for select biomolecules.

A small fraction of the POM synthesized in the euphotic zone reaches the seafloor. Most is in the form of large particles, such as fecal pellets, which have fast sinking rates. The POM that reaches the seafloor is subject to extensive degradation at the sediment–water interface. This causes a decrease in its carbon and nitrogen content, as well as an increase in C to N ratio. These same trends are exhibited to a lesser degree after burial as benthic organisms continue to decompose the organic matter. As a result, the organic matter content of the sediments is usually low, ranging from 1 to 10 percent.

DOM represents the second largest pool of carbon in seawater. Most is produced by the degradation of detrital POM as it sinks through the water column. Most of this DOM is low in molecular weight and relatively labile. These compounds are rapidly assimilated by bacteria, so their concentrations are usually less than 100's μg/L. The most abundant of the low-molecular-weight compounds are polypeptides and free amino acids. The relative abundance of the various compounds reflects their susceptibility to microbial attack, which is determined by their molecular structure.

Some of these compounds combine to form complex macromolecules, termed **humic substances**. Their chemical composition is variable and tends to be unreactive. As a result, their molecular structure has not been well characterized. Those that are insoluble in acid are termed **humic acids**, while those that are soluble at all pH's are termed **fulvic acids**. Most humic substances in seawater are fulvic acids.

Humic substances are also produced in terrestrial waters as a result of similar reactions that occur between molecular fragments derived from land plants. As a result, terrestrial humics are structurally distinct from marine forms and exhibit different chemical behavior. For example, a substantial fraction of terrestrial humic acids are removed in estuaries by precipitation reactions caused by increasing ionic strength.

In comparison to POM, DOM concentration profiles exhibit less vertical segregation. This is thought to reflect the presence of a large pool of refractory compounds that are primarily humic substances. This hypothesis may have to be modified in light of recent developments in the method by which DOM is measured. If valid, the results from this technique suggest the presence of a large and heretofore unrecognized pool of high-molecular-weight DOM that is also biologically labile.

The global distributions of POM, DOM, and sedimentary organic matter generally follow that of primary production, with highest concentrations in coastal waters and the lowest in open ocean areas, especially in the middle of geostrophic gyres. These patterns are reflected in the concentrations of organic matter in the underlying sediments. The relationship is not always direct, as other oceanographic factors, such as water depth and redox potential, also influence the degree to which POM is preserved prior to burial in the sediments.

CHAPTER 24

THE MARINE NITROGEN CYCLE

INTRODUCTION

As one of the biolimiting elements, nitrogen plays a central role in controlling biological productivity. Hence parts of the marine biogeochemical cycle of nitrogen are involved in feedback loops that regulate climate, formation of biogenic sediments, and seawater concentrations of some chemicals. Because nitrogen is naturally present in a variety of oxidation states, it tends to undergo redox reactions and as a result has a very complex biogeochemical cycle.

This complexity, compounded by great spatial and temporal variability, has made the marine nitrogen cycle difficult to study. As a result, our knowledge of global fluxes and reservoir sizes is still highly uncertain. The available data, which are presented in this chapter, suggest that either the marine biogeochemical cycle is not in a steady state or that some source of fixed nitrogen has been grossly underestimated.

Human activities have greatly increased some of the global nitrogen fluxes. Humans are presently fixing as much N_2 as the "natural" terrestrial biosphere. The rate at which fixed nitrogen is transported into the ocean has been greatly increased by sewage dumping and by agricultural activities. The latter causes soil erosion and fertilizer runoff. In some locations, this anthropogenic flux of nitrogen exceeds the natural riverine input and has led to the eutrophication of many estuaries. Nitrogen is also lost from the terrestrial biosphere by biomass burning, particularly in the tropical rainforests. This process, along with the burning of fossil fuels, has significantly increased atmospheric levels of nitrogen oxides. These air pollutants help create ozone in the troposphere, but cause its destruction in the stratosphere. In this chapter, the marine biogeochemistry of nitrogen is discussed along with some perturbations presently being caused by some of these anthropogenic inputs.

NITROGEN SPECIES

The marine chemistry of nitrogen is largely controlled by redox reactions that are mediated by phytoplankton and bacteria. As a result, nitrogen is present in seawater and the sediments in many oxidation states. The most abundant naturally occurring species and their oxidation states are given in Table 24.1. The inorganic species, NO_3^-, NO_2^-, and NH_4^+, are commonly referred to as DIN (dissolved inorganic nitrogen).

The oxidation number of nitrogen in all organic compounds is −III. The most abundant of these nitrogenous compounds are humic substances. Lesser amounts of organic nitrogen are also present as amino acids, nucleic acids, amino sugars, and urea, as well as their polymers (e.g., DNA, RNA, and chitin). Most marine organisms also contain alkyl and quaternary amines (Figure 24.1), which in some cases, are present at concentrations close to that of the most abundant amino acids.

The major function of these compounds is thought to be in osmoregulation. Thus they are present almost entirely in solution as part of the intracellular matrix. These amines are probably most important, and therefore abundant, in coastal marine organisms that are subject to fluctuations in salinity. They are released into seawater as DOM and are then degraded to low-molecular-weight species, such as methyl-, dimethyl-, and trimethylamines. Since these products are volatile, they are readily lost from the ocean by degassing across the air–sea interface. Their net global fluxes are as yet unquantified, so the relative importance of this process to the marine nitrogen budget is unknown.

THE GLOBAL BIOGEOCHEMICAL CYCLE OF NITROGEN

The nitrogen flows between the land, sea, atmosphere, and sediments are illustrated in Figure 24.2. Current estimates of the magnitudes of these fluxes are still highly uncertain and are discussed in the next-to-last section of this chapter. The sizes of the major reservoirs are given in Table 24.2. There is also some uncertainty associated with these data. For example, because of the effects of temporal variability, current estimates of the size of the atmospheric ammonia reservoir still vary by an order of magnitude.

TABLE 24.1
Common Species of Marine Nitrogen

Species	*Molecular Formula*	*Oxidation Number of Nitrogen*
Nitrate ion	NO_3^-	+V
Nitrite ion	NO_2^-	+III
Nitrous oxide gas	N_2O	+I
Nitric oxide gas	NO	+II
Nigrogen gas	N_2	0
Ammonia gas	NH_3	−III
Ammonium ion	NH_4^+	−III
Organic amine	RNH_2	−III

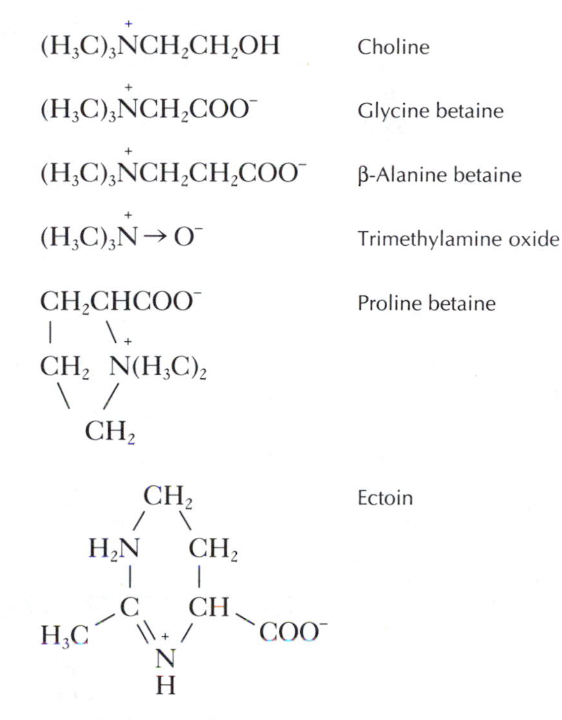

FIGURE 24.1. Structure of quarternary amines and related compounds of ecological significance.

As indicated in Table 24.2, most nitrogen is in the form of N_2 and thus is largely unutilizable because only nitrogen-fixing organisms can break its strong triple bond. Most of the marine fixed nitrogen is dissolved and is primarily in the form of deep-water nitrate and humic acids. The marine biota contain less than 0.002 percent of the marine nitrogen, which is equally distributed among plant and microbial (bacterial and protozoan) biomass. Although the land biota contains a larger fraction of the terrestrial nitrogen (2.6 percent), the vast majority is present in the form of plant biomass.

NITROGEN ASSIMILATION BY MARINE PHYTOPLANKTON

In the open ocean, primary production appears to be limited by nitrogen availability. A few marine photoautotrophs can fix nitrogen, but they appear to be restricted to anaerobic benthic environments and hence are found only in estuaries. Oceanic phytoplankton are not N_2 fixers and so must meet their nitrogen needs by absorbing, or **assimilating**, dissolved species such as nitrate, nitrite, ammonium, and urea.

After transport across the cell wall, the nitrogen is transformed into metabolites, such as proteins, by a series of anabolic reactions, as illustrated

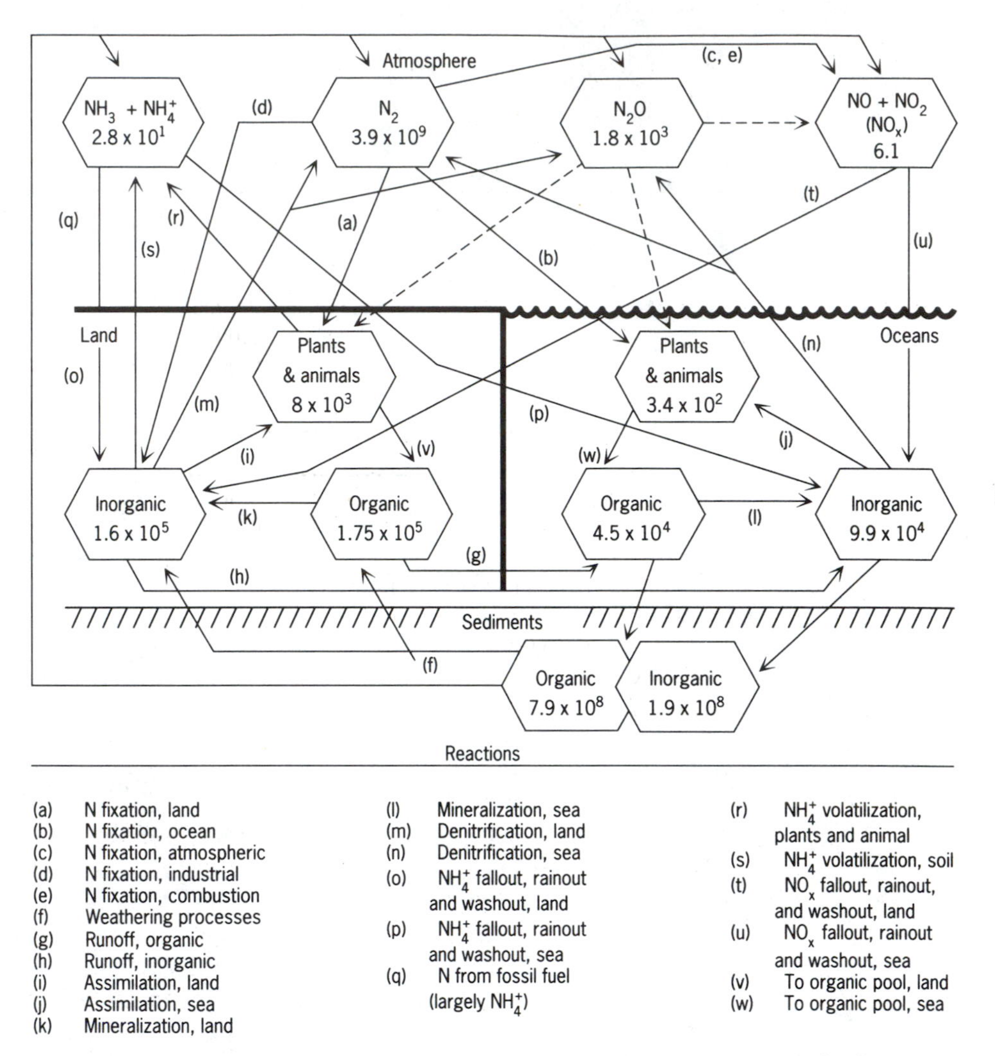

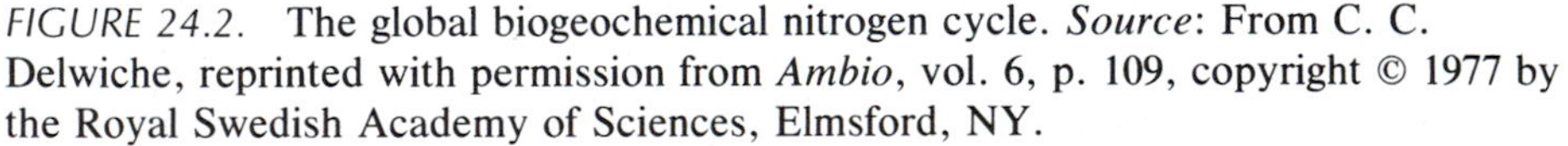

FIGURE 24.2. The global biogeochemical nitrogen cycle. *Source*: From C. C. Delwiche, reprinted with permission from *Ambio*, vol. 6, p. 109, copyright © 1977 by the Royal Swedish Academy of Sciences, Elmsford, NY.

in Eqs. 24.1 through 24.5. If nitrite is assimilated, the reaction sequence starts with Eq. 24.2; and if ammonium is assimilated, anabolism is initiated by the reaction given in Eq. 24.5.

$$NO_3^- + 2H^+ + 2e^- \rightarrow NO_2^- + H_2O \qquad (24.1)$$

$$2NO_2^- + 4H^+ + 4e^- \rightarrow N_2O_2^{2-} + 2H_2O \qquad (24.2)$$

$$N_2O_2^{2-} + 6H^+ + 4e^- \rightarrow 2NH_2OH \qquad (24.3)$$

$$NH_2OH + 2H^+ + 2e^- \rightarrow NH_3 + H_2O \qquad (24.4)$$

TABLE 24.2
Major Nitrogen Reservoirs (in units of 10^{15} g N)

Atmosphere		*% of total*
Dimolecular nitrogen	3,900,000	>99.999
Nitrous oxide	1.4	<0.0001
Ammonia	0.0017	<0.0001
Ammonium	0.00004	<0.0001
Nitric oxide + Nitrogen dioxide (NO_x)	0.0006	<0.0001
Nitrate	0.0001	<0.0001
Organic nitrogen	0.001	<0.0001
	Total	100
Ocean		
Plant biomass	0.30	0.001
Animal biomass	0.17	0.0007
Microbial biomass	0.02	0.00006
Dead organic matter (dissolved)	530	2.3
Dead organic matter (particulate)	3–240	0.01–0.1
Dimolecular nitrogen (dissolved)	22,000	95.2
Nitrous oxide	0.2	0.009
Nitrate	570	2.5
Nitrite	0.5	0.002
Ammonium	7	0.03
	Total	100
Pedosphere including biota		
Plant biomass	11–14	2.6
Animal biomass	0.2	0.04
Microbial biomass	0.5	0.1
Litter	1.9–3.3	0.5
Soil: Organic matter	300	63
Inorganic	160	34
	Total	100
Lithosphere		
Rocks	190,000,000	99.8
Sediments	400,000	0.2
Coal deposits	120	0.00006
	Total	100

Source: From *The Major Biogeochemical Cycles and Their Interactions*, T. Rosswall (ed.: B. Bolin and R. B. Cook), copyright © 1983 by John Wiley & Sons, Inc., New York, p. 48. Reprinted by permission. See Rosswall (1983) for data sources.

$$\underset{(\alpha\text{-ketoglutaric acid})}{HOOCCO(CH_2)_2COOH} + NH_3 + 2NADPH \rightarrow \underset{(\text{glutamic acid})}{HOOCCH(NH_2)CH_2CH_2COOH} + 2NADP + H_2O \quad (24.5)$$

$$\underset{(\text{pyruvic acid})}{CH_3COCOOH} + \underset{(\text{glutamic acid})}{HOOCCH(NH_2)CH_2CH_2COOH} \rightarrow \underset{(\text{alanine})}{CH_3CH(NH_2)COOH} + \underset{(\alpha\text{-ketoglutaric acid})}{HOOCCO(CH_2)_2COOH} \quad (24.6)$$

If nitrate or nitrite is assimilated, the nitrogen must be reduced to the −III oxidation state. Once in reduced form (NH_3), the nitrogen reacts with a carboxylic acid, which is α-ketoglutaric acid in the example given in Eq. 24.5. The product is an amino acid, which in this example is glutamic acid. As shown in Eq. 24.6, other amino acids are generated via the transfer of the amine group from the primary amino acid to the central carbon of another carboxylic acid. This chemical reaction is referred to as transamination. The overall process whereby nitrate or nitrite is reduced and incorporated into organic matter is termed assimilatory nitrogen reduction.

The nitrogen in nitrite is in a lower oxidation state than in nitrate and so requires less energy to be converted to organic form. Likewise, even less energy need be expended if phytoplankton assimilate ammonium or urea. To take advantage of these energy differences, phytoplankton have evolved transport mechanisms that favor the uptake of the reduced species. Thus phytoplankton fed a mixture of dissolved urea, ammonium, nitrite, and nitrate will assimilate the reduced species fastest. Urea uptake is significant in coastal regions due to its relatively large production rate in these waters.

LATITUDINAL VARIATIONS IN NITROGEN LIMITATION

The degree to which phytoplankton growth is limited by nitrogen availability is temporally and spatially variable. This is the result of seasonal shifts in nitrogen availability and variations in other growth-limiting factors such as light, temperature, micronutrients, and grazing pressure. The interaction of all these factors controls productivity at temperate latitudes, as illustrated in Figure 24.3.

At these latitudes, phytoplankton growth is light-limited during the winter. Heterotrophic activity continues, so organic detritus is remineralized, causing DIN concentrations to rise. As the intensity of insolation increases with the approach of spring, the phytoplankton growth rate increases. Due to the large size of the DIN pool, a phytoplankton bloom occurs. The increase in phytoplankton numbers stimulates an increase in the population of plant consumers (i.e., protozoans and zooplankton). As shown in Figure 24.4, the increased grazing pressure causes phytoplankton numbers to peak in mid spring.

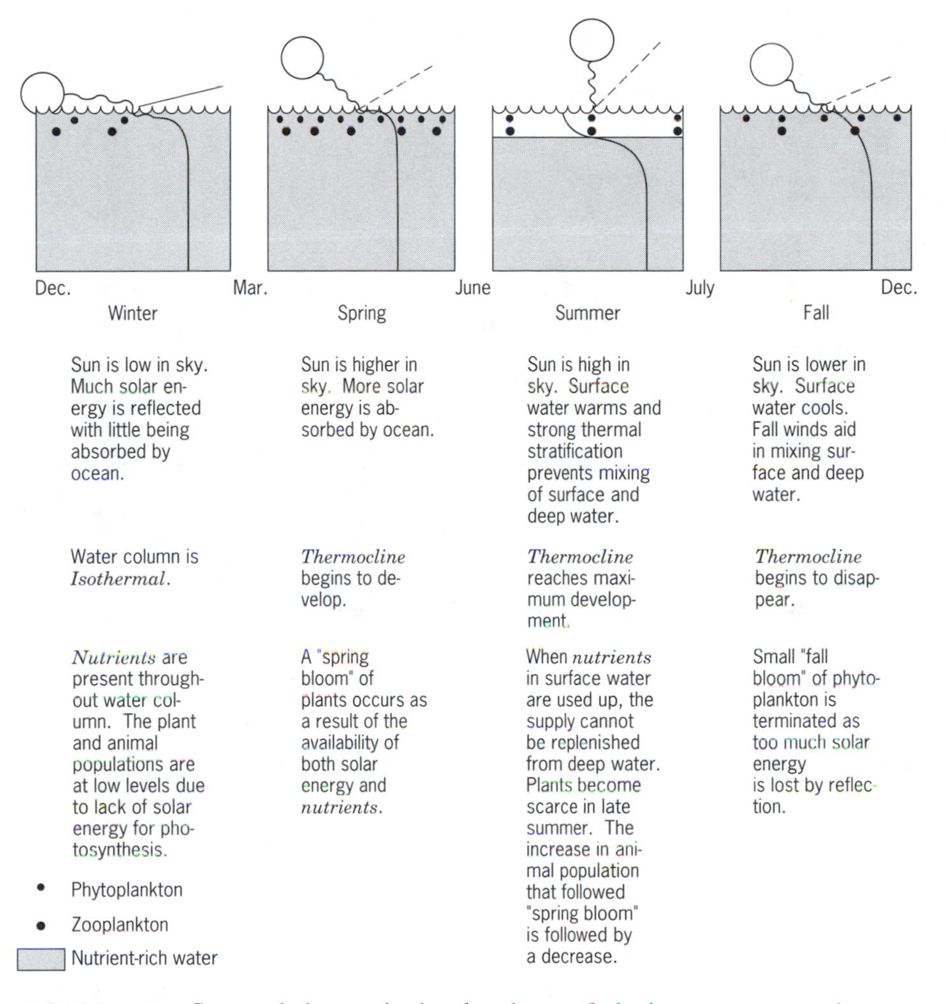

Dec.
Mar.
June
July
Dec.
Winter
Spring
Summer
Fall
Sun is low in sky. Much solar energy is reflected with little being absorbed by ocean.
Sun is higher in sky. More solar energy is absorbed by ocean.
Sun is high in sky. Surface water warms and strong thermal stratification prevents mixing of surface and deep water.
Sun is lower in sky. Surface water cools. Fall winds aid in mixing surface and deep water.
Water column is Isothermal.
Thermocline begins to develop.
Thermocline reaches maximum development.
Thermocline begins to disappear.
Nutrients are present throughout water column. The plant and animal populations are at low levels due to lack of solar energy for photosynthesis.
A "spring bloom" of plants occurs as a result of the availability of both solar energy and nutrients.
When nutrients in surface water are used up, the supply cannot be replenished from deep water. Plants become scarce in late summer. The increase in animal population that followed "spring bloom" is followed by a decrease.
Small "fall bloom" of phytoplankton is terminated as too much solar energy is lost by reflection.
Phytoplankton
Zooplankton
Nutrient-rich water

FIGURE 24.3. Seasonal changes in the abundance of plankton, grazers, and nutrients in temperate oceans. *Source*: From *Introductory Oceanography*, 5th ed., H. V. Thurman, copyright © 1988 by Merrill Publishing Company, Columbus, OH, p. 364. Reprinted by permission.

The consumers excrete DIN, which enables the plants to continue to grow rapidly despite intense grazing pressure. Thus nitrogen is rapidly recycled between the consumers and producers. This recycling is not wholly efficient. A significant amount is lost from the euphotic zone by the sinking of detrital PON. Eventually, this leak strips the surface waters of nitrogen. In the absence of DIN, plant growth rates and hence population numbers decline. This causes summer to be a period of relatively low productivity. The detrital PON that has sunk out of the euphotic zone is remineralized in the thermocline and deep zone. The return of these regenerated nutrients to the euphotic zone is inhibited by strong density stratification produced by warm summer temperatures. As

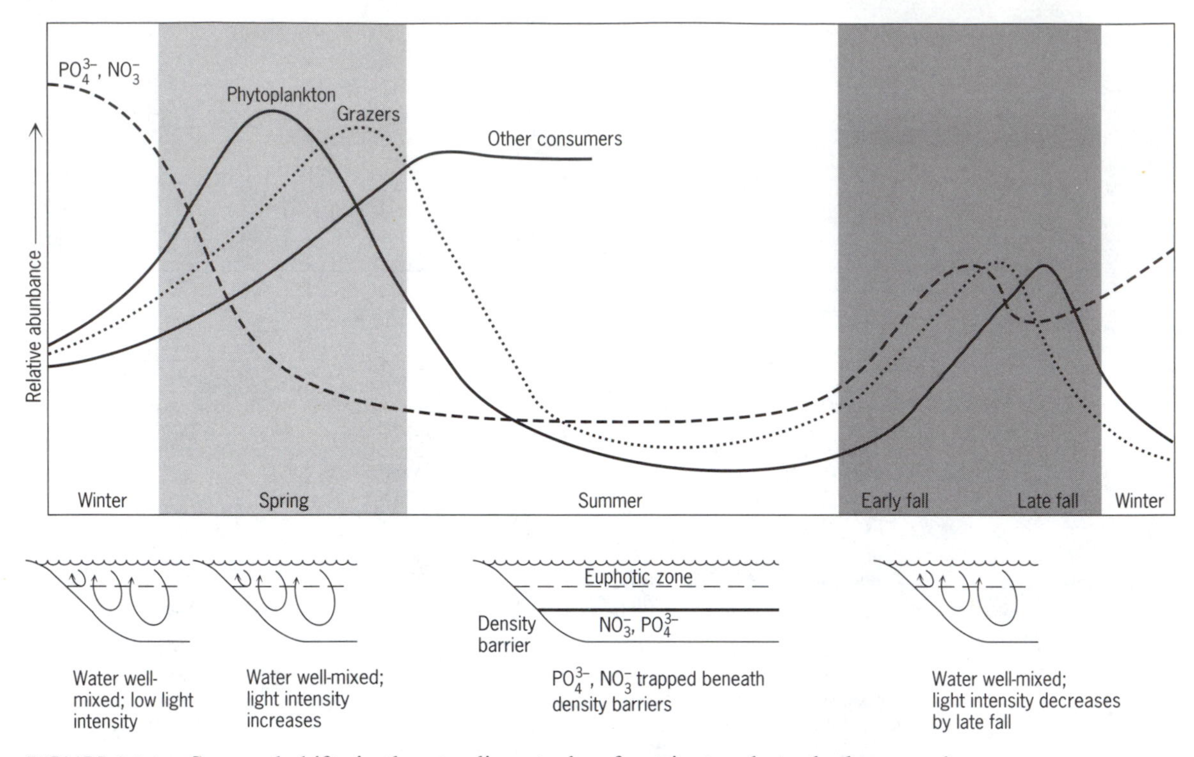

FIGURE 24.4. Seasonal shifts in the standing stocks of nutrients, phytoplankton, and the heterotrophic consumer community of bacteria, protozoa, and zooplankton. Also shown are seasonal changes in density stratification of the mixed layer. *Source*: From *Oceans and Coasts*, J. A. Black, copyright © 1986 by Wm. C. Brown Publishers, Dubuque, IA, p. 143. Reprinted by permission.

shown in Figure 4.4, this warming establishes a summer thermocline that inhibits vertical mixing, thereby greatly reducing eddy diffusion in the mixed layer.

In the fall, declining atmospheric temperatures and early winter storms destroy the summer thermocline. This causes some of the regenerated nitrogen to be mixed back up into the euphotic zone. Sunlight is still plentiful, so the injection of nitrogen fuels a second, though smaller, plankton bloom. As winter approaches, light levels decline, causing a decrease in plant growth rates. The resulting detrital PON is remineralized during the winter by protozoa and bacteria. Note that microbial regeneration of nitrogen occurs throughout the year, but only during the winter does its rate exceed that of phytoplankton assimilation.

As shown in Figure 24.5*b*, only one phytoplankton bloom occurs at subpolar latitudes, but is larger in amplitude than at mid latitudes. Plant growth in the subpolar region is very prolific because uniformly cold atmospheric temperatures suppress density stratification of the water column. Abundant winds ensure that the water column is well mixed and hence prevent vertical

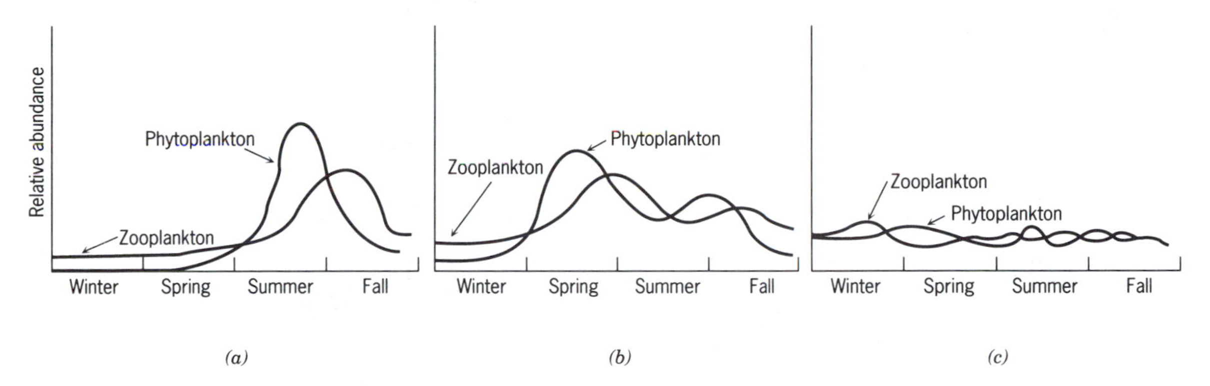

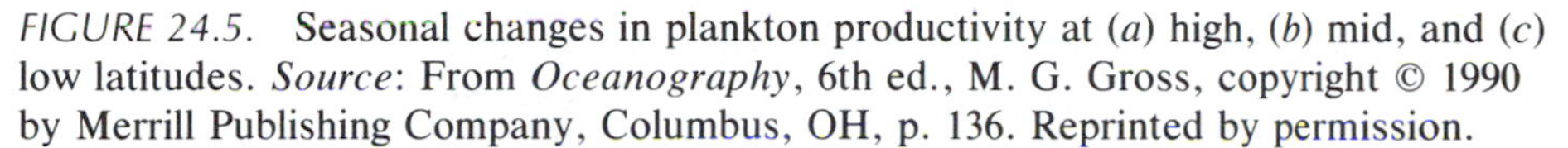

FIGURE 24.5. Seasonal changes in plankton productivity at (*a*) high, (*b*) mid, and (*c*) low latitudes. *Source*: From *Oceanography*, 6th ed., M. G. Gross, copyright © 1990 by Merrill Publishing Company, Columbus, OH, p. 136. Reprinted by permission.

segregation of DIN. Thus plant growth is never nutrient-limited. When insolation increases in early spring, a large bloom occurs and is sustained until light levels diminish at the end of the summer. Despite their limited growth period, the subpolar plankton are responsible for a significant amount of the global marine primary production.

Tropical and equatorial waters are characterized by uniformly low productivity (Figure 24.5*c*) because of permanent nutrient limitation. This condition is the result of a strongly stratified water column produced by high and constant atmospheric temperatures. As a result, the nutrients transported out of the euphotic zone by the sinking of biogenic detritus are not returned via eddy diffusion or convective overturn. Because the surface waters have very low DIN concentrations, the energy-expensive process of nitrogen fixation is ecologically favored at these latitudes.

HETEROTROPHIC TRANSFORMATIONS OF NITROGEN

The PON synthesized by phytoplankton has two possible fates. Either the plants die, their cells lyse, and the resulting DON is degraded by bacteria, or the cells are consumed by protozoa or zooplankton. The excretions and exudations of these consumers also supplies DON, as does lysis of their cells following death. Bacteria oxidize DON through a series of reactions, which are shown in the right-hand side of Figure 24.6. The oxidized products (DIN) can be reduced by phytoplankton as a consequence of nutrient assimilation. Alternatively, DIN can be reduced by heterotrophic bacteria that use DIN as an electron acceptor. This process is referred to as **dissimilatory nitrogen reduction** because some of the reduced nitrogen is released to seawater instead of being incorporated into bacterial biomass. These reduction processes are shown on the left-hand side of Figure 24.6 along with nitrogen fixation.

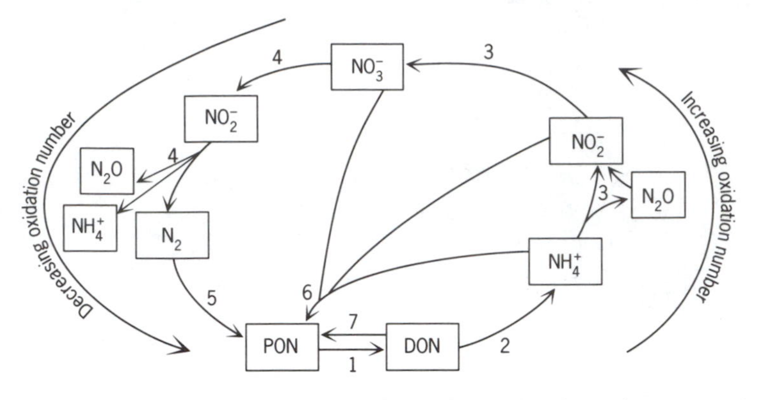

FIGURE 24.6. Schematic view of the biogeochemical nitrogen cycle: 1) remineralization, 2) ammonification, 3) nitrification, 4) denitrification (dissimilatory nitrate reduction), 5) nitrogen fixation, 6) assimilatory nitrogen reduction, and 7) assimilation of DON.

Oxidation Reactions

Remineralization and Ammonification

The term remineralization is strictly used to refer to the initial stage of PON decomposition during which solid nitrogen is converted to DON. The DON is then degraded by heterotrophic bacteria. This degradation is rapid due to the relative reactivity of the carbon-nitrogen bonds. Complete breakage of these bonds releases ammonia (NH_3), which then reacts with H^+ or H_2O to form ammonium (NH_4^+). This process is called **ammonification** and is depicted in Figure 24.7 on page 433 for the most abundant nitrogen-containing biomolecules, the proteins.

Nitrification

In oxic seawater, ammonium is readily oxidized to nitrite and then to nitrate by the marine bacteria *Nitrosomonas* and *Nitrobacter*, respectively. This process is termed **nitrification** and occurs in a stepwise fashion. This is illustrated in Figure 24.8 for the aerobic decomposition of detrital PON in a fixed volume of seawater in the dark. Initially, the degradation of PON produces ammonium, which stimulates the growth of *Nitrosomonas*. These bacteria oxidize the ammonium to nitrite, thereby causing ammonium concentrations to decline and nitrite concentrations to rise. The elevated nitrite levels stimulate the growth of *Nitrobacter*. These bacteria oxidize the nitrite to nitrate. Eventually all of the DIN is oxidized to nitrate. The remaining PON is composed of compounds that are so inert they cannot be degraded by aerobic marine bacteria.

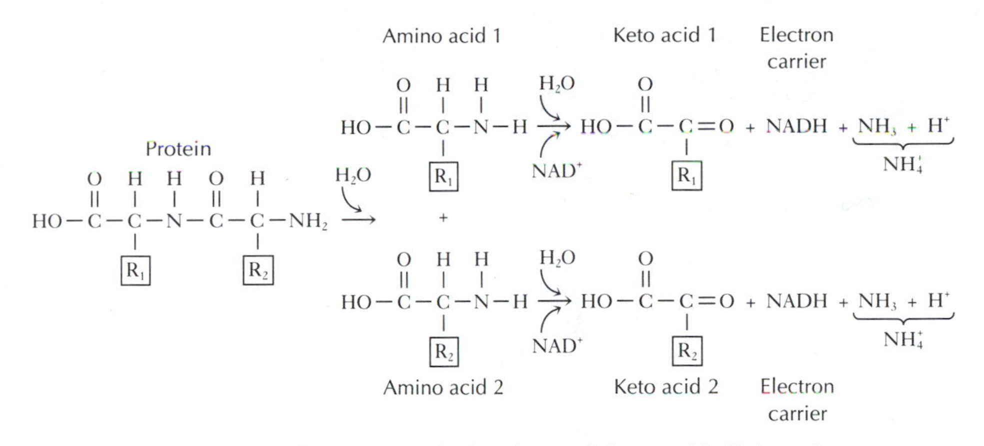

FIGURE 24.7. Ammonification. (1) The breakage of the peptide linkage between amino acids A and B. (2) the deamination of the amino acids and formation of NH_4^+. The type of electron carrier (e.g., NAD^+, $NADP^+$, etc.) involved in such reactions depends on the type of protein undergoing degradation and the species of organism mediating the reaction. These reactions are usually catalyzed by enzymes.

Reduction Reactions

Denitrification

In seawater that is undersaturated with respect to O_2, some species of heterotrophic bacteria respire organic matter by using nitrate as their electron acceptor. Some of the nitrate is reduced sequentially to nitrite and then to N_2 and hence is not incorporated into bacterial biomass. Since fixed nitrogen is lost by this process, it is termed **denitrification**. In addition to suboxic or anoxic conditions, large amounts of organic matter are required to fuel this process.

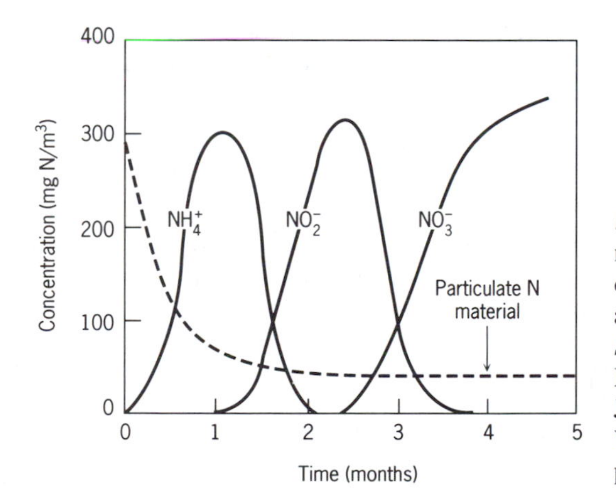

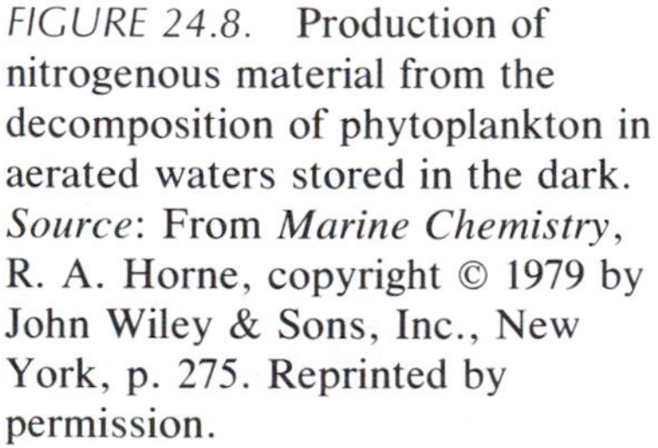
FIGURE 24.8. Production of nitrogenous material from the decomposition of phytoplankton in aerated waters stored in the dark. *Source*: From *Marine Chemistry*, R. A. Horne, copyright © 1979 by John Wiley & Sons, Inc., New York, p. 275. Reprinted by permission.

These conditions exist in coastal upwelling zones, such as the Peru upwelling area, as well as in regions of relative water stagnancy, such as the East Tropical North Pacific (Figure 8.3). Denitrification also occurs in coastal sediments and polluted estuarine waters due to the presence of very large amounts of organic matter.

Depth profiles from the East Tropical North Pacific (Figure 24.9) show the effects of nitrogen metabolism under O_2-deficient conditions. A sharp decline in O_2 concentrations coincides with the thermocline. This oxycline is produced by the respiration of organic matter under relatively stagnant conditions. As usual, the thermocline is also marked by sharp increases in nitrate and phosphate concentrations with depth. Below the oxycline, phosphate concentrations continue to increase, but at a slower rate. In contrast, nitrate concentrations decline and reach a mid-water minimum that coincides with a nitrite maximum. Both are the result of denitrification that occurs at these depths due to low O_2 concentrations and an abundant supply of organic matter. The size of the nitrite maximum is dependent on the relative rates of its production from NO_3^- and its loss from reduction to N_2.

As indicated in Eq. 8.7, ammonium is also released during dissimilatory nitrogen reduction. Since O_2 is limiting, this ammonium is not nitrified. Hence its concentrations are high in the denitrification zone. Some ammonium is also thought to be produced via the microbial reduction of nitrite, but the quantitative importance of this process is unknown.

Denitrification in the presence of small amounts of O_2 causes some of the nitrite to be reduced to N_2O. As suggested by its concentration maximum in the oxic zone, N_2O is also a minor by-product of nitrification (Figure 24.9*d*). As a result of production from nitrification and denitrification, the ocean as a whole is supersaturated with respect to N_2O. Thus there should be a net global flux of this gas out of the ocean across the air–sea interface. This flux has the potential to affect the marine nitrogen budget greatly, but its quantitative importance is unknown.

An ammonium maximum (Figure 24.9*c*) is present at the top of the oxycline and is the result of relatively rapid ammonification rates. Oxidation to nitrite is limited by low O_2 concentrations present at these depths. In comparison, the oxidation of nitrite is inhibited by high O_2 concentrations and light intensity. This causes the small nitrite maximum in the oxic zone. In areas where denitrification is occurring, eddy diffusion transports deep-water nitrite into the mixed layer. Because large amounts of O_2 are consumed during nitrification, the amount of nitrite that escapes from the denitrifying zone is important in determining the intensity of the mid-water O_2 minimum. The depths over which these different metabolic processes occur are summarized in Figure 24.10.

Nitrogen Fixation

Nitrogen is introduced into the ocean by river runoff, rainfall, diffusion from the sediments, and in situ **N_2 fixation**. During the latter, the triple bond in N_2 must be broken. The freed atoms are incorporated into reduced compounds,

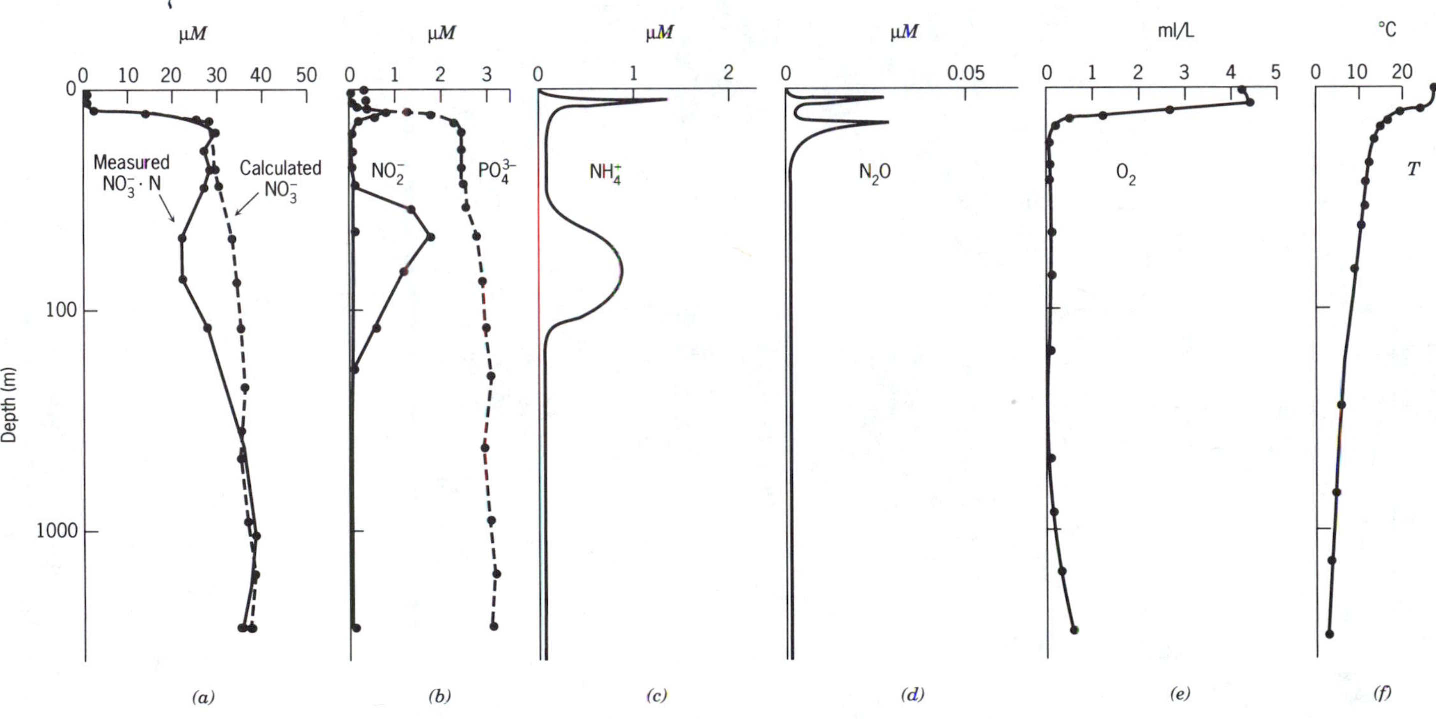

FIGURE 24.9. Vertical concentration profiles of (*a*) nitrate, (*b*) nitrite and phosphate, (*c*) ammonium[a], (*d*) N_2O[a], (*e*) O_2 and (*f*) temperature in the eastern tropical North Pacific. The calculated values for nitrate concentrations were determined by multiplying the Redfield–Richards Ratio of N/P (16) by the in situ phosphate concentrations. *Source*: After W. H. Thomas, reprinted with permission from *Deep-Sea Research*, vol. 13, p. 1110, copyright © 1966 by Pergamon Press, Elmsford, NY.
[a] This is an idealized profile showing concentrations that would be expected to occur at this site.

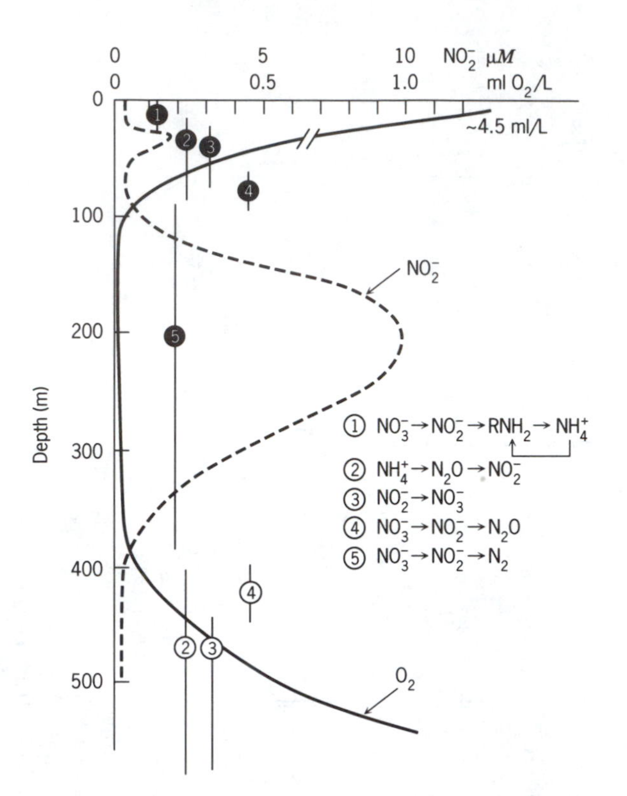

FIGURE 24.10. Nitrite and O_2 concentrations in a water column subject to nitrification in the surface waters and denitrification in the deep zone. Also shown are the depth distributions to the processes responsible for creating the nitrite profile. The white numbers are used to suggest that these reactions occur at different rates than those identified by the black numbers. (1) Rapid nitrogen assimilation and recycling in the photic zone. (2) Ammonium release and oxidation in water sufficiently well oxygenated to allow nitrite to be the major end product. (3) Nitrite oxidation is inhibited at high light and O_2 levels. (4) Nitrate reduction and denitrification in the presence of small amounts of O_2 favors the accumulation of some N_2O. (5) Nitrate reduction and denitrification at very low O_2 concentrations. *Source*: From L. A. Codispoti and J. P. Christensen, reprinted with permission from *Marine Chemistry*, vol. 16, p. 288, copyright © 1985 by Elsevier Science Publishers, B.V., Amsterdam, The Netherlands.

which are usually organic in nature. Since this is a very energy-expensive reaction, relatively few organisms can "fix" nitrogen. These organisms are either heterotrophic or phototrophic bacteria and are restricted to the few species listed in Table 24.3.

The only species which have free-living forms are the *Oscillatoria* ssp., which as photosynthetic cyanobacteria, are also considered to be a type of blue–green algae. These organisms are so prolific in some tropical coastal

TABLE 24.3
Nitrogen-Fixing Bacteria Isolated from Marine Environments:

Heterotrophs

Relation to O_2 and Genus	*Habitat*
1. Aerobes	
Azotobacter spp.	Black Sea sediments
	Seagrass sediments
	Macroalga *Codium*
	Intertidal sediments
	Estuarine sediments
	Saltmarsh sediments
2. Microaerophiles	
Azospirillum spp.	*Spartina* roots
	Zostera roots
	Seawater
Campylobacter spp.	*Spartina* roots
Beggiatoa spp.	Sediment surface
(?)	Shipworm
3. Facultative anaerobes	Beach sediment
Enterobacter spp.	Intertidal sediments
Klebsiella spp.	Estuarine sediments
	Halodule roots
	Beach sediment
	Mangrove bark
	Sea urchin
Vibrio spp.	Seawater
	Mangrove bark
4. Strict anaerobes	
Desulfovibrio spp.	Seagrass sediments
	Intertidal sediments
	Saltmarsh sediments
	Estuarine sediments
	Seagrass sediments
Clostridium spp.	Intertidal sediments
	Saltmarsh sediments
	Estuarine sediments

Phototrophs

Group and Genus	*Habitat*
1. Cyanobacteria	
A. Chroococcacean	
Synechococcus sp.	Snail shell, intertidal
	Tropical sediments
Gloeocapsa sp.	Microbial mat
B. Pleurocapsalean	
Dermocarpa sp.	Marine aquarium
Xenococcus sp.	Marine aquarium
Myxosarcina sp.	Snail shell, intertidal
Pleurocapsa sp.	Rock chip
C. Oscillatorian	
LPP[a] group	Snail shell, intertidal
Oscillatoria sp.	Intertidal mat
Microcoleus sp.	Marsh mat
Phormidium sp.	Microbial mat
D. Nostacalean	
Anabaena sp.	Algal mat
Calothrix sp.	Microbial mat
Nodularia sp.	Microbial mat
Nostoc sp.	Microbial mat
2. Prochlorales	
Prochloron[b]	Ascidians
3. Anoxyphotobacteria	
A. Rhodospirillaceae	
Rhodopseudomonas sp.	Harbour muds
B. Chromatiaceae	
Thiocapsa sp.	Anaerobic habitat

Source: From *Nitrogen Cycling Coastal Marine Environments*, D. G. Capone (eds.: T. H. Blackburn and J. Sørensen), copyright © 1988 by John Wiley & Sons, Inc., New York, pp. 88–89. Reprinted by permission. See Capone (1988) for data sources.

[a]Lyngbya/Plectonema/Phormidium group.

[b]Presumptive fixer, only active in association.

waters that they often form blooms. Nitrogen fixation is favored at such latitudes due to the presence of extreme nitrogen limitation. Because *Oscillatoria* ssp. are widespread and have high populations in tropical waters, they are commonly regarded as the most important oceanic nitrogen-fixing organisms. An estimate of their seasonal and annual contributions to the global marine nitrogen budget is given in Table 24.4.

The other nitrogen fixers are either members of the benthos and/or are symbionts. Most are heterotrophs, so they require large amounts of labile organic matter and thus are restricted to saltmarsh estuaries, mangrove swamps, and coral reefs. For example, symbiotic nitrogen fixers are common to the roots of *Spartina*, an abundant saltmarsh grass. The nitrogen fixed by the symbionts is excreted into the pore waters and rapidly assimilated by the roots of the host. Epiphytic nitrogen fixers have been observed on pelagic algae, such as Sargasso seaweed. Since this macroalgae lives in a region (the Sargasso Sea) that is also characterized by extreme nitrogen limitation, its growth is thought to be almost entirely supported by nitrogen supplied by its symbiotic N_2 fixers.

As shown in Table 24.5, most nitrogen fixation appears to occur in shallow benthic environments, particularly in saltmarsh sediments. In the open ocean, N_2 fixation below the euphotic zone is thought to occur in anoxic microzones, such as in sinking detrital PON. Due to lack of data, no quantitative assessment of this process has been made, but it is thought to supply an insignificant amount of fixed nitrogen to the ocean. The large standard errors in Table 24.5 indicate the great uncertainty associated with some of these data. As discussed below, global budget considerations suggest at least some of these rates have been grossly underestimated.

TABLE 24.4
Seasonal Estimates of Nitrogen Fixation by *Oscillatoria* ssp.

	Nitrogen Fixation (10^9g N/y)				
Oceanic Basin	*Spring*	*Summer*	*Autumn*	*Winter*	*Total*
Pacific	11	163	162	0.9	337
Atlantic	101	474	133	614	1322
Indian	1890	0.5	267	966	3124
South China and Arafuru seas	7	0.9	0.1	10.2	18
Total	2009	638	562	1591	4801

Source: From D. G. Capone and E. J. Carpenter, reprinted with permission from *Science*, vol. 217, p. 1141, copyright © 1982 by the American Association for the Advancement of Science, Washington, DC. See Capone and Carpenter (1982) for data sources.

TABLE 24.5
Estimates of the Total Annual Contribution of Combined Nitrogen to the Global Nitrogen Cycle by Nitrogen Fixation in Benthic Marine Environments[a]

Environment	*Area* ($km^2 \times 10^6$)	*N_2 Fixation* ($g\ m^{-2}y^{-1}$)	*N_2 Fixation* (Tg/y)
Depth			
>3000 m	272	0	0
2000–3000 m	31	0.0007	0.022
1000–2000 m	16	0.001	0.016
200–1000 m	16	0.01	0.16
0–200 m	27	0.1 ± 0.04	2.7
Bare estuary	1.08	0.4 ± 0.07	0.43
Sea grass	0.28	5.5	1.5
Coral reefs	0.11	25 ± 8.4	2.8
Salt marsh	0.26	24 ± 10.5	6.3
Mangroves	0.13	11	1.5
Total	363		15.4

Source: From D. G. Capone and E. J. Carpenter, reprinted with permission from *Science*, vol. 217, p. 1141, copyright © 1982 by the American Association for the Advancement of Science, Washington, DC. See Capone and Carpenter (1982) for data sources.
[a]Values for nitrogen fixation are annual averages ± the standard error.

GEOGRAPHIC DISTRIBUTION OF FIXED NITROGEN

The rate at which fixed nitrogen is supplied to the ocean is highest in coastal waters due to river runoff and benthic N_2 fixation. This supports high rates of primary production and a large flux of POM to the sediments. As a result, the most nitrogen-rich sediments are deposited on the continental margins, particularly in coastal upwelling areas (Figure 24.11). Oceanic upwelling does not produce similar enrichments as most of the PON is degraded before it reaches the seafloor.

High supply rates of nitrogen cause coastal zones to be characterized by short food webs. In these locations, nitrogen is rapidly recycled in seawater and the sediments. As shown in Table 24.6, the nitrogen supplied by benthic remineralization constitutes a significant fraction of the amount required to support local primary production.

Benthic nitrification is such a large O_2 sink that this process regulates the vertical position of the redox boundary in the sediments. The resulting suboxic

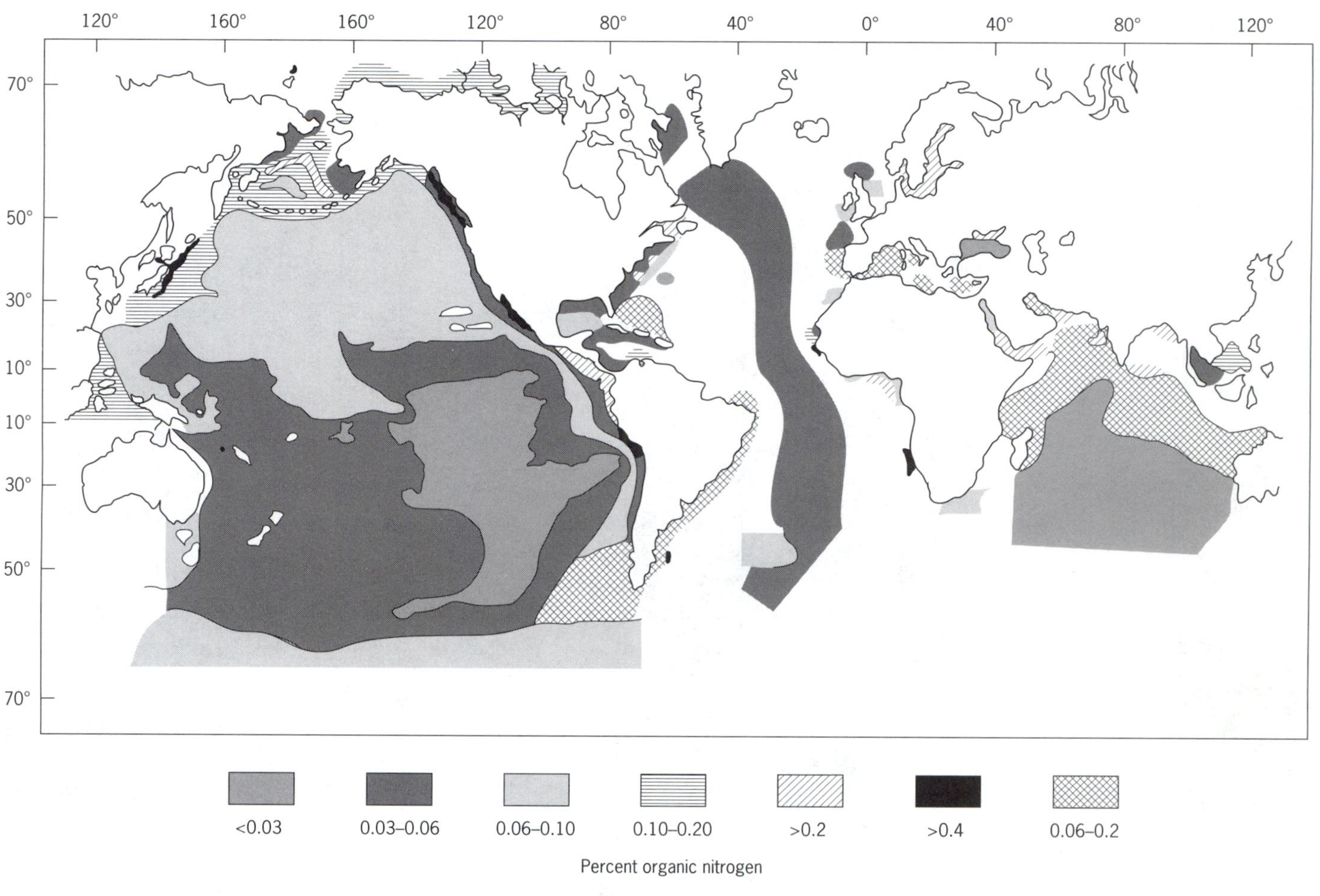

FIGURE 24.11. Distribution of organic nitrogen in the surficial sediments of the world ocean. *Source*: From E. T. Premuzic, C. M. Benkovitz, J. S. Gaffney, and J. J. Walsh, reprinted with permission from *Organic Geochemistry*, vol. 4, p. 66, copyright © 1982 by Pergamon Press, Elmsford, NY.

TABLE 24.6
Benthic Fluxes of Nitrogen Species and Their Relative Contribution to Local Primary Productivity

Environments	*N flux across sediment–water interface (mg N m⁻²d⁻¹)*			*Denitrification (mg N m⁻²d⁻¹)*		*N requirements of primary production (g N m⁻²y⁻¹)*	
	NH_4^+	NO_3^-	*Total*		*Percentage total N mineralization*		*Percentage supplied by benthic mineralization*
La Jolla Bight (CA, USA)	12	2	14	—	—	77	7
Loch Turnaig (UK)	13	0	13	—	—	29	17
Buzzards Bay (MA, USA)	18	1	19	—	—	10	70
Cap Blanc (Africa)	78	58	136	—	—	128	39
Belgian coastal zone	25	32	57	17	23	27	78
Southern Bight of the North Sea	13	17	30	5	15	28	38
Narraganset Bay (RI, USA)	34	4	38	33	16	—	—
South River (NC, USA)	38	0.5	38.5	—	—	48	29
Neuse River estuary (NC, USA)	76	1	77	—	—	107	26
Georgia Bight (USA)	55	3	59	—	—	132	16
Great Belt (DK)	10	7	17	49	16	17	36
W. Kattegat (DK)	13	5	18	69	22	15	42
E. Kattegat (DK)	20	5	25	28	8	17	53
Limfjord (DK)	25	13	38	188	26	26	55

Source: From *Nitrogen Cycling Coastal Marine Environments*, G. Billen and C. Lancelot, (eds.: T. H. Blackburn and J. Sørensen), copyright © 1988 by John Wiley & Sons, Inc., New York, p. 342. Reprinted by permission. See Billen and Lancelot (1988) for data sources.

conditions favor denitrification. Nitrification and denitrification are tightly coupled in the sediments due to redox heterogeneity. For example, organic-rich particles, such as fecal pellets, form anoxic microzones in the surface sediments. The irrigating activities of macrofauna produce oxic microzones below the redox boundary. Rates of denitrification and nitrification co-vary both seasonally and diurnally as a result of shifts in such factors as (1) the supply rate of POM, (2) the activity of benthic microbes, and (3) the supply rate of O_2 as determined by water circulation and production by benthic algae. These algae depress nitrification rates by competing for ammonium, excreting organic toxins, and elevating the pore-water pH.

Denitrification in the sediments is also associated with sulfate reduction. Redox heterogeneity causes both electron acceptors (nitrate and sulfate) to be used over the same depth range. A significant amount of the ammonium produced by these processes is adsorbed onto the exchange sites of clay minerals. The rest diffuses upward until it is either nitrified or assimilated by algae.

IS THE NITROGEN CYCLE IN A STEADY STATE?

Due to considerable uncertainty in global transport rates, it is not presently possible to determine whether the marine nitrogen cycle is in steady state. Transport rates are very difficult to determine because of small-scale temporal and spatial variability in processes such as phytoplankton growth and river input. In addition, a relatively large number of chemical species must be included in budget calculations. This requires a great deal of sampling and analytical effort. Furthermore, the nitrogen cycle is subject to variability over long time scales as a result of fluctuations in such environmental factors as thermohaline circulation and climate. Both of these influence the processes controlling the rates at which nitrogen is supplied to and removed from the ocean.

The existing set of observations are of insufficient duration to detect any such long-term trends. They are also too geographically limited to assess adequately whether a steady state exists even on short time scales. For example, the distribution of organic nitrogen in the surface sediments of one-

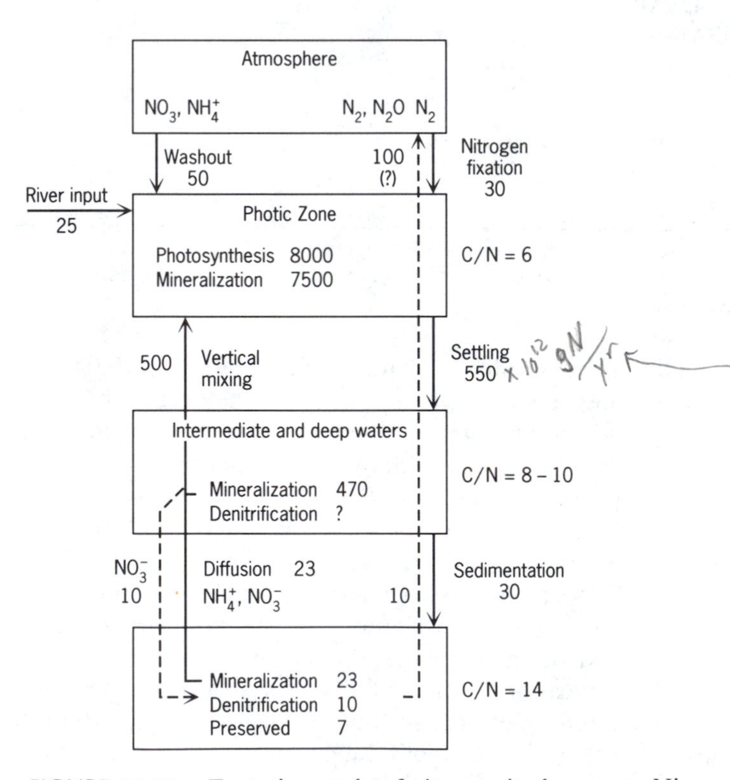

FIGURE 24.12. Tentative cycle of nitrogen in the ocean. Nitrogen values are given in units of 10^{12} g N/y. *Source*: From *Some Perspectives of the Major Biogeochemical Cycles*, R. Wollast (ed.: G. E. Likens), copyright © 1981 by John Wiley & Sons, Inc., New York, p. 134. Reprinted by permission.

third of the ocean are not known (hence the big gaps in Figure 24.11). As a result, the global burial rate of PON is difficult to assess. Likewise, global estimates of denitrification are limited by the lack of quantitative information on the importance of reduction in anoxic microzones, as well as of N_2O production and degassing from the sea surface.

Considerable uncertainty also exists in our knowledge of the rates at which nitrogen is supplied to the ocean. This is partially a result of problems associated with river sampling. Logistical problems make sampling of rain and aerosols very difficult, so estimates of aeolian input are still very uncertain. Estimates of global marine nitrogen fixation rates vary by two orders of magnitude because of analytical difficulties as well as our incomplete knowledge as to which organisms are capable of fixing nitrogen.

Slight changes in rate estimates can have a large impact on the global budget calculations. Thus much better precision in flux estimates is required before the existence of a steady state can be evaluated. Nevertheless, attempts have been made to construct global marine nitrogen budgets. As indicated in the two examples given in Figures 24.12 and 24.13, a steady state could exist if some supply has been grossly underestimated or some sink has been overestimated. Among these models, the missing nitrogen flux ranges from 2×10^{12} to 74×10^{12} g/y. This disparity is largely due to differences in the estimates of PON sedimentation rates and in the gaseous flux of N_2O out of

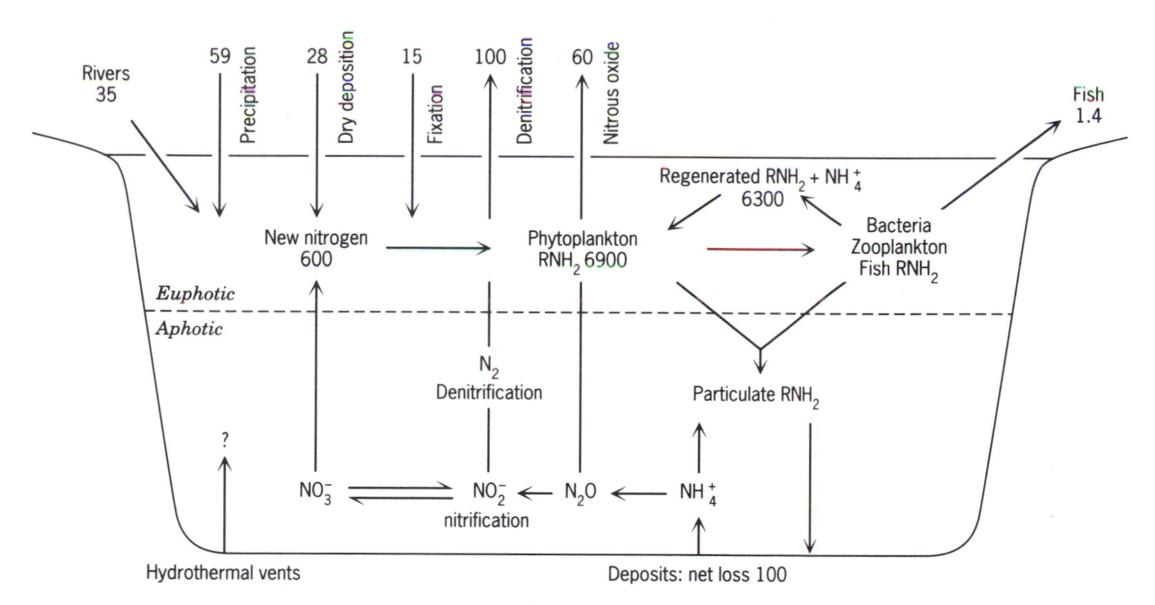

FIGURE 24.13. The oceanic nitrogen cycle. Estimates of nitrogen transport rates are given in units of 10^{12} g N/y. *Source*: From G. E. Fogg, reprinted with permission from the *Philosophical Transactions of the Royal Society of London*, vol. B296, p. 215, copyright © 1982 by the Royal Society of London, England. Data from G. E. Fogg, reprinted with permission from the *Stockholm Ecological Bulletin*, vol. 26, p. 12, copyright © 1978 by Ecological Bulletins, Stockholm, Sweden.

the sea surface. Most marine chemists hypothesize that the missing nitrogen is supplied by N_2 fixation, which apparently has been greatly underestimated.

ANTHROPOGENIC PERTURBATIONS OF THE MARINE NITROGEN CYCLE

Though the marine nitrogen cycle may have been in a steady state, human activities are likely to have caused a significant enough change in fluxes so as to perturb any natural balance. As indicated in Table 24.7, most nitrogen fixation is presently the result of human activities. Most of the industrial fixation is accomplished by the Born–Haber process, which is illustrated below.

$$N_2(g) + 3H_2(g) \xrightarrow{\text{catalyst}} 2NH_3(g) \qquad (24.7)$$

The resulting ammonia is converted to a form suitable for fertilizer or other use. Though some of this fixed nitrogen has resulted in an increase in biomass, much has been carried into rivers by groundwater and stormwater runoff. Poor agricultural practices have also increased the nitrogen load in rivers by accelerating rates of soil erosion. Combined with sewage input, these anthropogenic sources have overwhelmed the natural riverine flux of nitrogen in some estuaries. In Chesapeake Bay, decomposition of this organic matter causes periodic conditions of anoxia. Some of the excess nitrogen is then lost to denitrification. Changes in the relative abundance of the various forms of DIN and in redox levels have also altered the number and types of organisms found in heavily contaminated estuaries.

So much nitrogen has been introduced into the coastal ocean that nutrient concentrations appear to be increasing. This is thought to be stimulating coastal

TABLE 24.7
Comparison of Natural and Human Sources of Fixed Nitrogen (1 Tg = 10^{12} g)

Source	*Rate (Tg N/y)*
"Natural" (historic) biological	60
Atmospheric processes	7.4
Grain legumes	40.6
Hay and pasture legumes	28.4
Fossil fuel and other combustion	19.8
Industrial fixation	40

Source: From C. C. Delwhiche, reprinted with permission from *Ambio*, vol. 6, p. 108, copyright © 1977 by the Royal Swedish Academy of Sciences, Elmsford, NY.

productivity. Such an impact could be fortuitous as it would also consume CO_2, thereby mitigating some of the increase in atmospheric levels of this gas caused by the burning of fossil fuels and deforestation. The latter two processes are also a sink for fixed nitrogen as both cause organic nitrogen to be converted to nitrogen oxides. Nitrogen oxides contribute to acid rain and the formation of ozone in the lower atmosphere. Ozone is a component of photochemical smog and hence is a health hazard. In the upper atmosphere, nitrogen oxides contribute to the destruction of ozone. This is also a health hazard as stratospheric ozone absorbs ultraviolet light, thereby shielding Earth's surface from a form of mutagenic electromagnetic radiation. The increase in atmospheric levels of the nitrogen oxides should also increase their NAECs, thereby altering the degree to which the ocean acts as a net source or sink of these gases.

All of these anthropogenic inputs have the potential to alter marine productivity significantly. Positive feedback loops link biological production to the biogeochemical cycles of oxygen and carbon. Thus any of these perturbations could alter the redox level of seawater and thereby change the sedimentation rate of biogenic detritus. This in turn has the potential to affect global climate and hence circulation rates and the position of sea level.

SUMMARY

Though nitrogen is widespread in the ocean, it is largely inaccessible to marine life as most is in the form of N_2. As a result of its relatively low concentrations, fixed nitrogen limits biological productivity. In turn, its marine distributions are largely controlled by the activities of marine organisms. Thus the marine nitrogen cycle is dominated by biologically mediated processes, most of which involve redox reactions.

Marine algae meet their nitrogen needs by **assimilating** dissolved inorganic nitrogen (DIN). As less energy is required to metabolize ammonium, this species is assimilated more rapidly than nitrite and nitrate. At mid latitudes, seasonal variations in phytoplankton growth are caused by shifts in nutrient and light availability. As a result, these latitudes are characterized by a large spring plankton bloom and a smaller one in the fall. Similar blooms in consumer organisms also occur. Subpolar plankton are light-limited and as a result only experience one annual bloom. At low latitudes, the absence of convective mixing causes permanent nutrient limitation that results in uniformly low productivity year round.

Productivity is highest in coastal regions due to nitrogen input from rivers and to **nitrogen fixation**, which is favored in shallow benthic environments. Most nitrogen fixers appear to be heterotrophic bacteria, with the exception of the photosynthetic cyanobacteria *Oscillatoria* ssp., which are particularly abundant in the tropics.

PON synthesized by phytoplankton has two possible fates. Either the plants die, their cells lyse, and the resulting DON is degraded by bacteria, or the cells are consumed by protozoans and zooplankton. The excretions and exudations of the consumers are also a source of DON. The reduced nitrogen in these compounds is oxidized in a stepwise fashion. First, ammonium is produced by **ammonification**. It is then converted to nitrite and finally to nitrate. The latter two processes are termed **nitrification**. The oxidized products can be reduced by either plankton, during the process of nutrient assimilation, or by heterotrophic bacteria. The latter use DIN as their electron acceptor. This process is referred to as **denitrification** or **dissimilatory nitrogen reduction** because some of the reduced nitrogen is released to seawater instead of being incorporated into bacterial biomass.

Denitrification occurs in upwelling areas and other locations where seawater is relatively stagnant. The vertical profiles of the DIN species have maxima and minima reflecting the different depth distributions of the various metabolic processes. Denitrification also occurs in the sediments, though redox heterogeneity causes a direct coupling with nitrification. For denitrification to proceed, large supplies of organic matter must be present. Due to the decomposition of PON that occurs as it settles to the seafloor and prior to burial, the organic nitrogen composition of marine sediments is usually less than 1 percent. Since the highest concentrations are observed in estuarine and coastal sediments, these are the sites where most denitrification occurs.

Benthic respiration resolubilizes DIN, which diffuses into the water column. This flux represents a significant source of nutrients to shallow waters. Some of the ammonium produced during denitrification and sulfate reduction is also adsorbed onto the exchange sites of clay minerals.

Due to uncertainty in the estimates of global nitrogen fluxes, we are currently unable to establish whether the nitrogen cycle is in a steady state. Current estimates suggest that fixed nitrogen is being removed from the ocean faster than it is being supplied. Most geochemists hypothesize that we have grossly underestimated the global rate of marine N_2 fixation.

Though the marine nitrogen cycle may have been in a steady state, human activities are likely to have caused a significant enough change in fluxes so as to perturb any natural balance. For example, humans are now fixing by industrial processes more nitrogen than are "natural" processes. As a result of fertilizer runoff, sewage dumping, and soil erosion, the anthropogenic input of nitrogen has overwhelmed the natural riverine flux of nitrogen in some estuaries. This is thought to have caused an increase in the growth of coastal plankton that could be counteracting some of the anthropogenic greenhouse effect. The rate of atmospheric input of nitrogen also appears to have been increased as a result of nitrogen oxide emissions. These chemicals enter the ocean by diffusing across the air–sea interface or via rainfall.

All of these anthropogenic inputs have the potential to significantly alter marine productivity. Positive feedback loops link biological production to the biogeochemical cycles of oxygen and carbon. Thus any of these perturbations could alter the redox level of seawater and thereby change the sedimentation rate of biogenic detritus. This in turn has the potential to affect global climate and hence circulation rates and the position of sea level.

CHAPTER 25

THE MARINE CARBON CYCLE AND THE CARBON DIOXIDE PROBLEM

INTRODUCTION

Biogenic limestone is the single largest crustal reservoir of carbon. Sedimentary organic matter is the second largest and is composed primarily of carbon fixed by marine phytoplankton. Following deep burial, some of these deposits are converted to petroleum. As described in the next chapter, several million years are required for the formation of an economically attractive deposit. Hence petroleum is often referred to as a **fossil fuel**. In the absence of humans, geologic uplift eventually exposes these ancient marine deposits to atmospheric O_2. Chemical weathering ''unfixes'' the organic carbon by oxidizing it back to CO_2. This CO_2 is then taken up by plants, thereby closing the global biogeochemical cycle of carbon. The combustion of fossil fuel has greatly accelerated the rate at which organic carbon is unfixed and is causing atmospheric CO_2 levels to increase. Changes in land-use patterns, particularly deforestation, have also contributed to the atmospheric increase.

Any rise in the CO_2 content of the atmosphere is of great concern because this compound is a greenhouse gas. As such, CO_2 absorbs some of the solar radiation that is transmitted to this planet. The absorbed energy is partially dissipated as heat, thereby warming the atmosphere, land, and sea. A rise in the atmospheric level of CO_2 would increase the amount of heat on Earth's surface. If this increase is large enough, it could cause a rise in atmospheric temperatures and of sea level. The latter would result from the melting of glaciers and the thermal expansion of seawater. If this rise in sea level is large, it could cause the destruction of wetlands, the contamination of coastal freshwater aquifers, and eventually, the inundation of coastal cities. Rainfall patterns and storm activity are also likely to be altered, thereby affecting global agricultural activities, as well as the survival of undomesticated species.

The burning of fossil fuels has also mobilized a significant amount of nitrogen and sulfur, as evidenced by our growing acid rain problem. Thus our use of fossil fuels, as well as other activities, is causing a large-scale disruption of the global carbon, nitrogen, and sulfur cycles. These perturbations will undoubtedly result in some of the environmental changes described above. Due to our limited knowledge of transport rates and feedback mechanisms, we are unable to predict how large these changes will be and when they will occur.

Though widespread social, economic, and political impacts are likely, little or no effort has been made to alter our patterns of fossil fuel consumption or land use. This inaction is partially due to the uncertainty that accompanies current predictions of environmental impacts. It is also due to the perceived costs associated with the development and use of alternative energy sources. As a result, current attention is focused on removing the excess CO_2 from the atmosphere, rather than controlling its production. This has led to efforts at reforestation, as well as to a proposal that marine phytoplankton growth be stimulated by fertilizing the Southern Ocean. Some scientists believe that primary production in these nutrient-rich waters is limited by iron availability. They have estimated that fertilizing these waters with soluble iron will stimulate enough primary production to consume one-third to one-half of the anthropogenic CO_2 flux. Though some are leery of altering another facet of the global biogeochemical carbon cycle, supporters have described the artificial enhancement of primary production as a rapid method by which sedimentary organic matter can be recreated. Thus by increasing the rate at which marine organic matter is formed, we would be partially counteracting the effects of having accelerated its rate of destruction via the burning of fossil fuels.

Atmospheric levels of other greenhouse gases, such as freon, methane, and ozone, are also rising as a result of anthropogenic input. Thus geochemists find themselves in the midst of a global experiment in which several geochemical cycles can be studied by observing the effects of our perturbations. Insight into the workings of these cycles are also being obtained from paleoceanographic research. For example, observations from ice and sediment cores have demonstrated that large-scale fluctuations in atmospheric CO_2 levels have occurred over geologic time and are well correlated with global climate change. The likely causes for this relationship and the implications for future climate change are discussed in this chapter.

THE MARINE CARBON CYCLE

As shown in Figure 25.1, the global carbon cycle is largely controlled by biologically mediated processes. As with the nitrogen cycle, the sizes of some carbon reservoirs are still not well known. The large uncertainty associated with the terrestrial biomass reservoir is due to continuing changes in land-use patterns and vagueness in ecosystem classification. The results from recent developments in analytical methods suggest that the size of the marine DOC reservoir may be underestimated by as much as a factor of two.

Many transport rates are also not well known. The estimates of CO_2 fluxes across the air–sea interface and uptake by the terrestrial biota are so uncertain that the existence of a steady state cannot be established. For example, recent analytical advances suggest that global primary production has been significantly underestimated. The global burial rate of carbon in the sediments is best assessed by measuring biogenic particle fluxes with sediment traps.

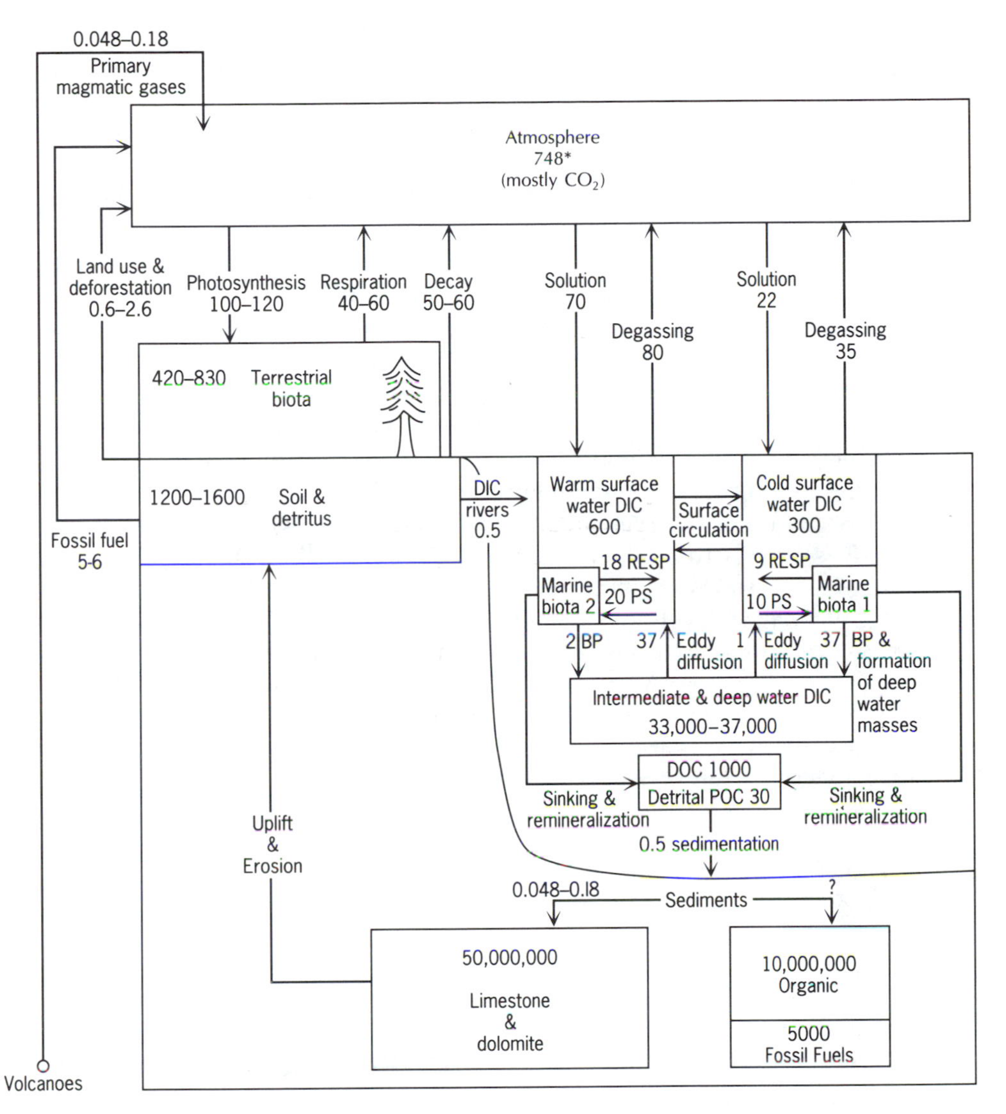

FIGURE 25.1. The biogeochemical cycle of carbon. Current estimates of the major reservoirs (in 10^{15} g C or billions of metric tons [BMT]) and fluxes (in 10^{15} g C/y or BMT/y). BP = transport of carbon to the deep sea by the biological pump. PS = conversion of DIC to POC by photosynthesis. RESP = conversion of organic carbon to DIC by respiration. *Source*: After W. M. Post, T.-H. Peng, W. R. Emmanuel, A. W. King, H. Dale, and D. L. DeAngelis, reprinted with permission from *American Scientist*, vol. 78, p. 314, copyright © 1990 by Sigma Xi, Raleigh, NC; and B. Moore and B. Bolin, reprinted with permission from *Oceanus*, vol. 29, p. 11, copyright © 1987 by Oceanus, Woods Hole, MA.

Since few such data have been collected, the global burial rate is also not well known but appears to be temporally variable as a result of seasonal plankton blooms.

The relatively small amount of carbon present in the atmosphere is primarily in the form of CO_2. As shown in Figure 25.2, the size of this reservoir oscillates seasonally. Concentrations are lowest in the summer due to an increase in plant biomass. During the winter, plant growth rates decline, but animals and microbes continue to respire organic matter. Thus atmospheric CO_2 levels rise. This effect is most pronounced at locations that experience the largest seasonal shifts in primary productivity (i.e., the mid latitudes). The oscillation is largest in the Northern Hemisphere due to the presence of greater amounts of terrestrial biomass. Since most atmospheric mixing is caused by the Trades, Westerlies, and Polar Easterlies, cross-latitudinal transport is inhibited. Thus regional differences in atmospheric CO_2 levels are maintained by relatively fast rates of local production and consumption.

Despite these seasonal oscillations, a long-term increase in the atmospheric CO_2 level is clearly evident. This is largely the result of fossil-fuel burning and changes in land-use patterns, such as deforestation. Data from sediments and ice cores suggest that large shifts in atmospheric CO_2 levels have occurred periodically over geologic time and are associated with climate change. A direct connection seems likely, as CO_2 is a greenhouse gas and thus affects atmospheric temperatures. Since the ocean is the single largest reservoir of biologically active carbon, it is thought to play a major role in controlling atmospheric CO_2 levels.

As shown in Figure 25.3, most oceanic carbon is in the form of bicarbonate. Plants assimilate this carbon and convert it to POC or calcium carbonate. As sinking particles, the carbon is transferred to the deep sea and sediments. This biological transfer connects the marine reservoir to the vast sedimentary deposits. As a result, most of the juvenile carbon emitted into the atmosphere as magmatic CO_2 has been buried in the sediments. Metamorphism converts the inorganic deposits into sedimentary rock, such as limestone and dolomite. Some of the organic carbon is transformed into petroleum. The carbon in these reservoirs is eventually recycled as a result of chemical weathering that occurs after geologic uplift. By burning fossil fuels, we have greatly accelerated the rate at which carbon is recycled in the crustal-ocean factory.

The biomass of land plants is much larger than that of marine phytoplankton. Because land plants tend to grow more slowly than phytoplankton, the rates of marine and terrestrial production are approximately equal. Most detrital organic matter on land is oxidized prior to incorporation in the soils. Thus the burial of marine POM and biogenic calcium carbonate is the main pathway by which carbon is removed from the biosphere. Since the DIC consumed by marine plants is readily replaced by gas exchange across the air–sea interface, the burial of biogenic carbon in marine sediments has the potential to remove excess carbon from the atmosphere. Hence this sink could consume some of the anthropogenic CO_2 that would otherwise contribute to an increased greenhouse effect.

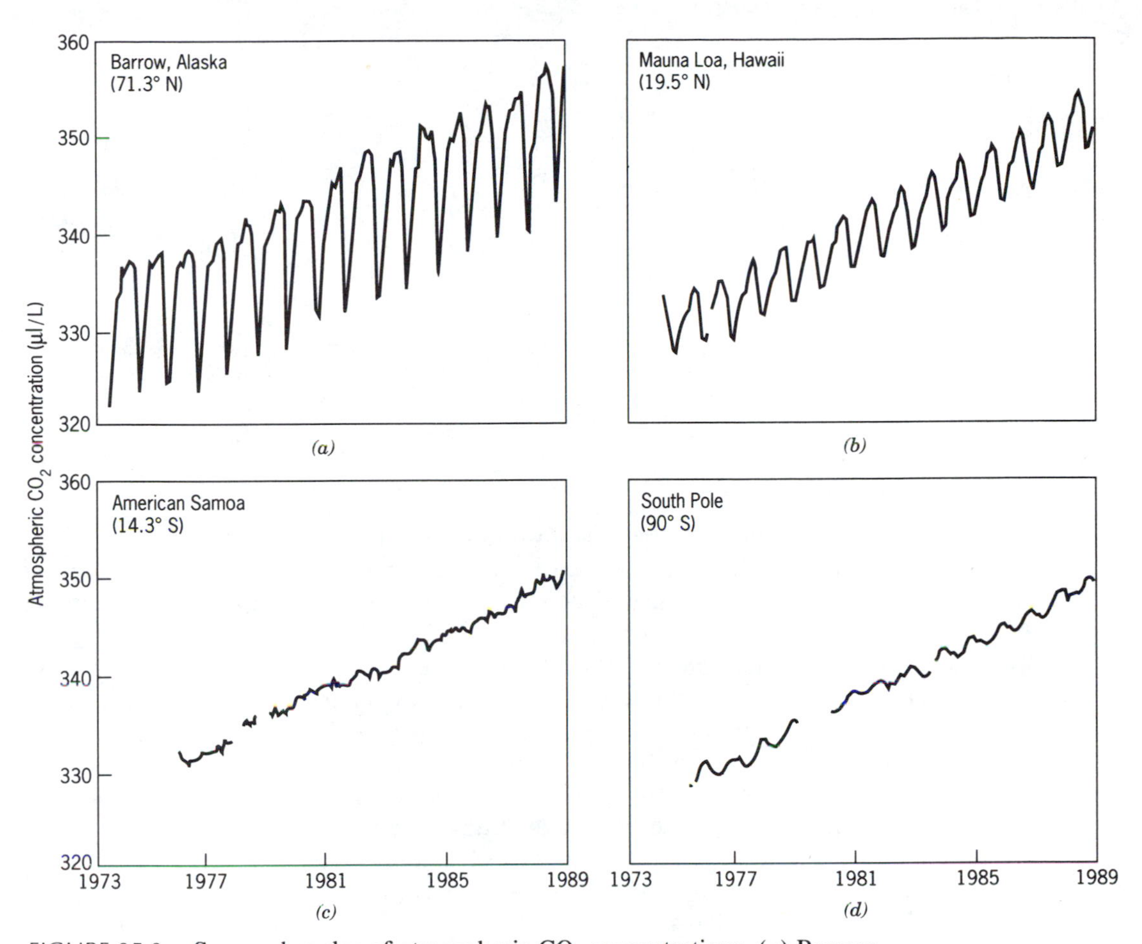

FIGURE 25.2. Seasonal cycles of atmospheric CO_2 concentrations. (*a*) Barrow, Alaska, (*b*) Mauna Loa, Hawaii, (*c*) American Samoa, and (*d*) the South Pole. *Source*: From W. M. Post, T.-H. Peng, W. R. Emmanuel, A. W. King, H. Dale, and D. L. DeAngelis, reprinted with permission from *American Scientist*, vol. 78, p. 325, 1990, copyright © 1990 by Sigma Xi, Raleigh, NC. Data sources: (*a*) from J. T. Peterson, W. D. Komhyr, L. S. Waterman, R. H. Gammon, K. W. Thoning, and T. J. Conway, reprinted with permission from the *Journal of Atmospheric Chemistry*, vol. 4, p. 497, copyright © 1986 by Reidel Publishing Company, Dordrecht, The Netherlands; (*b*) K. W. Thoning, P. P. Tans, and W. D. Komhyr, reprinted with permission from the *Journal of Geophysical Research*, vol. 94, p. 8559, copyright © 1989 by the American Geophysical Union, Washington, DC and W. D. Komhyr, T. B. Harris, L. S. Waterman, J. F. S. Chin, and K. W. Thoning, reprinted with permission from the *Journal of Geophysical Research*, vol. 94, p. 8546, copyright © 1989 by the American Geophysical Union, Washington, DC; (*c*) L. S. Waterman, D. W. Nelson, W. D. Komhyr, T. B. Harris, K. W. Thoning, and P. P. Tans, reprinted with permission from the *Journal of Geophysical Research*, vol. 94, p. 14828, copyright © 1989 by the American Geophysical Union, Washington, DC and (*d*) D. A. Gilette, W. D. Komhyr, L. S. Waterman, L. P. Steele, and R. H. Gammon, reprinted with permission from the *Journal of Geophysical Research*, vol. 92, p. 4233, coyright © 1987 by the American Geophysical Union, Washington, DC.

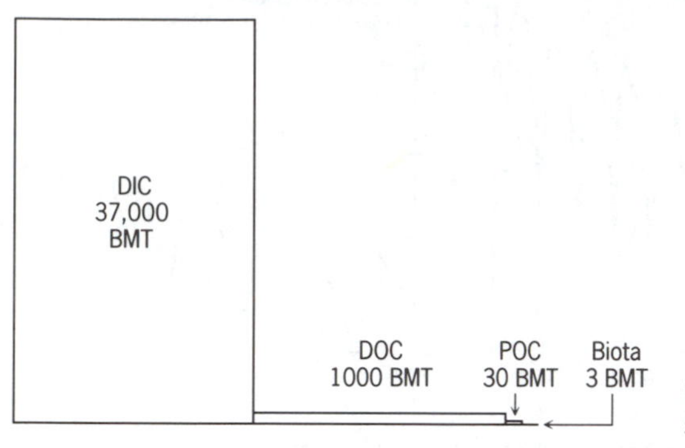

FIGURE 25.3. Sizes of the oceanic carbon reservoirs (1 BMT = 10^{15} g C) shown to scale.

THE GREENHOUSE EFFECT

The solar radiation received by Earth is predominantly of short wavelength. As shown in Figure 25.4, 30 percent is returned to outer space by reflection from atmospheric gases or Earth's surface. The majority of the incoming radiation is absorbed by the ocean and land (51%) or by the atmospheric gases (19%). The insolation absorbed by the ocean and land is returned to the atmosphere, and eventually to outer space, as either heat or long-wavelength radiation. The former is transmitted to the atmosphere by either (1) thermal conduction and convection or by (2) surface evaporation of water and condensation within clouds. The long-wavelength radiation emitted by the land and ocean is largely absorbed by the atmospheric gases.

Since the atmospheric gases absorb short-wavelength, as well as long-wavelength radiation, they retain a significant amount of solar radiation. Emission of this energy in the form of heat warms our atmosphere. Thus the steady-state temperature of our atmosphere is determined by its gas content. If gas levels were to rise, the rate at which solar energy is absorbed by the atmosphere would temporarily exceed its loss to outer space. Once the new gases had absorbed their requisite amount of radiation, steady state would be established, but at an elevated temperature. This is called the greenhouse effect because the Earth and its atmosphere convert the retained solar energy into heat. Without this natural greenhouse effect, the temperature of Earth's surface would be −4°C.

The majority of the long-wavelength radiation emitted from Earth's surface ranges in wavelength from 4 to 80 μm and hence is termed infrared radiation. As shown in Figure 25.5, atmospheric water and CO_2 can absorb radiation throughout this range except between 8 and 12 μm. While ozone can absorb radiation between 9 and 10 μm, its contribution is not very significant due to its relatively low concentration. Despite this absorption gap, 71 percent of the infrared radiation emitted from Earth's surface is absorbed by one of the atmospheric gases and then is released to the atmosphere as heat. Atmospheric temperatures would rise if the amount of infrared-absorbing gases were increased. Such gases include methane, nitrous oxide, and the chlorofluorocarbons.

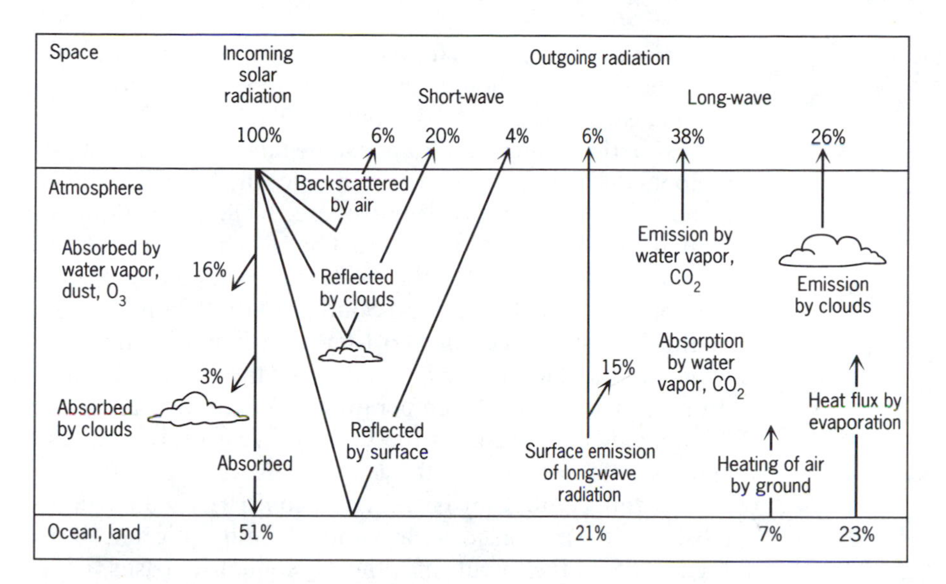

FIGURE 25.4. The steady-state radiation budget of the earth. *Source*: From M. O. Andreae, reprinted with permission from *Oceanus*, vol. 29, p. 32, copyright © 1987 by Oceanus, Woods Hole, MA.

ANTHROPOGENIC INPUT

Human activities have greatly increased the rate at which greenhouse gases are introduced into the atmosphere. As a result, their concentrations are rising and will eventually cause atmospheric temperatures to rise. Global climate is highly variable, especially over short time scales. Thus global temperatures will have to be monitored over several more decades before the onset of a global warming trend can be unequivocally substantiated. Selected historical

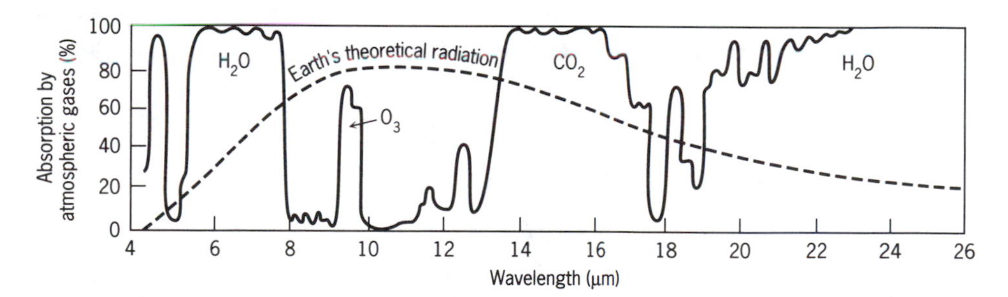

FIGURE 25.5. Absorbance spectrum for infrared radiation. The amount of radiation emitted by Earth is shown by the dashed line. The amount of radiation absorbed by the atmospheric gases is shown by the solid line. *Source*: From K. Bryan, reprinted with permission from *Oceanus*, vol. 29, p. 36, copyright © 1987 by Oceanus, Woods Hole, MA. After *Solar and Terrestrial Radiation*, K. L. Coulson, copyright © 1975 by Academic Press, Orlando, FL, pp. 265–267.

observations suggest that over the past century, atmospheric temperatures have been rising, though this trend appears to have been interrupted by a cooling spell which lasted from 1940 to 1970. Since the mid-1800s, global atmospheric temperatures appear to have increased by 0.5° to 1.0°C. Large regional differences exist. For example, the Southern Hemisphere has warmed faster than the Northern Hemisphere, while temperatures in the southeastern United States appear to have declined slightly over the past 30 y.

The warming trend of the 1980s stands out more clearly when temperatures are corrected for the El Niño/Southern Oscillation (Figure 25.6), which is one of the largest sources of short-term variability. These periodic shifts in the winds and surface-water circulation of the eastern Pacific cause a temporary drop in mean global temperatures. When the impact of these events is eliminated, 1989 becomes the warmest year on record (as of 1989), with 1988 and 1987 the second and third, respectively.

At this point, CO_2 is the greenhouse gas whose concentration has been most greatly increased as a result of anthropogenic input. But as shown in Figure 25.7, the input of other greenhouse gases is now significant and increasing. Since some are also more effective absorbers of infrared radiation, their contribution to the anthropogenic greenhouse effect has the potential to eventually supersede that of CO_2.

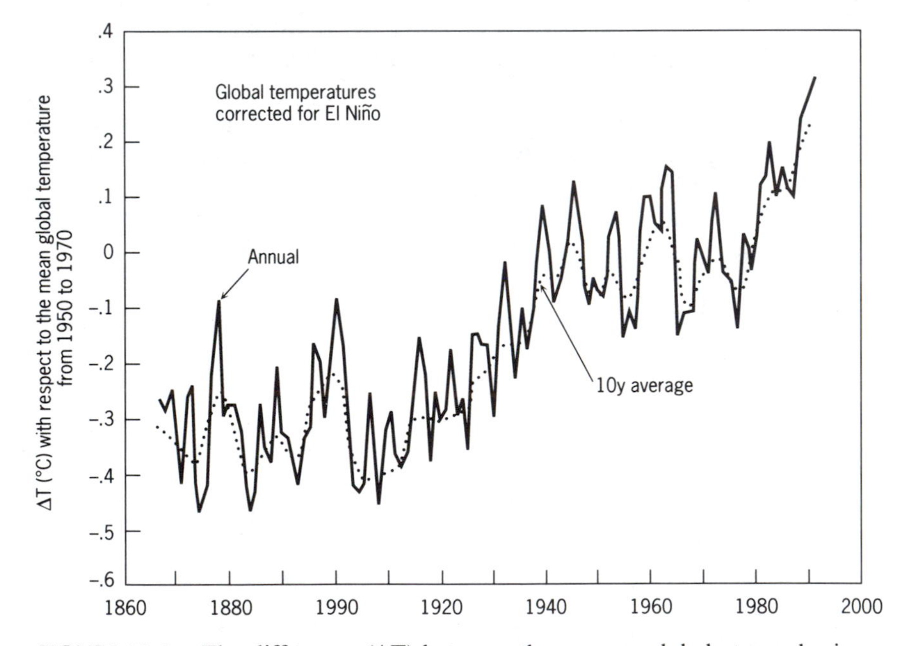

FIGURE 25.6. The difference (ΔT) between the average global atmospheric temperature and the average global temperature from 1950 to 1970. *Source*: From P. D. Jones and T. M. L. Wigley, reprinted with permission from *Scientific American*, vol. 263, p. 90, copyright © 1990 by Scientific American, New York.

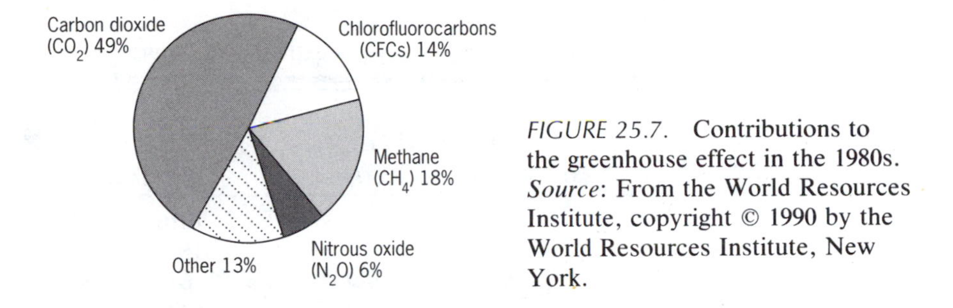

FIGURE 25.7. Contributions to the greenhouse effect in the 1980s. *Source*: From the World Resources Institute, copyright © 1990 by the World Resources Institute, New York.

Since the 1960s, fossil-fuel burning has been the single largest source of anthropogenic CO_2. As shown in Figure 25.8, the production rate of anthropogenic CO_2 accelerated in the late 1890s following the invention of the coal-fired steam engine which initiated the Industrial Revolution. The widespread use of petroleum-powered machines has also increased the atmospheric levels of O_3, CO, NO_x, and SO_x.

The United States is responsible for 23 percent of the CO_2 emissions generated by the use of fossil fuels. As shown below (Table 25.1), this large contribution is largely the result of our high per capita consumption rate. Nevertheless, our relative contribution is likely to decline as the rates of fossil-fuel consumption are rapidly increasing in developing third world countries.

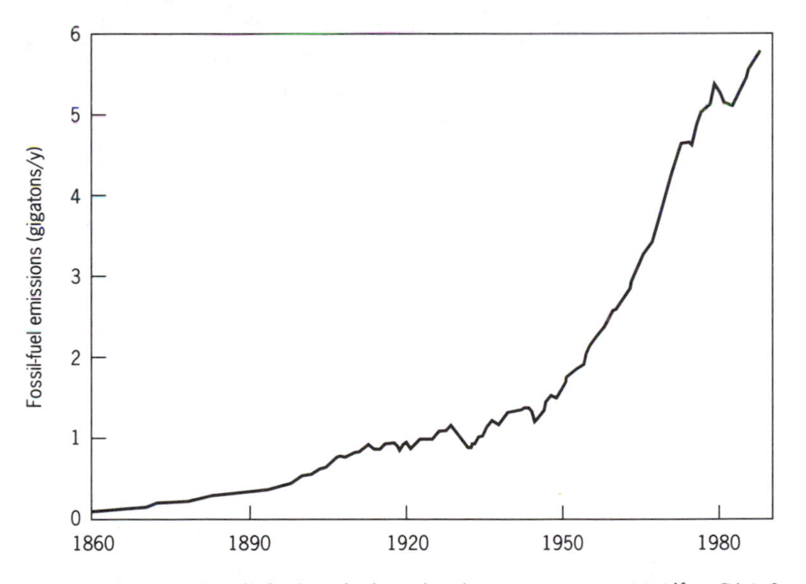

FIGURE 25.8. Fossil-fuel emissions in gigatons per year (10^{15} g C/y) from 1860 through the early 1980s. *Source*: From W. M. Post, T.-H. Peng, W. R. Emmanuel, A. W. King, H. Dale and D. L. DeAngelis, reprinted with permission from *American Scientist*, vol. 78, p. 313, copyright © 1990 by Sigma Xi, Raleigh, NC. Data from G. T. Marland, T. A. Boden, R. C. Griffin, S. F. Huan, P. Kancircuk, and T. R. Nelson, 1989, ORNL/CDIAC-25, Oak Ridge National Laboratory, Oak Ridge, TN.

TABLE 25.1
Carbon Dioxide Emissions from Fossil Fuels in 1987

Source	CO_2 *(million tons)*	CO_2 *per Capita (tons)*	CO_2 *per Dollar GNP (g)*
USA	4480	18.37	1010
USSR	3711	13.07	1563
W. Europe	2899	7.61	651
China	2031	1.90	6925
Japan	908	7.43	564
India	549	.70	2386
Canada	388	14.93	875
World	19,438	3.88	1138

Source: Courtesy of Greg Marland, Oak Ridge National Laboratory, Oak Ridge, TN.

As shown in Figure 25.9, the two largest uses of energy obtained from fossil-fuel combustion are the production of electricity and transportation. In the Middle East, large amounts of CO_2 are also produced by the flaring of natural gas at oilwell heads. A significant amount of CO_2 is also released as a result of changes in land use associated with agricultural and forestry activities (Figure 25.10). These include logging, the harvesting of fuel wood, deliberate burning, grazing, and conversion of undeveloped areas to croplands. Most of these changes are occurring in the tropics, where rainforests are being clear cut to create new range and croplands. Tropical rainforests, which are the most productive terrestrial ecosystems, originally covered 5 million square

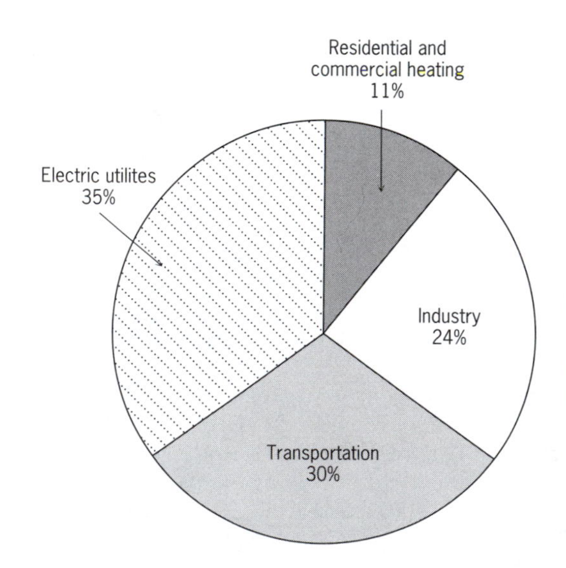

FIGURE 25.9. U.S. sources of carbon dioxide in 1987. *Source*: From M. Brower, reprinted with permission from *Nucleus*, vol. 10, p. 4, copyright © 1988 by the Union of Concerned Scientists, Cambridge, MA. Data from the World Resources Institute, copyright © 1988 by the World Resources Institute, New York.

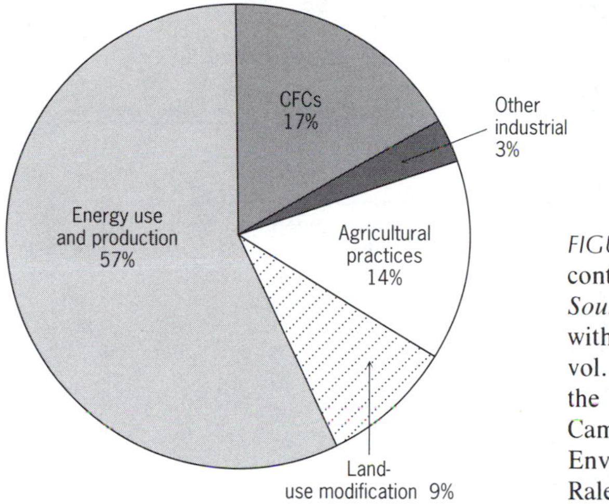

FIGURE 25.10. Sources contributing to global warming. *Source*: From A. Meyer, reprinted with permission from *Nucleus*, vol. 11, p. 4, copyright © 1989 by the Union of Concerned Scientists, Cambridge, MA. Data from the Environmental Protection Agency, Raleigh, NC.

miles of the planet. By 1900 A.D., 50 percent had been destroyed. If the current rate of deforestation is maintained (80,000 mi^2/y), the remainder will be gone by 2000 A.D.

Deforestation kills trees, which reduces the uptake rate of CO_2 but increases its production rate. The latter is caused by an acceleration in respiration rates supported by the increased amount of dead wood. As a result, the net flux of CO_2 to the atmosphere is increased. Clear cutting also increases soil erosion rates, which eventually leads to desertification. This has already occurred throughout large areas of China and Africa.

Much of the dead vegetation produced during deforestation and other agricultural activities is burned. As shown in Table 25.2, this practice is a significant and increasing source of greenhouse gases. Current estimates suggest that 5 percent of Earth's surface is burned each year.

As a result of these processes, atmospheric CO_2 concentrations are rising at a rate of about 1.5 ppm per year (Figure 25.11). This is equivalent to an annual addition of 2.0×10^{15} g C. Current estimates suggest that by the middle of the next century, the atmospheric CO_2 level will be double that which was present just prior to the start of the Industrial Revolution.

As shown in Figure 25.12, the atmospheric concentrations of other greenhouse gases are also increasing. In addition to biomass burning, a significant amount of methane is emitted by cattle and termites, which produce this gas during digestion of grass and wood, respectively. Emissions from rice paddies are also a significant source. In these locations, methane is produced by methanogenic bacteria that flourish in waterlogged, and hence anoxic, soils. A large amount of methane also is apparently being released from leaky natural gas pipelines in Eastern Europe.

Nitrous oxide is a by-product of fossil-fuel burning, industrial nitrogen fixation, fertilizer use, and denitrification. Denitrification is commonly conducted in sewage treatment plants as part of the water purification process.

TABLE 25.2
Biomass Burning as a Source of Greenhouse and Other Gases

Million Metric Tons per Year as C, N, or O	*All Combustion Sources*[a]	*% from Biomass*
Carbon dioxide (gross)	8700	40
Carbon dioxide (net)[b]	7000	26
Carbon monoxide	1100	32
Methane	380	10
Nonmethane hydrocarbons	100	24
Ozone	1100	38
Ammonia	44	12
Nitrogen oxides (NO_x)	40	21
Nitrous oxide (N_2O)	13	6

Source: Courtesy of Dr. M. O. Andrease, Max-Planck-Institut für Chemie, Mainz, Germany.
[a]Including fossil fuel and all biomass burning.
[b]Carbon dioxide released by all combustion minus that taken up by replacement vegetation.

Denitrification rates have also been increased in coastal waters that are subject to large sewage inputs.

Chlorofluorocarbons (CFCs) are synthetic gases composed of carbon and variable amounts of fluorine and chlorine. CFCs are used as refrigerants and aerosol propellants. In the upper atmosphere, ultraviolet light causes CFCs to react with ozone. This has resulted in a significant loss of stratospheric ozone, particularly in the polar regions. The loss of ozone is of great concern as this naturally occurring substance absorbs ultraviolet radiation, thereby shielding Earth's surface from a potent mutagenic agent. As a result, the United States has substantially reduced its use of CFCs and is seeking to coordinate an international agreement that would greatly limit production worldwide. It is somewhat ironic that ozone produced in the lower atmosphere by the burning of fossil fuel and biomass is photochemically converted into smog.

Several relatively nonindustrialized countries produce so much methane and chlorofluorocarbons that they can be considered as major sources of greenhouse gases. As shown in Table 25.3, Brazil contributes 11 percent of the annual input, primarily as a result of large-scale deforestation. India's large contribution is due to methane production from the raising of cattle and cultivation of rice.

The impact of these emissions on the atmospheric levels of greenhouse gases is difficult to assess since some are destroyed by natural photochemical processes. For example, some of the anthropogenic ozone is destroyed by methylhalides, such as methylchloride, methylbromide, and methyliodide, all of which appear to be synthesized by marine plants. Some anthropogenic ozone is probably also destroyed by reaction with CFCs. Anthropogenic sulfur

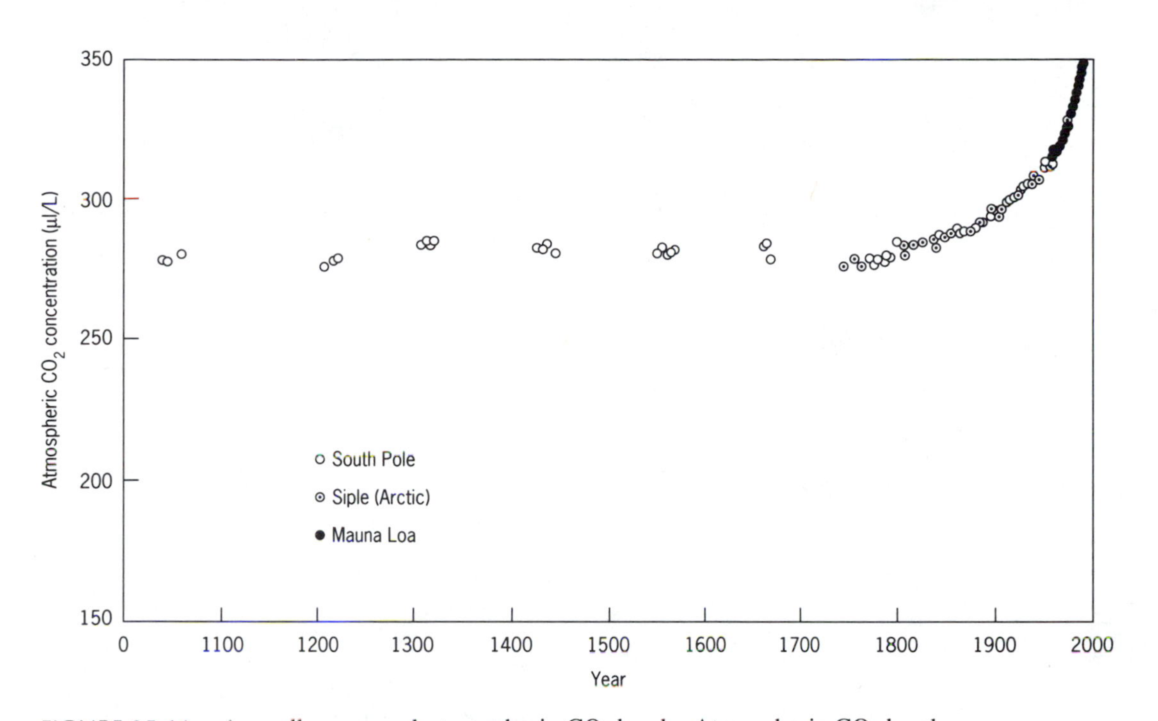

FIGURE 25.11. Annually averaged atmospheric CO_2 levels. Atmospheric CO_2 levels prior to 1958 were estimated from analyses of gas trapped in polar ice cores. *Source*: From W. M. Post, T.-H. Peng, W. R. Emmanuel, A. W. King, H. Dale, and D. L. DeAngelis, reprinted with permission from *American Scientist*, vol. 78, p. 314, copyright © 1990 by Sigma Xi, Raleigh, NC. Data from U. Siegenthaler, H. Friedli, H. Loetscher, E. Moor, A. Neftel, H. Oeschger, and B. Stauffer, reprinted with permission from *Annals of Glaciology*, vol. 10, p. 152, copyright © 1988 by the International Glaciological Society, Cambridge, England; H. Friedli, H. Lötscher, H. Oeschger, U. Siegenthaler, and B. Stauffer, reprinted with permission from *Nature*, vol. 324, p. 237, copyright © 1986 by Macmillan Journals, Ltd., London, England; and A. Neftel, E. Moore, H. Oeschger, and B. Stauffer, reprinted with permission from *Nature*, vol. 315, p. 45, copyright © 1985 by Macmillan Journals, Ltd., London, England.

oxides also appear likely to mitigate some of the anthropogenic greenhouse effect as they are photochemically converted to sulfate aerosols which reflect insolation.

WHERE DOES THE ANTHROPOGENIC CARBON DIOXIDE GO?

Based on our record of fossil fuel use (Figure 25.8) and changes in atmospheric concentrations (Figure 25.11), it appears that approximately half of the anthropogenic CO_2 emissions have remained in the atmosphere. Marine

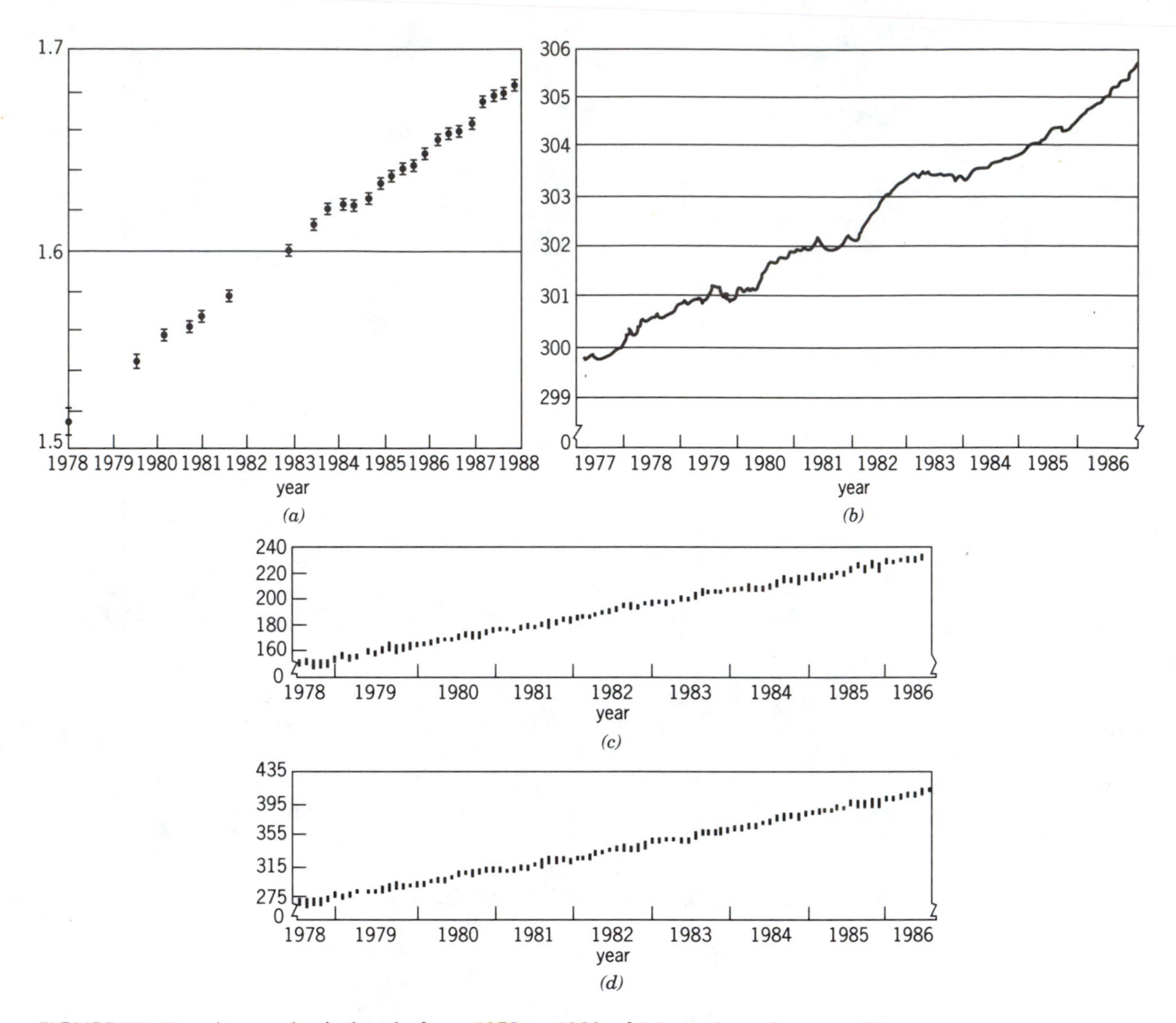

FIGURE 25.12. Atmospheric levels from 1978 to 1988 of (*a*) methane in ppm, (*b*) N_2O in ppb, (*c*) $CFCl_3$ in ppt, and (*d*) CF_2Cl_2 in ppt in the Northern Hemisphere. *Source*: From B. Hileman, reprinted with permission from *Chemical and Engineering News*, March 13, 1989, p. 31, copyright © 1989 by the American Chemical Society, Washington, DC. Methane data courtesy of D. Blake, University of California at Irvine, Irvine, CA. N_2O data courtesy of R. F. Weiss, Scripps Institute of Oceanography, La Jolla, CA. CFC graphs are from *The Changing Atmosphere*, R. G. Prinn (eds.: F. S. Rowland and I. S. A. Isaksen), copyright © 1988 by John Wiley & Sons, Inc., New York, p. 38. Reprinted by permission. See Prinn (1988) for data sources.

TABLE 25.3
Largest Contributors of Greenhouse Gases

Country	*Total Annual Greenhouse Gas Input (%)*
United States	18
USSR	12
Brazil	11
China	7
India	4

Source: Courtesy of Greg Marland, Oak Ridge National Laboratory, Oak Ridge, TN; and *World Resources 199–90*, copyright © 1990 by the World Resources Institute, New York.

chemists estimate that approximately 70 percent of the missing CO_2 has been absorbed by the ocean. The fate of the remaining 30 percent is not known, but some likely sinks are discussed below.

Absorption of Carbon Dioxide by the Ocean

The ocean has the potential to absorb large quantities of CO_2 for two reasons: First, gas solubility is enhanced by reaction with carbonate ($CO_3^{2-} + H_2O + CO_2 \rightarrow 2HCO_3^-$). The equilibrium constant for this reaction is so large that most of the CO_2 entering the ocean is rapidly converted into bicarbonate. Second, the amount of carbon that can be absorbed by the ocean is greatly augmented by the sinking of biogenic POM and calcium carbonate into the deep sea. As a result of this "biological pump", the bottom waters contain so much CO_2 that they are, on average, supersaturated by about 30 percent with respect to the NAEC. This disequilibrium is maintained by vertical density stratification that keeps remineralized CO_2 trapped in the cold, deep sea. The CO_2 cannot escape to the atmosphere until thermohaline circulation returns this deep water to the sea surface. As noted in Chapter 10, the efficiency of the biological pump is determined by the type and abundance of phytoplankton, as well as the pathways and rates of thermohaline circulation.

Once at the sea surface, a water mass undergoes gas exchange with the atmosphere. The direction of net exchange is determined by the difference in gas concentration that exists between the atmosphere and the surface waters. If the atmosphere has a higher partial pressure of CO_2, gas will diffuse into the sea surface. Net gas uptake will continue until the surface seawater acquires its NAEC or sinks below the mixed layer. Gas transfer rates depend on the magnitude of this difference in partial pressures, as well as on water temperature, wind velocity and sea state.

Gas transfer across the air–sea interface tends to be slower than other processes that affect the partial pressure of CO_2 in the surface waters. Thus most surface waters are not in gaseous equilibrium with the atmosphere. As shown in Figure 25.13, equatorial waters tend to be supersaturated. At these locations, Ekman transport causes the upwelling of cold, alkaline seawater. Warming at the sea surface causes this water to become supersaturated with respect to CO_2 and calcium carbonate. Thus upwelling is a net source of CO_2 to the atmosphere. At high latitudes, horizontal advection supplies warm waters, which become undersaturated upon cooling and hence absorb CO_2 from the atmosphere. The size of this sink is increased by photosynthesis, especially in subpolar waters, where nutrient concentrations are relatively high.

The gradients illustrated in Figure 25.13 can be converted to net CO_2 fluxes by multiplying by a transfer factor. These factors are functionally similar to the gas diffusion coefficients used in the gas exchange models presented in Chapter 6. The resulting fluxes are shown in Figure 25.14. Despite some regional variations, the ocean as a whole appears to be a net sink for CO_2. A similar result has been obtained from calculations based on radiocarbon distributions in deepwater DIC. Both calculations suggest that somewhere between 25 to 40 percent of the CO_2 produced by fossil-fuel burning has been absorbed by the ocean.

Due to upwelling, the equatorial Pacific is the single largest source region. This area accounts for a little over half of the CO_2 escaping from the ocean. During 1982 and 1983, this region became a net sink due to the occurrence of an intense El Niño/Southern Oscillation event, which halted upwelling. The subantarctic belt is the largest CO_2 sink, accounting for 35 percent of the total oceanic absorption.

Most of the gas that has been absorbed by the ocean is probably contained in the mixed layer. This limited penetration is the result of strong vertical density stratification at the latitudes where most of the anthropogenic CO_2 has been emitted. The only exception to this would be the CO_2 that has been transported by winds to high latitudes. At these locations, the sinking of deep water can transport CO_2 below the mixed layer. This process is likely to have absorbed only a small amount of CO_2 thus far, as its rate is limited by wind transport and the relatively small area over which it occurs.

In contrast, anthropogenic carbon can be transferred into the deep sea at all latitudes by sinking in the form of biogenic detritus. In this case, the ocean's capacity to absorb CO_2 would be limited by the efficiency with which biogenic particles are recycled in the surface waters and by nutrient availability. Thus on short time scales, most CO_2 transfer to the deep ocean would be accomplished by this biological pump.

On time scales greater than 1000 y, the ocean's capacity to absorb CO_2 will be considerably augmented by dissolution of sedimentary calcium carbonate. On these time scales, thermohaline circulation will transport CO_2-rich surface waters into the deep sea, bringing them into contact with the sediments. Some of this CO_2 will then react with $CaCO_3$, thereby lowering the partial pressure in the water mass. When the water mass returns to the sea surface, it will be able to absorb more CO_2. Thus a significant amount of time

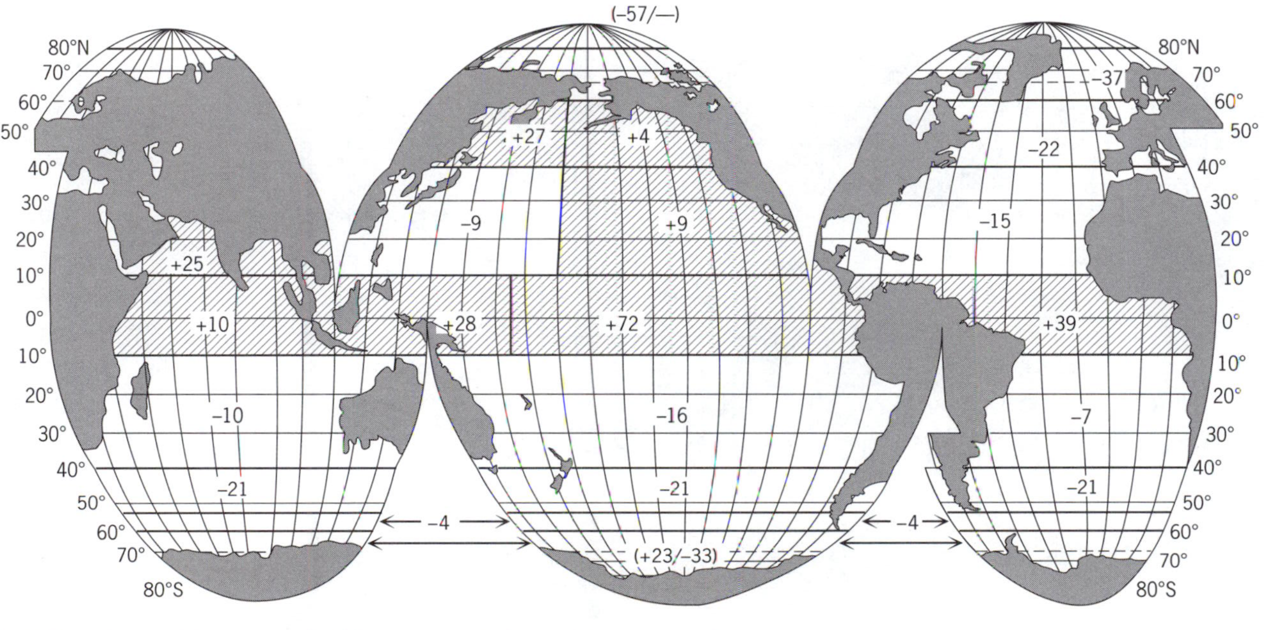

FIGURE 25.13. Mean annual differences in CO_2 partial pressures between surface seawater and the overlying air (ΔpCO_2). The values are given in microatmospheres. Positive values indicate that seawater is supersaturated and negative values indicate that seawater is undersaturated with respect to CO_2. *Source*: From T. Takahashi, reprinted with permission from *Oceanus*, vol. 32, p. 26, copyright © 1989 by Oceanus, Woods Hole, MA.

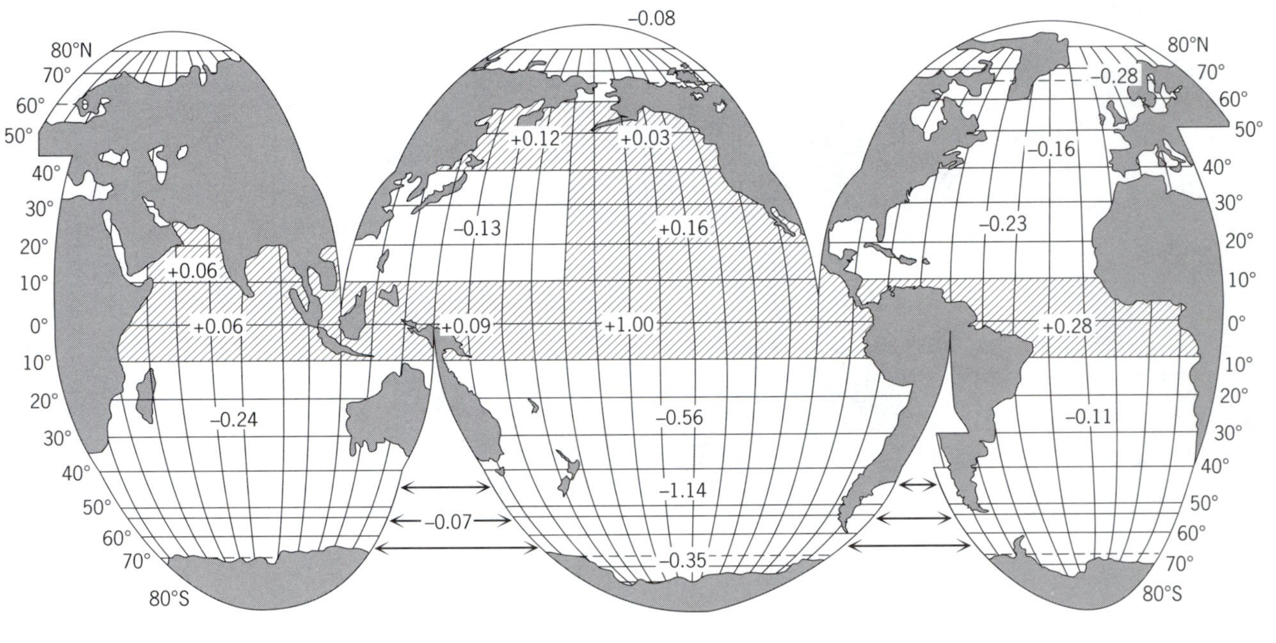

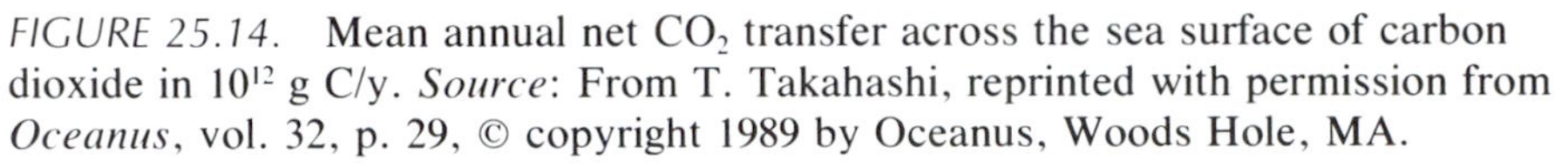
FIGURE 25.14. Mean annual net CO_2 transfer across the sea surface of carbon dioxide in 10^{12} g C/y. *Source*: From T. Takahashi, reprinted with permission from *Oceanus*, vol. 32, p. 29, © copyright 1989 by Oceanus, Woods Hole, MA.

will have to elapse before the ocean realizes its maximum carbon-absorbing potential. The magnitude of this potential is thought to be large, as the unperturbed ocean had a DIC reservoir 60 times bigger than the atmospheric pool of CO_2. Due to the considerable time lag in the ocean's response to increased atmospheric CO_2 levels, this ratio is not likely to be reestablished as CO_2 emission rates are continuing to increase.

The Missing Sink

The data in Figure 25.14 suggest that the ocean absorbs 1.6 gigatons of carbon per year. The annual emission rate of fossil-fuel derived CO_2 is 5.3 gigatons and the atmospheric increase of CO_2 is 3.0 gigatons. Thus 0.7 gigatons, or 13 percent of the annual emissions, are being consumed by some other sink in the global carbon cycle.

The missing sink represents only 4 percent of net annual terrestrial primary production or about 3 percent of the ocean–atmosphere exchange. Current measurements of these fluxes are so crude that this difference can not presently be resolved. Thus the missing carbon sink could just be the result of imprecision in flux estimates. If this sink is real, it is probably the result of an increase in terrestrial biomass. Though sewage dumping and runoff into the coastal ocean could be stimulating marine production, there is no direct evidence for such an increase.

Some scientists have interpreted an increasing amplitude in the seasonal oscillations of atmospheric CO_2 levels (1–2%/y) as evidence for an increase in plant activity. But this does not necessarily lead to increased carbon storage as any concurrent increase in temperature will also accelerate respiration rates. Increasing levels of atmospheric CO_2 are unlikely to enhance plant growth rates since other factors, such as temperature and nutrient availability, have much larger impacts. Thus, any increase in biomass is thought to have been caused by the intensive replanting of the northern temperate forests. These areas were almost completely harvested in the 1800s.

PAST VARIATIONS IN CLIMATE AND ATMOSPHERIC CARBON DIOXIDE LEVELS

With so much at stake for humanity, there is great political and scientific interest in predicting the effect of CO_2 emissions on the atmosphere and ocean. This is currently being done with sophisticated mathematical models. These models are tested by using them to "predict" global climate changes that occurred in the past. Evidence for these changes has been obtained from historical temperature records, tree rings, chemical properties in marine sediments and ice cores, as well as from marine and terrestrial fossils. The testing of these models has also provided great insight into the causes of these past climate changes.

In polar regions, partial melting of surface snow and ice occurs during the "summer" months. Refreezing in winter produces an ice layer that has a texture quite different from the underlying unmelted snow. As a result, the annual layers of snow and ice are readily identifiable and can be used to date the time of their deposition. Since snow is porous, small pockets of air are trapped in its pores when the surface layer of ice freezes. These air pockets represent a record of atmospheric gas composition. Though difficult to accomplish, ice cores can be obtained without exposing the ancient air pockets to the present atmosphere. Analyses of the gas in these cores have shown that the partial pressure of atmospheric CO_2 has undergone large oscillations over time. As shown in Figure 25.15*a*, these swings correlate closely with atmospheric temperature changes which have occurred over the past 160,000 y. In comparison, the relationship between atmospheric temperature and CO_2 levels has been far more variable over the past 100 y (Figure 25.15*b*).

Cold temperatures appear to be associated with decreased levels of atmospheric CO_2. These shifts are likely driven by external forces that affect the global heat budget. Over the past 1 million years, large climate changes have been the result of alternating periods of glaciation and deglaciation. The timing of these ice ages suggests that the external driving forces are the result of shifts in Earth's orbit relative to the sun and to sunspot activity. Over longer time scales, changes in the rate of volcanic activity and mantle convection are also likely to be important. These geological phenomena affect global climate by producing volcanic ash, which reflects sunlight, and by releasing CO_2 into the atmosphere.

Earth's distance from the sun is continuously altered by predictable shifts in the tilt and orientation of its spin axis, as well as in the eccentricity of its orbit. The interaction of these factors causes the intensity of insolation to vary over time. As a result, summertime insolation reaching the Northern Hemisphere has varied by as much as 20 percent over the past 1 million years. In addition, the intensity of the seasons has varied over time. Though periods of cooler summers should be accompanied by milder winters, the absence of very high temperatures is thought to have caused glaciers to grow. This appears to have triggered the ice ages.

Over the past 1 million years, global ice volume has peaked every 100,000 y, which is the period of the eccentricity variation. Smaller peaks have occurred at frequencies of 23,000 and 41,000 y, which match the periods of the precession and tilt variations. (These variations are termed **Milankovitch cycles** after their discoverer.) The amplitudes of these smaller peaks vary over time in a fashion that is predictable from interactions among the orbital effects. But the amplitude of the 100,000-y variation is much larger than predicted from the orbital motions. Due to its large amplitude, geochemists suspect that whatever drives the 100,000-y cycle also sets the fundamental frequency at which glaciation occurs.

The rate at which continental glaciers form has not been uniform, even over short time scales. The paleoceanographic record suggests that during an ice age, ice volumes tend to increase in an episodic fashion for approximately

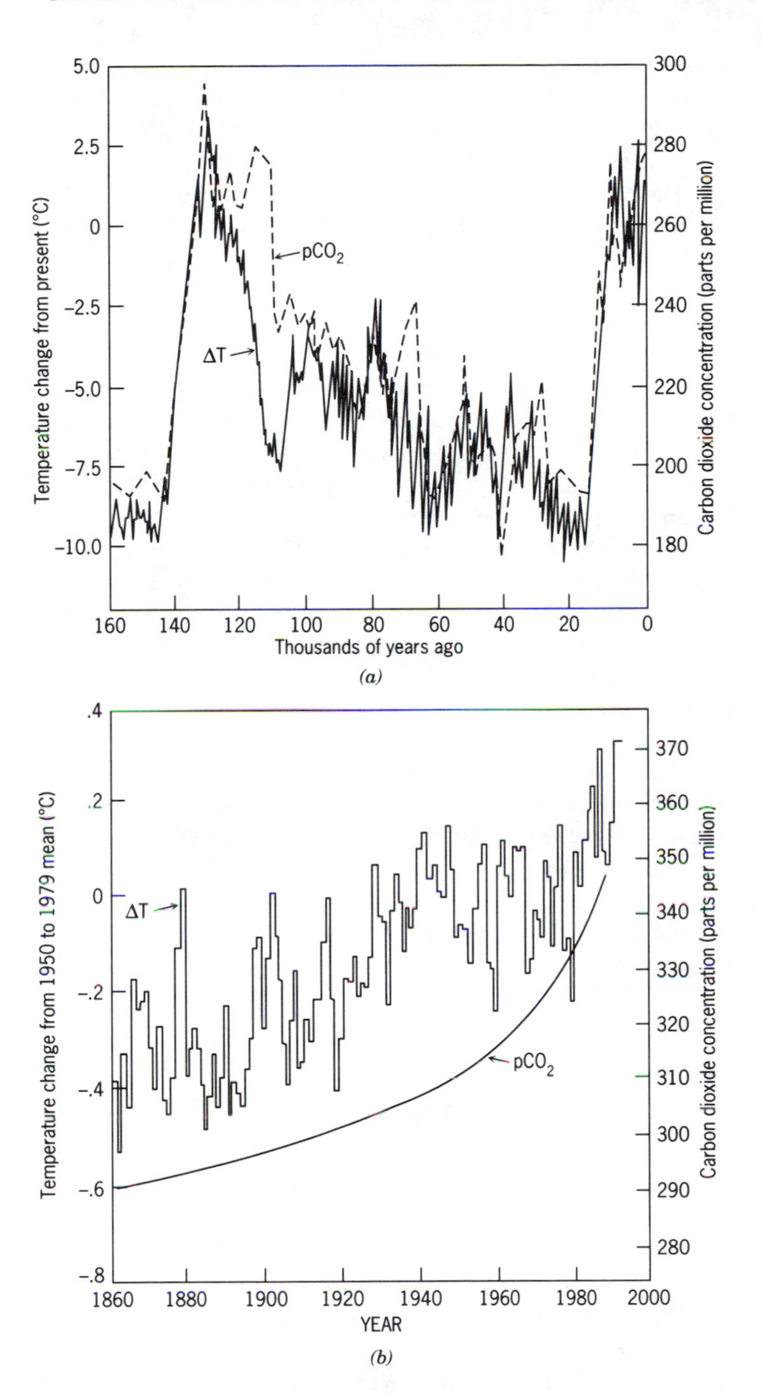

FIGURE 25.15. Changes in atmospheric temperature and carbon dioxide concentrations (*a*) over the past 160,000 years and (*b*) over the past 130 years. *Source*: From S. H. Schneider, reprinted with permission from *Scientific American*, vol. 261, p. 74, copyright © 1989 by Scientific American, New York.

100,000 y and then abruptly decline as a result of increasingly warmer summers. This effect is most pronounced in the Northern Hemisphere because the presence of a large land mass at mid- and subpolar latitudes supports the formation of large continental ice sheets. In the Southern Hemisphere, the amount of land is relatively small and limited to low latitudes. Hence, in the Southern Hemisphere, continental glaciers form only on mountaintops. Ice cores obtained from such glaciers in Hawaii, Columbia, New Guinea, and East Africa, suggest that their ice volumes have changed at about the same frequency as the northern ones. This evidence supports the occurrence of global climate change.

Other pieces of evidence indicate that changes in the sizes of these ice sheets did not have a large impact on the timing or severity of global climate change. Instead, shifts in the interaction between the atmosphere and ocean appear to have caused the large changes in ice volume which have occurred over the past 1 million years. Because of this interaction, the atmosphere and ocean behave as a coupled system that acts to transport and store heat. The operation of this system is dependent on the rates of air and water circulation. Shifts in these circulation patterns appear to cause global climate change. Over the past 1 million years, such shifts are thought to have caused the planet to switch abruptly from glacial to interglacial conditions. Although these global climate changes are ultimately driven by astronomical forces, they also appear to be amplified by feedbacks associated with the global hydrological and carbon cycles.

The rates of rainfall and evaporation at the sea surface are controlled by interactions between the ocean and atmosphere. Changes in these rates are likely to have nonlinear impacts on thermohaline circulation. For example, a shift in atmospheric circulation could affect the relative rates of precipitation and evaporation and thus alter surface-water salinities. If surface-water salinity is decreased, the rate of deep-water formation will decline because the maximum density that a water mass can attain is limited by its freezing point. Thus a decrease in salinity would slow the rate of thermohaline circulation. Such a shift appears to have interrupted the general warming trend of the past 18,000 y.

This shift occurred about 11,000 y before present and is thought to have been caused by a massive influx of freshwater from the melting of the North American Ice Sheet. (This hypothesis has been proposed by Dr. W. S. Broecker of the Lamont-Doherty Geological Observatory.) For the first 7000 y after the last ice age, melt waters appear to have flowed down the Mississippi River into the Gulf of Mexico. Eventually, the combined effects of the northward retreat of the glaciers and the overflow of meltwaters from existing lakes caused freshwater to start flowing across the Great Lakes and down into the St. Lawrence River. If this water entered the North Atlantic close to the present site of NADW formation, it would have reduced surface-water salinities. As a result, deep water would have stopped forming, despite winter cooling. This would have led to regional cooling, as the heat that would have otherwise been released by the sinking of deep water was instead retained by the low-salinity surface waters. The reestablishment of thermohaline circulation 1000 y later was probably caused by this cooling as it must have led to the regrowth of ice sheets and hence halted the flow of meltwater to the North Atlantic.

The temporary slowdown in thermohaline circulation must have had a global impact. For example, such a slowdown would have initially increased the efficiency of the biological pump and hence cause atmospheric CO_2 levels to decline. Evidence in support of this has been obtained from the cadmium content of forams. Foraminifera deposit cadmium in their shells in constant proportion to this chemical's relative abundance in seawater. The presence of cadmium in calcium carbonate is not unexpected as its charge and ionic radius are similar to that of calcium.

As shown in Figure 11.6*b*, seawater cadmium concentrations are directly related to phosphate and hence nitrate concentrations. Thus shifts in the cadmium content of the tests can be interpreted as a record of changes in seawater nutrient concentrations. The sedimentary foram record indicates that during ice ages, deep-water nutrient concentrations in the Atlantic were higher than in its mid waters. In addition, the horizontal deep-water nutrient gradient was smaller. Thus both lines of evidence indicate that NADW production was very slow or absent during the last ice age. Hence the bottom waters must have accumulated large amounts of biogenic detritus and thereby decreased atmospheric CO_2 levels.

This scenario suggests that ice ages were characterized by reduced rates of thermohaline circulation and lower atmospheric temperatures and CO_2 concentrations. Computer models indicate that the decrease in atmospheric CO_2 was insufficient to have caused the cooling that occurred during these periods. Instead, positive feedbacks probably contributed to the global decline in temperature. For example, cooling increases aridity and hence atmospheric dust levels. Dust increases the amount of reflected insolation, thereby enhancing the cooling effect. The drying of swamps would also cause a decrease in methane production. Other feedbacks are likely to involve changes in cloud and ice cover. In addition, a slowdown in thermohaline circulation should also have led to a decrease in deep-water O_2 concentrations. This change in redox level would have favored the preservation of sinking POM, thereby increasing its burial rate. The resulting removal of nutrients would have eventually limited the ocean's ability to remove atmospheric CO_2 via the biological pump.

Increased seasonality during interglacial times should have led to increased production of deep water and hence increased rates of thermohaline circulation. The ensuing decrease in efficiency of the biological pump would have decreased the storage of DIC in the deep water and caused atmospheric CO_2 levels to rise. This change in atmospheric composition would have enhanced the greenhouse effect, thereby amplifying the warming trend initiated by the Milankovitch cycles. Examples of other positive feedbacks that would have accelerated the warming trend include (1) faster evaporation rates, which would have increased the amount of water vapor in the atmosphere; (2) more melting of ice, which would have caused less insolation to be reflected; and (3) decreased CO_2 solubility caused by higher surface-water temperatures. Thus interglacial periods are thought to be characterized by relatively fast rates of thermohaline circulation and high atmospheric temperatures and CO_2 levels.

The hypothesis that global climate operates in "quantized" modes, such as the glacial and interglacial ones, is supported by the simultaneous shifts in several paleoclimatic indicators (e.g., the onset of glaciation, changes in plankton speciation, and shifts in atmospheric CO_2 levels). The timing of the mode shifts over the past 1 million years has not been exactly in phase with the external driving force (i.e., the Milankovitch cycles). This asynchrony is probably caused by nonlinear interactions among the positive feedbacks. In particular, the ocean–atmosphere system appears most susceptible to a mode shift once the ice sheets reach a critical size. This seems to have occurred with a frequency of 100,000 y over the past millennium.

The past millennium is not characteristic of the entire geologic history of this planet. The sedimentary record indicates that ice volumes did not peak every 100,000 y during the period from 1 to 2 million years before present. Evidently the response of global climate to external forces changes over long time scales. At present, anthropogenic emissions of CO_2 have caused atmospheric concentrations to reach levels that exceed the maximum attained during the height of deglaciation during the Pleistocene. Thus, it is possible that we are pushing the ocean–atmosphere system into an entirely new mode of climate regulation. The consequences of this could be severe, as mode shifts appear to have been accompanied by abrupt changes in environmental conditions.

CLIMATIC PREDICTIONS FOR THE NEXT CENTURY

Computer models indicate that the anthropogenic increase in greenhouse gases has altered the global radiation balance in a fashion roughly equivalent to a 1 percent increase in the sun's luminosity. This should have caused a 0.8° to 2.6°C increase in global mean temperatures. The observed increase is 0.5° to 1.3°C because the ocean has absorbed a considerable amount of the added heat. Thus the onset of climate change has been somewhat delayed by the oceanic absorption of heat. Since this sink is large, it will continue to moderate climate change for an extended period.

This heat absorption should cause sea level to eventually rise by 0.2 to 1.5 m as a result of the thermal expansion of seawater. Since we are currently in a warming period following a period of glaciation, sea level had already been rising at a natural rate of 2 mm per year. As shown in Figure 25.16, the melting of continental glaciers is also causing sea level to rise.

Computer models have also been used to predict the results of some "what-if" scenarios. Several such projections are given in Figure 25.17. In scenario A, our current rates of growth in greenhouse emissions, which average 1.5

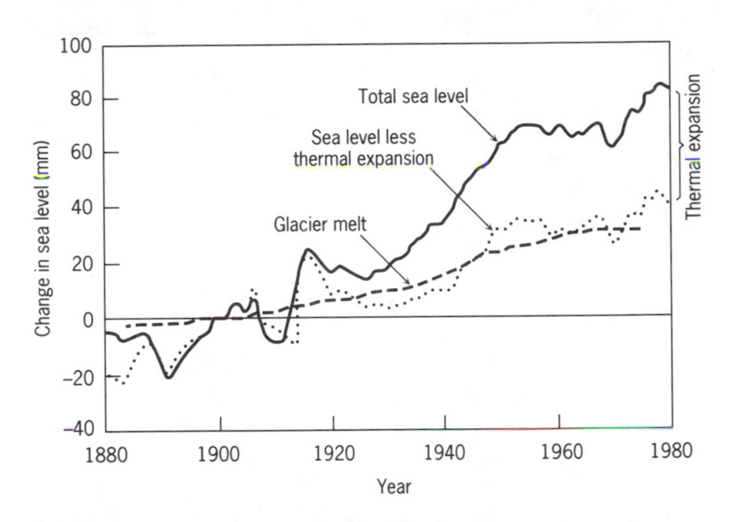

FIGURE 25.16. Role of thermal expansion and glacier melt in sea-level rise. The light line is an estimated sea-level curve predicted from natural rates of warming and the effect of increased atmospheric CO_2 levels. The rise in sea level caused solely by the melting of glaciers is shown by the heavy solid line. Subtraction of the glacial melt effect from the total sea-level curve yields the dotted line, which is interpreted as sea-level rise caused by thermal expansion. *Source*: From M. F. Meier, reprinted with permission from *Science*, vol. 226, p. 1420, copyright © 1984 by the American Association of Advancement of Science, Washington, DC.

percent annually, are continued. In scenario B, greenhouse emissions are maintained at our current rate of production. Scenario C illustrates the effect of cutting greenhouse gas production such as to halt any further increases in atmospheric levels by 2000 A.D. Note that even with drastic cuts, mean global atmospheric temperatures are predicted to remain almost 1°C higher, with larger increases at high latitudes.

If the present rate of growth in greenhouse gas emissions is continued, a 2° to 3°C increase in mean global atmospheric temperatures will probably occur. Temperatures are likely to increase by 4° to 6°C at high latitudes. This shift is similar in magnitude to the amount of warming that has occurred since the last ice age ended 18,000 y ago, but will take place 10 to 100 times faster. Though the remaining recoverable fossil fuel reservoir is relatively small, it is much larger than the atmospheric CO_2 reservoir and hence has the potential to wreak havoc if rapidly released. As noted above, rapid changes appear to cause positive feedbacks that result in abrupt shifts in global processes.

Most dramatic would be the loss of the Antarctic Ice Sheet, which comprises most of the polar ice. Partial melting could cause large chunks of this continental glacier to slide into the ocean, displacing large volumes of water. In 1987, the largest iceberg on record was observed floating in the Weddell Sea. Its surface area was equivalent to that of Rhode Island. If the entire Antarctic Ice Sheet were to slide into the ocean bit by bit over the next 500 y, sea level would rise by 5 m.

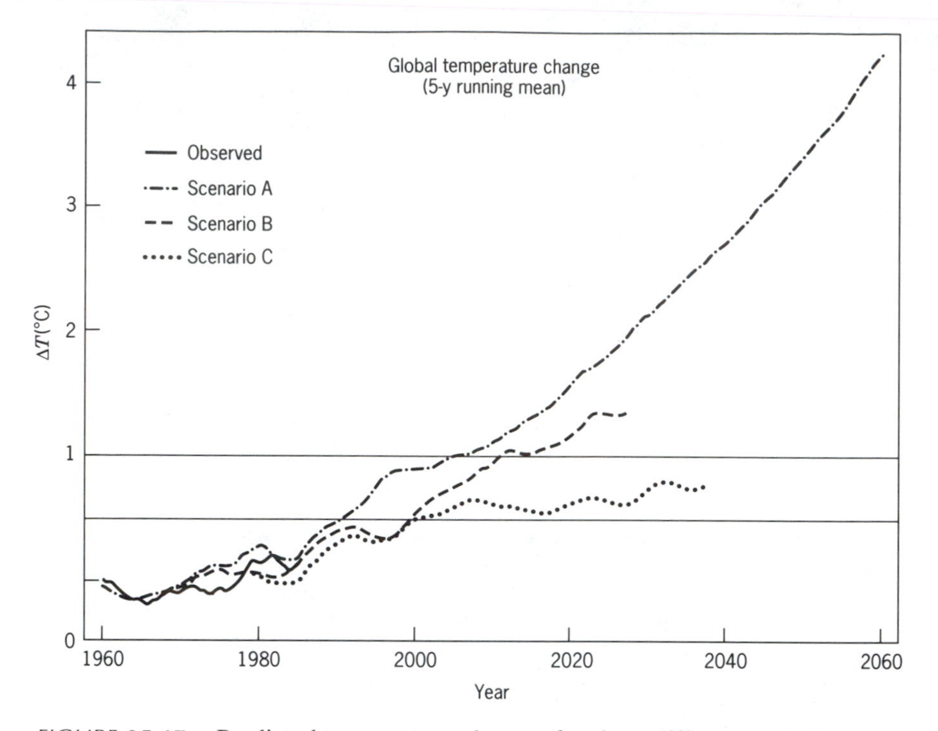

FIGURE 25.17. Predicted temperature changes for three different emission scenarios. The temperature is given as the change relative to the 1951–1980 global average. Scenario A: current rate of increased greenhouse gas production is continued. Scenario B: greenhouse gas production rate is maintained at current levels. Scenario C: greenhouse gas production is cut such that atmospheric levels stabilize by 2000 A.D. *Source*: Courtesy of J. E. Hansen, NASA Goddard Institute for Space Studies, New York.

The temperature rise is also likely to be amplified by a positive feedback caused by the warming of marine sediments. Such a warming would release vast quantities of methane due to the melting of methane–water clathrates. As shown in Figure 25.18, these clathrates are cagelike arrangements of water molecules which act to trap methane in the sediment. They contain enough methane to cause a 1° to 2°C rise in mean global atmospheric temperatures.

Ensuing changes in rainfall patterns would greatly affect the survival of plant species. After the last ice age, as temperatures rose 1° to 2°C every thousand years, the forest belts migrated poleward at a rate of 1 km per year. Since the current rate of temperature change is much faster, the forest belts will be forced to migrate faster. If they cannot, which seems likely, these forests will be destroyed and species driven to extinction. The resulting deforestation will also enhance global warming due to the increase in CO_2 and aridity.

At this point, no plausible policies are likely to prevent a 1° to 2°C warming. The impact of this warming will be largest at the temperate and subpolar

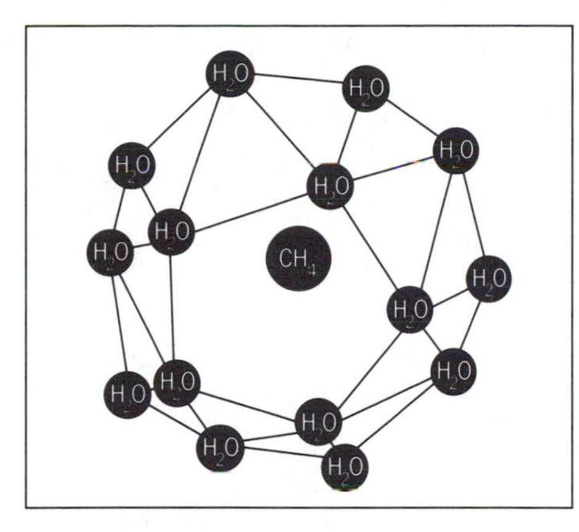

FIGURE 25.18. Molecular structure of methane-clathrate.

latitudes, causing changes in rainfall patterns and shifts in plant species. Some latitudes will benefit by an increase in the length of their growing season; others will be negatively impacted. In other words, a global redistribution of climate resources will occur with the climate and environment behaving in a fashion that is outside historical experience.

Our understanding of Earth's atmospheric and oceanic circulation is still too rudimentary to predict the rate and extent of these climatic and environmental changes. This is largely due to our poor knowledge of the terrestrial biosphere's and ocean's response to changes in the global carbon cycle. Another major problem is our inability to create a computer program detailed enough to accurately model small-scale climatic variability. Nevertheless, significant changes will occur. Hence the "CO_2 problem" cannot be ignored.

SUMMARY

The marine biogeochemical cycle of carbon is largely controlled by biologically mediated processes. As a result of photosynthetic carbon fixation and the deposition of calcareous shells, most of the juvenile carbon released into the atmosphere has been sequestered in marine sediments. Though the atmospheric reservoir of carbon is relatively small, it is biologically very active. It also plays an important role in regulating global climate because atmospheric carbon is primarily in the form of CO_2, which is a greenhouse gas. As a result of gas exchange across the air–sea interface, the marine cycle of carbon also plays an important role in regulating global climate.

The heat balance on this planet is maintained at a steady-state level determined by the amount of greenhouse gases in the atmosphere. Thus the CO_2 that is naturally present in the atmosphere helps keep Earth's surface at temperatures hospitable to life. An increase in greenhouse gases will cause atmospheric temperatures to rise. Such an increase is occurring as a result of CO_2 emissions from the burning of **fossil fuels** and deforestation, as well as from the production of other greenhouse gases, such as chlorofluorocarbons, methane, and nitrogen oxides.

Approximately half of the CO_2 that has been released as a result of fossil-fuel burning has remained in the atmosphere. The ocean appears to have absorbed 25 to 40 percent of the total emissions. The rest is thought to have been consumed by an increase in terrestrial biomass caused by the replanting of Northern Hemisphere forests. The ocean is able to absorb large amounts of CO_2 because the solubility of this gas is enhanced by the carbonate buffering reactions and dissolution of $CaCO_3$. On short time scales, the rate of CO_2 uptake is controlled by phytoplankton production as this determines the amount of biogenic POC and calcite that sink into the deep sea. The operation of this "biological pump" is affected by the rate of thermohaline circulation.

The past 1 million years of Earth's history have been characterized by periods of glaciation and deglaciation that appear to have driven by the **Milankovitch cycles**. These astronomical variations cause changes in Earth's distance from the sun. During times of cooler summers, ice sheets appear to experience net growth and trigger the occurrence of ice ages. These climate changes appear to be amplified by alterations in the marine carbon cycle that are driven by variations in the rate of thermohaline circulation. Because of nonlinear interactions between these and other positive feedbacks, atmospheric fluctuations in CO_2 levels and temperature have been slightly asynchronous over the past 1 million years.

The past millennium is not characteristic of the entire geologic history of this planet. The sedimentary record indicates that ice volumes have not always peaked every 100,000 y. Evidently the response of global climate to external forces changes over long time scales. At present, anthropogenic emissions of CO_2 have caused atmospheric concentrations to reach levels that exceed the maximum attained during the height of deglaciation in the Pleistocene. Thus, it is possible that we are pushing the ocean–atmosphere system into an entirely new mode of climate regulation. The consequences of this could be severe, as mode shifts appear to have been accompanied by abrupt changes in environmental conditions.

Current predictions suggest that our atmosphere will continue to warm and that sea level will rise. The impact of this warming will be largest at the temperate and subpolar latitudes, causing changes in rainfall patterns and shifts in plant species. Some latitudes will benefit by an increase in the length of their growing season; others will be negatively impacted. In other words, a global redistribution of climate resources will occur with the climate and environment behaving in a fashion that is outside historical experience.

Our understanding of Earth's atmospheric and oceanic circulation is still too rudimentary to predict the rate and extent of these climatic and environmental changes. This is largely due to our poor knowledge of the terrestrial biosphere's and ocean's response to changes in the global carbon cycle. Another major problem is our inability to create a computer program detailed enough to accurately model small-scale climatic variability. Nevertheless, significant changes will occur. Hence the "CO_2 problem" cannot be ignored.

CHAPTER 26

THE ORIGIN OF PETROLEUM IN THE MARINE ENVIRONMENT

INTRODUCTION

In coastal upwelling areas, the underlying sediments are enriched in organic matter due to high biological productivity and relatively shallow water depths. Given appropriate environmental conditions, diagenesis and catagenesis can convert this organic matter to petroleum. As noted in the previous chapter, these deposits contain less than one-third of the sedimentary carbon; the majority is buried as biogenic carbonate.

The global rate of petroleum formation is slow due to the limited size of upwelling areas. The carbon stored in these deposits is being rapidly returned to the atmosphere as a result of fossil-fuel burning. The accelerated rate at which these sedimentary deposits are being recycled has perturbed the biogeochemical carbon cycle. As noted in the preceding chapter, the resulting CO_2 emissions are a partial cause of the anthropogenic greenhouse effect. Nevertheless, petroleum remains the major source of energy fueling the world economy. Thus considered effort has been expended to determine how and where petroleum has formed. As discussed below, this information is used to determine likely sites of undiscovered deposits.

WHAT IS PETROLEUM?

Petroleum is a complex and variable mixture of organic compounds. Most of the low-molecular-weight compounds are hydrocarbons. Those with less than five carbons are gases at room temperature. These **natural gases** include methane, ethane, propane, and butane. The larger molecules are liquids and solids at room temperature. The liquid, or **crude oil**, portion of petroleum is usually separated into various fractions by distillation. These fractions are mixtures of compounds that are similar in structure and size, as these factors determine their boiling points.

As indicated in Figure 26.1, the **gasoline** fraction is composed primarily of **paraffins** and **naphthenes**. The former are saturated straight-chain, or aliphatic, hydrocarbons, which have five to ten carbons. The naphthenes are cycloparaf-

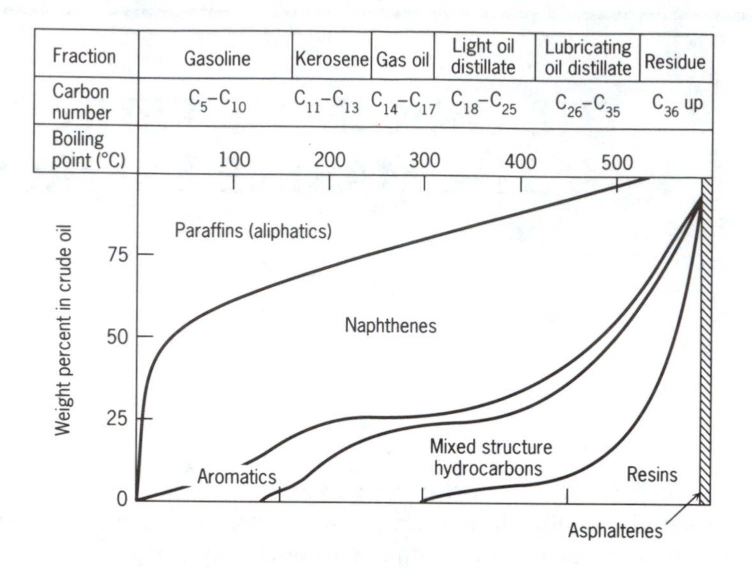

FIGURE 26.1. Distribution of hydrocarbon classes in a medium crude oil. *Source*: From *Fundamental Aspects of Petroleum Geochemistry*, M. A. Bestougeff (eds.: B. Nagy and U. Columbo), copyright © 1967 by Elsevier Science Publishers, Amsterdam, The Netherlands, p. 78. Reprinted by permission.

fins, such as cyclohexane. As shown in Figure 26.2, most paraffins have a relatively low molecular weight.

Continued distillation leaves a solid residue called **asphalt**, which is composed of **asphaltenes**. These compounds have very high molecular weights and are relatively inert as indicated by their high (~500°C) boiling point. Other solids present in marine petroleum deposits include kerogen and graphite. Deposits formed from terrestrial organic matter, such as peat, also contain lignite and coal. As shown in Table 26.1, the higher molecular weight compounds tend to have significant amounts of other elements, such as sulfur, nitrogen, oxygen, and trace metals. The relative abundances of these elements are variable and are the result of differences in organic matter source and maturation reactions as described below. In terms of fuel use, nitrogen and sulfur represent undesirable impurities. The burning of coal and oil with high nitrogen and sulfur contents produces nitrogen and sulfur oxides. These gases react with water to produce nitric and sulfuric acid, which cause acid rain.

CARBON SOURCES

Some hydrocarbons are synthesized by living organisms, but these biogenic compounds are not particularly abundant. They are also produced in relative abundances that differ from that which is present in petroleum. This dissimilarity suggests that most petroleum hydrocarbons are formed abiogenically by reactions that occur after burial. The organic compounds with the greatest likelihood of being transformed into petroleum hydrocarbons are the ones whose chemical composition is already similar to that of petroleum. This includes the biomolecules that have relatively few multiple bonds and are composed primarily of carbon and hydrogen. As shown in Table 26.2, the class of compounds closest in elemental composition to the paraffins and naphthenes are the lipids.

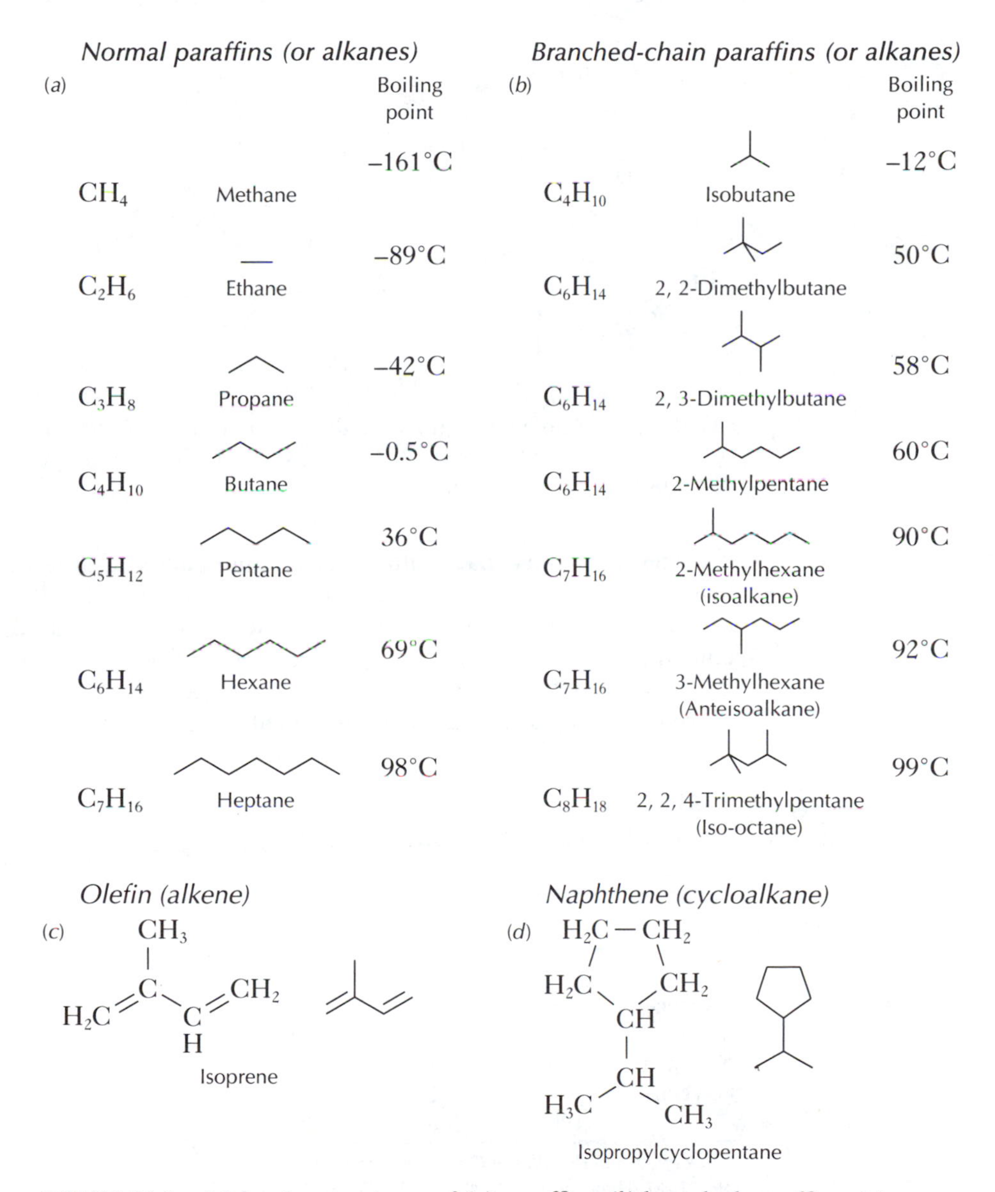

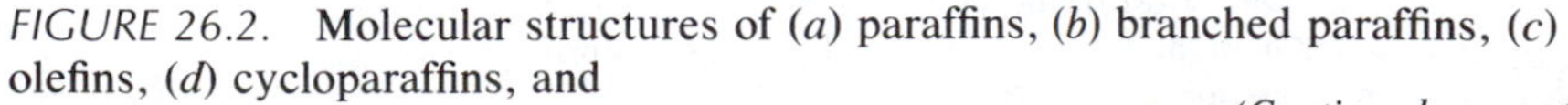

FIGURE 26.2. Molecular structures of (*a*) paraffins, (*b*) branched paraffins, (*c*) olefins, (*d*) cycloparaffins, and

(Continued on next page.)

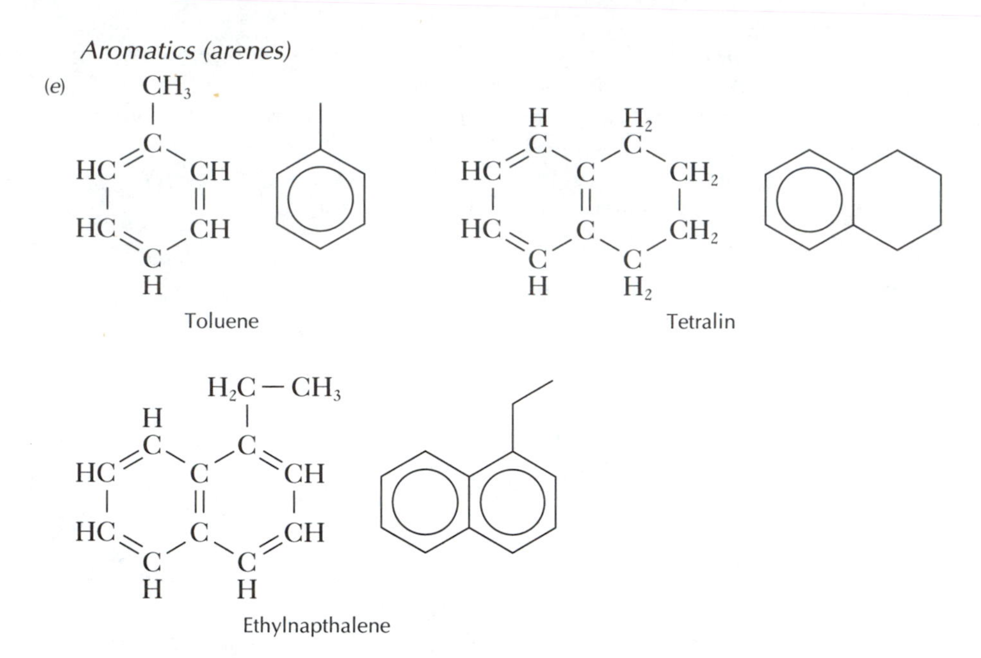

FIGURE 26.2. (Continued) (*e*) aromatics. *Source*: From *Petroleum Geochemistry and Geology*, J. M. Hunt, copyright © 1979 by W. H. Freeman and Co., New York, p. 32. Reprinted by permission.

The organisms that contain these compounds in greatest abundance are listed in Table 26.3. Terrestrial plants contain some lipids, but a significant amount of lignin is also present. These compounds, along with cellulose, constitute the rigid structural parts of terrestrial plants. They are relatively depleted in hydrogen and upon maturation yield gas, rather than oil or coal. Marine and aquatic plants do not synthesize these rather inert materials.

TABLE 26.1
Percent Elemental Composition of Oil, Asphalt, and Kerogen

	Oil	*Asphalt*	*Kerogen*
Carbon	84.5	84	79
Hydrogen	13	10	6
Sulfur	1.5	3	5
Nitrogen	0.5	1	2
Oxygen	0.5	2	8
	100	100	100

Source: From *Petroleum Geochemistry and Geology*, J. M. Hunt, copyright © 1979 by W. H. Freeman and Co., New York, p. 29. Reprinted by permission.

TABLE 26.2
Average Chemical Composition of Natural Substances

	Elemental Composition in Weight Percent				
	C	*H*	*S*	*N*	*O*
Carbohydrates	44	6	0	0	50
Lignin	63	5	0.1	0.3	31.6
Proteins	53	7	1	17	22
Lipids	76	12	0	0	12
Petroleum	85	13	1	0.5	0.5

Source: From *Petroleum Geochemistry and Geology*, J. M. Hunt, copyright © 1979 by W. H. Freeman and Co., New York, p. 85. Reprinted by permission.

Instead, their tissues contain high concentrations of lipids, which are relatively hydrogen-rich, and yield oil and coal, respectively, upon maturation.

DIAGENESIS

Maturation is the process by which biomolecules are converted into petroleum. As shown in Figure 26.3, maturation is initiated by diagenetic changes that occur under anoxic conditions at temperatures less than 50°C. Such conditions

TABLE 26.3
Composition of Living Matter

	Weight Percent of Major Constituents			
Substance	*Proteins*	*Carbohydrates*	*Lipids*	*Lignin*
Plants				
Spruce wood	1	66	4	29
Oak leaves	6	52	5	37
Scots-pine needles	8	47	28	17
Phytoplankton	23	66	11	0
Diatoms	29	63	8	0
Lycopodium spores	8	42	50	0
Animals				
Zooplankton (mixed)	60	22	18	0
Copepods	65	25	10	0
Oysters	55	33	12	0
Higher invertebrates	70	20	10	0

Source: From *Petroleum Geochemistry and Geology*, J. M. Hunt, copyright © 1979 by W. H. Freeman and Co., New York, p. 86. Reprinted by permission.

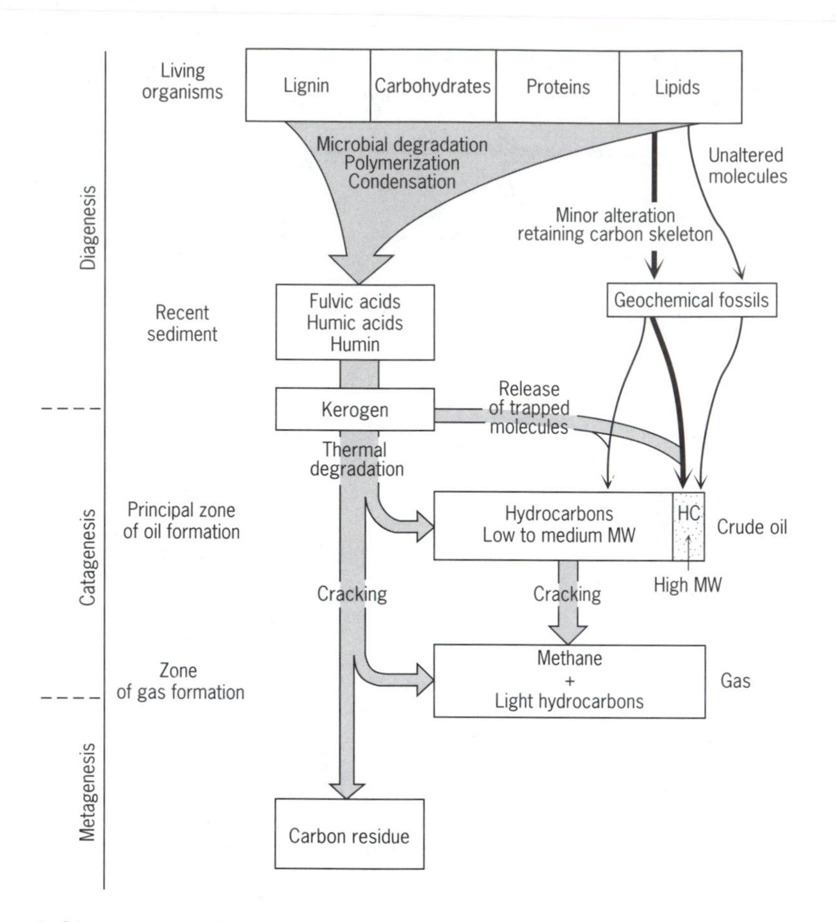

FIGURE 26.3. The petroleum maturation process. *Source*: From *Petroleum Formation and Occurrence: A New Approach to Oil and Gas Exploration*, B. P. Tissot and D. H. Welte, copyright © 1978 by Springer-Verlag, Heidelberg, Germany, p. 93. Reprinted by permission.

are a direct consequence of the large POM flux and lead to extensive amounts of denitrification, sulfate reduction, and methanogenesis. As a result of these metabolic reactions, oxygen is removed from the organic matter as water, nitrogen as N_2, and sulfur as H_2S. If iron is present in the deposit, pyrites precipitate. Long-term burial converts these iron sulfides into banded ironstones, which are common in shales. This mode of sulfur removal enhances the economic worth of the petroleum, as otherwise H_2S, or "sour gas," would remain in the deposit. Methanogenesis produces biogenic methane, which is termed dry gas.

Other reactions cause the rearrangement of some structures. For example, side chains separate from parent molecules, as illustrated for chlorophyll in Figure 26.4. The result of this diagenesis is a complex mixture of macromole-

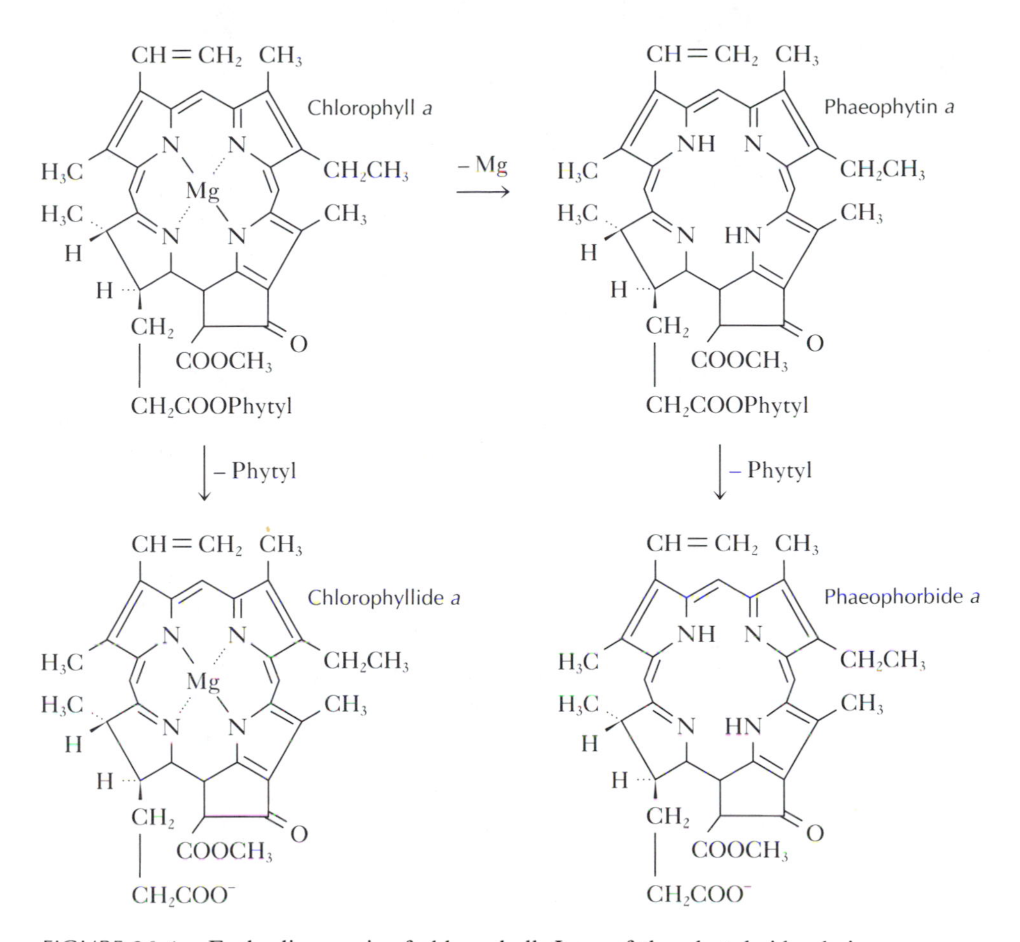

FIGURE 26.4. Early diagenesis of chlorophyll. Loss of the phytyl side chain can occur prior to or after the removal of the central magnesium atom. *Source*: From *Organic Geochemistry of Natural Waters*, E. M. Thurman, copyright © 1985 by Kluwer Academic Publishers, Dordrecht, The Netherlands, p. 266. Reprinted by permission.

cules. This mixture is termed a sapropelic kerogen if it is derived from marine biomolecules. The diagenesis of aquatic organic matter yields a mixture that is termed a humic kerogen. Due to differences in the chemical composition of their sources, the sapropelic compounds tend to be more reduced and have a higher degree of saturation than those in humic kerogen.

Some classes of compounds are not substantially altered during diagenesis and thus represent molecular fossils. For example, the C_{15} through C_{19} and C_{27} through C_{33} *n*-alkanes (or paraffins) are usually abundant in petroleum. These two classes are synthesized by water and terrestrial plants, respectively. As noted above, these compounds do not comprise a significant amount of petroleum.

CATAGENESIS

Most of the compounds that comprise petroleum are thought to be the products of abiogenic catagenetic reactions that occur at temperatures of 50° to 150°C. These temperatures are encountered at depths ranging from 1 to 3 km below the seafloor. Temperatures increase with depth beneath the seafloor for two reasons: First, primordial heat is still being conducted from Earth's interior. Second, the sedimentary overburden creates friction that is dissipated as heat.

During deep burial, the elevated pressures and temperatures cause **cracking reactions** in which short-chain hydrocarbons are broken off large parent molecules (Figure 26.5). This process creates most of the paraffins that are present in the gas and oil fraction. These reactions cause the residual parent material to become increasing depleted in hydrogen, thereby promoting the continued formation of aromatic structures. An example is given in Figure 26.6 in which napthalenes are formed from the closure of straight-chain unsaturated hydrocarbons.

As shown in Figure 26.7, catagenesis transforms sapropelic kerogens into oil, whereas humic kerogens are converted into coal. Both kerogens produce gas but differ in their yields and chemical composition. As shown in Figure 26.8, humic kerogens produce more carbon dioxide because their carbon source (terrestrial organic matter) is more oxidized than marine organic matter. The sapropelic kerogens generate more H_2S because the ready supply of sulfate in seawater supports large amounts of sulfate reduction in marine sediments.

Most oils produced from the maturation of marine organic matter are in the C_{18} to C_{40} size range, as shown in Figure 26.9. An extensive period of time

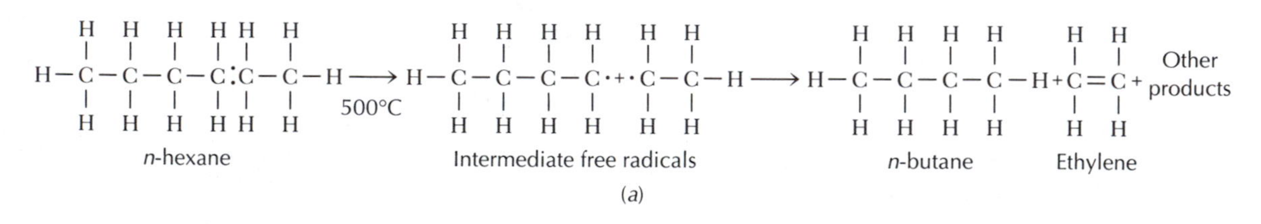

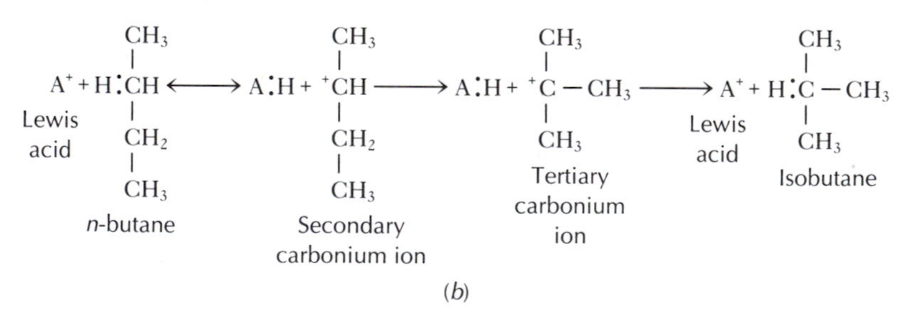

FIGURE 26.5. Mechanisms of cracking reactions: (*a*) Free-radical mechanism and (*b*) carbonium ion reaction. *Source*: From *Geochemistry of Marine Humic Compounds*, M. A. Rashid, copyright © 1985 by Springer-Verlag, Heidelberg, Germany, p. 206. Reprinted by permission.

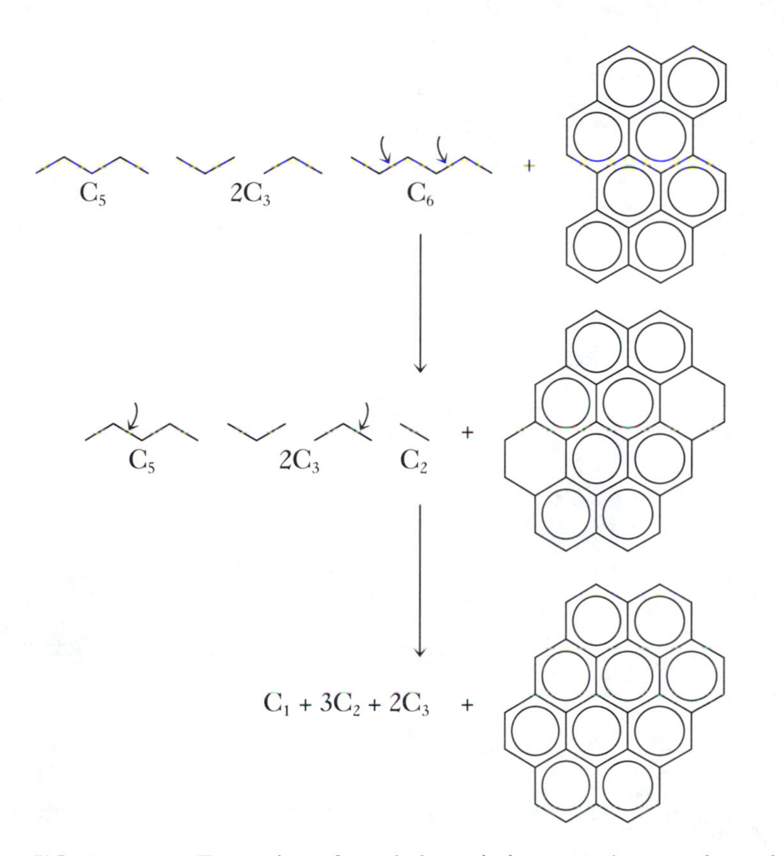

FIGURE 26.6. Formation of napthalenes in kerogen. Arrows show cleavage points. *Source*: From J. Connan, K. LeTran, and B. van der Weide, reprinted with permission from the *Proceedings of the Ninth World Petroleum Congress*, vol. 2, p. 172, copyright © 1975 by Applied Science Publishers, Ltd., Essex, England. Reprinted by permission.

is required for sediments to become deeply buried such that temperatures and pressures are high enough to promote catagenesis. An extended period of time is also required for the deeply buried deposits to mature and product significant quantities of gas and oil. As shown in Figure 26.10, the time required is on the order of 10 to several hundred million years. The sole exception to this are the petroleum hydrocarbons generated in relatively recent sediments as a result of heating near active plate boundaries, such as in the Guaymas Basin.

Since reaction rates increase with temperature, the time required for oil and gas production is shorter at higher temperatures. There is an upper limit to this effect as at very high temperatures and pressures, petroleum is converted into carbon dioxide, methane, and graphite. Thus deposits that have been subjected to metamorphism for extended periods of time are unlikely to have significant amounts of petroleum hydrocarbons. Even at lower temperatures and pressures, petroleum production will eventually cease as the source bed

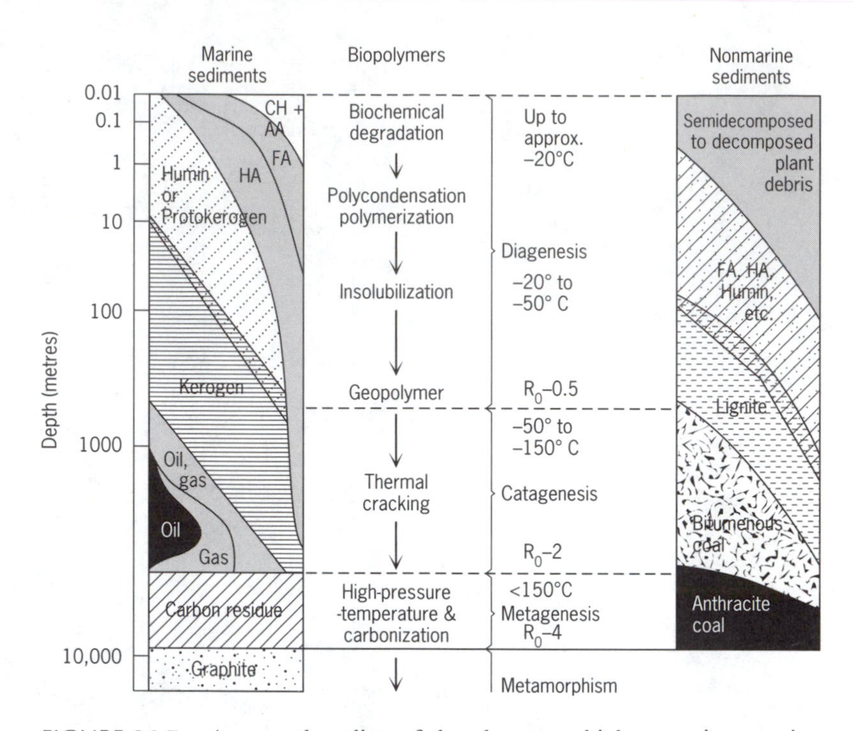

FIGURE 26.7. A general outline of the changes which occur in organic matter during diagenesis, catagenesis, and metamorphism. The relative abundances of the various forms of organic matter are shown on the x-axis. CH = chlorophyll; AA = amino acids; FA = fulvic acids, and HA = humic acids. *Source*: From *Geochemistry of Marine Humic Compounds*, M. A. Rashid, copyright © 1985 by Springer-Verlag, Heidelberg, Germany, p. 207. Reprinted by permission. After *Petroleum Formation and Occurrence: A New Approach to Oil and Gas Exploration*, B. P. Tissot and D. H. Welte, copyright © 1978 by Springer-Verlag, Heidelberg, Germany, p. 70. Reprinted by permission.

becomes depleted in labile organic carbon. The time–temperature window over which oil and gas are produced is illustrated in Figure 26.10.

Petroleum deposits can also be destroyed by exposure to surface water and O_2. The water contains bacteria that oxidize the hydrocarbons. Thus the formation of large petroleum deposits requires an intact, isolated source bed. As described below, the preservation of these deposits also requires an intact, isolated reservoir bed.

MIGRATION

At the completion of the maturation process, source beds contain a complex mixture of solids, liquids, and gases. In this state, the deposit is not economically attractive as a great deal of energy would have to be expended to separate the components into useful fractions. Fortunately, some oil and gas is naturally

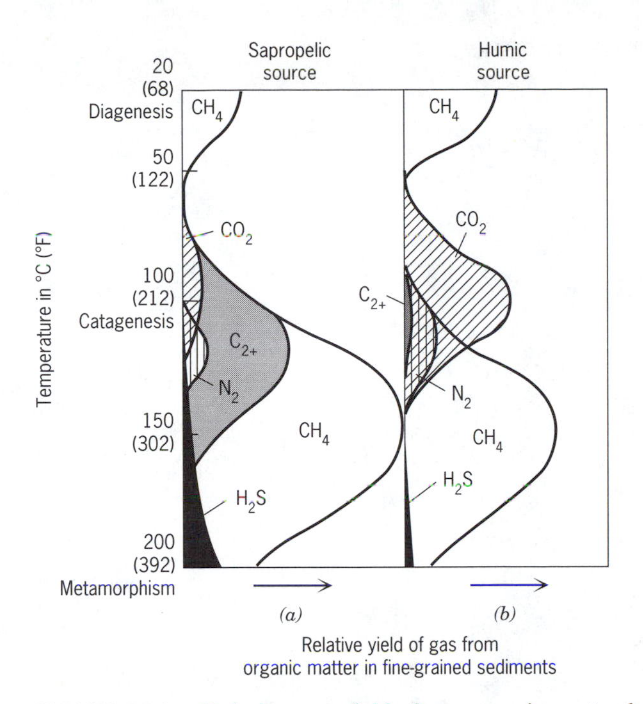

FIGURE 26.8. Relative gas yields from organic matter buried in fine-grained sediments: (*a*) Sapropelic source and (*b*) humic source. *Source*: From *Petroleum Geochemistry and Geology*, J. M. Hunt, copyright © 1979 by W. H. Freeman and Co., New York, p. 163. Reprinted by permission.

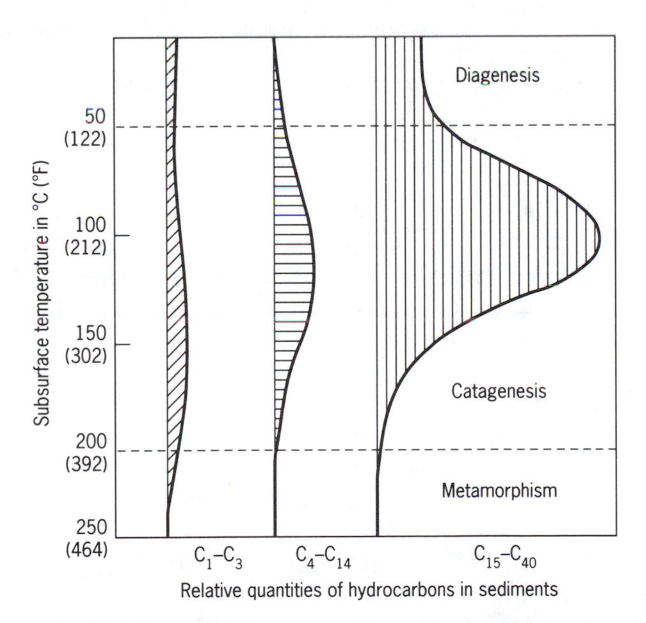

FIGURE 26.9. Relative quantities of hydrocarbons in sediments. *Source*: From J. Hunt, reprinted with permission from the *American Association of Petroleum Geologists Bulletin*, vol. 61, p. 103, copyright © 1977 by the American Association of Petroleum Geologists, Tulsa, OK.

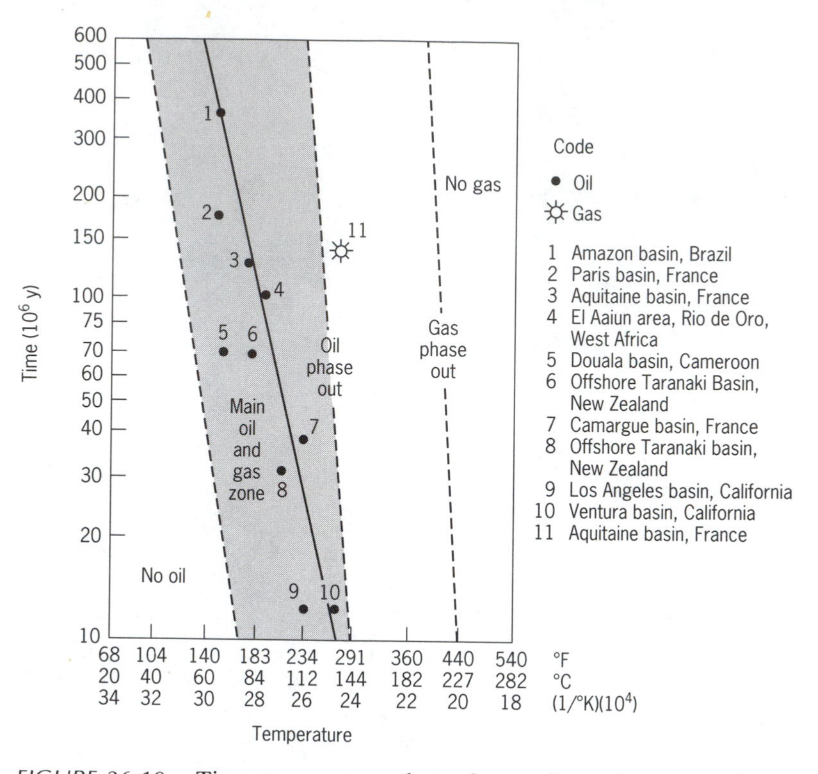

FIGURE 26.10. Time–temperature dependence of petroleum genesis. *Source*: From J. Connan, reprinted with permission from the *American Association of Petroleum Geologists Bulletin*, vol. 58, p. 2520, copyright © 1974 by the American Association of Petroleum Geologists, Tulsa, OK.

''purified'' by the process of **migration**. Due to the high overlying pressures, oil and gas migrate from deeply buried source beds into adjacent porous formations, called **reservoir beds**. The solid residues are left behind in the source beds. In terms of economic worth, sands and old coral reefs make the best reservoir beds as they are cleanest.

If migration continues, the oil and gas will become highly dispersed and the resulting deposits will be of little economic value. If the reservoir is partially surrounded by a relatively impermeable barrier, migration will be inhibited. The oil and gas will pool behind this barrier and form a concentrated deposit. The most effective barriers are provided by **stratigraphic** and **structural traps**, such as those illustrated in Figure 26.11. Stratigraphic traps are the result of differences in sediment porosity caused by the warping of sediment layers or by unconformities.

Diapirs are an example of a structural trap. These rocks are ancient evaporites that have flowed upward to fill cracks in sedimentary rocks. Evaporites are more easily deformed by overlying pressure than other sedimentary rocks. Thus they tend to form pillars, as illustrated in Figure 26.11*d*. Since

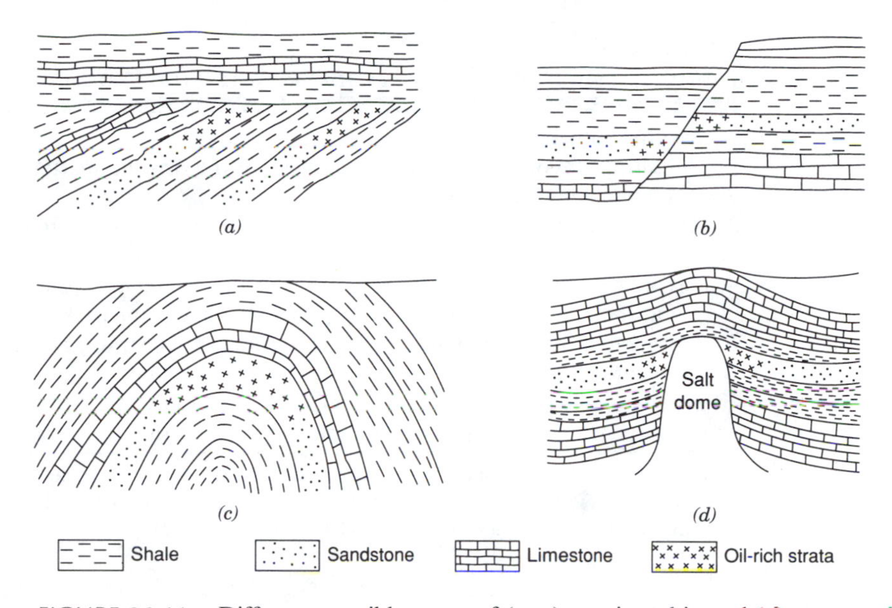

FIGURE 26.11. Different possible types of (*a–c*) stratigraphic and (*d*) structural oil traps. For simplicity, the reservoir beds are always indicated as sands. *Source*: From *Introduction to Oceanography*, 4th ed., D. A. Ross, copyright © 1988 by Prentice Hall, Inc., Englewood Cliffs, NJ, p. 350. Reprinted by permission.

diapirs are relatively nonporous, they are barriers to the migration of oil and gas and thus are sometimes referred to as a **cap rock**. Diapirs also act as barriers to sediment transport on continental margins. They prevent sediments from sliding down the continental slope and hence are a factor in shaping continental margins. Though not all diapirs have associated petroleum deposits, their frequent co-occurrence is probably due to a common origin under relatively stagnant conditions. Such conditions promote anoxia and net evaporation. The former favors the production of petroleum and the latter the deposition of evaporites.

PROSPECTING FOR MARINE PETROLEUM

The essential requirements for the formation of large petroleum deposits are listed in Table 26.4. Since all of these numerous requirements are rarely met, petroleum deposits are scarce.

As shown in Figure 26.12, most deposits of marine petroleum are located in marginal seas. These sites are conducive to the formation of sizable petroleum deposits because they are characterized by high rates of primary production, shallow water depths, and restricted circulation. As a result, marginal seas have large POM fluxes and anoxic sediments. Both favor the burial of large quantities of organic matter at rapid rates. Their close proximity to land ensures

TABLE 26.4
Essential Requirements for Favorable Petroleum Prospects

1. A sufficient source of the proper organic matter
2. Conditions favorable for the preservation of the organic matter—rapid burial or a reducing environment
3. An adequate blanket of sediments to produce the necessary temperatures for the conversion of the organic matter to fluid petroleum
4. Favorable conditions for the movement of the petroleum from the source rocks and the migration to porous and permeable reservoir rocks
5. Presence of accumulation traps, either structural or stratigraphic
6. Adequate cap rocks to prevent loss of petroleum fluids
7. Proper timing in the development of these essentials for accumulation and a postaccumulation history favorable for preservation

Source: From H. D. Hedberg, J. D. Moody, and R. M. Hedberg, reprinted with permission from the *American Association of Petroleum Geologists Bulletin*, vol. 63, p. 288, copyright © 1979 by the American Association of Petroleum Geologists, Tulsa, OK.

a supply of sand, which can form clean reservoir beds. Since marginal seas are likely to have been isolated from the ocean during periods of lower sea level, they tend to have diapirs. For a substantial amount of petroleum to have been produced in one of these locations also requires that enough time has elapsed for burial, maturation, and migration to have occurred. In the United States, the largest such deposit occurs in the Gulf of Mexico.

Abyssal plains are unlikely sites for large petroleum deposits as they lack an abundant source of POM and their sediments are too old. As indicated in Table 26.5, the continental margins that lie outside a marginal sea are also not promising. Due to rapid current motion, the sediments are more oxidizing at

TABLE 26.5
Estimated Undeveloped Total Offshore Petroleum Resources (Oil Plus Equivalent Amount of Natural Gas)

	Resources		
Ocean Area	*Billions of metric tons*	*Billions of barrels*	*%*
Undiscovered reserves			
Continental shelves, shallow seas	183	1370	70
Continental slopes	61	460	24
Continental rises	12	90	5
Deep-sea trenches and ridges	3.5	26	1
Total	260	1950	100
Proved reserves	19	143	—

Source: From *Mineral Resources and the Environment*, copyright © 1975 by the National Academy of Sciences, Washington, DC, p. 98. Reprinted by permission.

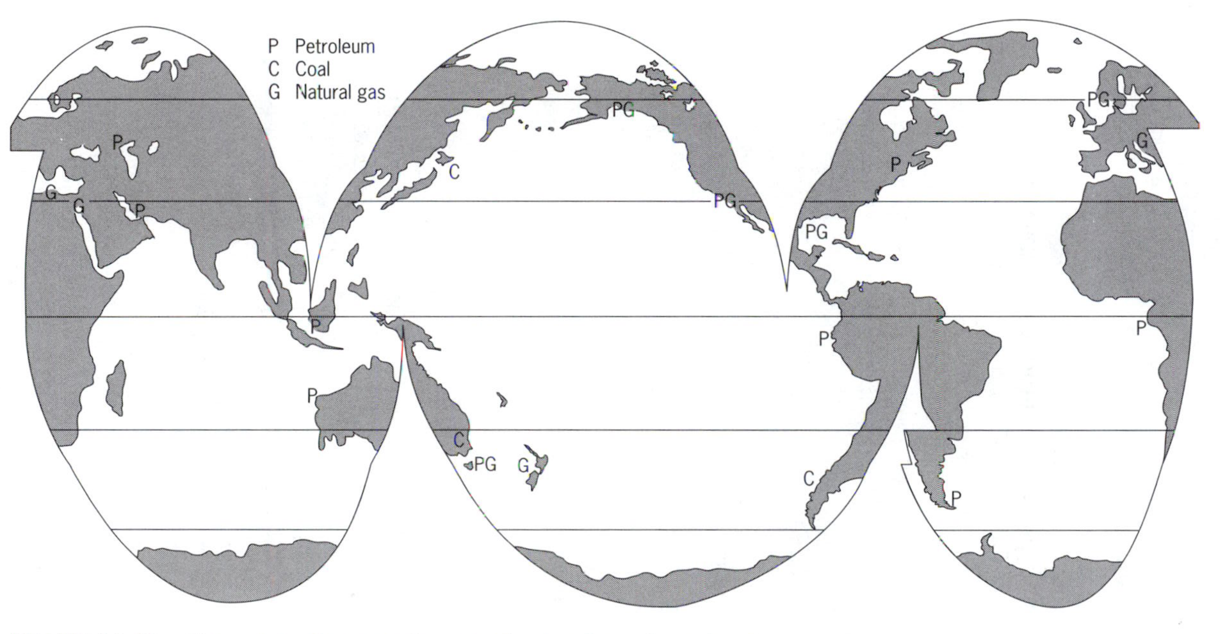

FIGURE 26.12. Known marine petroleum and natural gas deposits. *Source*: From *Oceanography*, 6th ed., M. G. Gross, copyright © 1990 by Merrill Publishing Company, Columbus, OH, p. 164. Reprinted by permission.

these sites and hence do not favor the burial of organic matter. Nevertheless, the United States government has subdivided the seafloor of its continental shelf and sold some leases to oil prospectors. Test wells drilled along the East Coast did not yield substantial amounts of petroleum. In the late 1980s, several large tanker spills occurred in coastal waters. As a result of concern over the environmental impact of increased ocean drilling, an open-ended moratorium on leasing was declared by the federal government in 1990.

The sole exceptions to the unpromising nature of most continental shelves are areas located beneath coastal upwelling areas. For example, the oil wells of Southern California are pumping petroleum created from sedimentary organic matter that accumulated beneath an ancient upwelling area. A similar and very large deposit is thought to lie beneath the Grand Banks. The intense upwelling that presently occurs at this site supports one of the world's largest fisheries. Though this area has been leased, environmentalists and fishermen have halted attempts to drill at this site.

SUMMARY

Petroleum is a complex mixture of organic compounds that can be present in the gas, liquid, or solid phase. The **natural gases** include methane, ethane, propane, and butane. The liquid fraction is termed a **crude oil**. The low-molecular-weight compounds in this fraction constitute **gasoline** and kerosene. They are primarily hydrocarbons, such as **paraffins** and **naphthenes**. The solid residue that remains following the distillation of a crude oil is termed **asphalt**. It is composed of **asphaltenes**, which are large, relatively inert compounds. They also tend to have significant amounts of sulfur, nitrogen, oxygen, and trace metals. In terms of fuel use, nitrogen and sulfur represent undesirable impurities. The burning of coal and oil with high nitrogen and sulfur contents produces nitric and sulfuric acid, both of which cause acid rain.

Most of the organic compounds in petroleum appear to be products of chemical reactions that occurred at high temperatures and pressures following the deep burial of sedimentary organic matter. The nature of the products is dependent on the chemical composition of the original sedimentary organic matter. Marine organic matter tends to be reduced and highly saturated due to the presence of lipids synthesized by phytoplankton. Maturation of this organic matter produces oil and gas. In contrast, aquatic organic matter is relatively oxidized and unsaturated, being composed primarily of cellulose and lignin. Maturation of this organic matter produces gas and coal.

Though a few compounds are well preserved following burial in the sediments, most undergo extensive diagenetic and catagenetic changes. The diagenetic changes occur at low temperatures and under anoxic conditions. Oxygen, nitrogen, and sulfur are removed from the sedimentary organic matter as a result of denitrification and sulfate reduction. The remaining hydrocarbons undergo reactions that result in rearrangements of their carbon skeletons.

Typically, side chains are removed and aromatic rings formed. Deep burial results in higher temperatures and pressures that cause **cracking reactions**. This catagenesis is responsible for the production of most oil and gas. As a result of the need for deep burial and extended periods of time for maturation, many millions of years are required for the production of large amounts of petroleum.

The maturation of organic matter is accelerated by increased temperatures and pressures. But extended periods of reaction at these environmental conditions will eventually convert petroleum to CO_2, CH_4, and graphite. Petroleum deposits can also be destroyed by contact with surface waters and O_2, which leads to microbial oxidation.

The petroleum produced from the maturation of organic matter is of little economic value unless it is naturally purified by the **migration** of oil and gas into clean **reservoir beds**. Migration is caused by high overlying pressures. If allowed to continue, migration will disperse the oil and gas, which also results in deposits of little economic value. Nonporous **structural** and **stratigraphic barriers** prevent migration and cause migrating gas and oil to collect in large pools. **Diapirs** are an example of a structural barrier commonly found on continental shelves, particularly in marginal seas.

Most petroleum appears to have been formed in the sediments of marginal seas or on continental shelves beneath upwelling areas. This reflects the need for a large source of organic matter, as well as good preservation in the sediments. The latter occurs in shallow-water depths, which, along with a large POM flux, ensure anoxic conditions. As noted above, neighboring clean reservoir beds must be present, as well as a **cap rock**, which stops migration. In addition, the formation of large petroleum deposits requires that sufficient time has elapsed for burial, maturation, and migration to have taken place. Finally, the reservoir bed must remain intact to ensure preservation. Due to these many requirements, petroleum deposits are relatively rare.

CHAPTER 27

DRUGS AND OTHER ORGANIC PRODUCTS FROM THE SEA

INTRODUCTION

Extracts from marine organisms have been used for medicinal purposes in China, India, the Near East, and Europe since ancient times. As late as the 1800s, many were a part of standard pharmacopoeias. For example, some seaweeds were used to treat dropsy, menstrual difficulties, gastrointestinal disorders, abscesses, and cancer. Hippocrates recorded that juices from various species of molluscs were commonly used as a laxative. Extracts from the sea hare, *Aplysia*, have been used as a depilatory and essences from gastropod opercula have been used in perfume and incense. Scientists have since come to appreciate that many of these remedies contain potent drugs. Substantial efforts are now being directed to the study of their active agents.

Marine organisms have long been recognized as likely to contain many potential novel drugs because of the environmental conditions that are unique to their habitat. In particular, adaptation to high ionic strengths, low light levels, cold temperatures, and pressures has led to the biosynthesis of unique compounds. For example, halides, such as bromine and iodine, are relatively abundant components in marine metabolites due to their high concentrations in seawater. In addition, many organisms synthesize bioluminescent compounds. Although scientific work has concentrated on highly visible and abundant marine organisms, such as seaweeds and large invertebrates, microorganisms are also likely to contain novel substances due to their unique metabolic pathways.

Recent advances in genetic engineering and chemical separation technology have facilitated the investigation and development of new marine drugs. But despite considerable effort, few new drugs have come to market as a result of these systematic studies of marine organisms. Progress in the development of marine drugs continues to be limited by problems associated with the collection, harvesting, and isolation of the pharmacologically active constituents, as well as in their clinical evaluation and processing. Most importantly, drug development requires an extensive period of time and financial commitment. Current estimates indicate that the complete process takes at least 10 y, mostly due to the requirements associated with governmental approval, and costs more than $50 million. Due to the risky nature of such a financial investment, few pharmaceutical companies are currently pursuing the development of marine drugs. As a result, only a dozen or so marine-derived drugs are in widespread use. In comparison, compounds from terrestrial organisms are the source of approximately half the pharmaceuticals currently is use; the rest are synthetic.

Many pharmacologically active marine compounds have no terrestrial analog and so represent wholly new classes of potential drugs. In the laboratory, some act as antibiotics, whereas others appear to have antitumorogenic, anticoagulant, antiviral, antiulcer, haemolytic, analgesic, antilipemic, cardioinhibitory, immunostimulant, or immunosuppressant effects. Some are currently used as fungicides and insecticides or as food and cosmetic additives. One interesting recent medical application has been the use of extracts from bivalve holdfasts to "glue" bones and false teeth. This application was suggested by the similar wet and salty environment of the human body and the sea.

The few marine compounds that are currently used as drugs, as well as some that are in various stages of development, are described below. Also discussed in this chapter is the general procedure used for marine drug development and potential applications of recent advances in biotechnology.

THE DISCOVERY AND DEVELOPMENT OF MARINE DRUGS

A potential natural product must go through many stages of research and testing before it can be brought to market. Since many thousands of marine organisms are available for study, a screening approach is used to assess those that are most likely to have pharmacologically active substances. The initial phase of screening typically involves environmental observations of the behavior and interactions among marine organisms in the natural environment.

For example, several groups of marine organisms appear defenseless but have few predators. These include soft-bodied sessile invertebrates, such as tunicates, soft coral, and certain sponges. Since these organisms are literally sitting targets, they have evolved chemical defenses that have since proven to be pharmacologically active. For example, shell-less molluscs, particularly the sea hares, have long been known for their toxins. These compounds are actually synthesized by red algae that are consumed by the sea hares. As the toxins are not metabolized, they become concentrated in the tissues of the sea hares, making the invertebrate unpalatable to potential predators. This reliance on exogenous compounds is common in the marine biota.

Unusual growth patterns in sessile organisms can also indicate the production of physiologically active compounds. For example, in cases where two organisms compete for the same substrate, one may defeat the other by exuding a chemical repellent. Since physiologically active substances provide a means of chemical communication, they are also associated with symbioses. Species-specific relationships tend to involve unique chemical exudates. Likewise, organisms that are free of bacterial infections or cancer are likely to contain a compound with antibiotic or antitumorogenic properties.

Once an organism has been targeted for research, the natural-products chemist must determine which organic compound is responsible for the behavior of interest. To do this, various classes of compounds are extracted from the organism. This is usually accomplished by subjecting the tissues to solvents of varying polarity. The extracts are then assayed for pharmacological activity. Some bioassays, such as that used to detect antibiotic activity, are fast, inexpensive, and can be done in the field. Others, such as those for anticancer activity, require sophisticated equipment, animal testing, and extended periods of time. More bioactivity is usually observed by testing pure compounds as compared to crude extracts. Thus extracts are usually separated into successively purer fractions, each of which is then tested. Though bioassays take much time and effort, potential drugs are unlikely to be overlooked if this methodical approach is used.

Bioassays have shown that more than 10 percent of marine organisms contain cytotoxic substances as compared with 2 to 3 percent of terrestrial species. These compounds are more likely to be found in marine invertebrates than in marine algae, in sessile rather than motile invertebrates, and in tropical rather than temperate or cold-water species. Though toxicity is obviously indicative of potent physiological activity, marine toxins are generally not very good drugs because they are usually too powerful and dangerous. Nevertheless, they have proven useful as model compounds in the study of biochemical mechanisms and for the synthesis of less toxic analogs.

If an extract demonstrates physiological activity, the next stage in drug development involves the isolation and identification of the active substance(s). This requires an extensive amount of extract and hence large field collections of marine organisms. This task can be complicated due to vagaries in taxonomic classification and the difficulty of field identification. As a result, screening is most successful if accomplished at the collection site. This approach is also ecologically superior, as only the "active" organisms need be collected in large amounts.

Compound isolation and identification are complicated tasks as most of these molecules have large and complex structures. Since many are unstable at high temperatures, these compounds must be isolated by techniques such as high-pressure liquid chromatography, rather than gas chromatography. Structural details are usually determined by mass spectrometry but must be confirmed by other analyses.

For ecological, as well as economic reasons, artificial means must be developed for the synthesis of these compounds. Usually a large number of slightly modified forms, called analogs, are also synthesized and tested in the hopes that they will prove to be more pharmacologically active than the natural product. These compounds can also provide insight into the mode of drug action. Indeed most marine compounds operate by mechanisms that differ from those of terrestrial and synthetic drugs. As a result, standard screening techniques and bioassays are often inappropriate for testing these potential drugs. On the other hand, information on the molecular basis of their activity has been used to develop more reliable ways to evaluate potential therapeutic applications.

All substances sold as drugs in the United States must be approved by the Federal Drug Administration. This requires that the substance be tested on animals. If no adverse side effects are observed and significant ameliorative effects are found, drug trials using human subjects are undertaken. This process can take years, as long-term effects, as well as the best method of drug administration and dosage, must also be determined.

If the substance receives approval, chemical engineers then develop economically effective methods by which the drug can be manufactured. Synthesis, rather than harvesting, ensures a reliable supply of pure drug and does not endanger marine life. The last stage in drug development is an assessment of the market potential of the drug. If the potential manufacturer is not able to realize a large enough profit, the drug will not be brought to market.

CARBOHYDRATES

The cell walls and mucilage of marine macroalgae are composed of carbohydrates. Some of these compounds are used for medicinal purposes and as food additives. They include agar, alginic acid, algin, laminarin sulfate, and carrageenan.

Agar, alginic acid, and alginates have extensive use as food stabilizers and emulsifiers. They are also used in the textile industries as adhesives and sizing agents. Agar is a mucilaginous substance isolated primarily from two species of red algae, *Gelidium* and *Gracilaria*. Frozen agar gels as it thaws due to the expulsion of water. This gel is used as a microbial culture medium because few bacteria can decompose it. Agar is also used in canned food because it can withstand low pH's and high sterilization temperatures. It is also used as a mild laxative and as a suspending and emulsifying agent in some ointments and cosmetics. Agar is a variable mixture of several polymers. The structures of the two most common repeating units, agarose and agaropectin, are shown below. A significant amount of the agarose also contains acidic residues, such as sulfate and pyruvate:

(a) Agarose (b) Agaropectin

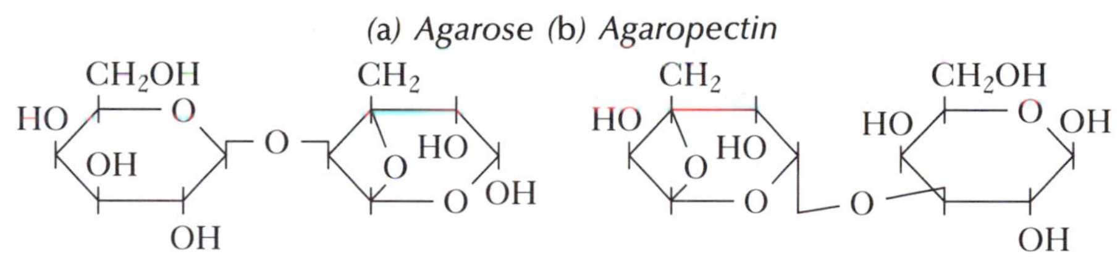

As illustrated below, alginic acid is a polymer of two sugar-like units, mannuronic and guluronic acid:

Partial Structure of Alginic Acid

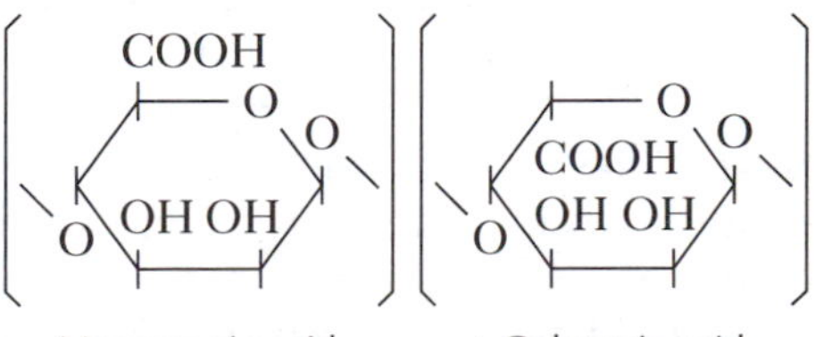

This polymer is obtained from brown seaweeds. Due to its coagulant properties, alginic acid has been incorporated into bandages. As with all the algal polysaccharides, alginic acid complexes with radioactive strontium and thus can be used as a detoxifying agent. It is also incorporated into drugs as a tablet-disintegrating agent. Brown seaweed has also been extensively used as a source of iodine, potassium-rich fertilizer, and soda ash (Na_2CO_3), which is an important component of hard soap. Laminarin sulfate is a sulfonated polysaccharide that has also been isolated from brown algae. It is an anticoagulant and prevents atherosclerosis by lowering serum cholesterol levels.

Algin is an extract obtained from the cell walls of kelp. Kelp itself has been used as food for livestock and poultry. Algin is presently used in a wide range of applications. It is added to foods, such as dairy products, beer, and salad dressings, as an emulsifier and stabilizer. In bakery products, from cake mixes to meringues, algin is used to improve texture and retain moisture. It is also used industrially in paper coatings and sizing, textile printing, and welding-rod coatings. In the pharmaceutical industry, algin is used in formulations of drug tablets, dental impressions, and antacids. For many of these applications, no substitutes exist. As a result, large quantities of kelp are harvested, primarily off the coast of California. By mowing the fronds, as much as 145,000 metric tons have been collected in a single year.

Carrageenan is another polymeric carbohydrate that has been isolated from red seaweeds. In food, it is used as an extender, stabilizer, and binder, particularly in cottage cheese and ice cream. It is also used to prevent ulcers and cholesterol absorption. Carrageenan complexes with metals such as iron and zinc, so it can be used to prevent poisoning if excess metal has been ingested. It also has anticoagulant and antiviral properties. As a result of the latter, carrageenan affords some protection against influenza. It also prolongs the activities of common analgesics and anticough medicines such as codeine and ethylmorphine. Carrageenan is also used to promote the rapid disintegration of drug tablets. The repeating units that make up this polymer have the following structure:

Partial Structure of Carrageenan

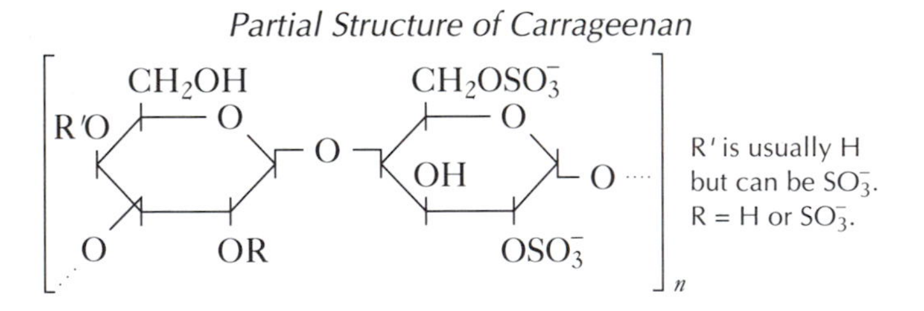

The locations and amounts of the sulfite residues are variable.

Chitin is a ubiquitous natural polymer that is composed of amino sugars (Figure 22.10) and is second in abundance to cellulose. Because it comprises most of the exoskeletons of crustaceans, chitin is an abundant component of shellfish wastes produced by the seafood industry. Controlled deacetylation produces a water-soluble material called Chitosan, which complexes metals and natural biomolecules very efficiently. Membranes made from Chitosan are currently being developed for use in ion transport, blood dialysis, and the sustained release of pharmaceuticals. Another derivative of chitin, called Kylan, is used to impart shrink resistance to wool. Chitin is also a source of glucosamine that is used as a precursor in drug synthesis. Because it is a selective binding agent, chitin is also used for isolating enzymes.

LIPIDS AND THEIR DERIVATIVES

Fatty Acids

Many fatty acids derived from macroalgae exhibit antibacterial activity. The strength of their antibiotic effect appears to be positively related to the degree of unsaturation in these fatty acids. An example, acrylic acid, is illustrated below. Due to its antibacterial properties, acrylic acid is used in the preservation of nonfood materials:

Acrylic Acid

$$CH_2 = CHCOOH$$

Polyunsaturated fatty acids isolated from marine fish oils have been observed to have antilipemic activity; that is, they lower blood cholesterol levels. The structure of one of these compounds, arachidonic acid is

Arachidonic Acid

$$CH_3(CH_2)_4(CH{=}CHCH_2)_4(CH_2)_2{-}COOH$$

Prostaglandins are very biologically active simple lipids. They have potential as a contraceptive or abortifacient because they induce labor. Prostaglandins also act as a central nervous system tranquilizer and have ameliorative effects on intestinal ulcers and renal problems. Other activities include depression of blood pressure, altered function of circulating platelets, alleviation of bronchoconstriction in asthmatics and nasal decongestion, as well as antacid, antidiarrheal, antihemorrhagic, and antitumoral effects.

The natural production of prostaglandins by land animals is too limited to provide sufficient quantities for medicinal use. In comparison, 1.5 percent of the dry weight of gorgonian corals is composed of prostaglandins. Although these prostaglandins lack significant biological activity, they can be chemically modified into mammalian forms as shown below (Figure 27.1). Scientists have determined that these coral will redevelop rapidly if at least 25 percent is left intact during harvesting. Thus gorgonian coral represent a large potential source of prostaglandins.

Steroids

As mentioned in Chapter 22, the synthesis of some sterols is species specific. For example, only sponges synthesize poriferasterol and spongosterol. Gorgonian corals are the sole source of gorgosterol. Green algae synthesize isofucosterol and sitosterol. Fucosterol and sargasterol are produced by several species of

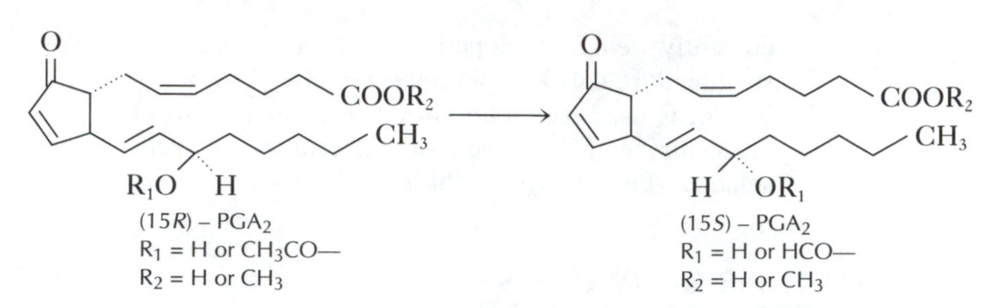

FIGURE 27.1. Transformation of gorgonian prostaglandins to mammalian prostaglandins. *Source*: From *Chemical Oceanography*, vol. 4, H. W. Youngken and Y. Shimizu (eds.: J. P. Riley and G. Skirrow), copyright © 1975 by Academic Press, Orlando, FL, p. 277. Reprinted by permission.

brown algae. Many of these sterols have hypocholesterolemic action. Some of their structures are illustrated in Figure 22.15.

Many insects, crayfish, and other crustaceans synthesize hormones that induce molting. These ecdysones are oxygenated cholestanes, a type of sterol. They have potential applications in mariculture as growth regulators and in the control of insect life cycles. The structure of a crustacean ecdysone, crustecdysone, is given below:

Crustecdysone

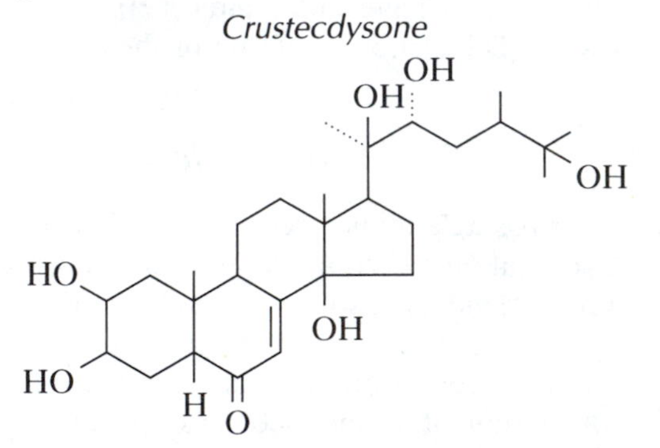

Holothurin is a steroidal saponin that has been isolated from sea cucumbers. (Saponins contain fatty acids.) This compound suppresses tumor growth and slows amoeboid movement. It also increases leukocyte phagocytosis (the destruction of foreign particles such as bacteria) by white blood cells. Holothurin also has antifungal and cardiotonic activity. Its molecular structure is

Holothurin

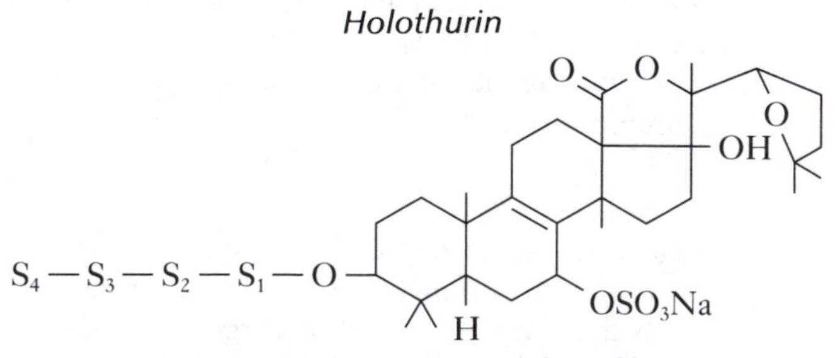

Variable fatty acid composition is indicated by S$_1$–S$_4$.

Sea cucumbers also synthesize the sterol stichopogenin A_3, which acts as an antitumorogenic agent. Its structure is

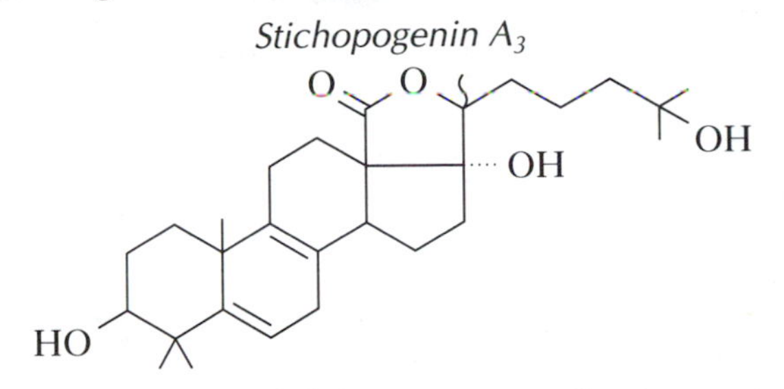

Steroids extracted from starfish have proven effective against influenza B. They also have antiinflammatory, antitumoral, and haemolytic activity. Starfish also synthesize steroids that are toxic to fish and molluscs. An example is given below. Some are used by the Japanese to kill fly larvae. This toxic effect is attributed to their surfactant nature which causes a disruption in the molting process.

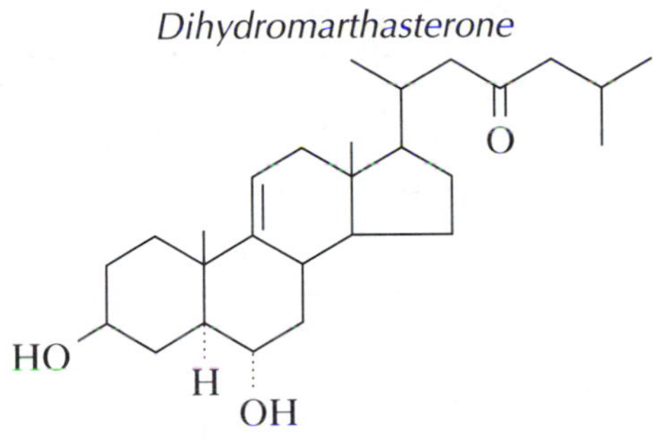

Terpenoids

The gorgonian and horny corals synthesize the terpenes, eunicin, and crassin acetate. Both are toxic to *Endamoeba hisolytica*, which causes dysentery in humans. These compounds also inhibit the growth of *Staphylococcus aureus* and *Clostridium*. The structure of eunicin is

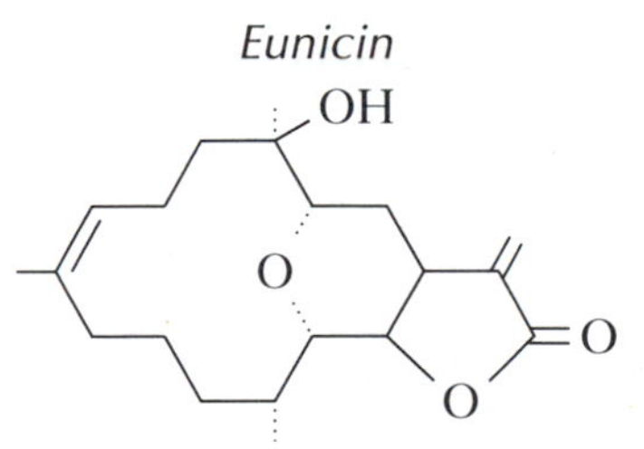

Sea hares are renowned for secreting paralytic neurotoxins. As shown in the following examples, bromine is present in some of these compounds and is likely the toxic agent:

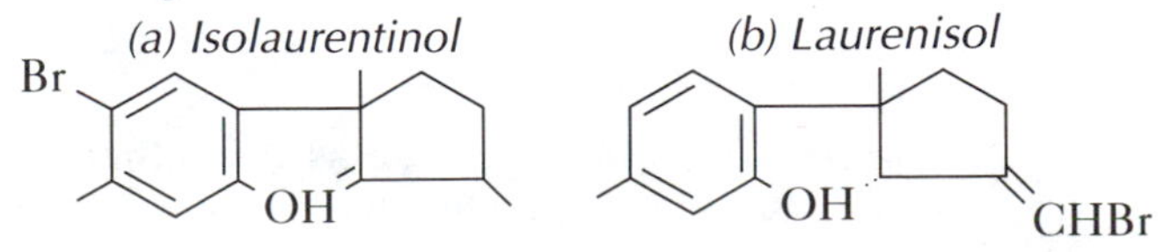

Some brown seaweeds synthesize the terpenes zonarol and isozonarol, both of which have antifungal activity. Their structures are

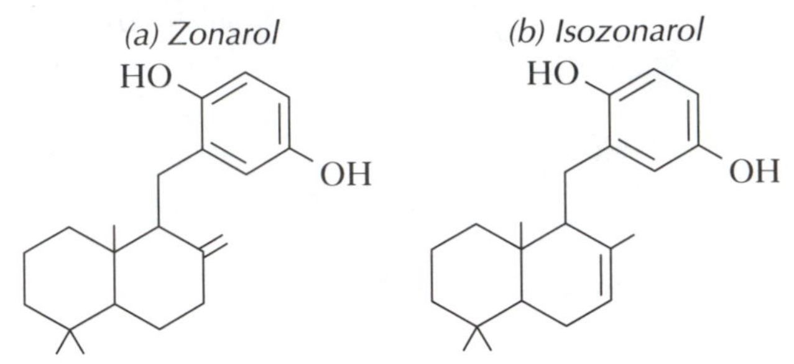

NITROGENOUS COMPOUNDS

Pharmacologically active substances often contain nitrogen, which is typically present as a simple amine or as a derivative of choline, betaine, creatine, or guanidine. Tetramine is an example of a simple amine that has been isolated from whelks and sea anemones, as well as other coelenterates. It has strong paralytic activity, acts as a curare-like muscle relaxant, and stimulates the parasympathetic nervous system. Its structure is

Tetramine

$$\left[\begin{array}{c} CH_3 \\ | \\ H_3C-N-CH_3 \\ | \\ CH_3 \end{array}\right]^{+} OH^{-}$$

Acrylcholine

$$H_2C{=}CHCO_2CH_2CH_2\overset{+}{N}(CH_3)_3$$

Acrylcholine has been isolated from gastropods. It affects the contraction of smooth muscle tissue and is a hypotensive. Its structure is

Acrylcholine

$$H_2C{=}CHCO_2CH_2CH_2N(CH_3)_3$$

Urocanylcholine has been isolated from the purple gland of the snail *Murex*. It is sold under the trade name Murexine as an insecticide. This compound is a strong paralytic. At low concentrations, it is used as a muscle relaxant, having nicotinic and curariform actions. Its structure is

Urocanylcholine

$$(\text{imidazol-4-yl})\text{—}CH{=}CH-CO_2CH_2CH_2\overset{+}{N}(CH_3)_3$$

Murex brandaris, as well as two other species of gastropod snail, were used as a source of the blue dye indigo and Tyrian purple by the Hebrews and

Romans during biblical times. These dyes were highly prized for their deep hues and colorfast nature. As a result, they were restricted to religious and government use. For example, only the Roman emperor was permitted to wear "true" purple as produced by Tyrian purple. As shown below, this compound is an azo dye:

Tyrian Purple

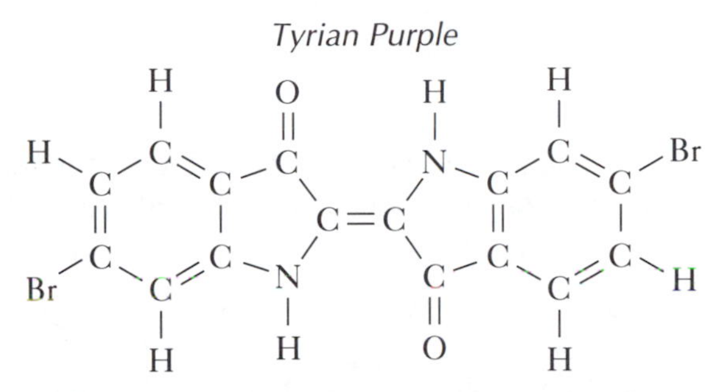

The snails emit precursors of these dyes as a clear fluid that also contains several toxic compounds, some of which probably repel predators. On exposure to O_2 and sunlight, enzymatic actions convert the precursors to a white, yellow, green, and finally blue or purple color. Variations in the color of the final products are related to differences in sex, species, and in situ environmental conditions, as well as in the method of dye processing. Knowledge of the latter was lost by 760 A.D. but rediscovered in 1856.

Eloidoisin, an 11 amino acid protein, has been isolated from the salivary gland of octopi. Like acrylcholine or histamine, eloidoisin is a vasodilator and hypotensive but is 50 times more potent. The amino acid composition of this protein is

Eloidoisin

H-Pyr-Pro-Ser-Lys-Asp(OH)-Ala-Phe-Ile-Gly-Leu-Met-NH_2

The salivary gland is also a source of octopamine, a long-chain fatty acid, which stimulates smooth muscle action and is also a strong hypotensive.

Dead annelid worms have long been used to repel and kill flies. The substance responsible for this insecticidal activity is nereistoxin. A synthetic derivative, padan, has been marketed under the trade name Cartap since 1967 as a substitute for dichlorodiphenyltrichloroethane (DDT) and hexachlorocyclohexane (BHC). The insecticide works by blocking the ganglionic action of the central nervous system. This is quite unlike the mode of action for the halogenated pesticides, such as DDT. As a result, padan is nontoxic to mammals and decomposes readily, making it an important addition to the antibug arsenal. It is used extensively in Asia and Japan against rice stem borers. The structures of nereistoxin and padan are

(a) Nereistoxin *(b) Padan*

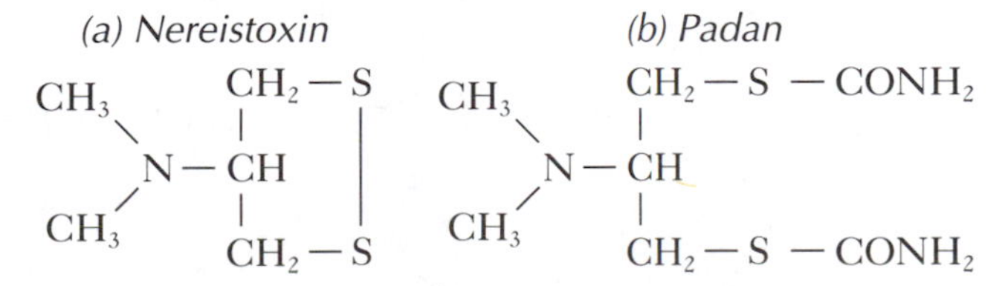

Anabaseine, another insecticidal compound, has also been isolated from marine polychaetes. Its structure is

Anabaseine

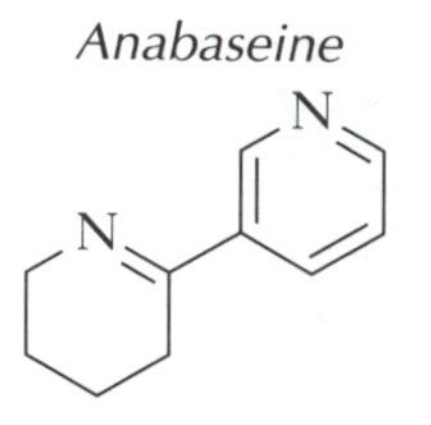

L-α kainic acid has been isolated from red seaweed and is used extensively in southern Japan as an antihelmintic. Its structure is

L-α Kainic Acid

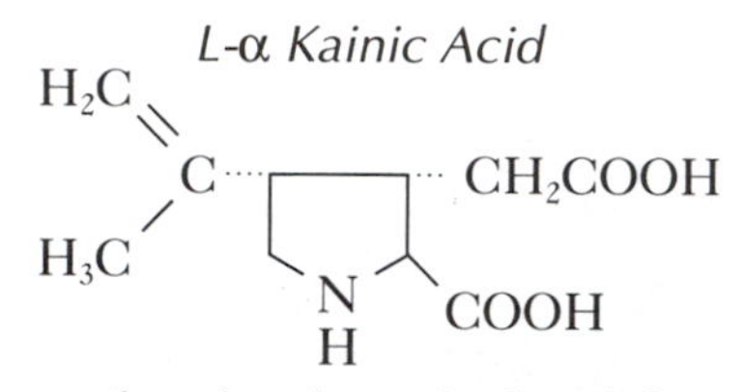

Laminine hydrogen oxalate has been isolated from brown seaweed. This compound has hypotensive activity comparable in strength to that of choline. It is also an anticoagulant and antilipemic. The structure of this oxalate salt is

Laminine Hydrogen Oxalate

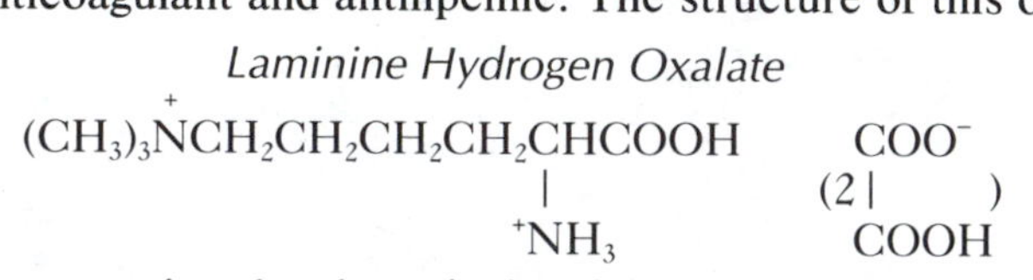

A protein, protamine, has been isolated from some marine fish. It is used as an antidote for heparin overdosage. By binding with the anticoagulant, protamine inactivates this drug. Protamine is also used in the preparation of protamine zinc insulin, which is a long-acting antidiabetic drug. Most protamine used in this application is isolated from trout and salmon. Protamine is also an antithromboplastic and antiprothrombic agent.

The great resistance of sponges to bacterial decomposition suggests that these creatures contain potent antibiotic compounds. Indeed, extracts from more than 100 species have given positive results as antibiotics effective against a wide spectrum of gram-positive and gram-negative bacteria. The compounds responsible for this activity are primarily brominated cyclohexadienes and polyhydroxybrominated phenols. The structure of another type of brominated compound, oroidin, is given below. The bromine atoms are the most likely cause of the antibiotic effect, functioning much like iodine in the antiseptic, tincture of iodine:

Oroidin

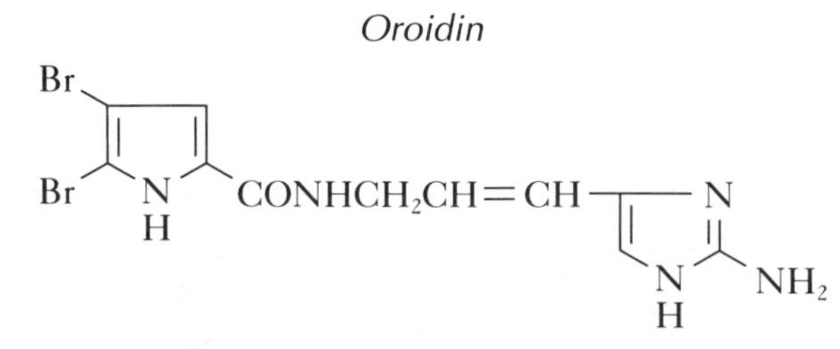

Other nitrogenous compounds synthesized by sponges are cytotoxic to human carcinoma cells. Examples include spongouridine and spongothymidine, as well as nucleosides, such as cytosine-arabinose. The structures of the first two are:

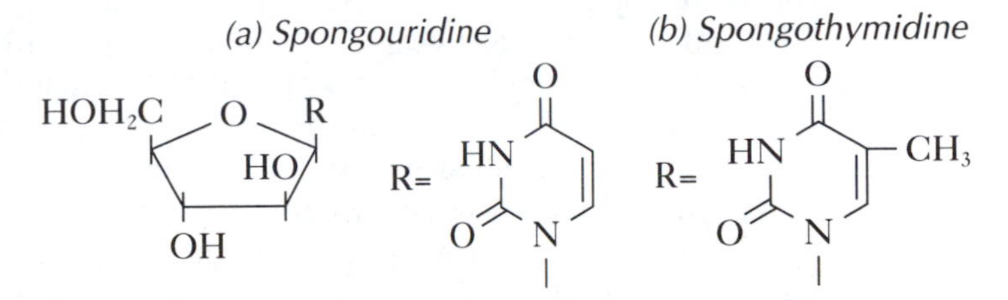

The most commonly used marine antibiotic was first isolated from the fungus *Cephalosporium*. The antibiotic effect of this marine fungus was discovered in the 1940s as a result of research conducted on the microbial flora that live near a sewage outfall in the Mediterranean Sea. This area was under study because scientists hypothesized that the self-purification properties of the water were due to the presence of antibacterial substances produced by the resident microorganisms. One antibiotic, cephalosporin C, was isolated from some of the ambient fungi and has proven to be effective against a variety of bacteria, including penicillin-resistant strains. It is now used widely in both oral and intravenous forms. Its structure is

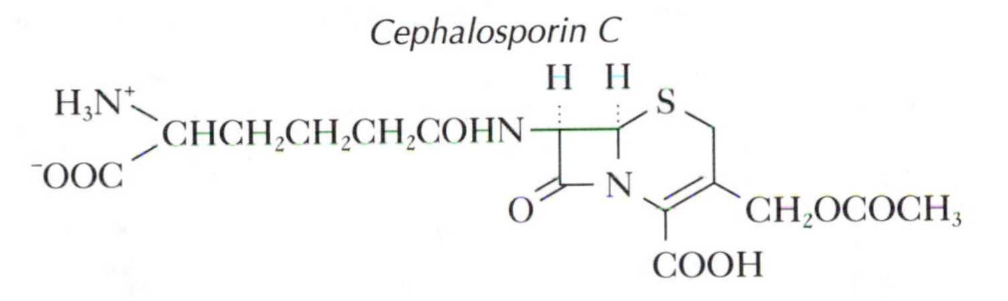

Aequorin, a chemical extracted from jellyfish, is used as a very sensitive bioassay for calcium because the organometallic complex is fluorescent. This has proven useful in diagnosing cardiac irregularities and metastatic carcinoma because both cause subtle changes in serum calcium levels. The structure of aequorin is

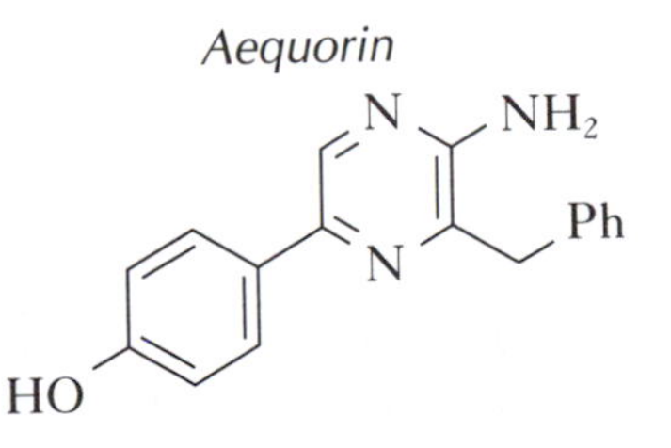

FISH AND SHELLFISH TOXINS

Due to the large-scale consumption of fish and shellfish by human, associated toxic effects are a matter of great concern. Such effects include muscle and respiratory paralysis that can be severe enough to cause death. Most physiological reactions are induced by algal toxins that are concentrated in the fish and shellfish tissues as a result of bioaccumulation (i.e., the selective retention of ingested compounds). This effect is prevalent in coral reef fish. The resulting disease, ciguatera, was recorded as early as the 1400s in the West Indies. The causative agent, ciguatoxin, is synthesized by dinoflagellates.

Other bioaccumulated toxins appear to be bacterial in origin. For example, the cnidarian coelenterates that belong to the genus *Palythoa* concentrate a very potent toxin, called palytoxin. This compound was first extracted from the species *Palythoa toxica*, which grows in only one tidal pool on the island of Maui. The extraordinarily toxic effect of this compound was known to the natives who used it on their spear tips and called the source organism *limu-make-o-Hana*, ''the deadly seaweed of Hana.''

This compound appears to be synthesized by a symbiotic bacterium of the genus *Vibrio*. As indicated below, its structure is unique and hence it represents an entirely new class of organic compounds. Palytoxin is one of the most poisonous substances in nature and is the most powerful nonproteinaceous toxin known. In rabbits, its LD_{50} is 25 ng/kg. (This is the dose required to kill 50 percent of the rabbits in a clinical trial.) The lethal effects are caused by cell lysis. Though palytoxin does not appear to have potential as a drug, it has important uses in medical and physiological research, particularly in the study of heart disease as it greatly perturbs ion fluxes within cells. Palytoxin also has potential in cancer research as it kills some tumors but causes others:

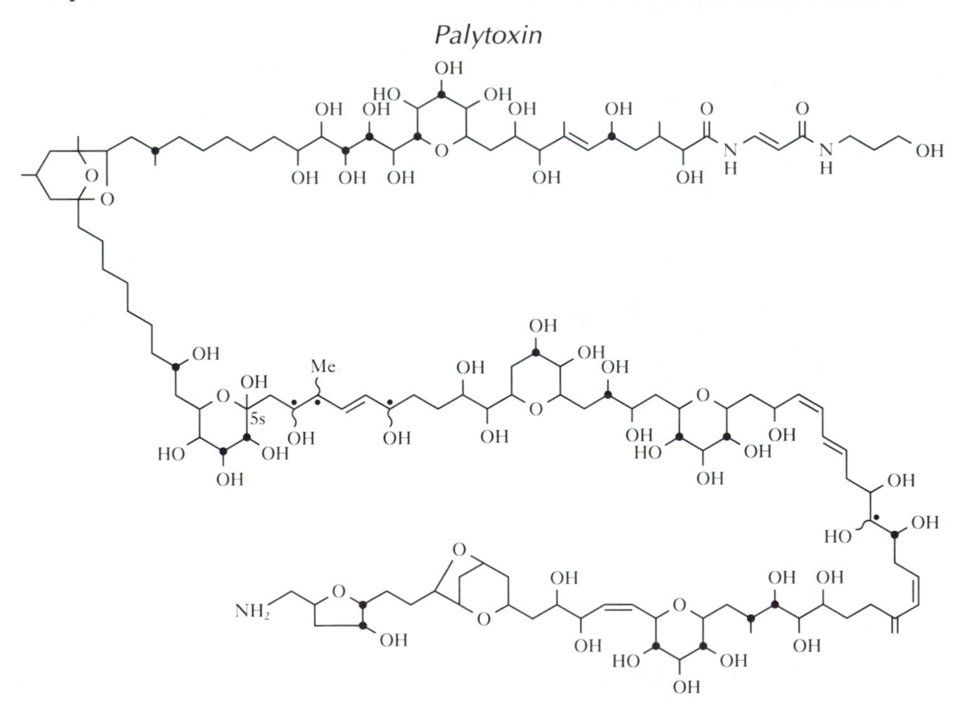

Puffers, ocean sunfish, and porcupine fish all contain tetrodotoxin, which is the most toxic low-molecular-weight poison known. These fish are eaten in

Japan and the United States. They must be prepared by licensed chefs who are trained to prevent the edible tissues from becoming contaminated with tetrodotoxin, which is concentrated in the liver. Cysteine is an effective antidote if consumed 10 to 30 minutes after ingestion of the tetrodotoxin. Biosynthesis of this toxin appears to be controlled by symbiotic bacteria.

Tetrodotoxin is commercially available in Japan and the United States. It is used as a muscle relaxant and as a pain killer in neurogenic leprosy and terminal cancer. In Japan, it is used as a local anesthetic. At low doses, its behavior is similar to that of cocaine, but it is 160,000 times more potent. Its structure is

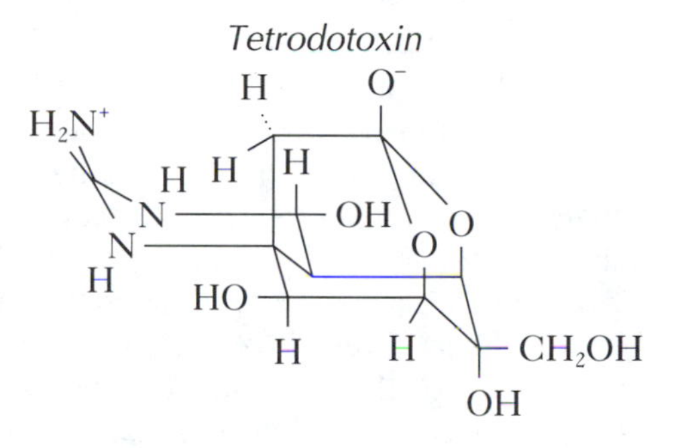

As shown below, saxitoxin is another large molecule that contains several nitrogen atoms in its rings. This compound is synthesized by the dinoflagellates, which cause some forms of red tide. Filter feeders, such as shellfish, consume these dinoflagellates. Saxitoxin is not poisonous to the shellfish and thus is bioaccumulated. The compound is toxic to humans and can cause paralytic shellfish poisoning, which is sometimes fatal:

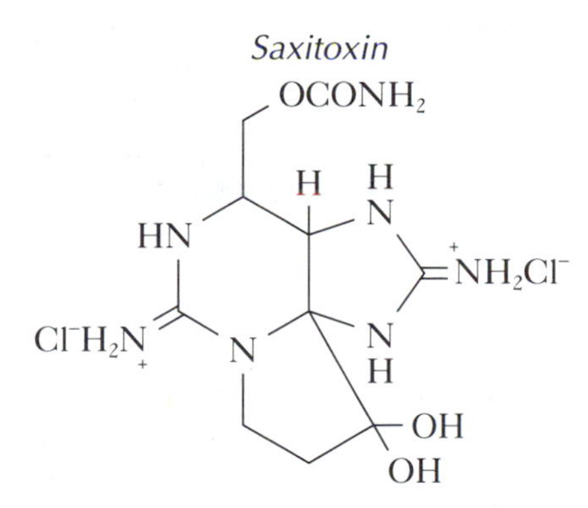

The dinoflagellate *Gonyaulax tamarensis* causes most of the red tides and shellfish contamination along the Atlantic shores of New England, Canada, and the North Sea. Its potent agent is not saxitoxin, but it also has a paralytic effect due to the blockage of membrane permeability.

Some toxins appear to be synthesized by fish. For example, pahutoxin has been isolated from the Hawaiian boxfish, Pahu. Its structure is given below.

Pahutoxin

$$CH_3(CH_2)_{12}CH(OAc)CH_2COOCH_2CH_2\overset{+}{N}(CH_3)_3$$

Similar toxins have been isolated from the mucus-producing glands of soapfish and Pacific bass. These toxins are excreted by the fish, probably as a defensive measure.

OTHER NATURAL PRODUCTS FROM FISH

Fish oil is rich in vitamins A (Figure 22.14*d*) and D, which promote the healing of wounds, burns, and abscesses. Cod and halibut liver oil are also used as a laxative. In addition, fish oils are known for their antilipemic activity.

Extracts of sturgeon and hake have been used as a source of gelatin and isinglass. The latter was produced in thin sheets for use as an inexpensive substitute for window glass.

Sharks are a source of immunoglobulins, which inhibit cancer lymphocyte activity. Shark liver oil is used as a bactericide and as an intermediate in the manufacture of pharmaceuticals. It contains squalene (Figure 22.14*c*), a terpene, which is used as a skin lubricant, an ingredient in suppositories, and as a carrier of oily drugs.

OTHER NATURAL PRODUCTS FROM THE SEA

Sponges contain iodine, which is used to treat tumors, goiter, dysentery, and diarrhea. Before the invention of synthetic sponges, natural ones were used for absorbent purposes. For example, anesthetics were delivered by impregnating a sponge with a soporific. Blood flow was stanched with sponges. Contraceptives, such as lemon juice and quinine, were held in place with sponges.

Bacteria continually shed fragments of their cell walls as they grow. These fragments are the cause of most fevers in humans and are termed endotoxins. Because endotoxins are difficult to degrade, they cannot be destroyed by simple sterilization. This has caused great problems in the preparation of pure drugs, such as vaccines and intravenous fluids. The presence of endotoxins is now monitored by the limulus amebocyte lysate (LAL) bioassay. Limulus amebocyte lysate is an extract from horseshoe crab blood that has a very high reactivity to gram-negative endotoxins. The crabs are thought to use these chemicals as a primitive immune system.

Due to the extreme sensitivity and ease of this bioassay (it only takes 15 minutes), drug and IV fluid purity has improved dramatically over the past 20 y. The bioassay is of limited utility in disease diagnosis because it is nonspecific. Instead, the LAL bioassay is most useful when fast results are essential, such as in the diagnosis of spinal meningitis. LAL is prepared from blood harvested from live horseshoe crabs. Up to 30 percent of the crab's total blood volume can be collected without ill effect.

POTENTIAL APPLICATIONS OF MARINE BIOTECHNOLOGY

Application of the new technologies associated with genetic engineering are likely to increase the rate of natural product development. They are also likely to improve the success of mariculture efforts and our attempts to deal with marine pollution. For example, genetically engineered bacteria could be used to produce unique biomolecules on a mass scale. In vitro manipulations, such as cloning, are also likely to provide major advances in mariculture by enabling the selection of traits such as increased hardiness, as well as fast growth and maturation rates. Genetic engineering could also be used to overcome fish diseases by aiding in the development of antisera and vaccines. Evolutionary relationships among species could be assessed by genetic mapping of DNA. Natural population structures could be studied by genetic tagging, which is done by the introduction of passive traits. Hybridoma technology could also prove useful in identifying, isolating, and characterizing physiologically active substances, such as toxins.

Many marine bacteria have unique metabolic pathways, such as those responsible for metal concentration, mineral deposition, photoautotrophy, and the synthesis of unusual pheromones. Thus these microorganisms contain a large collection of unique genes that could be used to develop novel organisms. In particular, the sulfide-oxidizing abilities of some marine bacteria might be useful in developing microbes that could aid in the decomposition of domestic sewage.

Likewise, genes of estuarine macroalgae could be used to develop salt-resistant land plants. Microorganisms have already been developed that can selectively decompose substances, as in wastewater treatment plants and oil spills. Thus it seems likely that genetic engineering could produce marine microorganisms capable of degrading wastes that are discharged into the ocean. It is also likely that genetic engineering could produce organisms capable of exuding algicides and other chemicals to inhibit larval settlement. Thus biofouling could be controlled on a molecular level, rather than by broad-spectrum toxins. Likewise, genetic engineering could produce organisms capable of dealing with specific problems of the seafood industry, such as the disposal of shellfish waste. For example, chitin is rich in amino acids and could be microbially converted to protein usable as a food additive.

Though our research results are still limited, marine organisms have clearly been identified as sources of unique compounds having pharmacological potential. It is highly likely that many more potentially useful, but as yet unknown, substances exist. As with the denizens of the tropical rainforests, there's no telling which species could hold a cure for diseases such as cancer or AIDS. For this reason, it behooves us to protect these ecosystems.

SUMMARY

Extracts from marine organisms have been used for medicinal purposes in China, India, the Near East, and Europe since ancient times. Scientists have long recognized marine organisms as likely to contain many potential novel drugs because of the environmental conditions unique to their habitat. Though research has been concentrated on highly visible and abundant marine organisms, such as seaweeds and large invertebrates, microorganisms are also likely to contain novel substances due to their unique metabolic pathways.

Recent advances in genetic engineering and chemical separation technology have facilitated the investigation and development of new marine drugs. Nevertheless, few new drugs have come to market. Progress continues to be limited by problems associated with the collection, harvesting, and isolation of the pharmacologically active constituents, as well as their clinical evaluation and processing. Most importantly, drug development requires an extensive period of time and financial commitment. Current estimates indicate that the complete process takes at least 10 y, mostly due to the requirements associated with governmental approval, and costs more than $50 million. Due to the risky nature of such a financial investment, few pharmaceutical companies are currently pursuing the development of marine drugs.

Many pharmacologically active marine compounds have no terrestrial analogs and so represent wholly new classes of potential drugs. In the laboratory, some act as antibiotics; others appear to have antitumorogenic, anticoagulant, antiviral, antiulcer, haemolytic, analgesic, antilipemic, cardioinhibitory, immunostimulant, or immunosuppressant effects. Some are currently used as fungicides and insecticides or as food and cosmetic additives. Recent advances in biotechnology are likely to have marine applications. For example, genetic engineering techniques could be used to improve the hardiness and speed growth rates of maricultured organisms. Likewise, genetically engineered bacteria could be developed to prevent biofouling and to degrade pollutants.

PROBLEM SET 4

Answer the following questions using the sediment trap data given in Table P4.1. All of the traps were identical in size and shape. They were all deployed for the same length of time.

1. Plot the data to help visualize what's going on.
2. Give some reasons as to why the concentrations of the fatty alcohols and sterols show mid-water maxima, rather than monotonically decreasing with increasing depth.
3. Compute degradation rates in mg/d for each compound between each sampling depth. Assume that the only source of particles to a trap is from downward vertical transport and that the sinking rate of the particles is 10 m/d. Report the results of your calculations in Table P4.2.
4. Discuss the likely causes for the differences in degradation rates for the different compounds.
5. Discuss the likely causes for the differences in degradation rates from depth zone to depth zone for each particular compound.

TABLE P4.1
Organic Geochemical Results for PARFLUX E (13.5°N 54°W) Sediment Traps

						Amount of Material per Trap											
Depth of 1.5 m^2 trap (m)	*Sample designation*	*Particle size (mm)*	*Particulate matter (g)*	*Organic matter (g)*	*Organic carbon (g)*	*Hexane-soluble lipids (mg)*	*Hydrocarbons (μg)*	*Total fatty acids (μg)*	*Free fatty acids (μg)*	*Wax esters (μg)*	*Steroid ketones (μg)*	*Amino acids (Lipid fraction) (μg)*	*Free fatty alcohols (μg)*	*Free sterols (μg)*	*Steryl esters (μg)*	*Triacylglycerols (μg)*
389	E6	<1	6.6	1.4	0.7	225	1200	39,100	4100	770	4720	300	20,500	19,100	170	390
	E2	>1	3.6	0.6	0.3	70	250	16,500	2600	530	130	90	4100	7400	60	90
988	E3	<1	6.7	1.2	0.5	170	1120	17,400	470	180	3300	43	1800	31,400	2300	2100
	E1	>1	0.5	0.1	0.04	20	170	5100	—	20	20	3.2	980	3360	200	110
3755	E4	<1	6.8	0.7	0.2	25	140	1280	—	16	530	13	6300	900	8	17
5068	E5	<1	6.9	0.7	0.2	13	97	910	200	49	590	14	760	550	9	12

Source: From S. G. Wakeham, J. W. Farrington, R. B. Gagosian, C. Lee, H. DeBaar, G. E. Nigrelli, B. W. Tripp, S. O. Smith, and N. M. Frew, reprinted with permission from *Nature*, vol. 286, p. 799, copyright © 1980 by Macmillan Journals, Ltd., London, England; and S. Honjo, reprinted with permission from the *Journal of Marine Research*, vol. 38, pp. 66–67, copyright © 1980 by the *Journal of Marine Research*, New Haven, CT.

TABLE P4.2

Computed Degradation Rates (mg/d)

	Depth Intervals (m)			
	389–988		*988–3755*	*3755–5068*
Compound	*<1 mm*	*>1 mm*	*<1 mm*	*<1 mm*
Organic matter				
Hydrocarbons				
Total fatty acids				
Free fatty alcohols				
Free sterols				
Amino acids				

SUGGESTED FURTHER READINGS

Biogeochemical Cycles of C, N, O, P, and S

BOLIN, B., and R. B. COOK (eds.). 1983. *The Major Biogeochemical Cycles and Their Interactions*. John Wiley & Sons, Inc., New York, 532 pp.

LASSERRE, P., and J.-M. MARTIN (eds.). 1986. *Biogeochemical Processes at the Land–Sea Boundary*. Elsevier Science Publishers, Amsterdam, The Netherlands, 214 pp.

LIKENS, G. E. (ed.). 1981. *Some Perspectives of the Major Biogeochemical Cycles*. John Wiley & Sons, Inc., New York, 175 pp.

Biogeochemistry of Organic Compounds

LEE, C., and S. G. WAKEHAM. 1989. Organic Matter in Seawater: Biogeochemical Processes. In: *Chemical Oceanography*, vol. 9 (ed.: J. P. Riley), Academic Press, Orlando FL, pp. 1–52.

SIMONEIT, B. R. 1978. The Organic Chemistry of Marine Sediments. In: *Chemical Oceanography*, vol. 7 (ed.: J. P. Riley), Academic Press, Orlando, FL, pp. 234–312.

The CO_2 Problem and Global Climate Change

F. P. BRETHERTON, et al. 1986/1987. Changing Climate and the Oceans. *Oceanus*, **29**:1–100.

BROECKER, W. S., and G. H. DENTON, 1990. What Drives Glacial Cycles? *Scientific American*, **262**:49–56.

HILLEMAN, B. 1989. Global Warming. *Chemical and Engineering News*, **March 13**:25–40.

JONES, P. D., and T. M. L. WIGLEY. 1990. Global Warming Trends. *Scientific American*, **263**:84–91.

POST, W. M., T.-H. PENG, W. R. EMANUEL, A. W. KING, V. H. DALE, and D. L. DeANGELIS. 1990. The Global Carbon Cycle. *American Scientist*, **78**:310–326.

STEELE, J. H., et al. 1989. The Oceans and Global Warming. *Oceanus*, **32**: 1–96.

SCHNEIDER, S. H. 1989. The Changing Climate. *Scientific American*, **261**: 70–79.

WHITE, R. M. 1990. The Great Climate Debate. *Scientific American*, **263**: 36–43.

Humic Substances

RASHID, M. A. 1985. *Geochemistry of Marine Humic Compounds*. Springer-Verlag, Heidelberg, Germany, 300 pp.

THURMAN, E. M. 1985. *Organic Geochemistry of Natural Waters*. Kluwer Academic Publishers Group, Dordrecht, The Netherlands, 497 pp.

Marine Drugs

COLWELL, R. R. 1984. The Industrial Potential of Marine Biotechnology. *Oceanus*, **27**:3–12.

FAULKNER, D. 1979. The Search for Drugs from the Sea. *Oceanus*, **22**: 44–50.

NOVITSKY, T. J. 1984. Discovery to Commercialization: The Blood of the Horseshoe Crab. *Oceanus*, **27**:13–18.

RUGGIERI, G. 1976. Drugs from the Sea. *Science*, **194**:491–497.

Nitrogen Biogeochemistry

BLACKBURN, T. H., and J. SORENSEN. 1988. *Nitrogen Cycling in Coastal Marine Environments*. John Wiley & Sons, Inc., New York, 451 pp.

DELWICHE, C. C. 1970. The Nitrogen Cycle. *Scientific American*, **223**: 137–146.

FOGG, G. E. 1982. Nitrogen Cycling in Sea Waters. *Philosophical Transactions of the Royal Society of London*, **B296**:511–520.

PART FIVE

The estimation of the paleotemperatures of the ancient oceans by measurement of the oxygen isotope distribution between calcium carbonate and water, as suggested by Harold Urey (1947), is one of the most striking and profound achievements of modern nuclear geochemistry.

Harmon Craig, 1965

ISOTOPE GEOCHEMISTRY

CHAPTER 28

MEASURING RATES AND DATES: THE USE OF RADIOISOTOPES IN THE STUDY OF MARINE PROCESSES

INTRODUCTION

Most of the elements in seawater and the sediments have several stable and radioactive isotopes. The relative oceanic abundances of these isotopes have been used to study biological, geological, physical, and chemical processes. This approach has arguably been the most productive in furthering our understanding of the crustal-ocean factory. For example, the radioactive isotopes, or **radionuclides**, have been used to determine rates of processes, such as those of sedimentation, photosynthesis, and various types of water motion. The stable isotopes have been used to establish the timing and intensity of climate changes, the structure of food chains, and the fate of terrestrial organic matter in the marine environment. Indeed, many hypotheses and conclusions presented in previous chapters are based on isotopic data. Some of this supporting evidence is presented in this chapter, which deals with the radionuclides. The use of stable isotopes is discussed in the next chapter.

MECHANISMS OF RADIOACTIVE DECAY

Radioisotopes are atoms that spontaneously lose nuclear material at a fixed rate. This process is termed **radioactive decay** and is an example of a nuclear reaction. Atoms with unstable nuclei spontaneously undergo radioactive decay and thereby attain a greater measure of stability. Nonradioactive atoms will undergo nuclear reactions, such as fusion, only if energy is applied either in the form of electromagnetic radiation or by collision with an energized particle.

In unstable nuclei, the repulsive forces are stronger than the attractive ones. Most nuclear repulsion is caused by electrostatic interactions among the positively charged **protons**. Thus the addition of protons to a nucleus causes a net increase in the repulsive forces. This instability can be counterbalanced by the addition of **neutrons**, which increases the attractive forces. The relative number of neutrons required to achieve a stable nucleus is shown in Figure 28.1. The relative number of neutrons needed to achieve stability increases

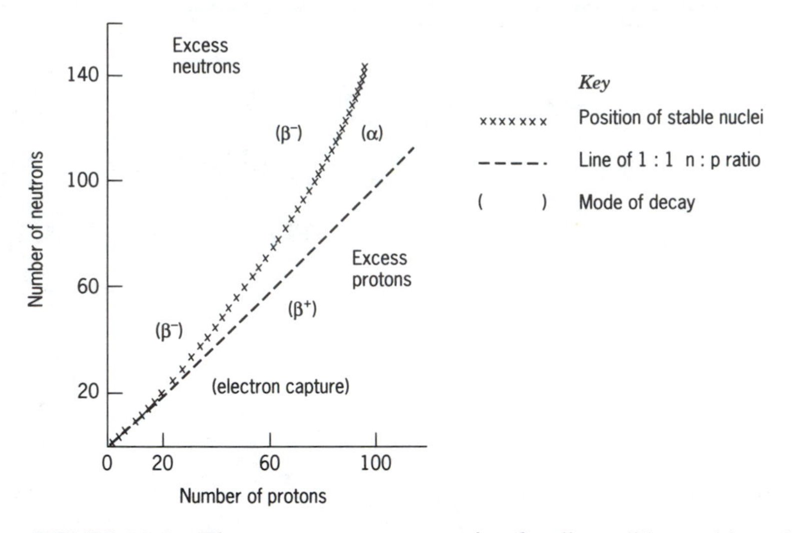

FIGURE 28.1. The neutron-to-proton ratio of radionuclides and its relationship to nuclear stability and radioactive-decay mechanisms.

with increasing mass number. Thus the neutron-to-proton ratio increases from 1:1 in the light nuclei ($Z \leq 17$) to 1.6:1 in the heavier elements. A nucleus with an excess of neutrons or protons spontaneously undergoes radioactive decay until a stable ratio is achieved. In many cases, this occurs through a series of radioactive decay steps, each of which progressively alters the neutron to proton ratio. Examples of such decay series are given later in this chapter.

If protons are in excess, stability is achieved either through emission of **α particles**, or positrons, or by electron capture. α particles are helium nuclei ($^{4}_{2}He$). Thus, their emission causes the radionuclide to lose two protons and two neutrons. This type of radioactive decay is characteristic of the larger radionuclides, as illustrated below:

$$^{238}_{92}U \rightarrow {}^{234}_{90}Th + {}^{4}_{2}He + Q \qquad (28.1)$$

Note that the neutron-to-proton ratio increases from 1.59 in the parent radionuclide, $^{238}_{92}U$, to 1.60 in its daughter isotope, $^{234}_{90}Th$. (Decay products are referred to as **daughters**.) Depending on the radionuclide, variable amounts of energy (Q) are released as gamma rays.

During positron emission, a proton spontaneously decomposes into a neutron and a positively charged particle ($^{0}_{1}e$) as follows:

$$^{1}_{1}p \rightarrow {}^{1}_{0}n + \beta^{+}({}^{0}_{1}e) + Q \qquad (28.2)$$

Positrons are also called β^{+} particles. They can be thought of as positively charged **electrons** and are destroyed as a result of collisions with electrons. Depending on the type of radionuclide undergoing positron emission, variable amounts of energy (Q) are released as neutrinos.

As shown in Figure 28.1, positron emission is characteristic of the radionuclides with intermediate atomic weights. These radionuclides are common products of nuclear reactions that occur in power plants and during detonation of atomic bombs. An example of positron emission is given below:

$$^{64}_{29}Cu \rightarrow {}^{64}_{28}Ni + {}^{0}_{1}e + Q \tag{28.3}$$

Note that the neutron-to-proton ratio of the nuclide increases from 1.21 to 1.29.

Excess protons are also destroyed through **electron capture**. In this reaction, an electron from the lowest energy level in the atom (1s) reacts with a proton, generating a neutron as illustrated below:

$$^{1}_{1}p + {}^{0}_{-1}e \rightarrow {}^{1}_{0}n \tag{28.4}$$

Due to their close proximity to the nucleus, the 1s electrons are most likely to be captured by the nucleus. They are replaced by higher-energy electrons. As these replacement electrons drop to the lower energy state, they emit short-wavelength radiation, called x-rays. This type of decay is characteristic of the low atomic weight isotopes.

^{40}K is unstable and spontaneously undergoes electron capture, as illustrated below:

$$^{40}_{19}K + {}^{-1}_{0}e \rightarrow {}^{40}_{18}Ar + Q \tag{28.5}$$

Only about 10 percent of ^{40}K decays by this mechanism. The majority (90%) decays to $^{40}_{20}Ca$ via β^- emission. **β^- particles** are high-energy electrons ($_{-1}^{0}e$). Their emission from the nucleus causes a neutron to be converted to a proton. Thus radionuclides with an excess of neutrons decay via this process. As shown below, variable amounts of energy (Q) are also released, either in the form of neutrinos or gamma rays, depending on the radionuclide.

$$^{1}_{0}n \rightarrow {}^{1}_{1}p + \beta^-({}^{0}_{-1}e) + Q \tag{28.6}$$

Tritium ($^{3}_{1}H$) also decays via β^- emission, as shown below:

$$^{3}_{1}H \rightarrow {}^{3}_{2}He + \beta^-({}^{0}_{-1}e) + Q \tag{28.7}$$

RADIOACTIVE DECAY LAW

The rate at which a radionuclide decays is directly proportional to the number of atoms present; that is,

$$-\frac{dN}{dt} = \lambda N \tag{28.8}$$

where N, the number of atoms (e.g., moles or grams) of radionuclide N, decreases over time (t) as a result of radioactive decay. The rate constant, λ, is a characteristic of the radionuclide. It is a statistical measure of the likelihood for an average atom to undergo decay in a specified unit of time.

If the original and final numbers of atoms of the radionuclide are known, the amount of time over which radioactive decay has occurred can be inferred from the **radioactive decay law**. To do this, Eq. 28.8 must be integrated from the original time and number (t_o and N_o) to the final time and number (t_f and N_f) as follows:

$$\int_{N_o}^{N_f} \frac{1}{N} dN = -\lambda \int_{t_o}^{t_f} dt \tag{28.9}$$

The solution of this integral is

$$\ln N \Big|_{N_o}^{N_f} = -\lambda t \Big|_{t_o}^{t_f} \tag{28.10}$$

which can be rewritten as

$$\ln N_f - \ln N_o = -\lambda(t_f - t_o) \tag{28.11}$$

Substituting in $t_o = 0$, $t_f = t$ and $N_f = N$ yields

$$\ln(N/N_o) = -\lambda t \tag{28.12}$$

or

$$\frac{N}{N_o} = e^{-\lambda t} \tag{28.13}$$

Thus radioactive decay causes radionuclide numbers to decrease exponentially over time, as illustrated in Figure 28.2 for ^{14}C.

If half of N_o has decayed, then $N_f/N_o = 0.5$ and Eq. 28.13 becomes

$$0.5 = e^{-\lambda t_{\frac{1}{2}}} \tag{28.14}$$

where $t_{\frac{1}{2}}$ is the **half-life**. This can also be written as

$$\frac{0.693}{\lambda} = t_{\frac{1}{2}} \tag{28.15}$$

Since ^{14}C has a **decay constant** of 1.22×10^{-4}/y, its half-life is 5680 y.

CLASSIFICATION OF MARINE RADIONUCLIDES

Marine radionuclides are classified according to their source. Some of the atoms that coalesced to form this planet were radionuclides. Many of those with long half-lives are still present. The most abundant of these primordial radionuclides are ^{235}U, ^{238}U, and ^{232}Th, which have half-lives of 0.71, 4.5, and 14 billion years, respectively. Since the earth is 4.6 billion years old, 1.1 percent of the primordial ^{235}U is still present, while 49 percent and 80 percent of the ^{238}U and ^{232}Th remain. As shown in Figure 28.3, these isotopes decay through a series of steps and produce a variety of daughters whose half-lives range

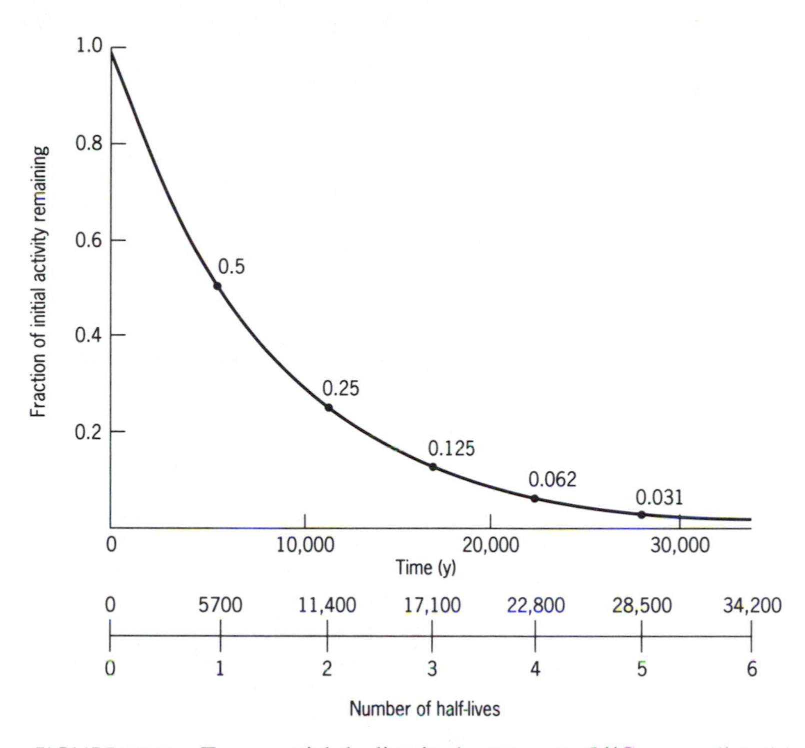

FIGURE 28.2. Exponential decline in the amount of ^{14}C as predicted by the radioactive decay law.

from fractions of seconds to thousands of years. The series end in stable isotopes of lead.

The primordial radionuclides decay by either α or β^- emission. The energy released during these nuclear reactions is partially dissipated as heat in the inner earth and makes a significant contribution to the geothermal gradient.

Cosmogenic radionuclides are formed by spallation reactions that occur in the atmosphere. In these reactions, gas nuclei are fragmented as a result of collisions with high-energy cosmic rays. Radiocarbon (^{14}C) is the cosmogenic radionuclide most commonly used by oceanographers. The artificial radionuclides are isotopes produced by humans. Most have been introduced into the ocean as a result of nuclear bomb testing and leakage from nuclear power plants. Due to differences in their sources and chemical behavior, the primordial, cosmogenic, and artificial radionuclides have distinct distributions and fates in the marine environment. As a result, they have proven useful for studying a multitude of marine processes. Some examples are described below.

THE PRIMORDIAL RADIONUCLIDES

All the **primordial radionuclides** are metals, with the exception of radon, which is a noble gas. As shown in Table 28.1, most are present at very low

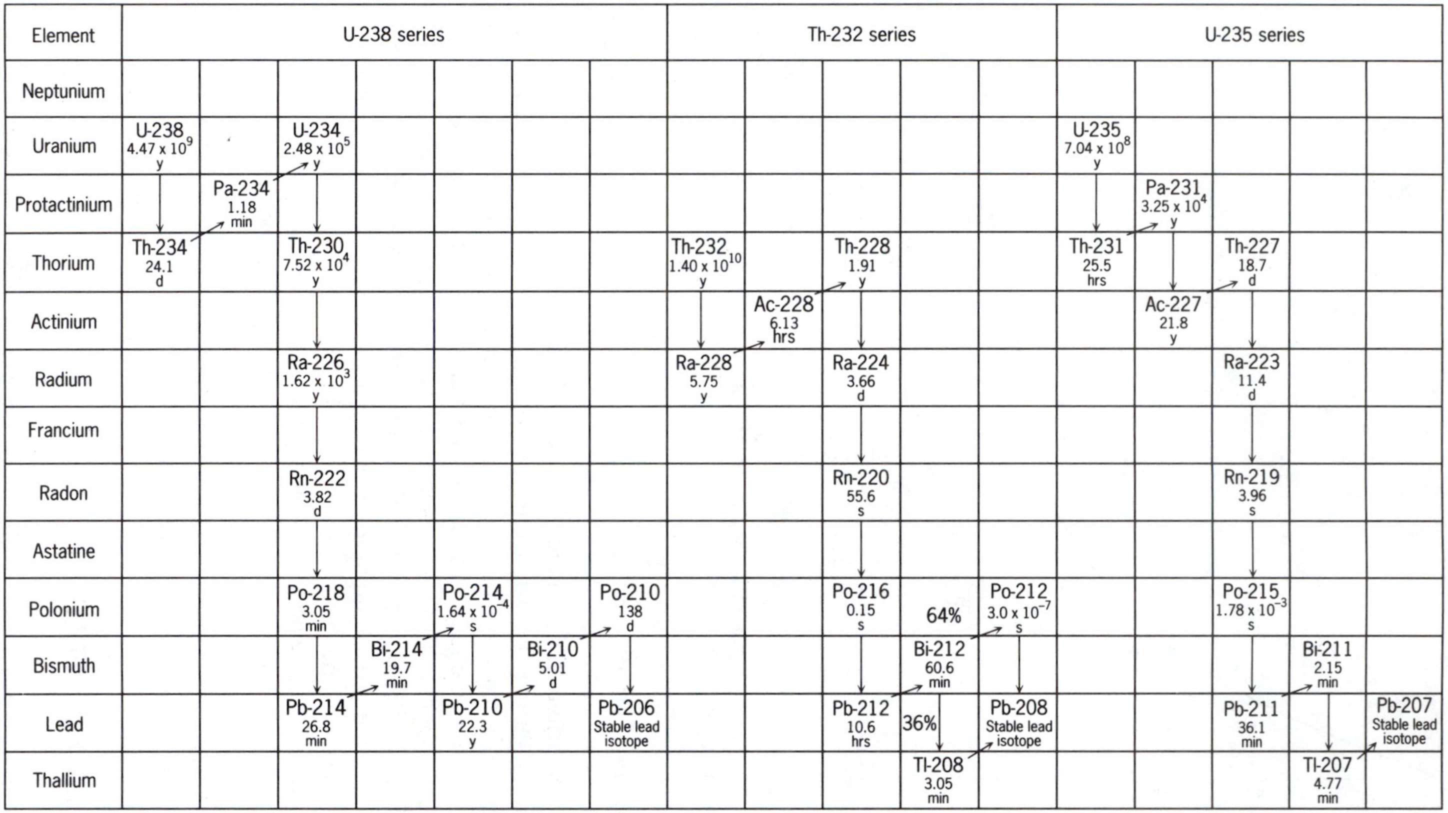

Element	U-238 series							Th-232 series					U-235 series				
Neptunium																	
Uranium	U-238 4.47×10^9 y		U-234 2.48×10^5 y										U-235 7.04×10^8 y				
Protactinium		Pa-234 1.18 min												Pa-231 3.25×10^4 y			
Thorium	Th-234 24.1 d		Th-230 7.52×10^4 y					Th-232 1.40×10^{10} y		Th-228 1.91 y			Th-231 25.5 hrs		Th-227 18.7 d		
Actinium									Ac-228 6.13 hrs					Ac-227 21.8 y			
Radium			Ra-226 1.62×10^3 y					Ra-228 5.75 y		Ra-224 3.66 d					Ra-223 11.4 d		
Francium																	
Radon			Rn-222 3.82 d							Rn-220 55.6 s					Rn-219 3.96 s		
Astatine																	
Polonium			Po-218 3.05 min		Po-214 1.64×10^{-4} s		Po-210 138 d			Po-216 0.15 s	64%	Po-212 3.0×10^{-7} s			Po-215 1.78×10^{-3} s		
Bismuth				Bi-214 19.7 min		Bi-210 5.01 d					Bi-212 60.6 min					Bi-211 2.15 min	
Lead			Pb-214 26.8 min		Pb-210 22.3 y		Pb-206 Stable lead isotope			Pb-212 10.6 hrs	36%	Pb-208 Stable lead isotope			Pb-211 36.1 min		Pb-207 Stable lead isotope
Thallium											Tl-208 3.05 min					Tl-207 4.77 min	

FIGURE 28.3. Chart showing the decay chain of the uranium and thorium series isotopes and the half-lives of each isotope. Alpha decays are shown by the vertical arrows and beta decays by the diagonal arrows. *Source*: From C. Giffin, A. Kaufman, and W. S. Broecker, reprinted with permission from the *Journal of Geophysical Research*, vol. 68, p. 1750, copyright © 1963 by the American Geophysical Union, Washington, DC.

TABLE 28.1
Average Oceanic and Sedimentary Concentrations of the Primordial Radionuclides

Isotope	*Estimated Average Concentration in Seawater (g/L)*	*Estimated Average Concentration in Surface Sediments (g/g)*	*Range of Concentration in Surface Sediments (g/g)*
^{238}U	3.0×10^{-6}	1×10^{-6}	$(0.4–80) \times 10^{-6}$
^{235}U	2.1×10^{-8}	7.1×10^{-9}	
^{234}U	1.6×10^{-10}	8.1×10^{-11}	
^{234}Pa	1.4×10^{-19}	4.7×10^{-20}	
^{231}Pa	$<2 \times 10^{-12}$	1×10^{-11}	$(0.08–9) \times 10^{-11}$
^{234}Th	4.3×10^{-17}	1.4×10^{-17}	
^{232}Th	$<2 \times 10^{-8}$	5.0×10^{-6}	$(1–16) \times 10^{-6}$
^{231}Th	8.6×10^{-20}	2.9×10^{-20}	
^{230}Th	$<3 \times 10^{-13}$	2.0×10^{-10}	$(0.3–20) \times 10^{-10}$
^{228}Th	4.0×10^{-18}	7×10^{-16}	
^{227}Th	$<7.0 \times 10^{-20}$	1.3×10^{-17}	
^{228}Ac	1.5×10^{-21}	2.4×10^{-19}	
^{227}Ac	$<1 \times 10^{-15}$	5.9×10^{-15}	
^{228}Ra	1.4×10^{-17}	2.3×10^{-15}	
^{226}Ra	1.0×10^{-13}	4.0×10^{-12}	$(0.3–40) \times 10^{-12}$
^{224}Ra	2.1×10^{-20}	3.4×10^{-18}	
^{223}Ra	$<4.4 \times 10^{-20}$	8.5×10^{-18}	
^{223}Fr	$<7.0 \times 10^{-24}$	1.4×10^{-21}	
^{222}Rn	6.3×10^{-19}	2.5×10^{-17}	
^{220}Rn	3.3×10^{-24}	5.4×10^{-22}	
^{219}Rn	$<1.7 \times 10^{-25}$	3.1×10^{-23}	
^{218}Po	3.4×10^{-22}	1.4×10^{-20}	
^{216}Po	1.0×10^{-26}	1.7×10^{-24}	
^{215}Po	$<8.1 \times 10^{-29}$	1.4×10^{-26}	
^{214}Po	3.0×10^{-28}	1.1×10^{-27}	
^{212}Po	1.2×10^{-32}	2.4×10^{-29}	
^{211}Po	$<6.8 \times 10^{-29}$	1.2×10^{-26}	
^{210}Po	2.2×10^{-17}	8.8×10^{-16}	
^{214}Bi	2.1×10^{-21}	8.8×10^{-20}	
^{212}Bi	2.2×10^{-22}	3.7×20^{-24}	
^{211}Bi	$<5.6 \times 10^{-24}$	1.0×10^{-21}	
^{210}Bi	7.8×10^{-19}	3.1×10^{-17}	
^{214}Pb	2.9×10^{-21}	1.2×10^{-19}	
^{212}Pb	2.4×10^{-21}	3.9×10^{-19}	
^{211}Pb	$<9.0 \times 10^{-23}$	1.6×10^{-20}	
^{210}Pb	1.1×10^{-15}	4.5×10^{-14}	
^{208}Tl	4.1×10^{-24}	6.7×10^{-22}	
^{207}Tl	$<1.2 \times 10^{-23}$	2.1×20^{-21}	

Source: From *Marine Chemistry*, R. A. Horne, copyright © 1969 by John Wiley & Sons, Inc., New York, p. 295. Reprinted by permission.

concentrations in both seawater and the sediments, making concentration measurements difficult.

The most sensitive technique for quantifying these radionuclides is measurement of their radioactivity. This is commonly done by first chemically separating the isotopes and then measuring the rate at which each produces either α or β^- emissions. This production rate is usually reported as counts per minute and is functionally equal to the decay rate, or **activity** of the radionuclide. Activities are usually reported as **disintegrations per minute (dpm)**. The activity of a radionuclide in a sample of seawater or sediment is reported in terms of dpm/L or dpm/g, respectively. These **specific activities** (A) are directly related to the concentration of the radionuclide as follows:

$$A = \lambda[N] \tag{28.16}$$

The average specific activities of some primordial radionuclides in various water masses are given in Table 28.2.

Secular Equilibrium

For most primordial radionuclides, the half-life of the **daughter** is much shorter than that of the parent. Thus the decay of the **parent** eventually becomes the rate-determining step that controls decay of the daughter, as shown in Figure 28.4. Due to slow decay rates, the concentration of parent and its specific

TABLE 28.2
Average Specific Activities of Some Primordial Radionuclides in Various Water Masses

Isotope	*Daughter Product Half-life (y)*	*Warm Surface Water (dpm/100 kg)*	*North Atlantic Bottom Water (dpm/100 kg)*	*Antarctic Bottom Water (dpm/100 kg)*	*North Pacific Bottom Water (dpm/100 kg)*
^{238}U	Parent	240	240	240	240
^{234}Th	0.066	230	240	240	240
^{234}U	248,000	280	280	280	280
^{230}Th	75,200	<0.02	—	—	0.15
^{226}Ra	1620	7	13	20	34
^{222}Rn	0.010	5	>13	>20	>34
^{210}Pb	22.3	20	8	10	16
^{210}Po	0.38	10	8	10	16
^{235}U	Parent	13	13	13	13
^{231}Pa	32,500	—	—	—	0.05
^{232}Th	Parent	<0.1	<0.1	<0.1	<0.1
^{228}Ra	5.8	3	0.4	—	0.4
^{228}Th	1.9	0.4	0.3	—	0.3

Source: From *Tracers in the Sea*, W. S. Broecker and T. -H. Peng, copyright © 1982 by the Lamont-Doherty Geological Observatory, Palisades, NY, p. 170. Reprinted by permission.

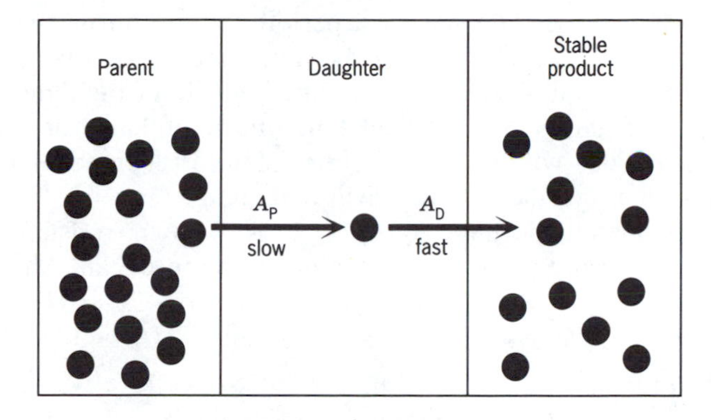

FIGURE 28.4. Secular equilibrium.

activity (A_P) remain relatively constant over time. In the absence of other processes, radioactive decay of these two isotopes will eventually reach a steady state. In this steady state, the rate at which the daughter is supplied from decay of parent (A_P) is matched by its rate of loss via its own decay (A_D). This condition, where $A_P = A_D$, is called **secular equilibrium**.

As shown in Figure 28.5, time is required for the attainment of secular equilibrium. In other words, introduction of the parent radionuclide to seawater

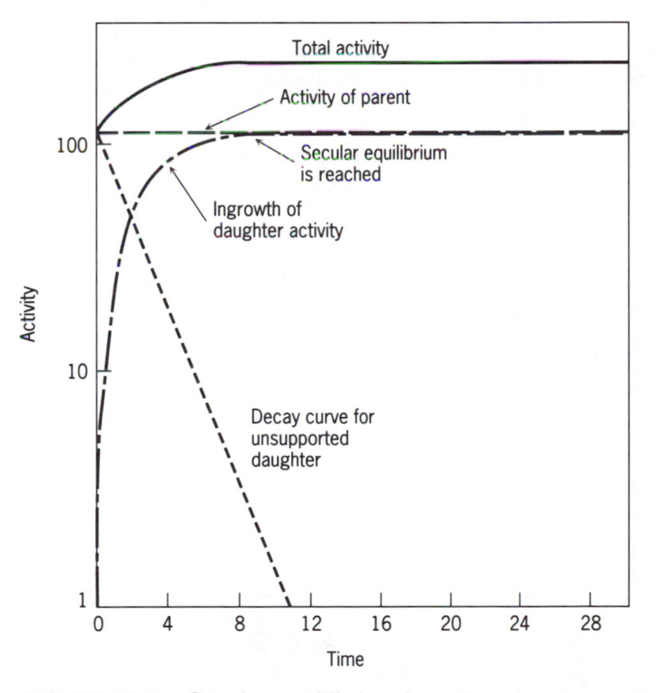

FIGURE 28.5. Secular equilibrium in radioactive decay. *Source*: From *Radioactivity in the Environment*, R. L. Kathren, copyright © 1984 by Harwood Academic Publishers, New York, p. 57. Reprinted by permission.

or sediment will be followed by a period of time during which A_D increases until it reaches a level equal to A_P. The amount of time required to reach secular equilibrium is determined by the half-life of the daughter, as shown in Table 28.3. Thus, given sufficient time and abundance of original parent, all the radionuclides within a decay series should reach secular equilibrium and hence all their specific activities will be equal.

At secular equilibrium, $A_D/A_P = 1$. As shown in Table 28.4, this is not observed for many parent–daughter pairs in the ocean. Most of these ratios

TABLE 28.3
Time Required for Various Daughters of ^{238}U and ^{232}Th To Achieve 99 and 95 Percent of A_P[a]

Nuclide	*99%*	*95%*
^{238}U	Parent	Parent
^{234}Th	160.13 d	104.16 d
^{234}Pa	7.84 min	5.10 min
^{234}U	1.661×10^6 y	1.081×10^6 y
^{230}Th	5.315×10^5 y	3.458×10^5 y
^{226}Ra	1.0777×10^4 y	7.011×10^3 y
^{222}Rn	25.41 d	16.53 d
^{218}Po	20.27 min	13.18 min
^{214}Pb	2.97 h	1.93 h
^{214}Bi	2.18 h	1.42 h
^{214}Po	1.09×10^{-3} s	7.09×10^{-4} s
^{210}Pb	146.18 y	95.09 y
^{210}Bi	33.22 d	21.61 d
^{210}Po	2.51 y	1.63 y
^{206}Pb	Stable (24.1% of natural Pb)	
^{232}Th	Parent	Parent
^{228}Ra	44.52 y	28.96 y
^{228}Ac	40.73 h	26.49 h
^{228}Th	12.62 y	8.21 y
^{224}Ra	24.19 d	15.73 d
^{220}Rn	6.02 min	3.93 min
^{216}Po	1.05 s	0.68 s
^{212}Pb	2.93 d	1.91 d
^{212}Bi	6.7 h	4.36 h
^{212}Po	1.99×10^{-6} s	1.3×10^{-6} s
^{208}Tl	20.6 min	13.4 min
^{208}Pb	Stable (52.3% of natural Pb)	

Source: From *Radioactivity in Geology*, E. M. Durrance, copyright © 1986 by Ellis Horwood, Ltd., Chichester, England, p. 286. Reprinted by permission.

[a]Since secular equilibrium is approached asymptotically, A_P is never quite attained.

TABLE 28.4
Typical Activity Ratios for Daughter–Parent Pairs in Various Water Types

	Estuaries	*Coastal*	*Surface Ocean*	*Deep Sea*
$^{210}Pb/^{226}Ra$	—	—	>1[a]	0.4–1.0
$^{230}Th/^{234}U$	—	—	$<3 \times 10^{-5}$	3×10^{-4}
$^{228}Th/^{228}Ra$	0.01	0.05	0.2	0.5–1.0
$^{234}Th/^{238}U$	0.2	0.6	>0.9	~ 1
$^{231}Pa/^{235}U$	—	—	—	2×10^{-3}
$^{210}Po/^{210}Pb$	—	—	0.5	1.0

Source: From *Tracers in the Sea*, W. S. Broecker and T.-H. Peng, copyright © 1982 by the Lamont-Doherty Geological Observatory, Palisades, NY, p. 173. Reprinted by permission.
[a]Although ^{210}Pb is being removed from surface water by particles, it has an additional source. Radon atoms escaping to the atmosphere from continental soils decay to ^{210}Pb. These ^{210}Pb atoms are incorporated into aerosols and are brought back to Earth's surface by rain and aerosol impact. The flux of these atoms to the sea surface exceeds by about a factor of 10 the in situ production by radiodecay of ^{226}Ra in the upper 200 m of the ocean.

are less than 1 due to preferential removal of daughter. This is usually the result of some chemical and/or physical process.

The net removal rate of the daughter can be calculated from a steady-state box model, as illustrated in Figure 28.6. The sole source of daughter is assumed to be from decay of the parent. Thus the supply rate of daughter is equal to A_P. At steady state, the supply rate must match the rate at which daughter is lost. The rate at which daughter is lost from its own decay is equal to A_D. Since nothing more is known about the chemical and/or physical processes that affect the daughter, its net nonradioactive removal rate is assumed to be directly proportional to its concentration, $[D]$. Thus

$$\text{Net nonradioactive removal rate} = k_D[D] \tag{28.17}$$

where k_D is the net removal rate constant.

This steady-state model is described by the following mass balance equation in which the supply rate of daughter is on the left and its losses are on the right: that is

$$A_P = A_D + k_D[D] \tag{28.18}$$

Since $A_D = \lambda_D[D]$, $[D]$ can be eliminated from Eq. 28.18 as follows:

$$A_P = A_D + \left[\frac{k_D A_D}{\lambda_D}\right] \tag{28.19}$$

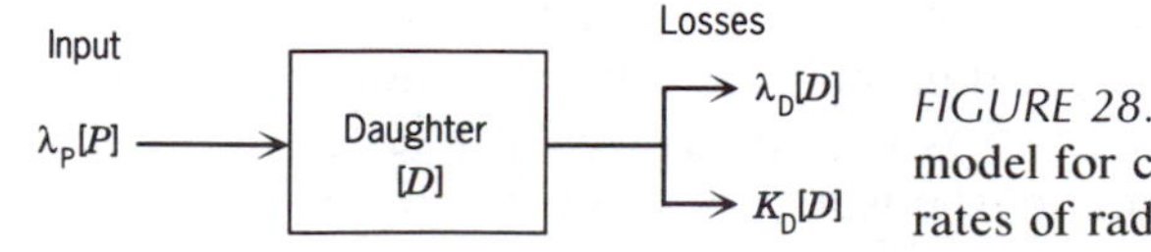

FIGURE 28.6. Steady-state box model for computing net removal rates of radionuclides.

Rearranging and solving for k_D yields

$$k_D = \left[\frac{A_P - A_D}{A_D}\right] \lambda_D \quad (28.20)$$

This equation can be rearranged to

$$k_D = \left[\frac{1 - \frac{A_D}{A_P}}{\frac{A_D}{A_P}}\right] \lambda_D \quad (28.21)$$

Thus without any knowledge of the actual removal process, its net rate constant can be inferred from A_D/A_P. Note that k_D includes the effects of all nonradioactive processes, whether they be additions or removals of daughter. Thus if daughter is also supplied by some nonradioactive process, its actual removal rate will be somewhat greater than the net removal rate. Further information is required to determine the mechanism by which daughter is removed (or added). But once this is done, the results can be generalized to chemically similar isotopes, such as cogeners or ions with similar charge densities. In this way, radionuclides are used as natural tracers of many marine processes. Several examples are given below.

Determination of Particle Scavenging Rates

The specific activity ratio of $^{228}Th/^{228}Ra$ ($A_{^{228}Th}/A_{^{228}Ra}$) is less than 1 because ^{228}Th is preferentially removed from seawater by adsorption onto sinking particles. The specific activity of ^{228}Ra is maintained by diffusion of this isotope from the sediments. The sediments contain radium because this element is incorporated into the tests of calcareous plankton. Biogenic uptake is favored by the chemical similarity of radium to calcium. ^{228}Ra is also produced from the in situ decay of ^{232}Th.

As shown in Figure 28.7, the effect of particle scavenging on ^{228}Th is most pronounced in coastal waters. This is due to higher particle concentrations and shallower water depths. Under these conditions, ^{228}Th has a greater chance of adsorbing onto particles or the surface sediments. Because of shallow water depths, diffusion of radium from the sediments has a relatively large impact on its specific activity in the water column. Thus relatively high levels of ^{228}Ra also contribute to the occurrence of low $A_{^{228}Th}/A_{^{228}Ra}$'s in coastal waters.

The net rate constant for the removal of ^{228}Th can be calculated by substituting the observed specific activity ratios into Eq. 28.21. This information is commonly reported as a scavenging turnover time in years ($1/k_D$). Indeed, this is how the scavenging turnover times were obtained for some of the trace metals listed in Table 11.3. This rate information is also commonly given as the chemical half-life, or half-scavenging time, of an isotope ($0.693/k_D$). This can be thought of as the time it would take for the nonradioactive processes to remove half of the isotope's specific activity. As shown in Figure 28.8, the

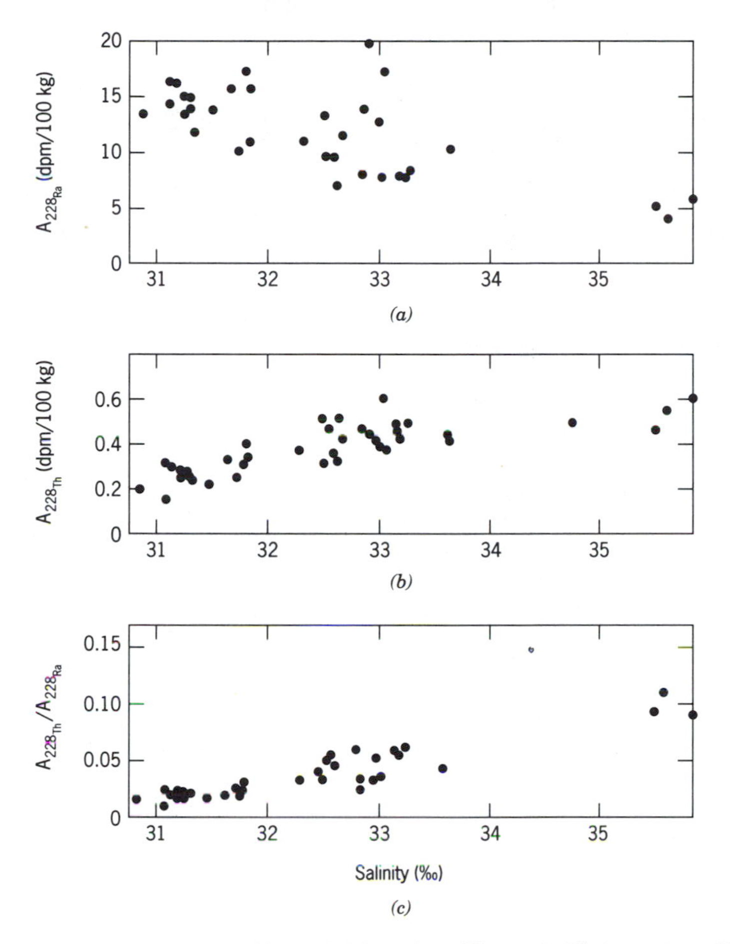

FIGURE 28.7. Specific activities of (*a*) ^{228}Ra, (*b*) ^{228}Th, and (*c*) ^{228}Th/^{228}Ra as a function of salinity in coastal samples taken adjacent to New York City. Due to the input of river water, the waters closest to the coast have the lowest salinity. Thus the increase in salinity is a measure of distance from the shore. *Source*: From Y.-H. Li, H. W. Feely, and P. H. Santschi, reprinted with permission from *Earth and Planetary Science Letters*, vol. 42, p. 20, copyright © 1979, Elsevier Science Publishers, Amsterdam, The Netherlands.

half-scavenging time for ^{228}Th ranges from approximately 0.1 to 1.0 y in the surface waters of the open ocean.

In the open ocean, A_{228Th}/A_{228Ra} increases with depth through the mixed layer and sometimes exceeds 1 at its base. This excess ^{228}Th is thought to be supplied by the remineralization of biogenic particles. In other words, marine organisms appear to bioaccumulate ^{228}Th.

The radioactive half-life of ^{234}Th is only 24 d. Thus its use as a tracer of scavenging is limited to coastal waters where high particle concentrations ensure rapid removal rates. At these locations, ^{234}Th distributions yield half-

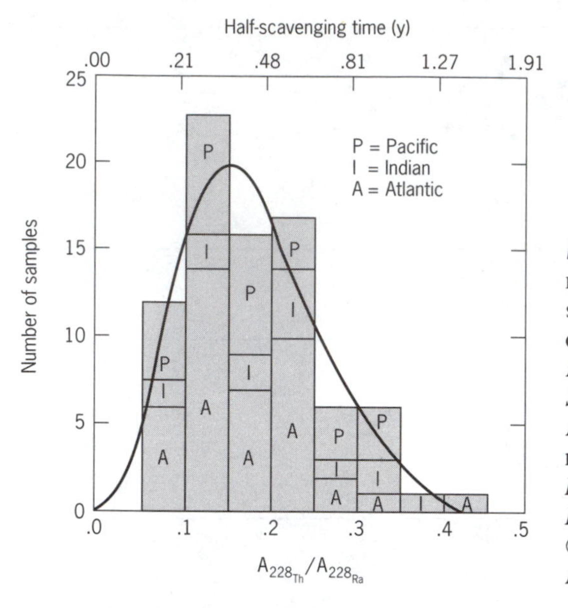

FIGURE 28.8. The specific activity ratio of $^{228}Th/^{228}Ra$ and the half-scavenging time for ^{228}Th in the open ocean surface waters of the Atlantic, Pacific, and Indian oceans. *Source*: From W. S. Broecker, A. Kaufman, and R. M. Trier, reprinted with permission from *Earth and Planetary Science Letters*, vol. 20, p. 41, copyright © 1973, Elsevier Science Publishers, Amsterdam, The Netherlands.

scavenging times similar to that of ^{228}Th. Because of its relatively short half-life, ^{234}Th cannot be used to measure scavenging rates in open ocean waters. In these waters, particle concentrations are so low that the ^{234}Th would decay before it had a chance to adsorb. As a result, A_{234Th}/A_{238U} in open ocean waters is close to 1. This illustrates an important factor that must be considered when selecting an isotope as a tracer for a particular geochemical process; its decay rate must be close to the rate of the process to be studied. As shown in Table 28.5, mismatches produce activity ratios too close to 1 or too small to be precisely measured.

In Case 1, $k_D >>> \lambda_D$, so the chemical half-life is much shorter than the radioactive half-life. Rapid removal keeps A_D very low and hence results in a very small specific activity ratio that is difficult to measure precisely. In Case

TABLE 28.5
Comparison of the Combined Effects of Chemical and Radioactive Removal on the Specific Activity Ratios of a Parent–Daughter Radionuclide Pair[a]

Case	k_D (y^{-1})	*Chemical* $t_{\frac{1}{2}}$ *(y)*	λ_D (y^{-1})	*Radioactive* $t_{\frac{1}{2}}$ *(y)*	*Calculated* A_D/A_P
1	0.69	1.0	0.0069	100	0.0099
2	0.0069	100	0.0069	100	0.50
3	0.0069	100	0.69	1.0	0.99

[a] A_D/A_P is inferred from the given net removal rate constants and radioactive half-lives using Eq. 28.21.

3, $k_D <<< \lambda_D$, so the chemical half-life is much longer than the radioactive half-life. Removal is too slow to cause secular disequilibrium and hence the specific activity ratio is close to, and difficult to precisely distinguish from, 1. In Case 2, $k_D = \lambda_D$, so the chemical and radioactive half-lives are equal. Here removal is fast enough to cause secular disequilibrium, but not so fast as to lower A_D greatly. In this situation, when A_D is significantly different from A_P, but not too different, k_D can be calculated with the greatest analytical precision.

Thus ^{230}Th, with a half-life of 75,000 y, is more suitable than ^{234}Th ($t_{\frac{1}{2}}$ = 24 d) or ^{228}Th ($t_{\frac{1}{2}}$ = 1.91 y) for determining half-scavenging times in the deep sea. As shown in Figure 28.9, the half-scavenging times of this isotope decrease with depth in the open ocean probably due to decreasing particle concentrations. Such data also indicate that the adsorption of ^{230}Th on sinking particles must be reversible.

The fate of pollutant lead in the ocean has also been investigated by inferring its scavenging rate from specific activity data. The burning of leaded gasoline injects lead aerosols into the atmosphere, some of which settle onto the sea surface. This aeolian input causes the surface-water lead enrichment ($A_D/A_P > 1$) shown in Figure 28.10. ^{210}Pb is also supplied to the ocean via decay of its parent, ^{226}Ra. The specific activity of ^{226}Ra increases with depth, because most is supplied by diffusion from the sediments. As mentioned above, radium is resolubilized in the sediments from biogenic calcium carbonate. ^{226}Ra is also produced from the decay of ^{230}Th transported to the sediments via particle scavenging.

At mid depths, $A_{^{210}Pb}/A_{^{226}Ra}$ is less than 1, suggesting that lead is removed faster than radium, probably by adsorption onto sinking particles. With these activity values, the chemical half-life of lead in the midwaters is calculated to be 100 y. Near the seafloor, the chemical half-life of lead is shorter (15 y) due

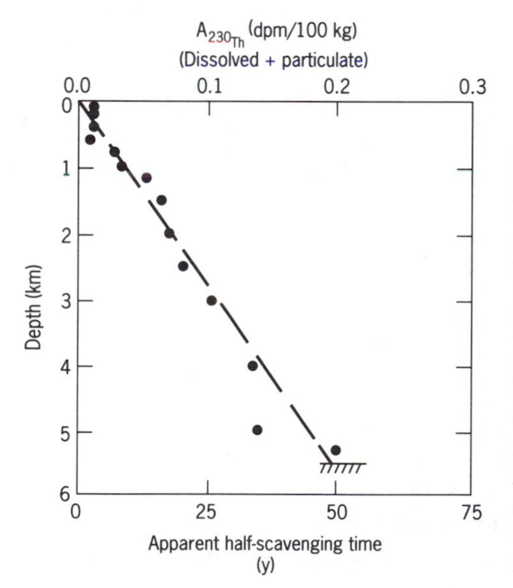

FIGURE 28.9. The specific activity of ^{230}Th and its apparent half-scavenging time as a function of depth at a site in the northwest Pacific. The specific activity of the parent, ^{234}U, at this site is 280 dpm/100 kg. The dissolved A_{230th} is ten times greater than the particulate activity. *Source*: From Y. Nozaki, Y. Horibe, and H. Tsubota, reprinted with permission from *Earth and Planetary Science Letters*, vol. 54, p. 213, copyright © 1981, Elsevier Science Publishers, Amsterdam, The Netherlands.

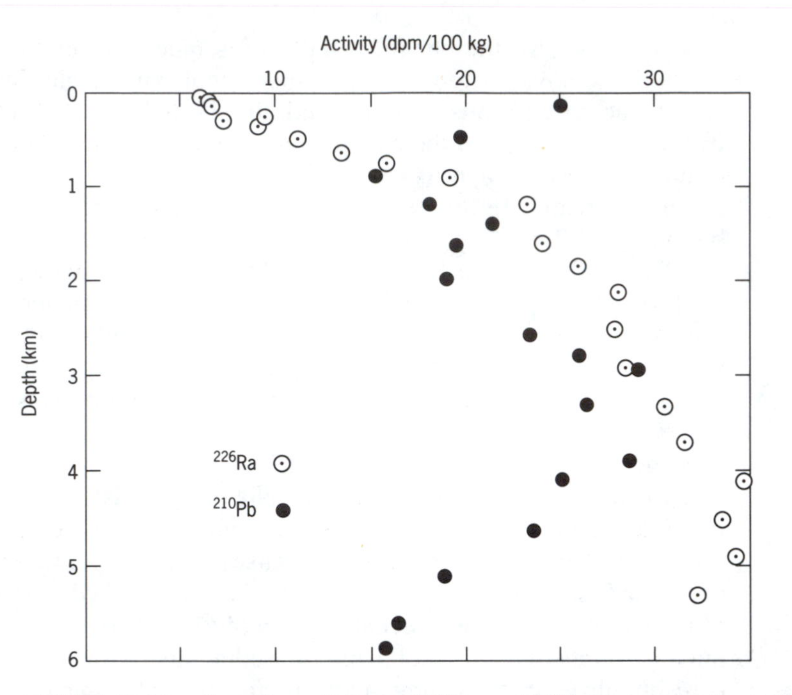

FIGURE 28.10. Specific activities of ^{226}Ra and ^{210}Pb versus depth at GEOSECS Station 214 (32°N 177°W) in the central north Pacific. *Source*: From *Tracers in the Sea*, W. S. Broecker and T.-H. Peng, copyright © 1982 by the Lamont-Doherty Geological Observatory, Palisades, NY, p. 193. Reprinted by permission. ^{226}Ra data from Y.-C. Chung and H. Craig, reprinted with permission from *Earth and Planetary Science Letters*, vol. 49, p. 271, copyright © 1980, Elsevier Science Publishers, Amsterdam, The Netherlands. ^{210}Pb data from Y. Nozaki, K. K. Turekian, and K. von Damm, reprinted with permission from *Earth and Planetary Science Letters*, vol. 49, p. 395, copyright © 1980, Elsevier Science Publishers, Amsterdam, The Netherlands.

to adsorption onto the surface sediments. Using estimates of anthropogenic lead input, the chemical half-life in the surface waters has been estimated to be less than 20 y. ^{210}Pb decays to ^{210}Po. Polonium is much more reactive than lead and has a very high enrichment factor in the tissues of marine organisms. As a result, its chemical half-life in surface waters is only about 0.4 y.

Dating Marine Sediments and Determining Sedimentation Rates

As indicated above, some daughters of the primordial radionuclides are preferentially transported to the seafloor via particle scavenging and/or by incorporation into biogenic particles. Since the rain rate of daughter exceeds that of its parent, the surface sediments have $A_D/A_P > 1$.

Once in the sediment, the parent will undergo decay and thus contribute to the A_D in the sediments. This contribution is termed the supported A_D. As noted above, preferential particle scavenging of the daughter causes the total A_D in the surface sediments to be somewhat larger than the supported A_D. This excess activity is termed the unsupported A_D and is equal to $A_D - A_P$. A_P is nearly constant with depth due to slow rates of radioactive decay. Hence the supported A_D is nearly constant with depth. In contrast, the sediment is cut off from its source of unsupported A_D following burial. Thus the unsupported A_D decreases with depth, as shown in Figure 28.11.

The time that has elapsed since particles, now at depth z, were last at the sediment surface can be calculated from the decrease in unsupported (or excess) A_D with depth. To do this, it must be assumed that the sedimentation rate (s) and the supply of unsupported A_D to the sediments has not varied over time. Since $s = z/t$, the radioactive decay law (Eq. 28.13) can be written as

$$\frac{A_{\text{DExcess}_z}}{A_{\text{DExcess}_{z=0\,\text{cm}}}} = e^{-\lambda_D \frac{z}{s}} \qquad (28.22)$$

where the radionuclide concentration, $[N]$, is given in terms of its specific

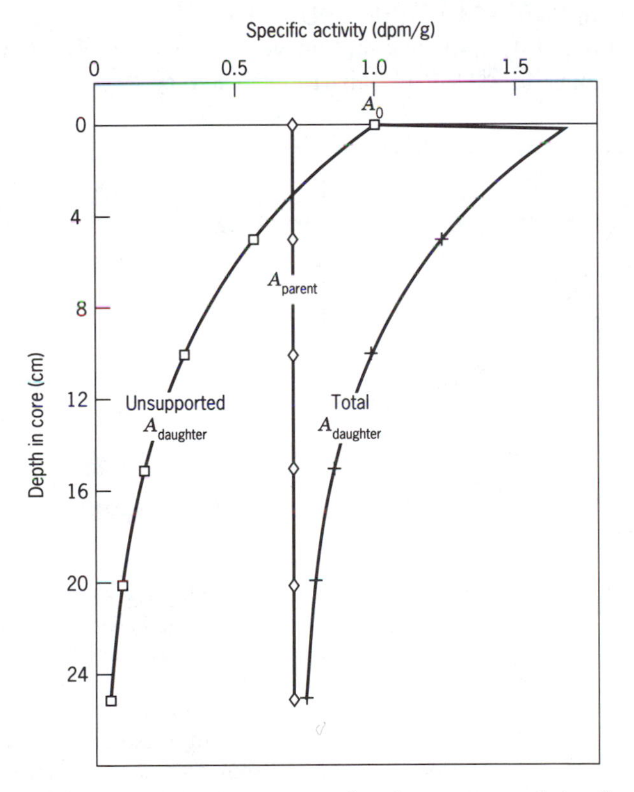

FIGURE 28.11. A_P, supported and unsupported A_D of a parent–daughter radionuclide pair in sediment core. The half-life of the daughter is 5700 y and the sedimentation rate is 1 cm/1000 y. Note that some time is required for the supported A_D to reach secular equilibrium with A_P. This causes A_D to increase with depth in the top few

activity, A_D (i.e., dpm/g). This can be rearranged to yield the equation of a straight line: that is,

$$\ln A_{\text{DExcess}_z} = -\frac{\lambda_D}{s} z + \ln A_{\text{DExcess}_{z=0\,\text{cm}}} \quad (28.23)$$

in which the slope is inversely proportional to the sedimentation rate. This is illustrated for an idealized data set in Figure 28.12*b*.

As shown in Figure 28.13, radionuclide distributions in sediments are noisy. Thus sedimentation rates must be inferred from "best-fit" equations determined by the method of linear regression. Alternatively, sedimentation rates are reported as the range of values that bracket the data. Some of this noise is the result of bioturbation. Some is also caused by variations in sedimentation rate and in the supply rate of daughter. Thus it is best to corroborate these inferred sedimentation rates using other radionuclide pairs that undergo decay and chemical transport over somewhat different time scales and by different mechanisms. For example, accretion rates of iron–manganese nodules have been inferred from several primordial radionuclide pairs. These radionuclides are incorporated into the nodules during the accretion process. As shown in Figure 28.14*a*, the ^{238}U–^{230}Th and ^{235}U–^{231}Pa pairs yield accretion rates of 4.0 and 4.3 mm per million years, respectively. This is close to that obtained from the ^{238}U–^{234}U pair (Figure 28.14*b*).

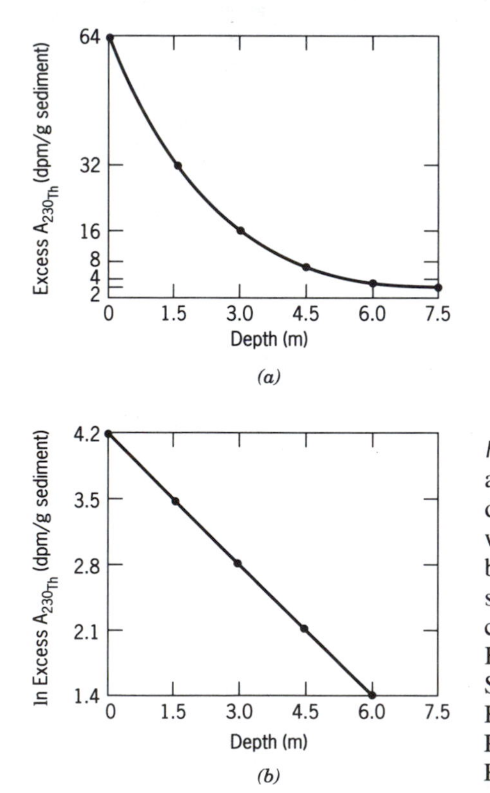

FIGURE 28.12. (*a*) Excess $A_{230\text{Th}}$ and (*b*) log Excess $A_{230\text{Th}}$ versus depth. This is for an ideal core in which the rate of ^{230}Th input has been constant over time and the sedimentation rate has also been constant at 2.0 cm/1000 y. *Source*: From *Chemical Oceanography*, W. S. Broecker, copyright © 1974 by Harcourt, Brace, and Javanovich Publishers, Orlando, FL, p. 78. Reprinted by permission.

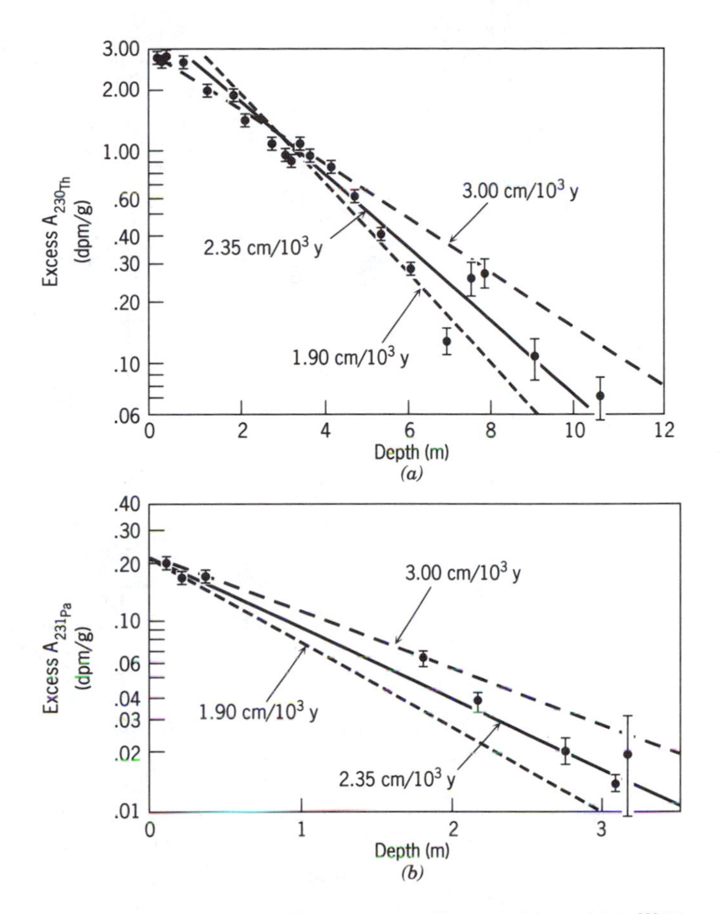

FIGURE 28.13. Actual excess specific activities of (*a*) ^{230}Th and (*b*) ^{231}Pa measured in a sediment core from the Caribbean Sea. The solid line is the best fit as determined by the method of linear regression. *Source*: From W. S. Broecker and J. van Donk, reprinted with permission from *Reviews of Geophysics and Space Physics*, vol. 8, p. 176, copyright © 1970 by the American Geophysical Union, Washington, DC.

Because the accretion rates are very slow, the activities of ^{231}Pa and ^{230}Th have been lowered to their detection limits within 1 mm of the nodule's surface. The half-life of ^{234}U is an order of magnitude longer; thus it can be detected 4 mm into the nodule. The time limits of application for other commonly used radionuclides are given in Figure 28.15. The datable time spans will expand as improvements in analytical technology lower detection limits.

Though not part of an extended decay series, ^{40}K is another primordial radionuclide. This isotope is ubiquitous in igneous rocks and decays to either ^{40}Ca or ^{40}Ar. Since the latter is a gas, it is not present in the rock at the time of solidification. Any ^{40}Ar now present in an igneous rock has been produced by in situ decay and can be used to determine the time that has elapsed since solidification. Due to the long half-life of ^{40}K (approximately 1 billion years),

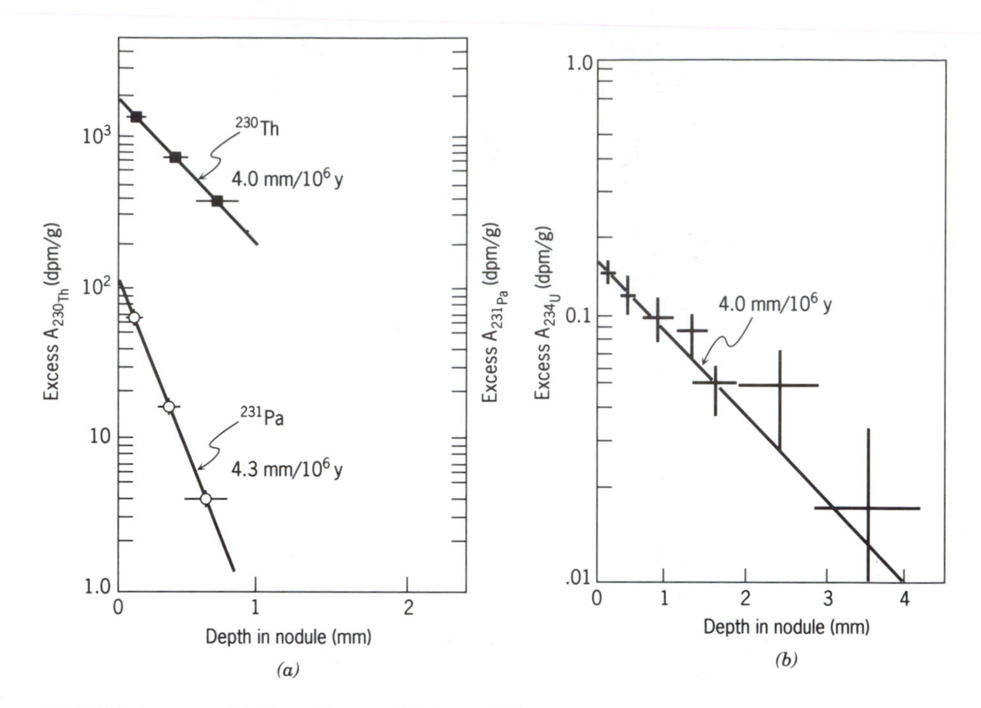

FIGURE 28.14. (*a*) Log Excess ^{230}Th and ^{231}Pa specific activities versus depth in a manganese nodule from the north Pacific Ocean. The computed accretion rates are given in mm per million years. (*b*) Log excess ^{234}U versus depth in the same manganese nodule. The line predicted from the ^{230}Th results is shown for reference. *Source*: From T. L. Ku and W. S. Broecker, reprinted with permission from *Earth and Planetary Science Letters*, vol. 2, pp. 319–320, copyright © 1967, Elsevier Science Publishers, Amsterdam, The Netherlands.

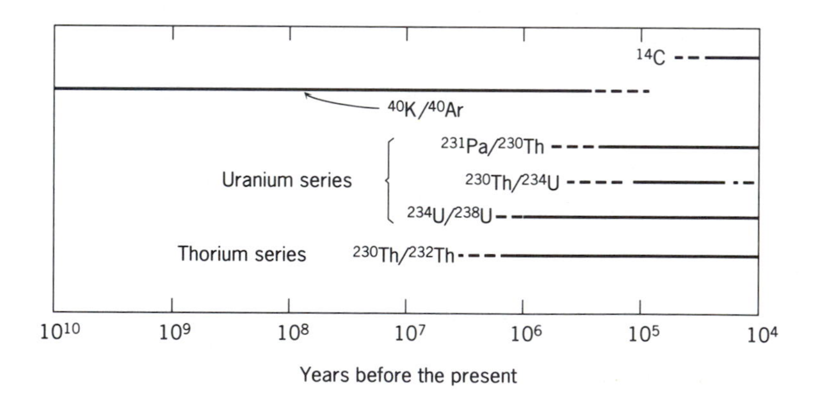

FIGURE 28.15. Time limitations on the application of radionuclides for determination of absolute ages of sediment deposits and times of rock formation. *Source*: From *Oceanography: A View of the Earth*, 4th ed., M. G. Gross, copyright © 1987 by Prentice Hall, Inc., Englewood Cliffs, NJ. Reprinted by permission.

400,000 y must pass before ^{40}Ar reaches measurable levels. Thus, the ^{40}K–^{40}Ar couple can date only those rocks older than 400,000 y.

Despite this limitation, the ^{40}K–^{40}Ar couple has been widely used as a dating tool. For example, it has been used to determine Earth's age and to date magnetic field reversals. The timing of magnetic field reversals has been used to support the theory of plate tectonics and to infer rates of seafloor spreading. ^{40}K–^{40}Ar has also been used to date volcanic ash layers, such as those at the Cretaceous–Tertiary boundary. Caution must be used in this application, as the date recorded by the isotopes is the time since the particles solidified. Due to resuspension and transport by bottom currents, ash layers can be composed of igneous material deposited at a time much later than its solidification. In such cases, potassium–argon dates will be too old. Spuriously young ages can be produced if the rocks have been subjected to elevated temperatures and pressures as this causes partial loss of argon.

Determination of Diffusion Rates and Diffusivity Constants

The primordial radionuclides and radiocarbon have been used to determine diffusion rates as well as the values of diffusivity coefficients. Two examples are given here. The first demonstrates how ^{222}Rn and ^{226}Ra have been used to study gas exchange at the air–sea interface. The second shows how rates of vertical diffusion in the deep sea have been measured from ^{228}Ra and ^{222}Rn.

As shown in Figure 28.16, the diffusion of ^{222}Rn across the air–sea interface causes its specific activity to be lower than that predicted from secular equilibrium with its parent, ^{226}Ra. This net flux of radon from the ocean to the atmosphere indicates that the surface waters are supersaturated with respect to this gas. The difference in specific activity between the parent and daughter ($A_{^{226}Ra} - A_{^{222}Rn}$) is a measure of how much ^{222}Rn has been degassed from the mixed layer.

The thermocline acts as a barrier that inhibits mixing between the deep and surface waters. Thus deep-water ^{222}Rn decays before it can diffuse through the density barrier into the mixed layer. This is why the depth of the mixed layer coincides with the lower boundary of the ^{222}Rn-deficient waters. Assuming that ^{222}Rn is present in a steady state in which its only supply to the mixed layer is via in situ decay of ^{226}Ra and its only losses are from decay and gas exchange, the following mass balance equation can be constructed:

$$hA_{^{226}Ra} = hA_{^{222}Rn} + F \qquad (28.24)$$

In this equation, h is the depth of the mixed layer and the activities are average values representative of the mixed layer. Solving for F, the gas flux, yields

$$F = h(A_{^{226}Ra} - A_{^{222}Rn}) \qquad (28.25)$$

This flux can also be inferred from the stagnant-film model as presented

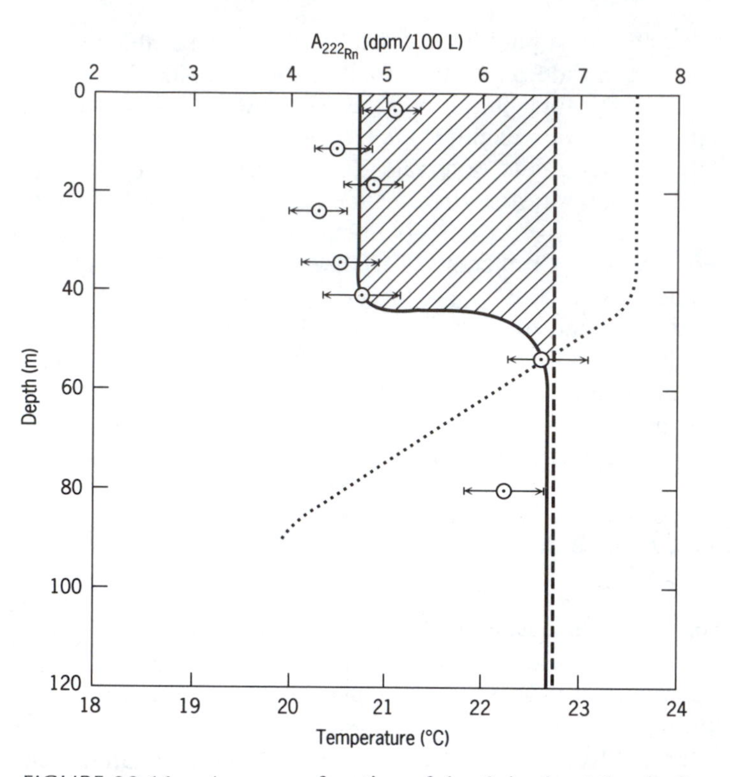

FIGURE 28.16. A_{222Rn} as a function of depth in the Atlantic Ocean at 24°S 35°W. The activity expected if no radon were escaping to the atmosphere is shown by the dashed line. The difference between this equilibrium value and the observed value is a measure of the amount of radon lost to the atmosphere and is shown by the shaded area. The dotted line is the temperature profile at this station which defines the depth of the mixed layer. *Source*: From W. S. Broecker and T.-H. Peng, reprinted with permission from *Tellus*, vol. 26, p. 30, copyright © 1974 by Munksgaard International Publishers, Ltd., Copenhagen, Denmark.

in Eq. 6.10. Since $[^{222}\mathrm{Rn}]_{\mathrm{atmosphere}}$ is approximately zero, this equation can be rewritten in terms of $A_{222\mathrm{Rn}}$ as follows:

$$F = D_{222\mathrm{Rn}} \frac{[^{222}\mathrm{Rn}]}{z} = D_{222\mathrm{Rn}} \frac{A_{222\mathrm{Rn}}}{z\,\lambda_{222\mathrm{Rn}}} \tag{28.26}$$

Assuming ^{222}Rn is in steady state, the fluxes inferred from Eqs. 28.25 and 28.26 should be constant and equal. Thus Eqs. 28.25 and 28.26 can be equated to yield:

$$z = \frac{D_{222\mathrm{Rn}}}{h\,\lambda_{222\mathrm{Rn}}} \left[\frac{1}{\left(\frac{A_{226\mathrm{Ra}}}{A_{222\mathrm{Rn}}} - 1 \right)} \right] \tag{28.27}$$

For the conditions illustrated in Figure 28.16, z = 28 μm. Results from other sites range from 0 to 120 μm and are similar to those obtained from CO_2 fluxes as measured by the gas' ^{14}C content. These results are currently being used to predict global ocean fluxes of anthropogenic gases such as CO_2.

Both ^{226}Ra and ^{228}Ra are produced in the sediments via decay from thorium parents. Both are also remobilized from the sediments as a result of the remineralization of biogenic calcium carbonate. Some of this radium diffuses into the deep ocean and some decays. The ^{222}Rn produced from decay of ^{226}Ra also diffuses into the bottom waters. Since ^{222}Rn is transported faster than ^{226}Ra, it is present in excess in the deep waters. As shown in Figure 28.17, this unsupported ^{222}Rn decreases with increasing height above the seafloor. A similar situation is observed for ^{228}Ra, whose parent is the insoluble ^{232}Th. ^{228}Ra penetrates higher into the deep waters due to its longer half-life and thus can be used to study mixing processes over a much larger depth range than ^{222}Rn.

On short time scales, vertical diffusion is the most important process transporting these isotopes. Thus their vertical distributions can be used to infer vertical "diffusion" rates, as well as the value of the vertical diffusivity coefficient, D_z. In this application, "diffusion" represents the combined effects

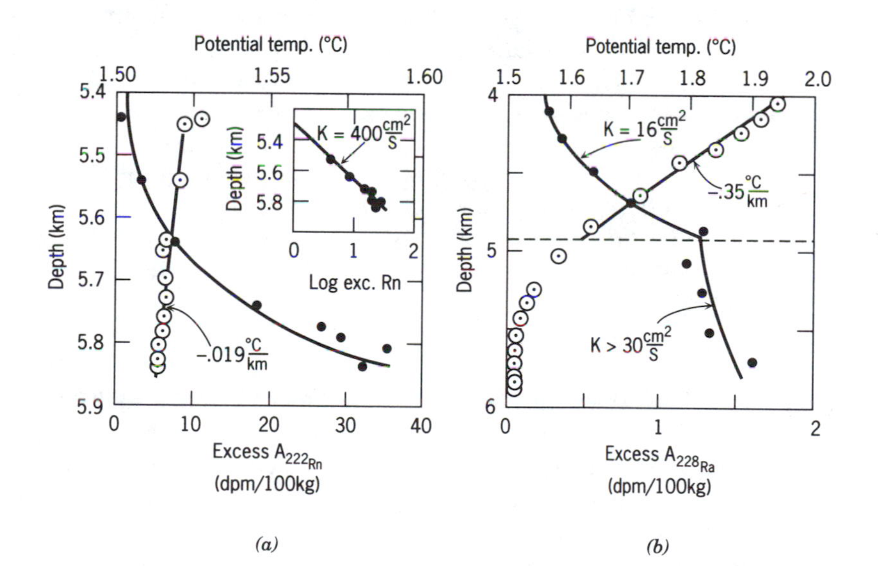

FIGURE 28.17. (*a*) Excess A_{222Rn} and (*b*) Excess A_{228Ra} versus depth in the deep water at 24°N 54°W in the north Atlantic. Also included are the potential temperature gradients. *Source*: From *Tracers in the Sea*, W. S. Broecker and T.-H. Peng, copyright © 1982 by the Lamont-Doherty Geological Observatory, Palisades, NY, p. 585. Reprinted by permission. After J. L. Sarmiento, H. W. Feely, W. S. Moore, A. E. Bainbridge, and W. S. Broecker, reprinted with permission from *Earth and Planetary Science Letters*, vol. 32, p. 362, copyright © 1976, Elsevier Science Publishers, Amsterdam, The Netherlands.

of eddy and molecular diffusion, where the former is likely to be much larger than the latter. If this diffusive mixing and radioactive decay are the sole processes affecting the radionuclides, their vertical distributions can be described by the following equation:

$$A_z = A_o e^{-\sqrt{\frac{\lambda}{D_z}}z} \tag{28.28}$$

In this equation, A_o and A_z are the excess A_D's present at the sediment surface and at some depth z above the seafloor, respectively.

Thus A_z should decrease exponentially with increasing depth above the seafloor. This behavior is verified by the linear relationship observed between log A_z and depth as shown in the insert in Figure 28.17*a*. The depth ($z_{\frac{1}{2}}$) at which A_z has declined to one-half its original value ($A_o/2$) is equal to $0.693\sqrt{D_z/\lambda}$. Thus $D_z = \lambda(z_{\frac{1}{2}}/0.693)^2$.

At this site in the North Atlantic, the $z_{\frac{1}{2}}$'s are such that the deep-water ^{222}Rn and ^{228}Ra distributions yield D_z's of 400 and 16 cm^2/s. These values lie within the range given in Figure 4.13 for oceanic eddy diffusion coefficients. The substantial dissimilarity in these estimates is probably due to differences in their chemical behavior and rates of radioactive decay. Since ^{228}Ra has a longer half-life, its distribution reflects the effects of mixing over longer time and space scales. In other words, the vertical distribution of ^{222}Rn is more influenced by short-term fluctuations in environmental conditions and water motion. The fairly large difference in these estimates of D_z suggests that mixing over long time and space scales is functionally distinct from that which occurs over short time and space scales.

As shown in Figure 28.17*b*, an abrupt change in the temperature gradient is present at 5 km, suggesting the presence of another water mass. The D_z in the overlying water mass is somewhat less than that in the lower one. This suggests that the rate of vertical diffusion is inversely related to the density gradient. In other words, the steeper the gradient, the slower the rate of vertical diffusion.

Though the half-life of ^{228}Ra is longer than that of ^{222}Rn, it is still too short (6 y) to use as a tracer of diffusive mixing on the time and space scales over which thermohaline circulation occurs. Since the half-life of ^{226}Ra is much longer (1600 y), marine chemists had hoped that this isotope could be used to measure some aspects of global oceanic circulation. As with ^{228}Ra, this isotope diffuses from the sediments, but its release is geographically variable. Until this variability is quantified, ^{228}Ra cannot be used as a tracer of large-scale oceanic circulation.

COSMOGENIC RADIONUCLIDES

Cosmic rays are charged particles that enter Earth's atmosphere from outer space. They are mostly protons (87%) and α particles (12%), though some larger nuclei (1%) are present. Their energies range up to 10^{14} MeV. Atmospheric gases are ionized as a result of collision with the low-energy cosmic rays. The high-energy cosmic rays can cause fragmentation of the gas nuclei. High-energy neutrons are also given off as a result of these **spallation reactions**. The energy of these neutrons is lowered by repeated collisions with gas atoms. When their energy has been sufficiently lowered, these neutrons can be captured by the nucleus of a gas atom. As shown in Figure 28.18, the capture of a neutron by an atom of $^{14}_{7}N$ is accompanied by the ejection of a proton, resulting in the formation of $^{14}_{6}C$.

As described below, oceanic **radiocarbon** measurements have been used to measure rates of water motion, sedimentation, bioturbation, and the timing of sea level changes. Other **cosmogenic radionuclides** that have provided information about marine processes include ^{3}H, ^{7}Be, ^{10}Be, ^{26}Al, and ^{32}Si. Their half-lives and global distributions are given in Table 28.6.

^{10}Be dissolves in rainwater and is adsorbed onto atmospheric particles. Following atmospheric fallout as rain or aerosols, ^{10}Be is transported to the seafloor by particle scavenging. Since this flux is relatively large, ^{10}Be can be used to measure sedimentation rates if its rain rate, as well as that of the other particles, has remained constant over time. Thus the atmospheric production rate of ^{10}Be must also have been constant. Changes in the flux of cosmic rays have caused variations in the ^{10}Be production rate and thus must be corrected for. Alternatively, these natural variations can be used as a record of changes in the cosmic ray flux if the sedimentation rate is independently known.

In addition to deposition in the sediment, some ^{10}Be is incorporated into manganese nodules. Due to its relatively long half-life (1.5×10^{6} y), ^{10}Be can be detected far deeper into the nodules than the primordial radionuclides. As shown in Figure 28.19, the accretion rates inferred from the ^{10}Be data range from 2.4 to 4.5 mm per million years, confirming the primordial radionuclide results.

Cosmogenic ^{14}C is incorporated into carbon dioxide gas and transferred from the atmosphere to the ocean via gas exchange across the air–sea interface.

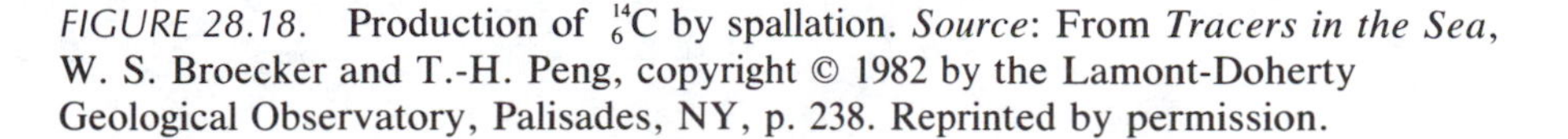

FIGURE 28.18. Production of $^{14}_{6}C$ by spallation. *Source*: From *Tracers in the Sea*, W. S. Broecker and T.-H. Peng, copyright © 1982 by the Lamont-Doherty Geological Observatory, Palisades, NY, p. 238. Reprinted by permission.

TABLE 28.6
Basic Information Concerning Nuclides Produced by Cosmic Rays

	Nuclide					
	^{3}H	^{7}Be	^{10}Be	^{14}C	^{26}Al	^{32}Si
Half-life (y)	12.3	0.145	2.5×10^{6}	5680	7.4×10^{5}	500
Production rate in total atmosphere (atom cm^{-2} s^{-1})	0.25	0.081	0.045	2.5	1.4×10^{-4}	1.6×10^{-4}
Fraction of total earth inventory in						
Atmosphere	0.072	0.71	3.9×10^{-7}	0.019	1.4×10^{-6}	2.0×10^{-3}
Land surface	0.27	0.08	0.29	0.04	0.29	0.29
Ocean—mixed layer	0.35	0.2	5.7×10^{-6}	0.022	1.4×10^{-5}	0.0035
Ocean–excluding mixed layer	0.3	0.002	10^{-4}	0.92	7×10^{-5}	0.68
Oceanic sediments	0	0	0.71	0.004	0.71	0.028
Average concentration in ocean (10^{-3} dpm kg $water^{-1}$)	36	—	10^{-3}	260	1.2×10^{-5}	2.4×10^{-2}
Average specific activity in ocean (dpm g $element^{-1}$)	3.3×10^{-4}	—	1600	10	0.0012	0.008
Global inventory (kg)	3.5	3.2×10^{-3}	4.3×10^{5}	7.5×10^{4}	1.1×10^{3}	1.4
Global inventory (MCi)	35	1.1	6.4	340	0.020	0.023

Source: From *Chemical Oceanography*, vol. 3, J. D. Burton (eds: J. P. Riley and G. Skirrow), copyright © 1975 by Academic Press, Orlando, FL, p. 140. Reprinted by permission. Data from *The Handbook of Physics, 2E*, E. U. Condon and H. Odishaw, copyright © 1967 by McGraw–Hill, Inc., New York, pp. 9.277, 9.285, 9.319. Reprinted by permission.

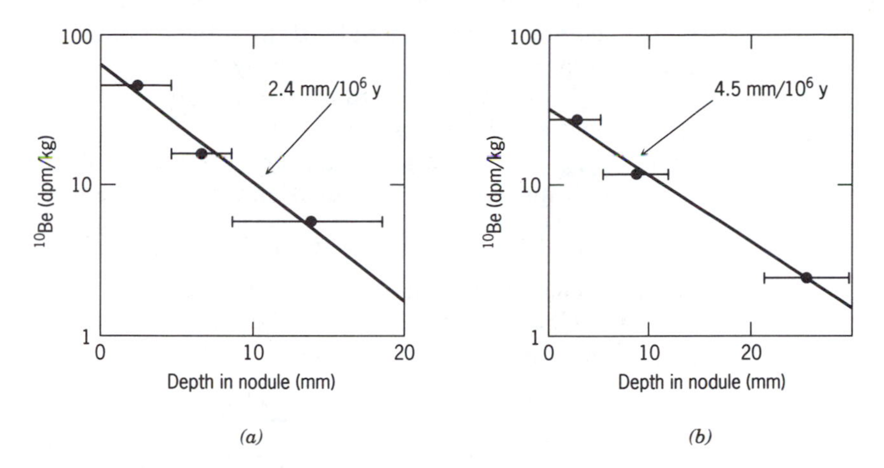

FIGURE 28.19. Log $A_{^{10}Be}$ as a function of depth in a manganese nodule from the (*a*) north Pacific Ocean (20°N 174°E) and (*b*) south Atlantic Ocean (28°S 41°W). *Source*: From *Tracers in the Sea*, W. S. Broecker and T.-H. Peng, copyright © 1982 by the Lamont-Doherty Geological Observatory, Palisades, NY, p. 265. Reprinted by permission. Data from (*a*) *Marine Geology and Oceanography of the Pacific Manganese Nodule Province*, T. L. Ku, A. Omura, and P. S. Chen (eds.: J. L. Bischoff and D. Z. Piper), copyright © 1979 by Plenum Press, New York, p. 801 (reprinted by permission); (*b*) K. K. Turekian, J. K. Cochran, S. Krishnaswami, W. A. Sanford, P. D. Parker, and K. A. Bauer, reprinted with permission from *Geophysical Research Letters*, vol. 6, p. 419, copyright © 1979 by the American Geophysical Union, Washington, DC.

Due to its relatively fast decay rate, radiocarbon is not homogeneously distributed through the ocean and exhibits significant vertical and horizontal segregation. In general, the oldest waters have the lowest radiocarbon content. Therefore radiocarbon tends to decrease with depth. In the deep waters, radiocarbon tends to decrease with increasing distance from the site of water mass formation. Due to continual input of "young" carbon via the remineralization of sinking biogenic particles, radiocarbon measurements cannot be directly converted into water-mass ages.

Radiocarbon gradients can be used to measure rates of ocean mixing. As illustrated below, oceanic radiocarbon distributions have been used to compute the rate of vertical mixing, v_{mix}, between the surface and deep reservoirs. This rate is used in the Broecker box model of vertical segregation.

In the lab, ^{14}C is commonly measured in terms of its abundance relative to the total carbon in a sample, (i.e., as $^{14}C/C$), which is hereafter represented as R. Thus the flux of radiocarbon into the deep reservoir of the Broecker box model is given by the sum of its transport via downwelling ($v_{mix}C_{surface}R_{surface}$) and by remineralization of sinking particulate carbon. ($C_{surface}$ is the total concentration of dissolved carbon [ΣCO_2] in the surface water.) Most particulate carbon is deposited by surface-dwelling organisms and thus has a ^{14}C content equal to that of surface-water DIC ($R_{surface}$). If the flux of remineralizable particulate carbon is represented by B, then the input of ^{14}C to the deep waters as a result of particle transport is given by BR_{deep}.

Assuming that the radiocarbon is in steady state, its inputs to the deep reservoir must equal its losses via upwelling ($v_{mix}C_{deep}R_{deep}$) and radioactive decay ($V_{deep}C_{deep}R_{deep}\ \lambda$) where V_{deep} represents the volume of the deep zone. Thus the following mass balance equation can be written for the deep reservoir:

$$v_{mix}C_{surface}R_{surface} + BR_{deep} = v_{mix}C_{deep}R_{deep} + V_{deep}C_{deep}R_{deep}\ \lambda \quad (28.29)$$

B can be eliminated from this equation using algebraic reasoning similar to that employed in Chapter 9 to replace P, the biogenic particle flux. The resulting equation can be rewritten as

$$v_{mix} = \frac{\lambda V_{deep}}{\dfrac{R_{surface}}{R_{deep}} - 1} \quad (28.30)$$

The volume of the deep water is equal to hA_{ocean}, where h is the average thickness of the deep reservoir (3 km) and A_{ocean} is the average surface area of the deep reservoir. Thus Eq. 28.30 can be written as:

$$v_{mix} = \frac{\lambda h A_{ocean}}{\dfrac{R_{surface}}{R_{deep}} - 1} \quad (28.31)$$

Since the average $R_{surface}/R_{deep}$ is 1.12, v_{mix}/A_{ocean} is equal to 300 cm/y. In other words, the annual amount of water exchanged between the surface and deep reservoirs is equal to the volume of water contained in a hypothetical layer that covers the entire ocean surface and is 300 cm thick. Since the deep reservoir has an average thickness of 3000 m, the residence time of water in this zone is given by

$$\text{Residence time of deep water} = \frac{3000\ \text{m}}{300\ \text{cm/y}} = 1000\ \text{y} \quad (28.32)$$

In reality, these reservoirs are not homogeneous. More sophisticated models of elemental cycling take this into account by partitioning the ocean into many more reservoirs. This approach is now being used to predict the rate at which anthropogenic CO_2 is being transported into and through the ocean.

This use of the Broecker box model also assumes that the amount of radiocarbon, and hence the $^{14}C/C$, in the ocean has remained constant over time. While this might have been true in the past, the "natural" ratio has been greatly altered as a result of two types of anthropogenic input. Since the 1850s, fossil-fuel burning has introduced "old" carbon into the atmosphere, thereby lowering the $^{14}C/C$ of CO_2. This is called the **Suess Effect**. The second input has been the result of ^{14}C production during bomb testing.

For the purposes of comparison, all $^{14}C/C$ ratios are now reported as $\Delta^{14}C$. This is essentially the difference (in parts per thousand, or "per mil") between a sample's $^{14}C/C$ and the ratio that was present in atmospheric CO_2 prior to the onset of these anthropogenic inputs. Since fossil-fuel burning lowers the atmospheric $^{14}C/C$, it has caused the $\Delta^{14}C$ of CO_2 to become negative. This change in isotopic composition has been recorded in trees that deposit atmospheric carbon in annual rings. Since these rings are datable, their relative radiocarbon content can be reported as a function of time, as shown in Figure 28.20. Fluctuations prior to 1850 were probably caused by changes in sunspot activity and in the strength of Earth's magnetic field, both of which affect the atmospheric flux of cosmic rays.

The relative radiocarbon content of coral reefs has also been used as a long-term record of changes in $\Delta^{14}C$. As a result of seasonal changes in growth rates, coral also deposit carbon in datable layers. But coral deposit carbon derived from surface-water DIC, so the radiocarbon composition of their calcium carbonate is not directly related to that which was concurrently present in the atmosphere. As shown in Figure 28.21, the radiocarbon content of newly deposited coral carbonate was relatively constant until the 1950s. At this time, a massive increase in the amount of ^{14}C occurred as a result of aboveground atomic bomb testing. This input declined abruptly in the 1960s as a result of the implementation of an international test ban. The input of radiocarbon was so large that it swamped the Suess Effect.

In the Northern Hemisphere, the anthropogenic input has increased the radiocarbon inventory by 50 percent. Unlike the Suess Effect, this addition of radiocarbon represents only a temporary perturbation as most will eventually be lost to decay. Thus radiocarbon ages of materials formed after 1950 will have to be corrected for the declining presence of this bomb-derived ^{14}C.

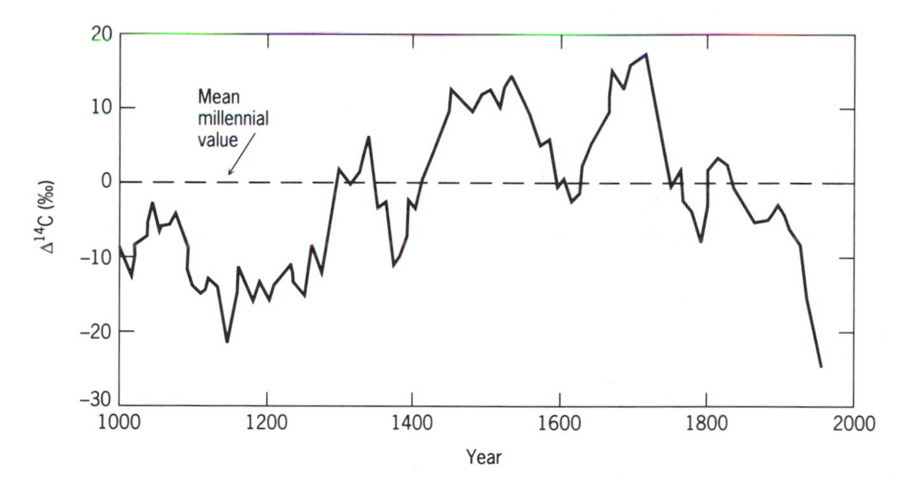

FIGURE 28.20. $^{14}C/C$ ratio (expressed as $\Delta^{14}C$) for atmospheric CO_2 over the last 1000 y as reconstructed from tree ring measurements. *Source*: From M. Stuiver and P. D. Quay, reprinted with permission from *Earth and Planetary Science Letters*, vol. 53, p. 354, copyright © 1981 by Elsevier Science Publishers, Amsterdam, The Netherlands.

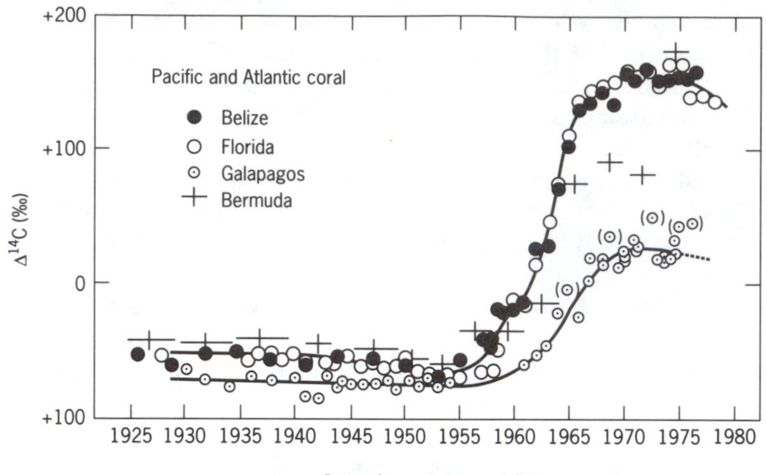

FIGURE 28.21. $\Delta^{14}C$ values obtained from corals. The Galapagos coral comes from an area of intense upwelling in the western equatorial Atlantic that has evidently supplied a significant amount of nonradiogenic DIC to the coral. *Source*: From E. M. Druffel and H. E. Suess, reprinted with permission from the *Journal of Geophysical Research*, vol. 88, p. 1273, copyright © 1983 by the American Geophysical Union, Washington, DC.

The coral isotope record has been used to estimate what the $^{14}C/C$ of DIC must have been prior to the testing of atomic bombs and the onset of the Suess Effect. This preindustrial value has been used to correct deep-water radiocarbon concentrations for atomic bomb and fossil-fuel contamination. As shown in Figure 28.22, the resultant deep-water distribution of radiocarbon is consistent with that of other water-mass tracers.

Some ^{14}C-labeled POC is transported to the seafloor. The radiocarbon content of this sedimentary organic matter can be used to determine sedimentation rates. Due to ^{14}C's relatively short half-life, this isotope is only of use in rapidly accumulating sediments. An example is given in Figure 28.23, where radiocarbon measurements have been converted to "ages."

The change in slope at mid depth in these cores is interpreted as the result of a change in sedimentation rate. The timing of the shift is coincident with the end of the last ice age. This suggests that rising sea level has reduced the rate at which sediment is supplied to this ocean site, probably by pushing the coastline inshore. Such records of changes in sedimentation rate are an important source of information regarding climate and other features of the crustal-ocean factory.

Bioturbation has homogenized the radiocarbon distributions in the top 10 cm of these cores, causing their surface sediments to have a constant "age." The absence of a radiocarbon gradient can be used to assess the depth to which bioturbation mixes the surface sediments. These data can also be used to infer

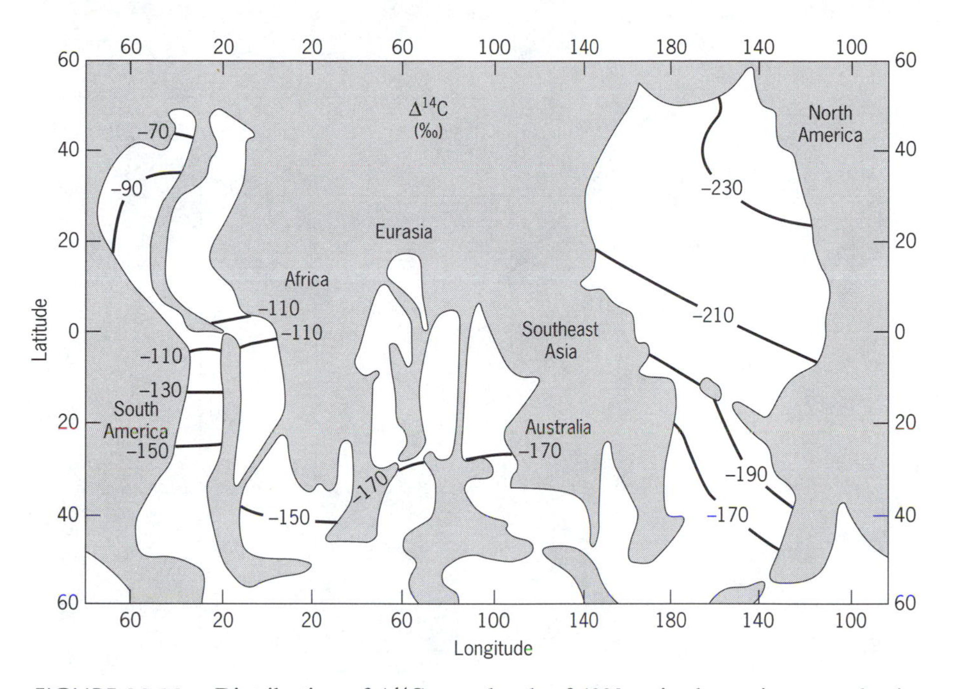

FIGURE 28.22. Distribution of $\Delta^{14}C$ at a depth of 4000 m in the main ocean basins. *Source*: From *Tracers in the Sea*, W. S. Broecker and T.-H. Peng, copyright © 1982 by the Lamont-Doherty Geological Observatory, Palisades, NY, p. 248. Reprinted by permission. Data from H. Stuiver and H. G. Östlund, reprinted with permission from *Radiocarbon*, vol. 22, p. 5, copyright © 1980 by the University of Arizona, Department of Geosciences, Tucson, AZ; H. G. Östlund and H. Stuiver, reprinted with permission from *Radiocarbon*, vol. 22, p. 29, copyright © 1980 by the University of Arizona, Department of Geosciences, Tucson, AZ, and H. G. Östlund, R. Oleson and R. Brescher, *Tritium Laboratory Data Report #9*, 1980, Rosenstiel School of Marine and Atmospheric Science, University of Miami, Miami, FL. Reprinted by permission.

bioturbation rates. On the other hand, this vertical transport of radiocarbon limits the resolution to which each sediment layer can be dated.

ARTIFICIAL RADIONUCLIDES

A large amount and great variety of radionuclides have been introduced into the marine environment as a result of human activities. The radioactivity of these isotopes is usually reported in picocuries (pCi). One Ci is equal to 2.200 $\times 10^{12}$ dpm, which is equivalent to the activity of 1 g of radium. Thus 1 pCi is equal to 2.200 dpm. Fallout from the testing of atomic bombs and intentional leakage from nuclear reactors are the two major sources of the artificial radionuclides now present in the ocean. The impact of nuclear reactors is discussed in Chapter 30.

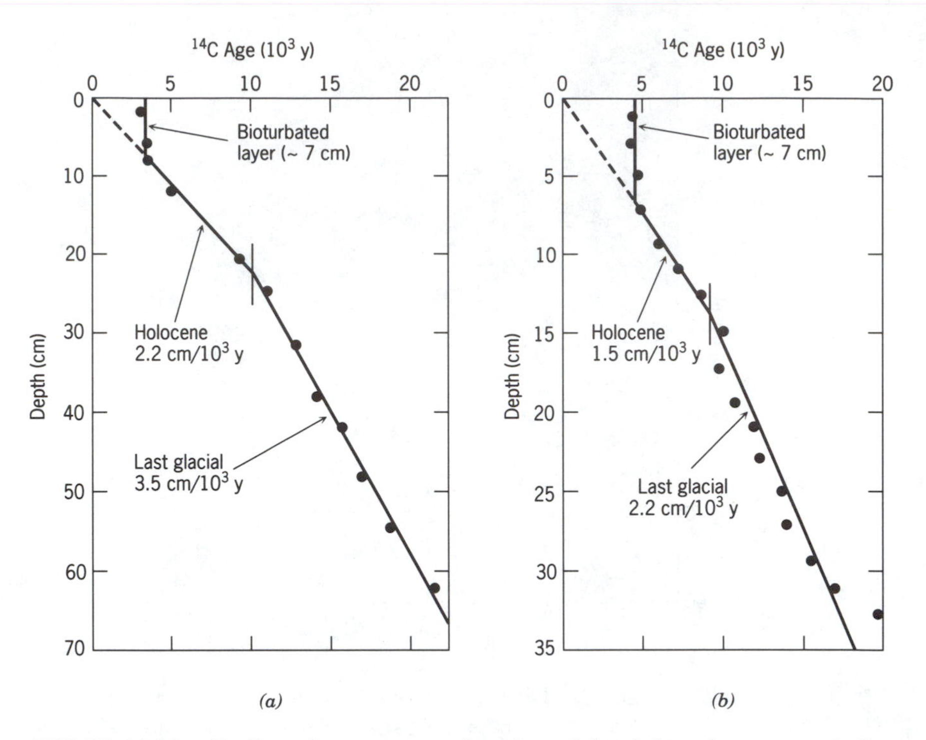

FIGURE 28.23. Radiocarbon ages as a function of depth in a deep-sea core from (*a*) the Indian Ocean (7°N 61°E) and (*b*) the Pacific Ocean (2°S 157°E). *Source*: From *Tracers in the Sea*, W. S. Broecker and T.-H. Peng, copyright © 1982 by the Lamont-Doherty Geological Observatory, Palisades, NY, p. 258. Reprinted by permission. Data from (*a*) *The Fate of Fossil Fuel CO_2 in the Ocean*, T.-H. Peng, W. S. Broecker, G. Kipphut, and N. Shackleton (eds.: N. Andersen and A. Malahoff), copyright © 1977 by Plenum Press, NY, p. 357 (reprinted by permission); (*b*) T.-H. Peng, W. S. Broecker, and W. H. Berger, reprinted with permission from *Quaternary Research*, vol. 11, p. 143, copyright © 1979 by Academic Press, Orlando, FL.

During the explosion of atomic bombs, radionuclides are produced directly by fission. High-energy neutrons are also produced. Some of these neutrons collide with the metallic bomb casings, earth, water, and atmospheric gases, thereby producing other **artificial radionuclides**. This process is termed neutron activation. In the first weeks after an explosion, radioactive decay is dominated by the short-lived radionuclides, such as ^{143}Pr, ^{140}Ba, ^{147}Nd, and ^{131}Ir. After 20 y, most of these isotopes have decayed, leaving ^{90}Sr and ^{137}Cs as the most abundant radionuclides. Other artificial radionuclides that are present in the marine environment are listed in Table 28.7. As indicated, some become strongly enriched in the tissues of marine organisms, much like the naturally occurring trace metals.

As shown in Figure 28.24, most of the aboveground bomb testing was done between 1958 and 1965 in the Northern Hemisphere. (The USA, USSR, and

TABLE 28.7
The Most Abundant Artificial Radionuclides in the Marine Environment[a]

Symbol	Element	Half-life	Radiation Type	Enrichment Factor	
				Mollusk	Fish
1. Fission products in nuclear reactions					
^{3}H	Tritium	12.3	β		
^{14}C	Carbon–14	5680	β		
^{85}Kr	Krypton–85	10.6	β	1	1
^{89}Sr	Strontium–89	0.14	β	1	0.2
^{90}Sr	Strontium–90	28.0	β	1	0.2
^{90}Y	Yttrium–90	0.007	β	15	10
^{91}Y	Yttrium–91	0.16	β	15	10
^{95}Nb	Niobium–95	0.10	β–γ	5	1
^{95}Zr	Zirconium–95	0.18	β–γ	5	1
^{103}Ru	Ruthenium–103	0.11	β-γ	10	1
^{106}Ru	Ruthenium–106	1.0	β–γ	10	1
^{131}I	Iodine–131	0.02	β–γ	50	10
^{137}Cs	Cesium–137	30.0	β–γ	10	10
^{144}Ce	Cerium–144	0.78	β–γ	400	3
2. Activation products					
^{32}P	Phosphorus–32	0.04	β	6000	3300
^{51}Cr	Chromium–51	0.08	K–γ	400	100
^{54}Mn	Manganese–54	0.86	K–γ	10,000	200
^{55}Fe	Iron–55	2.7	K	10,000	1500
^{57}Co	Cobalt–57	0.74	K–γ	500	80
^{60}Co	Cobalt–60	5.3	β–γ	500	80
^{65}Zn	Zinc–65	0.67	K–β–γ	10,000	1000
^{110}Ag	Silver–110	0.69	β–γ	10,000	—
^{134}Cs	Cesium–134	2.1	β–γ	10	10

Source: From *Marine Pollution, Diagnosis and Therapy*, S. A. Gerlach, copyright © 1981 by Springer-Verlag, Heidelberg, Germany, p. 108. Enrichment data from *Impingement of Man on the Oceans*, T. R. Rice and D. A. Wolfe (ed.: D. W. Hood), copyright © 1971 by John Wiley & Sons, Inc., New York, p. 351. Reprinted by permission.

[a]Their half-lives are listed in years. K emitters are radionuclides that generate gamma rays (γ) as a result of electron capture. The enrichment factors are given with respect to wet weight.

UK signed a test ban treaty in 1962.) Depending on the location of the bomb test, some radionuclides were introduced directly into the ocean; others were deposited on land or injected into the atmosphere. Wind mixing spread the atmospheric input across the globe. Eventually most of these radionuclides were deposited as fallout, either on the sea surface or land. Some of the radionuclides deposited on land were eventually transported to the ocean via river and groundwater runoff.

As is evident from Figure 28.24, the production and deposition of artificial radionuclides has been carefully monitored. Since the input rates of these

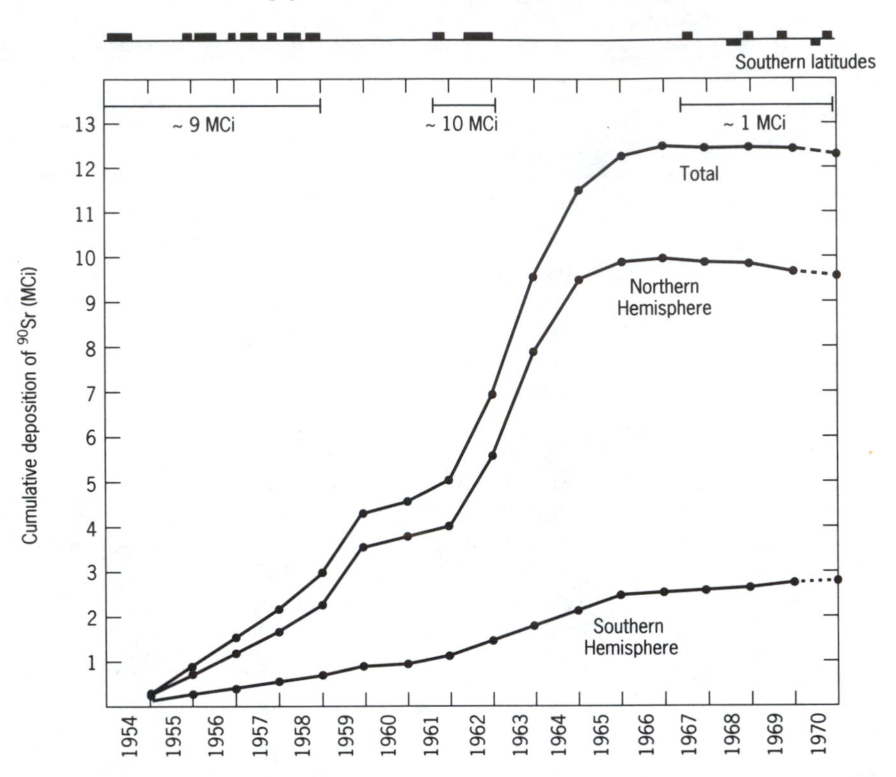

FIGURE 28.24. Cumulative deposition of ^{90}Sr in MCi (10^3 Ci). *Source*: From D. H. Pierson, reprinted with permission from *Nature*, vol. 234, p. 80, copyright © 1971 by Macmillian Journals, Ltd., London, England.

isotopes are very well known, they make excellent water-mass tracers. In particular, marine distributions of bomb-derived ^{90}Sr, tritium, and radiocarbon have been used to infer rates of water motion. Some of these applications are discussed below.

By the early 1970s, little of the bomb-derived ^{90}Sr had penetrated below the mixed layer as a result of mixing, at least at low and mid latitudes, as shown in Figure 28.25. The transport that had occurred at these sites was primarily the result of particle scavenging and incorporation into biogenic particles. At subpolar latitudes, the thermocline is weak and hence mixing would be expected to have caused a deeper injection of the artificial radionuclides. This has been observed for tritium, as described below.

Upon injection into the atmosphere, bomb-derived **tritium** was rapidly incorporated into gaseous water molecules. This caused rainwater to have very high tritium levels during the early 1960s (Figure 28.26). Thus bomb testing in the 1950s and 1960s caused anthropogenic tritium to be introduced into the ocean as a rapid, short-lived injection that was concentrated in the Northern

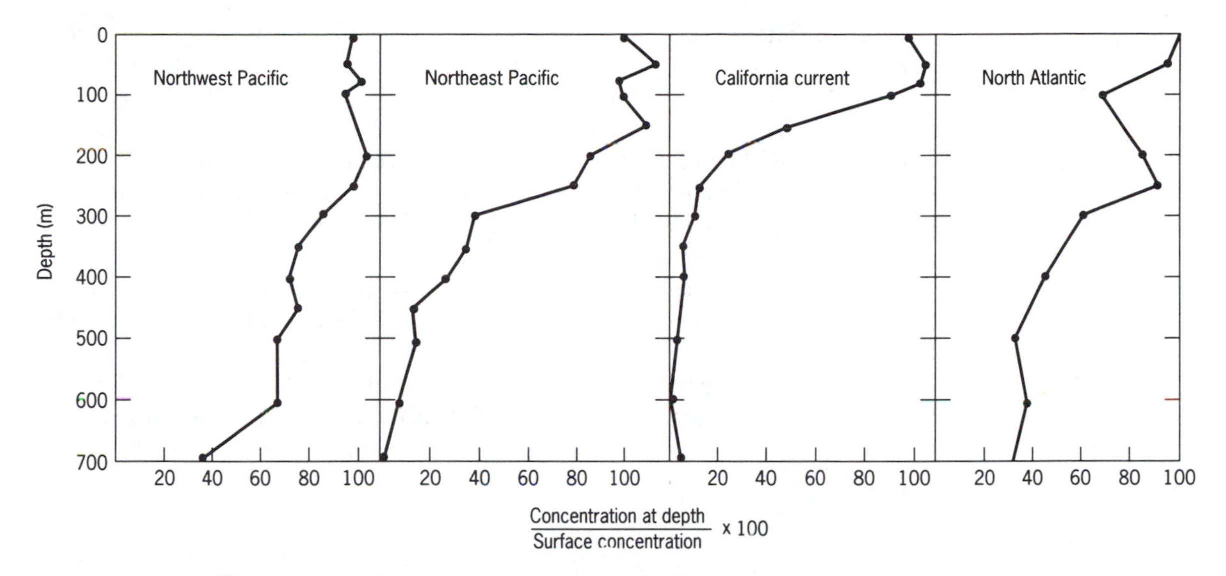

FIGURE 28.25. ^{90}Sr concentrations in the Atlantic and Pacific oceans given as the percentage of surface-water concentrations. *Source*: From *Radioactivity in the Marine Environment*, H. L. Volchok, V. T. Bowen, T. R. Folsom, W. S. Broecker, E. A. Schubert, and G. S. Bien (ed.: National Research Council), copyright © 1971 by the National Academy of Science, Washington, DC, p. 25. Reprinted by permission.

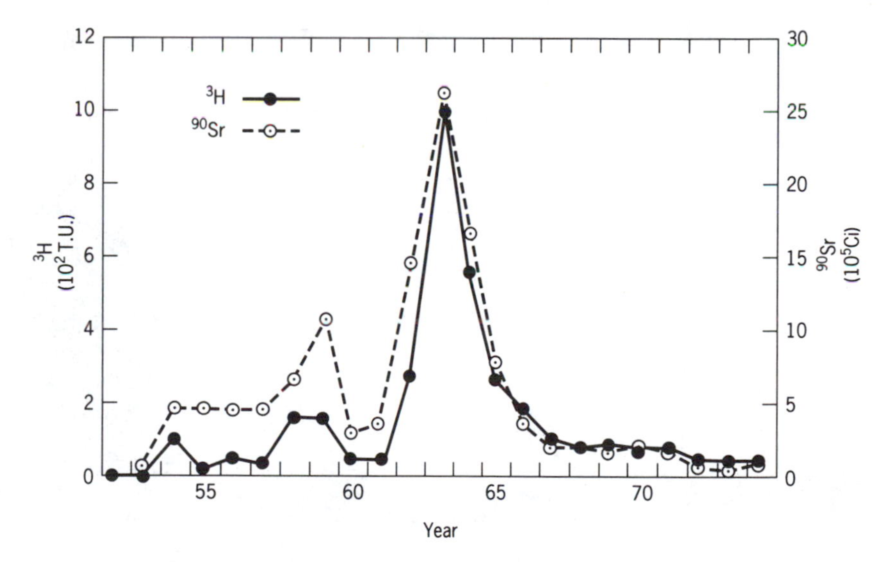

FIGURE 28.26. Plot of the ^3H content of rain at the west coast of Ireland, from 1952 to 1974 (solid circles). Tritium levels are reported in tritium units (T.U.), which are equal to $10^{18} \times (^3H/^1H)$. Also given is the total annual northern hemisphere ^{90}Sr deposition (open circles) in 10^5 Ci. *Source*: From E. Dreisigacker and W. Roether, reprinted with permission from *Earth and Planetary Science Letters*, vol. 38, p. 307, copyright © 1978 by Elsevier Science Publishers, Amsterdam, The Netherlands.

Hemisphere. The nature of this injection has made tritium a very useful tracer of water-mass motion. Since its half-life is only 12.5 y, much of the anthropogenic tritium has already decayed. Due to its rapidly decreasing activity, bomb-derived tritium will be usable as a tracer only for a few decades.

As shown in Figure 28.27, the penetration of anthropogenic tritium into the ocean has also been restricted to the mixed layer at low and mid latitudes. In contrast, tritiated water has already reached a water depth of 2 km in the subpolar North Atlantic. This is likely due to the formation of NADW via the sinking of surface waters at high latitudes. Although a small amount of tritium is produced by spallation in the atmosphere, rapid decay causes natural levels to be quite low.

This tritiated water has been followed as a coherent advective flow as far south as Florida. As shown in Figure 28.28, this deep-water current flows along the seafloor of the North Atlantic's western margin. It is thought to be

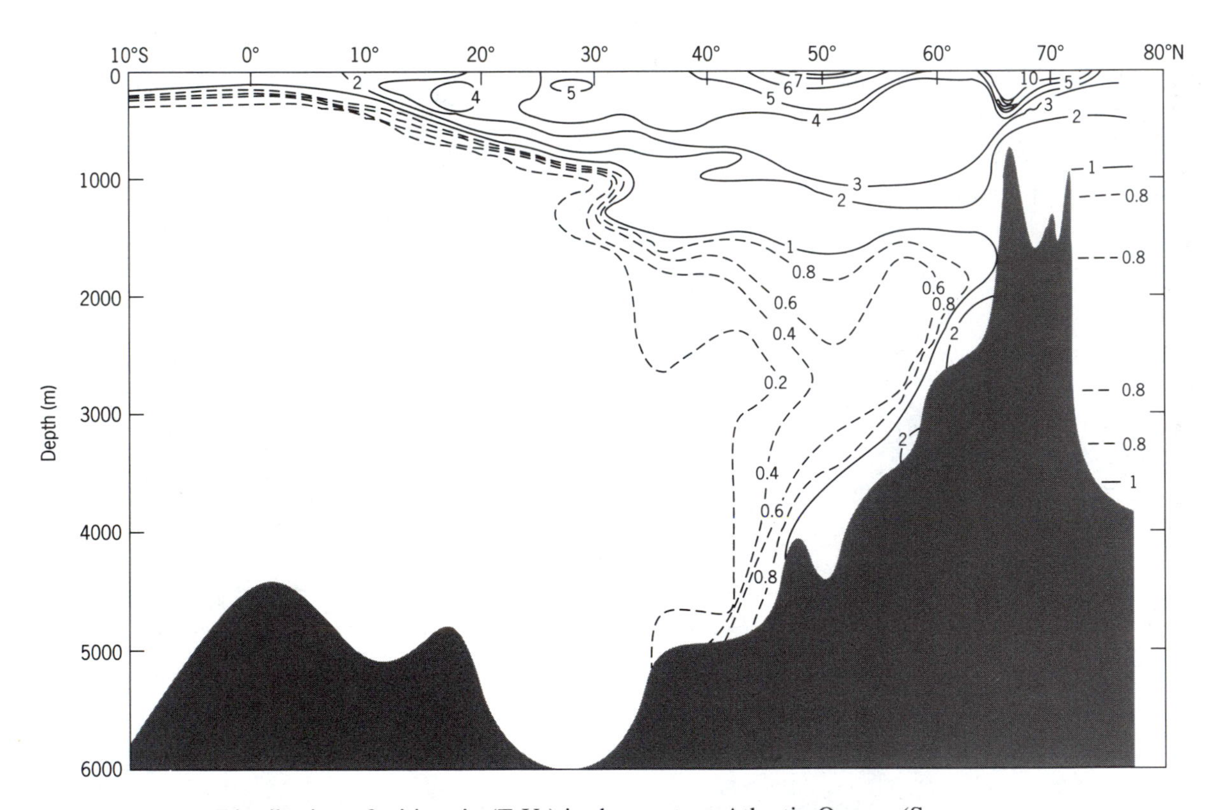

FIGURE 28.27. Distribution of tritium in (T.U.) in the western Atlantic Ocean. (See Figure 28.26 for unit definition.) This diagram indicates the extent to which bomb-produced material has penetrated the deep ocean over a period of about 10 y. *Source*: From H. G. Östlund and C. G. H. Rooth, reprinted with permission from the *Journal of Geophysical Research*, vol. 95, p. 20154, copyright © 1990 by the American Geophysical Union, Washington, DC.

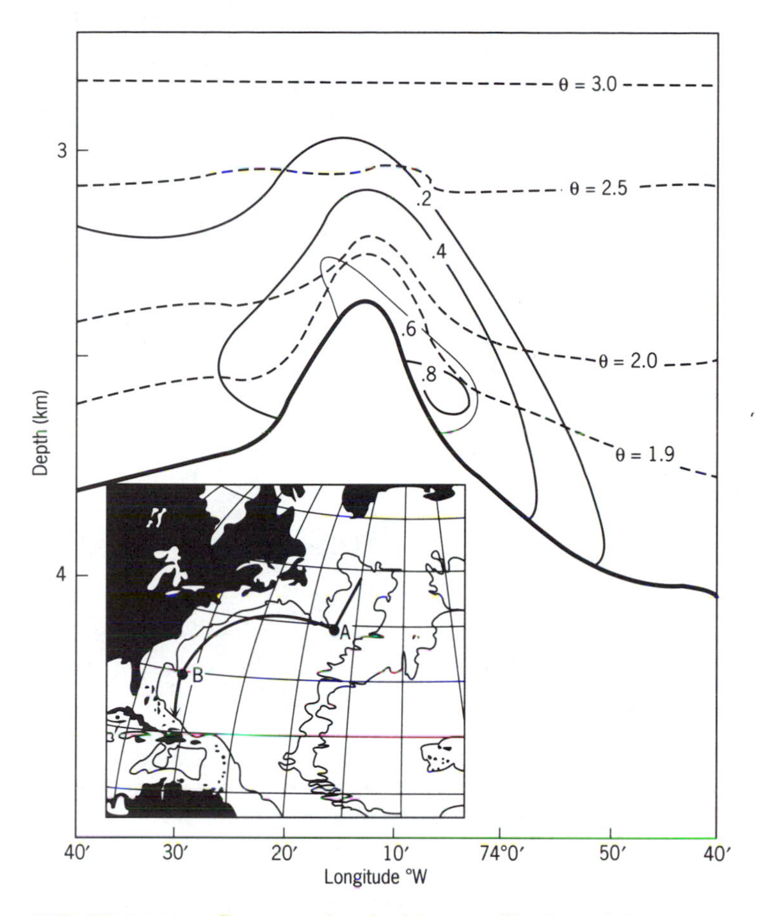

FIGURE 28.28. Cross-sectional tritium profile through bottom waters flowing along the Blake-Bahama Outer Rise. The site location is identified on the inset map as B. The solid contours are the relative tritium abundances in T.U.s (see Figure 28.26 for unit definition) and the dashed contours are for potential temperature. *Source*: From *Tracers in the Sea*, W. S. Broecker and T.-H. Peng, copyright © 1982 by the Lamont-Doherty Geological Observatory, Palisades, NY, p. 411. Reprinted by permission. Data from W. J. Jenkins and P. B. Rhines, reprinted with permission from *Nature*, vol. 286, p. 879, copyright © 1980 by Macmillan Journals, Ltd., London, England.

responsible for the transport of sediment that has been deposited on the Blake-Bahama Outer Rise as the huge delta over which the current now flows.

Tritium distributions have been used to study mixing processes in the surface waters and thermocline. For example, turnover times for the mixed layer and thermocline have been inferred from the application of the one-dimensional advection-diffusion model presented in Chapter 4 to vertical profiles of ^{3}H, salinity, and temperature. The results for temperate latitudes are summarized in Figure 28.29 as ventilation ages. This is the length of time

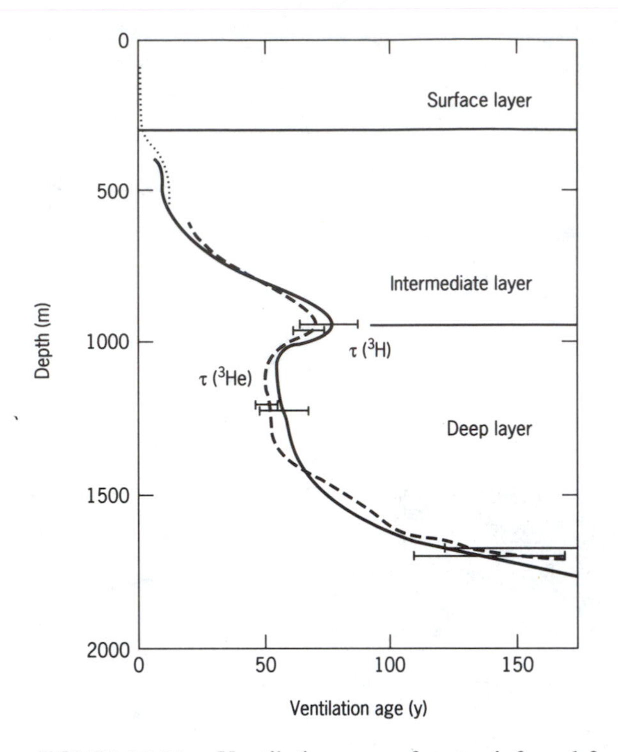

FIGURE 28.29. Ventilation ages of water inferred from tritium and ^3He distributions. The surface, intermediate, and deep layers are approximately analogous to the mixed layer, thermocline, and deep zone. *Source*: From *The Major Biogeochemical Cycles and Their Interactions*, M. E. Fiadeiro (eds.: B. Bolin and R. B. Cook), copyright © 1983 by John Wiley & Sons, Inc., New York, p. 464. Reprinted by permission. Data from W. J. Jenkins, reprinted with permission from the *Journal of Marine Research*, vol. 38, p. 599, copyright © 1979 by the Journal of Marine Research, New Haven, CT.

required for isopycnal processes to completely replace the water located at a particular depth. These processes are thought to involve convective overturn during the winter, followed by horizontal transport, as shown in Figure 28.30. Since the water masses start their journey at the sea surface, they are rich in atmospheric gases. Isopycnal transport carries this gas below the sea surface and hence acts to "ventilate" the ocean.

From these results, the water column at temperate latitudes appears to be occupied by three different types of water. The surface layer, which occupies the top 50 m, has a tritium "age" of less than 1 y. Thus this water mass can be characterized as well mixed and highly variable in composition on short time scales. Below this, the ventilation age of the water masses increases with depth to 1000 m. This coincides with the base of thermocline, suggesting that the waters in this intermediate layer are relatively stagnant and isolated as a result of slower isopycnal advection and density stratification. The underlying

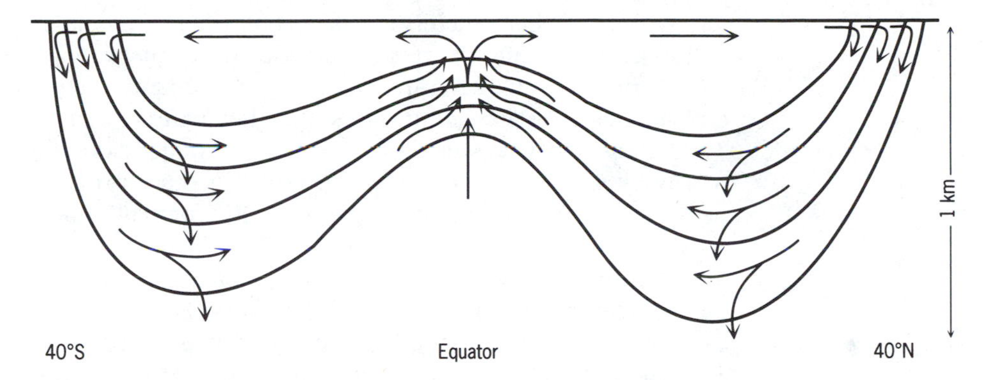

FIGURE 28.30. Isopycnal transport of water to depths below the sea surface. This process is thought to be responsible for the ventilation of the thermocline. *Source*: From *Tracers in the Sea*, W. S. Broecker and T.-H. Peng, copyright © 1982 by the Lamont-Doherty Geological Observatory, Palisades, NY, p. 440. Reprinted by permission.

water has a shorter ventilation age, reflecting the effects of relatively fast thermohaline-driven advection.

Due to its naturally high concentrations in seawater, bomb-derived radiocarbon does not make as "clean" a tracer of water-mass motion as tritium. Nevertheless, the pathway and degree of bomb radiocarbon penetration have also been used to validate ocean-mixing models. These models are presently being used to predict the fate of anthropogenic CO_2 in the ocean.

SUMMARY

Radioisotopes (or **radionuclides**) are atoms that spontaneously undergo nuclear transformations at a fixed rate and in doing so attain a greater measure of stability. **Radioactive decay** can occur through four types of nuclear reactions. If **protons** are in excess, stability is achieved either through emission of **α particles** ($^4_2$**He**) or **positrons** (**β^+**), or by **electron capture**. If **neutrons** are in excess, stability is achieved via the emission of **β^- particles**, which are high-energy **electrons** ($_{-1}^{0}e$).

The production rates of these high-energy particles are easily measured. They are functionally equivalent to the decay rate, or **activity**, of the parent radionuclide. Activities are usually reported as **disintegrations per minute (dpm)**. The activity of a radionuclide in a sample of seawater or sediment is reported in terms of dpm/L or dpm/g, respectively. These **specific activities** (A) are directly related to the concentration of the radionuclide by $A = \lambda[N]$, where λ is the **decay constant**. Thus the rate at which a radionuclide decays is directly proportional to its concentration. This decay rate is commonly expressed as the **radioactive decay law**: $\ln(N/N_o) = -\lambda t$ where N is the amount (e.g., grams or moles) of radionuclide N. Decay constants are usually given in terms of a **half-life**, which is the time required for radioactive decay to remove half of an initial amount of radionuclide. It is equal to $0.693/\lambda$.

Marine radionuclides are classified according to their source. The **primordial radionuclides** are isotopes that have been part of Earth since its formation. Thus they decay at very slow rates. In comparison, their **daughters** have relatively short half-lives. In the absence of other processes, the rapid decay of the daughter relative to its **parent** will eventually cause the pair to attain **secular equilibrium** in which $A_D = A_P$. Fortunately for marine chemists, many primordial radionuclides and their daughters are not present in secular equilibrium in the ocean. This is largely due to rapid removal of the daughter by adsorption onto sinking particles or incorporation into biogenic materials. The degree of disequilibrium can be used to compute the daughters' net removal rates. Secular disequilibrium has also been used to determine sedimentation rates, nodule accretion rates, and rates of water motion. These results are best confirmed with those from other isotopes. In doing so, it is important to recognize that the results reflect the half-lives of the isotopes. That is, the apparent behavior of the processes being measured is dependent on the time scales over which it is being observed.

Cosmogenic radionuclides are formed by **spallation reactions** that occur in the atmosphere. In these nuclear reactions, gas nuclei are fragmented as a result of collisions with high-energy **cosmic rays. Radiocarbon (^{14}C)** is the most commonly used cosmogenic radionuclide. Its marine distributions have been used to measure rates of water motion, sedimentation, bioturbation, and the timing of changes in sea level. Radiocarbon concentrations are usually measured relative to the amount of total carbon present in a sample ($^{14}C/C$). The relative abundance of radiocarbon has been greatly altered as a result of the introduction of "dead" carbon from the burning of fossil fuel. This is called the **Suess Effect** and had caused $^{14}C/C$ ratios to decline until the 1950s. At this time, radiocarbon concentrations were greatly increased as a result of atomic bomb testing. This input has been so large that it has, at least temporarily, swamped the Suess Effect.

Bomb testing, as well as intentional leakage from nuclear reactors, has introduced other **artificial radionuclides** into the ocean. For example, the natural inventory of **tritium** has also been greatly elevated. Both tritium and bomb radiocarbon have been used to obtain much information about the rates of thermohaline circulation and the processes by which waters in the mixed layer and thermocline are renewed.

CHAPTER 29

READING THE SEDIMENTARY RECORD: THE USE OF STABLE ISOTOPES IN THE STUDY OF PALEOCEANOGRAPHY

INTRODUCTION

Isotopes of an element react at different rates due to the slight difference in their atomic masses. Favorable energetics usually cause the lighter isotope to react faster and to a greater extent. In the case of the stable isotopes, this causes the reaction products to be relatively enriched in the light isotope. The degree of this enrichment is dependent upon such factors as (1) the reaction mechanism, (2) the degree to which the reaction has proceeded, (3) the isotopic composition of the reactants, and (4) environmental conditions, such as temperature and pressure. As a result, there is considerable spatial and temporal variability in the relative abundances of the naturally occurring stable isotopes.

The factors that control this variability have been quantified for some of the stable isotopes present in the ocean. This information has been used in a variety of applications, such as to (1) trace the fate or source of various materials in the ocean, (2) determine the type and extent of biogeochemical reactions that have acted on these materials, and (3) assess past environmental conditions in the ocean. As a result, the relative abundances of the naturally occurring stable isotopes have provided a vast amount of very important information on the biogeochemistry of the ocean. As with the radionuclides, many of the conclusions presented in earlier chapters are based on stable isotope data. Some examples are discussed below.

STABLE ISOTOPES

Most studies of the naturally occurring stable isotopes have focused on the elements carbon, hydrogen, oxygen, sulfur, and nitrogen. As shown in Table 29.1, carbon, nitrogen, and hydrogen have only two stable isotopes, while oxygen has three and sulfur has four. For each element, one isotope is most abundant, (i.e., ^{12}C, ^{1}H, ^{16}O, ^{14}N, and ^{32}S). The heavy isotope of H, ^{2}H, is called deuterium and is usually represented as D.

TABLE 29.1
Relative Abundances of Some Stable Isotopes[a]

Atomic Number	*Symbol*	*Mass Number*	*Abundance (%)*
1	H	1	99.99
		2	0.01
6	C	12	98.9
		13	1.1
		14[b]	10^{-10}
7	N	14	99.6
		15	0.4
8	O	16	99.8
		17	0.04
		18	0.2
16	S	32	95.0
		33	0.8
		34	4.2
		36	0.2

[a]These values are averages that are representative of Earth's crust, ocean, and atmosphere. They have been rounded to the nearest 0.1 percent, except for the very rare isotopes.

[b]Radioactive.

The relative abundances of the stable isotopes in Earth's crust, ocean, and atmosphere have been used to define the atomic weights of the elements. These atomic weights are a weighted average of the mass numbers of an element's naturally occurring isotopes. The weighting reflects the average natural abundance of each of the isotopes on the outer earth. The similarity of oxygen's atomic weight, 15.9994 amu, to the mass number of ^{16}O indicates that this is the most abundant naturally occurring isotope of oxygen on Earth's surface.

The relative abundances of the stable isotopes are usually expressed as ratios. As shown in Table 29.2, the most abundant isotope is in the denominator. Slight deviations from these average ratios are observed in many marine materials. For example, the $^{13}C/^{12}C$ ratio in most marine organic matter ranges between 0.000921 to 0.001098. These small differences are detected with a Nier-type ratio mass spectrometer, which measures only one isotope ratio at a time. (The mass spectra generated by the GC-MS systems are produced by mass spectrometers that sequentially measure the relative abundance of each mass fragment, or isotope.)

This type of mass spectrometer is very sensitive to changes in environmental conditions and the performance of various electronic components. As a result, absolute ratios cannot be determined with a high level of accuracy. On the

TABLE 29.2
Average Stable Isotope Ratios

$\frac{^{2}H}{^{1}H} = 0.0016$
$\frac{^{13}C}{^{12}C} = 0.001123$
$\frac{^{15}N}{^{14}N} = 0.00677$
$\frac{^{18}O}{^{16}O} = 0.00200$
$\frac{^{34}S}{^{32}S} = 0.0443$

other hand, excellent accuracy and precision are obtained by analyzing the sample and a standard as concurrently as possible. Thus the isotope ratio of a sample is always measured relative to a standard. This relative difference is reported as a **del value**, which is defined as

$$\delta \text{ in ‰} = \left[\frac{R_{\text{sample}} - R_{\text{standard}}}{R_{\text{standard}}}\right] \times 1000 \qquad (29.1)$$

where R is the isotope ratio with the most abundant isotope in the denominator. Since the ratio difference is multiplied by 1000, del values have units of parts per thousand, which are also called parts per mil (‰).

A del value of 0‰ means that the isotopic composition of the sample is equal to that of the standard. Positive del values indicate that the sample is enriched in the heavy (rare) isotope relative to the standard. Negative del values indicate isotope depletion; the sample has relatively less of the heavy (rare) isotope than the standard.

The internationally accepted stable isotope standards for C, H, N, O, and S are listed in Table 29.3. All meet the following criteria: (1) a reasonably abundant supply of this material exists; (2) the material is isotopically homogeneous; and (3) it is relatively easy to prepare for isotope analysis (e.g., the carbon and oxygen must be quantitatively converted to $CO_2(g)$, the hydrogen to $H_2(g)$, the nitrogen to $N_2(g)$ and the sulfur to $SO_2(g)$).

Most of these standards are materials that are active and abundant in the crustal-ocean factory. Thus del values represent a convenient comparison with a geochemically relevant benchmark. For example, a sample of limestone with a $\delta^{13}C$ value of +6‰ is relatively enriched in ^{13}C as compared to PDB, which is an ancient carbonate fossil of a marine invertebrate. The original fossils used as the PDB isotope standard have long since been used up. Thus other working standards, such as NBS-20 ($\delta^{13}C$[PDB] = −1.06‰), are presently used though the $\delta^{13}C$ results are still reported with respect to PDB.

TABLE 29.3
Internationally Accepted Stable Isotope Standards for Hydrogen, Carbon, Oxygen, Nitrogen, and Sulfur

Element	*Standard*	*Abbreviation*
H	Standard Mean Ocean Water	SMOW
C	*Belemnitella americana* from the Cretaceous Peedee formation, South Carolina	PDB
N	Atmospheric N_2	—
O	Standard Mean Ocean Water	SMOW
	Belemnitella americana from the Cretaceous Peedee formation, South Carolina	PDB
S	Troilite (FeS) from the Canyon Diablo iron meteorite	CD

Source: After *Stable Isotope Geochemistry*, J. Hoefs, copyright © 1980 by Springer-Verlag, Heidelberg, Germany, p. 19. Reprinted by permission.

FRACTIONATIONS

Variations in relative isotope abundance are caused by the preferential reaction or transport of one of the isotopes. Isotopic segregation, or **fractionation**, can occur during a variety of physicochemical processes, such as chemical reactions, phase changes, and molecular diffusion. For example, $H_2{}^{16}O(l)$ is more likely to evaporate than $H_2{}^{18}O(l)$. Thus the resulting water vapor will be relatively depleted in ^{18}O as compared to its parent liquid. The degree to which isotopes become segregated as a result of a particular process is usually expressed as a fractionation factor (α), which is defined as

$$\alpha = \frac{R_{\text{products}}}{R_{\text{reactants}}} \tag{29.2}$$

Since $\alpha \approx 1$, Eq. 29.2 can be rewritten as

$$(\alpha - 1)1000 \approx \delta_{\text{products}} - \delta_{\text{reactants}} \tag{29.3}$$

Fractionation is caused by the slight physicochemical differences that exist among the isotopes of an element. As discussed below, α's are dependent on the reaction mechanism and environmental conditions.

Although isotopes in a given compound form the same kinds of chemical bonds, their bond energies differ. This is due to the effect of the slight mass differences on the vibrational energy of molecules and their chemical bonds. Molecules that contain an atom of the light isotope have a higher vibrational energy than those that contain an atom of the heavier isotope. As a result, the chemical bonds in the lighter molecule are more apt to react, causing the products to become enriched in the light isotope.*

Most biologically mediated processes occur as a series of enzyme-catalyzed reaction steps. Each of these steps is reversible and thus has the potential to attain equilibrium. But when combined in a series, they cause an overall unidirectional transformation of reactants into products, such as in the bacterially mediated reduction of nitrate to N_2. In these processes, fractionations can arise from differences in the rates at which the isotopes are transformed from reactants into products. Since the light isotope reacts faster, it tends to become enriched in the products. In the case of denitrification, this causes N_2 to become relatively enriched in ^{14}N. This type of isotope segregation is rate dependent, so it is termed a **kinetic fractionation**. In many such reactions, the isotopes react at rates that are concentration dependent. Thus kinetic fractionation factors also tend to be concentration dependent.

Physical processes, such as diffusion and phase changes, can cause kinetic fractionations in chemical systems which are not at equilibrium. For example, the net evaporation of water produces vapor that is relatively depleted in ^{18}O as compared to its parent liquid.

The relative isotope composition of reactants and products can also differ even if equilibrium is achieved. The resulting isotope segregation is termed **thermodynamic fractionation** and can usually be described as a type of isotope exchange, such as illustrated below:

$$H_2{}^{18}O(g) + H_2{}^{16}O(l) \rightleftharpoons H_2{}^{16}O(g) + H_2{}^{18}O(l) \qquad (29.4)$$

In this reaction, the products are favored over the reactants at equilibrium because $H_2{}^{16}O$ requires less energy to maintain in the gas phase than $H_2{}^{18}O$. Equilibrium isotope exchange can also occur during chemical reactions as illustrated below.

$$H_2{}^{16}O(l) + C^{18}O^{16}O(aq) \rightleftharpoons H_2{}^{18}O(l) + C^{16}O^{16}O(aq) \qquad (29.5)$$

The fractionation factors for such simple systems can be calculated from theoretical principles. Less success has been achieved for processes that involve solids, as these fractionations are also dependent on differences in lattice energy. Kinetic fractionation factors for most naturally occurring chemical reactions are also difficult to calculate due to their multistep nature. As a result, most fractionation factors are determined experimentally.

All fractionation factors decrease with increasing reaction temperature. The addition of heat increases the total energy of the molecules, including their vibrational energy. At high enough temperatures, the vibrational energy of a

---

*A "reverse" isotope fractionation can occur during bond formation if the energy of the transition state complex is closer to that of the products than that of the reactants. Because the heavier isotope has a lower activation energy than the lighter one, less energy is required to form its new bond.

molecule is increased to a level such that differences due to dissimilar isotopic composition are insignificant. At these temperatures, $\alpha = 1$.

SOME APPLICATIONS OF THE STABLE ISOTOPES TO THE STUDY OF MARINE BIOGEOCHEMICAL PROCESSES

The Hydrogen and Oxygen Isotopes

The relative abundances of the hydrogen and oxygen isotopes in various natural materials are given in Figures 29.1 and 29.2, respectively. This type of diagram, in which the isotopic compositions of the various materials are reported as a range of del values, is called a Caltech Plot.

δD and $\delta^{18}O$ of Natural Waters

The range of del values seen in natural waters is largely the result of fractionations which occur during evaporation and condensation. The fractionation factor for this process (Eq. 29.4) is defined as

$$\alpha = \frac{R_{\text{liquid}}}{R_{\text{gas}}} \tag{29.6}$$

where $\alpha_D = (^2H/^1H)_{\text{liquid}}/(^2H/^1H)_{\text{gas}}$ and $\alpha_{18} = (^{18}O/^{16}O)_{\text{liquid}}/(^{18}O/^{16}O)_{\text{gas}}$. More energy

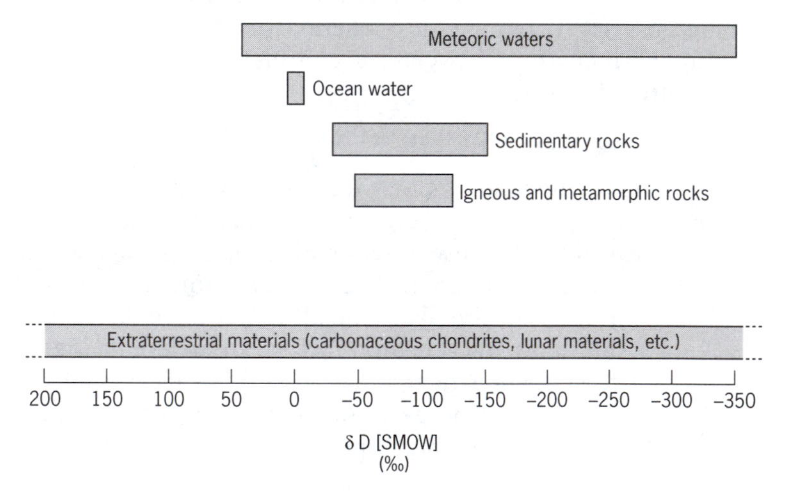

FIGURE 29.1. δD [SMOW] of some geologically important materials. Meteoric waters are produced by meteorological processes, such as rain and snow. They include all liquid and solid water on land and in underground aquifers. They do not include water in the ocean, sediments or below the crust. *Source*: From *Stable Isotope Geochemistry*, J. Hoefs, copyright © 1980 by Springer-Verlag, Heidelberg, Germany, p. 23. Reprinted by permission.

is required to keep the heavy isotope in the gaseous phase, so these fractionation factors are always greater than 1 and decrease with increasing temperature as shown below (Figure 29.3).

The del values of natural waters are also affected by the extent to which evaporation or condensation have occurred. As shown in Figure 29.4, an isotope distillation occurs as moisture-laden clouds produced by evaporation at low latitudes travel poleward and lose water via condensation.

At low latitudes, net evaporation of water from the ocean produces vapor which is depleted in ^{18}O. Atmospheric circulation transports this vapor poleward. The decline in temperature causes water vapor to condense en route. Since ^{18}O is more likely to condense than ^{16}O, the remaining water vapor becomes progressively depleted in ^{18}O. This causes the relative abundance of ^{18}O in the cloud to decrease, so that the ^{18}O content of any further condensate also declines. In other words, the $\delta^{18}O$ of the condensate is a function of the fraction of water vapor that remains in the atmosphere.

This process, in which the isotopic composition of the product varies as a result of the extent of reaction, is called a **Rayleigh Distillation**. Its effect on the isotopic composition of the remaining reactant is given by

$$\frac{R_f}{R_o} = f^{(\alpha - 1)} \tag{29.7}$$

where f is the fraction of remaining reactant, R_f is the del value of the reactant at some f and R_o is the del value when $f = 1$ (i.e., prior to the removal of any reactant). This expression can also be written in terms of del values as illustrated below for the Rayleigh Distillation of ^{18}O and ^{16}O in water vapor as it undergoes condensation.

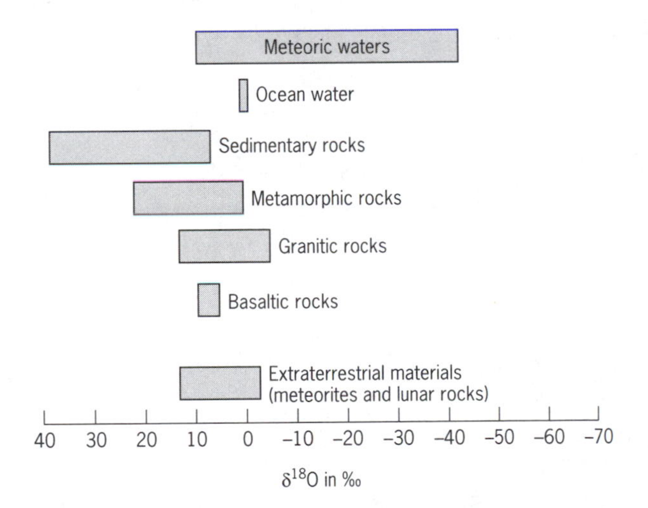

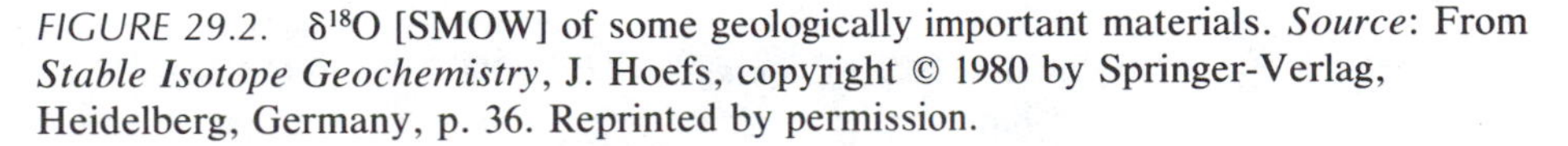

FIGURE 29.2. $\delta^{18}O$ [SMOW] of some geologically important materials. *Source*: From *Stable Isotope Geochemistry*, J. Hoefs, copyright © 1980 by Springer-Verlag, Heidelberg, Germany, p. 36. Reprinted by permission.

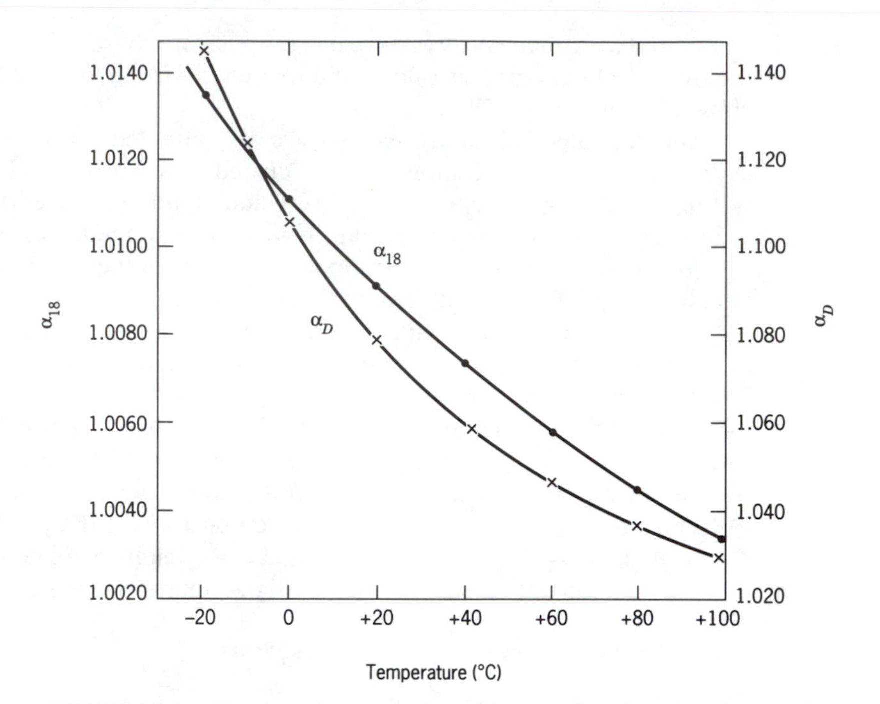

FIGURE 29.3. Temperature variation of isotope fractionation factors for the condensation of water where $\alpha_D = (^2H/^1H)_{liquid}/(^2H/^1H)_{gas}$ and $\alpha_{18} = (^{18}O/^{16}O)_{liquid}/(^{18}O/^{16}O)_{gas}$. These fractionation factors are for the case where isotopic equilibrium is achieved between the liquid and gas phases. *Source*: From *Principles of Isotope Geology*, 2nd ed., G. Faure, copyright © 1986 by John Wiley & Sons, Inc., New York, p. 433. Reprinted by permission. Data from W. Dansgaard, reprinted with permission from *Tellus*, vol. 16, p. 438, copyright © 1965 by Munksgaard International Publishers, Ltd., Copenhagen, Denmark.

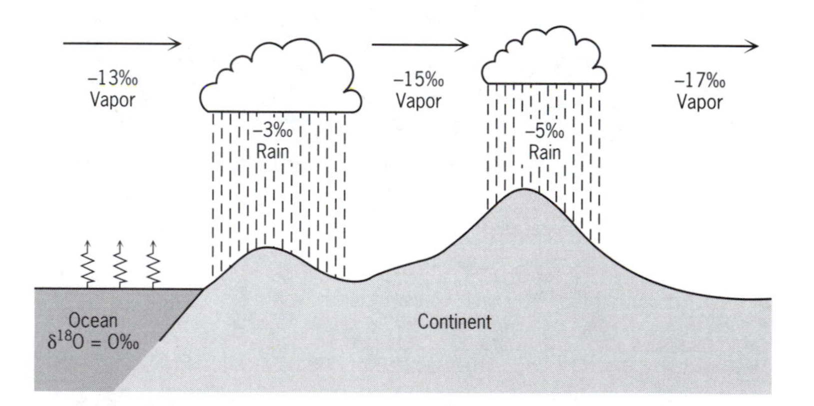

FIGURE 29.4. Schematic fractionation in the atmospheric water cycle. *Source*: From *Lectures in Isotope Geology*, U. Siegenthaler (eds.: E. Jager and J. C. Hunziker), copyright © 1979 by Springer-Verlag, Heidelberg, Germany, p. 266. Reprinted by permission.

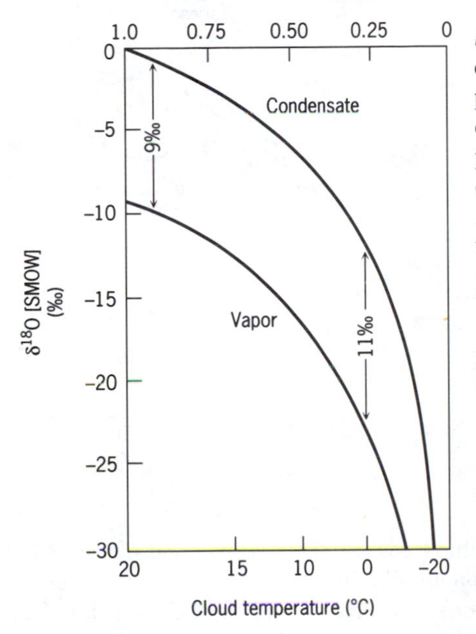

FIGURE 29.5. $\delta^{18}O$ [SMOW] in cloud vapor and condensate plotted as a function of the fraction of remaining vapor for a Rayleigh process. The temperature of the clouds is shown on the lower axis. As indicated, the increase in fractionation with decreasing temperature is taken into account. *Source*: From *Stable Isotope Geochemistry*, J. Hoefs, copyright © 1980 by Springer-Verlag, Heidelberg, Germany, p. 12. Reprinted by permission. After W. Dansgaard, reprinted with permission from *Tellus*, vol. 16, p. 440, copyright © 1964 by Munksgaard International Publishers, Ltd., Copenhagen, Denmark.

$$\delta^{18}O_f = [\delta^{18}O_o + 1000]f^{(\alpha - 1)} \tag{29.8}$$

A similar equation can be generated to describe the effect of this Rayleigh Distillation on the isotopic composition of the resulting condensate.

As clouds move from low to high latitudes, α increases due to decreasing temperature. The effect of this temperature change on the Rayleigh Distillation of water vapor is shown in Figure 29.5. Assuming that isotopic equilibrium is achieved in the clouds, the initial condensate will be enriched in ^{18}O by 9‰ relative to its parent vapor. As the remaining vapor moves to higher latitudes, the preferential removal of ^{18}O causes it to become progressively depleted in ^{18}O and hence its $\delta^{18}O$ becomes more negative. The $\delta^{18}O$ of the resulting condensate also decreases, but by less than would have been achieved at low latitudes due to the effect of temperature on the fractionation factor. For example, the condensate produced at 0°C is enriched in ^{18}O by 11‰ relative to its parent vapor. This increase in enrichment counters some of the depletion caused by the Raleigh Distillation Effect.

This process causes the $\delta^{18}O$ and δD of meteoric water* to be linearly related and generally decrease with increasing latitude, as illustrated in Figure 29.6. Deviations from these trends are caused by processes, such as an excess of evaporation over precipitation, that occur in some semi-isolated basins.

In comparison, the $\delta^{18}O$ and δD of seawater are not nearly as variable. The small differences that do exist are related to salinity, as shown in Figure 29.7. In the surface waters, evaporation causes a concurrent enrichment in salt

*See Figure 29.1 for a definition of meteoric water.

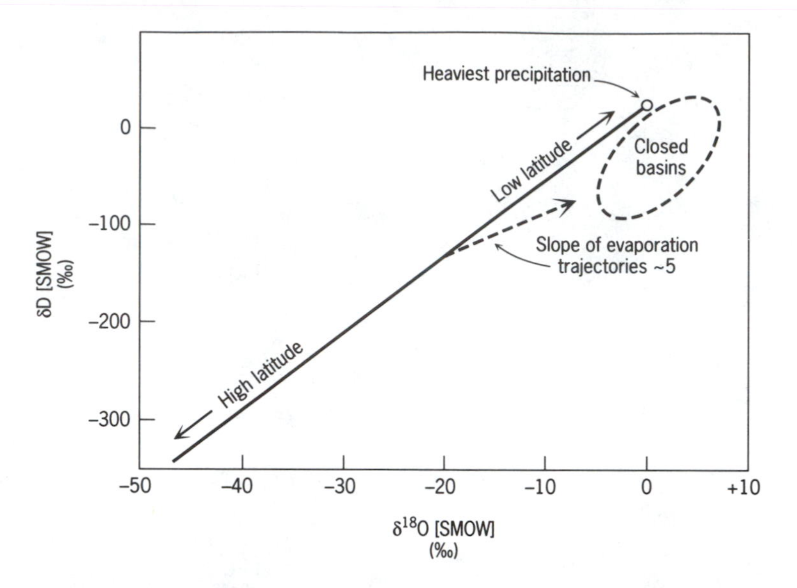

FIGURE 29.6. Relationship between δD [SMOW] and $\delta^{18}O$ [SMOW] in meteoric water. *Source*: From *Principles of Isotope Geology*, 2nd ed., G. Faure, copyright © 1986 by John Wiley & Sons, Inc., New York, p. 435. Reprinted by permission. After H. Craig, reprinted with permission from *Science*, vol. 133, p. 1702, copyright © 1961 by the American Academy of Sciences, Washington, DC.

and the heavy isotopes of oxygen and hydrogen. The latter is the result of the preferential evaporation of ^{16}O. Since deep water is created from the sinking of surface waters, the source of a deep-water mass can be determined from its isotopic composition and salinity. As shown in Figure 29.7, the isotopic composition and salinity of NADW is very different from that of AABW. The latter is a version of Weddell Sea water, whose salinity is increased by the removal of water as a result of freezing. The formation of sea ice has little impact on the isotopic composition of the remaining seawater.

The $\delta^{18}O$ and salinity of the waters that would be produced from the conservative mixing of AABW and NADW is shown by the dashed line. Since Indian and Pacific bottom water lie slightly off this trend, they appear to be produced by the mixing of AABW, NADW, and a small amount of some other water mass whose identity has not yet been established. This ''unknown'' water is likely some version of Weddell Sea water that has not been as intensely affected by freezing as pure AABW. These data support the conclusion that deep-water masses are currently being formed only in the North Atlantic and Southern oceans.

$\delta^{18}O$ of Igneous and Metamorphic Rocks

During the crystallization process, igneous and metamorphic rocks acquire oxygen from sources such as subterranean water. If the minerals solidify in

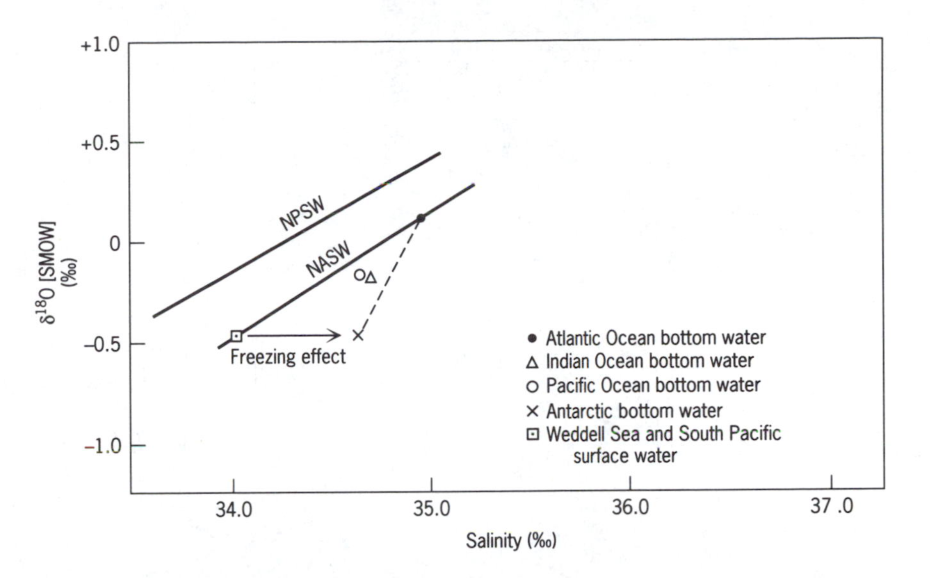

FIGURE 29.7. Relationship between $\delta^{18}O$ [SMOW] and salinity in the surface waters of the North Pacific (NPSW), the North Atlantic (NASW), and the major deep water masses. The formation of AABW from Weddell Sea water is shown by the arrow. The conservative mixing line between NADW and AABW is shown by the dashed line. *Source*: From *Principles of Isotope Geology*, 2nd ed., G. Faure, copyright © 1986 by John Wiley & Sons, Inc., New York, p. 440. Reprinted by permission. After *Stable Isotopes in Oceanographic Studies and Paleotemperatures*, H. Craig and L. I. Gordon (ed.: E. Tongiorgi), copyright © 1965 by Consiglio Nazionale delle Richerche, Laboratorio di Geologia Nucleare, Pisa, Italy, p. 39. Reprinted by permission.

isotopic equilibrium with this oxygen source, their $\delta^{18}O$ can be used to determine the temperature at which this process occurred. The minerals that can be used as geothermometers are ones that experience a significant amount of fractionation as a result of equilibrium isotope exchange with the oxygen source. The relationship between the degree of this fractionation and the solidification temperature is given by the semi-empirical equation

$$1000 \ln \alpha = A(10^6/T^2) + B \qquad (29.9)$$

where A and B are constants for a mineral that must be experimentally determined and $\alpha = R_{\text{rock}}/R_{\text{water}}$. Examples of geothermometer equations for various silicate minerals are plotted in Figure 29.8. Values of B are given by the y-intercept, and values of A are obtained from the slope of the line that best fits the calibration data as determined by the method of linear regression.

With values of A and B, the solidification temperature of a rock sample can be inferred from Eq. 29.9 if the isotopic compositions of the constituant mineral and its oxygen source are known. The requirement for the latter can be eliminated by combining geothermometer equations as follows. Consider two minerals whose fractionation factors are α_1 and α_2 and whose $\delta^{18}O$'s are δ_1 and δ_2, respectively. If the $\delta^{18}O$ of the oxygen source is represented by δ_o,

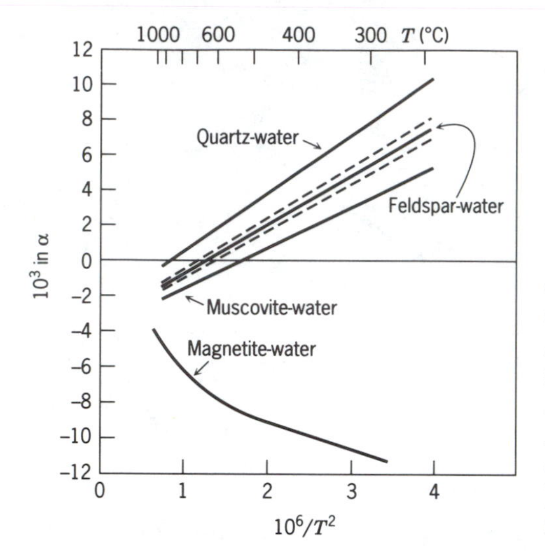

FIGURE 29.8. Schematic diagram of some experimentally determined oxygen isotope calibration curves. *Source*: From H. Friedrichsen, reprinted with permission from *Neues Jahrbuchfuer Mineralogie Monatshefte*, vol. 1, p. 29, copyright © 1971 by E. Schweizerbart'sche, Stuttgart, Germany.

then the fractionation factors can be written as follows, using the relationship given in Eq. 29.3:

$$1000\alpha_1 \approx \delta_1 - \delta_o \tag{29.10}$$

$$1000\alpha_2 \approx \delta_2 - \delta_o \tag{29.11}$$

Likewise, the geothermometer equations for minerals 1 and 2 are

$$1000 \ln \alpha_1 = A_1(10^6/T^2) + B_1 \tag{29.12}$$

$$1000 \ln \alpha_2 = A_2(10^6/T^2) + B_2 \tag{29.13}$$

α is rarely greater than 1.004 and can be thought of as $1.00X$, where $X \leq 4$. Since $\ln(1.004) = 0.00399$, $1000 \ln 1.00X \approx X$ or $1000 \ln \alpha \approx 1000(\alpha - 1)$. Thus Eqs. 29.10 and 29.12 can be equated to yield

$$\delta_1 - \delta_o = A_1(10^6/T^2) + B_1 + 1000 \tag{29.14}$$

and for mineral 2,

$$\delta_2 - \delta_o = A_2(10^6/T^2) + B_2 + 1000 \tag{29.15}$$

δ_o can be eliminated by subtracting Eq. 29.15 from Eq. 29.14 to yield

$$\delta_1 - \delta_2 = (A_1 - A_2)(10^6/T^2) + (B_1 - B_2) \tag{29.16}$$

In other words, when two solid phases have equilibrated with some common reservoir of oxygen, the difference in their δ values is a function of temperature. This assumes, of course, that no other process has altered their isotopic composition. In actuality, the isotopic compositions of most igneous and metamorphic rocks have been shown to change as they cool following crystallization. In these cases, cooling is so slow that the minerals continue to undergo isotope exchange with their oxygen sources. Thus temperatures inferred from their isotopic composition are underestimates of the crystallization temperature. As discussed below, the isotopic composition of minerals that equilibrated with oxygen sources at low temperatures can yield other useful information.

$\delta^{18}O$ and δD of Clay Minerals

Convincing evidence refuting the occurrence of reverse weathering in marine sediments has been obtained from the isotopic composition of clay minerals. During chemical weathering on land, clay minerals undergo isotopic exchange with meteoric waters. Thus the δD and $\delta^{18}O$ of a terrestrial clay mineral is determined by (1) its fractionation factor for isotope exchange, (2) the temperature at which weathering occurred, (3) the degree to which equilibrium with meteoric waters was achieved, and (4) the isotopic composition of the meteoric waters.

If the clay minerals reach isotopic equilibrium with meteoric waters, their δD and $\delta^{18}O$ should be linearly related in a fashion similar to that seen in meteoric waters (Figure 29.6). This linear relationship can be described as

$$\delta D = A\delta^{18}O + B \tag{29.17}$$

where A and B are constants for the mineral at a given temperature. Values of A and B have been determined for various terrestrial clay minerals, such as kaolinite, illite, and montmorillonite. The resulting equations, relevant for temperatures at which weathering occurs, are plotted in Figure 29.9 along with the meteoric water line.

Fractionation during equilibrium isotope exchange causes terrestrial clays to be enriched in ^{18}O relative to their meteoric waters, but depleted in δD. This fractionation is large and causes the mineral lines in Figure 29.9 to be considerably offset from the meteoric water relationship. By examining the effects of temperature on this fractionation, it appears that the clay minerals in modern soils were formed under somewhat warmer conditions than presently exist. Similar calculations can be done to determine what the isotopic composition of clay minerals should be if they are in isotopic equilibrium with seawater. Since the isotopic composition of seawater varies little, these predicted clay mineral values are represented by the small boxes in Figure 29.9.

The actual isotopic composition of clay minerals isolated from marine sediments is closer to that of terrestrial clays than of authigenic ones. This suggests that most of the clay minerals in marine sediments are detrital in origin.

$\delta^{18}O$ of Marine Carbonate Minerals

The $\delta^{18}O$ of sedimentary carbonates is one of the most widely used records of paleoclimate change. The basis for this application lies in the tendency of marine organisms to deposit calcite in isotopic equilibrium with ambient seawater. This isotopic equilibrium is the result of the following isotope exchange reaction

$$CaC^{16}O^{16}O^{16}O(s) + H_2^{18}O(l) \rightleftharpoons Ca^{18}O^{16}O^{16}O(s) + H_2^{16}O(l) \tag{29.18}$$

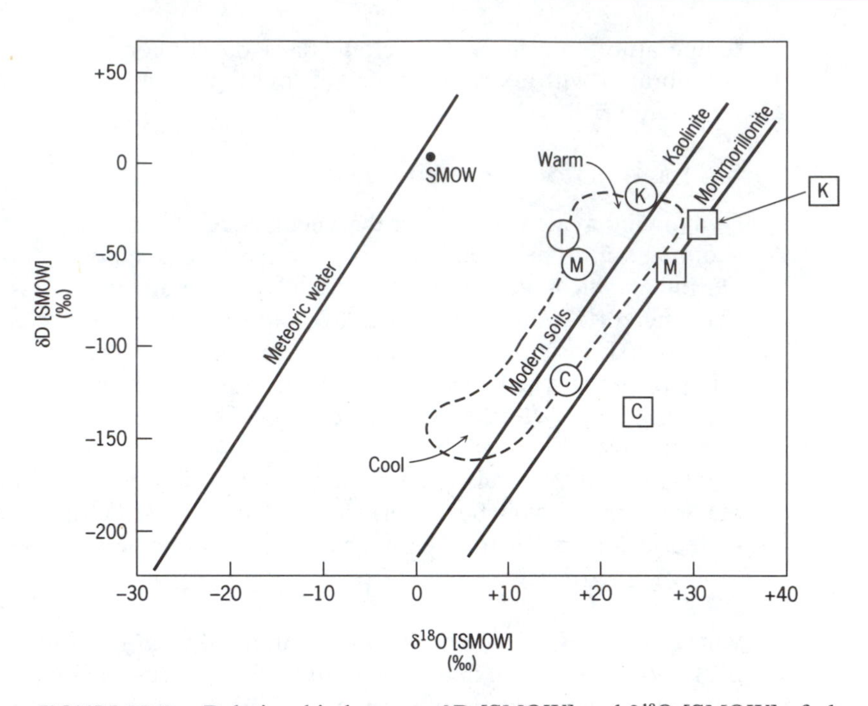

FIGURE 29.9. Relationship between δD [SMOW] and $\delta^{18}O$ [SMOW] of clay minerals. The lines represent the isotopic composition that would be achieved if the clay minerals reached isotopic equilibrium with meteoric waters at earth-surface temperatures. The boxes represent the isotopic composition that would be achieved if the clay minerals reached isotopic equilibrium with seawater at 0°C. The circles are the observed isotopic composition of clay minerals collected from marine sediment, where K = kaolinite; M = montmorillonite; C = chlorite; and I = illite. Also shown is the isotopic composition of meteoric waters. *Source*: From *Principles of Isotope Geology*, 2nd ed., G. Faure, copyright © 1986 by John Wiley & Sons, Inc., New York, p. 479. Reprinted by permission. After H. P. Taylor, Jr., reprinted with permission from *Economic Geology*, vol. 69, p. 857, copyright © 1974 by Economic Geology Publishing Co., El Paso, TX. Seawater Data from S. M. Savin and S. Epstein, reprinted with permission from *Geochimica et Cosmochimica Acta*, vol. 34, p. 55, copyright © 1971 by Pergamon Press, Elmsford, NY.

This isotope exchange causes pure calcite to be considerably enriched in ^{18}O at equilibrium. For example, calcite that has been isotopically equilibrated at 25°C is 28.6‰ enriched in ^{18}O as compared to the water. Thus calcite that is deposited in isotopic equilibrium with SMOW at this temperature will have $\delta^{18}O$ of +28.6‰ on the SMOW scale.

The $\delta^{18}O$ of marine carbonates is usually measured on the PDB scale, as this standard is also a fossil marine carbonate. The following equation relates the $\delta^{18}O$ of carbonates measured on the SMOW scale to the PDB scale.

$$\delta^{18}O[SMOW] = 1.03086\delta^{18}O[PDB] + 30.86 \quad (29.19)$$

The small correction factor (1.03086) compensates for fractionation that occurs during the preparation of marine carbonates for mass spectrometric analysis.

The fractionation factor for the equilibrium isotope exchange given in Eq. 29.18 increases with decreasing temperature. The effect of temperature on the $\delta^{18}O$ of isotopically equilibrated calcite (δ_c) has been experimentally established as

$$T(°C) = 16.9 - 4.2(\delta_c - \delta_w) + 0.13(\delta_c - \delta_w)^2 \quad (29.20)$$

where δ_w is the $\delta^{18}O$ of the water. Similar equations have been developed for other carbonate phases, such as aragonite. These relationships suggest that the isotopic composition of biogenic carbonates should reflect the water temperatures under which the shells were deposited. The use of $\delta^{18}O$ as a paleothermometer is complicated by several phenomena that are discussed below.

Biogenic calcites deviate slightly from the temperature relationship given in Eq. 29.20 as a result of several factors. These include (1) species differences in fractionation factors attributable to variations in mineralogy, (2) temporal changes in fractionation factors caused by variations in growth rate, (3) isotope exchange with respiratory CO_2, and (4) incomplete isotope exchange during shell deposition (i.e., isotopic disequilibrium). Some of these effects can be minimized by determining paleotemperatures from the isotopic composition of a single species. Foraminiferan tests are the biogenic carbonate of choice as they are geographically and temporarily widespread. In addition, these protozoans tend to live in narrowly defined depth ranges, with some species inhabiting the surface waters, while others live in the deep sea and on the seafloor.

In addition to the magnitude of the equilibrium fractionation factor, the $\delta^{18}O$ of biogenic calcite is also determined by the isotopic composition of ambient seawater. As described in the preceding section, the Rayleigh Distillation of atmospheric water vapor causes the $\delta^{18}O$ of surface seawater to vary geographically and is directly related to salinity. As a result, a 1‰ increase in salinity would be accompanied by a change in $\delta^{18}O$ that is equivalent to a 1°C decrease in the calculated water temperature.

The Rayleigh Distillation of water vapor also causes polar ice to be depleted in ^{18}O relative to seawater. Thus an increase in ice volume causes the ocean to become enriched in ^{18}O. During periods of maximum glaciation, the $\delta^{18}O$ of seawater increased to +0.90‰ [SMOW]. If all present-day continental glaciers melted, the $\delta^{18}O$ of seawater would decrease to −0.6‰ [SMOW]. Thus changes in temperature, ice volume, and the relative rate of local evaporation could have caused the $\delta^{18}O$ of forams to vary over time. Over the past 1 million years, the resulting variation has only amounted to 2‰.

Although the isotopic variations have been small, similar amplitudes and frequencies have been observed in cores from many different parts of the ocean (Figure 29.10*a* and *b*). Thus the causes of these fluctuations must have had global impact and are thought to be related to periodic episodes of glaciation and deglaciation. The increases in $\delta^{18}O$ are interpreted as records of ice ages,

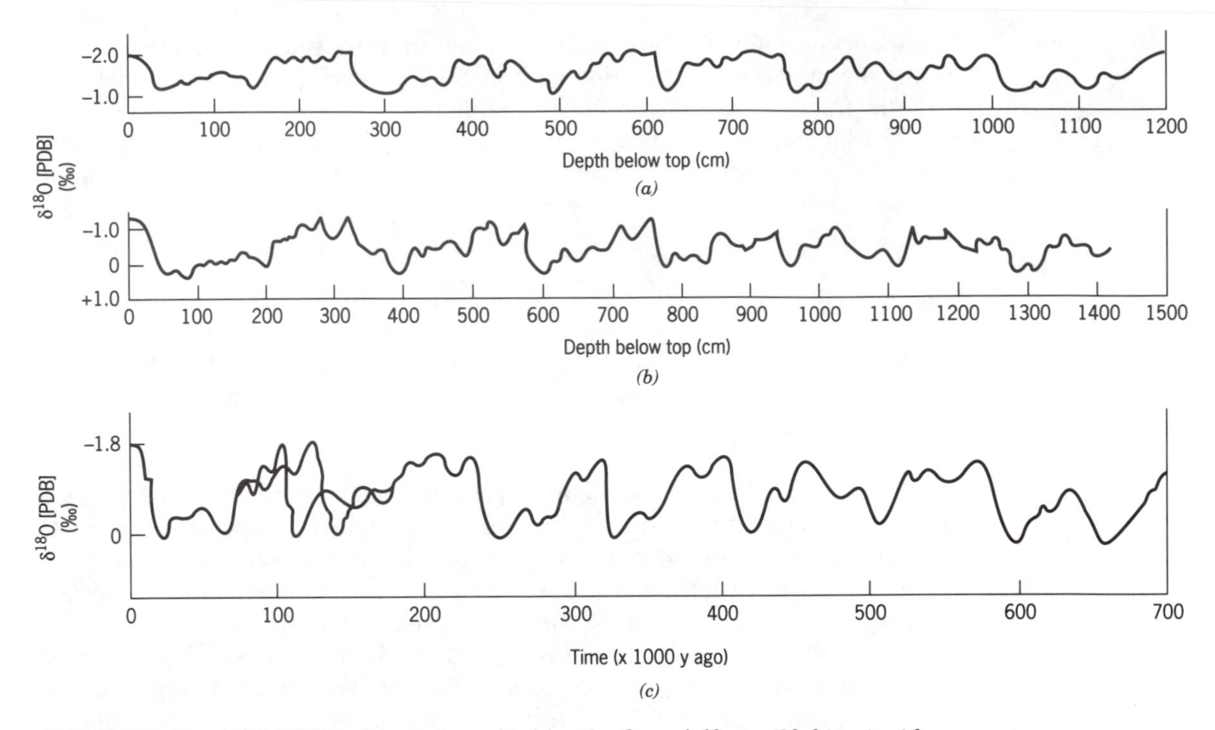

FIGURE 29.10. $\delta^{18}O$ [PDB] of tests deposited by the foraminiferan *Globigerinoides sacculifera* in cores collected from (*a*) the western equatorial Pacific and (*b*) the Caribbean Sea during the Brunhes Epoch. (*c*) Generalized isotope curve and time scales. The latter were obtained from radiometric dating. *Source*: From C. Emiliani and N. J. Shackelton, reprinted with permission from *Science*, vol. 183, pp. 511 and 513, copyright © 1974 by the American Association for the Advancement of Science, Washington, DC.

since ^{18}O enrichment is caused by both decreased temperatures and increased ice volume. Because these isotopic shifts were concurrently recorded in the sediments of all the ocean basins, they are also used as a stratigraphic marker. The timing of these events has been established by dating the sediments radiometrically, with either ^{14}C or $^{230}Th/^{231}Pa$. By averaging the isotope records from many cores, a generalized paleo-isotope curve has been constructed, as shown in Figure 29.10*c*.

The relative importance of changes in ice volume and water temperature in determining the $\delta^{18}O$ of biogenic calcite deposited over the past 1 million years has been assessed by comparing the isotopic composition of benthic and planktonic forams. The presence of polar ice caps during this period places a lower limit on the temperature of the bottom water. Since seawater can get no cooler than its freezing point, bottom-water temperatures could not have been lower than −2°C. Thus most of the variations in isotopic composition of the benthic species must have been caused by shifts in the $\delta^{18}O$ of seawater. Such shifts would have been caused by changes in ice volume.

If large changes in local surface-water temperatures occurred, the amplitude of isotopic variation in the planktonic forams should differ from that in the benthic species. For example, a greater cooling of the surface waters during an ice age would cause the planktonic species to experience a larger ^{18}O enrichment than seen in the benthic forams. In such cases, Eq. 29.20 can be used to infer δ_w from the δ_c of benthic plankton if a deep-water temperature is assumed. Since the ocean is well mixed on time scales much shorter than that of climate change, this δ_w can be used to compute a surface-water temperature from the δ_c of the planktonic tests.

As shown in Figure 29.11, the isotopic composition of the planktonic and benthic forams has fluctuated with virtually the same frequency and amplitude. This suggests that at this location, neither surface nor bottom-water temperatures varied much. This result appears applicable to most of the ocean and is supported by paleotemperatures estimated from changes in the relative abundances of surface-water species. These data indicate that surface-water temperatures varied by no more than 1.5°C. Thus the paleo-isotope curve for this time period is thought to be largely the result of changes in ice volume.

Since the surface waters are affected by short-term local variations in isotopic composition mostly as a result of meteorological events, the $\delta^{18}O$ of planktonic forams tends to be more variable than the isotopic composition of the benthic forams. Thus the $\delta^{18}O$ of benthic forams provides the least

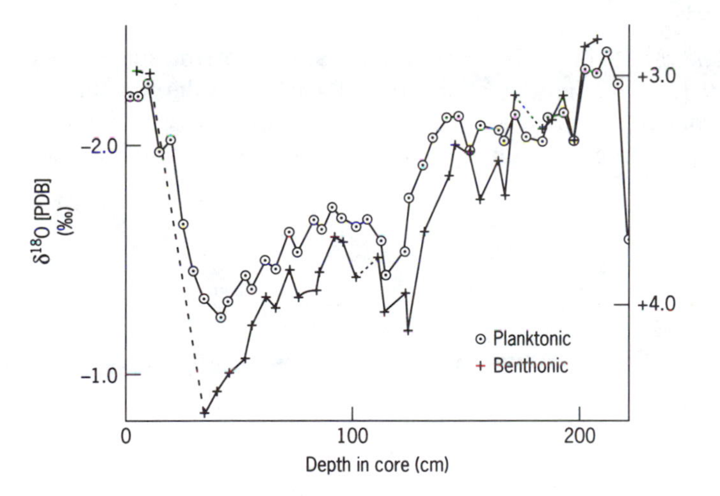

FIGURE 29.11. $\delta^{18}O$ [PDB] of planktonic and benthic foraminifera from DSDP core V28-238. The results are plotted to emphasize changes in the degree to which planktonic forams are isotopically different from the benthic species. To achieve this, the planktonic results are plotted with respect to the left vertical scale, whereas the benthic results are plotted with respect to the right vertical scale. Both are arranged such that the results overlap if the difference in their $\delta^{18}O$ is identical to that presently observed for these species (i.e., 5.3‰). *Source*: From N. J. Shackelton and N. D. Opdyke, reprinted with permission from *Quaternary Research*, vol. 3, p. 44, copyright © 1973 by Academic Press, Orlando, FL.

ambiguous record of climatic changes over the past 1 million years. Down-core variations in the $\delta^{18}O$ of benthic forams can also be interpreted as a record of paleosalinities because changes in ice volume affect the amount of water in the ocean, but not the amount of salt. The isotope data suggest that during the period of maximum glaciation, the mean salinity of the ocean was 3.5‰ higher than at present.

Biogenic calcite deposited prior to the formation of the polar ice caps must have been considerably depleted in ^{18}O relative to modern-day forams. Thus the formation of the present-day polar ice caps should have been recorded as a large increase in the $\delta^{18}O$ of benthic forams. As shown in Figure 29.12, this appears to have occurred 14 million years before present.

Interpretation of the ^{18}O record is complicated by the effects of shifts in oceanic circulation. In particular, changes in thermohaline circulation would affect bottom-water temperatures and the position of the CCD. In some cases, the impact of localized shifts can be assessed by comparing down-core variations in $\delta^{18}O$. For example, the isotope record in the subpolar North Atlantic is substantially different from that at midlatitudes and varies in such a way as to suggest that NADW was not formed in the Norwegian Sea during glacial times. Due to the variety of information obtained from the $\delta^{18}O$ record, its use as a paleoceanographic tool is considered to be the most important geochemical application of stable isotopes.

The Carbon Isotopes

The ranges of $\delta^{13}C$ in various naturally occurring substances are given in Figure 29.13. The marine chemistry of carbon is largely controlled by biological processes, many of which are accompanied by kinetic fractionations. The largest fractionation occurs during the photosynthetic fixation of carbon and

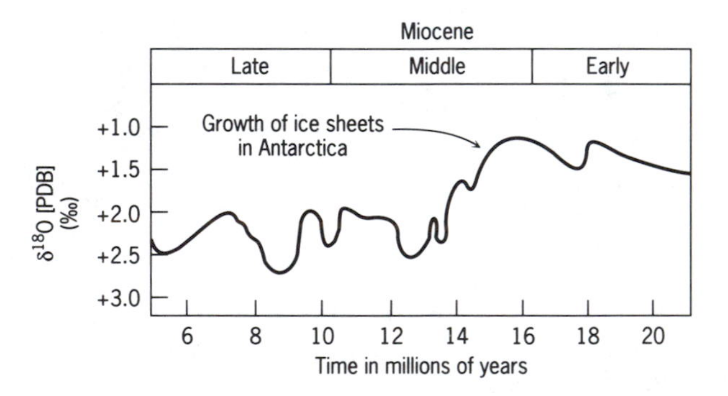

FIGURE 29.12. $\delta^{18}O$ [PDB] in the tests of the benthic foraminiferan *Cibicidoides* in a deep-sea core from the equatorial Pacific Ocean. *Source*: From *Principles of Isotope Geology*, 2nd ed., G. Faure, copyright © 1986 by John Wiley & Sons, Inc., New York, p. 446. Reprinted by permission. Data from R. Woodruff, S. M. Savin, and R. E. Douglas, reprinted with permission from *Science*, vol. 212, p. 666, copyright © 1981 by the American Association for the Advancement of Science, Washington, DC.

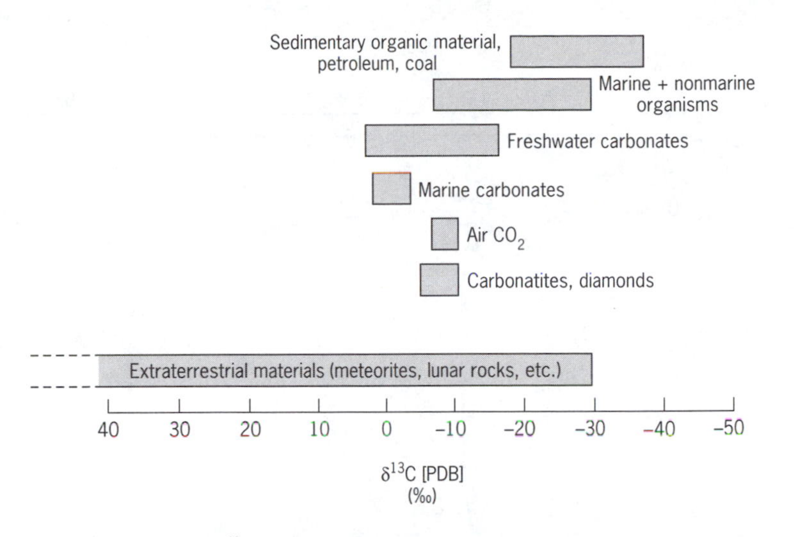

FIGURE 29.13. $\delta^{13}C$ [PDB] of some geologically important materials. *Source*: From *Stable Isotope Geochemistry*, J. Hoefs, copyright © 1980 by Springer-Verlag, Heidelberg, Germany, p. 32. Reprinted by permission.

causes most marine organic matter to be depleted in ^{13}C by approximately 20‰ relative to DIC. The magnitude of this fractionation is somewhat variable and depends on such factors as (1) the species of phytoplankton, (2) their growth rate, and (3) the temperature of ambient seawater.

As illustrated in Figure 29.14, the preferential uptake of ^{12}C by marine phytoplankton causes DIC in the euphotic zone to be relatively enriched in ^{13}C. Phytoplankton also produce O_2 that is depleted in ^{18}O. Following death, some of the ^{13}C-depleted phytoplankton biomass sinks below the euphotic zone and is remineralized. The addition of this ^{13}C-depleted CO_2 lowers the $\delta^{13}C$ of the ambient DIC. This remineralization proceeds via aerobic respiration, which involves the preferential uptake of ^{18}O-depleted O_2, thereby raising the $\delta^{18}O$ of the remaining gas.

The $\delta^{13}C$ of surface-water DIC is also influenced by isotopic equilibration with atmospheric CO_2. At 0°C, isotopic equilibration causes bicarbonate and carbonate to be 10.6‰ and 7.6‰ enriched in ^{13}C relative to atmospheric CO_2, respectively. At 30°C, bicarbonate and carbonate are 7.6‰ and 6.1‰ enriched, respectively. Isotopic equilibrium is often not achieved due to the relatively rapid preferential uptake of ^{12}C during photosynthesis. In comparison, little fractionation occurs during the deposition of biogenic calcite, so the $\delta^{13}C$ of biogenic calcite is close to that of its DIC source. Thus the $\delta^{13}C$ of biogenic calcite can be used to determine the depth or water mass in which the shell was deposited. As shown in Table 29.4, the isotopic composition of the foram tests is closest to that of surface-water bicarbonate. This isotopic similarity suggests that the tests were deposited at the surface from inorganic carbon obtained from the bicarbonate pool.

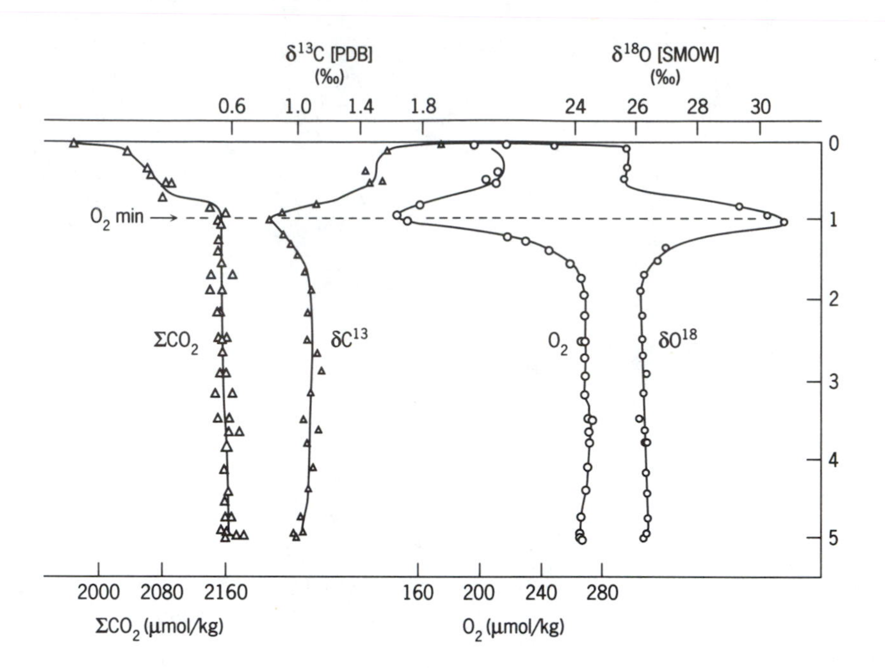

FIGURE 29.14. Vertical profiles of ΣCO_2, $\delta^{13}C$ [PDB] in DIC, dissolved O_2, and $\delta^{18}O$ (SMOW) in dissolved O_2 in the North Atlantic. *Source*: From P. Kroopnik, R. F. Weiss, and H. Craig, reprinted with permission from *Earth and Planetary Science Letters*, vol. 16, p. 106, copyright © 1972 by Elsevier Science Publishers, Amsterdam, The Netherlands.

TABLE 29.4
$\delta^{13}C$ [PDB] of DIC and a Carbonate Shell (The DIC values are for seawater at 15°C)

	$\delta^{13}C$(‰)
Total dissolved inorganic carbon	
At surface	+2.2
At 2.5 km (CO_2 maximum and O_2 minimum)	+0.27
In bottom water	+0.5
Foraminiferal tests	+2
Surface HCO_3^-	+2.5
Surface CO_3^{2-}	−0.5
Atmospheric CO_2	−6.5

Source: From H. Craig, reprinted with permission from the *Journal of Geophysical Research*, vol. 75, p. 693, copyright © 1970 by the American Geophysical Union, Washington, DC.

The $\delta^{13}C$ of sedimentary marine carbonates has not varied as much over time as has its $\delta^{18}O$, as illustrated in Figure 29.15. Since the deep waters are isolated from the atmosphere, changes in the carbon isotope composition of benthic plankton are thought to record shifts in the $\delta^{13}C$ of deep-water DIC. Such isotopic shifts have likely been caused by changes in the oceanic cycling of organic matter. Since the isotopic composition of biogenic calcite is close to that of DIC, changes in its rates of production and dissolution are unlikely to have affected the oceanic distributions of ^{12}C and ^{13}C.

On the other hand, an increase in the rate of POM remineralization in the deep sea should have caused the $\delta^{13}C$ of deep-water DIC to decline. This decline would then have been recorded in the biogenic calcite. As shown in Figure 29.15, this appears to have occurred during periods of glaciation. An increase in remineralization should also have caused deep-water phosphate concentrations to rise. Assuming that the remineralized POM had a C to P ratio of 106 to 16, the increase in deep-water phosphate concentrations can be estimated from the degree of ^{13}C depletion recorded by the forams. With this

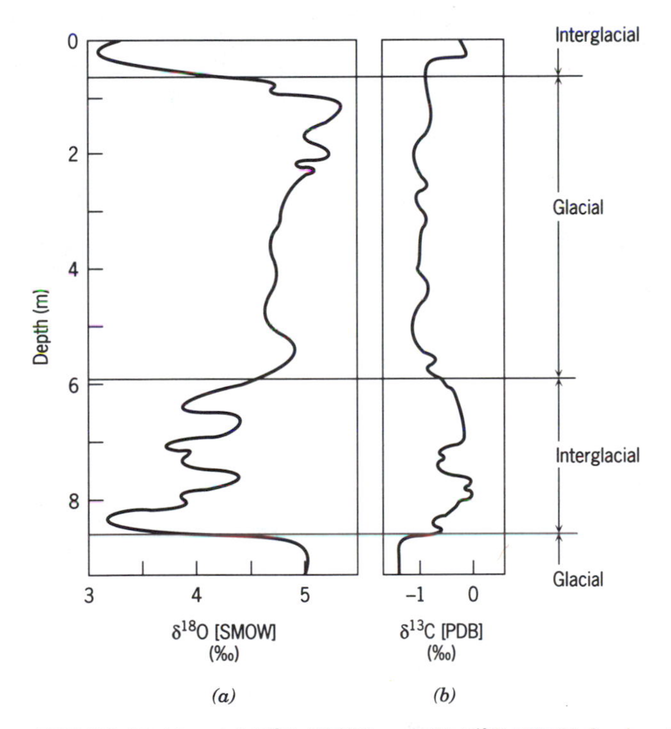

FIGURE 29.15. (*a*) $\delta^{18}O$ [PDB] and (*b*) $\delta^{13}C$ [PDB] for benthic forams as a function of depth in a core from the eastern margin of the north Atlantic Ocean (25°N 17°W). *Source*: From *Tracers in the Sea*, W. S. Broecker and T.-H. Peng, copyright © 1982 by the Lamont-Doherty Geological Observatory, Palisades, NY, p. 305. Reprinted by permission. Data from *The Fate of Fossil Fuel CO_2 in the Oceans*, N. J. Shackleton (eds.: N. R. Anderson and A. Malahoff), copyright © 1977 by Plenum Press, New York, p. 418. Reprinted by permission.

approach, the $\delta^{13}C$ of sedimentary calcite has provided information on how deep-water nutrient concentrations have changed over time.

The $\delta^{13}C$ of sedimentary organic matter also varies down core, but the causes of these variations are not well understood. Most sedimentary organic matter is derived from surface-water POM synthesized by phytoplankton. Thus the $\delta^{13}C$ of this POM is largely determined by the isotopic composition of the phytoplankton. The $\delta^{13}C$ of phytoplankton is determined by temperature, metabolic pathway, and the isotopic composition of the DIC pool. As shown in Figure 29.16, phytoplankton that grow at lower temperatures have larger ^{13}C depletions. This is primarily the result of an increase in the magnitude of the fractionation factor.

The magnitude of this kinetic fractionation is also dependent on the pathway of carbon metabolism in the plant. C_3 plants fractionate the carbon isotopes to a greater degree than do the C_4 (grasses) and CAM (succulents) plants. The $\delta^{13}C$ of plants is also determined by the isotopic composition of their carbon source. For land plants, this is atmospheric CO_2 ($\delta^{13}C$ [PDB] = −7‰) and for marine plants, carbon is assimilated as HCO_3^- ($\delta^{13}C$ [PDB] = ~0‰). As shown in Figure 29.17, this causes most terrestrial C_3 plants to have a lower

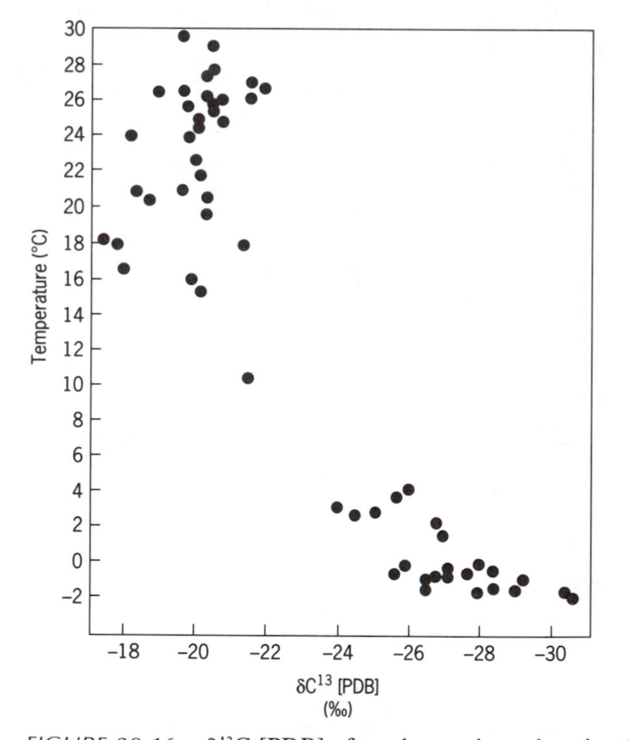

FIGURE 29.16. $\delta^{13}C$ [PDB] of total organic carbon in phytoplankton versus surface temperatures. *Source*: From *Advances in Organic Geochemistry, 1973*, W. M. Sackett, B. J. Eadie, and M. E. Exner (eds.: B. Tissot and F. Bienner), copyright © 1974 by Editions Technip, Paris, France, p. 668. Reprinted by permission.

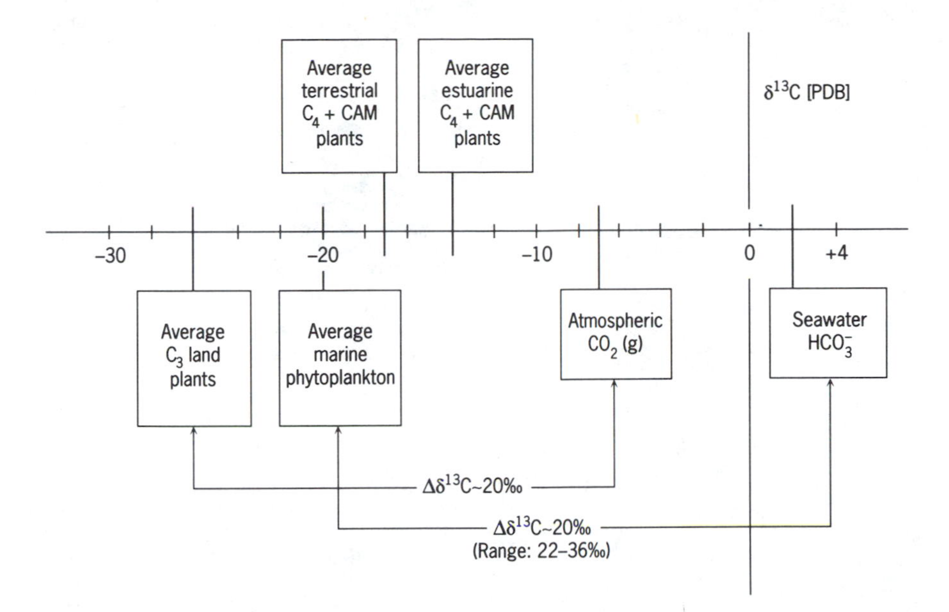

FIGURE 29.17. The influence of carbon source and kinetic fractionation on the average isotopic composition of marine and terrestrial plants. The isotopic composition of HCO_3^- represents that which would be present if isotopic equilibrium is achieved at 15°C.

$\delta^{13}C$ (−26‰) than most marine (−20‰) phytoplankton, which also use C_3 metabolism. Likewise, the average $\delta^{13}C$ of terrestrial C_4 and CAM plants (−17‰) is lower than that of their estuarine analogs (−12‰), which are primarily marsh grasses. It is important to note that variations in environmental conditions and biochemical behavior cause the isotopic composition of these plants to be quite variable. For example, the reported range in the $\delta^{13}C$ of marine plankton is −18 to −30‰.

Animals do not significantly fractionate the carbon isotopes as they pass organic matter up the food chain. As a result, the isotopic composition of an animal's tissues is similar to its source of dietary carbon. Thus the isotopic composition of the animal's tissues can be used to assess the source of their dietary carbon. This approach requires that all possible dietary sources be known. They must also be isotopically distinguishable from each other. If this is the case, the relative contribution of each carbon source can be assessed, as shown in Figures 29.18 and 29.19.

Diagenesis appears to cause a small ^{13}C depletion in sedimentary organic matter. This is thought to be the result of the preferential decomposition of compounds that happen to be enriched in ^{13}C. As shown in Figure 29.20, plant metabolites, such as proteins and carbohydrates, are enriched in ^{13}C relative to cellulose, lipids, and lignin. Since proteins and carbohydrates are more reactive, these compounds should degrade first. Since ^{13}C-enriched carbon is removed, the residual POM becomes depleted in ^{13}C.

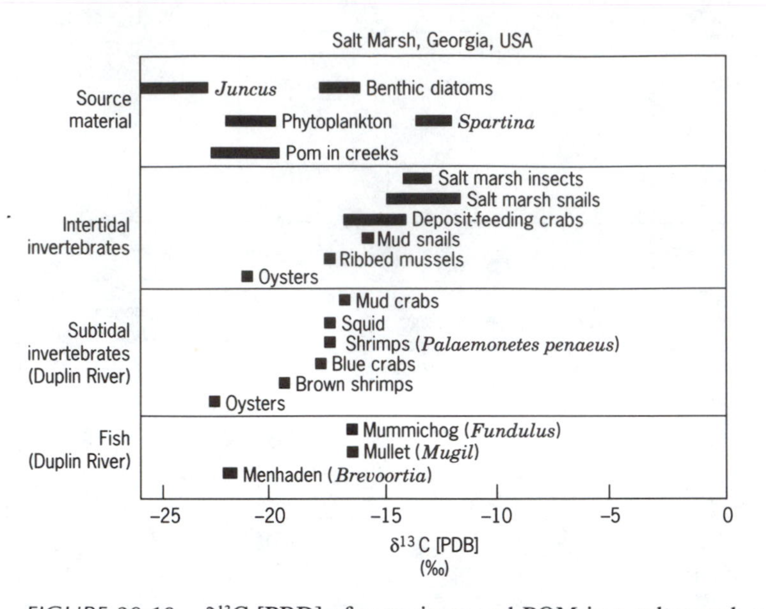

FIGURE 29.18. $\delta^{13}C$ [PBD] of organisms and POM in a salt-marsh estuary (Sapelo Island, GA). *Source*: From *Biogeochemical Processes at the Land-Sea Boundary*, K. H. Mann (eds.: P. Laserre and J. M. Martin), copyright © 1986 by Elsevier Science Publishers, Amsterdam, The Netherlands, p. 130. Reprinted by permission. Data from E. B. Haines and C. L. Montague, reprinted with permission from *Ecology*, vol. 60, p. 52, copyright © 1979 by the Ecological Society of America, Tempe, AZ; and E. B. Haines, reprinted with permission from *Estuarine and Coastal Marine Science*, vol. 4, p. 611, copyright © 1976 by Kluwer Academic Publishers, Dordrecht, The Netherlands.

The phenolic acid moieties (Figure 23.14*b*) are unique degradation products of woody plant detritus and so can be used as an unequivocal indicator of the presence of terrestrial organic matter. As shown in Figure 29.21, the $\delta^{13}C$ of coastal sediments is inversely related to its phenolic acid content. This relationship reflects the relative ^{13}C depletion of terrestrial organic matter as compared to marine POM, the latter being composed primarily of detrital plankton tissues. This relationship also indicates that even after early diagenetic alteration, terrestrial organic matter is still isotopically distinguishable from marine organic matter.

As with the carbonates, down-core variations in the $\delta^{13}C$ of sedimentary organic matter are thought to reflect long-term changes in the oceanic cycling of organic matter. The relationship is more complicated for organic matter because its isotopic composition is strongly influenced by short-term variability in environmental conditions and biological speciation, as well as by diagenesis. Nevertheless, the evolution of life must have caused a large readjustment in the sizes and isotopic composition of the global carbon reservoirs. Isotopic evidence suggesting the presence of life has been observed in rocks as old as 3.5 billion years. Though the isotopic record is somewhat compromised by the effects of metamorphosis, these rocks also contain structures that appear to

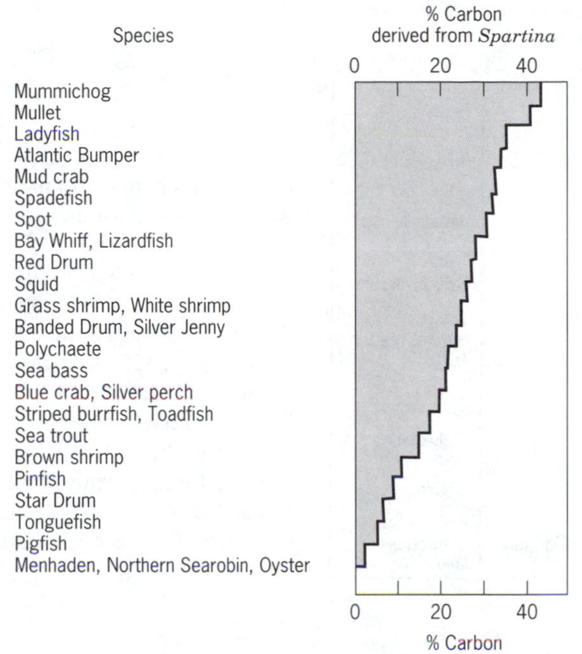

FIGURE 29.19. Percentage of carbon in several invertebrates and fish derived from *Spartina* as estimated from their $\delta^{13}C$. *Source*: From *Biogeochemical Processes at the Land–Sea Boundary*, K. H. Mann (eds.: P. Laserre and J. M. Martin), copyright © 1986 by Elsevier Science Publishers, Amsterdam, The Netherlands. Reprinted by permission. Data from E. H. Hughes and E. B. Scherr, reprinted with permission from the *Journal of Experimental Marine Biology and Ecology*, vol. 67, p. 239, copyright © 1983 by Elsevier Science Publishers, Amsterdam, The Netherlands.

be fossilized remains of stromatolites. Thus life appears to have evolved very early in this planet's history.

The Nitrogen Isotopes

The ranges in $\delta^{15}N$ of some naturally occurring substances are given in Figure 29.22. As with carbon, the marine chemistry of nitrogen is largely controlled by biological processes, many of which are accompanied by kinetic fractionations. The most notable exception to this is nitrogen fixation. As a result, the biomass of nitrogen fixers has a $\delta^{15}N$ value close to that of its metabolic substrate, atmospheric N_2 (0‰). As shown in Figure 29.23, nitrogen-fixing plankton are readily identifiable by their unique isotopic composition (Group III). Their metabolism is favored only under conditions of extreme nitrogen limitation, which accounts for their presence only in waters with low DIN concentrations.

Phytoplankton living in waters with higher DIN concentrations are able to achieve some kinetic fractionation during nutrient assimilation. As a result

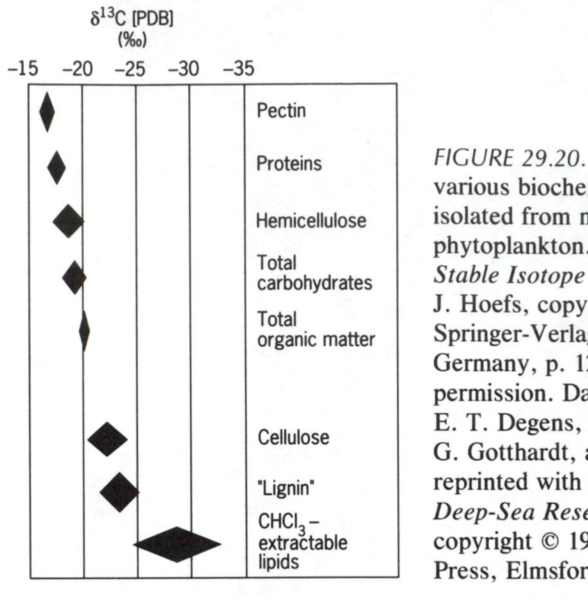

FIGURE 29.20. $\delta^{13}C$ [PDB] in various biochemical constituents isolated from marine phytoplankton. *Source*: From *Stable Isotope Geochemistry*, J. Hoefs, copyright © 1980 by Springer-Verlag, Heidelberg, Germany, p. 129. Reprinted by permission. Data from E. T. Degens, M. Behrendt, G. Gotthardt, and E. Reppmann, reprinted with permission from *Deep-Sea Research*, vol. 15, p. 14, copyright © 1968 by Pergamon Press, Elmsford, NY.

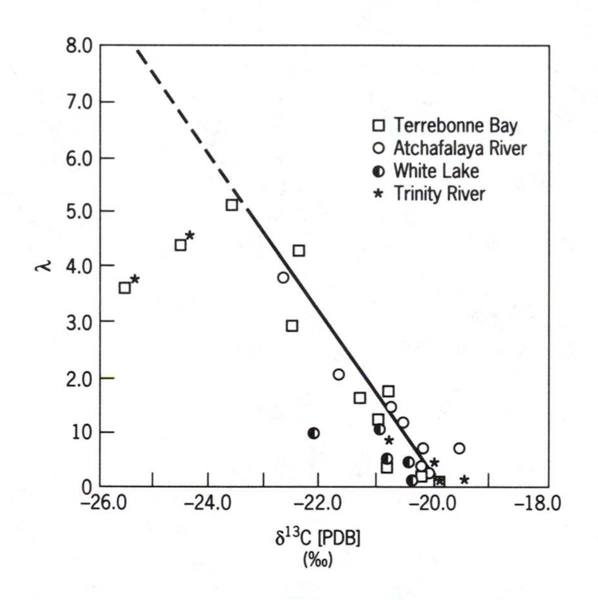

FIGURE 29.21. $\delta^{13}C$ [PDB] versus λ in coastal sediments. λ is the total weight in milligrams of two acidic phenols (vanillyl and syringyl) that are produced from the oxidation of 100 mg of sedimentary organic carbon. Since these phenols (Figure 23.14*b*) are ubiquitous and unique constituents of lignin, they are source tracers for the presence of terrestrial organic matter in marine sediments. *Source*: From J. I. Hedges and P. L. Parker, reprinted with permission from *Geochimica et Cosmochimica Acta*, vol. 40, p. 1026, copyright © 1976 by Pergamon Press, Elmsford, NY.

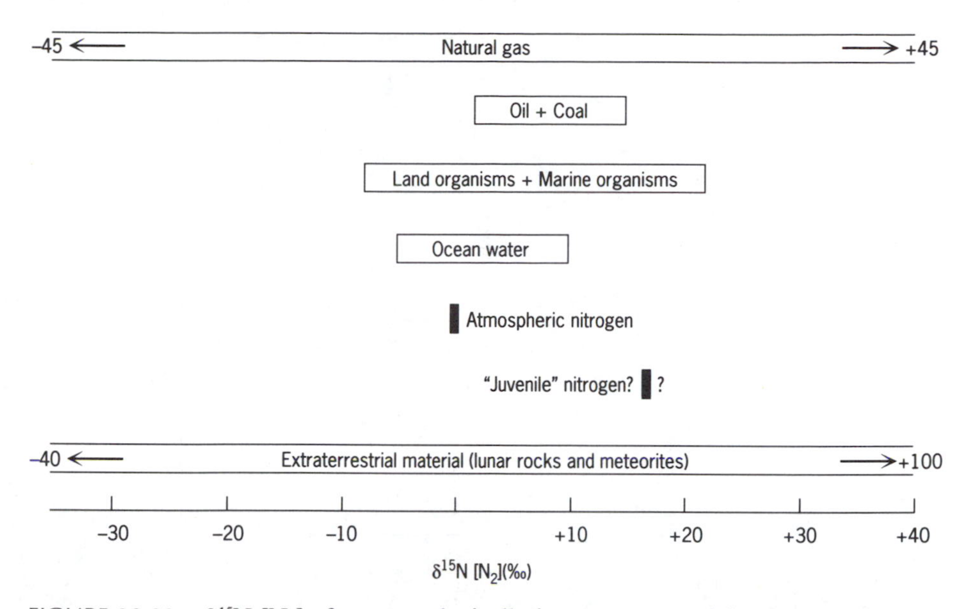

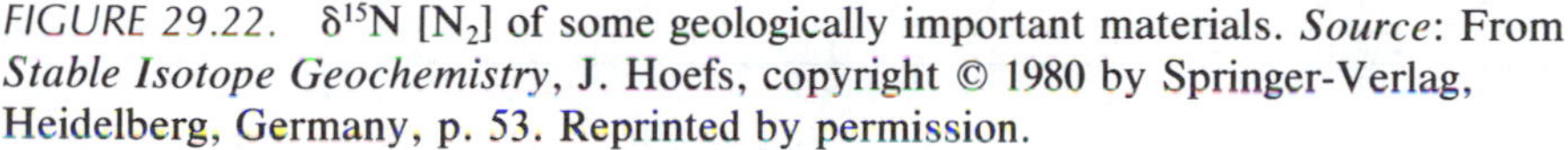

FIGURE 29.22. $\delta^{15}N\ [N_2]$ of some geologically important materials. *Source*: From *Stable Isotope Geochemistry*, J. Hoefs, copyright © 1980 by Springer-Verlag, Heidelberg, Germany, p. 53. Reprinted by permission.

their $\delta^{15}N$ is somewhat lower than that of the ambient DIN. By preferentially assimilating ^{14}N, phytoplankton uptake causes the remaining DIN pool to become increasingly ^{15}N enriched. Thus the $\delta^{15}N$ of DIN tends to increase with decreasing DIN concentration. Phytoplankton living in nutrient-deficient waters are less able to fractionate nitrogen because they extract nearly all of the DIN. Thus these phytoplankton tend to have the largest ^{15}N enrichments. Since nitrogen-limited growth is common to most of the open ocean, the average $\delta^{15}N$ of phytoplankton ranges from +5 to +7‰.

As shown in Figure 29.24, the $\delta^{15}N$ of marine organisms tends to increase with increasing trophic level. Apparently, animals preferentially excrete ^{15}N-depleted nitrogen, causing a biomagnification of ^{15}N in the food chain. A fairly constant enrichment of 1‰ to 2‰ has been observed to occur at each step. Thus, the $\delta^{15}N$ of an organism can be used to infer its trophic level.

This aspect of $\delta^{15}N$ has been used to determine the source of nitrogen that supports primary production at hydrothermal vents. As shown in Figure 29.25, the isotopic composition of the vent fauna suggests that the ultimate source of nitrogen to this ecosystem must have a $\delta^{15}N$ close to 0‰. Since the average $\delta^{15}N$ of marine DIN and PON ranges from +7 to +10 and +5 to +7‰, respectively, they are not likely sources. This suggests that in situ nitrogen fixation is the source of the ^{15}N-depleted nitrogen that supports the biological activity at hydrothermal vents.

Nitrogen fixation is also the ultimate source of nitrogen for terrestrial plants, whereas most marine phytoplankton assimilate DIN. Thus, terrestrial plant biomass should be isotopically distinguishable from that of marine plants.

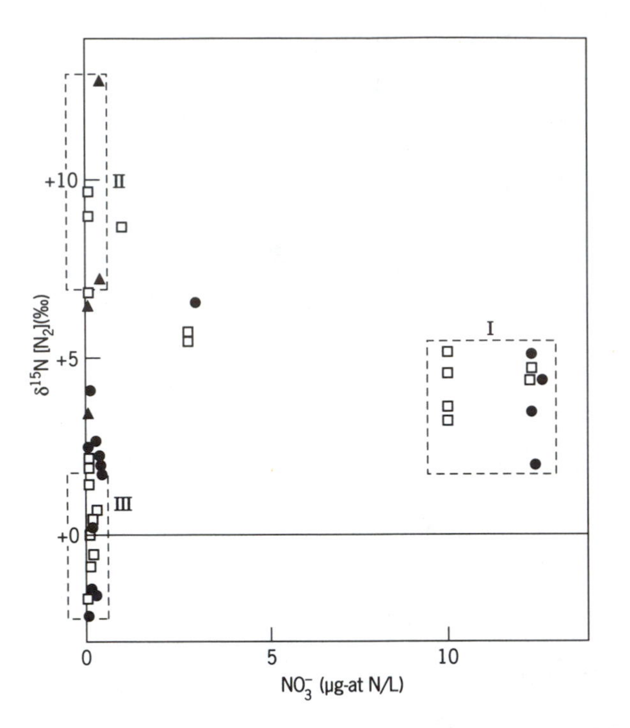

FIGURE 29.23. $\delta^{15}N$ [N_2] of phytoplankton versus DIN concentrations in the North Pacific. The phytoplankton in Group I are relatively ^{15}N-depleted due to kinetic fractionation during DIN assimilation. The phytoplankton in Group II are N_2 fixers and thus have $\delta^{15}N$ close to 0‰. The phytoplankton in Group III are relatively ^{15}N-enriched due to nitrogen limitation which prevents fractionation. *Source*: From *Isotope Marine Chemistry*, E. Wada (eds.: E. Goldberg, Y. Horibe, and K. Saruhashi), copyright © 1980 by Uchida Rokakuho Publishing Company, Ltd., Tokyo, Japan. Data from (·) E. Wada and A. Hattori, reprinted with permission from *Geochimica et Cosmochimica Acta*, vol. 40, p. 250, copyright © 1976 by Pergamon Press, Elmsford, NY; (▲) Y. Miyake and E. Wada, *Records of the Oceanographic Works in Japan*, vol. 9, p. 43, copyright © 1976 by Science Council of Japan, Tokyo, Japan; and (■) E. Wada (1980).

This is analogous to the situation with $\delta^{13}C$, where the source of carbon for terrestrial plants (CO_2) is depleted in ^{13}C relative to the DIC (HCO_3^-) that is assimilated by marine phytoplankton. Due to the large (approximately 20‰) fractionation that accompanies the photosynthetic fixation of carbon, the $\delta^{13}C$ of organic matter is substantially different from that of the substrate carbon. In comparison, little fractionation occurs during the assimilation of nitrogen.

The admixture of varying amounts of marine and terrestrial organic matter should produce a sediment whose $\delta^{15}N$ and $\delta^{13}C$ are linearly related as depicted by the line on the right-hand side of Figure 29.26. Since the isotopic composition of coastal sediments lies close to this theoretical mixing line, these deposits appear to be composed of variable amounts of marine and terrestrial organic

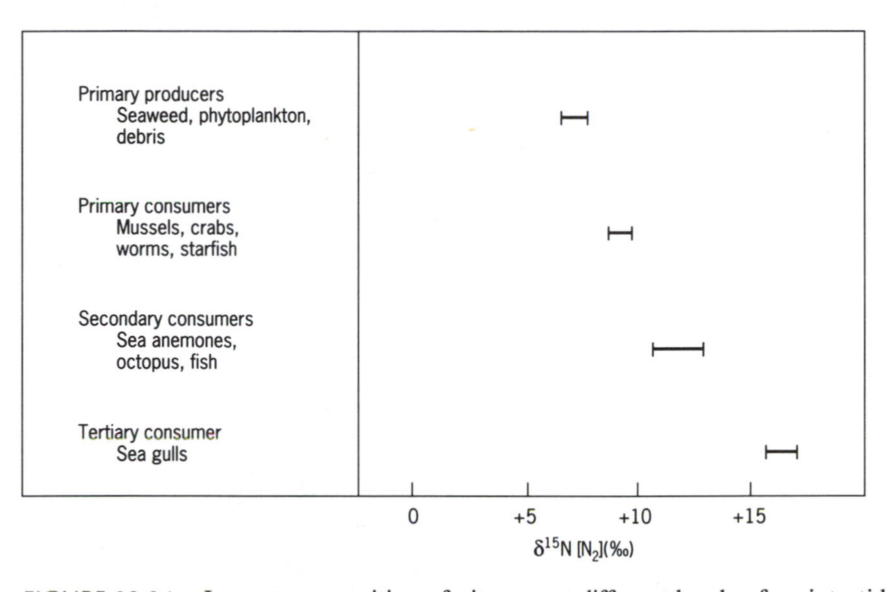

FIGURE 29.24. Isotope composition of nitrogen at different levels of an intertidal food web (Usujiri, Japan). *Source*: From *Principles of Isotope Geology*, 2nd ed., G. Faure, copyright © 1986 by John Wiley & Sons, Inc., New York, p. 518. Reprinted by permission. Data from M. Minagawa and E. Wada, reprinted with permission from *Geochimica et Cosmochimica Acta*, vol. 48, p. 1137, copyright © 1984 by Pergamon Press, Elmsford, NY.

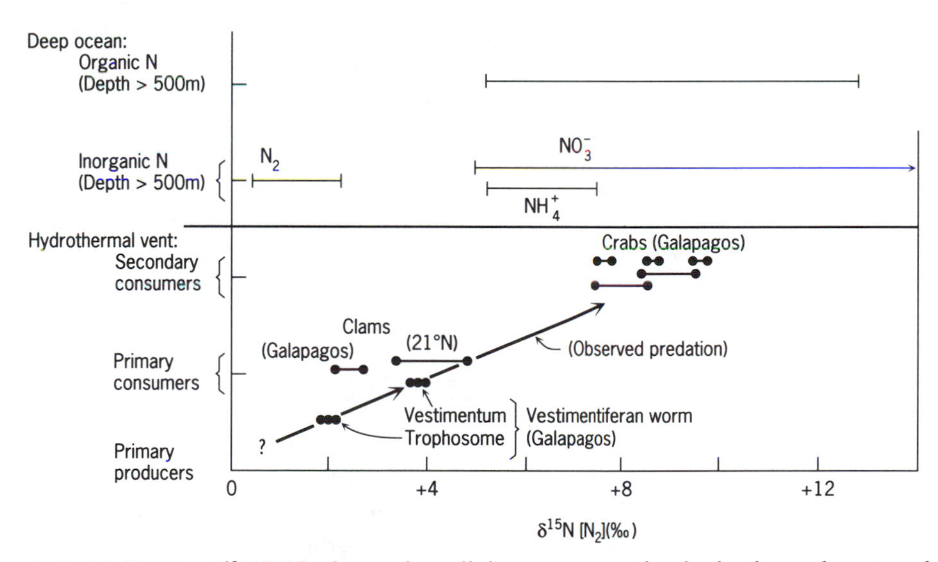

FIGURE 29.25. $\delta^{15}N$ $[N_2]$ of organisms living near an active hydrothermal vent and their potential nitrogen sources. *Source*: From G. H. Rau, reprinted with permission from *Nature*, vol. 289, p. 484, copyright © 1981 by Macmillan Journals, Ltd., London, England. See Rau (1981) for data sources.

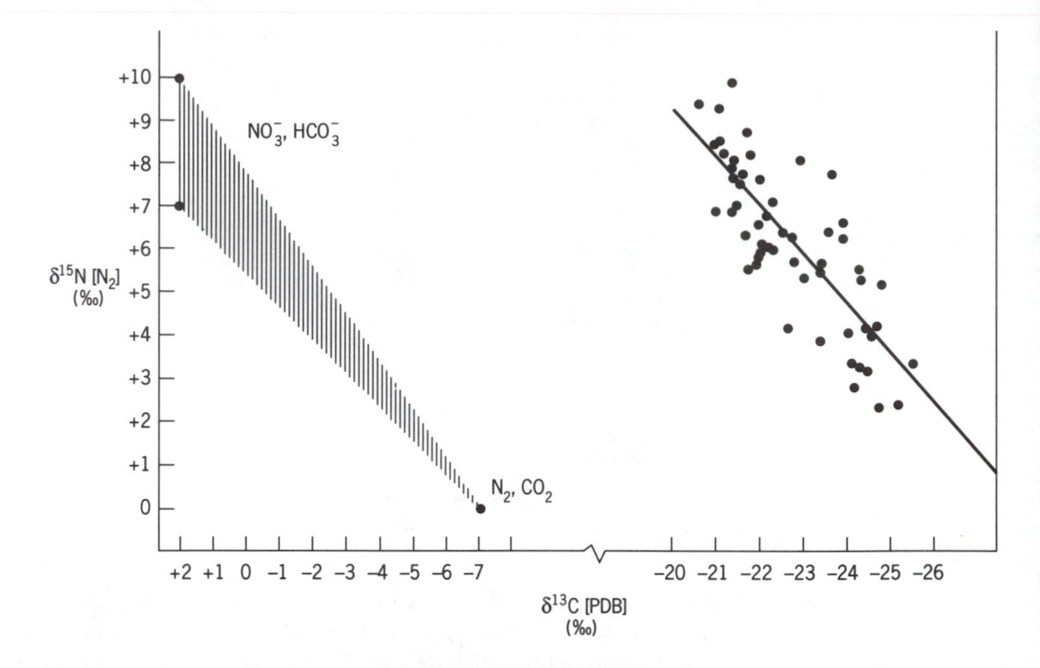

FIGURE 29.26. $\delta^{15}N$ [N_2] versus $\delta^{13}C$ [PDB] in coastal sediments. *Source*: From K. E. Peters, R. E. Sweeney, and I. R. Kaplan, reprinted with permission from *Limnology and Oceanography*, vol. 23, p. 602, copyright © 1978 by the American Society of Limnology and Oceanography, Seattle, WA.

matter. The unique isotopic signatures of the end members can be used to compute their relative contributions to the sediments. Sediments that do not follow this trend contain POM contributed by other sources, such as from sewage and nitrogen fixation.

Denitrification is accompanied by a fractionation that produces ^{15}N-depleted N_2. As shown in Figure 29.27, this causes the remaining nitrate to become enriched in ^{15}N. The degree of this enrichment is dependent on the fractionation factor and extent of nitrate depletion. Thus the largest enrichments are found in upwelling areas and regions of relative stagnancy. While residing in these regions, water masses acquire a very distinct isotopic signature that can be used to trace their pathway for fairly long distances through the ocean basins.

Denitrification is the only process that causes a preferential loss of ^{15}N-depleted nitrogen from the ocean. Thus it must be the underlying cause for the general ^{15}N enrichment of all marine nitrogen relative to atmospheric N_2. A more detailed understanding of the relative distributions of ^{15}N and ^{14}N in the ocean should provide unique information on the contribution of marine denitrification and nitrogen fixation to the global nitrogen budget.

The Sulfur Isotopes

The ranges in $\delta^{34}S$ of some naturally occurring substances are given in Figure 29.28. The $\delta^{34}S$ of sulfur is reported relative to a meteoritic standard that is

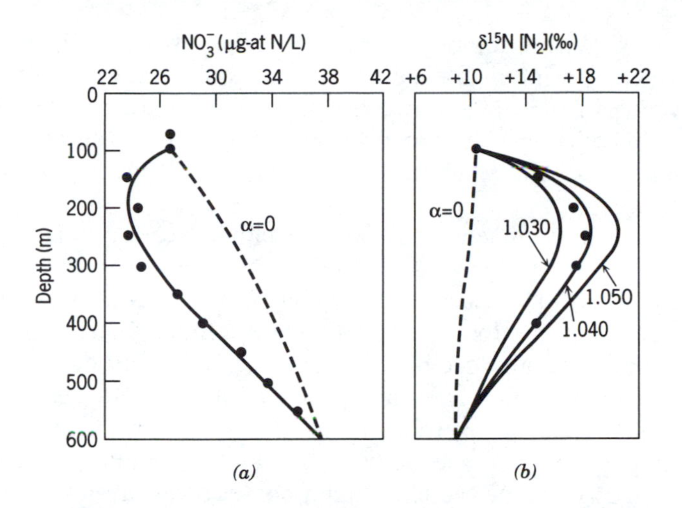

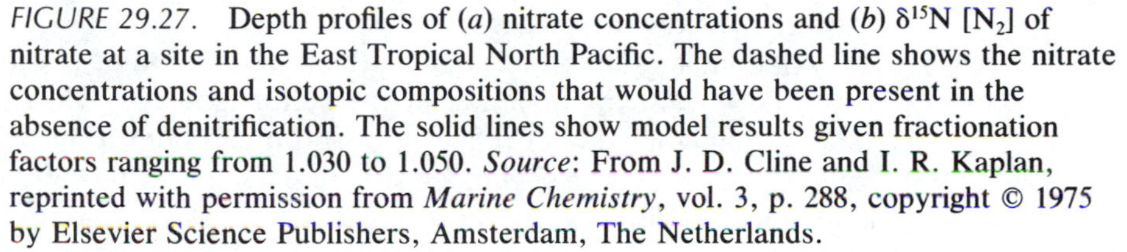

FIGURE 29.27. Depth profiles of (*a*) nitrate concentrations and (*b*) δ^{15}N [N$_2$] of nitrate at a site in the East Tropical North Pacific. The dashed line shows the nitrate concentrations and isotopic compositions that would have been present in the absence of denitrification. The solid lines show model results given fractionation factors ranging from 1.030 to 1.050. *Source*: From J. D. Cline and I. R. Kaplan, reprinted with permission from *Marine Chemistry*, vol. 3, p. 288, copyright © 1975 by Elsevier Science Publishers, Amsterdam, The Netherlands.

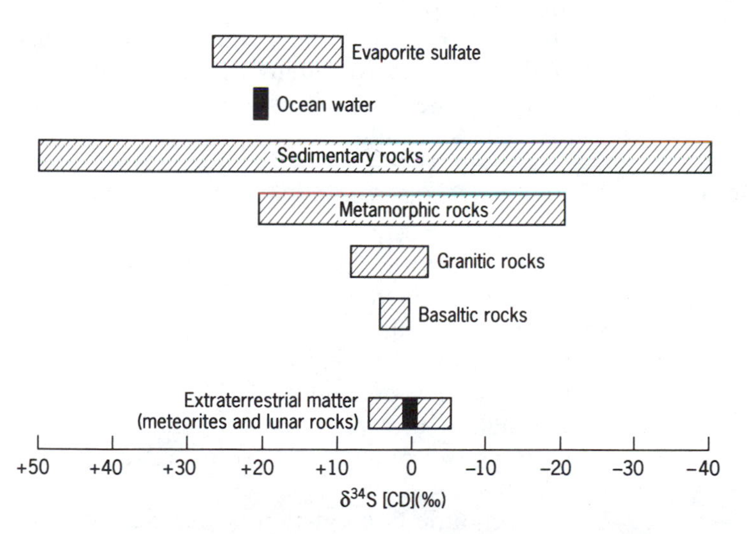

FIGURE 29.28. δ^{34}S [CD] of some geologically important materials. *Source*: From *Stable Isotope Geochemistry*, J. Hoefs, copyright © 1980 by Springer-Verlag, Heidelberg, Germany, p. 44. Reprinted by permission.

thought to have an isotopic composition similar to that of juvenile sulfur. Since modern-day marine sulfate has a $\delta^{34}S$ of +20‰, some process has preferentially removed ^{32}S from the ocean. This process is thought to be sulfate reduction as it is accompanied by kinetic fractionations that can be as large as 60‰. Precipitation of the resulting sulfide produces ^{34}S-depleted minerals, such as pyrite.

The oldest sedimentary rocks (3.7 billion years old) have a $\delta^{34}S$ approximately equal to 0‰, suggesting that sulfate reducers were not present on the very early earth. This seems reasonable as a substantial amount of time would have been required for seawater to acquire enough sulfate to support sulfate reduction. In the absence of O_2, sulfate was likely produced via the oxidation of abiogenic sulfide by green and purple photosynthetic sulfur bacteria. (These microorganisms are anaerobes.) The oldest sulfides that exhibit ^{34}S depletion were deposited sometime between 2.8 and 3.2 billion years before present. This suggests sulfate-reducing microbes evolved a short time prior to this during the Precambrian.

In comparison to the sulfides, the precipitation of gypsum is accompanied by little (<2‰) fractionation of the sulfur isotopes. Thus the $\delta^{34}S$ of marine sulfate minerals should reflect the isotopic composition of ambient seawater sulfate. As shown in Figure 29.29, the $\delta^{34}S$ of marine sulfates has varied over time but is difficult to interpret as a record of variations in the isotopic composition of sulfate for several reasons: First, most evaporites are deposited in isolated basins and thus are likely subject to Rayleigh Distillation. In addition, river input is often a source of ^{34}S-depleted sulfate. Nevertheless, the geologic variations in the $\delta^{34}S$ of marine sulfate minerals suggest that significant changes in oceanic sulfur fluxes have occurred.

Though the isotopic composition of oxygen in marine sulfate minerals has varied, the correlation with $\delta^{34}S$ is poor. As with sulfur, these variations are thought to reflect changes in oceanic oxygen fluxes. The use of marine sulfates as a paleoceanographic record requires a much better understanding of the complex set of interconnected processes that regulate the isotopic composition and fluxes of oxygen and sulfur.

The $\delta^{34}S$ of plants is largely determined by the isotopic composition of their dietary sulfur. In other words, little fractionation occurs during assimilation. Thus the $\delta^{34}S$ of terrestrial plants is close to that of rainwater sulfate (+2 to +8‰), while the $\delta^{34}S$ of phytoplankton is close to that of marine sulfate (+21‰). In estuaries, marsh grasses, such as *Spartina*, obtain most of their sulfur from sedimentary sulfides. Since these sulfides are produced by sulfate reduction, they tend to be greatly depleted in ^{34}S relative to marine sulfate. Uptake of this sulfide causes the $\delta^{34}S$ of *Spartina* to be intermediate in value between that of marine phytoplankton and upland plants. As shown in Figure 29.30*a*, some overlap in isotopic composition exists, mostly due to the variable ^{34}S depletion in sulfide.

As with carbon, little fractionation occurs during the transfer of sulfur up the food chain. Thus the isotopic composition of the animal is similar to the $\delta^{34}S$ of its diet. Because the isotopic signatures of marine, estuarine, and

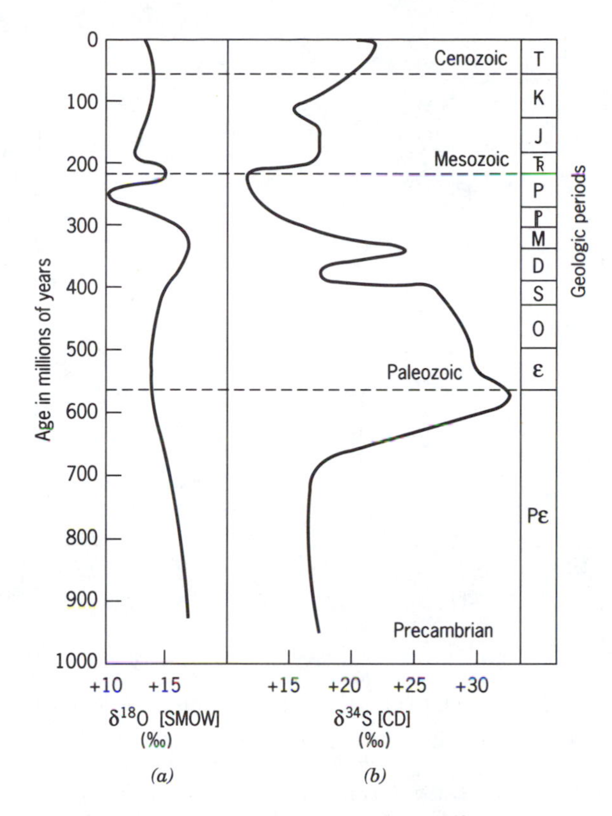

FIGURE 29.29. Variations in the (*a*) $\delta^{18}O$ [SMOW] and (*b*) $\delta^{34}S$ [CD] of marine sulfate minerals from the late Precambrian to the present. *Source*: From *Principles of Isotope Geology*, 2nd ed., G. Faure, copyright © 1986 by John Wiley & Sons, Inc., New York, p. 532. Reprinted by permission. Data from G. E. Claypool, W. T. Holser, I. R. Kaplan, H. Sakai, and I. Zak, reprinted with permission from *Chemical Geology*, vol. 28, p. 219, copyright © 1980 by Elsevier Science Publishers, Amsterdam, The Netherlands.

terrestrial plants are distinct, they can be used to assess the relative contributions of their organic matter to the biomass of estuarine animals. As shown in Figure 29.30*a*, *Spartina* and plankton appear to be equally important sources of organic matter to the estuarine animals. Their $\delta^{15}N$ also suggests that upland plants are not a significant source of dietary organic matter. The relative contributions of planktonic and *Spartina* nitrogen cannot be ascertained due to their similar isotopic composition.

SUMMARY

Most studies of the naturally occurring stable isotopes have focused on the elements carbon, hydrogen, oxygen, sulfur, and nitrogen. For these elements,

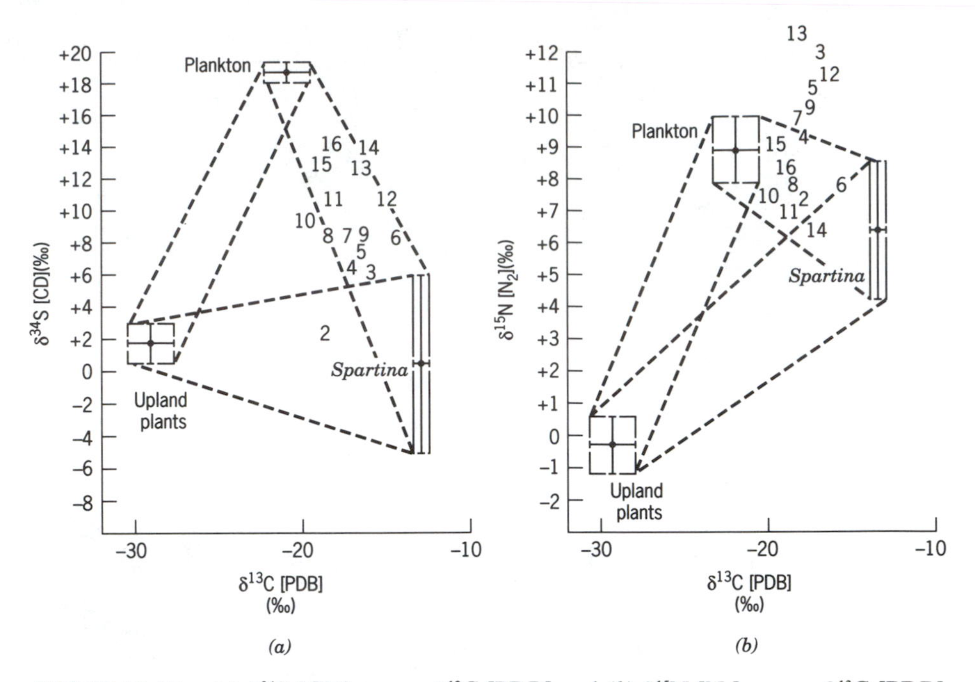

FIGURE 29.30. (*a*) $\delta^{34}S$ [CD] versus $\delta^{13}C$ [PDB] and (*b*) $\delta^{15}N$ [N_2] versus $\delta^{13}C$ [PDB] in various organic components of a salt-marsh ecosystem (Sapelo Island, GA). Key to consumers: 1) *Orchelimum fidicinium*; 2) *Mugil cephalus*; 3) *Fundulus heteroclitus*; 4) *Palaermonetes pugio*; 5) *Callinectes sapidus*; 6) *Littorina irrorata*; 7) *Penaeus setiferus*; 8) *Ilyanassa obsoleta*; 9) *Cairdiella chrysura*; 10) *Geukensia demissa*; 11) *M. cephalus*; 12) *M. cephalus*; 13) *Ocypode quadrata*; 14) *Uca pugnax*; 15) *Crassostria virginica*; 16) *C. virginica*. These organisms were collected from a variety of locations in the salt marsh. The dashed lines define the isotope signatures that would be produced by admixture of the two end members (e.g., the $\delta^{13}C$ and $\delta^{34}S$ in the tissues of an organism that consumes equal amounts of plankton and upland plants should range between −23 to −27 and +6 to 14‰, respectively). *Source*: From B. J. Peterson and R. W. Howarth, reprinted with permission from *Limnology and Oceanography*, vol. 32, pp. 1208–1209, copyright © 1987 by the American Society of Limnology and Oceanography, Seattle, WA.

the lightest isotope is most abundant. The relative abundances of the isotopes of these elements are usually reported as a **del value** relative to an internationally accepted standard.

Variations in the relative abundance of the naturally occurring stable isotopes are caused by the preferential reaction or transport of one of the isotopes. This isotopic segregation, or **fractionation**, can occur during a variety of physicochemical processes, such as chemical reactions, phase changes, and molecular diffusion. Such fractionations are caused by the slight physical and chemical differences which exist among the isotopes of an element. **Kinetic fractionations** arise from differences in the rates at which the isotopes undergo reaction. This type of fractionation is common to biologically mediated processes because on the macroscopic scale, these reactions tend to be unidirectional. Because the light isotope reacts faster, it becomes enriched in the products. **Thermodynamic fractionations** arise from isotope exchange reactions which attain equilibrium. Most fractionation factors are determined experimentally.

Fractionation factors depend on the reaction mechanism and tend to decrease with increasing temperature. The isotopic composition of a substance is thus determined by the type of reaction that it has undergone, as well as the temperature and extent of reaction. The selective reaction of one isotope causes the remaining reactant pool to become progressively depleted in the more reactive isotope. This also causes the product to become increasingly depleted in the more reactive isotope. This effect is referred to as a **Rayleigh Distillation**. Because of this effect, rainwater becomes progressively depleted in ^{18}O as clouds move poleward. This causes the water in polar ice caps to be depleted in ^{18}O relative to seawater. Thus seawater becomes enriched in ^{18}O during ice ages. This shift in the isotopic composition is recorded in biogenic carbonates because marine organisms tend to deposit their tests in isotopic equilibrium with seawater. Thus down-core variations in the $\delta^{18}O$ of biogenic carbonates provide a record of changes in ice volume and hence sea level.

The fractionation factor associated with this isotopic equilibrium is temperature dependent. Thus some of the down-core variability in $\delta^{18}O$ is the result of temperature changes that undoubtedly occurred as ice ages have come and gone. By comparing the isotopic composition of benthic and planktonic species, the ice volume effect can be separated from that caused by temperature changes. Thus the isotope record can, in some cases, be used to reconstruct paleotemperatures. Since changes in the rates of POM remineralization are likely to have affected the $\delta^{13}C$ of DIC, changes in the carbon isotope composition of the carbonate are thought to reflect changes in deep-water nutrient concentrations.

The oxygen isotope composition of crystalline rocks has been used to determine their solidification temperatures, while that of the clay minerals has been used to determine whether reverse weathering occurs in marine sediments. The naturally occurring stable isotopes of carbon, sulfur, and nitrogen isotopes have also been used to establish the structures of marine food webs and to trace the flow of these elements through ecosystems. Their isotopic composition in sedimentary organic matter has varied through time. The causes of these variations are not well understood because POM is not deposited in isotopic equilibrium. Instead, its isotopic composition is strongly affected by short-term local fluctuations in environmental conditions and biological speciation, as well as by diagenesis.

PROBLEM SET 5

1. Using a piece of graph paper, plot the unsupported ^{14}C activity against depth in the sediment for the data listed in Table P.5.1. Construct a second graph in which the log of ^{14}C activity is plotted against depth.

TABLE P.5.1
Unsupported ^{14}C Activity in a Hypothetical Deep-Sea Core

Depth Below the Sediment–Water Interface (cm)	*Unsupported ^{14}C Activity*
0	1.00
5	0.60
10	0.36
15	0.22
20	0.13
30	0.05
40	0.02

2. Explain the shapes of these curves.
3. If the sediment at the surface was formed today, how many years ago was the sediment at 10 cm formed? And at 20 cm?
4. What is the average rate of accumulation (in cm/1000 y) of sediment between 0 and 10 cm? And between 10 and 20 cm?
5. During preparation for radiocarbon analysis, a sediment sample that is 1×10^6 y old becomes contaminated by the introduction of some atmospheric CO_2 into its combustion ampule. If this contaminant CO_2 represents 10% of the total carbon now in the ampule, how old would the sample appear to be? Assume that atmospheric CO_2 has the same ^{14}C activity as the DIC in the surface ocean.
6. Explain why ^{14}C dates are considered to be minimum ages on very old samples.

SUGGESTED FURTHER READINGS

Radioisotope Geochemistry

BROECKER, W., and T.-H. PENG. 1982. *Tracers in the Sea*. Lamont-Doherty Geological Observatory, Palisades, NY, 690 pp.

DURRANCE, E. M. 1986. *Radioactivity in Geology: Principles and Applications*. John Wiley & Sons, Inc., New York, 441 pp.

TUREKIAN, K. K., and J. K. COCHRAN. 1978. Determining Marine Chronologies Using Natural Radionuclides. In: *Chemical Oceanography*, vol. 7 (ed: J. P. Riley). Academic Press, Orlando, FL, pp. 313–361.

Stable Isotope Geochemistry

BROECKER, W., and T.-H. PENG, 1982. *Tracers in the Sea*. Lamont-Doherty Geological Observatory, Palisades, NY, 690 pp.

FAURE, G. 1986. *Principles of Isotope Geology*. John Wiley & Sons, Inc., New York, 589 pp.

HOEFS, J. 1980. *Stable Isotope Geochemistry*. 2nd ed. Springer-Verlag, Heidelburg, Germany, 208 pp.

PART SIX

Don't dirty
the water
around you;
you may have
to drink it
someday.

A Mexican
Proverb

MARINE POLLUTION

CHAPTER 30

THE FATE OF POLLUTANTS IN THE COASTAL OCEAN

INTRODUCTION

The summers of 1987 and 1988 were bad ones for the coastal ocean. These were the summers when medical wastes washed up on the shores of recreational beaches in the northeastern United States. These were also the summers when hundreds of dolphin were fatally stranded on the eastern shoreline, when red tide decimated the fisheries in North Carolina, and when a garbage barge traveled up and down the East and Gulf coasts unable to find a landfill that would accept its load. Nearly half of the shellfish beds in the United States were closed due to the presence of pollutants. Across the ocean, 20,000 seals died in the North Sea as a result of a virus whose virulence was thought to have been intensified by the debilitating effects of pollution. Many interpreted these events as evidence that the seas were ''dying'' as a result of pollution.

Humans have always used the ocean as a waste disposal site. In addition, we have always and will continue to use it as a food and mineral resource, a transportation route, and for recreation. Since the ocean is vast, the effects of these conflicting uses were not evident until fairly recently. But the cumulative and increasing nature of our impacts have resulted in a significant alteration of some fluxes in the crustal-ocean factory. This has undoubtedly caused some of the catastrophes mentioned above and will inevitably lead to more.

At present, nearly 75 percent of the United States' population lives within 50 miles of a coastline. This percentage is increasing, as is our population. As a result, the anthropogenic impacts on the coastal ocean will continue to increase. This is doubly unfortunate, as these waters contain some of the most productive ecosystems on the planet, the salt-marsh estuaries. These ecosystems are particularly sensitive to pollution for several reasons: The mixing of salt and fresh water causes chemical changes that tend to precipitate pollutants and hence trap them in the sediments. These sediments have a large impact on the chemistry of estuarine waters due to the shallow depths. The pollutants are also actively cycled between sediment and water as a result of biological activity. Since estuaries are sites of great productivity, pollutants are readily transferred throughout food webs. Even though these processes tend to keep pollutants in estuaries, their impacts are transmitted to the open ocean by the many species of pelagic fish and shellfish that use these areas as breeding grounds.

Unfortunately, estuaries are the sites where most pollutants are introduced into the ocean. Thus the fate of pollutants in the marine environment is largely determined by the biogeochemical processes that occur in estuaries. In recognition of this, the subject of marine pollution is discussed herein from the perspective of the unique chemical processes that occur in estuaries. As described below, some of these chemical changes enhance the toxicity of the pollutants. Other chemical changes cause the degradation or immobilization of pollutants and, as a result, act to ''purify'' the waters.

Pollutants are introduced into the open ocean as a result of gas exchange and rainout at the air–sea interface, advection of continental runoff, an from offshore dumping. As a result, evidence of pollution can be found throughout the ocean but is greatest and most obvious in coastal waters. The first efforts at pollution control focused on banning the dumping of garbage and sewage sludge at sea. To this end, all ocean dumping by the United States will cease in 1991. Although important, ocean dumping has never been as large a threat to the ocean as chronic inputs from sewage outfalls and runoff. These sources have proven far more difficult to regulate and thus are currently being addressed via the adoption of water-quality standards. Because of the increasing shortage of landfill sites, the ban on ocean dumping is being reconsidered. Some proposals to emplace waste at ''low-risk'' sites are discussed at the end of this chapter, along with a short legal history of marine pollution control in the United States.

WHY HAS THE OCEAN BEEN USED FOR WASTE DISPOSAL?

Knowing what we know now, it is hard to understand why the ocean was ever, and still is, used as a waste-disposal site. The rationale has been the following—that the ocean is vast enough to accommodate our waste without undergoing an unacceptable amount of change. This view has been based on the assumption that any potentially toxic wastes would be diluted to innocuous levels and carried by currents far from our coastlines. Although this may have been true in the past, the assimilative capacity of the coastal ocean appears to have been exceeded. This is probably a result of the cumulative effects of our past activities, as well as of our ever-increasing rates of pollutant input.

WHAT IS MARINE POLLUTION?

Progress in controlling marine pollution has been impeded by the difficulty of defining it. Two international advisory groups (the United Nations Group of

Experts on the Scientific Aspects of Marine Pollution and the International Commission for Exploration of the Sea) have recommended the following:

> *Pollution is the introduction by man, directly or indirectly, of substances or energy to the marine environment resulting in such deleterious effects as harm to living resources; hazards to human health; hindrance of marine activities including fishing; impairment of the quality for use of seawater and reduction of amenities.*

Thus a **pollutant** can be a naturally occurring substance if its concentration is "above the natural background level for the area and for the organism." These pollutants are often referred to as **contaminants**. For example, microorganisms are considered as contaminants because their natural abundance in the ocean has been elevated by the introduction of sewage into seawater. Other pollutants, such as polychlorinated biphenyls (PCBs), most pesticides, and plastic, were never in the ocean until humans put them there.

WHY POLLUTION IS HARD TO MEASURE

Another major factor impeding progress in the control of marine pollution is the great difficulty associated with pollution measurement. Little data exist on the chemical composition of the unpolluted ocean due to the relatively recent development of the necessary analytical techniques. Thus, the "natural" levels of most contaminants are unknown. This prevents a determination of what levels constitute an elevated concentration.

Assessing the impact of a pollutant is also difficult as its effects can be complex and take a long time to develop. This is the result of the multitude of pathways that a pollutant can take in the marine environment, as illustrated in Figure 30.1.

Marine organisms tend to concentrate pollutants in their tissues by a process termed **bioaccumulation**. This pollutant enrichment is caused by either the passive adsorption of pollutants from seawater or active uptake followed by retention in tissues or hard parts as a result of nonexcretion. The degree of enrichment is variable and depends on factors such as (1) the chemical nature of the pollutant, (2) the type of organism, (3) its physiological state, (4) water temperature, and (5) salinity. Enrichment factors are greatest for metals, some of which are as high as 10^9. Though some pollutants are eventually excreted or degraded, the rates of these processes tend to be slow. Thus the consumption of tainted tissues causes pollutants to be passed up the food chain. If bioaccumulation occurs during each transfer to higher trophic levels, the concentration of pollutant increases. This process is termed **biomagnification**. Thus organisms at the top of the food chain tend to have the highest pollutant concentrations.

Because biomagnification and other transport processes take time, the harmful effect of many compounds may not become evident for decades. This

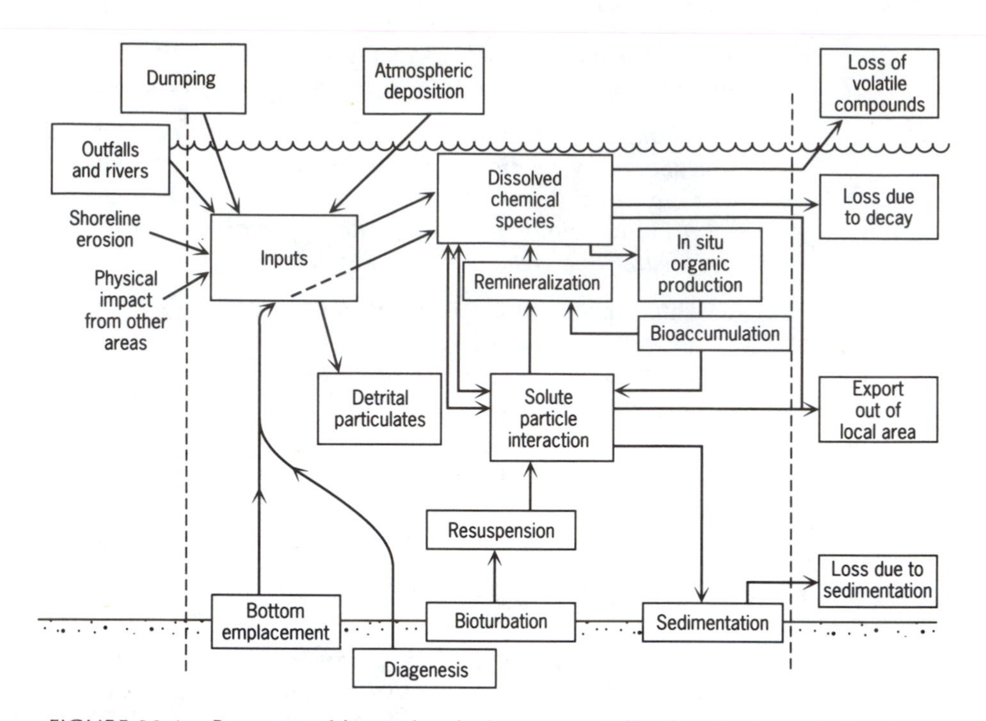

FIGURE 30.1. Important biogeochemical processes affecting the fate of chemical pollutants in marine ecosystems. *Source:* From *The Role of the Oceans as a Waste Disposal Option,* J. W. Farrington and J. Westall (ed.: G. Kullenberg), copyright © 1986 by Reidel Publishing Company, Dordrecht, The Netherlands, p. 363. Reprinted by permission.

makes direct causal relationships between specific pollutants and environmental change difficult to establish. Substantiating such relationships is further complicated by the complex network of positive and negative feedbacks that occur among most parts of the crustal-ocean factory.

The effects of pollution are also difficult to quantify because an essentially subjective judgment must be made as to what constitutes a "significant change." In most cases, the impact of a pollutant is determined by a laboratory bioassay in which its LD_{50} (LD for lethal dosage) is measured. This is the concentration that causes a 50 percent mortality in the test organisms during a designated exposure period. LD_{50}'s are used by regulatory agencies to establish the maximum concentrations of pollutants allowable in discharged effluents. They are also useful in determining the sensitivity of an organism to a particular pollutant, in comparing the relative toxicities of pollutants, and in developing new **water-quality criteria**. However, LD_{50}'s do not provide information on the effects of sublethal exposure. These effects can include significant physiological, behavioral, and ecological changes, as well as increased susceptibility to environmental stresses, such as disease. In other words, pollutants have impacts that extend over several levels of biological organization, as listed in Table 30.1.

TABLE 30.1
Summary of Response to Pollutant Stress

Level	*Adaptive Response*	*Destructive Response*	*Result at Next Level*
Biochemical–cellular	Detoxification		Adaptation of organism
		Membrane disruption Energy imbalance	Reduction in condition of organism
Organismal	Disease defense Adjustment in rate functions Avoidance		Regulation and adaptation of populations
		Metabolic changes Behavior aberrations Increased incidence of disease Reduction in growth and reproduction rates	Reduction in performance of populations
Population	Adaptation of organism to stress No change in population dynamics		No change at community level
		Changes in population dynamics	Effects on coexisting organisms and communities
Community	Adaptation of populations to stress		No change in community diversity or stability Ecosystem adaptation
		Changes in species composition and diversity	Deterioration of community
		Reduction in energy flow	Change in ecosystem structure and function

Source: From J. M. Capuzzo, reprinted with permission from Oceanus, vol. 24, p. 320, copyright © 1981 by Oceanus, Woods Hole, MA.

Environmental stress has a negative impact on growth and reproductive rates. Monitoring these characteristics in both individuals and populations provides an indication of the sublethal impacts of a pollutant. This evaluation, called the scope for growth, is particularly useful as it can be done in situ. But it does require knowledge of seasonal variations and behavior under unpolluted conditions. Other monitoring methods under development are ones based on the detection of biomolecules, such as enzymes, that are specific by-products of pollutant metabolism.

Marine pollution is also difficult to measure because humans introduce materials into the ocean through a great variety of mechanisms. The major pathways are listed in Table 30.2.

Pollutants emitted from a discrete source, such as a sewage outfall, are termed **point-source** pollutants. They are relatively easy to control as compared to **non-point-source** pollution, as they have a single readily identifiable emission source. Most non-point-source pollution is the result of stormwater runoff and ground-water seeps.

Ships are another important transport pathway. Though all dumping of garbage and sludge will be legally banned in U.S. coastal waters as of 1991, the introduction of dredge spoils will continue to be permitted. Most of this material is sediment displaced by dredging done to widen and deepen harbors. Approximately two-thirds of the dredge spoils are deposited in estuaries, with the rest split between the coastal and open ocean. The amount of sediment being redistributed is so large that it rivals the natural riverine flux of particles. Since pollutants tend to be trapped in estuarine sediments, dredging can remobilize them and create toxic conditions in the water column both at the dredge and spoil sites.

Finally, pollution is also difficult to measure due to the myriad number of compounds that humans introduce into the ocean. Though approximately 70,000 compounds are in daily use, thousands of new molecules are being synthesized every year. Special analytical strategies are required for monitoring the presence and effects of such a large number of potential pollutants. With so many pollutants present, synergistic effects often occur. In other words,

TABLE 30.2
Pathways of Pollutant Entry into the Ocean

Pathways
Direct outfalls
River inputs
Shipping
Offshore dumping of sewage and industrial wastes
Dredging spoil
Offshore industrial activities and accidents
Rain-out of airborne gases and particles
Ground-water seeps
Storm-water runoff

the pollutants interact such that their combined environmental impact cannot be predicted by simply summing their individual effects.

CONTAMINANTS

The naturally occurring substances whose seawater or sediment concentrations have been greatly elevated as a result of human activity are listed in Table 30.3.

Detrital Inorganic Particles

By accelerating the rate of continental erosion, humans have greatly increased the flux of particles to the ocean. We are presently supplying 3×10^9 tons/y, which amounts to about 10 percent of the global flux. This is arguably our single largest perturbation of the crustal-ocean factory. This accelerated rate is the result of dredging, deforestation, and other agricultural activities. The removal of plants destabilizes topsoil, which enhances the ability of storm waters to transport the soil into rivers. Storm-water runoff is also responsible for mobilizing soluble materials, such as fertilizers and pesticides, that are present in soil.

In addition to widening and deepening harbors, dredging is done as part of gravel-mining operations. At present, most particles that are dumped into the ocean are dredge spoils. In comparison, a relatively small amount of sewage and industrial sludge is being dumped into the ocean. A small amount of particles are also transported as a component of fly ash generated by waste incinerators.

Nutrients and Organic Matter

The introduction of sewage and fertilizers has caused local elevations in the rate at which nitrogen and phosphorus are introduced into the coastal ocean. This has enhanced algal productivity in some locations, as well as caused shifts in the species composition of phytoplankton communities. For example, elevated nutrient concentrations are thought to be the cause of the increasing number and intensity of red tides that occur along the coastlines of Japan and

TABLE 30.3
Marine Contaminants

Detrital inorganic particles
Nutrients and organic matter
Microorganisms
Petroleum hydrocarbons
Radionuclides
Trace metals
Polynuclear aromatic hydrocarbons

the United States. Red tides are phytoplankton blooms in which the dominant organisms are red-pigmented dinoflagellates. Some species synthesize substances that are passed up the food chain and have toxic effects on higher-order organisms. The increase in dolphin mortality in the late 1980s is partly attributed to the effects of an intense bloom of the dinoflagellate *Ptychodiscus brevi* in the coastal waters of the eastern United States.

Most nitrogen used as fertilizer now comes from industrial fixation. Fossil-fuel burning also mobilizes nitrogen by converting formerly buried organic nitrogen to nitrogen oxide. Some of these gases are rained out onto natural waters and thus contribute to the anthropogenic nitrogen flux. Most phosphorus that is used as fertilizer comes from the mining of ancient marine phosphorites. Phosphorous is also used in some detergents and thus becomes part of sewage effluent as laundry wastewater. These activities have mobilized nitrogen and phosphorus to such a large extent that their anthropogenic input to the coastal ocean is thought to rival the natural fluxes. Some marine chemists hypothesize that this increase in nutrient supply has caused a significant rise in the primary productivity of the coastal zone. Efforts to prove this have focused on detecting increases in the accumulation rate of sedimentary organic matter on the continental shelves.

The mariculture industry also has the potential to increase nutrient input to the coastal ocean. When seafood is raised in closed pens, decomposition of their wastes elevates the ambient levels of dissolved nitrogen and phosphorus. Also added to the water are drugs used in the treatment and prevention of disease, as well as growth regulation.

In extreme cases of nutrient loading, a mat of algae forms on the water surface. Respiration in the underlying waters lowers the O_2 concentrations as the algal mat inhibits gas exchange across the air–sea interface. The resulting O_2 depletion can cause fish kills. This process of eutrophication occurs most commonly in shallow or semi-isolated waters due to restricted water circulation. It is especially prevalent in the summer due to increased density stratification and decreased gas solubility. In an attempt to control eutrophication, most states in the United States have banned the use of phosphate in detergents. The most commonly used substitutes are linear alkyl sulfonates.

The introduction of heavy loads of organic matter, in the form of sewage sludge, can also cause anoxic conditions. As shown in Figure 30.2, the deep waters of the Mid-Atlantic Bight are subject to periods of anoxia. The O_2 depletion is caused by the respiration of organic matter and other reduced compounds that have been dumped in this area for decades. This has had a profound effect on some of the benthic communities. The Baltic and Mediterranean seas are also subject to periods of anoxia as a result of sewage input.

Microorganisms

Most microorganisms are introduced into the marine environment as a component of sewage. Included in this category are bacteria, viruses, and intestinal parasites. Most sewage treatment plants are not designed to remove these

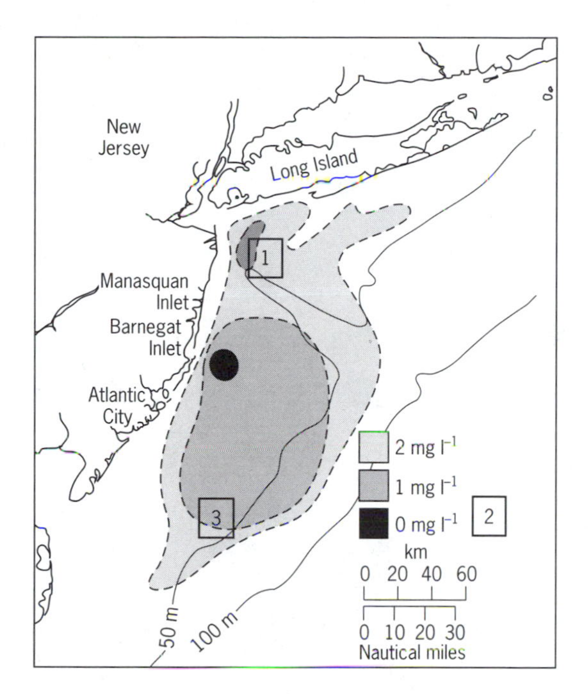

FIGURE 30.2. Dissolved O_2 concentrations (in mg O_2/L) in the bottom waters of the Mid-Atlantic bight during the summer of 1976. The numbered boxes mark the locations of active and historical dump sites. The New York bight sludge site (1) was used for the disposal of NYC's municipal wastes until 1981. The city then switched to the 106-mile dumpsite (2), which will be used until 1991. The Philadelphia sludge site (3) was used until 1981. *Source:* From F. W. Steimle and C. J. Sindermann, *Marine Fisheries Review,* vol. 40, p. 18, 1978, NOAA, Washington, DC.

organisms from waste waters as the required equipment is costly and complex. Thus even treated effluent is a source of these microorganisms. Marine filter feeders concentrate some of these organisms in their tissues. As shown in Table 30.4, some are pathogens. Since ingestion of even a single unit of virus can cause infection, shellfish are closely monitored for the presence of these pathogens.

Due to the high cost and time required, the concentrations of all pathogenic microorganisms are not routinely monitored. Instead, the level of microorganism contamination in seawater and seafood is estimated from the concentration of *Escherichia coli*, an intestinal bacterium indigenous to mammals. Since they do not survive long as free-living organisms, the presence of *E. coli* is indicative of the relatively recent input of substantial amounts of fecal material. While *E. coli* are nonpathogenic, their relative abundance suggests that other fecal-borne disease agents are likely to be present. Due to high *E. coli* counts, nearly half of the U.S. shellfish beds are presently closed for at least part of their annual harvesting season. Nevertheless, there are no laws requiring the inspection and grading of seafood, such as exist for poultry and meat.

TABLE 30.4
Principal Diseases in Humans Caused by Bacterial and Viral Contamination of Fish and Shellfish

	Etiological Agent	*Principal Aquatic Food Animals Involved as Source of Infection*	*Source of Infection for Aquatic Animal*	*Pathogenicity for Aquatic Food Animal*	*Mode of Transmission to Man*	*Diseases in Man and Most Common Manifestations*
Bacterial infection	*Salmonella* spp. (a) *S. typhimurium* *S. paratyphi* (b) other species (e.g., *S. typhimurium, S. enteritidis*)	Fish or shellfish secondarily contaminated through polluted waters or through improper handling	(a) Human feces and waters contaminated by human feces (b) Human and animal feces	Nonpathogenic	Ingestion of raw or insufficiently cooked contaminated fish or shellfish	(a) Typhoid and para-typhoid fever, septicemia (b) Salmonellosis: gastroenteritis
	Vibrio parahaemolyticus	Marine fish and shellfish	Organism occurs naturally in the marine environment	May cause death of shrimps and crabs: experimentally pathogenic for fish	Usually through consumption of raw or inadequately cooked fish or shellfish that has not been properly refrigerated	Diarrhea, abdominal pain
Bacterial intoxication	*Clostridium botulinum*	Fermented, salted, and smoked fish	Sediment, water, animal feces	Toxin can kill fish	Ingestion of improperly processed fish or shellfish	Botulism: neurological symptoms with high case fatality rate
	Staphylococcus aureus	Fish or shellfish secondarily contaminated through improper handling	Mannose and throat discharges, skin lesions	Nonpathogenic	Ingestion of fish or shellfish cross-contaminated after cooking	Staphylococcal intoxication: nausea, vomiting, abdominal pain, prostration

Bacterial intravital[a] intoxication	*Clostridium perfringens*	Fish or shellfish secondarily contaminated through polluted waters or through improper handling	Polluted waters, human and animal feces, sediment	Nonpathogenic	Ingestion of cooked fish or shellfish that has not been properly refrigerated	Diarrhea, abdominal pain
Bacterial skin infection	*Erysipelothrix insidiosa*	Fish, particularly spiny ones (e.g., sea robins, redfish)—organism is present in fish slime and meat		Nonpathogenic	Through skin lesions—usually an occupational disease	Erysipeloid—severe inflammation of superficial wounds
Viral infection	Virus of infectious hapatitis	Shellfish	Human feces and water polluted by human feces	Nonpathogenic	Ingestion of raw or inadequately cooked contaminated shellfish	Infectious hepatitis
Parasitic infestation	*Heterophyes heterophyes*	Freshwater or brackish-water fish	1st int. host[b]: snail 2nd int. host: fish Def. host: man, dog, cat, other fish-eating mammals, birds	Encyst in muscles and skin	Ingestion of raw or insufficiently cooked, infected fish (frequently salted or dried fish)	Heterophyiasis: abdominal pain, mucous diarrhea: eggs may be carried to the brain, heart, etc., causing atypical signs
Nematodes	*Anisakis matina*	Marine fish (e.g., cod, herring, mackerel)		Internal larvae infection	Usually from ingestion of raw or partially cooked, pickled or smoked herring	Anisakiiasis: eosinophilic enteritis

(Continued on next page.)

TABLE 30.4 (*Continued*)
Principal Diseases in Humans Caused by Bacterial and Viral Contamination of Fish and Shellfish

Etiological Agent	*Principal Aquatic Food Animals Involved as Source of Infection*	*Source of Infection for Aquatic Animal*	*Pathogenicity for Aquatic Food Animal*	*Mode of Transmission to Man*	*Diseases in Man and Most Common Manifestations*
Angiostrongylus cantonensis	Freshwater shrimp, land crab, possibly certain marine fish	1st. int. host[b]: snail, land snail Def. host: rat Paratenic hosts[c]: shrimp, land crab		Ingestion of raw or inadequately cooked shrimp or crabs (sometimes pickled)	Eosinophilic meningitis

Source: From IMCO/FAO/UNESCO/WMO/WHO/IAEA/UN Joint Group of Experts on the Scientific Aspects of Marine Pollution, reprinted with permission from *Reports and Studies of the Joint Group of Experts on the Scientific Aspects of Marine Pollution (GESAMP)* vol. 5, pp. 8–9, copyright © 1976 by the Food and Agriculture Organization of the United Nations, Rome, Italy.

[a]Intoxification by toxin produced in the body by bacteria present in heavily contaminated foods.

[b]The first intermediate host is an organism in which the parasites live as larva. In some cases, the larva must also pass to another host, called the second intermediate host, prior to reaching the adult stage in the definitive host.

[c]Paratenic hosts are hosts that are not necessary to the life cycle of a parasite. Since paratenic hosts can pass the parasite on to the definitive host, they can act as transport agents and thus not interrupt the parasite's life cycle.

Shellfish tend to bioaccumulate a variety of pollutants, such as metals and pesticides. For some chemicals, the degree of bioaccumulation is related to their seawater concentrations. In these cases, the chemical composition of shellfish tissues can be used as a long-term record of the average seawater concentration of that pollutant. One of the largest international monitoring efforts, Mussel Watch, uses sedentary bivalves as an indicator organism of marine pollution.

Petroleum

Petroleum is formed by natural processes that occur in marine sediments. In some locations, oil and gas seep from these deposits into the ocean. Therefore petroleum is regarded as a contaminant. Indeed, many of the compounds which compose petroleum, such as *n*-alkanes, are directly synthesized by marine plants and bacteria. Nevertheless, some components of petroleum, especially the aromatic compounds, are toxic if present at very high concentrations such as occur during an oil spill.

As shown in Table 30.5, the major sources of anthropogenic petroleum are chronic ones, such as tanker operations, sewage outfalls, and runoff. Though the atmospheric flux of petroleum into the ocean is not well known, it is thought to be significant. Most of the petroleum in runoff comes from streets, all of which are coated with oily residues. Storm waters wash these chemicals into rivers or ground waters, which eventually flow or seep into the ocean. This represents a non-point-source discharge of petroleum into the ocean and thus is very difficult to control. Tanker operations include such practices as bilge pumping, which involves the use of seawater as ballast. After a tanker offloads its oil, the bilge is filled with seawater to stabilize the ship during the trip back to home port. Just before arrival, the oily ballast water is pumped back into the ocean.

Shipping accidents represented only 12 percent of the oceanic input of petroleum during the early and mid 1980s. But largely as a result of these accidents and normal shipping operations, oil slicks are now commonly observed in coastal waters and in the open ocean of the North Atlantic as shown in Figure 30.3.

As shown in Figure 30.4, the fate of petroleum in these oil slicks is complicated and depends on such factors as (1) its chemical composition, (2) sea state, (3) wind speed, (4) temperature, (5) the geology of the seafloor and shoreline, and (6) local biological activity.

If the sea surface is calm, an oil spill will initially form a slick. The slick is subject to physical processes, such as advection and turbulence. The former tend to aid in dispersal, whereas the latter can form emulsions, often referred to as "chocolate mousse." Some of this material eventually forms tar balls. At the same time, the lower molecular weight compounds tend to evaporate or dissolve. Some are also photochemically degraded. As shown in Figure 30.5, these processes are largely complete within 1 d after the oil spill. On longer time scales, the altered petroleum is either transported horizontally by

TABLE 30.5
Input of Petroleum Hydrocarbons into the Marine Environment in Millions of Metric Tons per Year

	Input Rate (million metric tons/y)	
Source	*Best Estimate*	*Probable Range*
Natural sources		
Marine seeps	0.2	0.02–2.0[a]
Sediment erosion	0.05	0.0005–0.5
Offshore production	0.05	0.04–0.07
Transportation		
Tanker operations	0.7	0.4–1.5
Drydocking	0.03	0.03–0.05
Marine terminals	0.02	0.01–0.03
Bilge and fuel oils	0.3	0.2–0.6
Tanker accidents	0.4	0.3–0.4
Nontanker accidents	0.02	0.02–0.04
Atmosphere	0.3	0.05–0.5
Municipal and industrial wastes and runoff		
Refineries	0.2	0.1–0.6
Municipal wastes	0.7	0.4–1.5
Nonrefining industrial wastes	0.2	0.1–0.3
Urban runoff	0.03	0.01–0.2
River runoff	0.1	0.01–0.5
Ocean dumping	0.02	0.0005–0.02
Total	3.3	1.3–8.8

Source: From *Oil in the Sea: Inputs, Fates and Effects*, the Steering Committee for the Petroleum in the Marine Environment Update, copyright © 1985 by the National Research Council, Washington, DC, p. 82.
[a]May be as high as 6×10^6 ton/y.

currents or vertically by the sinking of tar balls. Some petroleum is also carried to the seafloor as a result of adsorption onto sinking particles or incorporation into POM as a result of biological uptake.

Tar balls are not rapidly degraded. Due to their inert nature and low density, tar balls are transported great distances before reaching the seafloor. This has caused them to become widely distributed throughout the ocean as shown in Figure 30.6. Many are eventually stranded on shorelines.

Once in the sediments, petroleum compounds will persist for long periods of time unless degraded by organisms. Since many of these compounds are toxic to the benthos, they are not readily degraded. This causes their impact on the biogeochemistry of the surface sediments to continue until the petroleum compounds are buried.

In addition to the fatal effects of some of its constituent compounds, petroleum harms marine life in other ways. For example, sublethal exposure

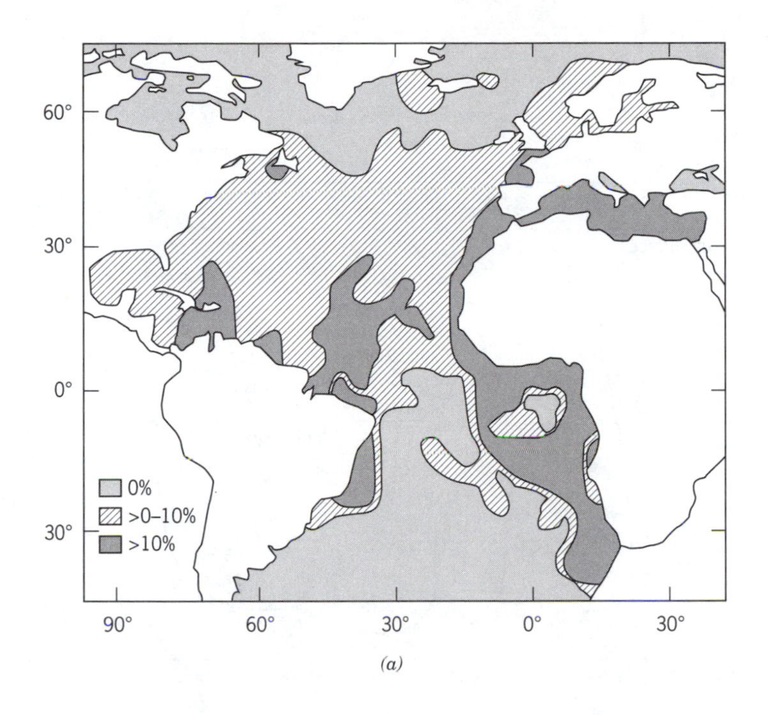

(a)

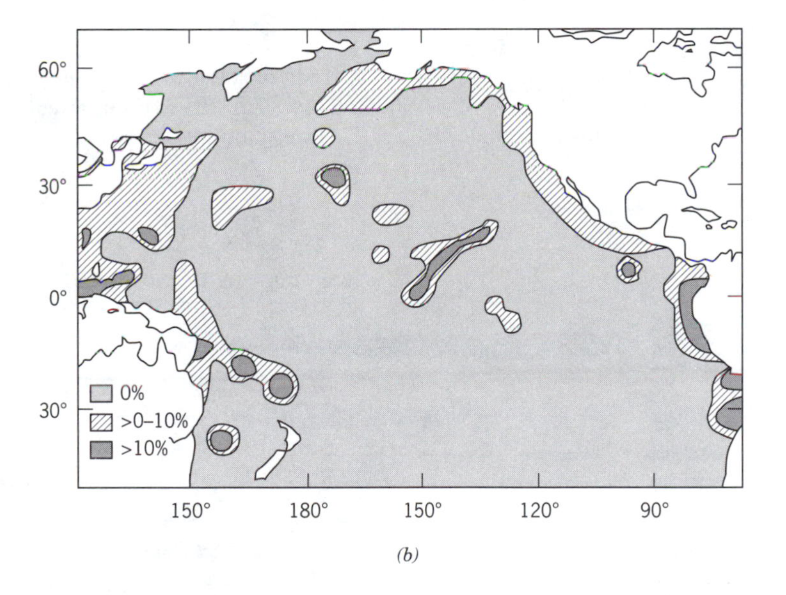

(b)

FIGURE 30.3. The geographical distribution of oil slicks on *(a)* the Atlantic Ocean and *(b)* the Pacific Ocean, as indicated by the percentage of positive sighting reports. *Source:* From *Chemical Oceanography,* vol. 9, M. R. Preston (ed.: J. P. Riley), copyright © 1988 by Academic Press, Orlando, FL., p. 73. Reprinted by permission. After *Global Oil Pollution,* E. M. Levy, D. Kohnke, E. Sobtchenko, T. Suzuoki, and A. Tokuhiro, copyright © 1981 by the Intergovernmental Oceanographic Commission, Paris, France, pp. 5–6. Reprinted by permission.

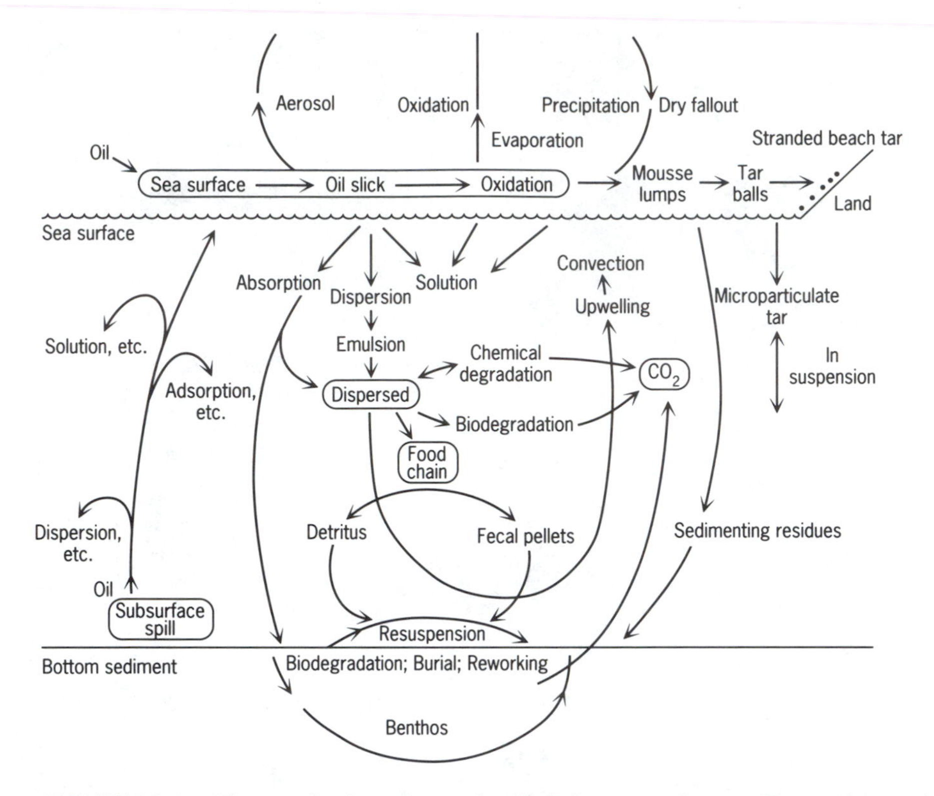

FIGURE 30.4. The weathering of a crude oil slick at sea. *Source:* From *Chemical Oceanography,* vol. 9, M. R. Preston (ed.: J. P. Riley), copyright © 1988 by Academic Press, Orlando, FL, p. 77. Reprinted by permission.

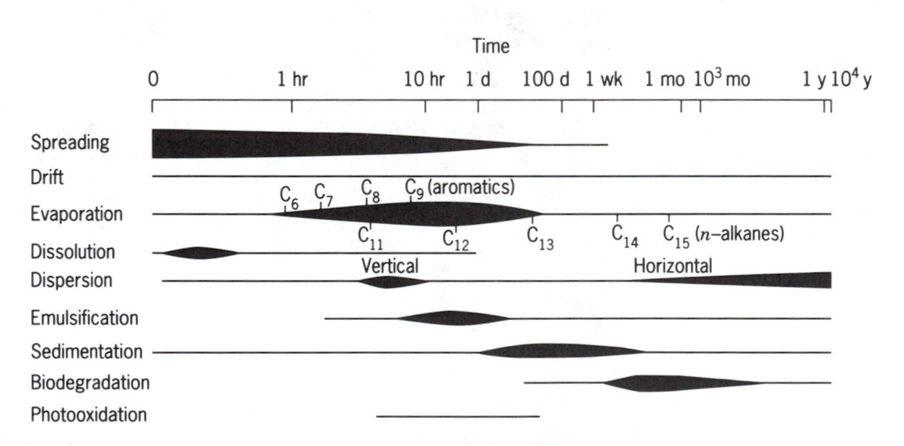

FIGURE 30.5. Time course of factors affecting oil spilled on the sea. *Source:* From K. J. Whittle, R. Hardy, P. R. Mackie, and A. S. McGill, reprinted with permission from the *Philosophical Transactions of the Royal Society of London,* vol. B297, p. 23, copyright © 1982 by the Royal Society of London, London, England.

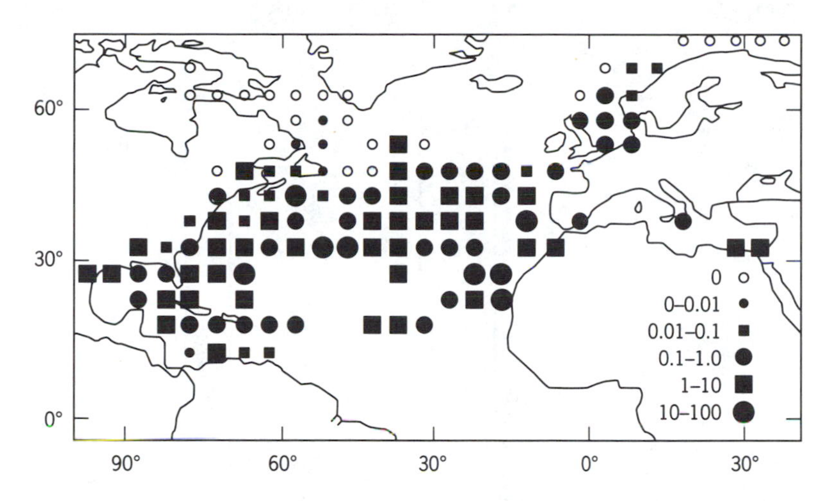

FIGURE 30.6. Tar-ball concentrations (mg/m²) in the North Atlantic. *Source:* From *Chemical Oceanography,* vol. 9, M. R. Preston (ed.: J. P. Riley), copyright © 1988 by Academic Press, Orlando, FL, p. 75. Reprinted by permission. After *Global Oil Pollution,* E. M. Levy, D. Kohnke, E. Sobtchenko, T. Suzuoki, and A. Tokuhiro, copyright © 1981 by the Intergovernmental Oceanographic Commission, Paris, France, p. 6. Reprinted by permission.

can interfere with metabolic processes, such as reproduction. Low levels of some compounds have also been observed to interfere with the perception of chemical stimuli, such as those that regulate reproduction. Petroleum also has negative physical effects. For example, a coating of oil will cause most sea birds to die from hypothermia or drown because the petroleum alters the insulating properties of their feathers and impairs their swimming ability.

Radioactivity

The total radioactivity in the world ocean is approximately 5×10^{11} Ci and consists mostly of ^{40}K. The oceanic inventory has been slightly increased by human activities. As shown in Table 30.6, the largest anthropogenic source has been from nuclear fallout produced since the 1950s from atom bomb tests conducted by the United States, the USSR, France, Britain, and China.

As a result of chronic emissions and accidents, the Sellafield Nuclear Fuel Reprocessing Plant has been the second major source of anthropogenic radionuclides to the sea. A small amount of low-level nuclear waste has also been introduced by ocean dumping at a site in the northeast Atlantic Ocean that is located in 4000 m of water. This dumpsite is regulated by the Nuclear Energy Agency of the Organization for Economic Cooperation and Development (NEA/OECD).

Smaller amounts of radionuclides have been introduced into the ocean as a result of accidents associated with the downing of nuclear-tipped missiles,

TABLE 30.6
Inventories of Artificial Radionuclides

	Plutonium–239, –240 (kCi)	*Cesium–137 (kCi)*	*Strontium–90 (kCi)*	*Carbon–14 (kCi)*	*Tritium (kCi)*
Total worldwide fallout by early 1970s	320	16,700	11,500	6,000	3,000,000
North Atlantic Ocean—early 1970s	63	3,300	2,300		650,000
Sellafield, 1957–78 discharge	14	830	130		370
	Total α-Emitters	*Total β/γ-Emitters (other than tritium)*			
The NEA dumpsite (1967–79)	8.3		258		262

Source: From G. T. Needler and W. L. Templeton, reprinted with permission from *Oceanus*, vol. 24, p. 62, copyright © 1981 by Oceanus, Woods Hole, MA.

as well as nuclear-powered submarines and satellites. In 1963 and 1968, the United States lost two nuclear submarines, the U.S.S. *Thresher* and *Scorpion*, when they sank off the New England coast and the Azores, respectively. The nuclear power plants in each submarine contain about 30,000 Ci of radionuclides, which will eventually be released into the ocean. In 1964, a nuclear power generator from a failed satellite reentered Earth's atmosphere and deposited 17 kCi of ^{238}Pu into the atmosphere. This event accounts for more than half of ^{238}Pu that has been deposited in marine sediments.

Other land-based emissions have resulted from several accidents at nuclear power plants: (1) the explosion of a nuclear waste dump in the Ural Mountains (USSR) in 1957, (2) the partial core meltdown at the Three-Mile Island plant (Harrisburg, PA) in 1979; (3) leakage from the Savannah River Nuclear Reactor Site (Aiken, SC) over the past four decades; and (4) the fire which occurred in the Chernobyl reactor (Kiev, USSR) in 1986. Some of these emissions have reached the ocean as a result of river and ground-water runoff, as well as atmospheric fallout.

The United States is responsible for the first intentional introduction of anthropogenic radionuclides into the ocean. Starting in the 1940s, radionuclides were directly discharged into the Columbia River from the Hanford Reservation, which is located 575 km upstream from the Pacific Ocean. During the peak period of operation in the 1950s and 1960s, nine plutonium production reactors were operating at this site. Their primary coolant fluid was river water, which following use, was discharged directly back into the river. As a result, thousands of curies of radioactivity were released into the Columbia River every day. This practice was stopped in 1971 and at present even the reactors have been shut down.

As shown in Figure 30.7, the entry of these radionuclides into the coastal ocean was traceable as a plume of water enriched in ^{51}Cr. Soluble chromium exists in seawater primarily as the chromate ion. The radioactive plume was traceable for 350 km downstream of the river mouth. In comparison, other radionuclides, such as ^{65}Zn and ^{60}Co, were rapidly transported to the sediments by adsorption onto sinking particles. Their distributions were used to trace the movement of sediments across the continental shelf.

As a result of the massive emissions of radionuclides from Sellafield, the Irish Sea has been recognized as the "most radioactive sea in the world." Sellafield is located on the midwestern coast of Great Britain (Figure 30.8*a*) and contains the world's largest nuclear reprocessing plants, as well as the oldest operating commercial nuclear power plant. Due to its extended and intense use, the Sellafield site is also thought to be the world's largest repository of stored radioactive materials.

Radionuclides have been emitted from Sellafield over the past 40 y as a result of accidents, such as the reactor fire that occurred in 1957, as well as from chronic discharges associated with plant-cleaning operations, fuel reprocessing, and the storage of nuclear wastes. The nuclear wastes are stored in ponds whose overflow water is presently discharged through two outfalls that extend about 2.5 km from the shoreline. The discharge contains low levels

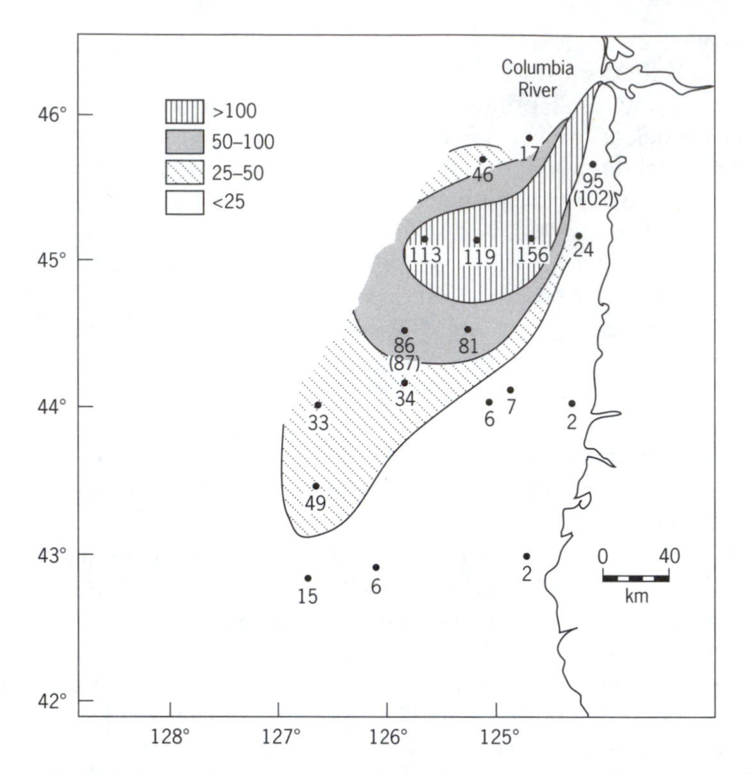

FIGURE 30.7. The Columbia River plume as defined by the specific activity of ^{51}Cr (dpm/100 L) of surface seawater. Numbers in parentheses are the results from duplicates. *Source:* From C. Osterberg, N. Cutshall, and J. Cronin, reprinted with permission from *Science,* vol. 150, p. 1586, copyright © 1965 by the American Association for the Advancement of Science, Washington, DC.

→

FIGURE 30.8. *(a)* The specific activity of ^{137}Cs (pCi/L) in the surface waters of the Irish and North Seas. Nuclear reactors are located at Sellafield, Dounreay, and Cape La Hague. *(b)* Estimated specific activity of ^{239}Pu and ^{240}Pu in the top 30 cm of sediments in the Irish Sea as of 1977/1978. *Source:* From *(a) Chemistry and Biogeochemistry of Estuaries,* U. Förstner (eds.: E. Olausson and I. Cato), copyright © 1980 by John Wiley & Sons, Inc., New York, p. 332. Reprinted by permission. Data from D. F. Jefferies, A. Preston and A. K. Steele, reprinted with permission from *Marine Pollution Bulletin,* vol. 4, p. 120, copyright © 1973 by Pergamon Press, Elmsford, NY. C. N. Murray and H. Kautsky, reprinted with permission from *Estuarine and Coastal Marine Science,* vol. 5, p. 324, copyright © 1977 by Kluwer Academic Publishers, Dordrecht, The Netherlands. *(b)* J. H. W. Hain, reprinted with permission from *Oceanus,* vol. 29, p. 23, copyright © 1986 by Oceanus, Woods Hole, MA. After R. J. Pentreath, D. S. Woodhead, P. J. Kershaw, D. F. Jeffries, and M. B. Lovett, reprinted with permission from *Rapports et Proc'es—Verbaux des Réunions, International Council for the Exploration of the Sea,* vol. 186, p. 66, copyright © 1986 by the International Council for the Exploration of the Sea, Copenhagen, Denmark.

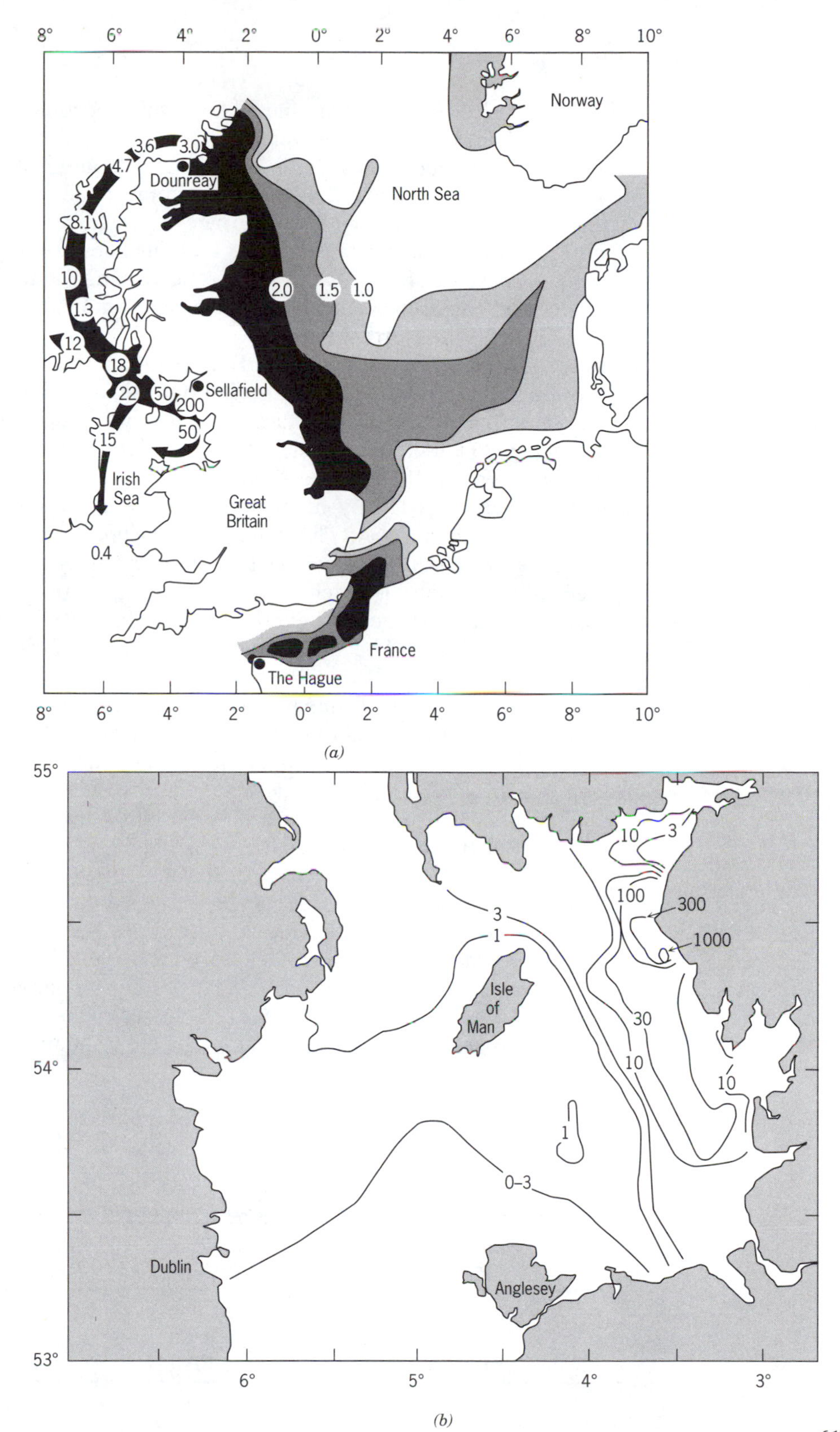

8°
6°
4°
2°
0°
2°
4°
6°
8°
10°
Norway
North Sea
3.6
3.0
4.7
Dounreay
8.1
10
1.3
12
18
22
50
200
50
15
Sellafield
2.0
1.5
1.0
Irish
Sea
Great
Britain
0.4
France
The Hague
(a)
55°
54°
53°
6°
5°
4°
3°
10
3
100
300
1000
3
1
Isle
of
Man
30
10
10
1
0–3
Dublin
Anglesey
(b)

of ^{137}Cs, ^{134}Cs, ^{106}Ru, ^{90}Sr, ^{239}Pu, and ^{241}Am, which account for 90 percent of the "normal" emissions from Sellafield.

The soluble radionuclides, such as ^{137}Cu, ^{134}Cs, and ^{90}Sr, have residence times in the Irish Sea of about 2 y. As shown in Figure 30.8*a*, the radionuclides are then transported north with the surface-water currents around Scotland and into the North Sea. From there, the radionuclides are transported by the Norwegian Coastal Current toward the Arctic Ocean. Dilution of this current with water from the open ocean has caused a small amount of the radionuclides to be carried into the North Atlantic.

In comparison, the isotopes of Pu, Am, and Ru have been rapidly transported to the sediments by adsorption onto sinking particles. As a result, more than 90 percent of the discharged Pu has been deposited in sediments located within 30 km of the outfall site (Figure 30.8*b*). Some of these radionuclides have been remobilized and returned to land by (1) bioaccumulation in shellfish that are ingested by humans and (2) resuspension of sediment by storms and bottom currents. The latter has transported radioactive muds up into estuaries and salt marshes. At low tide, the fine-grained particles are exposed to the air, during which time the winds can resuspend them, leading to inhalation. Humans are also exposed to radioactivity by close proximity and contact with the mud because some of the radionuclides emit high-energy gamma radiation.

As shown in Table 30.7, this high energy causes gamma radiation to be a very effective penetrator and thus can induce harm just through exposure. In comparison, α particles have relatively low energy and thus are not capable of deep penetration. But their large mass causes them to be very effective ionizing agents. Thus α particles are very harmful, but only if ingested. The same is true for β particles.

Due to improved safety practices, emissions from Sellafield have declined. Radioactive decay and surface currents have caused the specific activity of the waters to decrease. But the specific activity of the sediments is expected to increase over time as radioactive decay produces daughters that are stronger emitters. For example, the decay of ^{241}Pu produces relatively low-energy β particles and ^{241}Am. Decay of the latter releases higher-energy α particles. Therefore, the sediments are a repository of old radionuclides, as well as a source of new ones.

TABLE 30.7
Properties of Some Ionizing Radiation

Type	*Relative Mass Ratio*	*Relative Energy*	*Relative Penetration*	*Relative Ionization*
α	7380	Moderate	1	10^4
β	1	Low to high	10^2	10^2
γ	0	High	10^3–10^4	1

Trace Metals

Mining and fossil-fuel burning have greatly increased the rate at which some trace metals are introduced into the ocean. The mobilized metals are carried to the sea by atmospheric transport, continental runoff, and sewage outfalls. As shown in Table 11.1, anthropogenic input has greatly elevated the atmospheric fluxes of many trace metals, with the largest being the five-fold increase in Sn.

The metals whose atomic weights exceed 20 amu are termed **heavy metals**. They are of particular concern because most heavy metals are toxic at relatively low concentrations. The toxic heavy metals whose biogeochemical cycles have been most affected by human activity are As, Cd, Cu, Cr, Hg, Pb, Ni, Sb, Se, V, and Zn. Some, such as Zn, are also micronutrients. Thus their toxic effect is concentration dependent as shown in Figure 30.9. Elements that have a stimulatory effect on biological activity when present as low concentrations are termed essential elements. They become toxic only when concentrations reach a threshold level. Nonessential elements, such as Hg, have no stimulatory effect at any concentration and also provoke a toxic effect once a threshold concentration is exceeded. Unfortunately, heavy metals tend to have very high enrichment factors and slow clearance rates. As a result, biological uptake is a significant sink for these metals and also functions as a natural water-purification mechanism.

The toxic effect of a particular heavy metal is determined by the environmental and chemical factors listed in Table 30.8. In particular, the chemical speciation of the metal is important because not all forms are equally toxic. In general, methylated forms tend to be most toxic and are also the most greatly biomagnified. Methylation is thought to be the result of biological activity. Many methylated metals are also quite volatile.

Most heavy metals are introduced into seawater as components of the crystal lattice of mineral silicates. As indicated in Table 30.9, using the Amazon and Yukon rivers as examples, a lesser amount is present as a constituent of the hydrogenous oxyhydroxides and POM.

The metal composition of the silicate minerals is relatively constant. In addition to heavy metals, these minerals are also the largest source of particulate

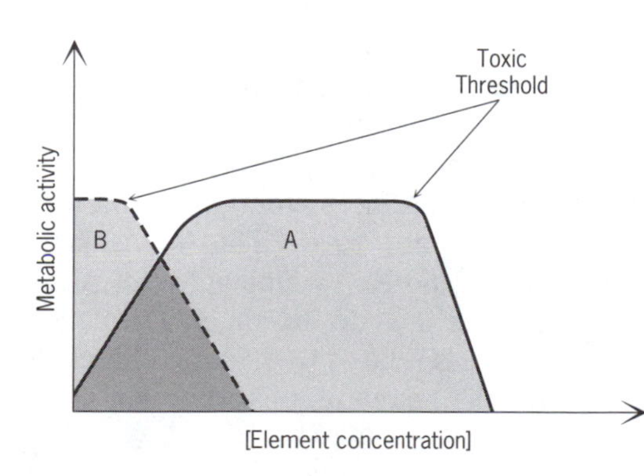

FIGURE 30.9. Metabolic activity versus concentration for an essential *(A)* and nonessential *(B)* element.

TABLE 30.8
Factors Influencing the Toxicity of Heavy Metals in Solution

Factor			
Form of metal in water	Inorganic Organic	Soluble	Ion Complex ion Chelate ion Molecule
		Particulate	Colloidal Precipitated Adsorbed
Presence of other metals or poisons	Synergy No interaction Antagonism		
Environmental Factors (influences on the physiology of organisms and possible form of metal in water)	Temperature pH Dissolved oxygen concentration Light intensity Salinity		
Condition of organism	Stage in life history (egg, larva, etc.) Changes in life cycle (e.g., moulting, reproduction) Age and size Sex Starvation Activity Additional protection (e.g., shell) Adaptation to metals		
Behavioral response			

Source: From *Marine Pollution*, G. W. Bryan (ed.: R. Johnston), copyright © 1976 by Academic Press, Orlando, FL, p. 257. Reprinted by permission.

Al to the sediments. Thus the heavy metal concentrations of a sediment should be directly proportional to its Al content. As shown in Figure 30.10, the Pb and Zn concentrations in the sediments of Biscayne and Pensacola bays exceed those that should be present if the silicate minerals were the major source of these metals. Anthropogenic input is a likely source of this excess metal.

A relatively small amount of metals are introduced into seawater as cations adsorbed onto clay minerals. Nevertheless, this pool of metal is biogeochemically important because it is very reactive. When river currents transport clay minerals into estuarine waters with salinities greater than 5‰, the exchangeable cations are desorbed. This is caused by the increased concentrations of Na^+, Mg^{2+}, and K^+, which act to displace the less abundant trace metals from the exchange sites on the clays. The degree to which a heavy metal undergoes ion exchange is related to its ionic charge density. This causes the order of desorption for the heavy metals to be Hg>Cu>Zn>Pb>Cr>As.

TABLE 30.9
Transport of Some Trace Metals by the Amazon and Yukon Rivers

Mechanism	*Fe*	*Ni*	*Co*	*Cr*	*Cu*	*Mn*
Amazon River						
1) In solution and organic complexes	0.7	2.7	1.6	10.4	6.9	17.3
2) Adsorbed	0.02	2.7	8.0	3.5	4.9	0.7
3) Precipitated and coprecipitated	47.2	44.1	27.3	2.9	8.1	50.0
4) In organic solids	6.5	12.7	19.3	7.6	5.8	4.7
5) In crystalline sediments	45.5	37.7	43.9	75.6	74.3	27.2
Yukon River						
1) In solution and organic complexes	0.05	2.2	1.7	12.6	3.3	10.1
2) Adsorbed	0.01	3.1	4.7	2.3	2.3	0.5
3) Precipitated and coprecipitated	40.6	47.8	29.2	7.2	3.8	45.7
4) In organic solids	11.0	16.0	12.9	13.2	3.3	6.6
5) In crystalline sediments	48.2	31.0	51.4	64.5	87.3	37.1

Source: From R. J. Gibbs, reprinted with permission from *Science*, vol. 180, p. 72, copyright © 1973 by the American Association for the Advancement of Science, Washington, DC.

The solubilized metals tend to complex with humic acids. The resulting colloids are destabilized by increasing salinities, causing some to flocculate. These colloids also tend to coprecipitate with metal oxyhydroxides. Because of the increase in alkalinity, the mixing of river water with seawater also promotes the formation of the oxyhydroxides. Most flocculation is complete by the time mixing has elevated salinities to 15‰. Biological uptake and incorporation into POM is also responsible for conversion of some dissolved metal into particulate form. As a result of all these processes, most dissolved metals exhibit **nonconservative** behavior in estuaries.

The occurrence of nonconservative behavior can be established by plotting dissolved metal concentrations against that of some conservative species, such as salinity. If the distribution of dissolved metal in an estuary is largely controlled by the mixing of river and seawater, its concentrations will be linearly related to that of salinity (Figure 30.11). The direction of the slope will be determined by the relative abundances of the metal in pure river water and seawater. Deviations from this simple relationship are observed in estuaries that receive freshwater from more than one river.

If the metal is removed by some chemical process during its transport through the estuary, its concentrations will plot below the theoretical mixing line. Conversely, if the metal is solubilized by some process that occurs during its transport through the estuary, the concentrations will fall above the theoretical mixing line. As shown in Figure 30.12, most DOC appears to behave conservatively in estuaries. In comparison, the small fraction composed of humic acids undergoes a significant removal. As mentioned above, this removal is the result of flocculation.

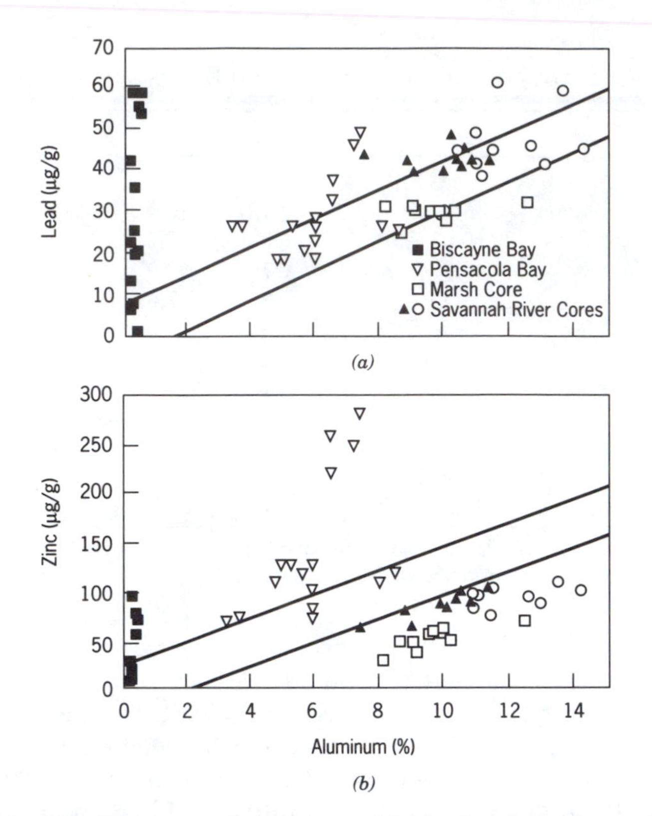

FIGURE 30.10. Comparison of *(a)* lead and *(b)* zinc concentrations in sediments from Biscayne Bay, Pensacola Bay, and the Savannah River. The straight lines represent the 95 percent confidence band for the metal composition of unpolluted sediments. *Source:* From H. L. Windom, S. J. Schropp, F. D. Calder, J. D. Ryan, R. G. Smith, L. C. Burney, F. G. Lewis, and C. H. Rawlinson, reprinted with permission from *Environmental Science and Technology,* vol. 23, p. 320, copyright © 1989 by the American Chemical Society, Washington, DC.

The speciation of heavy metals is strongly influenced by this decrease in humic acid concentration. As shown in Figure 30.13, the removal of this ligand causes the dissolved metals to form complexes with inorganic ligands.

Heavy metal speciation is also influenced by changes in pH and alkalinity, as well as by the presence of organisms as described in Table 30.10. The metals that are converted to particulate form tend to be retained by estuaries. This trapping effect is caused by the net riverward flow of bottom waters that occurs in all but the well-mixed estuaries (Figure 30.14). Since the particulate metals settle into these waters, they are carried landward. These particulate metals are either resolubilized in the bottom waters or settle onto the estuarine sediments.

As a result of this trapping effect, estuarine sediments are repositories of heavy metals and other pollutants. Dredging of these sediments resuspends

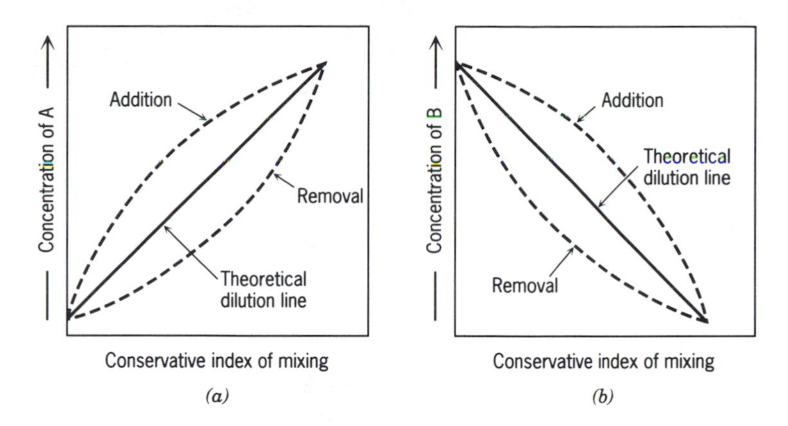

FIGURE 30.11. Idealized representation of the relationship between the concentration of a dissolved component and a conservative index of mixing in an estuary which has only one source of river and seawater. *(a)* Case I: the seawater concentration is greater than that in river water. *(b)* Case II: the seawater concentration is less than that in river water. *Source:* From *Estuarine Chemistry,* P. S. Liss (eds.: J. D. Burton and P. S. Liss), copyright © 1976 by Academic Press, Orlando, FL, p. 95. Reprinted by permission.

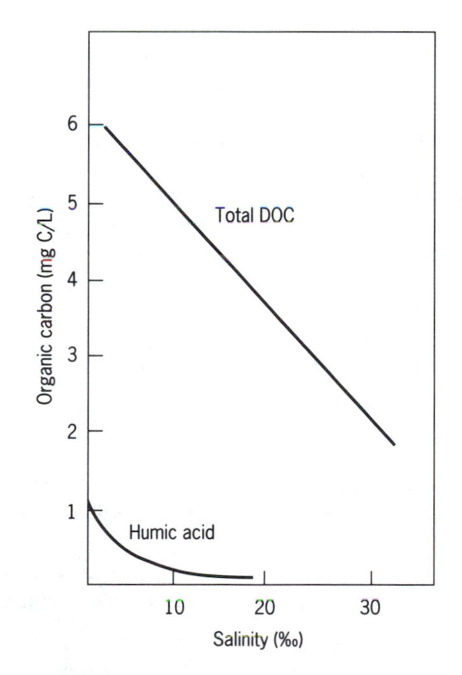

FIGURE 30.12. Typical changes in DOC and humic acid concentrations which occur in estuaries as a result of increasing salinities. *Source:* From *Humic Substances in Soil, Sediment and Water,* L. M. Mayer (eds.: G. R. Aiken, D. M. McKnight, R. L. Wershaw, and P. MacCarthy), copyright © 1985 by John Wiley & Sons, Inc., New York, p. 216. Reprinted by permission.

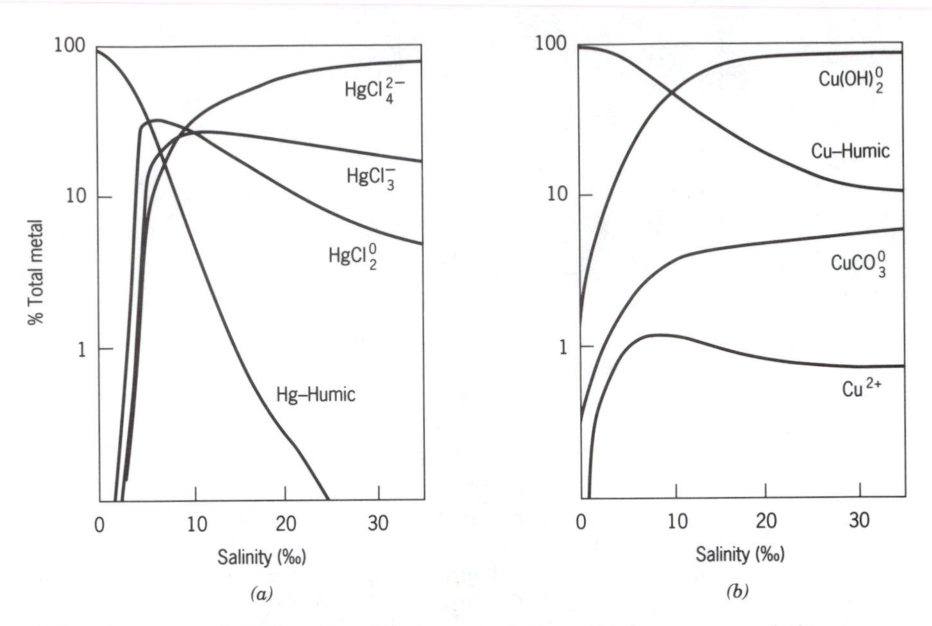

FIGURE 30.13. Calculated equilibrium speciation of *(a)* mercury and *(b)* copper during estuarine mixing of hypothetical river water with seawater. Hum = humic substance. *Source:* From *Humic Substances in Soil, Sediment and Water,* L. M. Mayer (eds.: G. R. Aiken, D. M. McKnight, R. L. Wershaw, and P. MacCarthy), copyright © 1985 by John Wiley & Sons, Inc., New York., p. 227. Reprinted by permission. After R. F. C. Mantoura, A. Dickson, and J. P. Riley, reprinted with permission from *Estuarine and Coastal Marine Science,* vol. 6, pp. 398 and 402, copyright © 1978 by Elsevier Science Publishers, Amsterdam, The Netherlands.

TABLE 30.10
Some Factors Affecting Chemical Speciation and Thus Toxicity of Trace Metals in Estuarine and Marine Organisms

Variables	*Explanation*
pH, alkalinity, organic and inorganic ligands	Changes species distribution and $[Me^{n+}]$; influences formation of hydroxo, carbonato, and other complexes; changes adsorbability of metal ions and resorptivity of cell organisms
Density of organisms	Reduces available $[Me]_T$ and changes species distribution because of adsorption on cell surfaces and/or by complexation by exudates of organisms
Concentration of particles and colloids	Metals are sequestered by particulate oxides of iron and manganese; Organic colloids are particularly effective
Redox potential	Affects oxidation state of metal; methylation often occurs more readily at low redox potentials

Source: From *Aquatic Chemistry*, W. S. Stumm and J. J. Morgan, copyright © 1981 by John Wiley & Sons, Inc., New York, p. 702. Reprinted by permission.

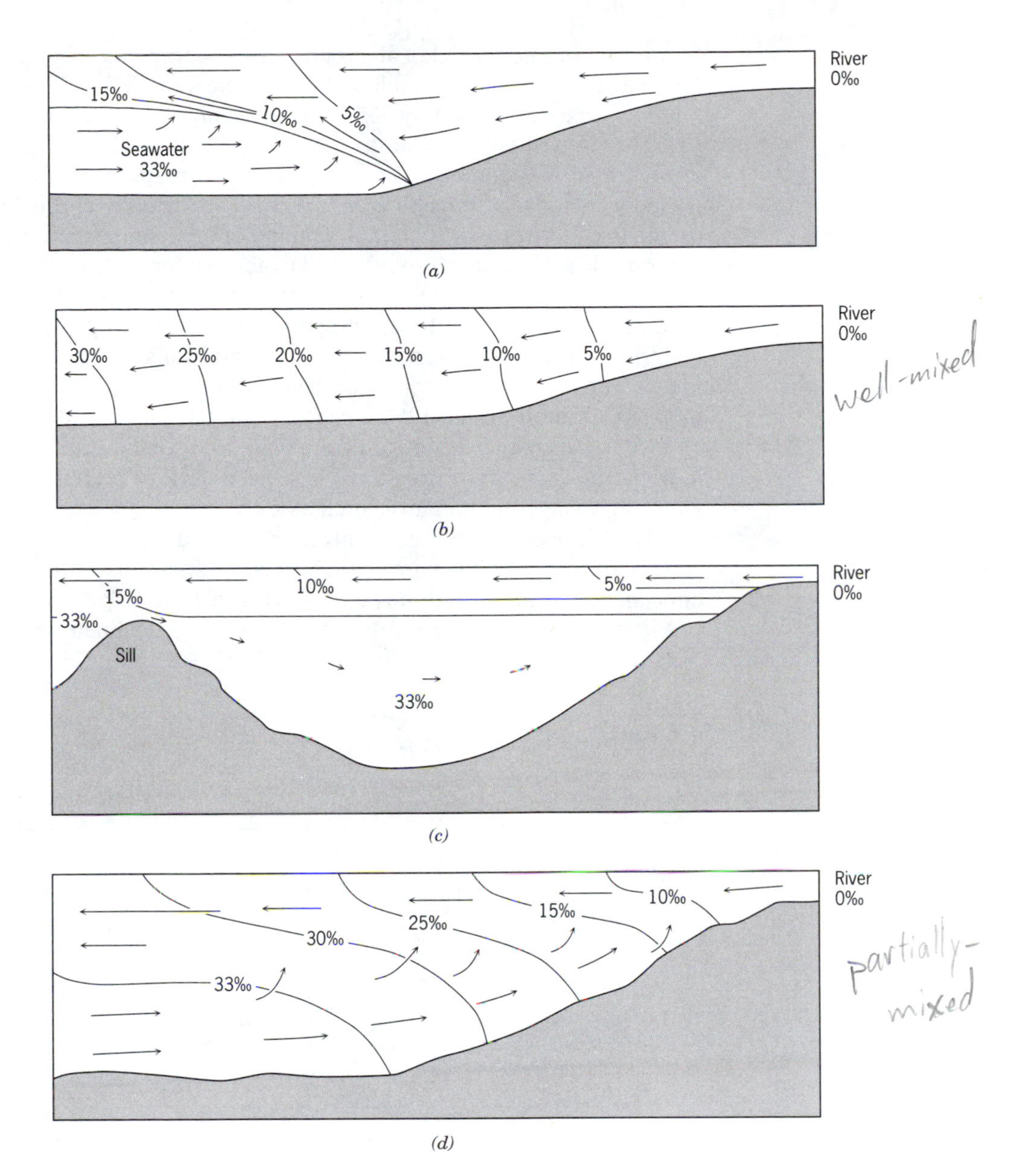

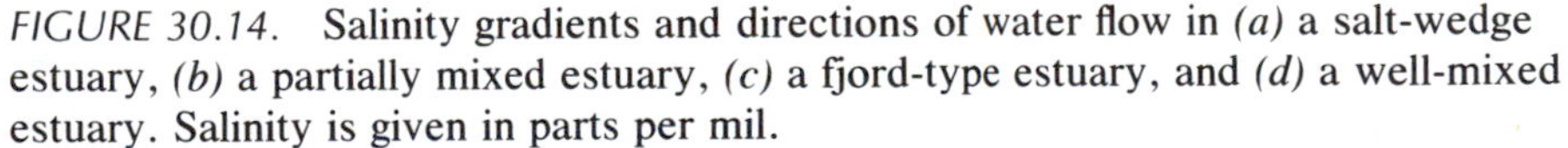

FIGURE 30.14. Salinity gradients and directions of water flow in *(a)* a salt-wedge estuary, *(b)* a partially mixed estuary, *(c)* a fjord-type estuary, and *(d)* a well-mixed estuary. Salinity is given in parts per mil.

the pollutants and causes significant water-quality problems. Estuarine circulation also acts to trap the biolimiting elements as biogenic particles are remineralized in the deep waters and the regenerated solutes are upwelled to the surface. This process supports high productivity in estuaries and amplifies the effects of nutrient loading from the discharge of sewage effluents. Incorporation into biogenic particles also enhances the trapping of pollutants that have high enrichment factors, such as heavy metals and pesticides.

The marine chemistry of anthropogenic lead and mercury has been particularly well studied due to their large inputs and toxic effects. As shown in Figure 30.15, human activities are presently responsible for most of the lead transported to the atmosphere and rivers. The majority of this lead is mobilized by the burning of fossil fuels, the refining of ores, and accelerated soil erosion.

Fossil-fuel burning converts lead to aerosol form. Though most is deposited relatively close to its site of input, some aerosol lead is transported to polar regions. The lead that settles onto polar ice is eventually buried and thus forms a record of changes in atmospheric concentrations. As shown in Figure 30.16, the rate of increase in lead concentration was moderate through the 1800s but increased sharply in the 1950s, marking the advent of leaded gasoline. Similar changes in atmospheric concentrations of other pollutants have been recorded in polar ice cores. Ice cores have also been used to reconstruct the "natural" levels of some contaminants, such as CO_2.

The highest seawater concentrations of lead have been observed in coastal waters adjacent to urban areas such as New York City. A depth profile of lead concentrations measured during the 1970s in Long Island Sound can be found in Figure 11.7. These concentrations have since declined as the use of leaded

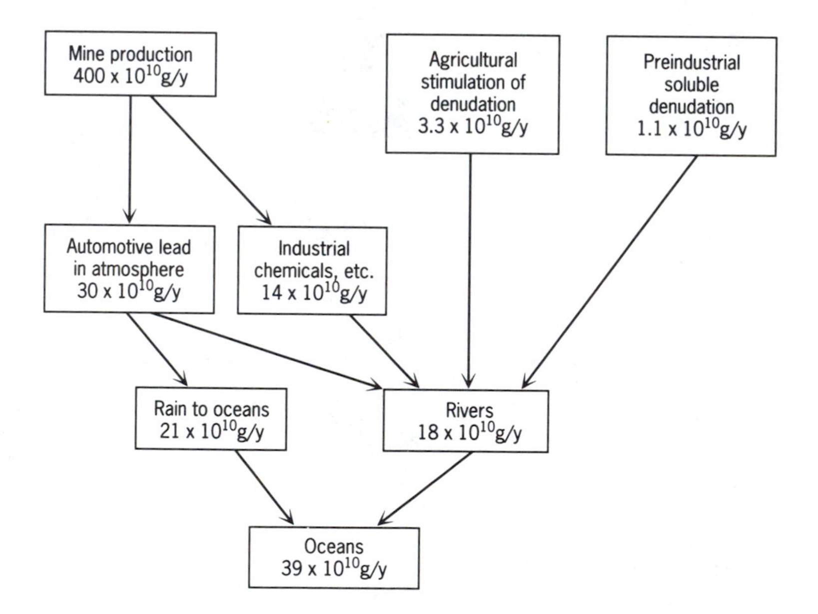

FIGURE 30.15. Lead input to the northern hemisphere ocean. *Source:* From *The Natural Environment and the Biogeochemical Cycles,* P. J. Craig (ed.: O. Hutzinger), copyright © 1980 by Springer-Verlag, Heidelberg, Germany, p. 191. Reprinted by permission. After *The Biogeochemistry of Lead in the Environment,* J. O. Nriagu (ed.: J. O. Nriagu), copyright © 1978 by Elsevier Science Publishers, Amsterdam, The Netherlands, p. 25. Reprinted by permission.

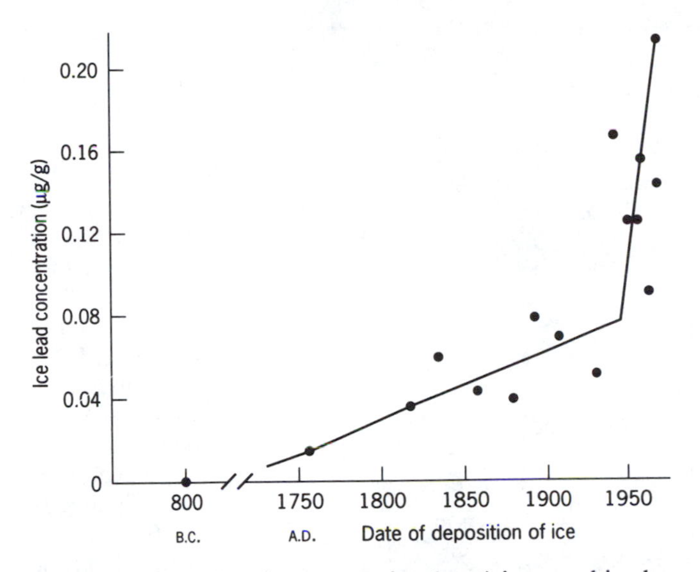

FIGURE 30.16. Lead concentration (ppm) in annual ice layers deposited on Greenland. *Source:* From M. Murozumi, T. J. Chow, and C. Patterson, reprinted with permission from *Geochimica et Cosmochimica Acta,* vol. 33, p. 1285, copyright © 1969, Pergamon Press, Elmsford, NY.

gasoline has been severely curtailed. Lead was also introduced to the waters of the New York Bight by the offshore dumping of acidic industrial wastes. This is the cause of the high sedimentary lead concentrations shown in Figure 30.17. Though industrial wastes are no longer being dumped, domestic sewage will continue to be legally permitted until 1991.

Because of mercury's toxicity, its marine biogeochemistry has been well studied. In contrast to lead, the natural fluxes of mercury are quite large due to its high volatility. Accurate methods for the measurement of environmental levels of mercury were developed in the 1970s. The first results suggested that most marine and aquatic fish contained levels of mercury that were high enough to represent a health threat. Initially, scientists assumed that these high levels were the result of pollution. But following analysis of museum specimens, it was determined that some species, such as tuna and swordfish, have naturally high mercury concentrations. In other words, these fish have very large enrichments factors for mercury.

Because of mercury's toxicity, health officials have recommended that its intake be limited to 0.03 mg/d. As shown in Figure 30.18, some swordfish have a high enough mercury concentration that consumption of reasonable amounts of this species (about 25 g/d) would cause the daily recommended intake of Hg to be exceeded.

In a few coastal areas, industrial effluents have caused mercury concentrations to reach toxic levels. The most extreme case occurred in Japan in the 1950s in Minamata Bay. The problem was first recognized when local inhabitants

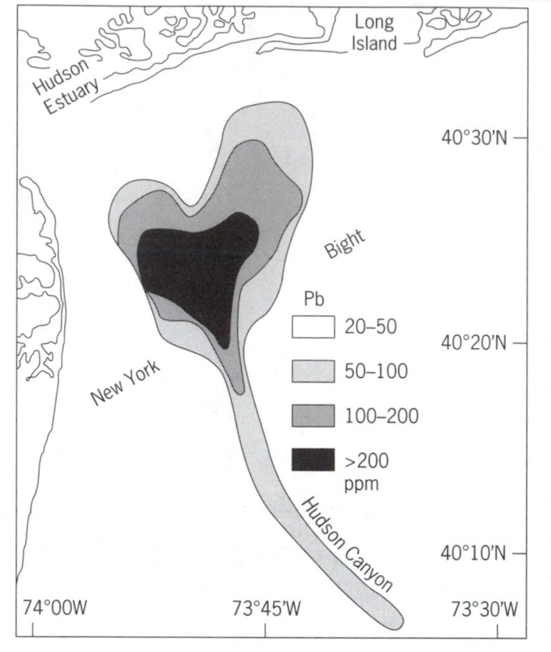

FIGURE 30.17. Distribution of lead (ppm) in bottom sediments from the New York bight. See Figure 30.2 for locations of dumpsites. *Source:* From *Chemistry and Biogeochemistry of Estuaries,* U. Förstner, (eds.: E. Olausson and I. Cato), copyright © 1980 by John Wiley & Sons, Inc., New York, p. 314. Reprinted by permission. After D. J. Carmody, J. B. Pearce, and W. E. Yasso, reprinted with permission from *Marine Pollution Bulletin,* vol. 4, p. 134, copyright © 1973 by Pergamon Press, Elmsford, NY.

developed severe neurological problems, including muscle spasms, loss of equilibrium, and impaired mental state. Birth defects and death were common. An important clue to the cause of these illnesses was the co-occurrence of these symptoms in the victims' pet cats. What the humans and the cats had in common was the ingestion of fish and shellfish from Minamata Bay. As shown in Figure 30.18, the fish had mercury concentrations well beyond any safe ingestion level. To make matters worse, the people of Minamata Bay had a very high rate of fish ingestion causing their tissues to become highly enriched in mercury as shown in Table 30.11.

The mercury was introduced as an industrial effluent from a chemical processing plant that made acetaldehyde using the metal as a catalyst. Once in the bay, the mercury was bioaccumulated by the phytoplankton and passed up the food chain until it was consumed by the bay's residents and their cats. Because the results of mercury poisoning are so extreme, its introduction into seawater is now strictly controlled in most westernized countries. This lesson has not been learned everywhere. For example, hundreds of tons of mercury are presently being dumped into the Amazon every year as effluent from the Brazilian gold-mining industry. Some scientists have predicted that this mercury will eventually cause a contamination of the aquatic food chain similar to that seen in Minamata Bay.

Little of the mercury passed through the food chain is in elemental form. Most is present as organometallic complexes, which are far more toxic. The most potent form, methyl mercury, is thought to be produced by bacterial activity, as shown in Figure 30.19. This process is termed biomethylation.

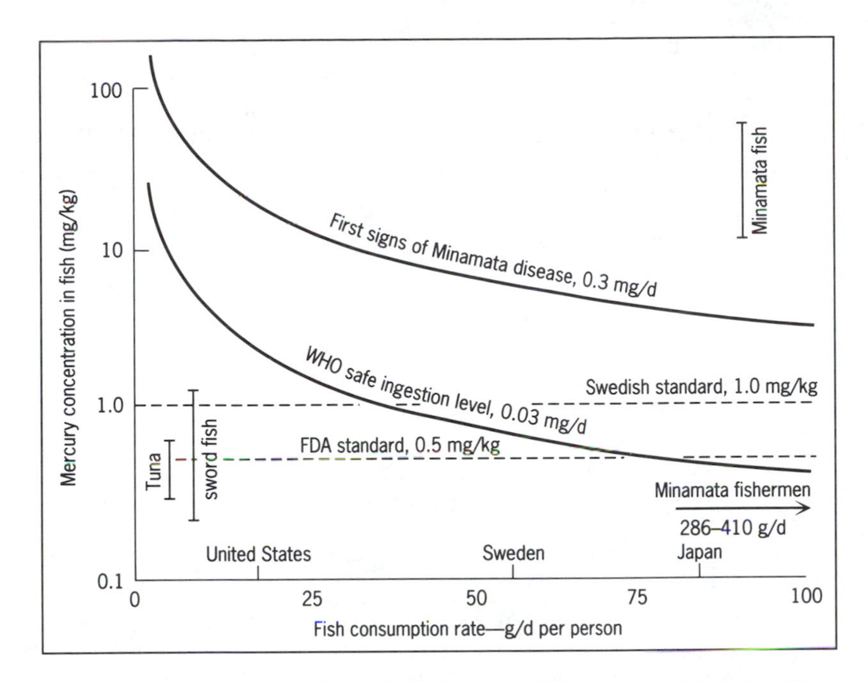

FIGURE 30.18. Mercury levels in fish versus fish consumption rates. The upper curve shows the ingestion rate of mercury at which symptoms of poisoning (Minamata disease) first become evident. The curve identifying the safe ingestion level was obtained by dividing the threshold curve by 10. This is considered to be a suitable safety factor. The average daily fish consumption rates in the United States, Sweden, and Japan are given on the x-axis. The range of mercury concentrations for tuna and swordfish are given near the y-axis. Since the average U.S. citizen consumes 20 g of fish per day, the safe ingestion level would be reached if the Hg concentrations in fish exceed (0.03 mg/d)/(20 g/d) or approximately 1 mg/kg. Due to an apparent arithmetic error, the FDA initially set its standard at 0.5 mg/kg, thus classifying a significant amount of the American fish supply as mercury contaminated. *Source:* From C. B. Officer and J. H. Ryther, reprinted with permission from *Oceanus,* vol. 24 p. 38, copyright © 1981 by Oceanus, Woods Hole, MA.

ARTIFICIAL SUBSTANCES

Humans are creating new compounds at a rate of several thousand per year. The ones that have been identified as problems in the marine environment can be classified as (1) high-molecular-weight aromatic compounds, such as pesticides, polychlorinated biphenyls (PCBs) and polynuclear aromatic hydrocarbons (PAHs); (2) low-molecular-weight, and hence volatile, organic compounds, such as freons; and (3) organometallic compounds, such as tributyltin. The latter are applied to marine structures to prevent biofouling. The others are used for agricultural purposes or are the by-product of industrial activities. They are introduced into seawater via atmospheric input and river

TABLE 30.11
Mercury Concentrations in Samples from Minamata, as well as from Nearby Areas Devoid of Mercury Pollution

	mg/kg Dry Weight	
Material	*Minamata Bay*	*Unpolluted Area*
Clams from the beach	11–39	1.70–6.00
Fish	10–55	0.01–1.70
Cats (from Minimata with mercury poisoning)		
Liver	40–145	0.64–6.60
Kidney	12–36	0.05–0.82
Brain	8–18	0.05–0.13
Man (from Minamata with mercury poisoning)		
Liver	22–70	0.07–0.84
Kidney	22–144	0.25–10.70
Brain	2–25	0.05–1.50
Hair	281–705	0.14–7.50

Source: From *Marine Pollution: Diagnosis and Therapy*, S. A. Gerlach, copyright © 1981 by Springer-Verlag, Heidelberg, Germany, p. 40. Reprinted by permission. Data from H. Tokuomi, reprinted with permission from *Revue Internationale Océanographie Médicale*, vol. 13, p. 23, 1969, copyright © 1969 by the Centre D'Études et de Recherches De Biologie et D'Océanographie Médicale, Nice, France.

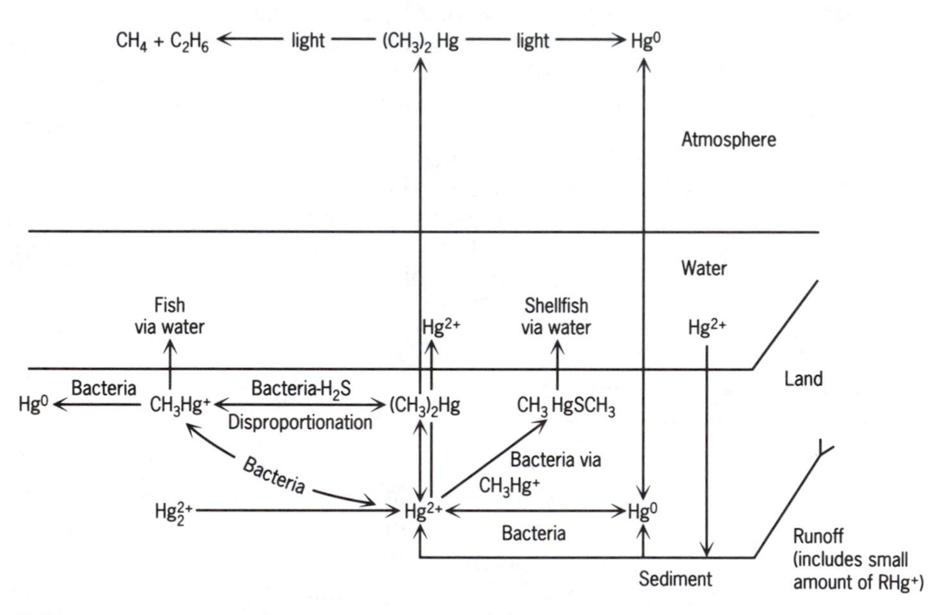

FIGURE 30.19. Transformations of mercury in the aquatic environment. *Source:* From *The Handbook of Environmental Chemistry*, P. J. Craig (ed.: O. Hutzinger), copyright © 1980 by Springer-Verlag, Heidelberg, Germany, p. 179. Reprinted by permission.

runoff. As shown in Figure 30.20, most are present in natural waters at very low concentrations. But due to their toxicity, some have a major impact on marine life.

Pesticides

Unsaturated carbon rings and halides confer toxicity to organic compounds. Synthetic organic chemists have used these substituents to develop potent pesticides, such as those illustrated in Figure 30.21. Another large class of pesticides are based on the toxic effect of phosphorus bound to sulfur as shown in Figure 30.21*k* and *l*. All of these pesticides are particularly effective as (1) they are not species specific and hence kill a wide spectrum of insects, and (2) they are relatively stable in the environment and hence remain active for a long time. Unfortunately, these characteristics also make pesticides toxic to nontarget organisms, such as birds and mammals.

These pesticides are applied to plants and soils. Due to their stability, many persist long enough to be transported by storm- and ground-water runoff into the ocean. Once in the ocean, these compounds enter the marine food web either via adsorption onto organisms or active uptake by phytoplankton. Some of these compounds are degraded by metabolic processes. For example, the most common degradation products of dichlorodiphenyltrichloroethane (DDT) are dichlorodiphenyl dichloroethane (DDD) and dichlorodiphenyl dichloroethylene (DDE). Thus the removal of pesticides from seawater is enhanced

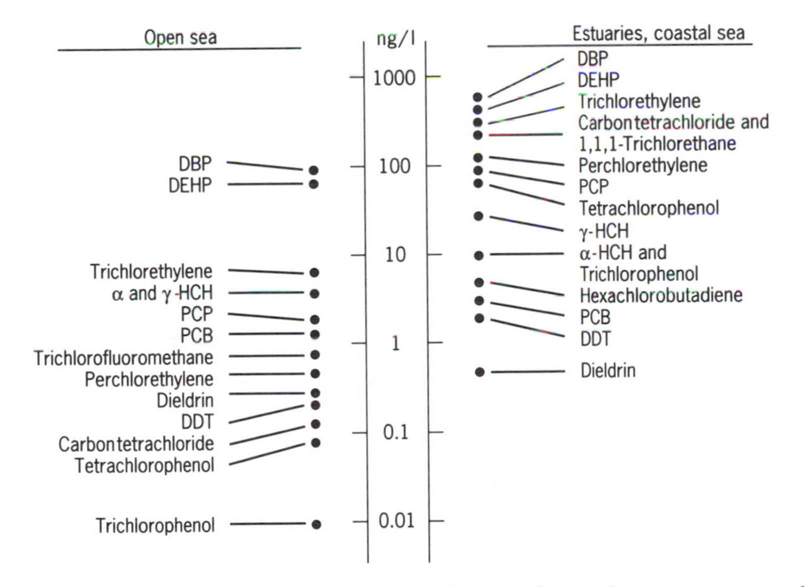

FIGURE 30.20. Concentrations of various pollutants in open ocean and estuarine waters. *Source:* From W. Ernst, reprinted with permission from *Helgol Meeresunters,* vol. 33, p. 302, copyright © 1980 by Biologische Anstalt Helgoland, Hamburg, Germany.

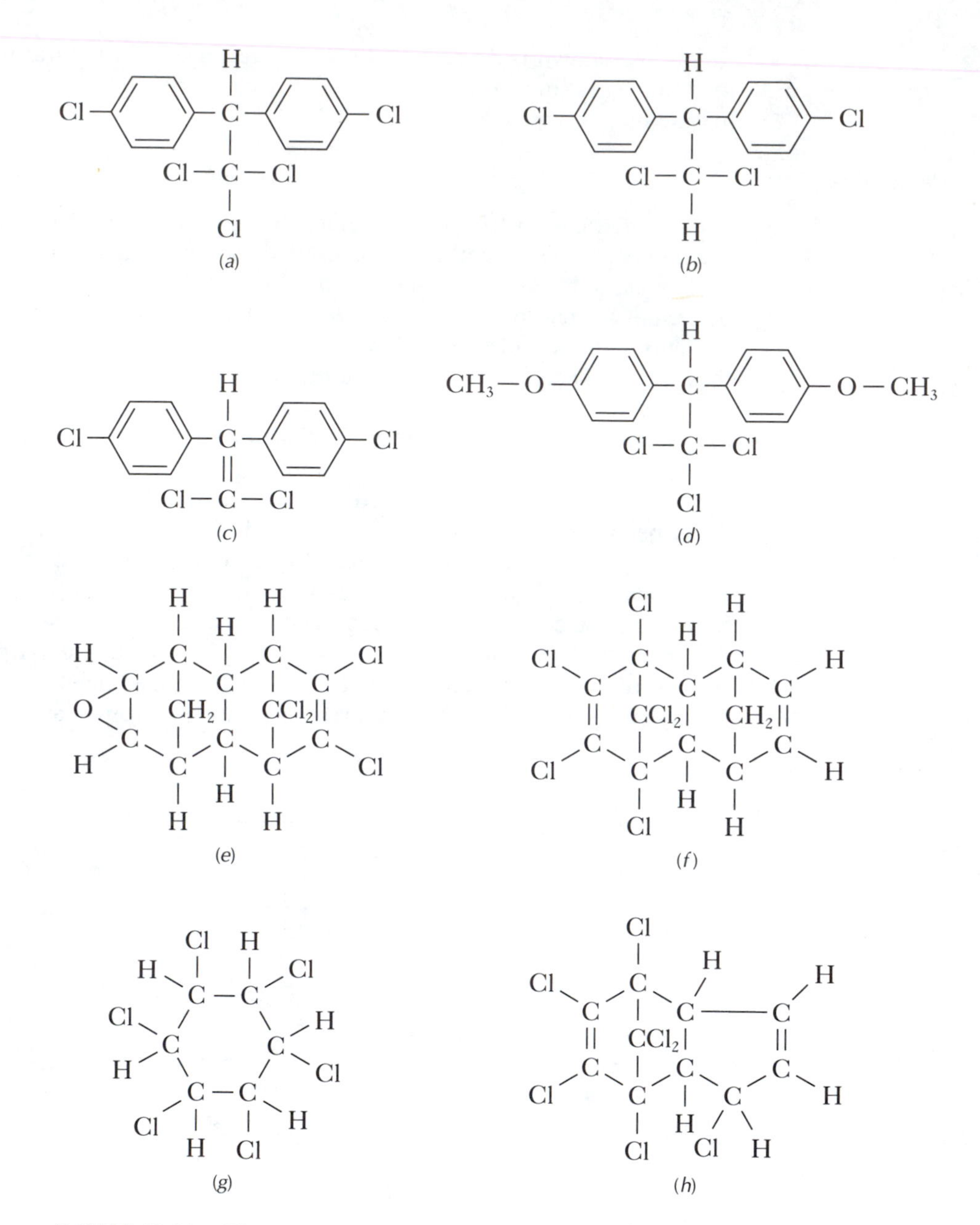

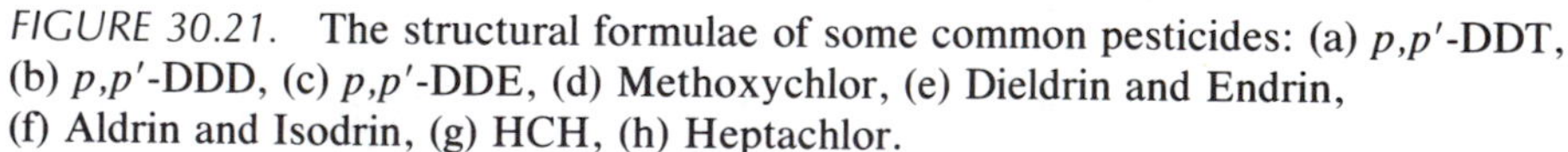
FIGURE 30.21. The structural formulae of some common pesticides: (a) *p,p′*-DDT, (b) *p,p′*-DDD, (c) *p,p′*-DDE, (d) Methoxychlor, (e) Dieldrin and Endrin, (f) Aldrin and Isodrin, (g) HCH, (h) Heptachlor.

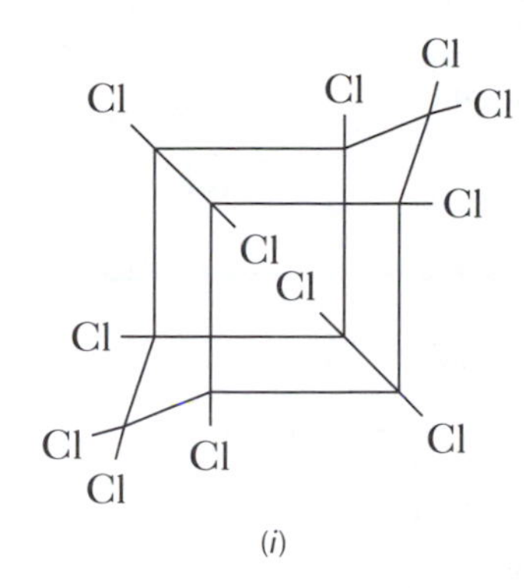

(i)

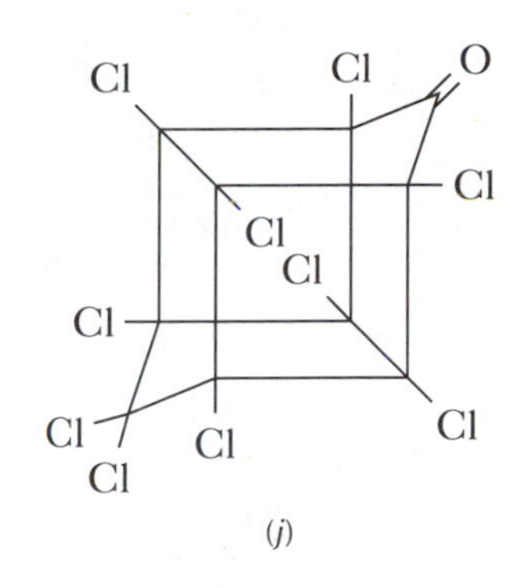

(j)

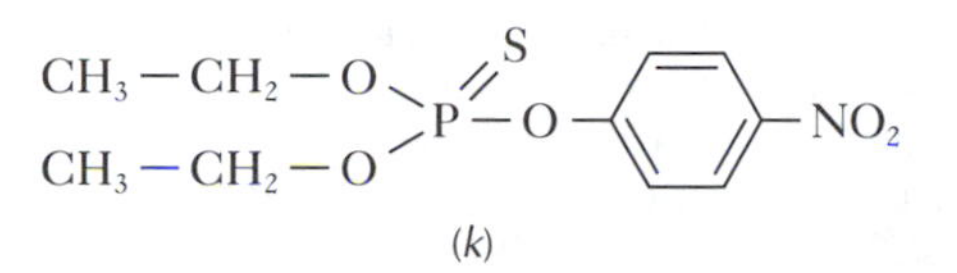

(k)

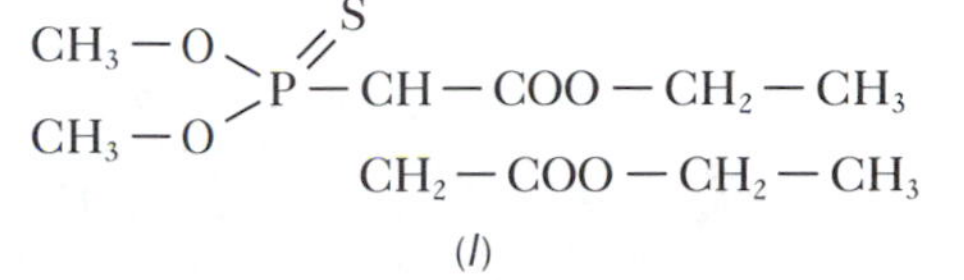

(l)

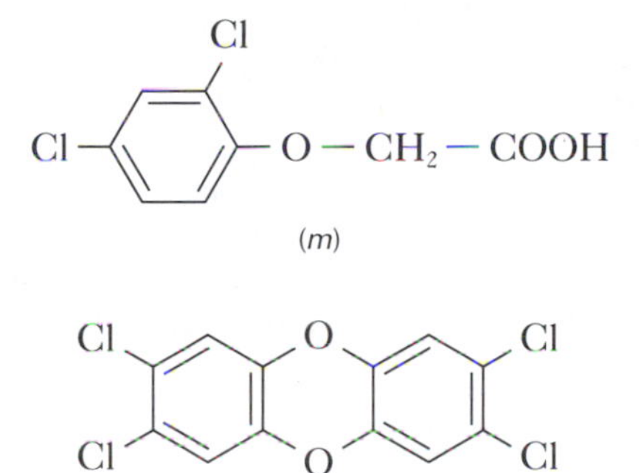

(m)

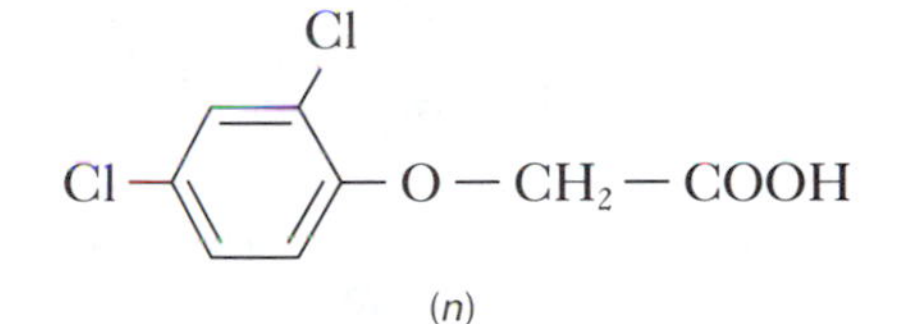

(n)

(o)

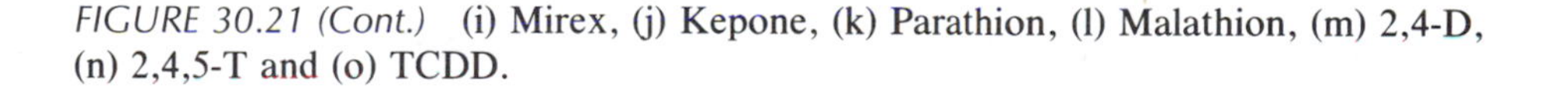

FIGURE 30.21 (Cont.) (i) Mirex, (j) Kepone, (k) Parathion, (l) Malathion, (m) 2,4-D, (n) 2,4,5-T and (o) TCDD.

by the presence of large amounts of particles and biological activity. As a result, pesticide residence times are longest in the open ocean and shortest in estuarine waters, as shown in Table 30.12.

As shown in Figure 30.22, biomagnification and bioaccumulation cause pesticides, such as DDT, to reach very high levels in organisms that are at the top of the food web. Such an enrichment occurred during the 1960s and 1970s in the pelicans that live on islands located off the coast of Southern California. As shown in Table 30.13, pelican numbers declined dramatically in the 1970s because of reproductive failure. The cause for this decline was traced to the DDT enrichment that led the females to lay very thin-shelled eggs. As a result, the eggs were crushed during gestation.

TABLE 30.12
Estimated Residence Times for Some Halogenated Hydrocarbons in the Euphotic Zone of Oligotrophic, Mesotrophic, and Eutrophic Waters[a]

Station	*Residence Time (y)*		
Area Type	*ΣHCH*[a]	*ΣDDT*[b]	*PCBs*[c]
Oligotrophic	5.1–10	0.19 –0.37	0.38 –0.76
Mesotrophic	4.3– 6.4	0.08 –0.12	0.082–0.12
Eutrophic	2.0– 3.4	0.031–0.052	0.070–0.12

Source: From *The Role of the Oceans as a Waste Disposal Option*, J. W. Farrington and J. Westall (ed.: G. Kullenberg), copyright © 1985 by Reidel Publishing Company, Dordrecht, The Netherlands. Reprinted by permission. After S. Tanabe and R. Tatsukawa, reprinted with permission from the *Journal of the Oceanographical Society of Japan*, vol. 39, p. 60, copyright © 1983 by the Oceanographical Society of Japan, Tokyo, Japan.
[a]ΣHCH = [α-HCH] + [β-HCH] + [α-HCH]
[b]ΣDDT = [*p, p*′-DDE] + [*p, p*′-DDT] + [*o, p*-DDT]
[c]The isomers of PCB are listed in Figure 30.23.

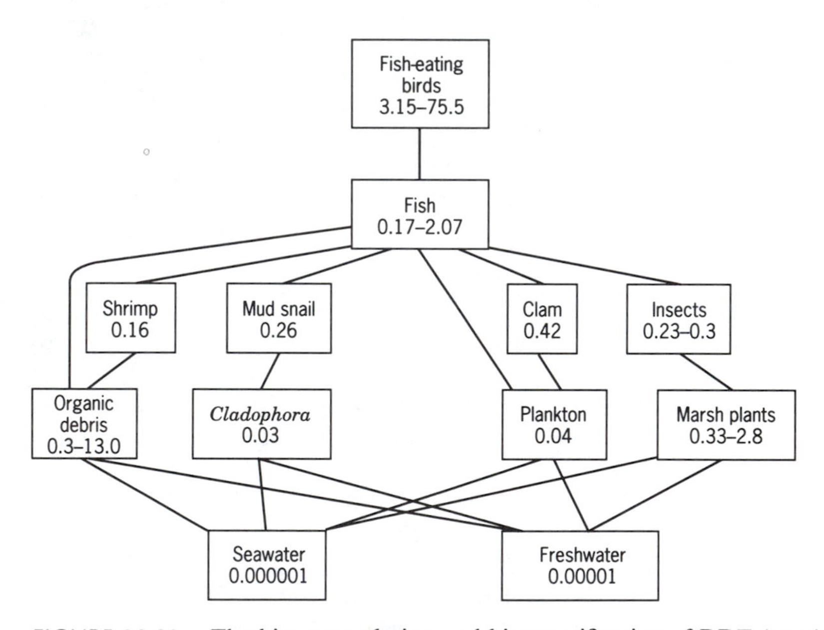

FIGURE 30.22. The bioaccumulation and biomagnification of DDT (ppm) in a marine and aquatic food web. *Source:* From *Persistent Pesticides in the Environment,* C. A. Edwards, copyright © 1973, CRC Press, Boca Raton, FL, p. 80. Reprinted by permission.

TABLE 30.13
ΣDDT Concentrations ([DDT] + [DDD] + [DDE]) in Sardines and Intact Pelican Eggs from 1969 to 1974[a]

Year	*Number of Nests*	*Number of Young Birds*	*DDT in Intact Eggs (mg/kg fat)*	*DDT in Sardines (mg/kg wet weight)*	*Abundance of Sardines (1000 shoals per sea area)*
1969	1125	4	907	4.27	140
1970	727	5		1.40	70
1971	650	42		1.34	80
1972	511	207	221	1.12	195
1973	597	134	183	0.29	275
1974	1286	1185	97	0.15	355

Source: From *Marine Pollution: Diagnosis and Therapy*, S. A. Gerlach, copyright © 1981 by Springer-Verlag, Heidelberg, Germany, p. 48. Reprinted by permission. Data from D. W. Anderson, J. R. Jehl, Jr., R. W. Risebrough, L. A. Woods, and W. G. Edgecomb, reprinted with permission from *Science*, vol. 190, p. 807, copyright © 1975 by the American Association for the Advancement of Science, Washington, DC.

[a]Also listed are numbers of pelican nests, young birds, and sardines living in the vicinity of Anacapa and Coronado Norte islands off the coast of Southern California.

The decline in the pelican population was probably worsened by a concurrent enrichment in PCBs and a concurrent decrease in sardine abundance. As shown in Table 30.13, the sardines also had large pesticide enrichments, but no evidence has been found to connect their decline with increasing DDT levels. Partly in response to the reproductive failure of this pelican population, the use of DDT was severely curtailed in 1970 and banned in 1971. This caused pelican numbers to rebound in the early 1970s as environmental levels of the pesticide declined rapidly.

Despite the ban on its use, DDT continues to be manufactured in the United States for use in other countries. Thus DDT is still introduced into the ocean. This illustrates the importance of international efforts in the control of marine pollution. In the early 1980s, DDT levels were observed to be increasing on land and in the coastal waters of the United States. This DDT is thought to be derived from either the illegal use of the pesticide or from the import of foodstuffs grown with its use.

Polychlorinated Biphenyls

Polychlorinated biphenyls (PCBs) are synthetic organic compounds used primarily as dielectrics in transformers and large capacitors because they are very stable at high temperatures. Hence they are also used as flame retardants. The base unit for PCBs is the biphenyl group shown in Figure 30.23. A wide variety of PCBs have been produced by various substitutions of chlorine onto this base unit.

(a)

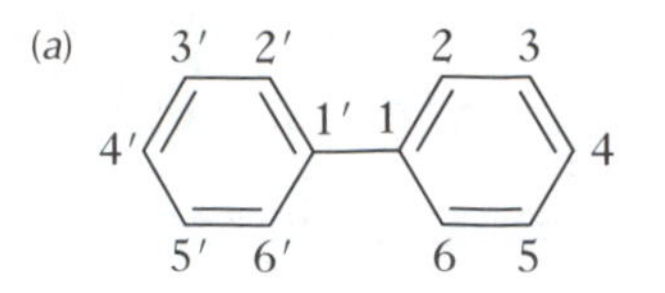

(b)

IUPAC No.	*Chlorine Substitution*
8	2, 4′
28	2, 4, 4′
29	2, 4, 5
44	2, 2′, 3, 5′
49	2, 2′, 4, 5′
52	2, 2′, 5, 5′
60	2, 3, 4, 4′
70	2, 3′, 4′, 5
86	2, 2′, 3, 4, 5
87	2, 2′, 3, 4, 5′
95	2, 2′, 3, 5′, 6
101	2, 2′, 4, 5, 5′
105	2, 3, 3′, 4, 4′
110	2, 3, 3′, 4′, 6
118	2, 3′, 4, 4′, 5
128	2, 2′, 3, 3′, 4, 4′
129	2, 2′, 3, 3′, 4, 5
137	2, 2′, 3, 3′, 6, 6′
138	2, 2′, 3, 4, 4′, 5
143	2, 2′, 3, 4, 5, 6′
153	2, 2′, 4, 4′, 5, 5′
156	2, 3, 3′, 4, 4′, 5
180	2, 2′, 3, 4, 4′, 5, 5′

FIGURE 30.23. *(a)* Base unit of PCBs and *(b)* molecular formulae of various chlorine substitutions.

PCBs have been introduced into seawater as a component of industrial effluent and by the dumping of electrical components. As with most pollutants, PCBs tend to become sequestered in the sediments close to their point of introduction as a result of adsorption and incorporation into biogenic POM. This causes these compounds to have relatively short residence times in seawater, as shown in Table 30.12. Like the halogenated pesticides, PCBs undergo biomagnification in the marine food chain, but due to their great stability and toxicity, they are not readily degraded. As a result, their residence time in the sediments is very long.

Because PCBs are human carcinogens, their use has been greatly restricted and even banned in some countries. Nevertheless, the sediments of many industrialized ports are already heavily contaminated with PCBs. This is a matter of some concern as any disturbance to the sediments can cause a release of these toxic compounds into the overlying waters. The sedimentary concentrations of PCBs are so high in the harbor of New Bedford, Rhode Island, that this area has been designated as a Superfund site. Though so targeted, no cleanup is presently under way due to lack of knowledge as to how to remove the toxic wastes without causing a water-quality problem.

Though PCBs are widely recognized as an environmental hazard, other halogenated aromatic compounds continue to be used. These compounds have structures and uses similar to that of PCBs and thus are likely to cause analogous environmental problems. Some examples are given in Table 30.14.

Dioxins are a particularly toxic class of halogenated aromatic compounds. One form, 2,3,7,8-tetrachlorodipenzol-*p*-dioxin (TCDD) (Figure 30.21*o*), is carcinogenic at such low levels that the **United States Environmental Protection Agency (USEPA)** has recommended that water concentrations not exceed 0.014 parts per quadrillion. Paper mills are presently the primary source of this dioxin to coastal waters. The compound is a by-product of the bleaching process used in the manufacture of white paper. Most mills are located on lakes, rivers, or estuaries into which their industrial wastes are discharged. The highest TCDD levels occur in Winyah Bay, South Carolina, as a result of the introduction of effluent from a mill operated by the International Paper Company. The enrichment factor for TCDD in fish exceeds 10^3. As a result, the USEPA has posted signs in Winyah Bay warning of the dangers of consuming local seafood.

Dioxins are also generated as a by-product during the manufacture of the herbicide 2,4,5-T, also known as Agent Orange. The incomplete incineration of hazardous wastes also produces dioxins, which are then introduced into natural waters as a result of atmospheric fallout and runoff. Due to the production of such air pollutants, incineration at sea is currently favored for the disposal of some hazardous wastes. In addition to the deposition of ash

TABLE 30.14
Production and Uses of Halogenated Aromatic Compounds

Compound	*Major Trade Names*	*World Production (1000 ton/y)*	*Major Uses*
Polychlorinated biphenyls (PCB)	Aroclor Phemoclor Kanechlor Clophen Fenclor Santotherm	70[a]	Capacitor dielectric, transformer coolant, hydraulic fluid, plasticizer, heat transfer fluid
Polychlorinated terphenyls (PCT)	Aroclor Kanechlor	5	Adhesives and sealants
Polychlorinated naphthalenes (PCN)	Halowax	5	Capacitor dielectric, oil additive
Polybrominated biphenyls (PBB)	Firemaster	5[b]	Fire retardant

Source: From *The Handbook of Environmental Chemistry*, vol. 3, Part B, C. R. Pearson (ed.: O. Hutzinger), copyright © 1982 by Springer-Verlag, Heidelberg, Germany, p. 90. Reprinted by permission.

[a]Peak annual rate for 1969–1970.

[b]Peak annual rate for 1974.

and aerosols, gaseous incinerator products are also likely to be introduced into the ocean as a result of diffusion across the air–sea interface.

Polynuclear Aromatic Hydrocarbons

Polynuclear aromatic hydrocarbons (PAHs) are truly contaminants, as they are a natural component of petroleum and are synthesized by some plants and bacteria. Despite this variety of sources, their "natural" concentrations are very low. Environmental concentrations have been greatly elevated as a result of oil spills, fossil-fuel burning and the use of petroleum tars as wood preservatives. Because of this atmospheric injection, PAHs have been transported across the globe. Evidence for the anthropogenic input of these compounds has been found in polar ice cores and is similar to that observed for lead.

The structures of some PAHs are given in Figure 30.24. Like most aromatic compounds, these substances are very potent human carcinogens. Like the PCBs, PAHs are fat soluble and hence tend to become concentrated in the livers of marine organisms. Both compounds are linked to the recent increase in cancers observed in coastal fish and are also thought to cause genetic mutations and impairment of the immune system. The health affects of consuming cancerous fish are unknown.

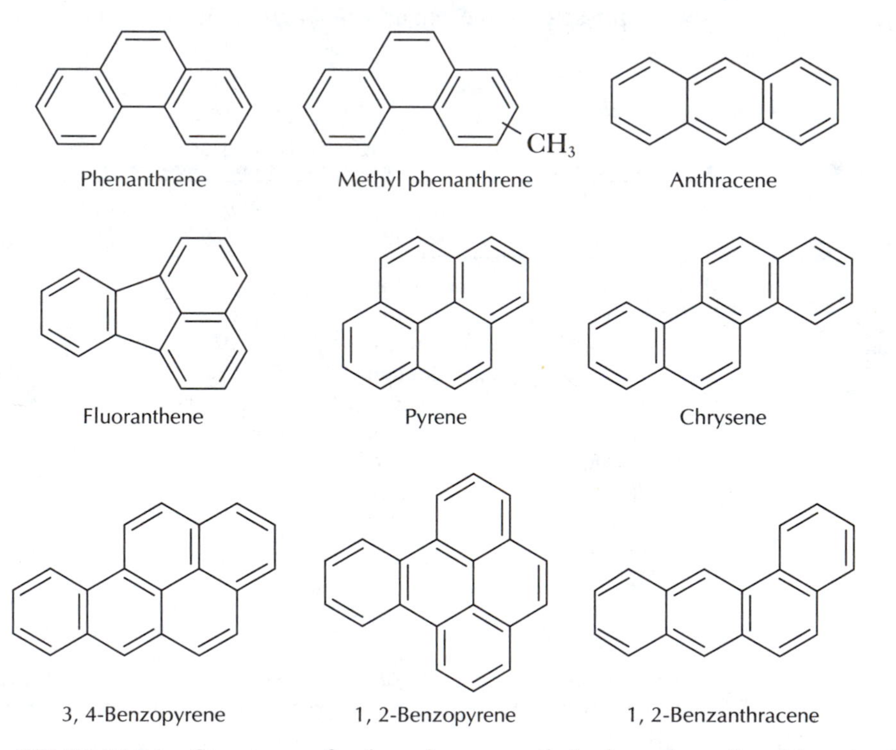

FIGURE 30.24. Structures of polynuclear aromatic hydrocarbons found in natural waters.

Volatile Organic Compounds

Low-molecular-weight molecules tend to have high vapor pressures and hence are also termed volatile organic compounds or VOCs. As shown in Figure 30.25, many are chlorinated. The primary use of VOCs, such as carbon tetrachloride and benzene, is as industrial solvents. Trichloroethylene is commonly used as dry-cleaning fluid. As a result of poor waste-disposal procedures, most of the ground water in the United States now contains some of these compounds. This is a matter of great concern as most VOCs are carcinogenic.

VOCs are introduced into the ocean via river runoff and gas exchange at the air–sea interface. The latter process is not well understood as some VOCs are thought to have a natural marine sources. The ocean appears to be a global sink for the synthetic ones. For example, 0.5 to 1 percent of the CCl_3F emissions have been dissolved into the ocean.

This compound is one of the chlorinated fluorocarbons, called freons, that are commonly used as refrigerants and flame retardants. Their use as aerosol propellants has been banned in the United States because these compounds have been shown to react with stratospheric ozone and cause its destruction. This loss is of concern as stratospheric ozone reacts with ultraviolet radiation, thereby shielding Earth's surface from highly mutagenic radiation. It is somewhat ironic that ozone is being added to the lower atmosphere (troposphere) as a result of fossil-fuel combustion. This ozone reacts with other air pollutants and sunlight to form photochemical smog, which is a health threat to humans.

Organometallic Paints

In 1975, more than $1 billion was spent to repair and protect marine surfaces from the effects of biofouling. The most common preventatives now in use are organometallic paints. Though copper-based compounds are still used, tributyltin (TBT) has become the major component of antifouling paints because it is 10 to 100 times more effective. In 1985, TBT was used on 20 to 30 percent of vessels worldwide and is also applied to docks and cooling towers. In the United States, 140,000 kg of TBT-containing antifouling paint is used each year on commercial and recreational boats and ships. Due to its broad-spectrum toxicity, tributyltin is also used as a fungicide, bactericide, and as insecticide on textiles, paper, leather, and electrical equipment.

TBT is leached from the paints and thereby introduced into seawater. This compound induces short-term acute effects in oysters and other nontarget molluscs at the ng/L level. The enrichment factor in oysters ranges from 2300 to 11,400. Due to slow clearance times, oysters make good sentinel organisms for monitoring changes in TBT concentrations which occur over periods of several months. Concentration data from the Mussel Watch program indicate that TBT concentrations are highest and increasing in harbors and marinas. This suggests that leaching from paints is the major source of TBT to the ocean.

Mono- and dichlorinated benzenes

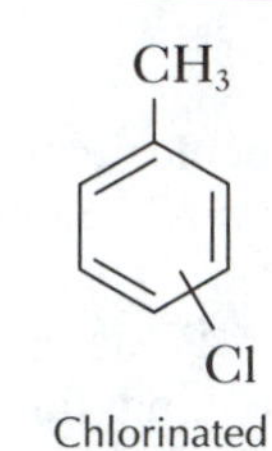

Chlorinated toluene

$CHCl_3$

Chloroform

$CHCl_2Br$ $CHClBr_2$ $CHBr_3$

Trihalomethanes

CH_3-CH_2Cl CH_2Cl-CH_2Cl

Chlorinated ethanes

$Cl_2C=CHCl$

Chlorinated ethenes

CCl_4

Carbon tetrachloride

$CH_3-CH_2-CH_2-Cl$

Chlorinated propanes

$CH_3-CH=CHCl$

Chlorinated propenes

$CHCl_3$ $CHBrCl_2$ $CHBr_2Cl$ $CHBr_3$

Halogenated methanes

Halogenated aromatics

$Cl_2C=CHCl$

Trichloroethane

$Cl_2C=CCl_2$

Tetrachloroethane

Halogenated alkenes

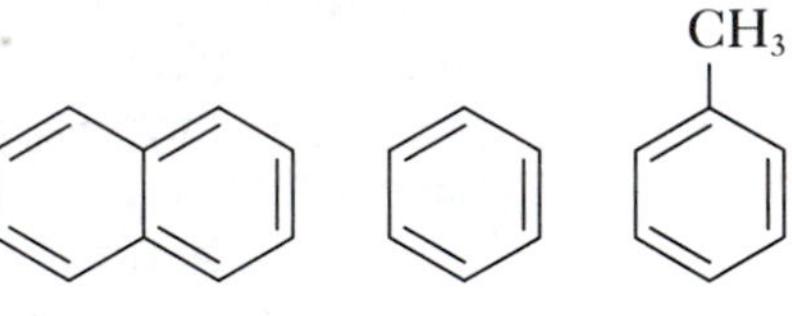

Aromatics

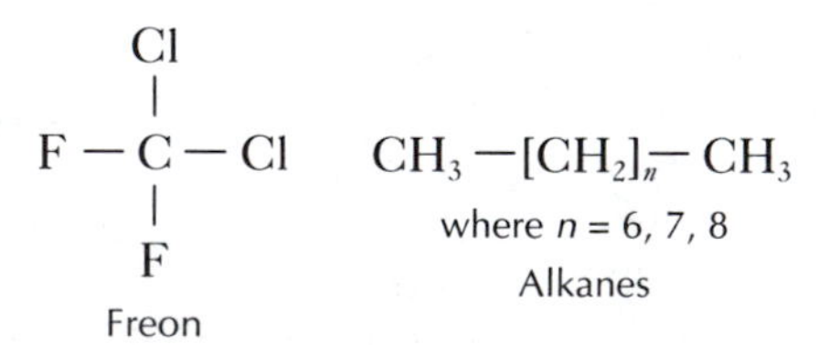

FIGURE 30.25. **Structures of some volatile organic compounds commonly found in natural waters.**

TBT is also bioaccumulated and biomagnified by fish, crabs, and microorganisms. The chemical half-life of TBT in harbor water is approximately 1 week and is thought to be controlled by this biological uptake and subsequent degradation. Most of the degradation appears to be the result of bacterial processes that convert TBT to dibutyltin (DBT) and monobutyltin (MBT). Both are somewhat less toxic than their parent compound. TBT also forms chloride, carbonate, and hydroxide complexes in seawater. Though TBT concentrations are higher in the sediments, degradation rates are slower than in the water column. This is thought to reflect the inhibitory effect of higher TBT concentrations on bacterial activity.

Due to the toxicity of TBT, Japan is planning to ban its use in 1991. Several substitutes have been developed, such as polysiloxane silicone polymers and epoxy urethane coatings, both of which create such a slippery surface that microorganisms are washed off if water flow rates exceed 5 ft/s. Also being developed are natural marine proteins, such as a coral extract called leptogorgia, to act as a supporting medium for copper and the antibiotic, tetracycline.

Marine Litter

An estimated 6 million tons of garbage, or *marine litter*, is disposed of at sea every year. As a result, floating debris is now a common feature of the surface ocean. The dumping of garbage is banned in most coastal waters; hence marine litter is largely the result of open-ocean dumping from passenger, recreational, military, and commercial ships. The United States is responsible for one-third of this input. The most common types of marine litter are listed in Table 30.15.

TABLE 30.15
Some Types of Marine Litter

Plastics	
Fishing gear	Nets, floats, ropes, traps, and fishing lines
Cargo-associated wastes	Plastic strapping and sheeting; debris from the offshore petroleum industry; and pellets of polyethylene, polypropylene, and polystyrene
Sewage-associated wastes	Condoms, tampon applicators, sanitary napkins, and disposable diaper shields
Domestic	Bags, bottles, lids, six-pack rings, and balloons
Rubber and metal fragments	
Glass bottles	
Wood and paper	
Metal drums containing synthetic wastes	
Munitions	

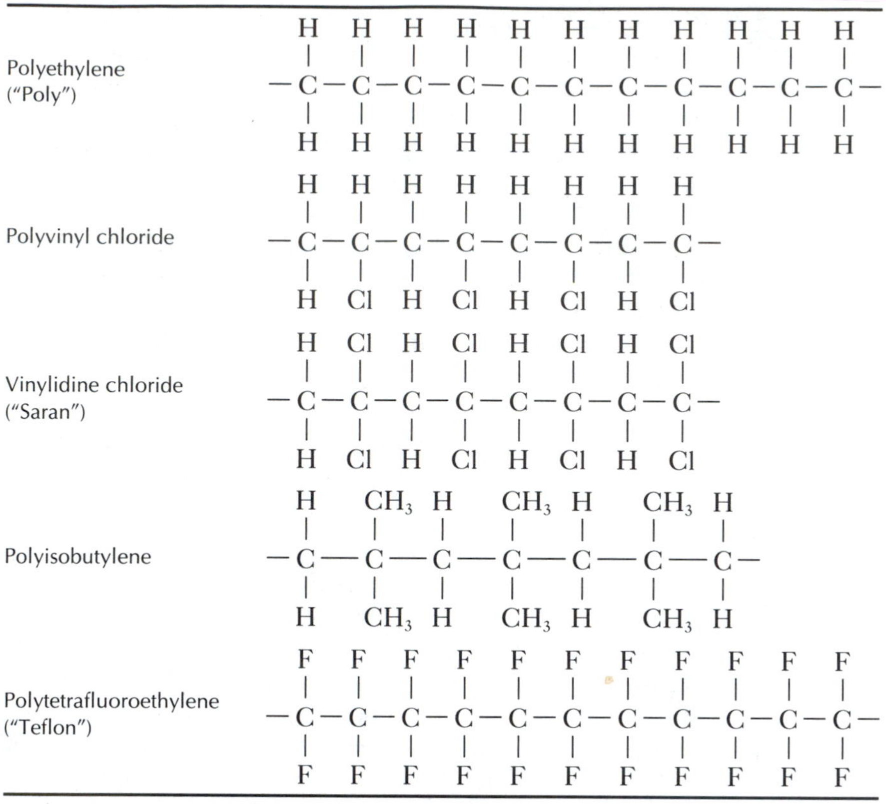

Polyethylene ("Poly")	H H H H H H H H H H H │ │ │ │ │ │ │ │ │ │ │ —C—C—C—C—C—C—C—C—C—C—C— │ │ │ │ │ │ │ │ │ │ │ H H H H H H H H H H H
Polyvinyl chloride	H H H H H H H H │ │ │ │ │ │ │ │ —C—C—C—C—C—C—C—C— │ │ │ │ │ │ │ │ H Cl H Cl H Cl H Cl
Vinylidine chloride ("Saran")	H Cl H Cl H Cl H Cl │ │ │ │ │ │ │ │ —C—C—C—C—C—C—C—C— │ │ │ │ │ │ │ │ H Cl H Cl H Cl H Cl
Polyisobutylene	H CH_3 H CH_3 H CH_3 H │ │ │ │ │ │ │ —C—C—C—C—C—C—C— │ │ │ │ │ │ │ H CH_3 H CH_3 H CH_3 H
Polytetrafluoroethylene ("Teflon")	F F F F F F F F F F F │ │ │ │ │ │ │ │ │ │ │ —C—C—C—C—C—C—C—C—C—C—C— │ │ │ │ │ │ │ │ │ │ │ F F F F F F F F F F F

FIGURE 30.26. Molecular structure of some plastic products. *Source:* From *The Chemistry of Our Environment,* R. A. Horne, copyright © 1969 by John Wiley & Sons, Inc., New York, p. 441. Reprinted by permission.

Though pieces of plastic constitute only about 1 percent of marine litter, they represent the majority of the floating debris due to their low density. Most plastics are polymers of polyethylene, polypropylene, or polystyrene. The molecular structures of the base units of some of these plastic polymers are illustrated in Figure 30.26.

In addition to dumping, plastics are also introduced into the ocean from (1) the degradation of flotation material in docks, (2) littering on beaches, (3) the use of balloons, and (4) loss during transportation. Though dumping is forbidden in coastal waters,* a significant amount of garbage is transported to landfills by barges. Much litter can be blown off the barges while in transit. A large amount of ''raw'' plastic, in the form of polystyrene spherules and

*The sole exception to this has been granted to NYC for the dumping of its municipal sewage until 1991.

polyethylene pellets, is also lost during shipment to manufacturing plants. This debris ranges in diameter from 0.1 to 5 mm. Densities as high as 10,000 and 34,000 particles/km^2 have been observed in the Sargasso Sea and Pacific Ocean, respectively. The pellets and spherules appear to be carried by currents until they are either degraded photochemically, consumed by marine organisms, or stranded on beaches. The pellets have been found embedded in tar balls, forming concretions that coat some Caribbean beaches.

Although plastics in the ocean represent an alternate substrate for some sessile organisms, they can cause considerable harm to other types of marine life. Sea turtles consume plastic bags, apparently mistaking them for jelly fish. The ingested bags eventually cause death because they take up so much space in the turtles' stomachs. Plastics are also a threat to marine life because of entanglement. This is particularly a problem with the plastic rings that hold beer and soda cans in six-packs. "Ghost fishing" by abandoned fishing nets is an even larger danger to marine life. Since most plastics are not readily degradable, these nets drift through the ocean for long periods of time, during which they continue to catch large numbers of marine organisms.

Because of the widespread distribution and increasing amount of marine litter, an international agreement has been reached restricting the dumping of garbage at sea. As indicated in Figure 30.27, ocean dumping of plastics is now strictly forbidden, but other types of garbage can be legally disposed of in offshore waters.

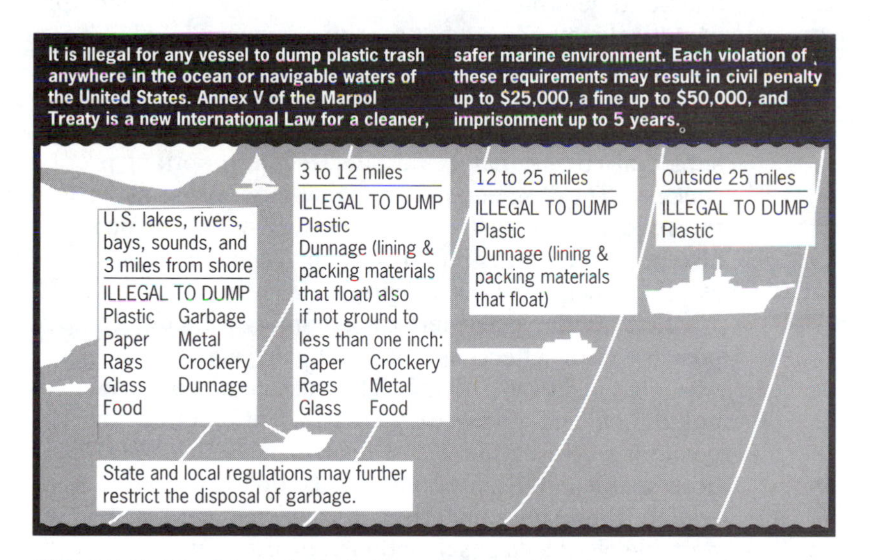

FIGURE 30.27. Marpol regulations on ocean dumping. *Source:* From *A Citizen's Guide to Plastics in the Ocean,* K. J. O'Hara, S. Iudicello, and R. Bierce, copyright © 1988 by the Center for Marine Conservation, Washington, DC, p. 37. Reprinted by permission.

THE FUTURE OF WASTE DISPOSAL AT SEA

The anthropogenic waste stream is now so large that we have become almost comically desperate in our search for dumpsites. For example, in 1987 a barge traveled up and down the eastern seaboard of the United States for six months in a futile attempt to find a place to dump the garbage it was carrying. Because no one wants to live next to a garbage dump, disposal at sea has always been an appealing option for waste disposal.

Despite our current problems with marine pollution, some scientists feel that with adequate management, the ocean can continue to be used as a site for the disposal of at least some wastes. In this view, the ocean is assumed to have some finite capacity to assimilate various compounds within which an unacceptable impact on living organisms or nonliving resources does not result. What constitutes an unacceptable impact would be largely defined by socioeconomic pressures. In other words, some amount of harm would be acceptable if the alternatives, such as contamination of drinking water, pose a larger threat to either the environment or to the quality of human life as evidenced by economic hardship. These scientists feel confident that specific indices could be developed to monitor the effects of pollutants and hence prevent "catastrophes."

Many scientists do not agree with this analysis. They feel that our lack of knowledge is so profound that a precautionary approach is warranted. That is, the benefit of any doubt should go to the ocean. For example, we are still unable to determine the toxic threshold levels of many pollutants because our analytical methods are not sensitive enough. In addition, most toxicity studies are based on short-term laboratory bioassays that investigate the effects of single compounds. These results do not provide any information from which to assess any long-term or synergistic effects. They also do not consider sublethal effects that could affect such important processes as reproductive success. These scientists also view monitoring as an unacceptable approach to preventing "catastrophes" due to the long response times involved. They also point out that careful management will have to be an international effort, as atmospheric and ocean currents distribute pollutants worldwide.

Due to the ever-increasing number of new compounds, pretesting for possible toxic effects is not practically achievable. Even if this information was obtained, some degree of uncertainty concerning the effects of marine pollution would always exist. In recognition of this uncertainty, a compromise position would be the use of management strategies that minimize the risk to both health and the environment. To do this would require an evenhanded and equally thorough comparison of environmental risks and benefits for all the possible waste disposal options. In other words, just because an option is cheap shouldn't cause it to be selected as a waste disposal site for which an appropriate implementation strategy is then designed. With this compromise approach, the ocean would be an acceptable waste disposal site only when land or atmospheric dissemination of pollutants becomes scientifically, economically, and socially unjustifiable.

Though the laws regulating marine pollution in the United States are probably the world's strictest, they have not been particularly successful, nor are they readily amenable to the compromise approach outlined above. Historically, the United States has attempted to regulate pollution by controlling the dumping and discharging of wastes. The most important pieces of legislation concerning this are the **Ocean Dumping Acts of 1972** and **1988** and the **Clean Water Acts of 1948** and **1977**. These laws contain the implicit assumption that ocean disposal is the least preferable method for dealing with pollutants. The approach taken has largely been the precautionary one: i.e., "If you don't know, don't dump."

The Clean Water Act also attempts to control pollution by establishing water-quality criteria for effluent. The USEPA has been charged with developing these water-quality criteria, but due to limited funding and the need for advances in analytical techniques, progress has been slow. This approach has come under attack for several reasons. For example, it ignores variations in the assimilative capacities of the receiving waters. The requirement that all regions use the same water-quality standard also ignores considerations of relative costs and risks. As a result, many municipalities have not met regulations because they do not have the funds to build the necessary secondary sewage treatment plants. This necessitated an amendment to the Clean Water Act that allows the USEPA to waive the requirement for secondary treatment on a case-by-case basis.

Marine pollution has also proven difficult to legally control because of the great many laws that exist. This is the product of a legislative process that responds to each pollutant as it becomes a particular problem. As a result, many pollutants are covered by multiple laws, some of which are contradictory. It has also necessitated that anyone proposing an activity that might alter the marine environment obtain a bewildering number of permits. Lack of enforcement and monitoring has also greatly diluted the original intent of this legislation.

Any long-term solution to the problem of waste disposal must involve a reduction in our waste stream. This can be achieved through conservation and recycling. What waste must be introduced should be as environmentally compatible as possible. But as landfill sites continue to close, the pressure to use the ocean as a dumpsite will increase. Thus it seems inevitable that marine pollution will also increase. In some cases, pollution has already caused changes in material fluxes, some of which have perturbed stabilizing feedbacks in the crustal-ocean factory. This has begun to cause ecological catastrophes, such as those described in the introduction to this chapter. To prevent further degradation of the marine environment requires that we be able to realistically and reliably predict the impacts of any future pollution.

To make such predictions requires a large increase in our understanding of the ocean. We need a better understanding of the kinds and amounts of pollutants that are being introduced, how they are being transported to the sea, and what happens to them after they get there. In particular we need to know more about (1) gas and particle exchange across the air–sea interface;

(2) the uptake, release, and transformation of chemicals by marine organisms; (3) particle scavenging and colloid flocculation; (4) the exchange of solutes and particles at the sediment-water interface; and (5) the factors that control marine productivity. Many of these require a better understanding of physical processes, such as turbulence and water-mass renewal, as well as geological processes, such as particle transport. It is also essential that we learn how these processes vary over time and space. To achieve this will require an enormous international effort, involving a large-scale commitment of time and resources. Obviously the need for new marine biogeochemists has never been greater or more important.

SUMMARY

A **marine pollutant** is defined as any substance that alters any natural feature of the ocean. Naturally occurring substances whose concentrations have been increased as a result of human activity are termed **contaminants**. Humans have always used the ocean as a site for waste disposal because they assumed the concentrations of all pollutants would be kept at low and innocuous levels as a result of dilution and dispersion.

Marine pollution is hard to study for the following reasons: (1) The natural levels of many contaminants are unknown due to the relatively recent development of analytical methods for their measurement; (2) Their transport through the crustal-ocean factory is very complex; (3) An ultimately subjective judgment is required to determine the toxicity of a pollutant; (4) A large number of input mechanisms exist; (5) The number of pollutants is large and increasing; and (6) These pollutants interact, often with synergistic results.

The marine contaminants include (1) detrital inorganic particles, (2) nutrients and organic matter, (3) microorganisms, (4) petroleum hydrocarbons, (5) radionuclides, and (6) **heavy metals**. The artificial pollutants can be classified as (1) high-molecular-weight aromatic compounds, such as pesticides, PCBs, and PAHs, (2) low-molecular-weight, and hence volatile, organic compounds, such as freons, and (3) organometallic compounds, such as tributyltin.

Most pollutants are introduced into the ocean as either **point-source** inputs, such as sewage outfalls and **non-point-source** inputs, such as ground-water runoff. Most of this occurs in estuaries. Pollutants tend to behave **nonconservatively** in estuaries. This is largely the result of (1) gradients in water chemistry, which affect speciation and solubility, (2) large amounts of biological activity, (3) shallow depths, which cause active cycling between sediments and water, and (4) circulation, which tends to move bottom waters landward. As a result, pollutants tend to become sequestered in the sediments. This is unfortunate as we are also in the habit of dredging estuaries. This causes a significant redistribution of pollutants with some undergoing redissolution and a significant fraction being dumped offshore.

Since estuaries are sites of great productivity, pollutants are readily transferred throughout food webs as a result of **bioaccumulation** and **biomagnifi-**

cation. Even though these processes tend to keep pollutants in estuaries, their impacts are transmitted to the open ocean by the many species of pelagic fish and shellfish which use these areas as breeding grounds.

Despite our current problems with marine pollution, some scientists feel that with adequate management, the ocean can continue to be used as a site for the disposal of at least some wastes. In this view, the ocean is assumed to have some finite capacity to assimilate various compounds within which an unacceptable impact on living organisms or nonliving resources does not result. Many scientists do not agree with this analysis. They feel that our lack of knowledge is so profound that a precautionary approach is warranted. A compromise position would be the use of management strategies that minimize the risk to both health and the environment.

Historically, the United States has attempted to regulate pollution by controlling the dumping and discharging of wastes. The most important pieces of legislation concerning this are the **Ocean Dumping Acts of 1972** and **1988** and the **Clean Water Acts of 1948** and **1977**. The latter also attempts to control pollution by establishing **water-quality criteria** for effluent. The **USEPA** has been charged with developing these water-quality criteria, but due to limited funding and the need for advances in analytical techniques, progress has been slow. Pollution has also proven legally difficult to control as many pollutants are covered by multiple laws, some of which are contradictory. Pollution legislation is also cumbersome in that a great number of permits are now required for any proposed environmental impact. Lack of enforcement and monitoring has also greatly diluted the original intent of pollution laws.

Any long-term solution to the problem of waste disposal must involve a reduction in our waste stream. This can be achieved through conservation and recycling. What must be introduced should be as environmentally compatible as possible. To prevent further degradation of the marine environment requires that we be able to realistically and reliably predict the impact of any future pollution. To make such predictions requires a large increase in our understanding of the ocean. This will require an international commitment of resources and the sustained efforts of many earth scientists.

PROBLEM SET 6

1. Using the data in Table P.6.1 and graph paper, determine whether nitrate is behaving conservatively. If nonconservative, postulate a mechanism by which nitrate removal or addition could occur.

TABLE P.6.1

Salinity (‰)	*Temperature* (°C)	$[O_2]$ (*ml/L*)	$[NO_3^-]$ (μM)
1	20	6.4	5.1
5	19	6.5	4.2
10	17	3.1	0.5
15	16	4.7	0.2
20	14	4.9	1.0
25	12	6.7	1.1
30	10	6.5	0.5

2. The following article appeared in the *Sun News* (Myrtle Beach, South Carolina) on September 24, 1989. Compute the enrichment factor for mercury in the whale's tissues.
3. Using the information in Figure 29.18, compute the maximum number of cans of tuna fish that you can ingest in 1 week without exceeding the WHO recommended limits of mercury intake. Assume that 1 can = 6.5 oz tuna.

Whale exposed to mercury

■ **VANCOUVER, British Columbia**—An autopsy on a killer whale found at Long Beach has revealed the highest mercury levels ever recorded in a marine mammal in British Columbia, said biologist Pam Stacey.

Stacey said Sunday night a report on the whale's liver from a provincial veterinary lab indicated the animal had a mercury level of 1,272 parts per million.

The amount was almost double the previous mercury level record found in another whale beached on the British Columbia coast last year.

"The amount of mercury in his liver is just phenomenal," Stacey said of the killer whale which was found last April at the Vancouver Island beach.

SUGGESTED FURTHER READINGS

Estuarine Chemistry

ASTON, S. R. 1978. Estuarine Chemistry. In: *Chemical Oceanography*, vol. 7 (ed: J. P. Riley). Academic Press, Orlando, FL, pp. 362–440.

LASSERRE, P., and J.-M. MARTIN. 1986. *Biogeochemical Processes at the Land-Sea Boundary*. Elsevier Science Publishers, Amsterdam, The Netherlands, 214 pp.

OLAUSSON, E., and I. CATO (eds). 1980. *Chemistry and Biogeochemistry of Estuaries*. John Wiley & Sons, Inc., New York, 452 pp.

Marine Pollution

CLARK, R. B. 1986. *Marine Pollution*. Clarendon Press, Oxford, England, 215 pp.

GOLDBERG, E. D. 1975. Marine Pollution. In: *Chemical Oceanography*, vol. 3 (ed: J. P. Riley). Academic Press, Orlando, FL, pp. 39–87.

GOLDBERG, E. D. (ed.). 1981. The Oceans as a Waste Space. *Oceanus*, **24**:1–65.

HAIN, J. H. W. 1986. Low-Level Radioactivity in the Irish Sea. *Oceanus*, **29**:16–27.

KAO, T. W., and J. M. BISHOP. 1985. Coastal Ocean Toxic Waste Pollution: Where Are We and Where Do We Go? *USA Today*, **July**:20–23.

O'HARA, K. J., S. IUDICELLO, and R. BIERCE. 1988. *A Citizen's Guide to Plastics in the Ocean: More than a Litter Problem*. Center for Marine Conservation, Washington, DC, 143 pp.

PRESTON, M. R. 1989. Marine Pollution. In: *Chemical Oceanography*, vol. 9 (ed: J. P. Riley). Academic Press, Orlando, FL, pp. 1–52.

RUSSELL, D. 1987. An Ecological Disaster as Gigantic as the Ocean. *In These Times*, **Sept. 30–Oct. 6**:6 pp.

SPENCER, D.W., et al. 1990. Ocean Disposal Reconsidered. *Oceanus*, **33**:96 pp.

APPENDIX I

THE PERIODIC TABLE OF THE ELEMENTS

New notation / Previous IUPAC form / CAS version

KEY TO CHART

Atomic Number →	50 +2	← Oxidation States
Symbol →	Sn +4	
1983 Atomic Weight →	118.71	
	18 18 4	← Electron Configuration

1 Group IA	2 IIA	3 IIIA IIIB	4 IVA IBV	5 VA VB	6 VIA VIB	7 VIIA VIIB	8 VIIIA VIII	9 VIIIA VIII	10 VIIIA VIII	11 IB	12 IIB	13 IIIB IIIA	14 IVB IVA	15 VB VA	16 VIB VIA	17 VIIB VIIA	18 VIIIA	Orbit
1 +1 −1 H 1.00794 1																	2 He 4.00260 2	K
3 +1 Li 6.941 2-1	4 +2 Be 9.01218 2-2											5 +3 B 10.81 2-3	6 +2 +4 −4 C 12.011 2-4	7 +1 +2 +3 +4 +5 −1 −2 −3 N 14.0067 2-5	8 −2 O 15.9994 2-6	9 −1 F 18.9984 2-7	10 0 Ne 20.179 2-8	K-L
11 +1 Na 22.9898 2-8-1	12 +2 Mg 24.305 2-8-2											13 +3 Al 26.9815 2-8-3	14 +2 +4 −4 Si 28.0855 2-8-4	15 +3 +5 −3 P 30.9738 2-8-5	16 +4 +6 −2 S 32.06 2-8-6	17 +1 +5 +7 −1 Cl 35.453 2-8-7	18 0 Ar 39.948 2-8-8	K-L-M
19 +1 K 39.0983 -8-8-1	20 +2 Ca 40.08 -8-8-2	21 +3 Sc 44.9559 -8-9-2	22 +2 +3 +4 Ti 47.88 -8-10-2	23 +2 +3 +4 +5 V 50.9415 -8-11-2	24 +2 +3 +6 Cr 51.996 -8-11-2	25 +2 +3 +4 +7 Mn 54.9380 -8-13-2	26 +2 +3 Fe 55.847 -8-14-2	27 +2 +3 Co 58.9332 -8-15-2	28 +2 +3 Ni 58.69 -8-16-2	29 +1 +2 Cu 63.546 -8-18-1	30 +2 Zn 65.39 -8-18-2	31 +3 Ga 69.72 -8-18-3	32 +2 +4 Ge 72.59 -8-18-4	33 +3 +5 −3 As 74.9216 -8-18-5	34 +4 +6 −2 Se 78.96 -8-18-6	35 +1 +5 −1 Br 79.904 -8-18-7	36 0 Kr 83.80 -8-18-8	-L-M-N
37 +1 Rb 85.4678 -18-8-1	38 +2 Sr 87.62 -18-8-2	39 +3 Y 88.9059 -18-9-2	40 +4 Zr 91.224 -18-10-2	41 +3 +5 Nb 92.9064 -18-12-1	42 +6 Mo 95.94 -18-13-1	43 +4 +6 +7 Tc (98) -18-13-2	44 +3 Ru 101.07 -18-15-1	45 +3 Rh 102.906 -18-16-1	46 +2 +4 Pd 106.42 -18-18-0	47 +1 Ag 107.868 -18-18-1	48 +2 Cd 112.41 -18-18-2	49 +3 In 114.82 -18-18-3	50 +2 +4 Sn 118.71 -18-18-4	51 +3 +5 −3 Sb 121.75 -18-18-5	52 +4 +6 −2 Te 127.60 -18-18-6	53 +1 +5 +7 −1 I 126.905 -18-18-7	54 0 Xe 131.29 -18-18-8	-M-N-O
55 +1 Cs 132.905 -18-8-1	56 +2 Ba 137.33 -18-8-2	57* +3 La 138.906 -18-9-2	72 +4 Hf 178.49 -32-10-2	73 +5 Ta 180.948 -32-11-2	74 +6 W 183.85 -32-12-2	75 +4 +6 +7 Re 186.207 -32-13-2	76 +3 +4 Os 190.2 -32-14-2	77 +3 +4 Ir 192.22 -32-15-2	78 +2 +4 Pt 195.08 -32-16-2	79 +1 +3 Au 196.967 -32-18-1	80 +1 +2 Hg 200.59 -32-18-2	81 +1 +3 Tl 204.383 -32-18-3	82 +2 +4 Pb 207.2 -32-18-4	83 +3 +5 Bi 208.980 -32-18-5	84 +2 +4 Po (209) -32-18-6	85 At (210) -32-18-7	86 0 Rn (222) -32-18-8	-N-O-P
87 +1 Fr (223) -18-8-1	88 +2 Ra 226.025 -18-8-2	89** +3 Ac 227.028 -18-9-2	104 +4 Unq (261) -32-10-2	105 Unp (262) -32-11-2	106 Unh (263) -32-12-2	107 Uns (262) -32-13-2												-N-O-P

*Lanthanides	58 +3 +4 Ce 140.12 -20-8-2	59 +3 Pr 140.908 -21-8-2	60 +3 Nd 144.24 -22-8-2	61 +3 Pm (145) -23-8-2	62 +2 +3 Sm 150.36 -24-8-2	63 +2 +3 Eu 151.96 -25-8-2	64 +3 Gd 157.25 -25-9-2	65 +3 Tb 158.925 -27-8-2	66 +3 Dy 162.50 -28-8-2	67 +3 Ho 164.930 -29-8-2	68 +3 Er 167.26 -30-8-2	69 +3 Tm 168.934 -31-8-2	70 +2 +3 Yb 173.04 -32-8-2	71 +3 Lu 174.967 -32-9-2	N O P
**Actinides	90 +4 Th 232.038 -18-10-2	91 +5 +4 Pa 231.036 -20-9-2	92 +3 +4 +5 +6 U 238.029 -21-9-2	93 +3 +4 +5 +6 Np 237.048 -22-9-2	94 +3 +4 +5 +6 Pu (244) -24-8-2	95 +3 +4 +5 +6 Am (243) -25-8-2	96 +3 Cm (247) -25-9-2	97 +3 +4 Bk (247) -27-8-2	98 +3 Cf (251) -28-8-2	99 +3 Es (252) -29-8-2	100 +3 Fm (257) -30-8-2	101 +2 +3 Md (258) -31-8-2	102 +2 +3 No (259) -32-8-2	103 +3 Lr (260) -32-9-2	O P Q

Numbers in parentheses are mass numbers of most stable isotope of that element

APPENDIX II

COMMON NAMES AND CHEMICAL FORMULAE

I. ELEMENTS PRESENT IN TISSUES

H Hydrogen
C Carbon
N Nitrogen
O Oxygen
S Sulfur
P Phosphorous

II. ELEMENTS PRESENT IN NODULES AND IRON–MANGANESE CRUSTS

Fe Iron
Mn Manganese
Cu Copper
Co Cobalt
Ni Nickel

III. CONSERVATIVE IONS

Na^+ Sodium
Cl^- Chloride
K^+ Potassium
Ca^{2+} Calcium
Mg^{2+} Magnesium
SO_4^{2-} Sulfate

IV. NATURALLY OCCURRING RADIONUCLIDES

U Uranium
Th Thorium
Pa Protactinium
Rn Radon
Ra Radium
Po Polonium
Pb Lead
He Helium
^3H Tritium

V. CARBONATE SYSTEM

CO_2 Carbon dioxide
H_2CO_3 Carbonic acid
HCO_3^- Bicarbonate
CO_3^{2-} Carbonate
$CaCO_3$ Calcium carbonate
$C(H_2O)$ Organic matter
ΣCO_2 Total dissolved inorganic carbon
C.A. Carbonate alkalinity

VI. GASES

Ne Neon
Ar Argon
Kr Krypton
O_2 Oxygen
N_2 Nitrogen
CO_2 Carbon dioxide
CH_4 Methane
H_2 Hydrogen
H_2S Hydrogen sulfide
N_2O Nitrous oxide

VII. NITROGEN COMPOUNDS

N_2 Nitrogen gas
NO_3^- Nitrate
NO_2^- Nitrite
N_2O Nitrous oxide
NH_4^+ Ammonium
NH_3 Ammonia

VIII. MISCELLANEOUS

PO_4^{3-} Phosphate
H_4SiO_4 Silicic acid
SiO_2 Silica
Si Silicon

APPENDIX III

METRIC UNITS AND EQUIVALENTS

METRIC PREFIXES

atto (a)	= 0.000000000000000001	= 10^{-18}
femto (f)	= 0.000000000000001	= 10^{-15}
pico (p)	= 0.000000000001	= 10^{-12}
nano (n)	= 0.000000001	= 10^{-9}
micro (μ)	= 0.000001	= 10^{-6}
milli (m)	= 0.001	= 10^{-3}
centi (c)	= 0.01	= 10^{-2}
deci (d)	= 0.1	= 10^{-1}
deca (da)	= 10	= 10^{1}
hecto (h)	= 100	= 10^{2}
kilo (k)	= 1000	= 10^{3}
mega (M)	= 1,000,000	= 10^{6}
giga (G)	= 1,000,000,000	= 10^{9}
tera (T)	= 1,000,000,000,000	= 10^{12}

UNITS OF LENGTH

1 ångström (Å)	= 0.0001 micron
1 nanometer (nm)	= 10^{-9} meter
1 micron (μm)	= 0.001 millimeter (or 10^{-3} mm)
1 millimeter (mm)	= 1000 microns 0.001 meter
1 centimeter (cm)	= 10 millimeters 0.394 inch
1 decimeter (dm)	= 0.1 meter
1 meter (m)	= 100 centimeters 3.28 feet
1 kilometer (km)	= 1000 meters 3280 feet 0.62 statute mile 0.54 nautical mile

1 inch	= 2.54 centimeters
1 foot	= 0.3048 meter
1 yard	= 3 feet 0.91 meter
1 fathom	= 6 feet 1.83 meters
1 statute mile	= 5280 feet 1.6 kilometers 0.87 nautical mile
1 nautical mile	= 6076 feet 1.85 kilometers 1.15 statute miles

UNITS OF AREA

1 square centimeter (cm^2)	= 100 square millimeters 0.155 square inch
1 square meter (m^2)	= 10^4 square centimeters 10.8 square feet
1 are (a)	= 100 square meters
1 square kilometer (km^2)	= 10^6 square meters 247.1 acres 0.386 square statute mile 0.292 square nautical mile
1 hectare (ha)	= 10,000 square meters

UNITS OF SPEED

1 knot	= 1 nautical mile/hour 1.15 statute miles/hour 1.85 kilometers/hour
Velocity of sound in water of salinity = 35‰	= 1507 meters/second

UNITS OF VOLUME

1 milliliter (ml)	= 0.001 liter 1 cm^3 (or 1 cc)
1 liter (L)	= 1000 cm^3 1.06 liquid quarts
1 gallon	= 3.785 dm^3 = 3.785 liters
1 cubic meter (m^3)	= 1000 liters

UNITS OF MASS

1 milligram (mg)	= 0.001 gram
1 kilogram (kg)	1000 grams
	2.2 pounds
1 tonne (or ton) (t)	= 1 metric ton
	10^6 grams
	1 Mg
1 pound (lb)	= 453.6 grams
1 long ton (UK)	= 1.016 Mg
1 short ton (US)	= 0.907 Mg

UNITS OF PRESSURE

1 atmosphere = 760 torr
12.5 psi

UNITS OT TEMPERATURE

°C = 5/9(°F − 32)

°K = 273.15 + °C

UNITS OF ENERGY

1 cal = 4.184 Joule

1 kcal = 1000 cal

UNITS OF CONCENTRATION

M	= moles of solute per liter of solution
m	= moles of solute per kilogram of solvent
μg-atom/L	= μM = mg-atoms/m^3
μg/L	= mg/m^3
% (parts per hundred)	= 1 g/100 mL
ppt (parts per thousand)	= g/1000 ml = g/L
(parts per mille)*	
ppm (parts per million)	= 10^{-3} g/1000 ml = mg/L
ppb (parts per billion)	= 10^{-6} g/1000 ml = μg/L
dpm	= disintegrations per minute

*mille is Latin for 1000.

APPENDIX IV

SYMBOLS, CONSTANTS, AND FORMULAE

SYMBOLS

a = Ionic activity
A = Radioactive activity
λ = Radioactive decay constant
γ = Ionic activity coefficient
K = Equilibrium constant
I = Ionic strength
G = Gibbs free energy
H = Enthalpy
ρ = Density
n = Moles
d = Finite increment
∂ = Infinitesimally small increment
z = Vertical dimension
x,y = Horizontal dimensions
C = Conservative ionic solute
S = nonconservative ionic solute
t = Time
S‰ = salinity
Cl‰ = chlorinity
P = Pressure
V = Volume

CONSTANTS AND FORMULAE

I. Ideal Gas Law

$$PV = nRT$$
$$R = 0.821 \text{ L atm } {}^\circ\text{K}^{-1} \text{ mol}^{-1}$$
$$T = 273.15 + {}^\circ\text{C}$$

1 mole of an ideal gas occupies 22.4 L at STP

II. Acidity

$$\text{pH} = -\log\{\text{H}^+\}$$
$$\text{pOH} = -\log\{\text{OH}^-\}$$
$$\text{pH} + \text{pOH} = 14 \text{ at } 25°\text{C and } I = 0$$

III. Redox

$$\text{F} = 96.485 \text{ coulombs mol electrons}^{-1} = 23.066 \text{ kcal volt}^{-1} \text{ mol electrons}^{-1}$$
$$\text{R} = 1.987 \times 10^{-3} \text{ kcal °K}^{-1} \text{ mol electrons}^{-1}$$

IV. Equilibrium

For $A + B \rightleftharpoons AB$,

$$K = \frac{[AB]}{[A][B]}$$

V. Logarithms

$$\log a + \log b = \log (ab)$$
$$\log a - \log b = \log (a/b)$$
$$\log 10^a = a$$

VI. O_2 Conversions

$$1 \text{ ml } O_2/\text{L} = 11.2 \times \text{mg-at } O_2/\text{L}$$
$$1 \text{ mg } O_2/\text{L} = 16.0 \times \text{mg-at } O_2/\text{L}$$

APPENDIX V

Area, Volume, and Mean Depth of Oceans and Seas

Body		*Area* *(10^6 km^2)*	*Volume* *(10^6 km^3)*	*Mean Depth* *(m)*
Atlantic Ocean	excluding adjacent seas	82.441	323.613	3926
Pacific Ocean	excluding adjacent seas	165.246	707.555	4282
Indian Ocean	excluding adjacent seas	73.443	291.030	3963
All oceans (excluding adjacent seas)		321.130	322.198	4117
Arctic Mediterranean		14.090	16.980	1205
American Mediterranean		4.319	9.573	2216
Mediterranean Sea and Black Sea		2.966	4.238	1429
Asiatic Mediterranean		8.143	9.873	1212
Large Mediterranean seas		29.518	40.664	1378
Baltic Sea		0.422	0.023	55
Hudson Bay		1.232	0.158	128
Red Sea		0.438	0.215	491
Persian Gulf		0.239	0.006	25
Small Mediterranean seas		2.331	0.402	172
All Mediterranean seas		31.849	41.066	1289
North Sea		0.575	0.054	94
English Channel		0.075	0.004	54
Irish Sea		0.103	0.006	60
Gulf of St. Lawrence		0.238	0.030	127
Andaman Sea		0.798	0.694	870
Bering Sea		2.268	3.259	1437
Okhotsk Sea		1.528	1.279	838
Japan Sea		1.008	1.361	1350
East China Sea		1.249	0.235	188
Gulf of California		0.162	0.132	813
Bass Strait		0.075	0.005	70
Marginal seas		8.079	7.059	874
All adjacent seas		39.928	48.125	1205
Atlantic Ocean	including adjacent seas	106.463	354.679	3332
Pacific Ocean	including adjacent seas	179.679	723.699	4028
Indian Ocean	including adjacent seas	74.917	291.945	3897
All oceans (including adjacent seas)		361.059	1370.323	3795

Source: From *An Introduction to Physical Oceanography*, J. A. Knauss, copyright © 1978 by Prentice Hall, Englewood Cliffs, NJ, p. 2. Reprinted by permission. Data from E. Kossinna, *Annalen für Geographisch-naturwissenschaft*, vol. 9, p. 70, 1921, published by the Institut für Meereskunde, Berlin University, Berlin, Germany.

APPENDIX VI

Important Rock-Forming Minerals

General Classification	Mineral	Empirical Formula	Crystal Structure
Sulfates	Barite	$BaSO_4$	Orthorhombic
	Gypsum	$CaSO_4 \cdot 2H_2O$	Monoclinic
	Anhydrite	$CaSO_4$	Orthorhombic; more stable than gypsum above 42°C
Carbonates	Calcite	$CaCO_3$	Trigonal. Mg- and Mn-calcite
	Rhodochrosite	$MnCO_3$	Similar to calcite
	Magnesite	$MgCO_3$	Similar to calcite
	Siderite	$FeCO_3$	Similar to calcite
	Dolomite	$MgCa(CO_3)_2$	One layer calcite combined with one layer magnesite
	Huntite	$Mg_3Ca(CO_3)_4$	
	Aragonite	$CaCO_3$	Orthorhombic
	Strontianite	$SrCO_3$	Similar to aragonite
Iron oxides	Goethite	α-FeOOH	Similar to diaspore
	Lepidocrocite	γ-FeOOH	Similar to boehmite
	Limonite	$FeOOH \cdot nH_2O$	Hydrated oxides of iron with poorly crystalline character
		α-Fe_2O_3	Trigonal, occurs in sediments; Spinel type
	Magnetite	Fe_3O_4	8 Fe^{2+} in 4 coordination; 16 Fe^{3+} in 6 coordination
Titanium oxide	Rutile	TiO_2	Tetragonal; band of octahedra
Magnesium hydroxide	Brucite	$Mg(OH)_2$	Trigonal, two sheets of OH parallel to basal plane with sheet of Mg ions between them
Phosphates	Apatite	$Ca_5(OH,F,Cl)(PO_4)_3$	Hexagonal
	Carbonate–apatite	$Ca_5(PO_4,OH,CO_3)_3(F,OH)$	
Silicon oxides	α-Quartz	SiO_2	Trigonal; densely packed arrayment of SiO_2 tetrahedra
	α-Tridymite	SiO_2	Orthorhombic, open structure
	α-Cristobalite	SiO_2	Tetragonal sheets of 6-membered rings of $[SiO_4]$ tetrahedra
	Opal	SiO_2	Hydrous, crypto crystalline form of cristobalite
Aluminum oxides	Corundum	α-Al_2O_3	Oxygen in hexagonal closest packing
	Gibbsite	$Al_2O_3 \cdot 3H_2O$	Monoclinic; a layer of Al ions sandwiched between two sheets of closely packed hydroxide ions

General Classification	*Mineral*	*Empirical Formula*	*Crystal Structure*
	Boehmite	γ-AlOOH	Orthorhombic; double sheets of octahedra with Al ions at their centers
	Diaspore	α-AlOOH	Orthorhombic Al^{3+} in octahedrally coordinated sites
Two-layer clays (kaolinites)	Kaolinite Halloysite	$Al_4[Si_4O_{10}](OH)_8$ $Al_4[Si_4O_{10}](OH)_8 \cdot 2H_2O$	Sheet consisting of two layers: (1) SiO_4 tetrahedra in a hexagonal array and (2) layer of Al in 6 coordination
Three-layer minerals	Micas Muscovite Biotite Glauconite Illite	 $K_2Al_4[Si_6Al_2O_{20}](OH,F)_4$ $K_2(Mg,Fe)_6[Si_6Al_2O_{20}](OH,F)_4$ $K_xAl_4[Si_{1-x}Al_xO_{20}](OH)_4$	A layer of octahedrally coordinated cations (usually Al) is sandwiched between two identical layers of $[(Si,Al)O_4]$ tetrahedra
Expandable three-layer clays	Montmorillonite Vermicullite	$(Na,K)_{x+y}(Al_{2-x}Mg_x)_2$ $[(Si_{1-y}Al_y)_8O_{20}](OH)_4 \cdot nH_2O$ $(Ca,Mg)(Mg_{3-x}Fe_x)_2$ $[(Si_6Al_2)_8O_{20}](OH)_4 \cdot 8H_2O$	Octahedral Al on Mg sheets, tetrahedral Si sheets. Al partially replaced by Mg and occasionally by Fe, Cr, Zn. In tetrahedral sheet occasional replacement of Si by Al
	Chlorite	$(Mg,Al)_{12}[(Si,Al)_8O_{20}](OH)_{16}$	
Sulfides	Pyrite	FeS_2	Cubic, octahedral coordination of Fe by S
	Marcasite	FeS_2	Orthorhombic, octahedral coordination of Fe by S
	Pyrrhotite	FeS	Monoclinic pseudohexagonal
	Galena	PbS	Cubic

Source: From *Aquatic Chemistry*, W. W. Stumm and J. J. Morgan, copyright © 1970 by John Wiley & Sons, Inc., New York, pp. 388–389. Reprinted by permission.

APPENDIX VII

GEOLOGIC TIME SCALES

The Geologic Time Scale for the Precambrian Eon

Eon	*Era*	*Sub-Era*		*Orogeny*	*Age (Ma)*[a]
	Paleozoic				~570
Proterozoic	Hadrynian	Late			~620
		Early		Avalonian	1000
	Helikian	Neo	Late	**Grenvillian**	1200
			Early	Elsevirian	1400
		Paleo	Late	Elsonian	~1500
			Early	Killarnian	~1750
	Aphebian	Late		**Hudsonian**	1870
		Middle		Moranian	2140?
		Early		Blezardian	~2510
Archean		Late		**Kenoran**	2670
		Late Middle		Laurentian	~2900
		Early Middle		Wanipigowan	~3400
		Early		Uivakian	

& Sons, Inc., New York, p. 85. Reprinted by permission. After C. H. Stockwell, reprinted with permission from the *Geological Survey of Canada*, Paper 80–19, p. 14, copyright © 1982 by the Geological Survey of Canada, Ottawa, Canada.

[a]1 Ma = 1 million years.

The Geologic Time Scale for the Phanerozoic Eon

Era	*Period*	*Epoch*	*Age*	*Ages of Boundaries (Ma)*		
				1	*2*	*3*
Cenozoic	Quaternary	Holocene				
					0.01	0.01
		Pleistocene	Calabrian			
				1.82	2.0	1.6
	Neogene (Tertiary)	Pliocene	Piacenzian			
						3.4
			Zanclean			
				5–5.5	5.1	5.3
		Miocene	Messinian			
						6.5
			Tortonian			
					11.3	11.2
			Serravallian			
						15.1
			Langhian			
						16.6
			Burdigalian			
						21.8
			Aquitanian			
				23	24.6	23.7
	Paleogene (Tertiary)	Oligocene	Chattian			
					32.8	30.0
			Rupelian			
				34	38.0	36.6
		Eocene	Priabonian			
					42.0	40.0
			Bartonian			
						43.6
			Lutetian		50.5	52.0
			Ypresian			
				53	54.9	57.8
		Paleocene	Thanetian (Selandian)			
						60.6
			Unnamed (Selandian)		60.2	63.6
			Danian			
				65	65	66.4
Mesozoic	Cretaceous	Late	Maastrichtian			
				95	97.5	97.5
		Early	Albian			
					113	113
			Aptian		119	119
					125	124
			Barremian			
					131	131
			Hauterivian			
					138	138
			Valanginian			
			Berriasian			
				130	144	144
	Jurassic	Late	Tithonian			
					150	152
			Kimmeridgian			
					156	156
			Oxfordian			
				150	163	163

The Geologic Time Scale for the Phanerozoic Eon *(Continued)*

Era	*Period*	*Epoch*	*Age*	*Ages of Boundaries (Ma)* *1*	*2*	*3*
		Middle	Callovian		169	169
			Bathonian		175	176
			Bajocian		181	183
			Aalenian	178	188	187
		Early	Toarcian		194	193
			Pliensbachian		200	198
			Sinemurian		206	204
			Hettangian	204	213	208
	Triassic	Late	Norian		225	225
			Carnian	229	231	230
		Middle	Ladinian		238	235
			Anisian	239	243	240
		Early	Scythian	245	248	245
Paleozoic	Permian	Late	Tatarian		253	253
			Kazanian			
			Ufimian		258	258
		Early	Kungurian		263	263
			Artinskian		268	268
			Sakmarian			
			Asselian	290	286	286
	Pennsylvanian (Late Carboniferous)		Gzelian			
			Kasimovian			296
			Moscovian			
			Bashkirian	320	320?	320
	Mississippian (Early Carboniferous)		Serpukhovian		333	333
			Visean		352	352
			Tournaisian	360	360	360
	Devonian	Late	Fammenian		367	367
			Frasnian	375	374	374

Era	*Period*	*Epoch*	*Age*	*Ages of Boundaries (Ma)*		
				1	*2*	*3*
		Middle	Givetian			
					380	380
			Eifelian			
				385	387	387
		Early	Emsian			
					394	394
			Siegenian			
					401	401
			Gedinnian			
				400	408	408
	Silurian	Late	Pridolian			
					414	414
			Ludlovian			
					421	421
		Early	Wenlockian			
					428	428
			Llandoverian			
				418	438	438
	Ordovician	Late	Ashgillian			
					448	448
			Caradocian			
					458	458
		Middle	Llandelian			
					468	468
			Llanvirnian			
					478	478
		Early	Arenigian			
					488	488
			Tremadocian			
				495	505	505
	Cambrian	Late	Trempealeauan			
			Franconian			
			Dresbachian			
					523	523
		Middle				
					540	540
		Early				
				530	590	570

Source: From *Principles of Isotope Geology*, 2nd ed., G. Faure, copyright © 1986 by John Wiley & Sons, Inc., New York, pp. 553–555. Reprinted by permission. Data sources: (1) *Numerical Data in Stratigraphy*, G. S. Odin, D. Curry, N. H. Gale, and W. J. Kennedy (ed.: G. S. Odin), copyright © 1982 by John Wiley & Sons, Inc., New York, p. 958. Reprinted by permission. (2) *A Geologic Time Scale*, W. B. Harland, A. V. Cox, P. G. Llewellyn, C. A. Pickton, A. G. Smith, and R. Walters, copyright © 1982 by Cambridge University Press, Cambridge, England, p. 12. Reprinted by permission. (3) A. R. Palmer, reprinted with permission from *Geology*, vol. 11, p. 504, copyright © 1983 by the Geological Society of America, Boulder, CO.

APPENDIX VIII

NAECs of O_2 and N_2

	SALINITY (‰)								
T(°C)	*0*	*10*	*20*	*30*	*34*	*35*	*36*	*38*	*40*
O_2									
− 1	—	—	9.162	8.553	8.321	8.264	8.207	8.095	7.984
0	10.218	9.543	8.913	8.325	8.100	8.045	7.990	7.882	7.775
1	9.936	9.284	8.676	8.107	7.890	7.836	7.783	7.679	7.575
2	9.666	9.037	8.449	7.898	7.689	7.637	7.586	7.484	7.384
3	9.409	8.801	8.232	7.699	7.496	7.446	7.397	7.298	7.202
4	9.163	8.574	8.024	7.509	7.312	7.264	7.216	7.121	7.027
5	8.927	8.358	7.825	7.327	7.136	7.089	7.043	6.950	6.860
6	8.702	8.151	7.635	7.152	6.967	6.922	6.877	6.787	6.699
8	8.280	7.763	7.278	6.824	6.650	6.608	6.565	6.481	6.398
10	7.891	7.406	6.950	6.522	6.359	6.319	6.279	6.199	6.121
12	7.534	7.077	6.647	6.244	6.090	6.052	6.014	5.939	5.865
14	7.204	6.773	6.367	5.987	5.841	5.805	5.769	5.698	5.628
16	6.898	6.491	6.108	5.748	5.610	5.576	5.542	5.475	5.409
18	6.615	6.230	5.868	5.527	5.396	5.363	5.331	5.268	5.205
20	6.352	5.987	5.644	5.320	5.196	5.166	5.135	5.075	5.015
22	6.106	5.761	5.435	5.128	5.010	4.981	4.952	4.895	4.838
24	5.878	5.550	5.241	4.949	4.836	4.809	4.781	4.727	4.673
26	5.664	5.353	5.058	4.780	4.673	4.647	4.621	4.569	4.518
28	5.464	5.168	4.887	4.623	4.521	4.496	4.471	4.421	4.372
30	5.276	4.994	4.727	4.474	4.377	4.353	4.329	4.282	4.235
32	5.099	4.830	4.576	4.335	4.242	4.219	4.196	4.151	4.106
34	4.932	4.676	4.433	4.203	4.114	4.092	4.070	4.027	3.984
36	4.775	4.530	4.298	4.078	3.993	3.972	3.951	3.910	3.869
38	4.627	4.393	4.171	3.960	3.878	3.858	3.838	3.799	3.760
40	4.486	4.262	4.050	3.848	3.770	3.750	3.731	3.693	3.656

	SALINITY (‰)								
T(°C)	0	10	20	30	34	35	36	38	40
N_2									
− 1	—	—	16.28	15.10	14.65	14.54	14.44	14.22	14.01
0	18.42	17.10	15.87	14.73	14.30	14.19	14.09	13.88	13.67
1	17.95	16.67	15.48	14.38	13.96	13.86	13.75	13.55	13.35
2	17.50	16.26	15.11	14.04	13.64	13.54	13.44	13.24	13.05
3	17.07	15.87	14.75	13.72	13.32	13.23	13.13	12.94	12.76
4	16.65	15.49	14.41	13.41	13.03	12.93	12.84	12.66	12.47
5	16.26	15.13	14.09	13.11	12.74	12.65	12.56	12.38	12.21
6	15.88	14.79	13.77	12.83	12.47	12.38	12.29	12.12	11.95
8	15.16	14.14	13.18	12.29	11.95	11.87	11.79	11.62	11.46
10	14.51	13.54	12.64	11.80	11.48	11.40	11.32	11.17	11.01
12	13.90	12.99	12.14	11.34	11.04	10.96	10.89	10.74	10.60
14	13.34	12.48	11.67	10.92	10.63	10.56	10.49	10.35	10.21
16	12.83	12.01	11.24	10.53	10.25	10.19	10.12	9.99	9.86
18	12.35	11.57	10.84	10.16	9.90	9.84	9.77	9.65	9.52
20	11.90	11.16	10.47	9.82	9.57	9.51	9.45	9.33	9.21
22	11.48	10.78	10.12	9.50	9.26	9.21	9.15	9.03	8.92
24	11.09	10.42	9.79	9.20	8.98	8.92	8.87	8.76	8.65
26	10.73	10.09	9.49	8.92	8.71	8.65	8.60	8.50	8.39
28	10.38	9.77	9.20	8.66	8.45	8.40	8.35	8.25	8.15
30	10.06	9.48	8.93	8.41	8.21	8.16	8.12	8.02	7.92
32	9.76	9.20	8.67	8.18	7.99	7.94	7.89	7.80	7.71
34	9.48	8.94	8.43	7.96	7.77	7.73	7.68	7.59	7.51
36	9.21	8.69	8.20	7.75	7.57	7.53	7.48	7.40	7.31
38	8.95	8.46	7.99	7.55	7.38	7.33	7.29	7.21	7.13
40	8.71	8.23	7.78	7.36	7.19	7.15	7.11	7.03	6.95

Source: From R. F. Weiss, reprinted with permission from *Deep-Sea Research*, vol. 17, p. 731, 1970, copyright © 1970 by Pergamon Press, Elmsford, NY.

APPENDIX IX

Sigma-T Values

S(‰)	T(°C) 0	1	2	3	4	5	6	7	8	9	10	11	12	13	14	15	16	17	18	19	20	21	22	23	24	25
0	−0.07	−0.01	0.03	0.05	0.06	0.05	0.03	−0.01	−0.06	−0.13	−0.21	−0.31	−0.41	−0.54	−0.67	−0.81	−0.97	−1.14	−1.32	−1.51	−1.71	−1.92	−2.14	−2.38	−2.62	−2.87
1	0.74	0.80	0.84	0.86	0.86	0.85	0.82	0.78	0.72	0.65	0.57	0.47	0.36	0.24	0.10	−0.04	−0.20	−0.37	−0.55	−0.75	−0.95	−1.16	−1.39	−1.62	−1.86	−2.12
2	1.56	1.61	1.64	1.66	1.66	1.65	1.62	1.57	1.51	1.44	1.35	1.25	1.14	1.01	0.88	0.73	0.57	0.39	0.21	0.02	−0.19	−0.40	−0.63	−0.87	−1.11	−1.37
3	2.37	2.42	2.45	2.46	2.46	2.44	2.41	2.36	2.30	2.22	2.13	2.03	1.91	1.79	1.65	1.50	1.33	1.16	0.97	0.78	0.57	0.35	0.13	−0.11	−0.36	−0.61
4	3.18	3.22	3.25	3.26	3.25	3.23	3.20	3.15	3.08	3.00	2.91	2.81	2.69	2.56	2.42	2.26	2.10	1.92	1.74	1.54	1.33	1.11	0.88	0.64	0.39	0.14
5	3.99	4.03	4.05	4.06	4.05	4.03	3.99	3.93	3.87	3.78	3.69	3.58	3.46	3.33	3.19	3.03	2.86	2.69	2.50	2.30	2.09	1.87	1.64	1.40	1.15	0.89
6	4.80	4.84	4.86	4.86	4.85	4.82	4.78	4.72	4.65	4.57	4.47	4.36	4.24	4.10	3.96	3.80	3.63	3.45	3.26	3.06	2.85	2.62	2.39	2.15	1.90	1.64
7	5.61	5.64	5.66	5.66	5.64	5.61	5.57	5.51	5.43	5.35	5.25	5.13	5.01	4.87	4.73	4.57	4.39	4.21	4.02	3.82	3.60	3.38	3.15	2.90	2.65	2.39
8	6.42	6.45	6.46	6.45	6.44	6.40	6.35	6.29	6.22	6.13	6.02	5.91	5.78	5.64	5.49	5.33	5.16	4.97	4.78	4.58	4.36	4.13	3.90	3.65	3.40	3.14
9	7.22	7.25	7.26	7.25	7.23	7.19	7.14	7.08	7.00	6.19	6.80	6.68	6.56	6.41	6.26	6.10	5.92	5.74	5.54	5.33	5.12	4.89	4.65	4.41	4.15	3.88
10	8.03	8.05	8.06	8.05	8.02	7.98	7.93	7.86	7.78	7.69	7.58	7.46	7.33	7.18	7.03	6.86	6.69	6.50	6.30	6.09	5.87	5.64	5.41	5.16	4.90	4.63
11	8.84	8.86	8.86	8.84	8.82	8.77	8.72	8.65	8.56	8.46	8.35	8.23	8.10	7.95	7.80	7.63	7.45	7.26	7.06	6.85	6.63	6.40	6.16	5.91	5.65	5.38
12	9.64	9.66	9.66	9.64	9.61	9.56	9.50	9.43	9.34	9.24	9.13	9.01	8.87	8.72	8.56	8.39	8.21	8.02	7.82	7.61	7.38	7.15	6.91	6.66	6.40	6.13
13	10.45	10.46	10.46	10.44	10.40	10.35	10.29	10.21	10.12	10.02	9.91	9.78	9.64	9.49	9.33	9.16	8.98	8.78	8.58	8.36	8.14	7.91	7.66	7.41	7.15	6.88
14	11.25	11.26	11.25	11.23	11.19	11.14	11.07	11.00	10.90	10.80	10.68	10.55	10.41	10.26	10.10	9.92	9.74	9.54	9.34	9.12	8.90	8.66	8.42	8.16	7.90	7.63
15	12.06	12.06	12.05	12.02	11.98	11.93	11.86	11.78	11.63	11.58	11.46	11.33	11.18	11.03	10.86	10.69	10.50	10.30	10.10	9.88	9.65	9.41	9.17	8.91	8.65	8.37
16	12.86	12.86	12.85	12.82	12.77	12.72	12.65	12.56	12.46	12.35	12.23	12.10	11.95	11.80	11.63	11.45	11.26	11.06	10.85	10.63	10.41	10.17	9.92	9.66	9.40	9.12
17	13.67	13.66	13.65	13.61	13.57	13.51	13.43	13.34	13.24	13.13	13.01	12.87	12.72	12.56	12.39	12.21	12.02	11.82	11.61	11.39	11.16	10.92	10.67	10.41	10.14	9.87
18	14.47	14.46	14.44	14.41	14.36	14.29	14.22	14.13	14.02	13.91	13.78	13.64	13.49	13.33	13.16	12.98	12.79	12.58	12.37	12.15	11.92	11.67	11.42	11.16	10.89	10.62
19	15.27	15.26	15.24	15.20	15.15	15.08	15.00	14.91	14.80	14.69	14.56	14.42	14.26	14.10	13.93	13.74	13.55	13.34	13.13	12.90	12.67	12.43	12.18	11.91	11.64	11.36
20	16.08	16.06	16.04	15.99	15.94	15.87	15.79	15.69	15.58	15.46	15.33	15.19	15.03	14.87	14.69	14.51	14.31	14.10	13.89	13.66	13.43	13.18	12.93	12.66	12.39	12.11
21	16.88	16.86	16.83	16.79	16.73	16.66	16.57	16.47	16.36	16.24	16.11	15.96	15.80	15.64	15.46	15.27	15.07	14.86	14.65	14.42	14.18	13.93	13.68	13.41	13.14	12.86
22	17.68	17.66	17.63	17.58	17.52	17.44	17.35	17.25	17.14	17.02	16.88	16.73	16.57	16.40	16.22	16.03	15.83	15.62	15.40	15.17	14.94	14.69	14.43	14.17	13.89	13.61
23	18.48	18.46	18.42	18.37	18.31	18.23	18.14	18.04	17.92	17.79	17.65	17.50	17.34	17.17	16.99	16.80	16.60	16.38	16.16	15.93	15.69	15.44	15.18	14.92	14.64	14.36
24	19.29	19.26	19.22	19.17	19.10	19.02	18.92	18.82	18.70	18.57	18.43	18.28	18.11	17.94	17.76	17.56	17.36	17.14	16.92	16.69	16.45	16.20	15.94	15.67	15.39	15.10
25	20.09	20.06	20.02	19.96	19.89	19.81	19.71	19.60	19.48	19.35	19.20	19.05	18.88	18.71	18.52	18.33	18.12	17.91	17.68	17.45	17.20	16.95	16.69	16.42	16.14	15.85
26	20.89	20.86	20.81	20.75	20.68	20.59	20.49	20.38	20.26	20.13	19.98	19.82	19.66	19.48	19.29	19.09	18.88	18.67	18.44	18.20	17.96	17.70	17.44	17.17	16.89	16.60
27	21.70	21.66	21.61	21.55	21.47	21.38	21.28	21.17	21.04	20.90	20.75	20.60	20.43	20.25	20.06	19.86	19.65	19.43	19.20	18.96	18.71	18.46	18.20	17.92	17.64	17.35
28	22.50	22.46	22.41	22.34	22.26	22.17	22.07	21.95	21.82	21.68	21.53	21.37	21.20	21.02	20.82	20.62	20.41	20.19	19.96	19.72	19.47	19.21	18.95	18.67	18.39	18.10
29	23.30	23.26	23.20	23.14	23.05	22.96	22.85	22.73	22.60	22.46	22.31	22.14	21.97	21.78	21.59	21.39	21.17	20.95	20.72	20.48	20.23	19.97	19.70	19.43	19.14	18.85
30	24.10	24.06	24.00	23.93	23.84	23.75	23.64	23.52	23.38	23.24	23.08	22.92	22.74	22.55	22.36	22.15	21.94	21.71	21.48	21.24	20.99	20.73	20.46	20.18	19.89	19.60
31	24.91	24.86	24.80	24.72	24.64	24.54	24.42	24.30	24.16	24.02	23.86	23.69	23.51	23.33	23.13	22.92	22.70	22.48	22.24	22.00	21.74	21.48	21.21	20.93	20.65	20.35
32	25.71	25.66	25.60	25.52	25.43	25.33	25.21	25.08	24.95	24.80	24.64	24.47	24.29	24.10	23.90	23.69	23.47	23.24	23.00	22.76	22.50	22.24	21.97	21.69	21.40	21.10
33	26.52	26.46	26.40	26.31	26.22	26.12	26.00	25.87	25.73	25.58	25.41	25.24	25.06	24.87	24.67	24.45	24.23	24.00	23.76	23.52	23.26	23.00	22.72	22.44	22.15	21.86
34	27.32	27.26	27.19	27.11	27.01	26.91	26.79	26.65	26.51	26.36	26.19	26.02	25.83	25.64	25.43	25.22	25.00	24.77	24.53	24.28	24.02	23.75	23.48	23.20	22.91	22.61
35	28.13	28.07	27.99	27.91	27.81	27.70	27.57	27.44	27.29	27.14	26.97	26.79	26.61	26.41	26.21	25.99	25.77	25.53	25.29	25.04	24.78	24.51	24.24	23.95	23.66	23.36

Source: From *A Manual of Chemical and Biological Methods for Seawater Analysis*, T. R. Parsons, M. Yoshiaki, and C. M. Lalli, copyright © 1984 by Pergamon Press, Elmsford, NY, pp. 172–173. Reprinted by permission.

APPENDIX X

Equilibrium Constants for Ion Pairs and Complexes[a]

	OH^-		CO_3^{2-}		SO_4^{2-}		Cl^-		Br^-		F^-		NH_3		$B(OH)_4^-$	
H^+			HL	10.33	HL	1.99	—		—		HL	3.2	HL	9.24	HL	9.24
	HL·w	14.00	H_2L	16.68									L·g	−1.8	HL_3	10.4
			H_2L·g	18.14											H_2L_3	20.4
															H_2L_4	21.0
															H_4L_5	38.8
Na^+	—		NaL	1.27	NaL	1.06	—									
			NaHL	10.08												
K^+	—		—		KL	0.96	—									
Ca^{2+}	CaL	1.15	CaL	3.2	CaL	2.31	—				CaL	1.1	—			
			CaHL	11.59	CaL_2·s	4.62					CaL_2·s	10.4				
	CaL_2·s	5.19	CaL·s	8.22												
			CaL·s	8.35												
Mg^{2+}	MgL	2.56	MgL	3.4	MgL	2.36	—		—		MgL	1.8	—			
	Mg_4L_4	16.28	MgHL	11.49							MgL_2·s	8.2				
	MgL_2·s	11.16	MgL·s	4.54												
			MgL·s	7.45												
Sr^{2+}	—		SrL·s	9.0	SrL	2.6					SrL_2·s	8.5				
					SrL·s	6.5										
Ba^{2+}	—		BaL	2.8	BaL	2.7	—				BaL_2·s	5.8				
			BaL·s	8.3	BaL·s	10.0										
Cr^{3+}	CrL	10.0			CrL	3.0	CrL	0.1			CrL	5.2				
	CrL_2	18.3									CrL_2	9.2				
	CrL_3	24.0									CrL_3	12.0				
	CrL_4	28.6														
	Cr_3L_4	47.8														
	CrL_3·s	30.0														
Al^{3+}	AlL	9.0														
	AlL_2	18.7									AlL	7.0				
	AlL_3	27.0									AlL_2	12.6				
	AlL_4	33.0									AlL_3	16.7				
	Al_3L_4	42.1									AlL_4	19.1				
	AlL_3·s	33.5														
Fe^{3+}	FeL	11.8			FeL	4.0	FeL	1.5	FeL	0.6	FeL	6.0				
	FeL_2	22.3			FeL_2	5.4	FeL_2	2.1			FeL_2	10.6				
	FeL_4	34.4									FeL_3	13.7				
	Fe_2L_2	25.0														
	FeL_3·s	42.7														
	FeL_3·s	39.5														

	OH^-		CO_3^{2-}		SO_4^{2-}		Cl^-		Br^-		F^-		NH_3		$B(OH)_4^-$	
Mn^{2+}	MnL	3.4	MnHL	12.1	MnL	2.3	MnL	0.6			MnL	1.9	MnL	1.0		
	MnL_2	5.8	MnL·s	9.3									MnL_2	1.5		
	MnL_3	7.2														
	MnL_4	15.7														
	MnL_2·s	12.8														
Fe^{2+}	FeL	4.5	FeL·s	10.7	FeL	2.2					FeL	1.4				
	FeL_2	7.4														
	FeL_3	11.0														
	FeL_2·s	15.1														
Co^{2+}	CoL	4.3	CoL·s	10.0	CoL	2.4	CoL	0.5			CoL	1.0	CoL	2.0		
	CoL_2	5.1											CoL_2	3.5		
	CoL_3	10.5											CoL_3	4.4		
	CoL_2·s	15.7											CoL_4	5.0		
Ni^{2+}	NiL	4.1	NiL·s	6.9	NiL	2.3	NiL	0.6			NiL	1.1	NiL	2.7		
	NiL_2	9.0											NiL_2	4.9		
	NiL_3	12.0											NiL_3	6.6		
	NiL_2·s	17.2											NiL_4	7.7		
													NiL_5	8.3		
Cu^{2+}	CuL	6.5	CuL	6.7	CuL	2.4	CuL	0.5			CuL	1.5	CuL	4.0		
	CuL_2	11.8	CuL_2	10.2	$Cu_4(OH)_6L$·s	68.6							CuL_2	7.5		
	CuL_4	16.4	CuL·s	9.6									CuL_3	10.3		
	Cu_2L_2	10.4	$Cu_2(OH)_2L$·s	33.8									CuL_4	11.8		
	CuL_2·s	19.3	$Cu_3(OH)_2L_2$·s	46.0												
	CuL_2·s	20.4														
Zn^{2+}	ZnL	5.0	ZnL·s	10.0	ZnL	2.1	ZnL	0.4			ZnL	1.2	ZnL	2.2		
	ZnL_2	11.1			ZnL_2	3.1	ZnL_2	0.0					ZnL_2	4.5		
	ZnL_3	13.6					ZnL_3	0.5					ZnL_3	6.9		
	ZnL_4	14.8					$Zn_2(OH)_3L$·s	26.8					ZnL_4	8.9		
	ZnL_2·s	15.5														
	ZnL_2·s	16.8														
Pb^{2+}	PbL	6.3			PbL	2.8	PbL	1.6	PbL	1.8	PbL	2.0				
	PbL_2	10.9	PbL·s	13.1	PbL·s	7.8	PbL_2	1.8	PbL_2	2.6	PbL_2	3.4				
	PbL_3	13.9					PbL_3	1.7	PbL_3	3.0	PbL_2·s	7.4				
	PbL_2·s	15.3					PbL_4	1.4	PbL_2·s	5.7						
							PbL_2·s	4.8								
Hg^{2+}	HgL	10.6	HgL·s	16.1	HgL	2.5	HgL	7.2	HgL	9.6	HgL	1.6	HgL	8.8		
	HgL_2	21.8			HgL_2	3.6	HgL_2	14.0	HgL_2	18.0			HgL_2	17.4		
	HgL_3	20.9					HgL_3	15.1	HgL_3	20.3			HgL_3	18.4		
	HgL_2·s	25.4					HgL_4	15.4	HgL_4	21.6			HgL_4	19.1		
							HgOHL	18.1	HgL_2·s	19.8						
Cd^{2+}	CdL	3.9	CdL·s	13.7	CdL	2.3	CdL	2.0	CdL	2.1	CdL	1.0	CdL	2.6		
	CdL_2	7.6			CdL_2	3.2	CdL_2	2.6	CdL_2	3.0	CdL_2	1.4	CdL_2	4.6		
	CdL_2·s	14.3			CdL_3	2.7	CdL_3	2.4					CdL_3	5.9		
							CdL_4	1.7					CdL_4	6.7		
Ag^+	AgL	2.0	Ag_2L·s	11.1	AgL	1.3	AgL	3.3	AgL	4.7	AgL	0.4	AgL	3.3	AgL	0.6
	AgL_2	4.0			Ag_2L·s	4.8	AgL_3	5.3	AgL_2	6.9			AgL_2	7.2	$AgHL_2$·s	22.9
	AgL·s	7.7					AgL_3	6.4	AgL_3	8.7						
							AgL·s	9.7	AgL_4	9.0						
									AgL·s	12.3						

	SiO_3^{2-}		S^{2-}		$S_2O_3^{2-}$		PO_4^{3-}		$P_2O_7^{4-}$		$P_3O_{10}^{5-}$		CN^-	
H^+	HL	13.1	HL	13.9	HL	1.6	HL	12.35	HL	9.4	HL	9.3	HL	9.2
	H_2L	23.0	H_2L	20.9	H_2L	2.2	H_2L	19.55	H_2L	16.1	H_2L	18.8		
	H_2L_2	26.6	$H_2L\cdot g\cdot$	21.9			H_3L	21.70	H_3L	18.3	H_3L	21.3		
	H_4L_4	55.9							H_4L	19.7	H_4L	22.3		
	H_6L_4	78.2												
	$H_2L\cdot s$	2.7												
Na^2					NaL	0.5	NaHL	13.5	NaL	2.3	NaL	2.7		
									Na_2L	4.2	NaHL	11.6		
									NaHL	10.8				
K^+					KL	1.0	KHL	13.4	KL	2.1	KL	2.8		
Ca^{2+}	CaL	4.2			CaL	2.0	CaL	6.5	CaL	6.8	CaL	8.1		
	CaHL	14.1					CaHL	15.1	CaHL	13.4	CaHL	14.1		
	CaH_2L_2	29.9					CaH_2L	21.0	CaOHL	8.9	CaOHL	10.4		
							$CaHL\cdot s$	19.0	$Ca_2L\cdot s$	14.7				
Mg^{2+}	MgL	5.3			MgL	1.8	MgL	4.8	MgL	7.2	MgL	8.6		
	MgHL	14.3					MgHL	15.3	MgHL	14.1	MgHL	14.5		
	MgH_2L_2	30.8					MgH_2L	20.8	MgOHL	9.3	MgOHL	11.0		
							$Mg_3L_2\cdot s$	25.2						
							$MgHL\cdot s$	18.2						
Sr^{2+}					SrL	2.0	SrL	5.5	SrL	5.4	SrL	7.2		
							SrHL	14.5	SrOHL	7.7	SrHL	13.6		
							SrH_2L	20.3	$Sr_2L\cdot s$	12.9	SrOHL	9.3		
							$SrHL\cdot s$	19.3						
Ba^{2+}					BaL	2.3	$BaHL\cdot s$	19.8			BaL	6.3		
					$BaL\cdot s$	4.8					BaHL	12.9		
											$Ba_2L\cdot s$	16.1		
Cr^{3+}														
Al^{3+}														
Fe^{3+}	FeHL	22.7			FeL	3.3	FeHL	22.5					FeL_6	43.6
							FeH_2L	23.2						
							$FeL\cdot s$	26.4						
Mn^{2+}			$MnL\cdot s$	10.5	MnL	2.0					MnL	9.9		
				13.5							MnHL	14.8		
Fe^{2+}			$FeL\cdot s$	18.1			FeHL	16.0					FeL_6	35.4
							FeH_2L	22.3						
							$Fe_3L_2\cdot s$	36.0						
Co^{2+}			$CoL\cdot s$	21.3	CoL	2.1	CoHL	15.5	CoL	7.9	CoL	9.7		
			$CoL\cdot s$	25.6					CoHL	14.1	CoHL	14.8		
Ni^{2+}			$NiL\cdot s$	19.4	NiL	2.1	NiHL	15.4	NiL	7.7	NiL	9.5	NiL	7.7
			$NiL\cdot s$	24.9					NiHL	14.4	NiHL	14.7	NiL_4	30.9
			$NiL\cdot s$	26.6									$NiHL_4$	36.7
													NiH_2L_4	41.4
													NiH_3L_4	44.0
Cu^{2+}			$CuL\cdot s$	36.1			CuHL	16.5	CuL	9.8	CuL	11.1	CuL_2	16.3
							CuH_2L	21.3	CuHL	15.5	CuHL	15.5	CuL_3	21.6
									CuL_2	12.5			CuL_4	23.1
									CuH_2L	19.2				

	SiO_3^{2-}		S^{2-}		$S_2O_3^{2-}$		PO_4^{3-}		$P_2O_7^{4-}$		$P_3O_{10}^{5-}$		CN^-	
Zn^{2+}			ZnL	16.6	ZnL	2.4	ZnHL	15.7	ZnL	8.7	ZnL	10.3	ZnL	5.7
			ZnL·s	24.7	ZnL_2	2.5	ZnH_2L	21.2	ZnL_2	11.0	ZnHL	14.9	ZnL_2	11.1
					ZnL_3	3.3	Zn_3L_2	35.3	ZnOHL	13.1	ZnOHL	13.6	ZnL_3	16.1
					Zn_2L_2	7.0							ZnL_4	19.6
													ZnL_2·s	15.9
Pb^{2+}			PbL·s	27.5	PbL	3.0	PbHL	15.5	PbL	9.5				
					PbL_2	5.5	PbH_2L	21.1	PbL_2	10.2				
					PbL_3	6.2	Pb_3L_2·s	43.5						
					PbL_4	7.3	PbHL·s	23.8						
Hg^{2+}			HgL	7.9	HgL_2	29.2			HgOHL	18.6			HgL	17.0
			HgL_2	14.3	HgL_3	30.6							HgL_2	32.8
			HgOHL	18.5									HgL_3	36.3
			HgL·s	52.7									HgL_4	39.0
			HgL·s	53.3									HgOHL	29.6
Cd^{2+}			CdL	19.5	CdL	3.9			CdL	8.7	CdL	9.8	CdL	6.0
			CdHL	22.1	CdL_2	6.3			CdOHL	11.8	CdHL	14.6	CdL_2	11.1
			CdH_2L_2	43.2	CdL_3	6.4					CdOHL	12.6	CdL_3	15.7
			CdH_3L_3	59.0	CdL_4	8.2							CdL_4	17.9
			CdH_4L_4	75.1	Cd_2L_2	13.0								
			CdL·s	27.0										
Ag^+			AgL	19.2	AgL	8.8	Ag_3L·s	17.6					AgL_2	20.5
			AgHL	27.7	GaL_2	13.7							AgL_3	21.4
			$AgHL_2$	35.8	AgL_3	14.2							AgOHL	13.2
			AgH_2L_2	45.7	Ag_2L_4	26.3							AgL·s	15.7
			Ag_2L·s	50.1	Ag_3L_5	39.8								
					Ag_6L_8	78.6								

	Ethylene-diamine		*NTA*[b]		*EDTA*[c]		*CDTA*[d]		*IDA*[e]		*Picolinate*		*Cysteine*		*Desferr-ferrioxamine B*	
H^+	HL	9.93	HL	10.33	HL	11.12	HL	13.28	HL	9.73	HL	5.39	HL	10.77	HL	10.1
	H_2L	16.78	H_2L	13.27	H_2L	20.01	H_2L	22.63	H_2L	12.63	H_2L	6.40	H_2L	19.13	H_2L	19.4
			H_3L	14.92	H_3L	21.04	H_3L	23.98	H_3L	14.51			H_3L	20.84	H_3L	27.8
			H_4L	16.02	H_4L	23.76	H_4L	26.62								
					H_5L	24.76	H_5L	28.34								
Na^+			NaL	1.9	NaL	2.5			NaL	0.8						
K^+					KL	1.7										
Ca^{2+}			CaL	7.6	CaL	12.4	CaL	15.0	CaL	3.5	CaL	2.2			CaL	3.5
					CaHL	16.0					CaL_2	3.8				
Mg^{2+}	MgL	0.4	MgL	6.5	MgL	10.6	MgL	12.8	MgL	3.8	MgL	2.6			MgL	5.2
					MgHL	15.1					MgL_2	4.0				
Sr^{2+}			SrL	6.3	SrL	10.5	SrL	12.4	SrL	3.1	SrL	1.8			SrL	3.1
					SrHL	14.9					SrL_2	3.0				
Ba^{2+}			BaL	5.9	BaL	9.6	BaL	10.5	BaL	2.5	BaL	1.6				
					BaHL	14.6	BaHL	17.8								
Cr^{3+}					CrL	26.0			CrL	12.2						
					CrHL	28.2			CrL	23.2						
					CrOHL	32.2										

Al^{3+}			AlL	13.4	AlL	18.9	AlL	22.1	AlL	9.9						
			AlOHL	22.1	AlHL	21.6	AlHL	24.3	AlL_2	17.5						
					AlOHL	26.6	AlOHL	28.1								
					$Al(OH)_2L$	30.0										
Fe^{3+}			FeL	17.9	FeL	27.7	FeL	32.6	FeL	12.5	FeL_2	13.9			FeL	31.9
			FeL_2	26.3	FeHL	29.2	FeOHL	36.5			$FeOHL_2$	24.9			FeHL	32.6
					FeOHL	33.8										
					$Fe(OH)_2L$	37.7										
Mn^{2+}	MnL	2.8	MnL	8.7	MnL	15.6	MnL	19.2			MnL	4.0	MnL	5.6		
	MnL_2	4.9	MnL_2	11.6	MnHL	19.1	MnHL	22.4			MnL_2	7.1				
	MnL_3	5.8									MnL_3	8.8				
Fe^{2+}	FeL	4.3	FeL	9.6	FeL	16.1	FeL	20.8	FeL	6.7	FeL	5.3			FeHL	17.7
	FeL_2	7.7	FeL_2	13.6	FeHL	19.3	FeHL	23.9	FeL_2	11.0	FeL_2	9.7			FeH_2L	23.1
	FeL_3	9.7	FeOHL	12.6	FeOHL	20.4					FeL_3	13.0				
					$Fe(OH)_2L$	23.7										
Co^{2+}	CoL	6.0	CoL	11.7	CoL	18.1	CoL	21.4	CoL	7.9	CoL	6.4			CoL	11.2
	CoL_2	10.8	CoL_2	15.0	CoHL	21.5	CoHL	24.7	CoL_2	13.2	CoL_2	11.3			CoHL	18.0
	CoL_3	14.1	CoOHL	14.5							CoL_3	15.8			CoHL	23.6
Ni^{2+}	NiL	7.4	NiL	12.8	NiL	20.4	NiL	22.1	NiL	9.1	NiL	7.2	NiL	10.7	NiL	11.8
	NiL_2	13.6	NiL_2	17.0	NiHL	24.0	NiHL	25.4	NiL_2	15.2	NiL_2	13.2	NiL_2	20.9	NiHL	18.3
	NiL_3	17.9	NiOHL	15.5	NiOHL	21.8					NiL_3	17.9			NiH_2L	23.8
Cu^{2+}	CuL	10.5	CuL	14.2	CuL	20.5	CuL	23.7	CuL	11.5	CuL	8.4			CuL	15.0
	CuL_2	19.6	CuL_2	18.1	CuHL	23.9	CuHL	27.3	CuL_2	17.6	CuL_2	15.6			CuHL	24.1
	CuOHL	11.8	CuOHL	18.6	CuOHL	22.6									CuH_2L	27.0
Zn^{2+}	ZnL	5.7	ZnL	12.0	ZnL	18.3	ZnL	21.1	ZnL	8.2	ZnL	5.7	ZnL	10.1	ZnL	11.0
	ZnL_2	10.6	ZnL_2	14.9	ZnHL	21.7	ZnHL	24.4	ZnL_2	13.5	ZnL_2	10.3	ZnL	19.1	ZnHL	17.5
	ZnL_3	13.9	ZnOHL	15.5	ZnOHL	23.4					ZnL_3	13.6	ZnHL	16.4	ZnH_2L	22.9
Pb^{2+}	PbL	7.0	PbL	12.6	PbL	19.8	PbL	22.1	PbL	8.3	PbL	5.0	PbL	12.5		
	PbL_2	8.5			PbHL	23.0	PgHL	25.3			PbL_2	8.6				
Hg^{2+}	HgL	14.3	HgL	15.9	HgL	23.5	HgL	26.8	HgL	12.6	HgL	8.1	HgL	15.3		
	HgL_2	23.2			HgHL	27.0	HgHL	30.3			HgL_2	16.2				
	HgOHL	24.2			HgOHL	27.7	HgOHL	29.7								
	$HgHL_2$	28.0														
Cd^{2+}	CdL	5.4	CdL	11.1	CdL	18.2	CdL	21.7	CdL	6.6	CdL	5.0			CdL	8.8
	CdL_2	9.9	CdL_2	15.1	CdHL	21.5	CdHL	25.1	CdL_2	11.1	CdL_2	8.8			CdL	8.2
	CdL_3	12.7	CdOHL	13.4							CdL_3	11.4			CdH_2L	22.7
Ag^{+}	AgL	4.7	AgL	5.8	AgL	8.2	AgL	9.9			AgL	3.6				
	AgL_2	7.7			AgHL	14.9					AgL_2	6.1				
	AgHL	11.9														

	Glycine		*Glutamate*		*Acetate*		*Glycolate*		*Citrate*		*Malonate*		*Salicylate*		*Phthalate*	
H^{+}	HL	9.78	HL	9.95	HL	4.76	HL	3.83	HL	6.40	HL	5.70	HL	13.74	HL	5.41
	H_2L	12.13	H_2L	14.47					H_2L	11.16	H_2L	8.55	H_2L	16.71	H_2L	8.36
			H_3L	16.70					H_3L	14.29						
Na^{+}									NaL	1.4	NaL	0.7			NaL	0.7
K^{+}									KL	1.3						

	Glycine		*Glutamate*		*Acetate*		*Glycolate*		*Citrate*		*Malonate*		*Salicylate*		*Phthalate*	
Ca^{2+}	CaL	1.4	CaL	2.1	CaL	1.2	CaL	1.6	CaL	4.7	CaL	2.4	CaL	0.4	CaL	2.4
									CaHL	9.5	CaHL	6.6				
									CaH_2L	12.3						
Mg^{2+}	MgL	2.7	MgL	2.8	MgL	1.3	MgL	1.3	MgL	4.7	MgL	2.9				
									MgHL	9.2	MgHL	7.1				
Sr^{2+}	SrL	0.9	SrL	2.3	SrL	1.1	SrL	1.2	SrL	4.1	SrL	2.1				
											SrHL	6.5				
Ba^{2+}	BaL	0.8	BaL	2.2	BaL	1.1	BaL	1.1	BaL	4.1	BaL	2.1	BaL	0.2	BaL	2.3
									BaHL	9.0						
									BaH_2L	12.4						
Cr^{3+}					CrL	5.4					CrL	9.6				
					CrL_2	8.4										
					CrL_3	11.2										
Al^{3+}					AlL	2.4							AlL	14.2	AlL	5.0
													AlL_2	24.0	AlL_2	8.7
													AlL_3	32.8		
Fe^{3+}	FeL	10.8	FeL	13.8	FeL	4.0	FeL	3.7	FeL	13.5	FeL	9.3	FeL	17.6		
					FeL_2	7.6	FeOHL	16.7	$Fe_2(OH)_2L_2$	36.			FeL_2	28.6		
					FeL_3	9.6	$FeOHL_2$	19.4					FeL_3	37.2		
							$FeOHL_3$	20.9								
Mn^{2+}	MnL	3.2			MnL	1.4	MnL	1.6	MnL	5.5	MnL	3.3	MnL	6.8	MnL	2.7
									MnHL	9.4			MnL_2	10.7		
Fe^{2+}	FeL	4.3	FeL	4.6	FeL	1.4	FeL	1.9	FeL	5.7			FeL	7.4		
									FeHL	9.9			FeL_2	12.1		
Co^{2+}	CoL	5.1	CoL	5.4	CoL	1.5	CoL	2.0	CoL	6.3	CoL	3.7	CoL	7.5	CoL	2.8
	CoL_2	9.0	CoL_2	8.7			CoL_2	3.0	CoHL	10.3	CoL_2	5.1	CoL_2	12.3	CoHL	7.2
	CoL_3	11.6							CoH_2L	12.9	CoHL	7.0				
Ni^{2+}	NiL	6.2	NiL	6.5	NiL	1.4	NiL	2.3	NiL	6.7	NiL	4.1	NiL	7.8	NiL	3.0
	NiL_2	11.1	NiL_2	10.6			NiL_2	3.4	NiHL	10.6	NiL_2	5.8	NiL_2	12.6	NiHL	7.6
	NiL_3	14.2					NiL_3	3.7	NiH_2L	13.4	NiHL	7.2				
Cu^{2+}	CuL	8.6	CuL	8.8	CuL	2.2	CuL	2.9	CuL	7.2	CuL	5.7	CuL	11.5	CuL	4.0
	CuL_2	15.6	CuL_2	15.0	CuL_2	3.6	CuL_2	4.7	CuHL	10.7	CuL_2	8.2	CuL_2	19.3	CuHL	7.7
							CuL_3	4.7	CuH_2L	13.9	CuHL	8.3				
									CuOHL	16.4						
									Cu_2L_2	16.3						

Zn^{2+}	ZnL	5.4	ZnL	5.8	ZnL	1.6	ZnL	2.4	ZnL	6.3	ZnL	3.8	ZnL	7.5	ZnL	2.9
	ZnL_2	9.8	ZnL_2	9.5	ZnL_2	1.8	znL_2	3.6	ZnL_2	8.6	ZnL_2	5.4			ZnL_2	4.2
	ZnL_3	12.3	ZnL_3	9.8			znL_3	3.9	ZnHL	13.5	ZnHL	7.1				
									ZnH_2L	12.9						
Pb^{2+}	PbL	5.5			PbL	2.7	PbL	2.5	PbL	5.4	PbL	4.0				
	PbL_2	8.9			PbL_2	4.1	PbL_2	3.7	PbL_2	8.1	PbL_2	4.5				
							PbL_3	3.6	PbHL	13.5						
									PbH_2L	13.1						
Hg^{2+}	HgL	10.9			HgL	6.1			HgL	12.2						
	HgL_2	20.1			HgL_2	10.1										
					HgL_3	14.1										
					HgL_4	17.6										
Cd^{2+}	CdL	4.7	CdL	4.8	CDL	1.9	CdL	1.9	CdL	5.1	CdL	3.2	CdL	6.4	CdL	3.4
	CdL_2	8.4			CdL_2	3.2	CdL_2	2.7	CdL_2	7.2	CdL_2	4.0				
	CdL_3	10.7							CdHL	12.7	CdHL	6.9				
									CdH_2L	12.6						
Ag^{+}	AgL	3.5			AgL	0.7	AgL	0.4								
	AgL_2	6.9			AgL_2	0.6	AgL_2	0.5								

Source: From *Principles of Aquatic Chemistry*, F. M. M. Morel, copyright ©1983 by John Wiley & Sons, Inc., New York, pp. 242–249. Reprinted by permission. See Morel (1983) for data sources.

[a]Given as the logarithm of stability constants for the formation of complexes and solids from metals and ligands at $I=0$ and 25°C.
[b]Nitrilotriacetic acid
[c]Iminodiacetic acid
[d]Ethylenedinitrilotetraacetic acid
[e]trans–1, 2–Cyclohexylenedinitrilotetraacetic acid

APPENDIX XI

RULES FOR ASSIGNING OXIDATION NUMBERS

1. The sum of all the oxidation numbers of the atoms in an electrically neutral chemical substance must be zero. In H_2S, for example, each H atom has an oxidation number of $+1$, and the oxidation number of the S atom is -2. The sum is $2(+1) + (-2) = 0$.
2. In polyatomic ions, the sum of oxidation numbers must equal the charge on the ion. For example, in NH_4^+, each H atom has an oxidation number of $+1$, and the oxidation number of the N atom is -3, making the total $4(+1) + (-3) = +1$, which is the charge on the ion. In SO_4^{2-}, each O atom has an oxidation number of -2 (see Rule 5). The sum of oxidation numbers for the ion must be -2, which is the charge on the ion. Since the sum of the oxidation numbers of the four O atoms is -8, the oxidation number of the S atom must be $+6$.
3. The oxidation number of an atom in a monatomic ion is its charge. In Na^+, sodium has an oxidation numbe of $+1$. In S^{2-}, sulfur has an oxidation number of -2.
4. The oxidation number of an atom in a single-element neutral substance is 0. Thus, the oxidation number of every sulfur atom in S_2, S_6, and S_8 is 0; the oxidation number of chlorine in Cl_2 is 0; the oxidation number of oxygen in O_2 or O_3 is 0. Every element has at least one form in which its oxidation state is 0.
5. Some elements have the same oxidation number in all or nearly all of their compounds. When F combines with other elements, its oxidation number is always -1. The halogens Cl, Br, and I have the oxidation number -1 except when they combine with oxygen or a halogen of lower atomic number. Oxygen usually has the oxidation number -2, except when it combines with F or with itself in such compounds as the peroxides or superoxides. Hydrogen always has the oxidation number $+1$ when combined with nonmetals and -1 when combined with metals. A metal in group IA of the periodic table always has an oxidation number of $+1$. A metal in group IIA always has an oxidation number of $+2$.
6. Metals almost always have positive oxidation numbers.
7. A bond between identical atoms in a molecule makes no contribution to the oxidation number of that atom because the electron pair of the bond is divided equally. In hydrogen peroxide, H_2O_2, for example, the two O

atoms are bonded to one another. We can calculate the oxidation number of O by determining the contribution of the two H atoms, each of which has an oxidation number of +1. Since the sum of the oxidation numbers of the H atoms is +2 and the molecule is neutral, the sum of the oxidation numbers of the two O atoms is −2, giving each an oxidation number of −1.

8. In a primary alcohol or hydrocarbon, the R—C bond is between two carbon atoms and is assigned zero polarity. Each C—H bond and the O—H bond are assigned polarities with the negative end at carbon or oxygen and the positive end at hydrogen. The C—O bond is assigned a polarity with the positive end at carbon and the negative end at oxygen. The same is true for each bond of the double bond in the carboxyl group of the acid.

$$
\begin{array}{ccccccc}
 & & & \mathrm{H} & & & \\
 & & +1 & | & & & \\
 & & -1 & | & +1\ \ -1 & & -1\ \ +1 \\
\mathrm{R} & \text{——} & & \mathrm{C} & \text{——} & \mathrm{O} & \text{——}\ \mathrm{H} \\
0 & & 0 & | & -1 & & \\
 & & & | & +1 & & \\
 & & & \mathrm{H} & & &
\end{array}
\qquad
\begin{array}{ccccccc}
 & & & \mathrm{O} & & & \\
 & & -1 & \| & -1 & & \\
 & & +1 & \| & +1 & & \\
\mathrm{R} & \text{——} & & \mathrm{C} & \text{——} & \mathrm{O} & \text{——}\ \mathrm{H} \\
0 & & 0 & & +1\ \ -1 & & -1\ \ +1
\end{array}
$$

The algebraic sum of the charges on any atom is its polar number or oxidation number. Thus the polar number of carbon in a primary alcohol group is $0 - 1 + 1 - 1 = -1$, and that in a carboxyl group is $0 + 1 + 1 + 1 = +3$.

Source: After *Chemical Principles*, 2nd ed., R. S. Boikess and E. Edelson, copyright © 1981 by Harper and Row, New York, p. 241. Reprinted by permission.

APPENDIX XII

Solubility Products of Various Minerals at 25°C and $I = 0$

Mineral	*Reaction*	*Log* K_{sp}
Chlorides		
Halite	$NaCl(s) = Na^+ + Cl^-$	1.54
Sylvite	$KCl(s) = K^+ + Cl^-$	0.98
Chlorargyrite	$AgCl(s) = Ag^+ + Cl^-$	−9.74
Sulfates		
Gypsum	$CaSO_4{\cdot}2H_2O(s) = Ca^{2+} + SO_4^{2-} + 2H_2O$	−4.62
Celestite	$SrSO_4(s) = Sr^2 + SO_4^{2-}$	−6.50
Barite	$BaSO_4(s) = Ba^{2+} + SO_4^{2-}$	−9.96
Oxyides and Hydroxides		
Calcium hydroxide	$Ca(OH)_2(s) = Ca^{2+} + 2OH^-$	−5.19
	$CaOH^+ = Ca^{2+} + OH^-$	−1.15
Brucite	$Mg(OH)_2(s) = Mg^{2+} + 2OH^-$	−11.1
	$MgOH^+ = Mg^{2+} + OH^-$	−2.6
	$Mg_4(OH)_4^{4+} = 4Mg^{2+} + 4OH^-$	−16.3
Gibbsite	$Al(OH)_3(s) = Al^{3+} + 3OH^-$	−33.5
	$AlOH^{2+} = Al^{3+} + OH^-$	−9.0
	$Al(OH)_2^+ = Al^{3+} + 2OH^-$	−18.7
	$Al(OH)_3 = Al^{3+} + 3OH^-$	−27.0
	$Al(OH)_4^- = Al^{3+} + 4OH^-$	−33.0
	$Al_2(OH)_2^{4+} = 2Al^{3+} + 2OH^-$	−20.3
	$Al_3(OH)_4^{5+} = 3Al^{3+} + 4OH^-$	−42.1
Manganous	$Mn(OH)_2(s) = Mn^{2+} + 2OH^-$	−12.8
Hydroxide	$MnOH^+ = Mn^{2+} + OH^-$	−3.4
Ferrous hydroxide	$Fe(OH)_2(s) = Fe^{2+} + 2OH^-$	−15.1
	$FeOH^+ = Fe^{2+} + OH^-$	−4.5
	$Fe(OH)_2 = Fe^{2+} + 2OH^-$	−7.4
	$Fe(OH)_3^- = Fe^{2+} + 3OH^-$	−11.0
	$Fe(OH)_4^{2-} = Fe^{2+} + 4OH^-$	−10.0
Goethite	$\alpha{\cdot}FeOOH(s) + H_2O = Fe^{3+} + 3OH^-$	−41.5
Ferric hydroxide	$am{\cdot}Fe(OH)_3(s) = Fe^{3+} + 3OH^-$	−38.8
Hematite	$\frac{1}{2}\alpha{\cdot}Fe_2O_3(s) + \frac{3}{2}H_2O = Fe^{3+} + 3OH^-$	−42.7
	$FeOH^{2+} = Fe^{3+} + OH^-$	−11.8
	$Fe(OH)_2^+ = Fe^{3+} + 2OH^-$	−22.3
	$Fe(OH)_4^- = Fe^{3+} + 4OH^-$	−34.4
	$Fe_2(OH)_2^{4+} = 2Fe^{3+} + 2OH^-$	−25.1
	$Fe_3(OH)_4^{5+} = 3Fe^{3+} + 4OH^-$	−49.7
Tenorite	$CuO(s) + H_2O = Cu^2 + 2OH^-$	−20.4
Cupric hydroxide	$Cu(OH)_2(s) = Cu^{2+} + 2OH^-$	−19.4
	$Cu(OH)_4^{2-} = Cu^{2+} + 4OH^-$	−16.4
	$Cu_2(OH)_2^{2+} = 2Cu^{2+} + 2OH^-$	−17.6
Litharge (red)	$PbO(s) + H_2O = Pb^{2+} + 2OH^-$	−15.3

Mineral	*Reaction*	*Log* K_{sp}
Massicot (yellow)	$PbO(s) + H_2O = Pb^{2+} + 2OH^-$	−15.1
	$PbOH^+ = Pb^{2+} + OH^-$	−6.3
	$Pb(OH)_2 = Pb^{2+} + 2OH^-$	−10.9
	$Pb(OH)_3^- = Pb^{2+} + 3OH^-$	−13.9
	$Pb_2(OH)^{3+} = 2Pb^{2+} + OH^-$	−7.6
	$Pb_3(OH)_4^{2+} = 3Pb^{2+} + 4OH^-$	−32.1
	$Pb_4(OH)_4^{4+} = 4Pb^{2+} + 4OH^-$	−35.1
	$Pb_6(OH)_8^{4+} = 6Pb^{2+} + 8OH^-$	−68.4
Carbonates		
Aragonite	$CaCo_3(s) = Ca^{2+} + CO_3^{2-}$	−8.22
Calcite	$CaCO_3(s) = Ca^{2+} + CO_3^{2-}$	−8.35
Magnesite	$MgCO_3(s) = Mg^{2+} + CO_3^{2-}$	−7.46
Nesquehonite	$MgCO_3{\cdot}3H_2O(s) = Mg^{2+} + CO_3^{2-} + 3H_2O$	−4.67
Dolomite	$CaMg(CO_3)_2(s)$ = Calcite + Magnesite	−1.70
Disordered dolomite	$CaMg(CO_3)_2(s)$ = Calcite + Magnesite	−0.08
Strontianite	$SrCO_3(s) = Sr^{2+} + CO_3^{2-}$	−9.0
Rhodochrosite	$MnCO_3(s) = Mn^{2+} + CO_3^{2-}$	−10.4
Siderite	$FeCO_3(s) = Fe^{2+} + CO_3^{2-}$	−10.7
Malachite	$Cu_2CO_3(OH)_2(s) = 2Cu^{2+} + CO_3^{2-} + 2OH^-$	−33.8
Azurite	$Cu_3(CO_3)_2(OH)_2(s) = 3Cu^{2+} + 2CO_3^{2-} + 2OH^-$	−46.0
Cerussite	$PbCO_3(s) = Pb^{2+} + CO_3^{2-}$	−13.1
Phosphates		
Brushite	$CaHPO_4{\cdot}2H_2O(s) = Ca^{2+} + HPO_4^{2-} + 2H_2O$	−6.6
Hydroxylapatite	$Ca_5(PO_4)_3OH(s) = 5Ca^{2+} + 3PO_4^{3-} + OH^-$	−55.6
Newberyite	$MgHPO_4{\cdot}3H_2O(s) = Mg^{2+} + HPO_4^{2-} + 3H_2O$	−5.8
Bobierrite	$Mg_3(PO_4)_2{\cdot}8H_2O(s) = 3Mg^{2+} + 2PO_4^{3-} + 8H_2O$	−25.2
Vivianite	$Fe_3(PO_4)_2{\cdot}\ 8H_2O(s) = 3Fe^{2+} + 2PO_4^{3-} + 8H_2O$	−36.0
Strengite	$FePO_4{\cdot}2H_2O(s) = Fe^{3+} + PO_4^{3-} + 2H_2O$	−26.4
Berlinite	$AlPO_4(s) = Al^{3+} + PO_4^{3-}$	−20.6
Sulfides		
Pyrrhotite	$FeS(s) = Fe^{2+} + S^{2-}$	−18.1
Pyrite	$FeS_2(s)$ = Pyrrhotite + S^0	−10.4
Alabandite	$MnS(s) = Mn^{2+} + S^{2-}$	−13.5
Covellite	$CuS(s) = Cu^{2+} + S^{2-}$	−36.1
Galena	$PbS(s) = Pb^{2+} + S^{2-}$	−27.5
Chalcopyrite	$CuFeS_2(s)$ = Covellite + Pyrrhotite	−6.0
Silicates		
Quartz (α + β)	$SiO_2(s) + H_2O = H_2SiO_3$	−4.00
Cristobalite (α + β)	$SiO_2(s) + H_2O = H_2SiO_3$	−3.45
Amorphous silica	$SiO_2(s) + H_2O = H_2SiO_3$	−2.71
Fosterite	$\frac{1}{4}[2MgO{\cdot}SiO_2](s) + H^+ = \frac{1}{2}Mg^{2+} + \frac{1}{4}H_2SiO_3 + \frac{1}{4}H_2O$	7.11
Fayalite	$\frac{1}{4}[2FeO{\cdot}SiO_2](s) + H^+ = \frac{1}{2}Fe^{2+} + \frac{1}{4}H_2SiO_3 + \frac{1}{4}H_2O$	4.21
Enstatite	$\frac{1}{2}[MgO{\cdot}SiO_2](s) + H^+ = \frac{1}{2}Mg^{2+} + \frac{1}{2}H_2SiO_3$	5.82
Wollastonite	$\frac{1}{2}[CaO{\cdot}SiO_2](s) + H^+ = \frac{1}{2}Ca^{2+} + \frac{1}{2}H_2SiO_3$	6.82
Hedenbergite	$\frac{1}{4}[CaO{\cdot}FeO{\cdot}2SiO_2](s) + H^+ = \frac{1}{4}Ca^{2+} + \frac{1}{4}Fe^{2+} + \frac{1}{2}H_2SiO_3$	4.60

Mineral	*Reaction*	*Log* K_{sp}
Diopside	$\frac{1}{4}[CaO \cdot MgO \cdot 2SiO_2](s) + H^+ = \frac{1}{4}Ca^{2+} + \frac{1}{4}Mg^{2+} + \frac{1}{2}H_2SiO_3$	5.30
Tremolite	$\frac{1}{14}[2CaO \cdot 5MgO \cdot 8SiO_2 \cdot H_2O](s) + H^+ = \frac{1}{7}Ca^{2+} + \frac{5}{14}Mg^{2+} + \frac{4}{7}H_2SiO_3$	4.46
Aluminosilicates		
Gibbsite	$Al(OH)_3(s) + H_2SiO_3 = \frac{1}{2}$ Kaolinite $+ \frac{3}{2}H_2O$	4.25
Phlogopite	$\frac{1}{14}[K_2O \cdot 6MgO \cdot Al_2O_3 \cdot 6SiO_2 \cdot 2H_2O](s) + H^+ = \frac{1}{14}$ Kaolinite $+ \frac{1}{7}K^+ + \frac{3}{7}Mg^{2+} + \frac{2}{7}H_2SiO_3 + \frac{3}{14}H_2O$	5.01
Muscovite	$\frac{1}{2}[K_2O \cdot 3Al_2O_3 \cdot 6SiO_2 \cdot 2H_2O](s) + H^+ + \frac{3}{2}H_2O = \frac{3}{2}$ Kaolinite $+ K^+$	3.51
Anorthite	$\frac{1}{2}[CaO \cdot Al_2O_3 \cdot 2SiO_2](s) + H^+ + \frac{1}{2}H_2O = \frac{1}{2}$ Kaolinite $+ \frac{1}{2}Ca^{2+}$	9.83
Albite	$\frac{1}{2}[Na_2O \cdot Al_2O_3 \cdot 6SiO_2](s) + H^+ + \frac{5}{2}H_2O = \frac{1}{2}$ Kaolinite $+ 2H_2SiO_3 + Na^+$	−0.68
K-feldspar	$\frac{1}{2}[K_2O \cdot Al_2O_3 \cdot 6SiO_2](s) + H^+ + \frac{5}{2}H_2O = \frac{1}{2}$ Kaolinite $+ 2H_2SiO_3 + K^+$	−3.54
Na-montmorrillonite	$\frac{1}{2}[Na_2O \cdot 7Al_2O_3 \cdot 22SiO_2 \cdot 6H_2O](s) + H^+ + \frac{15}{2}H_2O = \frac{7}{2}$ Kaolinite $+ 4H_2SiO_3 + Na^+$	−9.1
Ca-montmorillonite	$\frac{1}{2}[CaO \cdot 7Al_2O_3 \cdot 22SiO_2 \cdot 6H_2O](s) + H^+ + \frac{15}{2}H_2O = \frac{7}{2}$ Kaolinite $+ 4H_2SiO_3 + \frac{1}{2}Ca^{2+}$	−7.7

Source: From *Principles of Aquatic Chemistry*, F. M. M. Morel, copyright © 1983 by John Wiley & Sons, Inc., New York, pp. 184–187. Reprinted by permission.

APPENDIX XIII

Oceanic Residences Times and Elemental Concentrations in River Water and Seawater[a]

Atomic No.	Element	Mean Concentration (10^{-6} mol/kg) River	Sea	τ (y)
1	H (as H_2O)	—	5.4×10^{7}	—
2	He	—	1.8×10^{-3}	—
3	Li	1.7	2.5×10^{1}	5.7×10^{5}
4	Be	—	6.5×10^{-5}	—
5	B	1.7	4.2×10^{2}	9.6×10^{6}
6	C (inorganic)	—	2.3×10^{3}	—
	(organic)	—	4	—
7	N (dissolved N_2)	—	5.8×10^{2}	—
	(NO_3^-)	—	3.0×10^{1}	—
8	O (as H_2O)	—	5.4×10^{7}	—
	(dissolved O_2)	—	2.2×10^{2}	—
9	F	5.3	6.8×10^{1}	5.0×10^{5}
10	Ne		7.5×10^{-3}	
11	Na	2.2×10^{2}	4.7×10^{5}	8.3×10^{7}
12	Mg	1.6×10^{2}	5.3×10^{4}	1.3×10^{7}
13	Al	1.9	3×10^{-2}	6.2×10^{2}
14	Si	1.9×10^{2}	1.0×10^{2}	2.0×10^{4}
15	P	1.3	2.3	6.9×10^{4}
16	S	—	2.8×10^{4}	—
17	Cl	—	5.5×10^{5}	—
18	Ar	—	1.5×10^{1}	—
19	K	3.4×10^{1}	10.2×10^{3}	1.2×10^{7}
20	Ca	3.6×10^{2}	10.3×10^{3}	1.1×10^{6}
21	Sc	8.9×10^{-5}	1.5×10^{-5}	
22	Ti	2.1×10^{-1}	$<2.0 \times 10^{-2}$	3.7×10^{3}
23	V	2.0×10^{-2}	2.3×10^{-2}	4.5×10^{4}
24	Cr	1.9×10^{-2}	4×10^{-3}	8.2×10^{3}
25	Mn	1.5×10^{-1}	5×10^{-3}	1.3×10^{3}
26	Fe	7.2×10^{-1}	1×10^{-3}	5.4×10^{1}
27	Co	3.4×10^{-3}	3×10^{-5}	3.4×10^{2}
28	Ni	3.8×10^{-2}	8×10^{-3}	8.2×10^{3}
29	Cu	1.6×10^{-1}	4×10^{-3}	9.7×10^{2}
30	Zn	4.6×10^{-1}	6×10^{-3}	5.1×10^{2}
31	Ga	1.3×10^{-3}	3×10^{-4}	9.0×10^{3}
32	Ge	—	7×10^{-5}	—
33	As	2.3×10^{-2}	2.3×10^{-2}	3.9×10^{4}
34	Se	2.5×10^{-3}	1.7×10^{-3}	2.6×10^{4}

Atomic No.	Element	Mean Concentration (10^{-6} mol/kg)		
		River	Sea	τ (y)
35	Br	2.5×10^{-1}	8.4×10^{2}	1.3×10^{8}
36	Kr	—	3.4×10^{-3}	—
37	Rb	1.8×10^{-2}	1.4	3.0×10^{6}
38	Sr	6.9×10^{-1}	8.7×10^{1}	5.1×10^{6}
39	Y	7.9×10^{-3}	1.5×10^{-4}	7.4×10^{2}
40	Zr	—	3×10^{-4}	—
41	Nb	—	$<5 \times 10^{-5}$	—
42	Mo	5.2×10^{-3}	1.1×10^{-1}	8.2×10^{5}
43	Tc	—	—	—
44	Ru	—	—	—
45	Rh	—	—	—
46	Pd	—	—	—
47	Ag	2.8×10^{-3}	2.5×10^{-5}	3.5×10^{2}
48	Cd	—	7×10^{-4}	—
49	In	—	1×10^{-6}	—
50	Sn	—	4×10^{-6}	—
51	Sb	8.2×10^{-3}	1.2×10^{-3}	5.7×10^{3}
52	Te	—	—	—
53	I	5×10^{-2}	4.4×10^{-1}	3.4×10^{5}
54	Xe	—	5.0×10^{-4}	—
55	Cs	2.6×10^{-4}	2.2×10^{-3}	3.3×10^{5}
56	Ba	4.4×10^{-1}	1.0×10^{-1}	8.8×10^{3}
57	La	3.6×10^{-4}	3×10^{-5}	3.2×10^{3}
58	Ce	5.7×10^{-4}	2×10^{-5}	1.4×10^{3}
59	Pr	5.0×10^{-5}	4×10^{-6}	3.1×10^{3}
60	Nd	2.8×10^{-4}	2×10^{-5}	2.8×10^{3}
61	Pm	—	—	—
62	Sm	5.3×10^{-5}	4×10^{-6}	2.9×10^{3}
63	Eu	6.6×10^{-6}	9×10^{-7}	5.3×10^{3}
64	Gd	5.1×10^{-5}	6×10^{-6}	4.6×10^{3}
65	Tb	6.3×10^{-6}	9×10^{-7}	5.6×10^{3}
66	Dy	3.0×10^{-4}	6×10^{-6}	7.8×10^{2}
67	Ho	6.1×10^{-6}	2×10^{-6}	1.3×10^{4}
68	Er	2.4×10^{-5}	5×10^{-6}	8.1×10^{3}
69	Tm	5.9×10^{-6}	8×10^{-7}	5.2×10^{3}
70	Yb	2.3×10^{-5}	5×10^{-6}	8.5×10^{3}
71	Lu	5.7×10^{-6}	9×10^{-7}	6.2×10^{3}
72	Hf	—	$<4 \times 10^{-5}$	—
73	Ta	—	$<1.4 \times 10^{-5}$	—
74	W	1.6×10^{-4}	6×10^{-4}	—
75	Re	—	2×10^{-5}	—
76	Os	—	—	—
77	Ir	—	—	—
78	Pt	—	—	—
79	Au	1.0×10^{-5}	2.5×10^{-5}	9.7×10^{4}
80	Hg	3.5×10^{-4}	5×10^{-6}	5.6×10^{2}

Atomic No.	*Element*	*Mean Concentration (10^{-6} mol/kg)* River	Sea	τ *(y)*
81	Tl	—	6×10^{-5}	—
82	Pb	4.8×10^{-3}	1×10^{-5}	8.1×10^{1}
83	Bi	—	1×10^{-4}	—
90	Th		$<3 \times 10^{-6}$	
92	U	$\sim 1 \times 10^{-3}$	1.3×10^{-2}	$\sim 5 \times 10^{5}$

Source: From *Tracers in the Sea*, W. S. Broecker and T.-H. Peng, copyright © 1982 by the Lamont-Doherty Geological Observatory, Palisades, NY, pp. 26–27. Reprinted by permission. See Broecker and Peng (1982) for data sources.

[a]Also given are the corresponding mean ocean residence times.

GLOSSARY

For definitions of units, chemical composition of minerals, chemical formulae of molecules and time spans of geological periods, see the Appendices.

AABW, Antarctic bottom water The densest water mass in the open ocean.

Abiogenic Formed without the intervention or support of organisms.

Abiotic Without life.

Absorb To become incorporated in.

Abyssal Of the deep sea.

Accrete The process by which inorganic bodies grow larger by the addition of fresh material to the outer surface.

Acidic Having a pH less than 7 (i.e., the hydrogen ion concentration is greater than the hydroxide ion concentration).

Activity (1) The effective concentration of a solute. Solute activity is significantly lower than total concentration in solutions of high ionic strength. (2) Rate of radioactive decay, usually measured in disintegrations per minute (dpm).

Adsorb To adhere to the exterior surface.

Advection The large-scale mass transport of matter.

Aeolian Of the wind.

Aerobic respiration The biological process during which animals and some bacteria oxidize organic carbon to carbon dioxide to yield cellular energy. The oxidizing agent is O_2.

Air–Sea interface The boundary between the surface ocean and atmosphere.

Alkaline Having a pH greater than 7 (i.e., the hydrogen ion concentration is less than that of the hydroxide ion concentration).

Alkalinity The concentration of negative charge in a solution that can be titrated by a strong acid. In seawater, HCO_3^- and CO_3^{2-} contribute most of this negative charge. The units of measurement are meq/L.

Alpha (α) particles Helium nuclei (^4_2He). A common by-product of the radioactive decay of primordial radionuclides.

Aluminosilicates Minerals composed primarily of the elements aluminum, oxygen, and silicon.

Amino acids Biochemicals found in all organisms. They are the building blocks of proteins. Each contains at least one amine and one carboxylic acid group.

Ammonification The microbial process by which dissolved organic nitrogen is converted to ammonium.

Anabolism Metabolic reactions in which the resultant molecules are larger than those of the reactants.

Anaerobes Organisms that live in the absence of O_2.

Analog A molecule with a fundamentally similar structure but subtly different, such as in the chemical composition of a side chain.

Anhydrite A mineral composed of calcium sulfate. It is a common hydrogenous mineral deposited in hydrothermal systems.

Anion A negatively charged ion.

Anoxic Without O_2.

Anthropogenic Caused by humans.

AOU, apparent oxygen utilization The difference in O_2 concentration between that in a deep-water sample and its NAEC. It is a measure of the amount of O_2 consumed via the respiration of organic matter since a deep-water mass was last at the sea surface.

Aragonite A mineral form of calcium carbonate. It is more soluble than calcite. Some marine organisms, such as pteropods, deposit shells composed of aragonite. This mineral is also a common component of evaporites.

Aromaticity The number of benzene groups in an organic molecule.

Artificial radionuclides Radionuclides produced by the explosion of atomic bombs and nuclear reactors.

Asphalt A solid residue left after distillation has removed all other compounds from petroleum. The residue is composed of asphaltenes.

Asphaltenes A class of compounds that are part of the high-molecular-weight component of petroleum. They are relatively inert, have high boiling points, and have high degrees of unsaturation.

Assimilation The update of nutrients and other dissolved materials by phytoplankton.

Assimilatory nitrate reduction The reduction of nitrate to organic nitrogen compounds that constitute the tissues of marine organisms. Plankton and some bacteria assimilate nitrogen via this process.

Asynchrony Not in phase; out of sync.

ATP, adenosine triphosphate A biomolecule that acts to transport energy within cells by enabling electron transfer.

Authigenic Created by inorganic processes in the ocean.

Autotrophs Organisms that can use inorganic substances as their energy source.

Barite A mineral form of barium sulfate that is most common beneath surface waters of high productivity. The barium is primarily biogenic in origin.

Basalts The igneous extrusive rocks that constitute the oceanic crust.

Basic Of a solution that has a greater concentration of hydroxide (OH^-) ions as compared to hydrogen (H^+) ions.

Benthic Of the organisms that live in or on the seafloor.

Bioaccumulation An enrichment of a particular chemical caused by either passive adsorption from seawater or active uptake followed by retention in living tissues or hard parts as a result of nonexcretion.

Bioassay A technique of concentration measurement that relies on the measured response of an organism, or some part of an organism, to an analyte.

Bioavailibility In a form that is readily consumed or assimilated by organisms.

Bioflocculation Flocculation caused by living organisms (e.g., the formation of large clumps of phytoplankton during a bloom).

Biofouling The sequential colonization of marine surfaces by microbes and higher order organisms.

Biogenic Of marine organisms.

Biogenic silica A mineral form of silica that is amorphous in structure and deposited by marine organisms such as diatoms and radiolaria. Also called opal or opaline silica.

Biogenous sediments Sediments that are composed of hard or soft parts, such as shells and tissues, that were synthesized by marine organisms.

Biogeochemical cycle The transport of materials on the earth as a result of biological, chemical, and geological processes.

Biogeochemistry The science that studies the biological, chemical, and geological aspects of environmental processes.

Biointermediate element An element whose marine distributions are controlled by both physical and biogeochemical processes.

Biolimiting element An element whose marine distributions are controlled primarily by biogeochemical processes. Such elements are characterized by low concentrations in the surface waters and tend to limit the growth of phytoplankton.

Biomagnification The concentration of a chemical with increasing trophic level. This is caused by bioaccumulation that occurs during each transfer of the chemical to higher trophic levels. Biomagnification causes organisms at the top of the food chain to have the highest pollutant concentrations.

Biomarkers Organic compounds that are synthesized by specific organisms and thus can be used as an indicator of the current or past presence of those organisms.

Biomass The quantity of living particulate organic matter.

Biomolecules Organic compounds synthesized by organisms.

Biopolymers Organic molecules synthesized by organisms. They are composed of repeating units.

Biota All of the organisms in an ecosystem.

Bioturbation The physical mixing of sediments caused by the burrowing and feeding activities of benthic organisms.

Biounlimited element An element whose marine distributions are controlled primarily by physical processes and are relatively uninfluenced by biogeochemical phenomena. These elements demonstrate conservative behavior.

Box model A conceptual representation of a geochemical cycle in which reservoirs are depicted as boxes and transports by arrows. No details as to the processes that occur in the boxes are provided.

Bunsen solubility coefficient (α_A) The term that relates the concentration of a gas in seawater to its partial pressure in the atmosphere. It is dependent on temperature and salinity.

Cap rock A relatively impermeable rock that retards the migration of petroleum through marine sediments. Many are composed of ancient evaporites, also called diapirs.

Carbohydrates A class of biopolymers whose building blocks are composed of simple sugars such as glucose and fructose. These compounds contain only carbon, hydrogen, and oxygen.

Carbonate alkalinity The concentration of negative charge in seawater contributed by bicarbonate (HCO_3^-) and carbonate (CO_3^{2-}). It is usually reported in units of meq/L.

Carbon fixation reactions Biochemical reactions conducted by plants and some bacteria in which inorganic carbon is incorporated into organic molecules.

Catabolism The biochemical process by which organic compounds are degraded. This provides cellular energy.

Catagenesis Geochemical reactions that occur in the sediments following burial on time scales greater than 1000 years and at temperatures of 50 to 150°C.

Cation A positively charged ion.

Cation exchange The displacement of one cation for another on the surface of a negatively charged solid, such as a clay mineral.

CCD, calcite compensation depth The depth at which the rate of calcite supply from the sinking of detrital shells is equal to their rate of loss from dissolution.

CEC, cation exchange capacity A measure of the amount of cations that will adsorb to the negatively charged surface of a clay mineral. It is usually measured in units of meq of charge per 100 g of clay mineral. This adsorption is reversible.

Cell potential (E_h) A measure of how far a redox reaction is from equilibrium. It is reported in units of volts. The higher the E_h, the greater the driving force.

Chelation The complexation of a metal cation by an organic ligand. Some ligands have multiple binding sites. Those with two sites are called bidentate ligands.

Chemoautolithotrophs Organisms that use inorganic chemicals as their source of electrons, energy and carbon.

Chlorinity The grams of halides expressed as the number of grams of chloride that are present in 1000 g of seawater. This concentration is directly proportional to salinity.

Chlorite A magnesium-rich clay mineral produced by terrestrial weathering under polar and subpolar conditions.

Circumpolar Surrounding either the north or south poles.

Clay (1) A grain whose diameter is less than 4 μm. Most are inorganic silicates. (2) A sedimentary deposit that is composed of more than 70 percent by mass clay-sized grains.

Clay mineral A layered aluminosilicate, such as kaolinite, illite, chlorite, and montmorillonite. Most are formed by chemical weathering of rocks on land.

Clean Water Acts of 1948 and 1977 Federal legislation enacted to regulate marine pollution by controlling the dumping and discharge of wastes into the ocean. These acts also include water-quality criteria for effluent.

Coastal upwelling The upward advection of water from the base of the mixed layer toward the sea surface caused by Ekman Transport. This water motion brings nutrient-rich water to the sea surface.

Coccolithophorids Phytoplankton that deposit calcareous plates called coccoliths.

Cogener An element that is in the same periodic group.

Colligative property Any property of a solution that depends only on the number of solute particles, rather than on the nature of the particles.

Colloid A form of matter that ranges in size from 0.001 to 0.1 μm. This is in between that of a solute and a solid.

Compaction Compression. This causes a reduction in sediment porosity.

Complexation A chemical bond between a cation and a ligand that ranges in behavior from polar to nonpolar covalent.

Compressibility The degree to which the volume of a given quantity of material can be reduced as a result of an increase in pressure.

Condensation The phase change during which a gas is transformed into a liquid.

Conduction (1) The transport of heat via radiation of energy. (2) The transport of electrons causing an electrical current to flow.

Conservative Chemical behavior that is controlled by physical processes. In other words, physical transport is much faster than any chemical processes that can either supply or remove the chemical. All of the major ions exhibit conservative behavior. The concentrations of conservative species are directly proportional to each other and to salinity.

Contaminant Naturally occurring substances whose concentrations have been increased as a result of human activity.

Continental crust The thickened part of the crust that comprises the continents and consists primarily of granitic rock.

Continental margin The zone separating the emergent continents from the deep-sea bottom; it generally consists of a continental shelf, slope, and rise.

Continental rise The large sedimentary deposit that lies at the foot of the continental slope.

Continental shelf The sea floor adjacent to a continent, extending from the low-water line to the change in slope, usually at 180 m water depth, where the continental shelf and slope join.

Continental slope A declivity that extends from the outer edge of the continental shelf to the continental rise. The angle is approximately 4 to 5°.

Contour currents Water currents that flow along the foot of the continental slope.

Convection The transport of heat as a result of the physical movement of a carrier, such as air, water, or magma. Convection occurs spontaneously due to density differences.

Coordination complex A molecule composed of one or more metal atoms, each of which is covalently bonded to more than one ligand or electron donor.

Coprecipitate To precipitate together. Due to their chemical similarity, most metals tend to precipitate together into solids that are thus composed of a large and variable amount of these elements (e.g., polymetallic sulfides and oxyhydroxides).

Coriolis effect The apparent force acting on moving particles that results from Earth's rotation. The Coriolis Effect causes freely moving bodies to be deflected to the right of their direction of motion in the northern hemisphere and to the left in the southern hemisphere.

Cosmic rays Protons and α particles that enter Earth's atmosphere from outer space. When they collide with atoms or molecules of atmospheric gas, high-energy neutrons can be given off. These neutrons can then undergo nuclear reactions with other atmospheric gas atoms and molecules.

Cosmogenic radionuclides See **Spallation reactions**.

Cosmogenous From outer space.

Covalent bond A chemical bond in which the bonding electrons are shared between the bonded atoms. If the sharing is equal, the bond is termed nonpolar covalent. If one atom is more electronegative, the electrons are not equally shared. See **Polar covalent bond**.

Cracking reactions Chemical reactions that occur during catagenesis and metagenesis in marine sediments and sedimentary rocks. During this process, petroleum compounds are formed as hydrocarbons are broken off heteroatomic macromolecules.

Crude oil The liquid portion of petroleum. It contains various fractions, such as kerosene, gasoline, and napthenes, that can be isolated by distillation.

Crustal Of Earth's crust, which is the outer shell of the planet. It is composed of sedimentary, metamorphic, and igneous rocks and ranges in thickness from 5 to 35 km.

Crustal-ocean factory The conceptual model that describes the material flows between the crust, ocean, atmosphere, and upper mantle.

Crystal lattice The regular, repeating framework created by atoms, ions, or molecules constituting a crystalline solid.

Cyanobacteria These bacteria are photosynthetic and hence are sometimes referred to as blue-green algae.

Cytotoxic Toxic to cells.

Daughters The radionuclides produced by the decay of primordial radionuclides.

Debouches Exits from an orifice.

Decay constant The constant that describes the rate at which a radionuclide decays. It is equal to $0.693/\tau$, where τ is the half-life of the radionuclide.

Deep water or zone That portion of the water column from the base of the permanent thermocline or pycnocline to the ocean floor.

Degas The process by which gas is lost from a solution.

Del value A measure of how different the stable isotope composition of a sample is relative to a standard. Samples that are enriched in the rare stable isotope relative to the standard have positive del values. Samples that are depleted in the rare stable isotope relative to the standard have negative del values.

Denitrification The conversion of nitrate to N_2 gas. This is achieved by heterotrophic bacteria that use nitrate as an electron acceptor under suboxic and anoxic conditions.

Density Mass per unit volume of a substance, usually expressed in units of g/cm^3.

Density stratification Gradients in the density of seawater caused by the presence of different water masses. In a stable density configuration, density increases with increasing depth.

Deposit An accumulation of crustal solids.

Desertification The process by which fertile land is turned into a desert. This is usually caused by removal of vegetation and/or climate change.

Desorb The reverse of adsorb; to be released from a surface.

Detrital (1) Nonliving; (2) from land.

Diagenesis Geochemical reactions that occur in the sediments following burial on time scales less than 1000 years and at temperatures ranging from 0 to 150°C.

Diagenetic remobilization The solubilization of materials from sedimentary particles after their accumulation on the seafloor.

Diapir An ancient evaporite that has become buried in marine sediments. The overlying pressure causes the rock to flow like toothpaste out of a tube, thereby forming vertical pillar and domal structures.

Diatoms Microscopic phytoplankton that deposit siliceous tests. These are the most abundant plants in the ocean.

DIC, dissolved inorganic carbon CO_2, H_2CO_3, CO_3^{2-}, and HCO_3^-.

Diel Daily.

Diffusion The transfer of matter or heat as a result of molecular motion. This motion causes net transport from regions of high concentration (or heat) to regions of lower concentration. In the absence of gradients, this motion is random and causes no net transport.

Dinoflagellates Phytoplankton that have flagella. These organisms do not synthesize hard parts, though some have exoskeletons composed of chitin.

Disequilibria Not in thermodynamic equilibrium and thus liable to spontaneous reactions.

Disintegrations per minute (dpm) The unit measurement of radioactive decay rates.

Disproportionation A reaction in which a molecule reacts with itself.

Dissimilatory nitrate reduction Denitrification.

Divalent Having two units of electrical charge.

Divergence Horizontal flow of water in different directions from a common center or zone that results in upwelling. This occurs in the open ocean at the equator and 60°S as result of surface currents driven by the Trades and Westerlies, respectively.

DOC, dissolved organic carbon The fraction of the organic carbon pool that is dissolved in seawater. In practical terms, this includes all organic compounds that pass through a filter with a 0.4 μm pore size.

DOM, dissolved organic matter The fraction of the organic matter pool that is dissolved in seawater. In practical terms, this includes all organic compounds that pass through a filter with a 0.4 μm pore size.

Domal Having a dome-like shape.

DON, dissolved organic nitrogen The fraction of the organic nitrogen pool that is dissolved in seawater. In practical terms, this includes all organic nitrogen compounds that pass through a filter with a 0.4 μm pore size.

Downwelling The downward advection of water as a result of water-mass formation.

Eddy diffusion Transport as a result of vertical water turbulence.

Ejecta Small fragments spewed out of volcanoes during an eruption. Most are pieces of ash or glassy fragments of basalt.

Ekman transport The advection of water in the mixed layer caused by the winds and the Coriolis Effect. The latter causes net water transport to occur at a direction that is at a 90° angle relative to the wind direction.

Electron A negatively charged subatomic particle that constitutes an insignificant amount of the mass of an atom.

Electron activity (pe) $\{e^-\}$; A measure of how far a redox reaction is from equilibrium. The larger the pe, the greater the driving force toward spontaneous reaction.

Electron capture A nuclear reaction in which an electron from the lowest energy level of an atom (1s) reacts with a proton, thereby generating a neutron and causing the emission of x-rays.

Electron carriers Biomolecules, such as NAD and NADP, that act to transport energy within cells by enabling electron transfers.

Electronegativity The degree to which an atom can attract electrons when it is bonded to other atoms.

Electroneutrality Having no net electrical charge.

Electrostriction The drawing together of water molecules in the presence of ionic solutes as a result of specific and nonspecific interactions.

Elute To remove a solute as a result of transport in an advecting solution. This usually involves desorption from a solid support.

Embayment An indentation in the coastline.

Empirical Determined through observation and measurement rather than from theoretical principles.

Empirical formulae The molar combining ratio of a compound that has a variable size, such as a crystalline solid.

Endosymbiotic Of a close biological association in which an organism lives inside its host.

Enrichment factor (E.F.) The degree to which a marine organism is enriched in a particular chemical with respect to the seawater concentration, e.g., E.F. = [metal concentration in biogenic material]/[metal concentration in seawater].

Equation of state of seawater The semiempirical equation that relates the density of seawater to its salinity, temperature, and pressure.

Equilibrium The chemical state in which the concentrations of reactive species do not change over time.

Equilibrium Constant An expression that relates the equilibrium concentrations (or pressures) of reactive species to one another. Its value is dependent on temperature.

Estuarine Of estuaries, which are coastal areas where freshwater mixes with salt water. Most are semienclosed bodies subject to tidal motions, such as the seaward end of a river valley.

Euphotic Of the depth zone through which light penetrates. The bottom of the euphotic zone is defined as the depth at which less than 1 percent of the incident solar radiation remains, the rest having been either absorbed or reflected.

Eutrophic Of waters that have a great abundance of marine life, usually due to high nutrient levels.

Eutrophication The overgrowth of algae in marine and fresh waters caused by an overabundance of nutrients. The algae form a mat at the water surface that retards gas exchange across the air–sea interface. This can lead to O_2 depletion in the underlying waters and, hence, fish kills.

Evaporation The phase change during which a liquid is transformed into the gaseous state.

Evaporite A suite of minerals formed as a result of the evaporation of seawater.

Exothermic Releases heat and is therefore the result of a spontaneous reaction.

Extracellular Occurs outside the cell.

Exudates Materials that are slowly discharged from various pores and orifices of organisms.

Facies The physical characteristics of a rock, which usually reflect its conditions of origin.

Fatty acids Biochemicals that are composed of hydrocarbons attached to a terminal carboxylic acid group. They vary in the number of carbons and degree of multiple bonding and branching in the hydrocarbon.

Fecal pellet Organic excrement generally ovoid in form and less than 1 mm in length. Produced primarily by marine invertebrates.

Feedback loop A set of geochemical processes that influence each other. In a negative feedback, an alteration in the rate of one process is at least partially compensated for by changes in the rates of the other interconnected processes. In a positive feedback, an alteration in the rate of one process is amplified by accompanying changes in the rates of the other interconnected processes.

Filter feeding The filtering or trapping of edible particles from seawater. This feeding mode is typical of many zooplankton and other marine organisms of limited mobility.

Floc Clotlike masses composed of small particles.

Flocculation The process by which small particles aggregate into clumps.

Fluidize To cause material to flow by increasing the motion of individual particles.

Flux The transport of matter or energy through a given surface area or volume in a given unit of time.

Foraminifera Protozoans that deposit tests, called forams, composed of calcite.

Fossil fuel Refined petroleum products that are burned in automobiles and factories to provide energy.

Fractionation The segregation of isotopes of a particular element that can occur during a variety of physicochemical processes. This causes the products to have a different isotopic composition than that of the reactants. See **Kinetic and Thermodynamic Fractionations**.

Free energy change (ΔG) The total energy of a chemical system that would have to be expended or absorbed to reach the equilibrium state.

Free water Water that is not contained within the crust.

Freezing The phase change during which a liquid is transformed into a solid.

Fulvic acids Humic substances that are soluble at all pHs.

Gas chromotagraph A scientific instrument used to identify and quantify organic compounds of intermediate boiling points

Gasoline The fraction of petroleum that has a boiling point less than 180°C. Most of the compounds are low-molecular-weight, straight-chain hydrocarbons.

Geomorphological Of the general configuration of Earth's surface.

Geostrophic current The advection of water resulting from the balance between gravity, wind stress, and the Coriolis Effect.

Geothermometers Minerals whose chemical composition can be used to determine their temperature of crystallization.

Gibbs free energy (G) The total chemical energy of a system.

Glacioeustatic Of changes in sea level caused by variations in continental ice loading.

Glauconite A hydrogenous mineral commonly found in carbonate sediments of continental shelves located in the tropics. This iron-rich, greenish, hydrous silicate contains fossilized biogenic detritus such as fecal pellets and siliceous tests. It is found as nodules and encrustations. In some locations, concentrations are high enough to give the sediments a greenish cast and hence are referred to as green muds.

Glycolysis The chemical reactions during which glucose is catabolized to generate cellular energy.

Graded bedding A type of sediment stratification in which each stratum displays a gradation in grain size from coarse below to fine above.

Gradient The rate of decrease (or increase) of one quantity with respect to another.

Granite A crystalline intrusive igneous rock consisting of alkaline feldspar and quartz.

Gravimetric Of the chemical analyses that involve mass measurements.

Greenhouse effect The warming of Earth's atmosphere as a result of the retention of solar radiation. This retention is possible because insolation absorbed by the land and ocean is radiated back to the atmosphere as UV energy. This energy is absorbed by atmospheric gases and then radiated from them as heat.

Groundwater That part of the subsurface water that is in the soil or flows through and around terrestrial crustal rocks.

Gypsum A mineral form of calcium sulfate. Gypsum is a common component of evaporites.

Gyres A set of four interlocking geostrophic currents that move water in each ocean basin. Northern hemisphere gyres move surface waters clockwise, while southern hemisphere gyres move water counterclockwise.

Half-cell reaction A conceptual representation of electron transfer in which the number of electrons gained by a molecule or atom is indicated. For example, the half-cell reduction of Mn^{4+} to Mn^{2+} is written as: $Mn^{4+} + 2e^- \rightarrow Mn^{2+}$.

Half-life The time required for some physical or chemical process to remove half of the original amount of a substance.

Halite A mineral form of sodium chloride. Halite is a common component of evaporites.

Halmyrolysis The processes that alter the chemical composition of terrestrial clay minerals during their first few months of exposure to seawater.

Halogenated Having one or more atoms of a halide (e.g., chloride, bromide, or iodide).

Heavy metal Metals whose atomic weights exceed 20 amu.

Heat capacity The amount of heat required to raise the temperature of a substance by a given amount.

Henry's Law The mathematical formula that relates the concentration of gas in a solution to its partial pressure in the overlying atmosphere.

Heterotrophs Organisms that require organic matter as their carbon source.

Horizontal segregation The horizontal gradient in biogenic materials, such as nutrients and O_2, that is established by the interaction between the biogeochemical cycling of particulate organic matter and thermohaline circulation.

Humic acids Humic substances that are insoluble at acidic pHs.

Humic substances High-molecular-weight organic compounds that are variable in composition, have complex structures, and are relatively inert. They comprise most of the DOM. The very large ones are components of soils and sediment.

Hydration The binding of water molecules by adsorption onto solid surfaces or through electrostatic attractions to ions or molecules.

Hydrocarbons Organic compounds composed entirely of carbon and hydrogen atoms.

Hydrogen bond The relatively weak intermolecular interaction that occurs between oppositely charged ends of adjacent water molecules.

Hydrogenous Same as **Authigenic**.

Hydrographic Pertaining to seawater.

Hydrological cycle The global water cycle involving the transport of this substance between the atmosphere, hydrosphere, and lithosphere.

Hydrolysis A chemical reaction involving water.

Hydrosphere The water portion of the earth, as distinguished from the solid (lithosphere) and gaseous (atmosphere) parts. This includes water in lakes, ponds, streams, rivers, glaciers, icebergs, the ocean, pore waters, and that which is trapped in crustal rocks.

Hydrothermal Of the hot-water systems that are present at active mid-ocean spreading centers.

Hydroxyl A chemical group composed of an oxygen atom bound to a hydrogen atom (i.e., $-OH$).

Hypersaline Water with a salinity in excess of that at which halite will spontaneously precipitate.

IAP, ion activity product If the product of the ion activities exceeds the K_{sp} of a mineral, it will spontaneously precipitate.

Ice The solid form of water.

Iceberg A large mass of detached land ice that either floats in the sea or is stranded in shallow water.

Ice rafting The transport of lithogenous material by icebergs. The rafted material is rock eroded from the continents as the icebergs flowed seaward.

Igneous rock A rock formed by the solidification of magma.

Illite A potassium-rich clay mineral.

Inert Unreactive.

In situ Occurring or formed in place.

Insolation Solar radiation received at Earth's surface.

Intercalibration A standardization process in which scientists compare measurements of a single substance as produced by different equipment or methods at different times and locations.

Interface A surface separating two substances of different properties, such as different densities, salinities, or temperatures (e.g., the air–sea interface or the sediment–water interface).

Interstitial water Water trapped in the sediments between the particles.

Ion An atom that is electrically charged due to a surplus or deficiency of electrons. The former is termed an anion and the latter a cation.

Ionic bond A chemical bond in which electron(s) have been transferred from the less electronegative atom to the more electronegative atom. The difference in these electronegativities is somewhat greater than that in a polar covalent bond.

Ionic solid A solid held together by ionic bonds between monatomic ions or complex ions e.g., $NaCl(s)$.

Ionic strength The total concentration of positive and negative charges in a solution contributed by ionic solutes, usually given in units of molality.

Ion exchange See **Cation exchange**.

Ion pair A weak electrostatic attraction that occurs between solutes in a highly concentrated solution. This is an example of a specific interaction.

Ion pairing Electrostatic attractions that arise between cations and anions in solutions of high ionic strength. These attractions are not as strong as ionic bonds. Nevertheless, their presence can influence the physical and chemical behavior of the solution.

IR, infrared radiation This type of electromagnetic radiation has a wavelength that ranges from 75 to 1000 μm.

Iron–manganese oxides Hydrogenous precipitates composed of iron and manganese oxides. Most of the metals are hydrothermal in origin.

Isomorphic substitution The replacement of some of the aluminum and silicon in aluminosilicate minerals by cations of similar ionic charge and radius. This usually occurs as a result of chemical weathering.

Isopycnal A line on a chart or map that connects points of constant seawater density.

Isostacy The condition of equilibrium that is attained by lithospheric units as a result of adjustments in their positions as they float in the asthenosphere.

Isotope Nuclides that have the same number of protons in their nucleus, and hence belong to the same element, but have different numbers of neutrons.

Iteration Repetition.

Juvenile Of material that has directly issued from Earth's interior, i.e., it has not been recycled through the sediments.

Kaolinite A clay mineral produced by intense weathering under tropical conditions. As a result, it is depleted in all cations except for silicon and aluminum.

Kinetic energy Energy of motion. This energy causes particles to more with a velocity which increases with temperature and decreases with increasing mass.

Kinetic fractionation Isotope segregation that is dependent on the rates at which the isotopes react. Since the lighter isotope reacts faster, the products tend to be enriched in the light isotope relative to the reactants.

Kerogen The complex mixture of solid organic compounds formed from the diagenesis and catagenesis of soils and marine sediments.

K_{sp} The equilibrium constant describing the solubility of minerals.

Labile Reactive.

Lagoonal Of lagoons, which are semiisolated bodies of seawater trapped between coral reefs and volcanic islands or between the mainland and barrier islands. Seawater in lagoons tends to be hypersaline.

Latent heat of evaporation The amount of heat required to transform 1 g of liquid water into steam or the amount of heat that must be removed to transform 1 g of steam into liquid water.

Latent heat of fusion The amount of heat required to transform 1 g of ice into liquid water or the amount of heat that must be removed to transform 1 g of liquid water into ice.

LD_{50} The lethal dose at which a chemical causes the mortality of 50 percent of the test organisms during a specified period of exposure.

Ligand An electron donor. Ligands from complexes with cations (i.e., electron acceptors).

Lignin A large polymeric macromolecule synthesized only by woody plants. The degradation of lignin is a unique source of phenolic acids.

Lipids A class of organic compounds composed of carbon, hydrogen, and oxygen atoms. Complex lipids contain fatty acids attached to a backbone molecule such as glycerol. Simple lipids, such as carotene, are polymers of terpene. Lipids are used in organisms for energy storage.

Liquid water Water in its liquid state.

Lithogenous Of the continental or oceanic crust.

Lithosphere The outer, solid portion of Earth, including the crust and upper mantle.

Lysis The breakage of cell membranes.

Lysocline The depth at which shell dissolution starts to have a detectable impact on the calcium carbonate content of the surface sediments.

Macroalgal Of large marine plants such as seaweed and marsh grasses.

Macrofauna Benthic animals that are larger than 0.5 mm.

Magma Mobile, usually molten rock material generated within the earth. It can intrude into crevices and fissures as it upwells through the crust. Alternatively, it can upwell through fissures to be extruded onto the crust's surface. Solidification produces igneous rock.

Mantle The bulk of the Earth. It is the layer that lies between the crust and core.

Marcet's Principle See the **Rule of Constant Proportions**.

Marginal sea A semienclosed body of water adjacent to, widely open to, and connected with the ocean at the water surface but bounded at depth by submarine ridges.

Mariculture Human efforts at the cultivation and propagation of seafood.

Marine pollutant Any substance introduced into the ocean by humans that alters any natural feature of the marine environment.

Marine snow Large, loosely aggregated solids composed of biogenous and lithogenous particles. The organic material is often colonized by microbes.

Mass balance equation An equation that accounts for the total amount of material in a given system.

Mass spectrometer A scientific instrument used to identify organic compounds. Other types of mass spectrometers are used to measure the relative abundance of stable isotopes in a sample.

Melting The phase transition during which a solid is transformed into a liquid.

Mesotrophic Seawater that has biological productivity intermediate between that of eutrophic and oligotrophic areas.

Metabolites Compounds produced or required by the reactions that occur within cells.

Metagenesis Geochemical reactions that occur in the sediments following burial on time scales greater than 1 million years and temperatures in excess of 150°C.

Metalliferous Metal rich.

Metalliferous sediments Metal-rich sediments. Most are found around active spreading centers because hydrothermal activity is the primary source of the metals.

Metal sulfides Hydrogenous sulfides, such as FeS_2 and CuS_2. Metal sulfides are common components of metalliferouus sediments; most of the constituents are hydrothermal in origin.

Metamorphic rock Rocks formed in the solid state as a result of changes in temperature, pressure, or chemical environment.

Meteoric waters Waters produced by meteorological processes, such as rain and snow. They include all liquid and solid water on land and in underground aquifers. They do not include water in the ocean, in sediments, or below the crust.

Meteorological Of changes in the temperature, humidity, or air motion of the atmosphere.

Methanogenesis The anaerobic microbial process that produces methane.

Microlayer The thin layer present at the sea surface, also referred to as a "sea slick." This layer tends to have high solute and particle concentrations.

Micronutrient Elements that are required by organisms in smaller amounts than nitrogen and phosphorus. Most are trace metals.

Microtektites Small meteorites.

Microzone A small volume of water or solid matter in which the redox environment is considerably different from that in the surrounding sediment or seawater.

Migration The movement of petroleum through marine sediments and sedimentary rock as a result of overlying pressure.

Milankovitch cycles Changes in global ice volume that reach maxima every 23,000, 41,000, and 100,000 years. They are thought to be related to changes in astronomical alignments that have similar periods.

Mineral An inorganic substance that occurs naturally in the Earth and has distinctive physical properties. Its chemical composition can be expressed as an empirical formula that shows the molar combining ratios of the constituent elements.

Mixed layer Near-surface waters down to the pycnocline that are isohaline and isothermal as a result of wind mixing.

Molecular diffusion The random motion of molecules in a solution. In the absence of external forces, solutes spontaneously undergo net diffusion from regions of higher concentration to lower concentration. This continues until a homogeneous distribution of the solute is achieved.

Molluscs Soft, unsegmented animals usually protected by a calcareous shell and having a muscular foot for locomotion; includes snails, clams, and octopuses.

Montmorillonite An iron-rich clay mineral that has a very high cation-exchange capacity. Unlike the other clay minerals, a significant amount of sedimentary montmorillonite is hydrothermal in origin.

Mud A sedimentary deposit composed of 70 percent or more by mass clay-sized grains; also called deep-sea muds.

Mutagenic A form of energy or chemical that can alter the molecular structure of the genes.

NAD, nicotine adenine dinucleotide A biochemical that functions as a cellular electron carrier.

NADP, nicotine adenine dinucleotide phosphate A biochemical that functions as a cellular electron carrier.

NADW, North Atlantic Deep Water This is a deep water mass formed in the North Atlantic.

NAEC, normal atmospheric equilibrium concentration The gas concentration that a water mass would attain if it reached equilibrium with the atmosphere. The NAEC of a gas is a function of water temperature and salinity as well as the partial pressure of the gas in the atmosphere.

Napthenes Cycloparaffins, such as cyclohexane. They are a common component of gasoline.

Natural gases Hydrocarbons that have fewer than 5 carbons, i.e., methane, ethane, propane, and butane.

Nepheloid layer Deep and bottom waters that have high concentrations of resuspended sediment.

Neritic Of the coastal ocean (i.e., over the continental margin).

Nernst equation The mathematical equation that relates the cell potential of a redox reaction to the temperature and concentrations of the reacting chemicals, e.g., $E_{\text{cell}} = E^0_{\text{cell}} - ((RT/nF) \ln Q)$.

Neutron An electrically neutral particle found in the nucleus of atoms.

Nitrification The microbial oxidation of ammonium to nitrite and then to nitrate.

Nitrogen fixation The process by which some bacteria and phytoplankton are able to convert N_2 into organic nitrogen. The energy required is large because a triple bond must be broken.

Nodules Mineral precipitates, such as iron–manganese oxides and glauconites, that form as roundish lumps.

Nonconservative Chemical behavior that is largely controlled by biogeochemical reactions. The concentrations of nonconservative substances are not directly proportional to salinity.

Nonideal Physicochemical behavior that does not conform to ideal thermodynamic predictions.

Nonpelagic Sediments that accumulate at rates in excess of 1 cm/1000 y.

Non-point-source inputs Pollutants introduced into the ocean from widely dispersed sources, such as ground-water seeps.

Nonpolar covalent bond See **Covalent bond**.

Nucleic acids A class of organic compounds that contain carbon, hydrogen, oxygen, and nitrogen atoms. They are one of the building blocks of DNA and RNA.

Nucleus The central, positively charged part of an atom, composed primarily of protons and neutrons.

Nuclide A species of atom uniquely identified by the number of protons and neutrons in its nucleus.

Nutrient An inorganic or organic solute necessary for the nutrition of primary producers.

Nutrient regeneration The process whereby particulate organic nitrogen and phosphorous are transformed into dissolved inorganic species, such as nitrate and phosphate. Microorganisms are largely responsible for this process.

Ocean Dumping Acts of 1972 and 1988 Federal legislation enacted to regulate marine pollution by controlling the dumping and discharging of wastes into the ocean.

Oceanic Of the water or sediment beyond the continental margins.

Oceanic crust The mass of basaltic material, approximately 5–7 km thick, that underlies the ocean basins.

Oligotrophic Waters with very low biological productivity, such as those of the mid-ocean gyres.

Oolite A hydrogenous precipitate commonly found in carbonate sediments of continental shelves located in the tropics. They are composed primarily of calcium carbonate and are thought to be an abiogenic precipitate formed from warm seawater supersaturated with respect to calcite and aragonite.

Ooze A sediment that contains greater than 30 percent detrital biogenic hard parts by mass.

Opal (or opaline silica) An amorphous silicate formed through the polymerization of silicic acid molecules. Though most is biogenic in origin, some forms as a result of diagenesis.

Open ocean That part of the ocean seaward of the continental margin.

Ophiolites Ancient pieces of oceanic crust that have been thrust up onto the continents as a result of geologic uplift.

Organometallic Compounds that contain both organic structures and metals.

Osmotic pressure The pressure exerted across a semipermeable membrane separating two solutions that differ in their concentrations of a dialyzable solute.

Outfalls Discrete locations where large quantities of water and/or human wastes are introduced into rivers or the ocean.

Oxic Waters that contain O_2 concentrations close to that of their NAEC.

Oxidation The loss of electrons.

Oxidizing agent A chemical that gets reduced, thereby causing some other reactant to become oxidized.

Oxycline The depth over which O_2 concentrations decrease rapidly. This usually coincides with the thermocline.

Oxygen free radical An atom or molecule of oxygen that contains extra electrons and hence is negatively charged. These species are very strong oxidizing agents.

Oxyhydroxides Amorphous precipitates of oxides and hydroxides that form in alkaline solutions such as seawater. These precipitates usually contain a variety of cations, such as trace metals.

Paleoceanography The study of the ancient oceans that seeks to reconstruct past environmental conditions by examining chemical and fossil records left in the sediments and polar ice cores.

Paraffins Saturated straight-chain or aliphatic, hydrocarbons that have five to ten carbons. A common component of gasoline.

Partial pressure The pressure exerted by a particular gas in a liquid or gaseous solution.

Particulate matter Any solid.

pe Electron activity, i.e., $\{e^-\}$.

Pelagic sedimentation Sedimentation that occurs at rates less than 1 cm/1000 y. This is characteristic of sediment on the abyssal plains and mid-ocean ridges.

pH The negative log of the hydrogen ion activity. Solutions with pH greater than 7 are acidic and those with pH less than 7 are alkaline. A solution is neutral if its pH equals 7.

Phosphorite A hydrogenous mineral commonly found in sediments underlying surface waters of high biological productivity. It is composed primarily of calcium phosphate that is biogenic in origin.

Phosphorylation The biochemical reactions in which a phosphate group is attached to a biomolecule. The attachment and removal of such groups is used as a means of energy transfer in cells.

Photic Same as **Euphotic**.

Photoautolithotrophs Organisms that use solar radiation as their energy source and inorganic substances as their carbon and electron sources.

Photodissociation The decomposition of a molecule as a result of the input of solar energy.

Physicochemical Of a process that involves either physical and/or chemical change.

Phytoplankton Single-celled plants that have weak swimming ability.

Pillow basalts Large mounds of basalt that form when lava is extruded onto the seafloor at active spreading centers.

Plankton Organisms that drift passively or are weak swimmers. Includes mostly microscopic plants (phytoplankton) and protozoans, as well as larval animals and small filter feeders (zooplankton).

Plate tectonics The theory of crustal plate movement caused by seafloor spreading.

POC, particulate organic carbon The fraction of the organic carbon pool that is not dissolved in seawater. In practical terms, this includes all organic compounds that do not pass through a filter with a 0.4 μm pore size.

Point-source inputs Inputs of pollutants into the ocean from discrete sources, such as sewage outfall pipes.

Polar covalent bond Chemical bonds in which electrons are not equally shared due to the greater electronegativity of one of the atoms. As a result, the more electronegative atom acquires a small net negative charge relative to the less electronegative one. The difference in electronegativities is somewhat smaller than that in an ionic bond.

Polymetallic Containing a variety of different metals.

Polymetallic sulfides Hydrogenous metal sulfides that form from the precipitation of metals and sulfides that are hydrothermal in origin. The metals (Fe, Cu, Co, Zn, and Mn) tend to coprecipitate to form heterogeneous sulfide deposits.

Polynuclear Of an organometallic molecule that contains more than one atom of metal.

Polyprotic Of an acid that contains more than one proton that it can donate to a base.

Polysaccharide A biopolymer composed of two or more simple sugars, such as glucose. They are used for energy storage in organisms.

POM, particulate organic matter The fraction of the organic matter pool that is not dissolved in seawater. In practical terms, this includes all organic compounds that do not pass through a filter with a 0.4 μ pore size.

PON, particulate organic nitrogen The fraction of the organic nitrogen pool that is not dissolved in seawater. In practical terms, this includes all organic nitrogen compounds that do not pass through a filter with a 0.4 μ pore size.

Ponding The slumping of abyssal sediments into depressions. This process tends to smooth out irregular topographic features on the seafloor of the deep ocean.

Poorly sorted sediments Sediments that contain a wide range of different size class particles, such as clays, sands, and pebbles.

Pore waters Same as **Interstitial waters**.

Porosity A measure of the open space between grains in a sediment.

Positron Positively charged electrons ($^{0}_{1}e$). Also called β^{+} particles.

Postdepositional Of a change that occurs in a sediment following its accumulation on the seafloor.

Precipitation (1) Rainfall. (2) Formation of solids from dissolved materials.

Primary See **Juvenile**.

Primary productivity The amount of organic matter synthesized by organisms from inorganic substances per unit time in a unit volume of water.

Primordial radionuclides Long-lived radionuclides that were present at Earth's formation. The most abundant are ^{235}U, ^{238}U, and ^{232}Th. These isotopes decay to radioactive daughters and hence form a series of decay steps ending in a stable nuclide of lead.

Proteins Biopolymers composed of amino acids that are nitrogen-rich organic compounds.

Proton A nuclear particle that contains one unit of positive electrical charge.

Protozoans Heterotrophic single-celled animals. Benthic and pelagic forms are present in the ocean.

Pteropods Free-swimming gastropods in which the foot is modified into fins; both shelled and nonshelled forms exist.

Pycnocline The depth zone over which density increases rapidly with depth. This usually coincides with the thermocline.

Radioactive decay See **Radioactivity**.

Radioactive decay law The mathematical description of how the amount of radioactive material diminishes over time as a result of radioactive decay.

Radioactivity The spontaneous breakdown of the nucleus of an atom that results in the emission of radiant energy either as particles or waves.

Radiocarbon The radioactive isotope of C, ^{14}C, which has a half-life of approximately 5700 y.

Radioisotope See **Radionuclide**.

Radiolaria A group of protozoans that deposit siliceous tests.

Radionuclide A radioactive isotope.

Rain rate The rate at which particulate matter settles to the seafloor. Since some of the particles can dissolve prior to permanent burial, the rain rate can be less than the accumulation rate.

Rayleigh distillation A distillation process in which the isotopic composition of the products changes over time. This is the result of isotope fractionation during the process that causes the isotope composition of the remaining reactant pool to change over time. Typically, the reactant pool becomes progressively enriched in the rare isotope, thereby causing the products to become progressively enriched.

Recycling efficiency The degree to which a biolimiting element is remineralized prior to removal from the ocean via sedimentation.

Redfield–Richards Ratio The elemental ratio of C to N to P that is present in average phytoplankton (i.e., 106:16:1).

Redox reactions Chemical reactions that involve changes in the oxidation numbers of some of the participating species.

Red tide A red or red-brown discoloration of surface waters caused by high concentrations of certain microorganisms, particularly dinoflagellates. Most common in coastal waters.

Reducing agent A chemical that is oxidized, thereby causing some other reactant to become reduced.

Reduction Gain in electrons.

Refractory Inert; unreactive.

Relict sediment Sediments that are no longer forming. Most of the sediments on the continental margins are presently relict. Some are even eroding.

Remineralization The dissolution of hard parts or the degradation of POM that leads to the solubilization of nutrients.

Reprecipitate The formation of solids from solutes that were introduced into seawater as a result of some dissolution or remineralization process.

Reservoir The biogeochemical form and/or location of a chemical in the crustal-ocean factory.

Reservoir beds Marine sediments that are porous enough for large amounts of petroleum to migrate into.

Residence time The average amount of time that a chemical species spends in the ocean or sediment, assuming the species is at a steady state.

Resolubilization Same as **Remineralization**.

Reverse weathering Chemical reactions that are theorized to occur in the sediments. In these reactions, seawater is thought to react with clay minerals and bicarbonate producing secondary clays and consuming alkalinity and some cations. This process is approximately the reverse of chemical weathering on land that produces clay minerals.

Riverine Of rivers.

River runoff The transport of water, solutes, and lithogenous particles from the continents to the ocean as a result of riverine input.

Rule of Constant Proportions The relative abundances of the six major cations (Na^+, K^+, Ca^{2+}, Mg^{2+}, Cl^-, and SO_4^{2-}) is constant regardless of the salinity of seawater.

Sabkha Intertidal mudflats in which modern-day evaporites form. Most are located in the Middle East.

Salinas Coastal salt lakes in which modern-day evaporites form. Most are located in the Middle East and Western Australia.

Salinity The grams of inorganic ionic solutes that are present in 1000 g of seawater.

Salinometer The device used to measure the salinity of seawater by determining its electrical conductivity.

Salt A molecule composed of one or more cations and anions that are held together by ionic bonds, e.g., KCl and Na_2SO_4.

Saltmarsh A relatively flat area of the shore where fine-grained sediment is deposited and salt-tolerant grasses grow. One of the most biologically productive ecosystems on Earth's surface.

Sand Particles that range in diameter from 1/16 to 1 mm.

Saturated (1) Of a solution containing the equilibrium concentrations of solutes dictated by the solubility of a particular solid. As a result, the mass of the solid in solution will remain constant over time. (2) Of a solution that contains the equilibrium concentration of a gas. This concentration is determined by the temperature and ionic strength (salinity) of the solution.

Saturation horizon The depth range over which seawater is saturated with respect to calcium carbonate, i.e., $D = 1$. At depths below the saturation horizon ($D < 1$), calcium carbonate will spontaneously dissolve if exposed to the seawater for a sufficient period of time.

Scavenging The removal of dissolved materials (such as trace metals) from seawater by adsorption onto the surfaces of sinking particles.

Scavenging turnover time The amount of time required to remove the entire inventory of a solute from seawater via scavenging.

Sea state A description of the ocean surface with respect to the height of the wind waves.

Secondary Of the material that has been recycled through the sediments.

Secular equilibrium A steady state in which the decay rate of a parent radionuclide is equal to that of its radioactive daughter. Secular equilibrium can only be established when (1) the decay rate of the parent is much slower than that of the daughter, and (2) the only process influencing the amount of daughter is its production from the parent and its loss to radioactive decay.

Sediment Particulate inorganic or organic particles that accumulate in loose, unconsolidated form.

Sedimentary rock A rock resulting from the consolidation of loose sediment.

Sedimentation The process by which sediments accumulate on the seafloor.

Sessile Of benthic organisms that live attached to the seafloor.

Seuss Effect The lowering of the $^{14}C/C$ in atmospheric CO_2 as a result of the burning of fossil fuels.

SHE, standard hydrogen electrode The electrode used as a standard against which all other half-cell potentials are measured. The following reaction occurs at the platinum electrode when immersed in an acidic solution and connected to the other half of an electrochemical cell: $2H^+(aq) + 2e^- \rightarrow H_2(g)$. The half-cell potential of this reaction at 25°C, 1 atm and 1 *M* concentrations of all solutes is agreed, by convention, to be 0 V.

Sigma-T A measure of the density of seawater (i.e., $\sigma_t = (\rho - 1) \times 1000$). It is a function of temperature, salinity, and hydrostatic pressure.

Silica Silicon oxide ($SiO_2(s)$), which can occur in crystalline form, e.g., quartz, or noncrystalline form, e.g., opal.

Silicic acid H_4SiO_4.

Sill Shallow portion of the seafloor that partially restricts water flow, usually at the mouth of a marginal sea or estuary.

Silt Particles that range in diameter from 1/256 to 1/32 mm.

Sink A reservoir that is the recipient of material transport.

Solubility A measure of the maximum quantity of a substance that can dissolve in a given quantity of solvent under a specified set of conditions.

Solubilized Dissolved.

Solution A state in which a solute is dissolved in a solvent.

Solute A substance that is dissolved in a solvent.

Solvent The most abundant substance in a solution.

Source The reservoir that is the origin of material undergoing transport, e.g., water is transported from lakes to the ocean by river runoff. Thus the lakes are the source and the ocean the sink.

Spallation reactions Nuclear reactions that occur in the atmosphere as a result of the collision of cosmic rays with atoms and molecules of atmospheric gases. The resulting high-energy neutrons can then react with atmospheric gases to form cosmogenic radionuclides.

Spatial Of space.

Speciation The chemical form(s) that an atom is in.

Specific activity The radioactive decay rate of a specified amount of sample, usually expressed as disintegrations per liter (dpm/L) or disintegrations per gram (dpm/g).

Spherules Small spheres.

Spicules Minute, needle-like spines produced by some of the microorganisms that deposit calcareous and/or siliceous hard parts.

Steady state The absence of change over time caused by equal rates of production and removal. Equilibrium is a special case in which the processes producing and removing the chemicals are reversible.

Steam The gaseous phase of water.

Stoichiometry The relative abundance of atoms in a molecule or in a reaction, given in terms of molar quantities.

Stormwater Surface water produced by heavy rains that either percolates into the soil or flows directly into rivers and the coastal ocean.

STP—standard temperature and pressure The temperature is 273.15°K and the pressure is 1 atm.

Stratographic trap Changes in the porosity of marine sediments that retard the migration of petroleum and hence cause large pools to accumulate.

Stratosphere That part of Earth's atmosphere that lies between the troposphere and the upper layer (ionosphere).

Stromatolites The fossilized remains of microorganisms that grew in such a fashion as to create domal structures.

Structural trap Geological structures, such as diapirs, that retard the migration of petroleum through marine sediments and sedimentary rock.

Subaqueous Below the air–sea interface.

Suboxic O_2 concentrations significantly below that of the NAEC.

Subpolar Latitudes ranging from 40 to 60°.

Substituents Atoms or groups of atoms that are attached to a molecular framework or backbone.

Substrate The base upon which an organism lives and grows.

Subtidal At depths below the lowest reach of the low tide.

Supersaturated (1) Of a solution containing concentrations of solutes that exceed equilibrium levels for a particular solid. As a result, the ions will spontaneously precipitate to form that solid. (2) Of a solution that contains greater than the equilibrium concentration of a gas. If the solution is in contact with the atmosphere, a net flux of gas out of the solution will occur.

Supratidal That part of the shoreline that lies above the highest reach of the high tides. This zone receives seawater only as a result of storm surges and wind transport.

Surface tension The force that exists among water molecules at the air–liquid interface. This force is the result of hydrogen bonding between water molecules.

Surfactant A chemical that has soap-like behavior (i.e., it is able to cause the dissolution of nonpolar solutes in a polar solvent).

Surficial Of the surface of a solid.

Symbiosis A close relationship between two species upon which the survival of at least one member is dependent.

Taxonomic Of the system used to classify organisms (e.g., kingdom, phylum, class, species, etc.).

Tectonism Earthquake, volcanic, and other crustal motions associated with the process of plate tectonics.

Temporal Of time.

Terrigenous Of the land.

Tests Microscopic shells of marine plankton.

Thermocline The depth range over which temperature decreases rapidly with depth. At mid latitudes this zone usually spans depths ranging from 300 to 1000 m.

Thermodynamic equilibrium See **Equilibrium**.

Thermodynamic fractionation Isotope segregation that accompanies attainment of chemical equilibrium.

Thermohaline circulation Deep-water circulation caused by density differences created in the surface waters of polar regions. Cooling increases the density of the surface waters, which sink and then advect horizontally throughout the deep ocean. The water is returned to the sea surface by eddy diffusion.

Titrimetry The analytical technique of concentration measurement that involves reaction of a solution of analyte with a known quantity of standard (e.g., titration).

Trace element An element whose dissolved concentration is between 50 and 0.05 μmol/kg. Most are metals and are referred to as trace metals.

Transamination The reaction in which an amine group is transferred from an amino acid to a carboxylic acid, thereby creating a new amino acid.

Tritium The radioactive isotope of H, ^3H, which has a half-life of approximately 12.5 y.

Trophic level The relative position occupied by an organism in a food chain or web.

Troposphere The portion of the atmosphere closest to Earth's surface in which temperature generally decreases with altitude, clouds form, and convection is active.

Turbidite The sedimentary deposit created by turbidity currents. The latter are underwater mudslides common to the continental slope and rise.

Turbidity current An underwater mudslide common to the continental slope. The particles deposit at the foot of the slope to form the continental rise. The resulting deposit is often referred to as a turbidite.

Turbulence The random motion of water molecules.

Turnover time The time required for a specific process or set of processes to remove all of a substance from a reservoir.

Unconsolidated Composed of small particles or grains, e.g., a sandy beach.

Undersaturated (1) Of a solution containing concentrations of solutes that are less than those required to reach equilibrium with a particular solid phase. As a result, that solid phase will spontaneously dissolve. (2) Of a solution that contains less than the equilibrium concentration of a gas. If the solution is in contact with the atmosphere, a net flux of gas into the solution will occur.

Universal solvent Water, which is so named due to its ability to dissolve at least a small amount of all substances.

Unsorted sediments Same as **Poorly sorted sediments**.

USEPA, United States Environmental Protection Agency The federal agency charged with developing water-quality criteria for effluents discharged into marine waters.

UV, ultraviolet radiation This form of electromagnetic radiation has wavelengths that range from 1 to 400 nm.

Vertical segregation The vertical gradient in biogenic materials, such as nutrients and O_2, that is established by the interaction between the biogeochemical cycling of particulate organic matter and the vertical density stratification of the water column at mid and low latitudes.

Volatiles Compounds with such high vapor pressures under environmental conditions that a significant fraction is present in the gas phase.

Volcanism Volcanic eruptions during which magma from the mantle is extruded into or onto Earth's crust. Once the magma has escaped from the mantle, it is termed lava.

Viscosity A measure of the ability of fluid to flow.

Water mass A water body that has the same temperature and salinity throughout.

Water-quality criteria Legal limits that set the permissible levels of pollutants in marine and fresh waters.

Weathering Destruction or partial destruction of rock by thermal, chemical and/or mechanical processes.

Well-sorted sediments Sediments composed of one size class of particles, e.g., a deep-sea clay.

Zooplankton The animal and protozoan members of the plankton.

INDEX